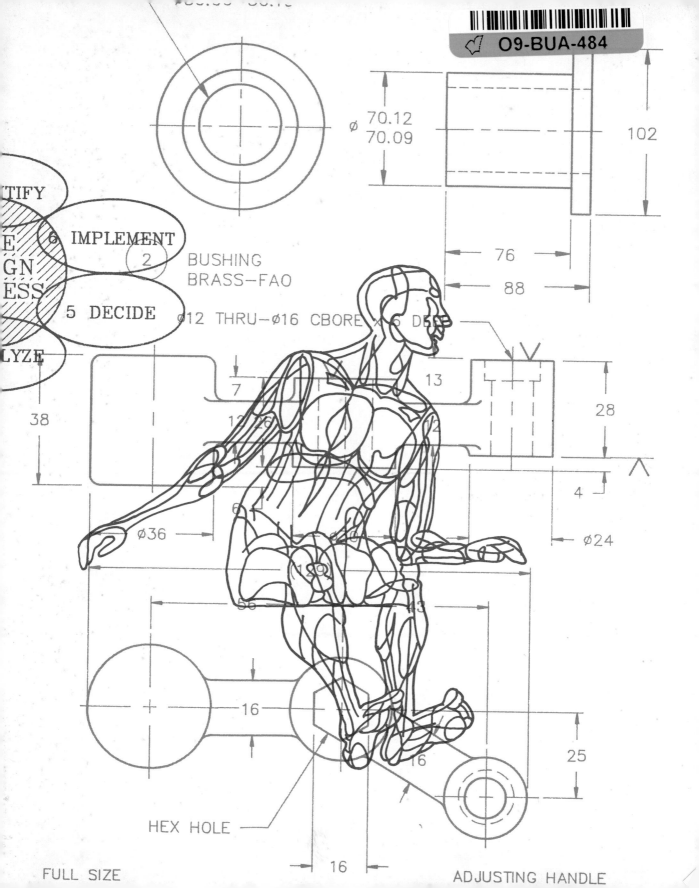

ø 70.12
70.09

102

76

88

TIFY

6 IMPLEMENT
2

5 DECIDE

LYZE

BUSHING
BRASS—FAO

ø12 THRU—ø16 CBORE X6 D5

7

13

38

28

4

ø36

ø24

16

HEX HOLE

FULL SIZE

25

16

ADJUSTING HANDLE

Engineering Design Graphics
Fifth Edition

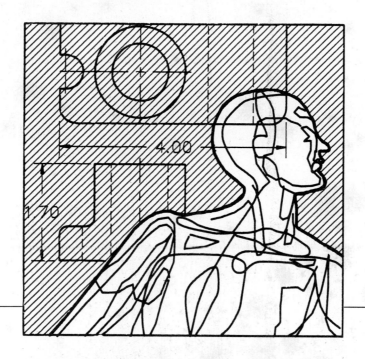

JAMES H. EARLE

Texas A&M University

Addison-Wesley Publishing Company

Reading, Massachusetts • Menlo Park, California • New York
Don Mills, Ontario • Wokingham, England • Amsterdam • Bonn
Sydney • Singapore • Tokyo • Madrid • Bogotá
Santiago • San Juan

Sponsoring Editor: Donald Fowley
Production Supervisor: Laura Skinger
Production Coordinator: Ezra C. Holston
Text Designer: Designworks
Illustrator and Cover Designer: James H. Earle
Layout Artist: Lorraine Hodsdon
Editorial Assistant: Laurie McGuire
Cover Photograph: Honeywell Information Systems

Library of Congress Cataloging-in-Publication Data

Earle, James H.
  Engineering design graphics.

  Includes index.
  1. Engineering design. 2. Engineering graphics.
I. Title.
TA174.E23 1987   620'.00425   86-22269
ISBN 0-201-11641-3

Reprinted with corrections April, 1988

EFGHIJ-HA-898

Dedicated to my father,
Hubert Lewis Earle,
October 25, 1900–October 22, 1967

The following supplements are compatible with the following textbooks from Addison-Wesley Publishing Co.:

Drafting Technology,

Engineering Design Graphics,

Design Drafting,

Graphics for Engineers,

Geometry for Engineers

DRAFTING TECHNOLOGY PROBLEMS is a problem book designed to cover basic graphics, descriptive geometry, and specialty drafting areas. It is designed to accompany *Drafting Technology*, that is available from Addison-Wesley Publ. Co., Reading, Mass. 01867. 131 pages.

BASIC DRAFTING (With Computer Graphics) is a problem book that covers the basics for a one-semester high school drafting course. 67 pages.

CREATIVE DRAFTING (With Computer Graphics) is a problem book that covers mechanical drawing and architectural drafting for the high school. 106 pages.

DRAFTING & DESIGN (With Computer Graphics) is a problem book for mechanical drawing for a high school or college course. 106 pages.

DRAFTING FUNDAMENTALS 1 is a problem book for a one-year high school course in mechanical drawing. 94 pages.

DRAFTING FUNDAMENTALS 2 is a second version of **DRAFTING FUNDAMENTALS 1** for the same level and content. 94 pages.

TECHNICAL ILLUSTRATION is a problem book for a course in pictorial drawing for the college or high school. 70 pages.

ARCHITECTURAL DRAFTING is a problem book for a first course in architectural drafting. 71 pages.

GRAPHICS FOR ENGINEERS 1 (With Computer Graphics) is a problem book for a first course in engineering graphics for the college student. 100 pages.

GRAPHICS FOR ENGINEERS 2 (With Computer Graphics) is a problem book for a first course in engineering graphics for the college student. 100 pages.

GRAPHICS FOR ENGINEERS 3 (With Computer Graphics) is a problem book for a first course in engineering graphics for the college student. 108 pages.

GEOMETRY FOR ENGINEERS 1 (With Computer Graphics) is a problem book for a college-level descriptive geometry course. 100 pages.

GEOMETRY FOR ENGINEERS 2 (With Computer Graphics) is a problem book for a college-level descriptive geometry course. 100 pages.

GEOMETRY FOR ENGINEERS 3 (With Computer Graphics) is a problem book for a college-level descriptive geometry course. 100 pages.

GRAPHICS & GEOMETRY 1 (With Computer Graphics) is a problem book for a college course in graphics and descriptive geometry. 119 pages.

GRAPHICS & GEOMETRY 2 (With Computer Graphics) is a problem book for a college course in graphics and descriptive geometry. 121 pages.

GRAPHICS & GEOMETRY 3 (With Computer Graphics) is a problem book for a college course in graphics and descriptive geometry. 138 pages.

Creative Publishing Co.
BOX 9292   COLLEGE STATION, TEXAS   77840
PHONE 409-775-6047

Preface

Engineering Design Graphics covers the principles of engineering drawing, computer graphics, descriptive geometry, design, and problem solving for a college-level course. The fifth edition is a major revision of the previous edition, but it retains the same class-tested format and general sequence of topics.

Computer graphics

Two chapters, 10 and 38, are devoted to computer graphics. Chapter 10 provides a general introduction to the use of computer graphics in design and drafting and covers the fundamentals of AutoCAD Version 2.15 and its usage with microcomputers. Chapter 38 covers updated features of Version 2.5 that was released in late summer of 1986 and other applications of AutoCAD. Additionally, computer graphics techniques have been integrated throughout other chapters to offer an easy and natural transition from traditional graphics with a pencil to graphics by a computer.

AutoCAD software was selected as the example program to use because it is the most widely sold microcomputer software in the computer graphics field. Therefore, it is likely that more students will encounter AutoCAD in the industrial employment environment than any other software. The IBM XT/AT computer (and their compatibles) were chosen as the microcomputer because they too dominate the market.

Essentially all computer graphics principles have been presented in illustrations that use a two-step format and are supplemented with the prompts from AutoCAD that would be seen on the screen and the response from the user. By this means, the teacher and student will be able to cover the principles of computer graphics with the greatest ease.

Major topics

Other major areas in this text are communications, working drawings, descriptive geometry, and design concepts. Each of these areas is important to the engineer, technologist, and technician.

Communications is covered as it relates to the presentation of ideas, three-dimensional concepts, data analysis, and pictorials. The methods of making working drawings are given in accordance with ANSI standards and include materials, tolerances, welding specifications, pipe drafting, and electric/electronics drafting. The design process is covered in Chapters 2 through 9 to illustrate the importance of engineering graphics in developing creative solutions to technical problems. Descriptive geometry is covered in Chapters 27 through 32. Vectors, nomography, empirical equations, and graphical calculus are other specialty areas receiving emphasis.

Objectives

The objective of this book is to support courses in which the student learns the various standards and techniques of preparing engineering drawings and specifications, learns how to read and interpret drawings, learns how to supervise the preparation of drawings by others, and learns how to solve three-dimensional technical problems that require the application of descriptive geometry and graphical analysis. Once the principles of preparing drawings correctly have been mastered, the student should be able to utilize these techniques in preparing drawings made by computer graphics methods. Above all, this textbook is designed to help students expand their creative talents and communicate their ideas in an effective manner.

Format

Engineering Design Graphics can be used as a self-instructional book enabling the student to work independently. Many problems are presented in a step-by-step sequence that shows the progressive steps in solving a problem and a second color is used to emphasize important points of construction. There are over 600 problems, many of which are new. These problems offer a range of assignments that can be solved by students to aid them in grasping principles covered in each chapter.

Revision features

The fifth edition of *Engineering Design Graphics* has been updated to reflect important developments in the field, including the use of new tools and technologies, recent changes in governing standards, and others. In addition this revision has benefited in numerous ways from the close and continuous scrutiny of the author, colleagues at the author's and other institutions, and students. The opportunity to use the book over the course of several years has enabled us to refine and improve the material in hundreds of ways.

Some of the major revision features are

- the revision of several hundred illustrations to enhance clarity and student learning
- the inclusion of several hundred new drawings
- Chapter 10, "The Computer in Design and Drafting," replaces Chapter 37 from the fourth edition. This chapter is much changed in content: it provides a general introduction to the area of computer-aided design and drafting and then gives a detailed description of the AutoCAD software program and its application to engineering design graphics
- the insertion of computer graphics techniques in all appropriate chapters
- the addition of a new chapter, Chapter 38, which details the updated features of AutoCAD's Version 2.5 and offers some applications of these features to engineering design graphics
- the major rewriting of the chapter on tolerances (Chapter 22)
- the addition of many new problems
- the editing of existing material in all chapters, including: extensive updating of references, illustrations, and photographs; the enlargement of many problem illustrations from the fourth edition for easier student use; correction of minor inaccuracies that were brought to our attention by students, colleagues, and reviewers

Supplementary problems

Seventeen different problem books and teachers' guides are available to be used with this textbook. A listing of these books and their source is given on the page preceding the Preface. Twelve of the manuals have computer graphics problems on the backs of the problem sheets that allow the coverage of the problems by both computer and traditional methods with the same problem books.

Acknowledgments

We are grateful for the assistance of many who have influenced the development of this volume. We have been aided by the use of illustrations developed by the late William E. Street and Carl L. Svensen for their earlier publications. Many industries have furnished photographs and drawings that have been acknowledged in the legends where they are used. The Engineering Design Graphics staff of Texas A&M University has been helpful in making suggestions for the preparation of this book. Professor Tom Pollock provided valuable information on various metals and their designations.

We are indebted to Mary Ann Zadfar, Josef Woodman, and Joseph Oakey of Autodesk Inc. for their assistance with AutoCAD. We appreciate the assistance and cooperation of Karen Kershaw of MEGA CADD, INC. After our association with these individuals and their companies, it is easy to understand why they are leaders in their respective fields.

We are appreciative of the many institutions who have thought enough of our publications to adopt them for classroom use. It is an honor for one's work to be accepted by his colleagues. We are hopeful that this textbook will fill the needs of engineering and technology programs. As always, comments and suggestions for improvement and revision of this book will be appreciated.

College Station, Texas   J. H. E.

Brief Contents

Contents

CONTENTS

CONTENTS

CONTENTS

CONTENTS

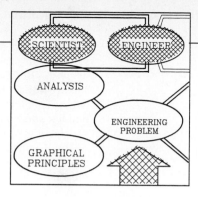

CHAPTER 1

Introduction to Engineering and Technology

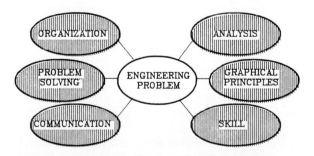

FIG. 1.1 Problems in this text require a total engineering approach with the engineering problem as the central theme.

1.1
Introduction

This book is devoted to the introduction of elementary design concepts of engineering and to the application of engineering graphics to the design process. Examples are given that have an engineering problem at the core and that require organization, analysis, problem-solving graphical principles, communication, and skill (Fig. 1.1).

Creativity and imagination are essential ingredients of the engineering profession. Albert Einstein said, ''Imagination is more important than knowledge, for knowledge is limited, whereas imagination embraces the entire world . . . stimulating progress, or, giving birth to evolution'' (Fig. 1.2).

FIG. 1.2 "Imagination is more important than knowledge."

1.2
Engineering graphics

Engineering graphics is the total field of graphical problem solving and includes two areas of specialization: descriptive geometry and working drawings. Other areas that can be used for a wide variety of applications are nomography, graphical mathematics, empirical equations, technical illustration, vector analysis, data analysis, and computer graphics. Graphics is one of the designer's primary methods of thinking, solving problems, and communicating ideas.

Descriptive geometry

Gaspard Monge (1746–1818) is the "father of descriptive geometry" (Fig. 1.3). While a military student in

FIG. 1.3 Gaspard Monge, the "father of descriptive geometry."

France, young Monge used this graphical method to solve design problems related to fortifications and battlements. He was scolded by his headmaster for not solving problems of this type by the usual (long and tedious) mathematical process. Only after long explanations and comparisons of both methods' solutions was he able to convince the faculty that his graphical methods solved problems in considerably less time. Descriptive geometry was such an improvement upon the mathematical method that it was kept a military secret for fifteen years before it was allowed to be taught as part of the technical curriculum. Monge became a scientific and mathematical aide to Napoleon during his reign as emperor of France.

Descriptive geometry is the projection of three-dimensional figures onto a two-dimensional plane of paper in a manner that allows geometric manipulations to determine lengths, angles, shapes, and other descriptive information about the figures.

1.3
The technological team

So rapidly has the scope of technology broadened that it has become necessary for professional responsibilities to be performed by people with specialized training. Thus technology has become a team effort involving the scientist, engineer, technologist, technician, and craftsman (Fig. 1.4).

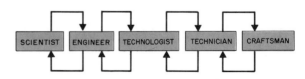

Fig. 1.4 The technological team.

Scientist

The scientist is a researcher who seeks to establish new theories and principles through experimentation and testing (Figs. 1.5 and 1.6). Since they are primarily interested in isolating significant relationships, scientists often show little concern for the application of specific principles. Scientific discoveries are the basis for development of practical applications that may not actually exist until years after discovery.

FIG.1.5 Scientists conduct research to establish fundamental relationships. This chemist is trying to determine the best combination of ingredients to yield a better tire. (Courtesy of Uniroyal, Inc.)

FIG. 1.6 A researcher solders fine wires to each strain gauge in an experimental model. (Courtesy of the Bureau of Reclamation, U.S. Department of the Interior.)

Engineer

Engineers are trained in science, mathematics, and industrial processes to prepare them to apply the basic principles discovered by the scientist (Fig. 1.7). They are concerned with converting raw materials and power sources into needed products and services. Creatively applying these principles to new products or systems is the process of design, the engineer's most

unique function. Generally, the engineer practices the art of using known principles and available resources to achieve a practical end at a reasonable cost.

FIG. 1.7 Engineers at Ford Motor Company use computers to design, build, and test vehicles to provide more efficient and durable cars and trucks. (Courtesy of Ford Motor Co.)

Technologist

The technologist assists the engineer at a semiprofessional level slightly below the engineer's and above the technician's. Whereas the engineer is concerned with the application of theoretical concepts, the technologist is more concerned with the routine, practical aspects of engineering, whether at the planning or production stage (Fig. 1.8). Several technologists

FIG. 1.8 Engineering technologists combine their knowledge of production techniques with the design talents of the engineers to produce a product. (Courtesy of Omark Industries, Inc.)

working with an engineer will greatly increase the engineer's productivity. Likewise, the technologist can coordinate the activities of several technicians, thereby sharing supervisory responsibility with the engineer. The technologist usually has a four-year college degree in engineering technology.

Technician

Technicians assist the engineer and technologist at a level below the technologist. In general, they work as liaisons between the technologist and the craftsman (Fig. 1.9). Their work may vary from conducting routine laboratory experiments to supervising craftsmen in manufacturing or construction. Usually the technician is required to have two years of technical training beyond high school.

FIG. 1.10 This craftsman is a welder skilled in joining metal parts in accordance with prescribed specifications. (Courtesy of Texas Eastern: *TE Today*, photo by Bob Thigpen.)

FIG. 1.9 Technicians are skilled in performing routine jobs requiring technical knowledge and skill. This factory technician is operating computerized controls. (Photo by John J. Krieger, Jr./The Picture Cube.)

Craftsman

Craftsmen are responsible for implementing the engineering design by producing it according to the engineer's specifications. Craftsmen may be machinists who fabricate various components of the product or electricians who assemble electrical components. The craftsman's ability to produce a part according to design specifications is as necessary as the engineer's ability to design the part. Craftsmen include electricians, welders, machinists, fabricators, drafters, and

members of many other occupational groups (Fig. 1.10).

Designer

The designer may be an engineer, an inventor, or a person who has special talents for devising creative solutions. Designers often do not have an engineering background, especially in newer technologies where there may be little precedent for the work involved. Thomas A. Edison (Fig. 1.11) had little formal education, but he had an exceptional ability to design and implement some of the world's most useful inventions.

FIG. 1.11 Thomas A. Edison had essentially no formal education, but he gave the world some of its most creative designs.

FIG. 1.12 The stylist is more concerned with the outward appearance of a product than the functional aspects of its design. (Courtesy of Ford Motor Corp.)

Stylist

Concerned with the outward appearance of a product rather than the development of a functional design (Fig. 1.12), stylists may design an automobile body or the configuration of an electric iron. An automobile stylist considers the functional requirements of the body, driver vision, enclosure of passengers, space for a power unit, and so forth. However, the stylist is not involved with the design of internal details of the product, such as the engine or steerage linkage. The stylist must have a high degree of aesthetic awareness plus a feel for the consumer's acceptance of designs.

1.4
The engineering fields

Significant recent changes in engineering include the emergence of the technologist and the technician and the growing number of women pursuing engineering careers. Indeed, more than 15% of today's freshman engineering class is composed of women.

Aerospace engineering

Aerospace engineering deals with all aspects (all speeds and altitudes) of flight. Aerospace engineering assignments range from complex vehicles traveling 350 million miles to Mars to hovering aircraft used in deep-sea exploration. In the space exploration branch of this profession, aerospace engineers work on all types of aircraft and spacecraft—state-of-the-art missiles and rockets, as well as conventional propeller-driven and jet-powered planes.

Second only to the auto industry in sales, the aerospace industry contributes immeasurably to national defense and the economy. Specialized areas include (1) aerodynamics, (2) structural design, (3) instrumentation, (4) propulsion systems, (5) materials, (6) reliability testing, and (7) production methods. Aerospace engineers may also specialize in a particular product, such as conventional-powered planes, jet-powered military aircraft, rockets, satellites, or manned space capsules.

Aerospace engineers can be divided into two major areas: research engineering and design engineering. The research engineer investigates known principles in search of new ideas and concepts. The design engineer translates these new concepts into workable applications for the improvement of the existing state of the art. This approach has elevated the field of aerospace engineering from the Wright brothers' first flight at Kittyhawk, North Carolina, in 1903 to the penetration of outer space with an unlimited future (Fig. 1.13).

FIG. 1.13 The space shuttle *Columbia*, first launched in April 1981, introduced a new era for aerospace engineering—the era of the reusable spaceship. (Courtesy of NASA.)

The professional society for aerospace engineers is the American Institute of Aeronautics and Astronautics (AIAA).

Agricultural engineering

Agricultural engineers are trained to serve the world's largest industry—agriculture. Agricultural engineering problems deal with the production, processing, and handling of food and fiber.

MECHANICAL POWER The agricultural engineer who works with manufacturers of farm equipment is concerned with gasoline and diesel engine equipment such as pumps, irrigation machinery, and tractors. Machinery must be designed for the electrical curing of hay, milk processing and pasturizing, fruit processing, and heating environments in which to raise livestock and poultry. Farm machinery designed by agricultural engineers has been largely responsible for the increased productivity in agriculture.

FARM STRUCTURES The construction of barns, shelters, silos, granaries, processing centers, and other agricultural buildings requires specialists in agricultural engineering. Engineers must understand heating, ventilation, and chemical changes that might affect the storage of crops.

ELECTRICAL POWER Agricultural engineers design electrical systems and select equipment that will provide efficient operation to meet the requirements of the specific situation. They may serve as consultants or designers for manufacturers or processors of agricultural products.

SOIL AND WATER CONTROL The agricultural engineer is responsible for devising systems for improving drainage and irrigation systems, resurfacing fields, and constructing water reservoirs (Fig. 1.14). These activities may be performed in association with the U.S. Department of Agriculture, the Department of Interior, state agricultural universities, consulting engineering firms, or irrigation companies.

Most agricultural engineers are employed in private industry, especially by manufacturers of heavy farm equipment and specialized lines of field, barnyard, and household equipment, electrical service companies, and distributors of farm equipment and

FIG. 1.14 Agricultural engineers improve production by designing irrigation systems. California's Coachella Valley was a desert (as shown at the right) that was converted into 52,000 fertile acres of farmland by diverting water from the Colorado River. (Courtesy of the Bureau of Reclamation, photo by E. E. Hertzog.)

supplies. Although few agricultural engineers live on a farm, it is helpful if they have a clear understanding of agricultural problems, farming, crops, animals, and farmers themselves. Thus, agricultural engineering has helped increase the farmer's efficiency: today one farmer produces enough food for eighty people, whereas 100 years ago one farmer was able to feed only four other people.

The professional society for agricultural engineers is the American Society for Agricultural Engineers.

Chemical engineering

Chemical engineering involves the design and selection of equipment that facilitates the processing and manufacturing of large quantities of chemicals. Chemical engineers design unit operations, including fluid transportation through ducts or pipelines, solid material transportation through pipes or conveyors, heat transfer from one fluid or substance to another fluid or substance through plate or tube walls, absorption of gases by bubbling them through liquids, evaporation of liquids to increase concentration of solutions, distillation under carefully controlled temperatures to separate mixed liquids, and many other similar chemical processes. They may employ chemical reactions of raw products such as oxidation, hydrogenation, reduction, chlorination, nitration, sulfonation, pyrolysis, and polymerization (Fig. 1.15).

Process control and instrumentation are important specialities in chemical engineering. With the measurement of quality and quantity by instrumentation, process control is a fully automated operation.

Chemical engineers develop and process chemicals such as acids, alkalies, salts, coal-tar products, dyes, synthetic chemicals, plastics, insecticides, fungicides, and many others for industrial and domestic uses. They help develop drugs and medicine, cosmetics, explosives, ceramics, cements, paints, petroleum products, lubricants, synthetic fibers, rubber, and detergents. They also design equipment for food preparation and canning plants.

Approximately 80% of chemical engineers work in the manufacturing industries—primarily the chemical industry. The remainder work for government agencies, independent research institutes, and as independent consulting engineers. New fields requiring chemical engineers are nuclear sciences, rocket fuels, and environmental pollution.

The professional society for chemical engineers is the American Institute of Chemical Engineers (AICE).

Civil engineering

Civil engineering, the oldest branch of engineering, is closely related to practically all of our daily activities. The buildings we live and work in, the transportation we use, the water we drink, and the drainage and sewage systems we rely on are all the results of civil engineering. Civil engineers design and supervise the construction of roads, harbors, airfields, tunnels, bridges, water supply and sewage systems, and many other types of structures.

CONSTRUCTION ENGINEERS manage the resources, workers, finances, and materials necessary for

FIG. 1.15 The efforts of scientists, chemical engineers, and others are realized in the construction and operation of a refinery that produces vital products for our economy. (Courtesy of Houston Oil and Refining Co.)

construction projects, which may vary from the erection of skyscrapers to the movement of concrete and earth.

CITY PLANNERS develop plans for the future growth of cities and the systems related to their operation. Street planning, zoning, and industrial site development are problems that city planners solve.

STRUCTURAL ENGINEERS design and supervise the erection of buildings, dams, powerhouses, stadiums, bridges, and other structures.

HYDRAULIC ENGINEERS work with the behavior of water from its conservation to its transportation. They design wells, canals, dams, pipelines, drainage systems, and other methods of controlling and using water and petroleum products (Fig. 1.16).

TRANSPORTATION ENGINEERS develop and improve railroads and airlines in all phases of their operations. Railroads are built, modified, and maintained under the supervision of civil engineers. Design and construction of airport runways, control towers, passenger and freight stations, and aircraft hangars are done by civil engineers.

HIGHWAY ENGINEERS develop the complex network of highways and interchanges for moving automobile traffic. These systems require the design of tunnels, culverts, and traffic control systems.

FIG. 1.16 This photograph shows several civil engineering projects, including highways, waterways, bridges, and structures. (Courtesy of Los Angeles Dept. of Highways.)

SANITARY ENGINEERS assist in maintaining public health through the purification of water, control of water pollution, and sewage control. Such systems involve the design of pipelines, treatment plants, dams, and related systems.

Many civil engineers find positions in administration and municipal management. Most civil engineers are associated with federal, state, and local government agencies and the construction industry. Many work as consulting engineers for architectural firms and independent consulting engineering firms. The remainder work for public utilities, railroads, educational institutions, steel industries, and other manufacturing industries.

The professional society for civil engineers is the American Society of Civil Engineers (ASCE). Founded in 1852, it is the oldest engineering society in the United States.

Electrical engineering

Electrical engineers are concerned with the use and distribution of electrical energy. The two main divisions of electrical engineering are (1) power, which deals with the control of large amounts of energy used by cities and large industries, and (2) electronics, which deals with the small amounts of power used for communications and automated operations that have become part of our everyday lives. These two divisions of electrical engineering have many areas of specialization, a few of which are given below.

POWER GENERATION poses many electrical engineering problems from the development of transmission equipment to the design of generators for producing electricity (Fig. 1.17).

POWER APPLICATIONS are numerous in homes, where toasters, washers, dryers, vacuum cleaners, and lights are used. Of total energy consumption, only about one quarter is used in the home. Industry uses about half of all energy for metal refining, heating, motor drives, welding, machinery controls, chemical processes, plating, electrolysis, and so forth.

ILLUMINATION is required in nearly every area of modern life. Improving the efficiency and economy of illumination systems is a challenging area of study for the electrical engineer.

COMPUTERS are a gigantic industry that is the domain of the electrical engineer. Used with industrial electronics, computers have changed industry's man-

FIG. 1.17 Electrical engineers designed and supervised this power plant to convert waterpower into electricity. (Courtesy of the Bureau of Reclamation, photo by E. E. Hertzog.)

ufacturing and production processes, resulting in greater precision and fewer employees.

COMMUNICATIONS is devoted to the improvement of radio, telephone, telegraph, and television systems, which are the nerve centers of most industrial operations.

INSTRUMENTATION is the study of systems of electronic instruments used in industrial processes. Extensive use has been made of the cathode-ray tube and the electronic amplifier in industry and atomic power reactors. Increasingly, instrumentation is applied to medicine for diagnosis and therapy.

MILITARY ELECTRONICS is used in practically all areas of military weapons and tactical systems from the walkie-talkie to the distant radar networks for detecting enemy aircraft. Remote-controlled electronic systems are used for navigation and interception of guided missiles.

More electrical engineers are employed than any other type of engineer. The increasing need for electrical equipment, automation, and computerized systems is expected to contribute to the growth of this field.

> The professional society for electrical engineers is the Institute of Electrical and Electronic Engineers (IEEE). Founded in 1884, the IEEE is the world's largest technical society.

Industrial engineering

Industrial engineering, one of the newer engineering professions, is defined by the National Professional Society of Industrial Engineers as follows:

> Industrial engineering is concerned with the design, improvement, and installation of integrated systems of men, materials, and equipment. It draws upon specialized knowledge and skill in the mathematical, physical, and social sciences together with the principles and methods of engineering analysis and design to specify, predict, and evaluate the results to be obtained from such systems.

Industrial engineering differs from other branches of engineering in that it is more closely related to people

FIG. 1.18 Industrial engineers lay out and design complex industrial facilities to provide efficient production. This plant produces and assembles automobiles. (Courtesy of Chrysler Corp.)

and their performance and working conditions (Fig. 1.18). Often the industrial engineer is a manager of people, machines, materials, methods, money, and the markets involved.

The industrial engineer may be responsible for plant layout, the development of plant processes, or the determination of operating standards that will improve the efficiency of a plant operation. They also design and supervise systems for improved safety and production.

Specific areas of industrial engineering include management, plant design and engineering, electronic data processing, systems analysis and design, control of production and quality, performance standards and measurements, and research. Industrial engineers are also increasingly involved in the implementation of automated production systems.

People-oriented areas include the development of wage incentive systems, job evaluation, work measurement, and the design of environmental systems. Industrial engineers are often involved in management-labor agreements that affect the operation and production of an industry.

More than two thirds of industrial engineers are employed in manufacturing industries. Others work for insurance companies, construction and mining firms, public utilities, large businesses, and governmental agencies.

> The professional society for industrial engineers is the American Institute of Industrial Engineers (AIIE), which was organized in 1948.

Mechanical engineering

The mechanical engineer's major areas of specialization are power generation, transportation, aeronautics, marine vessels, manufacturing, power services, and atomic energy. Each of these areas is outlined below.

POWER GENERATION requires that prime movers (machines that convert natural energy into work) be developed to power electrical generators that will produce electrical energy. Mechanical engineers design and supervise the operation of steam engines, turbines, internal combustion engines, and other prime movers (Fig. 1.19).

TRANSPORTATION including trucks, buses, automobiles, locomotives, marine vessels, and aircraft are designed and manufactured by mechanical engineers.

AERONAUTICS requires mechanical engineers to develop aircraft engines, as well as aircraft controls and environmental systems.

MARINE VESSELS powered by steam, diesel, or gas-generated engines are designed by mechanical engineers, as are power services throughout the vessel such as light, water, refrigeration, and ventilation.

FIG. 1.19 This hydroelectric power plant at Hoover Dam was completed by mechanical engineers. (Courtesy of the Bureau of Reclamation.)

MANUFACTURING requires mechanical engineers to design new products and factories in which to build them. Economy of manufacturing and achieving uniform product quality are major functions of manufacturing engineers. The professional society for manufacturing engineers is the Society of Manufacturing Engineers (SME).

POWER SERVICES include the movement of liquids and gases through pipelines, refrigeration systems, elevators, and escalators. These mechanical engineers must have a knowledge of pumps, ventilation equipment, fans, and compressors.

ATOMIC ENERGY needs mechanical engineers for the development and handling of protective equipment and materials. The mechanical engineer plays an important role in the construction of nuclear reactors.

> The professional society for mechanical engineers is the American Society of Mechanical Engineers (ASME).

Mining and metallurgical engineering

Mining engineers are responsible for extracting minerals from the earth and for preparing them for use by manufacturing industries. Working with geologists to locate ore deposits, which are exploited through the construction of tunnels and underground operations, mining engineers must have an understanding of safety, ventilation, water supply, and communications. Two main areas of metallurgical engineering are (1) extractive metallurgy, the extraction of metal from raw ores to form pure metals, and (2) physical metallurgy, the development of new products and alloys.

Many metallurgical engineers work on the development of machinery for electrical equipment and in the aircraft and aircraft parts industries. The development of new lightweight, high-strength materials for space flight vehicles, jet aircraft, missiles, and satellites will increase the need for metallurgical engineers (Fig. 1.20). Mining engineers who work at mining sites are usually employed near small, out-of-the-way communities, whereas those in research and consulting often work in metropolitan areas.

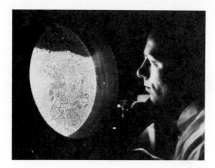

FIG. 1.20 A metallograph shows the structure of an alloy that may be used in the construction of a refinery unit. Materials are specially developed for specific applications by metallurgical engineers. (Courtesy of Exxon Corp.)

The professional society for mining and metallurgical engineers is the American Institute of Mining, Metallurgical, and Petroleum Engineering (AIME).

Nuclear engineering

The earliest work in the nuclear field has been for military and defense purposes. However, nuclear power for domestic needs is being developed for the medical profession and other areas.

Peaceful applications of nuclear engineering are divided into two major areas—radiation and nuclear power reactors. Radiation is the propagation of energy through matter or space in the form of waves. In atomic physics, radiation includes fast-moving particles (alpha and beta rays, free neutrons, etc.), gamma rays, and x rays. Nuclear science is closely allied with botany, chemistry, medicine, and biology.

The production of nuclear power in the form of mechanical or electrical power is a major peaceful use of nuclear energy (Fig. 1.21). For the production of electrical power, nuclear energy is used as the fuel for producing steam that will drive a turbine generator in the conventional manner.

Although nuclear engineering degrees are offered at the bachelor's level, advanced degrees are recommended for this area of engineering. Most training of the nuclear engineer centers on the design, construction, and operation of nuclear reactors. Other areas of study include the processing of nuclear fuels, thermonuclear engineering, and the use of various nuclear by-products.

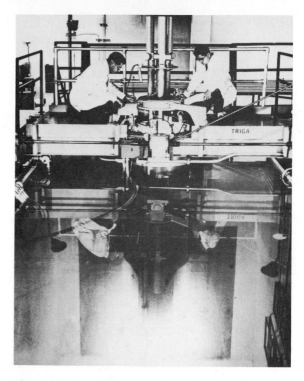

FIG. 1.21 This pulsing reactor, operated by a nuclear engineer, tests the effect of extremely high radiation on delicate equipment. (Courtesy of General Dynamics Corp.)

The professional society for nuclear engineers is the American Nuclear Society.

Petroleum engineering

The recovery of petroleum and gases is the primary concern of petroleum engineers; however, they must also develop methods for the transportation and separation of various products. Moreover, they are responsible for the improvement of drilling equipment and its economy of operation (Fig. 1.22). In the exploration for petroleum, the petroleum engineer is assisted by the geologist and by instruments like the airborne magnetometer, which indicates uplifts on the earth's subsurfaces that could hold oil or gas.

Oil well drilling is supervised by the petroleum engineer, who also develops the equipment to most

FIG. 1.22 The petroleum engineer supervises the operation of offshore platforms used in the exploration for oil. (Courtesy of Marathon Oil Co.)

efficiently remove the oil. When oil is found, the petroleum engineer must design piping systems to remove and transport the oil to its next point of processing. Processing itself is a joint project with chemical engineers.

The Society of Petroleum Engineers is a branch of AIME, which includes mining and metallurgical engineers and geologists.

1.5
Technologists and technicians

As technology has become more complex two new members of the technological team have emerged: the technologist and the technician. The primary mission of both is to assist the engineer at a technical level below that of the engineer and about that of the craftsman.

THE TECHNOLOGIST works under the supervision of an engineer but has considerable responsibility,

thereby freeing the engineer for more advanced applications. Most technologists have a four-year college background in a specialty area of engineering technology that enables them to perform semiprofessional jobs with a high degree of skill. Their interest in the practical aspects of engineering qualifies them to offer advice to the engineer about production specifications and on-the-site procedures when new projects are being designed.

THE TECHNICIAN performs tasks less technical than those performed by the technologist. Most technicians have graduated from a two-year technical pro-

FIG. 1.23 These technicians are making a cartographic survey. Data from the tellurometer are being recorded in a notebook. (Courtesy of the U.S. Forest Service.)

FIG. 1.24 Technicians check parts within thousandths of an inch to ensure that the part conforms to the specifications. (Photo by John J. Krieger, Jr./The Picture Cube.)

gram that enables them to be repairers, inspectors, production specialists, or surveyors (Fig. 1.23). They may work under the supervision of an engineer, but ideally they are supervised by a technologist. In turn, the technician coordinates the activities between the technologist and the craftsman (Fig. 1.24). These levels of responsibility ensure that the proper skills and qualifications are available throughout the chain of command.

1.6
Drafters

Today drafters carry a great responsibility in assisting the engineer and designer. The experienced drafter may be involved in preparing complex drawings, selecting materials, detailing designs, and writing specifications.

DESIGN AND CONSTRUCTION DRAWINGS are made by drafters to explain how to fabricate, build, or erect a project or product in fields such as aerospace engineering, architecture, machine design, mechanical engineering, and electrical/electronic engineering.

TECHNICAL ILLUSTRATION is a type of graphics that is usually prepared as three-dimensional pictorials to illustrate a project or product as realistically as possible. Technical illustration is the most artistic area of engineering design graphics.

MAPS, GEOLOGICAL SECTIONS, AND HIGHWAY PLATS are used for locating property lines, physical features, strata, right-of-way, building sites, bridges, dams, mines, utility lines, and so forth. Drawings of this type are usually prepared as permanent ink drawings.

There are three levels of certification for drafters: drafters, design drafters, and engineering designers.

FIG. 1.25 These drafters are working at a computer using AutoCAD sofware to generate technical drawings. (Courtesy of Autodesk, Inc.)

DRAFTERS are graduates of a two-year post–high-school curriculum in engineering design graphics.

DESIGN DRAFTERS complete two-year programs in an approved junior college or technical institute.

ENGINEERING DESIGNERS are drafters who have completed a four-year college course in engineering design graphics. Graduates of these programs can become certified as technologists.

Computerized systems are being adopted by industry to improve the drafter's productivity. Computer graphics systems do not lessen the need for drafters or engineers to know fundamental graphical principles; rather they offer a different medium of expression (Fig. 1.25). Chapter 10 discusses computer graphics systems and their uses in engineering graphics, and further demonstration of the use of the computer in engineering graphics problems can be found throughout the remainder of the text.

Problems

1. Write a report that outlines the specific duties and relationships between the scientist, engineer, craftsman, designer, and stylist in an engineering field of your choice. For example, explain this relationship for an engineering team involved in an aspect of civil engineering. Your report should be

supported by factual information obtained from interviews, brochures, or library references.

2. Write a report that investigates the employment opportunities, job requirements, professional challenges, and activities of your chosen branch of engineering or technology. Illustrate this report with charts and graphs where possible for easy interpretation. Compare your personal abilities and interests with those required by the profession.

3. Arrange a personal interview with a practicing engineer, technologist, or technician in the field of your interest. Discuss with him or her the general duties and responsibilities of the position to gain a better understanding of this field. Summarize your interview in a written report.

4. Write to the professional society of your field of study for information concerning this area. Prepare a notebook of these materials for easy reference. Include in the notebook a list of books that would provide career information for the engineering student.

Addresses of professional societies

Publications and information from these societies were used as the basis for the content of this chapter.

American Ceramic Society
65 Ceramic Drive, Columbus, Ohio 43214

The American Institute of Aeronautics and Astronautics
1290 Avenue of the Americas, New York, N.Y. 10019

American Institute of Chemical Engineers
345 East 47th Street, New York, N.Y. 10017

American Institute for Design and Drafting
3119 Price Road, Bartlesville, Okla. 74003

The American Institute of Industrial Engineers
345 East 47th Street, New York, N.Y. 10017

American Institute of Mining, Metallurgical, and Petroleum Engineering
345 East 47th Street, New York, N.Y. 10017

American Nuclear Society
244A East Ogden Avenue, Hinsdale, Ill. 60521

American Society of Agricultural Engineers
2950 Niles Road, St. Joseph, Mich. 49085

American Society of Civil Engineers
345 East 47th Street, New York, N.Y. 10017

American Society for Engineering Education, Technical Institute Division
11 DuPont Circle, Suite 200, Washington, D.C. 20036

American Society of Mechanical Engineers
345 East 47th Street, New York, N.Y. 10017

The Institute of Electrical and Electronic Engineers
345 East 47th Street, New York, N.Y. 10017

National Society of Professional Engineers
2029 K Street, N.W., Washington, D.C. 20006

Society of Petroleum Engineers (AIME)
6300 North Central Expressway, Dallas, Tex. 75206

Society of Women Engineers
United Engineering Center, Room 305, 345 East 47th Street, New York, N.Y. 10017

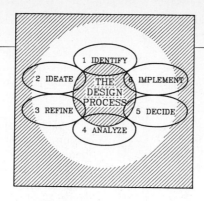

CHAPTER 2

The Design Process

2.1
Introduction

Engineering graphics and descriptive geometry provide methods of solving technical problems. An engineer working on a design solution must make many sketches and drawings to develop preliminary ideas before effective communication with associates is possible.

2.2
Types of design problems

Most design problems fall into one of two categories: *systems design* and *product design*.

Systems design

Systems design deals with the arrangement of available products and components into a unique combination that yields the desired result. A residential building is a complex system made up of components and products: a heating-cooling system, a utility system, a plumbing system, a gas system, an electrical system, and other systems that form the overall system (Fig. 2.1). The electrical system involves wiring, insulation, light bulbs, meters, controls, switches, and other electrical components (Fig. 2.2).

The engineer who develops a traffic system will need assistance from other professionals, for the project will also involve legal problems, economics, historical data, human factors, social considerations, scientific principles, and politics (Fig. 2.3). By applying engineering principles, the engineer can design a driving

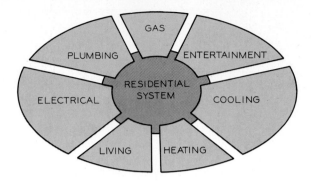

FIG 2.1 The typical residence is a system composed of many component systems.

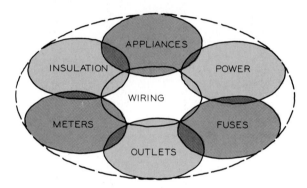

FIG. 2.2 The electrical system of a residence is a composite of related components.

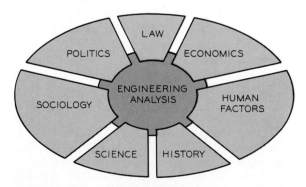

FIG. 2.3 An engineering project may involve the interaction of many professions, with engineering analysis the central function.

surface, drainage system, overpasses, and other components of the traffic system; however, adhering to budgetary constraints is equally necessary, and the budget is closely related to legal and political requirements.

Traffic laws, zoning ordinances, right-of-way possession, and liability clearances are other legal areas that must be considered. Past trends and historical data also should be reviewed, as should human factors including driver characteristics and safety features that would affect the function of the traffic system. Social problems may arise if heavily traveled highways attract commercial establishments, shopping centers, and filling stations. Finally, scientific principles developed through research can be applied in building more durable roads, cheaper bridges, and a more functional system.

SYSTEMS DESIGN PROBLEM The following problem illustrates the various steps of the systems design process.

PARKING AREA Select a building on your campus that needs an improved parking lot. This may be a dormitory, office building, or classroom building. Design a combination traffic and parking system for the building. The solution of this problem must adhere to regulations and policies of your campus.

Product design

Product design is concerned with the design, testing, manufacture, and sale of an item that will be mass-produced such as an appliance, a tool, or a toy (Fig. 2.4). The primary function of an automotive system is to provide transportation. However, the automobile

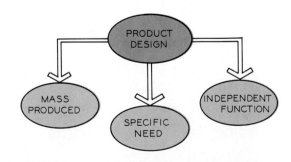

FIG. 2.4 Product design is more limited in scope and specific in application than systems design.

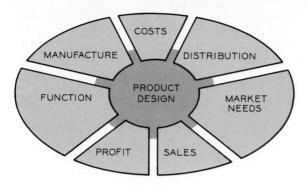

FIG. 2.5 Areas associated with product design are related to the manufacture and sale of completed products.

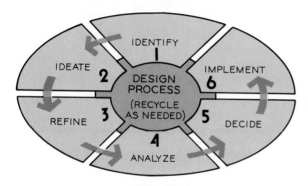

FIG. 2.6 The steps of the design process. Each step can be recycled when needed.

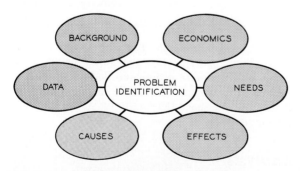

FIG. 2.7 Problem identification requires accumulating as much information about the problem as possible before attempting a solution.

must also provide communication, illumination, comfort, and safety, which seemingly would classify it as a system. Nevertheless, since it is mass produced for a large consumer market the automobile is regarded as a product. A petroleum refinery, on the other hand, is definitely a system. Although all refineries have certain processes in common, no two are entirely alike.

Product design is related to current market needs, production costs, function, sales, distribution methods, and profit predictions (Fig. 2.5). This concept may encompass a complex system, for example the automobile has expanded to a system of highways, service stations, repair shops, parking lots, drive-in businesses, residential garages, traffic enforcement, and endless other related components.

PRODUCT DESIGN PROBLEM The following problem is an example of a product design problem.

HUNTING SEAT Many hunters, especially deer hunters, hunt from trees to obtain a vantage point. Sitting in a tree for several hours can be uncomfortable and hazardous to the hunter. Design a hunting seat that provides hunters with comfort and safety and that meets the requirements of economy and the limitations of hunting.

2.3
The design process

Design, the responsibility that most distinguishes the engineer from the scientist and technician is the act of devising an original solution to a problem by a combination of principles, resources, and products.

This book emphasizes a six-step design process: (1) problem identification, (2) preliminary ideas, (3) problem refinement, (4) analysis, (5) decision, and (6) implementation (Fig. 2.6). Designers work sequentially from step to step, but they may recycle to previous steps as they progress.

Problem identification

Most engineering problems are not clearly defined at the outset, so they must be identified before an attempt is made to solve them (Fig. 2.7). A prominent concern in our society is air pollution, but before this problem can be solved, we must identify what air pollution is and what causes it. Is pollution caused by

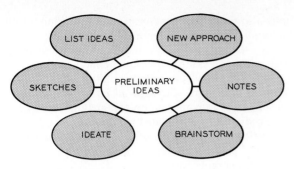

FIG. 2.8 Preliminary ideas are developed after the identification process has been completed. All ideas should be listed and sketched to give the designer a broad selection to work from.

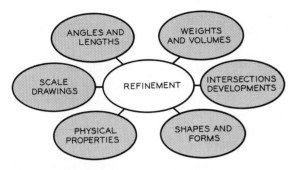

FIG. 2.9 Refinement begins with the construction of scale drawings of the better preliminary ideas. Descriptive geometry and graphical methods are used to find necessary geometric characteristics.

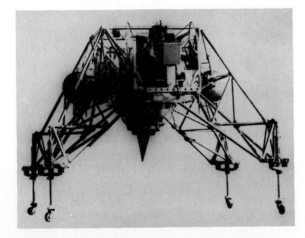

FIG. 2.10 The refinement of the lunar vehicle required using descriptive geometry and other graphical methods. (Courtesy of Ryan Aeronautical Co.)

automobiles, factories, atmospheric conditions that harbor impurities, or geographic features that contain impure atmospheres? When you enter a street intersection where traffic is unusually congested, do you identify the reasons for the congestion? Are there too many cars? Are the signals poorly synchronized? Are there visual obstructions?

Problem identification requires a good deal more study than just a simple statement like "solve air pollution." You will need to gather data of several types: field data, opinion surveys, historical records, personal observations, experimental data, and physical measurements and characteristics (Fig. 2.7).

Preliminary ideas

The second step is to accumulate as many ideas for solving the problem as possible (Fig. 2.8). Preliminary ideas should be broad enough to allow for unique solutions that could revolutionize present methods. Many rough sketches of preliminary ideas should be made and retained to generate original ideas and stimulate the design process. Ideas and comments should be noted on the sketches.

Problem refinement

Next, several of the better preliminary ideas are selected for refinement to determine their true merits. Rough sketches are converted to scale drawings that will permit space analysis, critical measurements, and the calculation of areas and volumes affecting the design (Fig. 2.9). Consideration is given to spatial relationships, angles between planes, lengths of structural members, intersections of surfaces and planes.

Designing the landing gear of the lunar vehicle (Fig. 2.10) required descriptive geometry to determine this type of information. The configuration of the landing gear was drawn to scale in the descriptive views of the landing craft. The length of each leg of the landing apparatus and the angles between the members at the point of junction had to be found to design a connector, and the angles the legs made with the body of the spacecraft had to be known to design these joints.

Analysis

The step of the design process where engineering and scientific principles are most often used is analysis (Fig. 2.11)—the evaluation of the best designs to de-

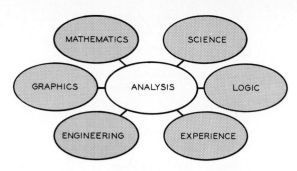

FIG. 2.11 In the anlysis phase of the design process all available technological methods, from science to graphics, are used to evaluate the refined designs.

termine the comparative merits of each with respect to cost, strength, function, and market appeal. Graphical methods of analysis are means of checking a solution. Data that are difficult to mathematically interpret can be graphically analyzed. Models constructed at reduced scales are also a valuable analytical tool. They help to establish relationships of moving parts and outward appearances and to evaluate other design characteristics.

Decision

At this stage a decision must be made. A single design must be selected as the solution of the design problem (Fig. 2.12). Often, the final design is a compromise that offers many of the best features of several designs. The decision may be made by the designer alone, or it may be made by several associates. Regardless of who decides, graphics is a primary means of presenting the

proposed designs for a decision. The outstanding aspects of each design usually lend themselves to graphical presentations that compare manufacturing costs, weights, operational characteristics, and other data that would be considered before arriving at the final decision.

Implementation

The final design concept must be presented in a workable form. Working drawings and specifications are usually used as the instruments for fabrication of a product, whether it is a small piece of hardware or a huge bridge (Fig. 2.13). Workers must have detailed instructions for the manufacture of each part, measured to a thousandth of an inch to ensure its proper manufacture. Working drawings must be sufficiently explicit to provide a legal contractual basis for the contractor's bid on the job.

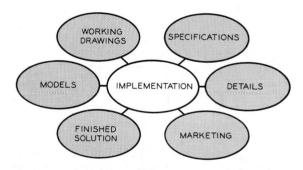

FIG. 2.13 During implementation, the final step of the design process, drawings and specifications are prepared from which the final product can be constructed.

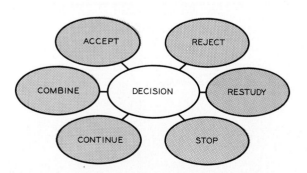

FIG. 2.12 Decision is the selection of the best design or design features to be implemented.

2.4
Application of the design process to a simple problem

To illustrate the steps of the design process as they would be applied to a simple design problem, the following example is given.

Swing-set anchor problem

A child's swing set is unstable during the peak of the swing. The momentum of the swing causes the

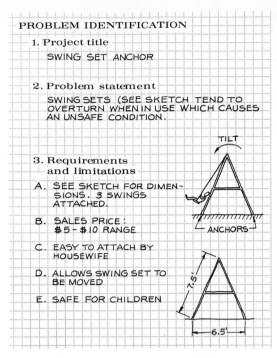

PROBLEM IDENTIFICATION

1. Project title

   SWING SET ANCHOR

2. Problem statement

   SWING SETS (SEE SKETCH TEND TO OVERTURN WHEN IN USE WHICH CAUSES AN UNSAFE CONDITION.

3. Requirements and limitations

   A. SEE SKETCH FOR DIMEN-SIONS. 3 SWINGS ATTACHED.

   B. SALES PRICE: $5 - $10 RANGE

   C. EASY TO ATTACH BY HOUSEWIFE

   D. ALLOWS SWING SET TO BE MOVED

   E. SAFE FOR CHILDREN

FIG. 2.14 A worksheet showing a portion of problem identification.

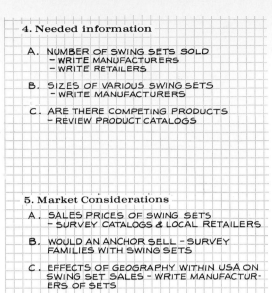

4. Needed information

   A. NUMBER OF SWING SETS SOLD
      - WRITE MANUFACTURERS
      - WRITE RETAILERS

   B. SIZES OF VARIOUS SWING SETS
      - WRITE MANUFACTURERS

   C. ARE THERE COMPETING PRODUCTS
      - REVIEW PRODUCT CATALOGS

5. Market Considerations

   A. SALES PRICES OF SWING SETS
      - SURVEY CATALOGS & LOCAL RETAILERS

   B. WOULD AN ANCHOR SELL - SURVEY FAMILIES WITH SWING SETS

   C. EFFECTS OF GEOGRAPHY WITHIN USA ON SWING SET SALES - WRITE MANUFACTUR-ERS OF SETS

   D. WHERE ARE SWING SETS SOLD? CHECK YELLOW PAGES VISIT STORES

FIG. 2.15 The information needed to identify the problem is listed. This information needs to be gathered to complete this part of the design process.

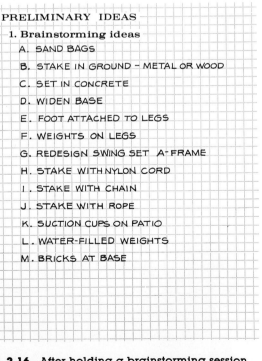

PRELIMINARY IDEAS

1. Brainstorming ideas

   A. SAND BAGS

   B. STAKE IN GROUND - METAL OR WOOD

   C. SET IN CONCRETE

   D. WIDEN BASE

   E. FOOT ATTACHED TO LEGS

   F. WEIGHTS ON LEGS

   G. REDESIGN SWING SET A-FRAME

   H. STAKE WITH NYLON CORD

   I. STAKE WITH CHAIN

   J. STAKE WITH ROPE

   K. SUCTION CUPS ON PATIO

   L. WATER-FILLED WEIGHTS

   M. BRICKS AT BASE

FIG. 2.16 After holding a brainstorming session, all ideas are listed.

A-frame to tilt with a possibility of overturning and causing injury. The swing set can accommodate three children at a time. Design a device that eliminates this hazard and has market appeal for owners of swing sets of this type.

PROBLEM IDENTIFICATION   First, the designer writes down the problem statement (Fig. 2.14) and a statement of need. The limitations and desirable features are listed, along with necessary sketches, to enable the designer to better understand the problem requirements (Figs. 2.14 and 2.15). Much of the information in the problem identification step may be obvious to the designer, but making sketches and writing statements about the problem help the designer get off "dead center," a common difficulty at the beginning of the creative process.

PRELIMINARY IDEAS   A worksheet is used to list brainstorming ideas (Fig. 2.16). The better ideas and design features are then summarized on the worksheet (Fig. 2.17). These verbal ideas are then translated into rapidly drawn freehand sketches (Fig. 2.18). The de-

2. Description of best ideas

   A. STAKE IDEAS SEEM TO BE

      BEST SUITED FOR DEVELOPMENT

   B. EXTENSION FEET OK

3. Attach sketches

FIG. 2.17 The better ideas are selected from the brainstorming list to be developed as preliminary ideas.

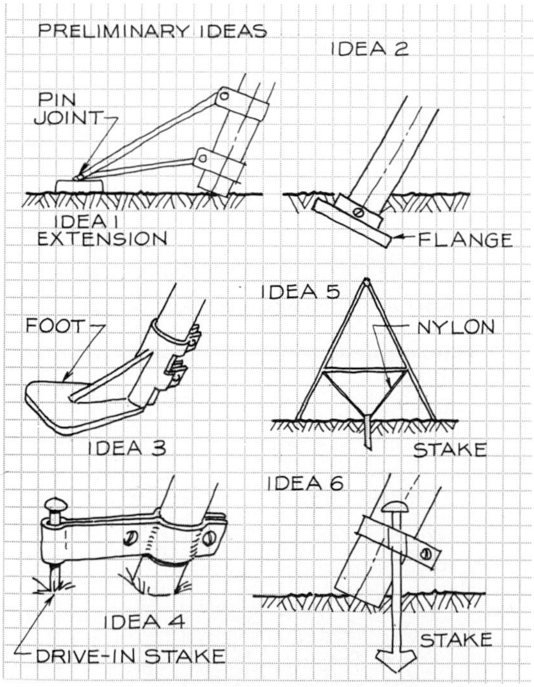

FIG. 2.18 Preliminary ideas are sketched. This is the most creative step of the design process.

REFINEMENT

1. Description of design

  A. STAKE

    1. METAL OR WOOD STAKE – 6 IN – 10 IN

    2. ATTACHED WITH CHAIN, CORD, OR

      CABLE – ABOUT 10 IN LONG

    3. ATTACH TO A-FRAME WITH COLLAR

    4. NEED 4

  B. EXTENSION FOOT

    1. ATTACH TO A-FRAME LEGS

    2. MUST DETERMINE LENGTH OF FOOT

      NECESSARY TO PREVENT TILTING

    3. MUST BE EASY TO ATTACH TO A-FRAME

    4. NEED 4 FEET

    5. NO HEAVIER THAN 1 POUND EACH

2. Attach scale drawings

FIG. 2.19 Refining preliminary ideas begins by giving verbal descriptions of the selected designs.

signer should attempt to develop as many ideas as possible.

PROBLEM REFINEMENT A verbal description of the design features of one or more of the preliminary ideas is listed on a worksheet (Fig. 2.19) for comparison. The better designs are drawn to scale in preparation for their analysis (Fig. 2.20). Only a few dimensions need to be given at this stage.

Orthographic projection, working drawings, and descriptive geometry may be used. In this example (Fig. 2.20), orthographic views with auxiliary views depict the two designs.

ANALYSIS An analysis worksheet is used to analyze the tubular stake design (Figs. 2.21, 2.22, and 2.23). If several solutions are being considered each design should be analyzed.

The force F at the critical angle can be measured or estimated (Fig. 2.24) and used as the basis of a vector polygon to determine the magnitude of R at the base of the swing that must be resisted by the anchor. Again, graphics is used as a design tool.

21

REFINEMENT

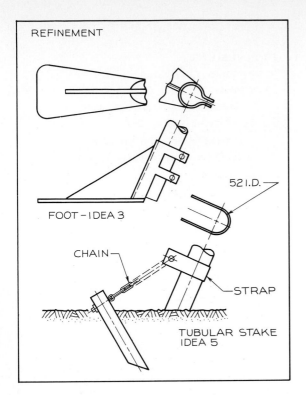

FOOT – IDEA 3

52 I.D.

CHAIN

STRAP

TUBULAR STAKE
IDEA 5

FIG. 2.20 Scaled refinement drawings are used to develop two or more of the designs. Only a few dimensions are needed.

ANALYSIS

1. Function
   A. ANCHORS SWING SET
   B. ATTACHES TO GROUND – NOT CONC. SLAB
   C. PREVENTS OVERTUNING

2. Human engineering
   A. EASY TO INSTALL
   B. REQUIRES NO SPECIAL TOOLS
   C. PROVIDES SAFETY, FOR CHILDREN

3. Market & consumer acceptance
   A. KEEP PRICE UNDER $15 FOR SWING SET
   B. SHOULD BE SOLD WITH SWING SET
   C. CAN BE SOLD AS AN ACCESSORY TO THE SWING SET.
   D. NEED ESTIMATE OF SWING SETS SOLD PER YEAR.

FIG. 2.21 An analysis worksheet.

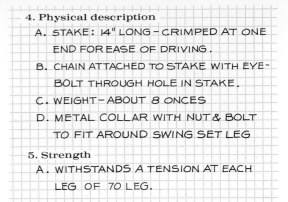

4. Physical description
   A. STAKE: 14" LONG – CRIMPED AT ONE END FOR EASE OF DRIVING.
   B. CHAIN ATTACHED TO STAKE WITH EYE-BOLT THROUGH HOLE IN STAKE.
   C. WEIGHT – ABOUT 8 ONCES
   D. METAL COLLAR WITH NUT & BOLT TO FIT AROUND SWING SET LEG

5. Strength
   A. WITHSTANDS A TENSION AT EACH LEG OF 70 LEG.

FIG. 2.22 Continuing the analysis step of the design process.

6. Production procedures
   A. USE 15 GAGE GALVANIZED IRON PIPE (∅ 40 OD) FOR STAKE. CUT ON DIAGONAL TO CRIMP & POINT STAKE
   B. 15 GAGE METAL COLLAR TO BE FORMED INTO U-SHAPE & DRILLED FOR ∅ 5 BOLT
   C. STAKE DRILLED ∅8 FOR ∅5 BOLT FOR ATTACHING CHAIN

7. Economic analysis
   A. COSTS
      1. CHAIN $0.30
      2. STAKE   .20
      3. COLLAR   .20
      4. EYE BOLT & NUT .10
      5. COLLAR BOLT   .10
                                .90
   B. LABOR   .60
   C. PACKAGING   .30
   D. PROFIT   1.15
   E. WHOLESALE PRICE   2.95
   F. RETAIL PRICE $4.95

FIG. 2.23 Continuing the analysis step of the design process.

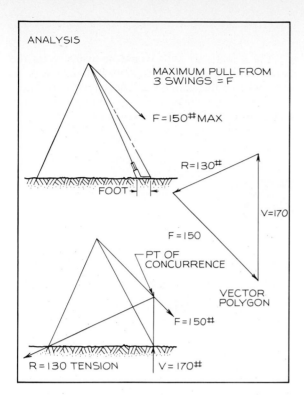

ANALYSIS

MAXIMUM PULL FROM
3 SWINGS = F

F=150#MAX

R=130#

FOOT

V=170

F=150

PT OF
CONCURRENCE

VECTOR
POLYGON

F=150#

R=130 TENSION   V=170#

FIG. 2.24 Descriptive geometry and graphics can be used to analyze a solution. Here, the force (F) is calculated graphically.

DECISION Two designs, the stake and chain and the extension foot, are compared in the decision table (Fig. 2.25). Factors for analysis are assigned points for a total value of ten points. The factors for each design are evaluated to determine which has the best overall score.

Conclusions are then given: a summary of the recommended design and the product's profit outlook in the marketplace (Fig. 2.26). A decision is made to implement the stake and chain solution.

IMPLEMENTATION The tubular stake design is presented in a working drawing, where each individual part is detailed and dimensioned. All principles of graphical presentation are used, including a freehand sketch illustrating how the parts will be assembled (Fig. 2.27).

Standard parts—the nuts, bolts, and chain—should be noted but not drawn, since they will not be specially fabricated. With this working drawing, the design has been implemented as far as can be done without actually building a prototype or model.

DECISION
1. Decision table for evaluation

Design 1: STAKE & CHAIN
Design 2: EXTENSION FOOT
Design 3:
Design 4:
Design 5:

Maximum value	Factors for analysis	1	2	Designs 3	4	5	6	7
2	Function	2	1					
2	Human Factors	1.5	1.5					
1	Market analysis	.5	.2					
1	Strength	1	1					
1	Production procedures	1	.5					
1	Cost	.5	.2					
2	Profitability	1.5	1.0					
0	Appearance	0	0					
10	TOTALS	8	5.9					

FIG. 2.25 A worksheet with a decision table that is used to evalute the final designs.

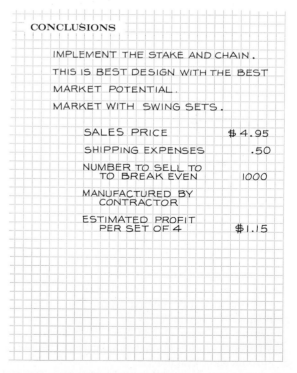

CONCLUSIONS

IMPLEMENT THE STAKE AND CHAIN.
THIS IS BEST DESIGN WITH THE BEST
MARKET POTENTIAL.
MARKET WITH SWING SETS.

SALES PRICE	$4.95
SHIPPING EXPENSES	.50
NUMBER TO SELL TO TO BREAK EVEN	1000
MANUFACTURED BY CONTRACTOR	
ESTIMATED PROFIT PER SET OF 4	$1.15

FIG. 2.26 The final decision and other conclusions are summarized on this worksheet.

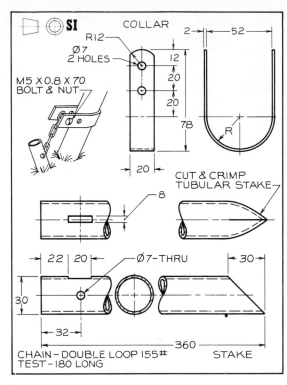

FIG. 2.27 A working drawing is made of the design that is to be implemented.

The actual part as it would be available for the market is shown in Fig. 2.28. A market analysis or an evaluation of the commercial prospects of the item would be another requirement for implementing any device produced for general market consumption.

FIG. 2.28 The completed swing-set anchor.

Problems

Most problems are to be solved on 8½-by-11-inch paper, using instruments or drawing freehand as specified. The paper can be printed with a ¼-in. grid to assist in laying out the problems, or plain paper can be used with the layout made with a 16-scale (architect's scale). The grid of the given problems in later chapters represents ¼-in. intervals that can be counted and transferred to a like grid paper or scaled on plain paper.

Each problem sheet should be endorsed and should include the student's name and seat number, the date, and the problem number. All points, lines, and planes should be lettered using ⅛-in. letters with guidelines in all cases.

Essay problems should have their answers lettered, using approved, single-stroke Gothic lettering, as introduced in Chapter 12.

1. List engineering achievements that have demonstrated a high degree of creativity in the following areas: (a) the household, (b) transportation, (c) recreation, (d) educational facilities, (e) construction, (f) agriculture, (g) power, (h) manufacture.

2. Outline your plan of activities for the weekend. Indicate areas in your plans that you feel display a degree of creativity or imagination. Explain why.

3. Write a short report on the engineering achievement or person that you feel has exhibited the highest degree of creativity. Justify your selection by outlining the creative aspects of your choice. Your report should not exceed three typewritten pages.

4. Test your creativity in recognizing needs for new designs. List as many improvements for the typical automobile as possible. Make suggestions for implementing these improvements. Follow this same procedure in another area of your choice.

5. List as many systems as possible that affect your daily life. Separate several of these systems into component parts or subsystems.

6. Subdivide the following systems into components: (a) a classroom, (b) a wrist watch, (c) a movie theater, (d) an electric motor, (e) a coffee percolator, (f) a golf course, (g) a service station, (h) a bridge.

7. Indicate which of the items in Problem 6 are systems and which are products. Explain your answers.

8. Make a list of new products introduced within the last five years with which you are familiar.

9. Make a list of products and systems that you would anticipate for life on the moon.

10. Assume you have been assigned the responsibility for organizing and designing a skate board installation on your campus. This must be a self-supporting enterprise. Write a paragraph on each of the six steps of the design process explaining how the steps would be applied to the problem. For example, what action would you take to identify the problem?

11. You are responsible for designing a motorized wheelbarrow to be marketed for home use. Write a paragraph on each of the six steps of the design process explaining how the steps would be applied to the problem. For example, what action would you take to identify the problem?

12. List and explain a sequence of steps that you feel would be adequate for, yet different from, the design process given in this chapter. Your version of the design process may contain as many of the steps discussed here as you desire.

13. Can you design a simple device for holding a fishing pole in a fishing position while you are rowing a boat? Make sketches and notes to describe your design.

14. Design a car jack that would be more serviceable than present models. Review the six steps of the design process given in Section 2.3 and outline what you would do to apply these steps to your attempt to design a jack. Write the sequential steps and the methods that you would use to carry out each step.

15. Design a doorstop that could be used to prevent a door from slamming into a wall. Make sketches and notes as necessary to give tangible evidence that you have proceeded through the six steps, and label each step. Your work should be entirely freehand and rapid. Do not spend more than thirty minutes on this problem. Indicate any information you would need in a final design that may not be accessible to you now.

16. List areas that you must consider during the problem identification phase of a design project for the following products: a new skillet design, a bicycle lock, a handle for a piece of luggage, an escape from prison, a child's toy, a stadium seat, a desk lamp, an improved umbrella, a hot-dog stand.

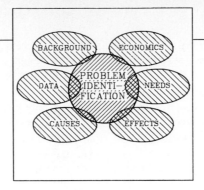

CHAPTER 3

Problem Identification

3.1

Introduction

Problem identification, the initial step that a designer takes to solve a problem, can be either of two general types: (1) identification of a need or (2) identification of design criteria (Fig. 3.1).

IDENTIFICATION OF A NEED is the beginning of the process. Recognizing a problem, a defect, or a shortcoming in an existing product or situation, the designer investigates the causes and effects, which may result in the development of a new product or solution. The need may be for an improved automobile safety belt, a solution to air pollution, or a special hunting seat.

IDENTIFICATION OF DESIGN CRITERIA is that part of the problem where the designer identifies the specifications that must be met by a new design.

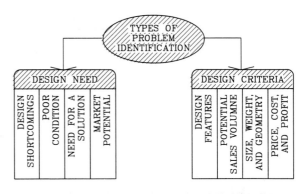

FIG. 3.1 The two types of problem identification.

3.2

Design worksheets

Throughout the design process, designers must make numerous notes and sketches as they search for a solution. Periodically reviewing earlier ideas and notes is

26

important to avoid overlooking previously identified concepts. Moreover, written record of a designer's work will help establish ownership to patentable ideas.

MATERIALS The following materials will aid designers in maintaining permanent records for their design activities.

1. Sketchpad (8½ by 11 inches). A sketchpad can be either grid-lined or plain, depending on the person's preference. Sheets should be punched for insertion in a notebook or file.

2. Pencils. A medium-grade pencil is adequate for most papers.

3. Binder or envelope. All worksheets should be kept in a binder or envelope for reference.

FORMAT Each sheet should have the following information (Fig. 3.2):

1. Name of Project.

2. Name of Designer.

3. Date.

4. Page number of each sheet.

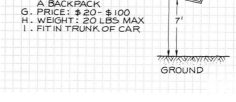

FIG. 3.2 The format of a worksheet for recording problem identification information about a design project.

3.3
The problem identification process

Problem identification requires the designer to analyze requirements, limitations, and other background information without becoming involved with the solution to the problem. The following steps should be used in problem identification (Fig. 3.3).

1. *Problem statement*. Write down the problem statement to begin the thinking process.

2. *Problem requirements*. List factors that the design must meet. Some of these statements may be questions that will be answered later when data have been gathered.

3. *Problem limitations*. List factors that will help determine the design specifications. For example, cannot weigh more than twenty-five pounds; must fit in the trunk of a car.

4. *Sketches*. Make sketches, with notes and dimensions, of the problem's physical characteristics.

5. *Data collection*. Gather data on population trends, related designs, physical characteristics, sale records, market studies, and so on. Once collected,

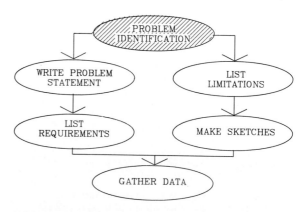

FIG. 3.3 The steps suggested for identifying a problem.

FIG. 3.4 The design of a new automobile demonstrates the problem identification step of the design process. (Courtesy of Ford Motor Co.)

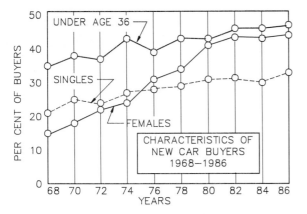

FIG. 3.5 The data plotted here were gathered to identify the changing characteristics of new-car buyers.

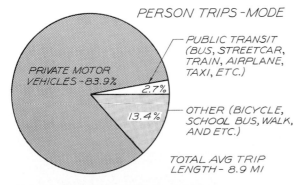

FIG. 3.6 When a car is to be designed, it is important to know what kinds of trips it will be used for. This pie chart shows that 84 percent of person trips are made in private automobiles.

the data should be graphed for easy interpretation.

The designer should review the identification of the problem throughout the design process. Any new information about the problem should be added to the notes.

3.4
Automobile design—problem identification

Suppose you were involved in the development of a new model automobile that would have a broad market appeal (Fig. 3.4). It would be necessary you to consider the prevailing climate of the marketplace: need for fuel economy, rigorous emission controls, expensive safety standards, and everchanging lifestyles. Knowing as much as possible about the consumer is also part of problem identification.

The data graphed in Fig. 3.5 were gathered in a national study about the backgrounds of new-car buyers. As the graph shows, most car buyers are under the age of thirty-six. The most rapid growth has been among female car buyers. Single people buy an increasingly larger portion of new cars. The data in Fig. 3.6 reveal that 84% of all "person trips" are made in private cars, and the remaining 16% is divided among other forms of transportation. Other data show that Americans drive more than a trillion miles per year: 42% for business; 33% for recreation and vacations; 19% for errands; and 5% for education and civics. Figure 3.7 reflects the boom in the number of men, women, and children participating in outdoor sports. How are the data interpreted? What does it mean? Experience as well as instinct are used to answer these questions.

Ford Motor Company concluded that a car was needed that was economical, fun to drive, lively but safe, and sporty but functional. Since younger people and singles were a greater part of the market, there were things the car did not have to be. It did not need a lot of room for passengers; a one- or two-seater with room for luggage was adequate. Likewise, the car did not have to be "hot" to be sporty. Instead, the car needed to be lively and attractive and give good overall performance. The collection of this data, and the knowledge gained from it, resulted in the production of the Ford 1986½ EXP (Fig. 3.8).

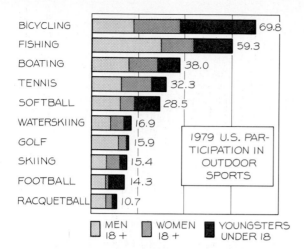

BICYCLING	69.8
FISHING	59.3
BOATING	38.0
TENNIS	32.3
SOFTBALL	28.5
WATERSKIING	16.9
GOLF	15.9
SKIING	15.4
FOOTBALL	14.3
RACQUETBALL	10.7

1979 U.S. PARTICIPATION IN OUTDOOR SPORTS

☐ MEN 18 +   ☐ WOMEN 18 +   ■ YOUNGSTERS UNDER 18

FIG. 3.7 Survey data showing the recreational activities the public participates in are presented in this bar graph. These findings suggest that a car should be designed to meet the needs of consumers active in outdoor sports.

FIG. 3.8 Ford introduced the 1986½ EXP with features and characteristics identified by the problem identification phase of their design process. (Courtesy of Ford Motor Co.)

3.5
Hunting seat—problem identification

The following example is used to illustrate the problem identification step of designing a hunting seat.

Hunting seat

Many hunters hunt from trees to obtain a better vantage point. Design a seat that provides hunters with comfort and safety while hunting from a tree and that meets the requirements of economy and the limitations of hunting.

WORKSHEET COMPLETION The worksheet in Fig. 3.9 is typical of the information the designer needs to understand the background of the problem.

TITLE AND PROBLEM STATEMENT The title of the project is recorded along with a brief problem statement.

REQUIREMENTS AND LIMITATIONS The requirements and limitations are listed along with any sketches that would aid in understanding the problem. You might have to list some requirements as questions for the time being.

After further investigation, you should list the limits, for example, must cost between $60 and $100. It is better to give a range of prices or weights rather than to attempt to be exact. Catalogs offering similar products is one source of information on sales prices, weights, and sizes.

NEEDED INFORMATION How many hunters are there? How many hunt from trees or elevated blinds? This and similar information is available from your state game office.

What is the average income of the hunter? How much do they spend on their hobby per year? Sporting-goods dealers could help you by sharing their experiences, and perhaps they could direct you to other sources for answers to these questions.

MARKET CONSIDERATIONS The designer must think about cost control even in the problem identification step (Figs. 3.9 and 3.10).

Figure 3.11 shows how an item is priced from wholesale to retail. Percentages vary by product: There is less profit in retail food sales than in furniture sales. If an item retails for $50, the production cost cannot be more than about $20 to maintain the necessary margins.

Another method of collecting information is surveying hunters about the merits of introducing a hunting seat to the market (Fig. 3.10). The data show that twenty out of fifty people gave the market potential of the seat a high ranking.

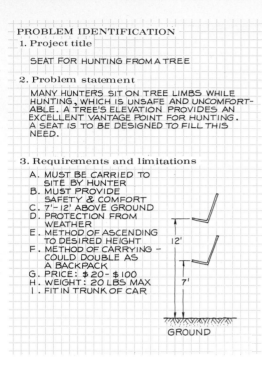

PROBLEM IDENTIFICATION

1. Project title

SEAT FOR HUNTING FROM A TREE

2. Problem statement

MANY HUNTERS SIT ON TREE LIMBS WHILE HUNTING, WHICH IS UNSAFE AND UNCOMFORTABLE. A TREE'S ELEVATION PROVIDES AN EXCELLENT VANTAGE POINT FOR HUNTING. A SEAT IS TO BE DESIGNED TO FILL THIS NEED.

3. Requirements and limitations

A. MUST BE CARRIED TO SITE BY HUNTER
B. MUST PROVIDE SAFETY & COMFORT
C. 7'-12' ABOVE GROUND
D. PROTECTION FROM WEATHER
E. METHOD OF ASCENDING TO DESIRED HEIGHT
F. METHOD OF CARRYING — COULD DOUBLE AS A BACKPACK
G. PRICE: $20-$100
H. WEIGHT: 20 LBS MAX
I. FIT IN TRUNK OF CAR

4. Needed information

A. NUMBER OF HUNTERS WHO HUNT FROM TREES? IN STATE? NATION? — CONTACT STATE GAME COMMISSION.
B. WHAT HAPPENS ON A TYPICAL HUNTING TRIP? SURVEY HUNTERS
C. HOW DO HUNTERS HUNT FROM TREES WITHOUT SEATS? INTERVIEW HUNTERS
D. LAWS CONCERNING HUNTING FROM TREES — WRITE STATE GAME COMMISSION
E. HOW MANY BOW & ARROW HUNTERS ARE THERE? CHECK LIBRARY FOR SOURCES
F. HOW LONG & WHEN ARE THE VARIOUS HUNTING SEASONS? CONTACT GAME COMMISSION
G. EQUIPMENT CARRIED BY HUNTER?

5. Market Considerations

A. WOULD SPORTING GOODS RETAILERS LIKE A TREE SEAT? INTERVIEW DEALERS
B. WHAT IS COMPETITION? REVIEW ADS IN HUNTING MAGAZINES
C. WHAT IS BEST PRICE RANGE? — INTERVIEW DEALERS
D. HOW MUCH DO HUNTERS SPEND PER SEASON? INTERVIEW HUNTERS
E. FEATURES DESIRED BY HUNTERS? — INTERVIEW HUNTERS
F. POSSIBLE MARKET OUTLETS? VISIT STORES

FIG. 3.9 A worksheet for the problem identification step for the hunting seat design.

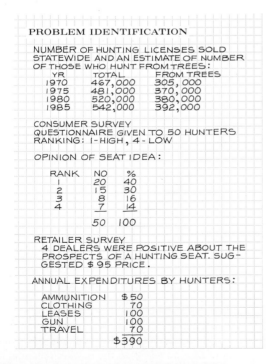

PROBLEM IDENTIFICATION

NUMBER OF HUNTING LICENSES SOLD STATEWIDE AND AN ESTIMATE OF NUMBER OF THOSE WHO HUNT FROM TREES:

YR	TOTAL	FROM TREES
1970	467,000	305,000
1975	481,000	370,000
1980	520,000	380,000
1985	542,000	392,000

CONSUMER SURVEY
QUESTIONNAIRE GIVEN TO 50 HUNTERS
RANKING: 1-HIGH, 4-LOW

OPINION OF SEAT IDEA:

RANK	NO	%
1	20	40
2	15	30
3	8	16
4	7	14
	50	100

RETAILER SURVEY
4 DEALERS WERE POSITIVE ABOUT THE PROSPECTS OF A HUNTING SEAT. SUGGESTED $95 PRICE.

ANNUAL EXPENDITURES BY HUNTERS:

AMMUNITION	$50
CLOTHING	70
LEASES	100
GUN	100
TRAVEL	70
	$390

FIG. 3.10 A worksheet for collecting data.

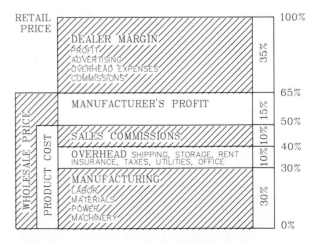

FIG. 3.11 A model showing the breakdown of expenses and costs involved in arriving at the retail price of a product.

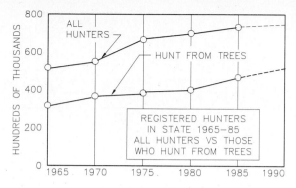

FIG. 3.12 The survey data plotted in this graph describes the population trends of potential customers for the hunting seat.

GRAPHS Data is easier to interpret if presented graphically. For example, the number of hunters is compared with those who hunt from trees in Fig. 3.12. Thorough problem identification includes graphs, sketches, and schematics that improve the communication of the findings and the conclusions of the designer.

The problem identification of this problem is not yet complete. By following the method in this example, you should be able to incorporate your own innovations to arrive at a more thorough problem identification.

3.6
Organization of effort

A schedule of the activities that must be performed in achieving a solution should be prepared immediately after the identification of a need has been sufficiently established. One technique of scheduling project work is Project Evaluation and Review Technique (PERT), which was developed by industries participating in certain governmental projects where coordination of many activities was essential to the successful completion of the project. PERT provides a means of scheduling the activities in their appropriate sequence and reviewing the progress being made toward their completion.

The *critical-path method* of scheduling project activities evolved from, and is used with, PERT. Since some jobs cannot be performed until others have been

completed, the critical-path method determines the chain of events that depend on other activities. The critical path is the sequence of tasks requiring the *longest* time and the least flexibility before the project can be completed. Other activities, not in the critical path, can be scheduled to receive secondary emphasis.

3.7
Planning design activities

The flowchart in Fig. 3.13 suggests steps for achieving the best results in planning your project. These steps are (1) list the jobs that must be performed on a form called a Design Schedule and Progress Record; (2) prepare an Activities Network to place the jobs in sequence, and (3) prepare an Activity Sequence Chart that graphs the jobs in the same sequence shown in the network.

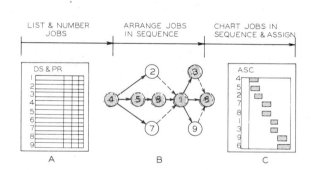

FIG. 3.13 The three steps of planning and scheduling a project.

DESIGN SCHEDULE AND PROGRESS RECORD Figure 3.14 suggests a layout for this form. Each job is broken into reasonably small tasks, which are listed on the Design Schedule and Progress Record (DS&PR) form with no concern for their sequence. When these are all listed, estimate the time required for each job in the second column. The sum of the times for each job should be adjusted to approximate the total time allotted for the project. Extra time may be left unassigned for emergencies. Each job is given a number to identify it.

ACTIVITIES NETWORK The jobs listed on the DS&PR must now be arranged in proper sequence (Fig. 3.15). The note form (Fig. 3.15A) lists the events

Design Schedule & Progress Record

TEAM 7 PROJECT PRODUCT DESIGN

WORK PERIODS 13 MAN-HOURS 200 FINISHING DATE 12-10 SECTION 500

JOB	ASSIGNMENT	ESTIMD HOURS	ACTUAL HOURS	PER CENT COMPLETE 0 50 100
1	WRITE REPORT-ALL	15		
2	BRAINSTORM-ALL	0.5		
3	WRITE LETTERS-JHE	2		
4	GRAPH DATA-BLW	2		
5	MARKET ANAL-ALL	2		
6	COL. DATA-JWS	2		
7	ORGANIZATION-ALL	1		

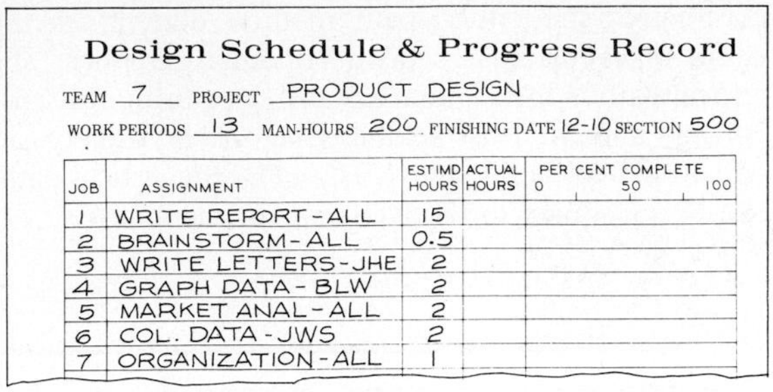

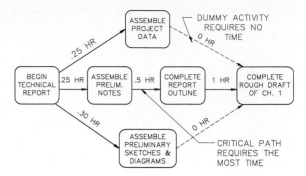

FIG. 3.14 A Design Schedule and Progress Record showing typical entries to identify example design tasks.

FIG. 3.16 The critical path is the path that requires the most time to arrive at the last event of a project.

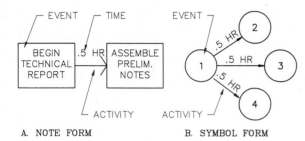

FIG. 3.15 Two ways of preparing an Activities Network, which graphically places the jobs in sequence.

by name, and the symbol form (Fig. 3.15B) uses the job numbers as the events. An *event* is some specific point in the Activities Network. Events do not require time but are milestones along the sequence of activities.

Events are connected by arrows that represent the *activities* that consume time from one event to the next. For example, in Fig. 3.15A, it requires one-half hour to assemble preliminary notes after beginning the technical report. The time is marked on the activity arrow.

Dummy activities indicate a connection between activities even though they require no time. All events must be connected by activity arrows coming to and going from the event. Two dummy activities are used in the Activities Network in Fig. 3.16 to show a sequential connection but no expenditure of time. Dashed lines are used for dummy activities.

The critical path marked in the portion of the Activity Network shown in Fig. 3.16 is the longest time path from the first event to the completion of the project; one and three-quarters hours are required to complete a draft of the first chapter of the report.

A more sophisticated method of estimating the activity times of an Activity Network is shown in Fig. 3.17. Each activity is given an optimistic time, a most likely time, and a pessimistic time. These three times are written on each activity arrow and are used in arriving at critical paths based on each of these estimates.

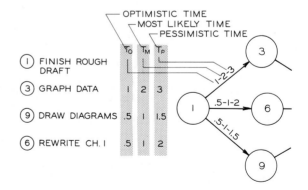

FIG. 3.17 To show a tolerance in the schedule, an Activities Network can be constructed with optimistic, most likely, and pessimistic times on the activity arrows.

ACTIVITY SEQUENCE CHART The jobs of the Activities Network are listed on the Activity Sequence Chart (ACS) (Fig. 3.18). The job numbers are taken from the Network, listed in sequence, and assigned to members of the team. A bar graph shows the time for each job. For example, the first half hour of the project

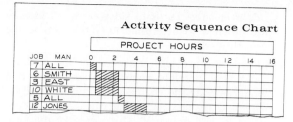

FIG. 3.18 The jobs listed on the DS&PR are converted into an Activities Network and then graphed on an Activity Sequence Chart (ASC), which assigns each job to a fixed time schedule in the proper order.

will be used to perform job 7, which is "organization" on the DS&PR. If your team has four members, this is equivalent to two man-hours of work. Continue until all the jobs are listed. The project hours across the top of the ASC can be used to schedule the events and to estimate when each event should be reached during the project.

As progress is made, the status of each assignment can be graphed on the DS&PR (Fig. 3.19). When a job is completed, the hours it required and the completion date are listed in the last two columns. If more time was required than was scheduled, adjustments can be made in subsequent jobs to compensate for this loss.

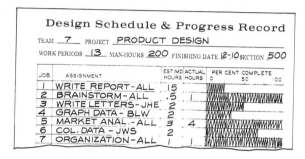

FIG. 3.19 The Design Schedule and Progress Record is completed by graphing the actual time required for each job.

Problems

Problems should be presented on 8½-by-11-inch paper, grid or plain. All notes, sketches, drawings, and graphical work should be neatly prepared in keeping with good practices. Written matter should be typed or lettered using ⅛-in. guidelines.

General

1. Identify a need for a design solution that could be used as a short design problem for a class assignment (less than three man-hours for complete solution). Submit a proposal outlining this need and your general plan for solution. Limit the proposal to two typewritten pages.

2. Assume you were marooned on a deserted island with no tools, supplies, or anything. Identify major problems that you would be required to solve. List factors that would identify the problem in detail. For example, need for food: Determine (a) available sources of food on island, (b) method of storing food supply, (c) method of cooking, (d) method of hunting and trapping. Although you are not in a position to gather data or supply answers to these questions, list factors of this type that would need to be answered before a solution could be attempted.

3. While walking to class, what irritations or discomforts did you recognize? Using worksheets, identify the problem causing these irritations; write down the problem statement, the recognition of a need, the requirements, a d the limitations.

4. Apply the criteria given in Problem 3 to your living quarters, classroom, recreation facilities, dining facilities, and other environments with which you are familiar.

Product design problems

5. Assume you are attempting to reconstruct the designer's approach to the development of the self-opening can. Even though the problem has been solved and completed, follow the identification steps with which the designer was concerned. Using the procedure outlined in Section 3.3, list these on worksheets. After identifying the problem, do you feel that the solution is the most ap-

propriate one, or does your identification suggest other designs?

6. Follow the same procedure outlined in Problem 5 in identifying the problem of designing a travel iron for pressing clothes.

7. Identify the problems of designing a motorized wheelbarrow. List your ideas on worksheets.

8. You have noticed the need for a device that, attached to a bicycle, would allow the bicycle to ride over street curbs to sidewalk level. Identify the problem to determine its application to the general market.

9. Identify the problem and need for the development of a portable engineering travel kit that would provide the engineer with on-the-road facilities to make engineering calculations, notes, sketches, and drawings. This may take the form of a case that includes a calculator, drawing instruments, paper, reference material, and so on.

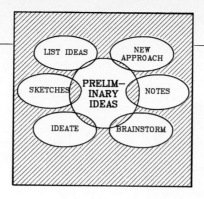

CHAPTER 4

Preliminary Ideas

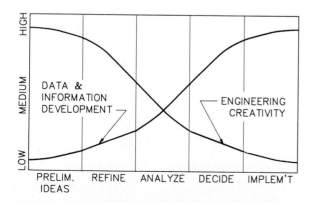

FIG. 4.1 Engineering creativity is highest during the initial stages of the design process, whereas data and information development increase during the final sages.

4.1
Introduction

The relationship between creativity and accumulating information during the design process is shown in Fig. 4.1. You have no limitations on generating preliminary ideas; be as creative as possible, as wild as you like. During the later stages of the design process, the need for creativity diminishes while the need for information increases.

The CAR-FFEE beverage maker (Fig. 4.2) is an example of an advanced design idea using available heat in an automobile and applying it to a hot beverage maker. The CAR-FFEE heats up to a gallon of water for beverages by using the wasted heat from a car's engine. Hot water is then available at the dashboard for coffee, tea, soups, and hot chocolate.

FIG. 4.2 The CAR-FFEE hot beverage maker uses wasted automobile engine heat to maintain a supply of hot water for beverages and soups beneath the dashboard of a car. (Courtesy of Arel Industries, Inc.)

4.2
Individual versus team

Designers work as individuals and as members of design teams. Each approach is discussed below.

Individual approach

Designers who work alone must keep notes and make sketches to communicate not with others but with themselves. Obtaining as many ideas as possible is their primary goal for the better ideas will more likely come from a long list than from a short list.

Possible solutions are sketched and accompanied by notes and schematics. These are not working drawings, but rapid freehand sketches used to retain ideas that might otherwise be lost.

Team approach

The more people involved in a project, the more problems of management and human relations. A team must overcome personality differences and ego problems. Teams perform best if the members select a leader to be responsible for moderating and guiding the team's activities.

Design teams should alternate between individual and group work. For example, members could individually develop preliminary ideas, which would then be discussed with the group.

4.3
Plan of action

In gathering preliminary ideas for a design problem, the following sequence of steps is suggested: (1) hold brainstorming sessions, (2) prepare sketches and notes, (3) research existing designs, and (4) conduct surveys (Fig. 4.3). Periodically reviewing your notes and worksheets from the previous step, problem identification, will ensure that your efforts are directed toward the "target."

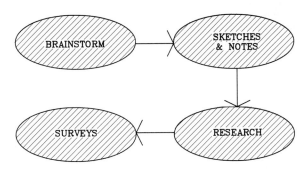

FIG. 4.3 A suggested plan of action for gathing preliminary ideas for a design solution.

4.4
Brainstorming

Brainstorming is a problem-solving technique in which all members of a group spontaneously contribute ideas.

Rules of brainstorming

The fundamental guidelines of a brainstorming session are as follows:*

1. *Criticism is ruled out.* Adverse judgment of ideas must be withheld until later.

*From Alex F. Osborn, *Applied Imagination.* New York: Scribner, 1963, p. 156.

2. *"Free-wheeling"* is welcomed. The wilder the idea, the better; it is easier to tame down than to think up.

3. *Quantity is wanted.* The greater the number of ideas, the more likelihood of useful ideas.

4. *Combination and improvement are sought.* Participants should seek ways of improving the ideas of others.

Organization of a brainstorming session

The organizational steps of a brainstorming session are (1) selection of the panel, (2) preliminary group work, (3) selection of a moderator and a recorder, (4) the session, and (5) the follow-up (Fig. 4.4).

PANEL SELECTION The optimum number of participants in a brainstorming session is twelve. To encourage a variety of ideas, the group should include people with and people without knowledge of the subject. Supervisors often restrict a flow of ideas, so it is desirable that panels be composed of people with a similar professional status.

PRELIMINARY WORK A one-page outline of information about the session should be given to panel members a couple of days beforehand the session to allow ideas to incubate.

THE PROBLEM The problem to be brainstormed should be concisely defined. For example, instead of presenting the problem as "how to improve our campus," present it as "how to improve student parking."

THE MODERATOR AND RECORDER A moderator is selected to be in charge of the session. The recorder is assigned the job of recording ideas as they are verbalized by the panel.

THE SESSION The moderator tosses out the problem and recognizes the first member holding up a hand. That person responds with an idea in as few words as possible. The moderator then recognizes the next person who holds up a hand, and the process continues in this manner.

Ideally, a suggestion made by one member stimulates ideas in other members, who would then hold up their hands and snap their fingers signifying they want to "hitchhike" on the previous idea. This kind of idea development is central to the brainstorming session.

The moderator should not permit individuals to give long responses that would slow down the flow of ideas.

LENGTH A session should move at a brisk pace and should be called to a halt when ideas slow to an unproductive rate. An effective session can last from a few minutes to an hour, but twenty-minutes is considered about the best length.

FOLLOW-UP The recorder should reproduce the list of ideas gathered during the session for distribution to the participants. Approximately 100 ideas are usually gathered during a twenty-five-minute session. The list should be pared down to ten or twelve ideas believed to have the most merit.

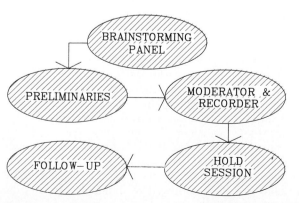

FIG. 4.4 The organizational steps of planning a brainstorming session.

FIG. 4.5 Designers communicate with themselves and others through sketches and notes to develop their preliminary ideas. (Courtesy of Ford Motor Co.)

4.5
Sketching and notes

> Sketching is the designer's most useful medium for developing preliminary ideas (Fig. 4.5).

By sketching, the designer's ideas take form as three-dimensional pictorials (Fig. 4.6) or as two-dimensional views. When properly used, sketching is a rapid, visual extension of the thinking process. The sketch shown in Fig. 4.6 was used in completing the styling features of an automobile. Additional sketches

FIG. 4.6 Sketching is used by the designer and stylist to develop the exterior design of an automobile. (Courtesy of Ford Motor Co.)

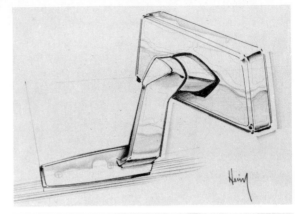

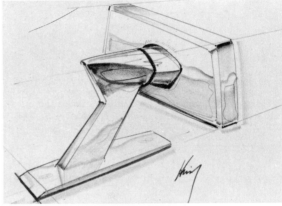

FIG. 4.7 Preliminary sketches of a rearview mirror design. (Courtesy of Ford Motor Co.)

were used to develop preliminary designs for a rearview mirror for an automobile (Fig. 4.7).

The Transportable Uni-Lodge (Fig. 4.8) illustrates the value of sketches in developing and communicating ideas. The Uni-Lodge could be transported by helicopter to previously unreachable areas. Retractable legs with pontoons permit it to float on water. Most ideas are indicated on the sketches by notes.

Sketching was also used to develop the concept shown in Fig. 4.9, a check-out system for a grocery store. The shopper sets the machine in operation by inserting a credit card in a slot. After the card is scanned, the customer receives an order number tag, and the conveyor moves the items under a scanner that totals the prices. If the customer has any questions, the machine can be stopped for communication with the store's employees by lifting the phone. The items are then conveyed to a unit where they are

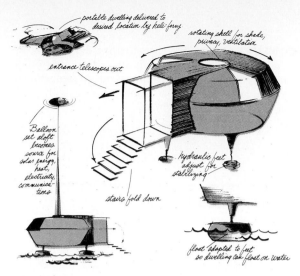

FIG. 4.8 Preliminary sketches of a Transportable Uni-Lodge, a mobil dwelling of the future. (Courtesy of Lippincott and Margulies, Inc., and Charles Bruning Co.)

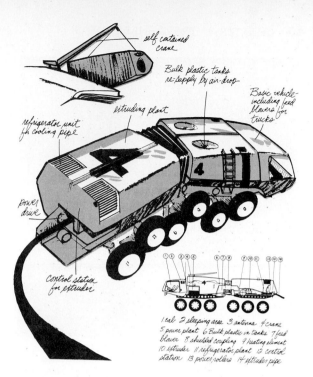

FIG. 4.10 A designer's preliminary design of a self-contained pipelayer, which is intended to lay pipe for the irrigation of large tracts of desert. (Courtesy of Donald Desky Associates, Inc., and Charles Bruning Co.)

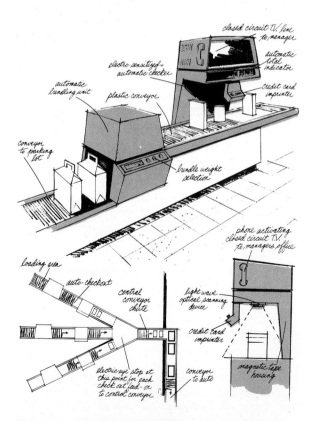

FIG. 4.9 The conceptualization of an automatic checkout/packaging unit for a supermarket is shown in a sequence of sketches. (Courtesy of Lester Beall, Inc., and Charles Bruning Co.)

packaged in plastic containers marked with the customer's number. Large orders are transported by a central conveyor to an exterior pick-up near where the customer is parked.

Another concept is a self-contained pipelayer (Fig. 4.10) that can be used to lay pipe to provide irrigation of desert areas. The first unit is the tractor, which includes a cab, sleeping accommodations, radio equipment, power plant, and bulk storage tanks for plastic. The second unit, the van, consists of an extrusion machine, a refrigeration unit, and a control station. The pipelayer is capable of transporting bulk plastic and machinery to extrude and lay approximately two miles of plastic pipe from each pair of storage tanks. The tanks are discarded when empty and are replaced by new tanks that are air-dropped into the area.

After much freehand sketching has been done, computer graphics can be used to modify and develop a number of preliminary ideas (Fig. 4.11).

FIG. 4.11 After a number of freehand sketches have been made, these preliminary ideas can be developed further by computer graphics techniques. (Courtesy of Ford Motor Co.)

4.6
Research methods

One way of gathering preliminary ideas is to research similar products and designs in use. Applying known principles to new applications and designs is called *synthesis.* Some reference sources available for this research are suggested below.

TECHNICAL MAGAZINES Articles in technical magazines often give detailed explanations of unique designs, complete with sketches and photographs. Advertisements in these magazines can furnish information on materials and innovations that may be helpful.

GENERAL MAGAZINES Significant design developments are often reported in popular periodicals. Magazines that are several years old can also be helpful in finding ideas.

PATENTS Patents from the U.S. Patent and Trademark Office can be used to good advantage by the designer. Patents are available to anyone for $1.50 each.

CONSULTANTS A comprehensive design may require a team of specialists in structures, electronics, power systems, and instrumentation. Manufacturers' representatives are also available to assist with problems related to their products.

4.7
Survey methods

Survey methods are used to gather opinions and reactions to a preliminary or completed design. This is especially important when a product is being designed for the general market.

Opinions

Designers need to know consumers' attitudes toward their preliminary designs. Are potential buyers excited about the design, or do they show little interest in it? Before conducting the survey, the particular group whose opinions would be most valuable should be targeted for survey by (1) personal interview, (2) telephone interview, or (3) mail questionnaire.

THE PERSONAL INTERVIEW should be organized to provide reliable, unbiased opinions. Questions should be true-false or multiple choice to make conclusions easier to determine.

Interviewers should introduce themselves, mention the purpose of the interview, and ask for permission to proceed. They should then tabulate the responses and thank the interviewee for participating.

THE TELEPHONE INTERVIEW can be conducted by randomly picking names from the directory if the opinions of a cross section of the public are desired. If the opinions of a select group—sporting goods retailers—would be more valuable, a Business-to-Business Yellow Pages would supply a list of prospects.

THE MAIL QUESTIONNAIRE is an economical method of contacting large numbers of people in a wide range of locations. To identify questions that need to be revised, the questionnaire should first be sent to a small group as a pilot test. Send at least three times as many questionnaires as the number of responses that you wish to receive. A self-addressed,

PRELIMINARY IDEAS
1. Brainstorming ideas
    A. USE LAWN CHAIR
    B. CHAIR ON STILTS
    C. INFLATABLE CHAIR
    D. PROVIDE FOOTREST
    E. ROOF FOR RAIN
    F. PADDED SEAT
    G. HEADREST
    H. RIFLE REST
    I. AMMUNITION COMPARTMENT
    J. REFRESHMENT COMPARTMENT
    K. ENTERTAINMENT COMPARTMENT
    L. TV ACCESSORY
    M. RADIO
    N. CB RADIO
    O. HOIST SYSTEM
        1. PULLEYS & CABLES
        2. TREE CLIMBER
        3. LADDER
        4. STEPS
    P. PLATFORM FOR STANDING
    Q. PLATFORM FOR SLEEPING
    R. LIGHTS FOR NIGHT
    S. SAFETY BELT
    T. SAFETY BELT FOR ARCHER
    U. SAFETY BELT FOR RIFLEMAN
    V. DOUBLES AS BACKPACK
    W. DOUBLES AS CAMP CHAIR
    X. DOUBLES AS TENT
    Y. CARRYING CASE FOR SEAT
    Z. SEAT ON WHEELS
    A1. MOTORIZED SEAT
    A2. TELESCOPE MOUNT
    A3. HEATING SYSTEM

FIG. 4.12 A worksheet listing the brainstorming ideas recorded by a design team.

2. Description of best ideas

    A. SEAT WITH FOOTREST

    B. NEED METHOD OF ASCENDING – CABLE AND PULLEY SYSTEM

    C. ACCESSORIES

        1. STANDING PLATFORM

        2. RIFLE RACK

        3. BOW RACK

        4. CLIMBING ACCESSORY

    D. SAFETY BELT

    E. FOLDING SEAT

    F. WEIGH UNDER 10 LBS

3. Attach sketches

FIG. 4.13 A description of the better ideas selected from the original brainstorming ideas.

stamped envelope should be included to improve response to the inquiry.

What types of opinions would be helpful to you? First is there a need for the product? What features do consumers like or dislike? What price would they willingly pay for it? What do retailers think it will sell for? Does size matter? Color?

Where appropriate, findings and data should be graphed for study and analysis.

4.8
Hunting seat—preliminary ideas

The design of the tree-borne hunting seat is used to illustrate the preliminary ideas step of the design process. The problem is restated below.

HUNTING SEAT Many hunters hunt from a sitting position in trees to obtain a better vantage point. Design a seat that provides comfort and safety and that meets the requirements of economy and the limitations of hunting.

BRAINSTORMING IDEAS are gathered from a brainstorming session with classmates. All ideas are listed on a worksheet (Fig. 4.12). Remember, wild ideas are encouraged. The better ideas are then selected and their features described on the worksheet (Fig. 4.13). You may list more features than would be possible to include in a single design, and be sure that no ideas are forgotten or lost at this state.

SKETCHES OF PRELIMINARY IDEAS are drawn on worksheets, using rapid freehand techniques. Orthographic views and pictorial methods are both used. Thoughts or questions that come to mind during this sketching should be noted on the drawings. Lettering and sketching techniques need not be highly detailed or precisely executed.

In Fig. 4.14, notice how ideas have been adapted from various types of chairs, lawn chairs in particular. Each idea is numbered for identification. Another worksheet (Fig. 4.15) shows a fourth idea, and idea #2 is modified to include a footrest and to suggest a method of guying the seat while suspended.

Many other ideas of this type need to be developed and sketched before leaving the preliminary ideas step of the design process.

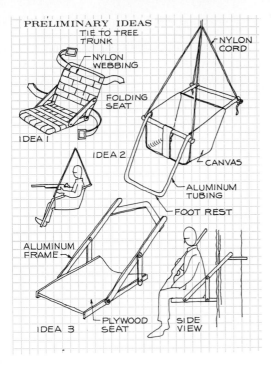

FIG. 4.14 A worksheet for presenting preliminary ideas on the development of a hunting seat.

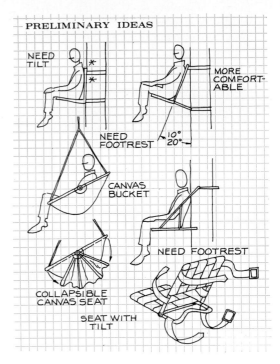

FIG. 4.15 Additional preliminary ideas illustrate design concepts for the hunting seat problem.

Problems

Problems should be presented on 8½-by-11-inch paper, grid or plain. Each grid square represents ¼ in. All notes, sketches, drawings, and graphs should be neatly prepared in keeping with the practices introduced in this volume. Written matter should be legibly lettered using ⅛-in. guidelines.

1. Select one or several of these items and list as many uses as possible: empty vegetable cans (3 in. by 5 in.), two thousand sheets of 8½-by-11-inch bond paper, one cubic yard of dirt, three empty oil drums (24 in. DIA by 36 in.), a load of egg cartons, twenty-five bamboo poles (10 ft. long), ten old tires, or old newspapers.

2. If you were going to select an ideal team to develop an engineering problem, what would you look for? List the characteristics with your explanations.

3. What are the advantages and disadvantages of working independently on a project? What are the advantages and disadvantages of working as a member of a design team? List your reasons and give examples of the types of problems where each approach would have the greater advantage.

4. Research the materials available to you to accumulate information on one of the following design problems or on one of your own selection. You are concerned with costs, methods of construction, dimensions, existing models, estimates of need, and other information of a general nature that will assist you in understanding the problem and deciding whether it is a feasible project. List the references you used. Prepare your research in a presentable form that could be reviewed by your instructor. The design problems are a one-person canoe, a built-in car jack, an automatic blackboard eraser, a built-in coffee maker for an automobile, a self-opening door to permit a pet to leave or enter the house, an emergency fire escape for a two-story building, a rain protector for persons attending outdoor spectator activ-

ities, a new household appliance, a home exerciser.

5. List and describe the type of engineering or professional consulting services that would be required in the following design projects: a zoning system for a city of twenty thousand people, a shopping center, a go-cart, a water-purification facility, a hydroelectrical system, a nuclear fallout disaster plan, a processing plant for refining petroleum products, a drainage system for residential and rural areas.

6. Develop a questionnaire that could be given to the general public to determine its attitude toward a particular product. Select a product, and prepare the questionnaire to measure response to its unique features. Indicate how you would tabulate the information received from your questionnaire.

7. Review the brainstorming techniques, and then organize a group of associates to brainstorm a selected problem or to determine a problem in need of solution. List all ideas as they are suggested.

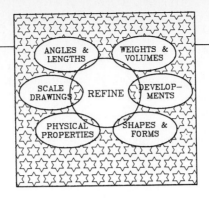

CHAPTER 5

Design Refinement

5.1
Introduction

Descriptive geometry has its greatest application in the refinement step of the design process, for in this step it is necessary to make scale drawings with instruments to check the critical dimensions that cannot accurately be shown in sketches (Fig. 5.1).

> Refinement is the first departure from unrestricted creativity and imagination. Practicality and function must now be given primary consideration.

5.2
Physical properties

One of the important concerns of the design's refinement is determining the physical properties of the proposed solutions. An example of a refinement drawing

FIG. 5.1 The designer's first step in the refinement step of the design process is to draw preliminary ideas as scale drawings with instruments. (Courtesy of Chrysler Corp.)

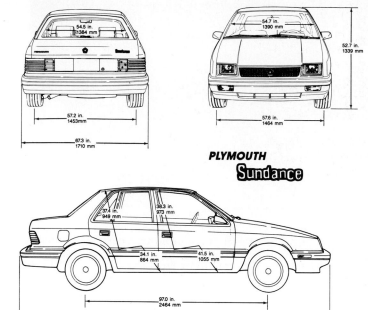

PLYMOUTH
Sundance

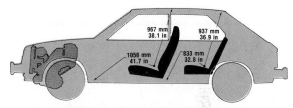

HEAD ROOM AND LEG ROOM DIMENSIONS

FIG. 5.2 These orthographic views were drawn to illustrate the overall appearance and to give the dimensions of the Plymouth Sundance. (Courtesy of Chrysler Corp.)

FIG. 5.3 This scale drawing gives the dimensions and clearances necessary for a comfortable automobile interior. (Courtesy of Chrysler Corp.)

of the exterior of an automobile with the appropriate overall dimensions is shown in Fig. 5.2. Additional scale drawings show dimensions and functions essential to good automobile design, such as seating space (Fig. 5.3).

5.3
Application of descriptive geometry

Descriptive geometry is the study of points, lines, and surfaces in three-dimensional space. The calculation of practically any given properties begins with these geometric elements.

Before descriptive geometry can be applied, orthographic views must be drawn to scale from which auxiliary views can be projected. An example problem has been solved in Fig. 5.4, where descriptive geometry determined the clearance between a hydraulic cylinder and the fender of an automobile.

The design of a surgical light (Fig. 5.5) involves the application of geometry. The light fixture had to be designed to provide the maximum of light on the operating area. A scaled refinement drawing (Fig.

5.5A) shows the converging beams of light that are emitted from the reflectors. The beams are very narrow at their centers and are positioned at shoulder level to minimize shadows cast by the surgeon's shoulders, arms, and hands.

Additional scale drawings (Fig. 5.5B) show the geometry of the fixture in detail. From these scale drawings, measurements, angles, areas, and other geometry can be determined (Fig. 5.6). Figure 5.7 is a refinement drawing that shows critical dimensions of the surgical lamp. Eventually a working model will be constructed to confirm the geometry of the refinement drawings.

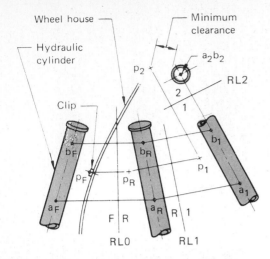

FIG. 5.4 Descriptive geometry is an effective means of determining clearances between components. Shown here is the clearance between a hydraulic cylinder and a fender. (Courtesy of General Motors Corp.)

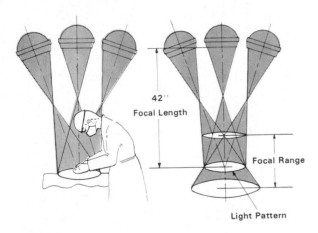

FIG. 5.5 A well-adapted surgical lamp emits light that passes around the surgeon's shoulders with the minimum of shadow. The focal range of this surgical lamp is between 30 and 60 inches. (Courtesy of Sybron Corp.)

5.4
Refinement considerations

Advanced designs, such as the development of a new model automobile, have numerous features that must be improved and optimized at the refinement stage (Fig. 5.8). The overall dimensions of the automobile are shown in the orthographic view in Fig. 5.9. In Fig.

FIG. 5.6 By using scale drawings developed in the refinement step of the design process, the geometry of a surgical lamp can be studied. (Courtesy of Sybron Corp.; photo by Brad Bliss.)

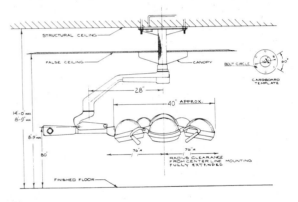

FIG. 5.7 The overall dimensions of the final design of the surgical lamp are shown in this refinement drawing. (Courtesy of Sybron Corp.)

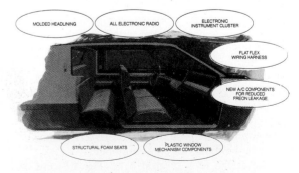

FIG. 5.8 The various features of an automobile's interior are indicated in this illustration. (Courtesy of Ford Motor Co.)

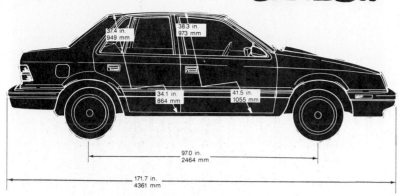

FIG. 5.9 The side view of the Dodge Shadow that shows its significant dimensions. (Courtesy of Chrysler Corp.)

5.10 the arrangement of the exhaust system is shown pictorially. Thus, many refinement drawings were required before the general pictorials were possible; for example, descriptive geometry had to be used to determine the bend angles and clearances in the exhaust pipe to fit a particular chassis (Fig. 5.11).

Computer graphics is a powerful tool that can be used to refine a preliminary idea. In Fig. 5.12, the designer is using a system that yields three-dimensional representations of his designs.

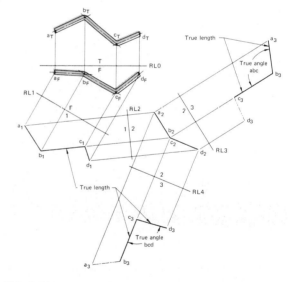

FIG. 5.11 An application of descriptive geometry is shown in this example, which finds the lengths and angles between exhaust pipe segments. (Courtesy of General Motors Corp.)

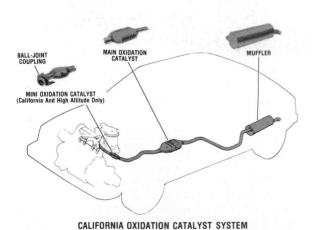

CALIFORNIA OXIDATION CATALYST SYSTEM

FIG. 5.10 An automobile's exhaust system must be developed by using descriptive geometry to determine the lengths and angles necessary to clear the structural members of the chassis. (Courtesy of Chrysler Corp.)

The Space Shuttle Orbiter was also refined using a combination of orthographic drawings and notes. A profile view of the Orbiter (Fig. 5.13) gives the overall size of the craft, and a series of notes show its geometry and specifications.

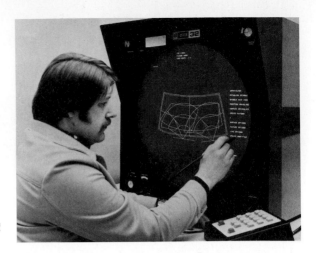

FIG. 5.12 A windshield system can be refined by plotting it on a cathode-ray tube, using computer graphics. (Courtesy of Ford Motor Co.)

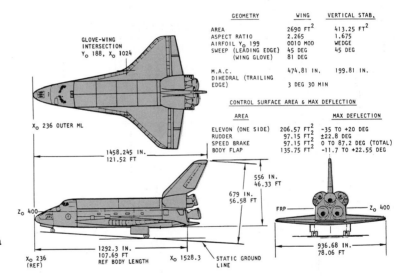

FIG. 5.13 Additional information and specifications for the orbiter vehicle are given in this refinement drawing. (Courtesy of NASA.)

GEOMETRY	WING	VERTICAL STAB.
AREA	2690 FT^2	413.25 FT^2
ASPECT RATIO	2.265	1.675
AIRFOIL Y_O 199	0010 MOD	WEDGE
SWEEP (LEADING EDGE)	45 DEG	45 DEG
(WING GLOVE)	81 DEG	
M.A.C.	474.81 IN.	199.81 IN.
DIHEDRAL (TRAILING EDGE)	3 DEG 30 MIN	

CONTROL SURFACE AREA & MAX DEFLECTION

	AREA	MAX DEFLECTION
ELEVON (ONE SIDE)	206.57 FT^2	-35 TO +20 DEG
RUDDER	97.15 FT^2	±22.8 DEG
SPEED BRAKE	97.15 FT^2	0 TO 87.2 DEG (TOTAL)
BODY FLAP	135.75 FT^2	-11.7 TO +22.55 DEG

GLOVE-WING INTERSECTION Y_O 188, X_O 1024

X_O 236 OUTER ML

1458.245 IN.
121.52 FT

Z_O 400

1292.3 IN.
107.69 FT
REF BODY LENGTH

X_O 236 (REF)

X_O 1528.3

STATIC GROUND LINE

556 IN.
46.33 FT

679 IN.
56.58 FT

FRP

Z_O 400

936.68 IN.
78.06 FT

5.5

Hunting seat—problem identification

The hunting seat problem illustrates the method of refining a preliminary product design.

HUNTING SEAT Many hunters hunt from trees to obtain a better vantage point. Design a seat that provides the hunter with comfort and safety while hunting from a tree and that meets the requirements of economy and the limitations of hunting.

REFINEMENT Preliminary ideas for this design were developed in Chapter 4. The features to be incorporated into the design are listed on a worksheet (Fig. 5.14).

Idea #2 is refined in Fig. 5.15, where a scale drawing of the seat is shown orthographically. Tubular parts, like the separator bars, are blocked in to expedite the drawing process; also, some hidden lines are omitted.

REFINEMENT

1. Description of design

    A. SUPPORT 350 LBS

    B. WEIGH UNDER 10 LBS

    C. FIT TREES 6 IN TO 12 IN DIA

    D. HAS CLIMBING DEVICE

    E. HAS SAFETY BELT

    F. COMFORTABLE SEAT

    G. FOLDS UP FOR EASY STORAGE

    H. PLATFORM FOR STANDING

2. Attach scale drawings

FIG. 5.14 A design's specifications and desirable features are listed on a worksheet of this type. In this example, the hunting seat is refined.

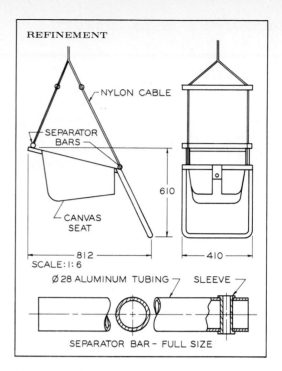

REFINEMENT

NYLON CABLE

SEPARATOR BARS

CANVAS SEAT

610

812

410

SCALE: 1:6

Ø 28 ALUMINUM TUBING   SLEEVE

SEPARATOR BAR - FULL SIZE

FIG. 5.15 A refinement drawing of idea #3 for a hunting seat. Note that only general dimensions are given on the scale drawing.

It is important that refinement drawings be made to scale to give an accurate proportion of the design and to serve as a basis for finding angles, lengths, shapes, and other geometric specifications. Only overall dimensions are given on the drawing, but specific details are shown for the separator bar to explain an idea for a sleeve to protect the nylon cord from being cut. Another design concept is refined in Fig. 5.16. Once again, only the major dimensions are given on these scaled refinement drawings made with instruments. These worksheets do not represent a complete refinement of the design; they are merely examples of the type of drawings required in this step of the design process.

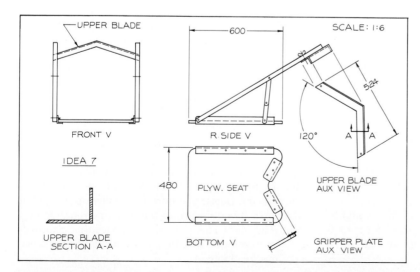

UPPER BLADE

FRONT V

IDEA 7

UPPER BLADE
SECTION A-A

600

R SIDE V

PLYW. SEAT

480

BOTTOM V

SCALE: 1:6

120° A A

524

UPPER BLADE
AUX VIEW

GRIPPER PLATE
AUX VIEW

FIG. 5.16 Another design concept for a hunting seat is shown as a refinement drawing.

Problems

Problems should be presented on 8½-by-11-inch paper, grid or plain. Each grid square represents ¼ in. All notes, sketches, drawings, and graphical work should be neatly prepared in keeping with the practices covered in this book. Written matter should be legibly lettered using ⅛ in.-guidelines.

1. When refining a design for a folding lawn chair, what physical properties would a designer need to determine? What physical properties would be needed for the following items: a TV-set base, a golf cart, a child's swing set, a portable typewriter, an earthen dam, a shortwave radio, a portable camping tent, a warehouse dolly used for moving heavy boxes?

2. Why should scale drawings rather than freehand sketches be used in the refinement of a design? Explain.

3. List five examples of problems involving spatial relationships that could be solved by the application of descriptive geometry. Explain your answers.

4. Make a freehand sketch of two oblique planes that intersect. Indicate by notes and algebraic equations how you would mathematically determine the angle between these planes.

5. What is the difference between a working drawing and a refinement drawing? Explain your answer and give examples.

6. In the refinement step, how many preliminary designs should be refined? Explain.

7. Prepare refinement drawings of Problems 5–15 of Chapter 3. It may be necessary to postpone preparing these refinement drawings until you read some of the succeeding chapters and understand sufficient theory. Keep all these drawings together as they accumulate throughout the design process.

8. Make a list of refinement drawings that would be needed to develop the installation and design of a 100-foot radio antenna. Make rough sketches indicating the type of drawings needed with notes to explain their purposes.

9. Make a list of refinement drawings that would be needed to refine a preliminary design for a rearview mirror that will attach to the outside of an automobile. Refer to Fig. 4.7.

10. After a refinement drawing has been made, the design is found to be lacking in some respects so that it is eliminated as a possible solution. What should be the designer's next step? Explain.

11. Would a pictorial be helpful as a refinement drawing? Explain your answer.

12. List several design projects that an engineer or technician in your particular field would probably be responsible for. Outline the type of refinement drawings that would be necessary.

13. For the hunting seat design discussed in Section 4.8, what refinement drawings are necessary that were not given on the worksheets? Make freehand sketches of the necessary drawings with notes to explain what they would reveal.

14. A seating layout is shown in Fig. 5.17 for continuous pews, which are shown in detail. Prepare the necessary refinement drawings to determine the following by descriptive geometry:
   a) The angle between the seats of row B.
   b) The angle between the pew backs of row B.
   c) The miter angles of the seats and the backs.
   d) A scale drawing of each seat and back.

Refinement problems

The following sketches show products that a designer has developed as preliminary ideas. Prepare scaled refinement drawings are made with a instruments of each to better understand how the products are to be made. Types of refinement drawings you can make are

a) Orthographic views of each individual part of the design.
b) Orthographic views of the products' assemblies and subassemblies.
c) Pictorials to explain relationships between parts.

You must develop and design the details as they are drawn since the preliminary sketches are just that—preliminary. Devise solutions that will work by adding your own inventiveness to the refinement drawings.

15. (Fig. 5.18) Sit-up device. A device that clamps to an interior door to hold one's feet while doing sit-ups for fitness.

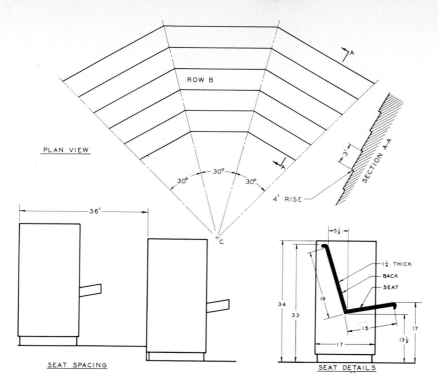

PLAN VIEW

ROW B

SECTION A-A

4" RISE

SEAT SPACING

SEAT DETAILS

BACK

SEAT

FIG. 5.17 Problem 14. Auditorium seating refinement.

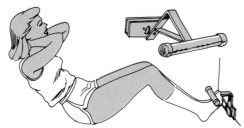

FIG. 5.18 Problem 15. Sit-up device refinement.

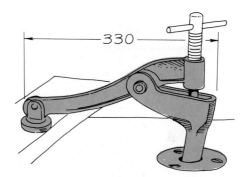

FIG. 5.20 Problem 17. Woodworking clamp refinement.

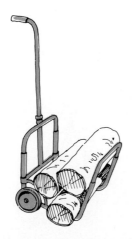

FIG. 5.19 Problem 16. Fireplace caddy refinement.

16. (Fig. 5.19) Fireplace caddy. A small handcart for carrying firewood decorative enough to be kept indoors as a holder for the wood.

17. (Fig. 5.20) Woodworking clamp. A clamp permanently attached to a workbench for holding wood up to 6 inches (150 mm) thick. The design involves refining the mechanism and the collar that attaches to the workbench to permit height adjustment and easy removal.

18. (Fig. 5.21) Hold-down clamp. A clamp designed to attach to a workbench by drilling and counter-

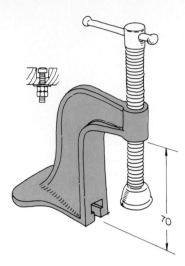

FIG. 5.21   Problem 18. Hold-down clamp refinement.

FIG. 5.23   Problem 20. Sharpener guide refinement.

boring a hole for a mounting bolt that fits in the T-slot of the clamp. When the clamp is removed, the bolt will drop below the surface of the work-bench.

19. (Fig. 5.22) Pointer mount. A mount that connects on the top of a drawing table to hold a rotational pencil pointer.

20. (Fig. 5.23) Sharpener guide. A device for holding a chisel cutting edge at a constant angle while sharpening it on a whetstone.

21. (Fig. 5.24) Luggage carrier. A portable cart for carrying luggage that will fold up to as small a size as possible.

FIG. 5.22   Problem 19. Pointer mount refinement.

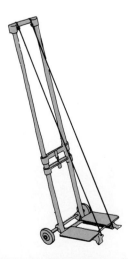

FIG. 5.24   Problem 21. Luggage carrier refinement.

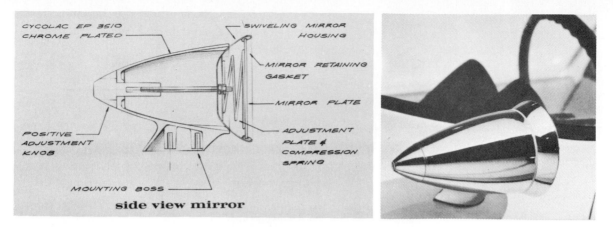

FIG. 5.25 Problem 22. Sideview mirror refinement.

22. (Fig. 5.25) Sideview mirror. A fully adjustable mirror to be mounted on the side of an automobile.

23. (Fig. 5.26) Rotary pump. A rotary pump that operates by squeezing liquid through a flexible tube.

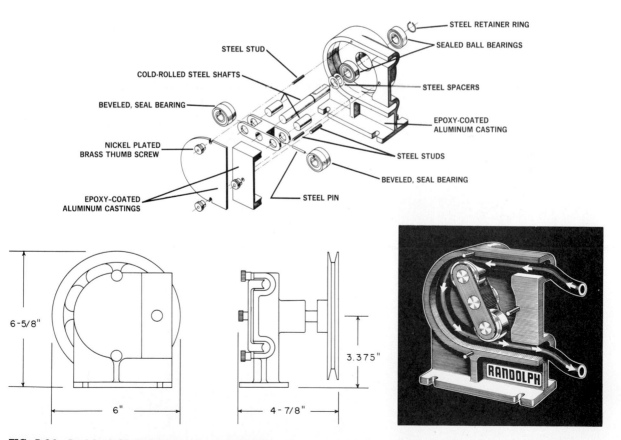

FIG. 5.26 Problem 23. Rotary pump refinement.

24. (Fig. 5.27) Safety hook. A self-locking lifting hook.

25. (Fig. 5.28) Ball display. A system for stacking inflated balls at sporting goods stores.

FIG. 5.27 Problem 24. Safety hook refinement.

FIG. 5.28 Problem 25. Ball display refinement. (Courtesy of AMF/Voit.)

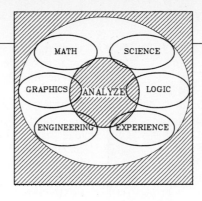

CHAPTER 6

Design Analysis

6.1
Introduction

Design analysis is the process most commonly associated with traditional engineering courses. For example, a bridge design must be analyzed to select the proper structural materials and components for it to be strong, economical, and functional. Analysis is the evaluation of a proposed design. This stage is characterized by objective thinking and the application of technical knowledge. Less creativity is employed during this stage than during the previous stages of the design process.

6.2
Types of analysis

The general areas of analysis are

1. Functional Analysis.
2. Human Engineering Analysis.

3. Market and Product Analysis.
4. Specifications Analysis.
5. Strength Analysis.
6. Economic Analysis.
7. Model Analysis.

FUNCTIONAL ANALYSIS Functional analysis is the most important characteristic of a design; if a design does not function, it is a failure regardless of its other desirable features (Fig. 6.1). A doorknob that will not open a door is an unacceptable design even if it is attractive, strong, and economic.

HUMAN ENGINEERING All designs must serve the humans who use the product, travel on or in it, or profit from its existence. Therefore the designer must consider the physical, mental, and emotional characteristics of the user of the product.

MARKET AND PRODUCT ANALYSIS The market for which a product is designed is studied during the early stages of the products development (Chapters 3 and 4) and before its production. To de-

FIG. 6.1 Experimental automobiles designs are analyzed and tested to arrive at the most functional and efficient performance. (Courtesy of Pontiac Motor Div.)

termine consumer attitude toward a proposed concept, an initial market survey is done.

PHYSICAL SPECIFICATIONS ANALYSIS   In the refinement stage, various specifications, such as lengths, areas, shapes, and angles were obtained. In the analy-

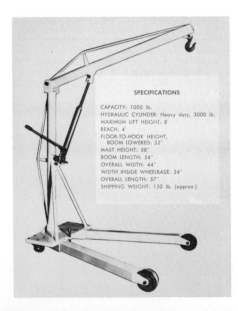

SPECIFICATIONS

CAPACITY: 1000 lb.
HYDRAULIC CYLINDER: Heavy duty, 3000 lb.
MAXIMUM LIFT HEIGHT: 8'
REACH: 4'
FLOOR-TO-HOOK HEIGHT,
   BOOM LOWERED: 32"
MAST HEIGHT: 58"
BOOM LENGTH: 54"
OVERALL WIDTH: 44"
WIDTH INSIDE WHEELBASE: 34"
OVERALL LENGTH: 57"
SHIPPING WEIGHT: 150 lb. (approx.)

FIG. 6.2 Designs must be analyzed to determine their physical properties, including dimensions, weights, and capacities. (Courtesy of Air Technical Industries.)

sis stage, other specifications, such as weights, volumes, capacities, velocities, and ranges of operation, must also be obtained (Fig. 6.2).

STRENGTH ANALYSIS   A proposed design must be strong enough to support the maximum design load that can be anticipated. Strength is closely associated with funciton, since a weak design is not a functional design.

ECONOMIC ANALYSIS   Designs that are unduly expensive have little chance of being profitable in a competitive marketplace. As the design nears the completion stage, the designer must consider economy and type of fabrication.

MODEL ANALYSIS   A proposed design is seldom produced before a model or prototype has been constructed for analysis and evaluation. Extensive tests may be run on a functional design to gather data to support its acceptance or rejection.

6.3
Graphics and analysis

Engineering graphics and descriptive geometry are valuable in the analysis step of the design process. Empirical data obtained from laboratory experiments and field data can be transformed into algebraic equations by graphical techniques (Fig. 6.3). Mathematical evaluation can also be handled graphically.

When the data does not fit an algebraic equation, but instead plots as an irregular curve, graphical calculus must be used. Examples of graphical analysis include analyzing clearances between parts and linkage systems (Fig. 6.4) and presenting and analyzing market surveys, populations, trends, and technical data (Fig. 6.5).

6.4
Functional analysis

Functional analysis requires applying judgment and experience in order to balance the economy, durability, appearance, and marketability of a product's design. For example, in times of high fuel costs, a need for an economical automobile exists, but simply designing a car that gets good mileage is not enough.

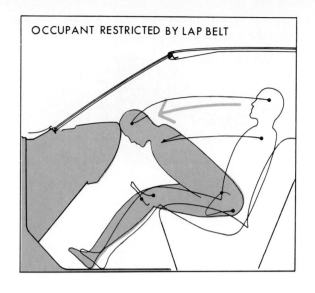

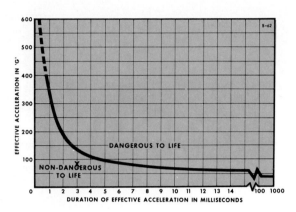

IMPACT ACCELERATION - TIME
TOLERANCE FOR THE HUMAN BRAIN IN FOREHEAD
IMPACTS AGAINST PLANE, UNYIELDING SURFACES

FIG. 6.3 Empirical data obtained from laboratory experiments can be analyzed more efficiently by graphical techniques than by mathematical methods. (Courtesy of General Motors Corp.)

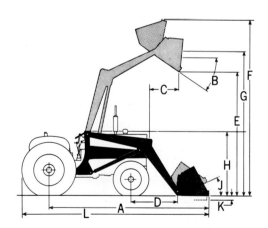

FIG. 6.4 Clearance between functional parts and linkage systems can be analyzed efficiently by using graphical methods. (Courtesy of Navistar International.)

Most buyers are willing to give up some luxuries, but few buyers would give them all up just for better gasoline economy.

Certain functional characteristics can be analytically evaluated when function is the only considera-

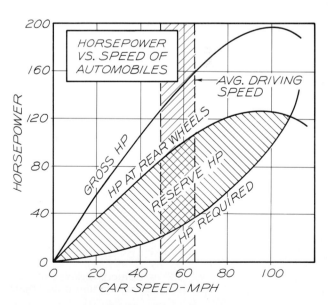

FIG. 6.5 An automobile's power system is analyzed by plotting experimental data on a graph of this type.

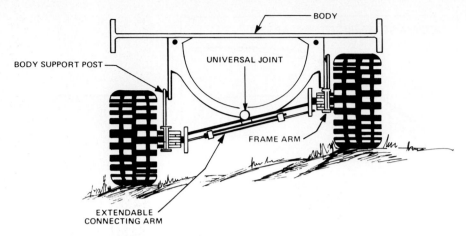

FIG. 6.6 This linkage ensures that an all-terrain vehicle will remain level on an irregular surface. (Courtesy of *Design News.*)

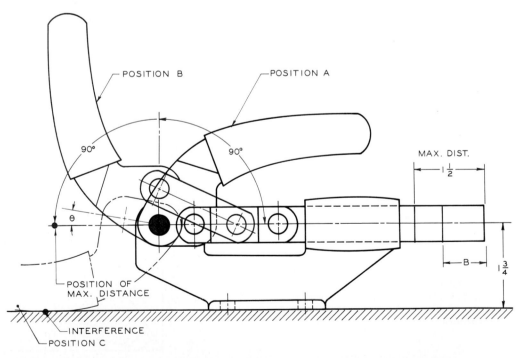

FIG. 6.7 Graphical analysis of a hand-operated clamping device of this type is an efficient means of determining operating limits.

tion. The all-terrain vehicle was designed and its function analyzed graphically in Fig. 6.6. Similarly, the function of a hand-operated clamp is analyzed for its function of clamping a part in position while the part is machined or drilled (Fig. 6.7). In cases like these, function is limited to a narrow range of requirements.

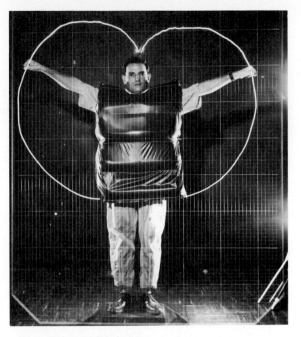

FIG. 6.9 Body dimensions and movements are measured to analyze human mobility that is restricted by a radiation protection vest used by astronauts in space travel. (Courtesy of General Dynamics Corp.)

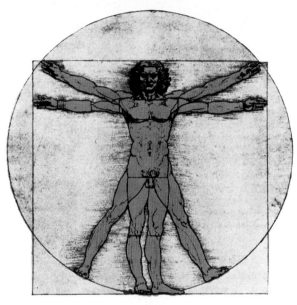

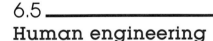

FIG. 6.8 Leonardo da Vinci analyzed body dimensions and proportions more than four hundred years ago.

6.5
Human engineering

Human engineering is defined by Woodson* as follows:

> The design of human tasks, man-machine systems, and specific items of man-operated equipment for the most effective accomplishments of the job, including displays for presenting information to the human senses, controls for human operation, and

*Woodson, W. E., *Human Engineering Guide for Equipment Designer*, Berkeley, CA.: University of California Press, 1954.

complex man-machine systems. In the design of equipment, human engineering places major emphasis upon efficiency, as measured by speed and accuracy, of human performance, in the use and operation of equipment. Allied with efficiency are safety and comfort of the operator.

Leonardo da Vinci analyzed body dimensions and proportions in about 1473 (Fig. 6.8). A close similarity can be seen in Fig. 6.9, where human factors are measured to determine the restriction of human mobility by a radiation protection garment used by astronauts.

Human factors are very important to a successful space program. In a weightless atmosphere even the simplest, most familiar tasks require training and getting used to. What's more, the spacecraft's cabin must be designed to provide the proper life systems as well

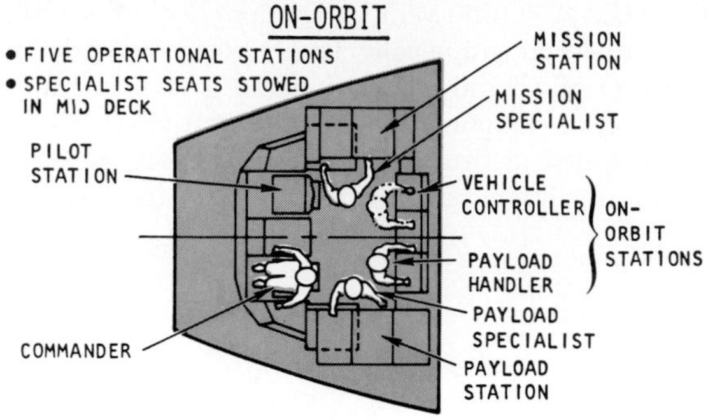

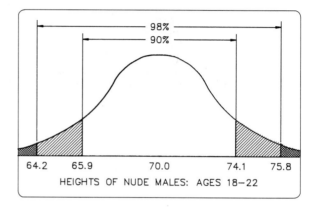

FIG. 6.10 Human factors and living environments had to be analyzed when the crew cabin was designed for the space shuttle. (Courtesy of Rockwell International.)

as a work area in which the on-orbit assignments can be performed (Fig. 6.10).

DIMENSIONS A design must take into account the dimensions, ranges of manipulations, and senses of the person who will be using the finished product (Fig. 6.11). Variations in physical characteristics tend to conform to the normal distribution curve shown in Fig. 6.12.

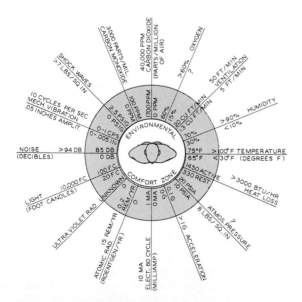

FIG. 6.11 The inner circle is the environmental comfort zone, and the outer circle is the bearable limit zone of human environment. (Courtesy of Henry Dreyfuss, *The Measure of Man*, New York: Whitney Library of Design, 1967.)

FIG. 6.12 The distribution of the average heights in inches of American males from 18 to 22 years of age. (Based on data gathered by B. D. Corpinos, *Human Biology* 30:292.) Fifty percent of American males in this age range are taller than 70 inches, and 50 percent are shorter.

Body dimensions of the average American male are shown in Figs. 6.13 and 6.14. Collected by Henry Dreyfuss, the measurements are used for industrial designs. The average female body dimensions are given in Figs. 6.15 and 6.16.

Two car seats are shown in Fig. 6.17. The safety bar of one seat can easily be raised to clear the child's head and thus permits her to dismount from the seat with a minimum of assistance. The other car seat has a safety bar that cannot be lifted easily, which makes it better to protect a child riding in the back seat.

MOTION The study of body motion includes the amount of space required to function comfortably,

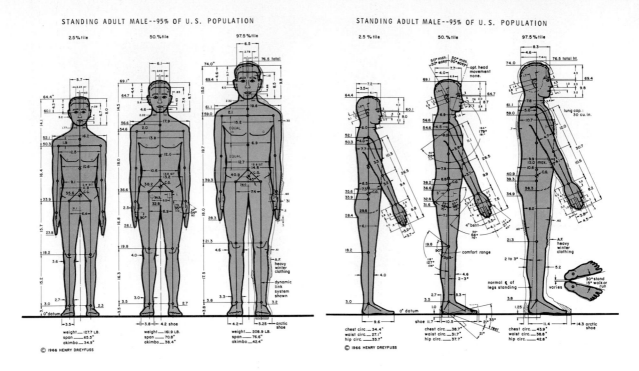

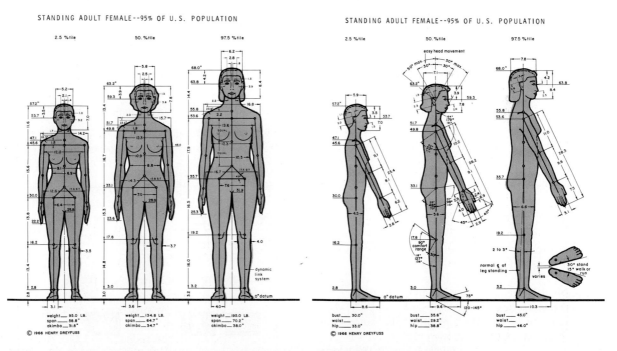

FIGS. 6.13—6.16 Front and profile body measurements of the adult male and female. These measurements describe 95 percent of the U.S. adult population. (Courtesy of Henry Dreyfuss, *The Measure of Man*, New York: Whitney Library of Design, 1967.)

FIG. 6.17 This child's seat has a safety bar that encircles the child's body. The safety bar lifts over the child's head with generous clearance.

FIG. 6.18 The design of an automobile is based on the dimensions of the consumers who will use it. This is an application of human engineering. (Courtesy of General Motors Corp.)

SOUND To be audible, sound must be within specified frequencies. Studies indicate that sound affects a person's stress level and, consequently, productivity.

ENVIRONMENT Working environments may include the entire layout of an industrial plant, the conditions in a particular workstation, or a specialized location, such as the cockpit of an airplane. Environment comprises temperature, lighting, color scheme, sound control, and comfort of operation.

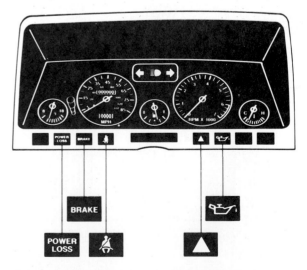

safely, and efficiently. A great deal of analysis is devoted to the interior of an automobile (Fig. 6.18) in order to adapt it to the human body.

VISION Designs with gauges and controls are developed to take advantage of the most visually effective means of aiding the operator. The automobile dashboard configuration shown in Fig. 6.19 is very similar to those of other models. Lights make it easy for the driver to manipulate the controls and to obtain information from the instruments.

FIG. 6.19 The design of efficient, safe automobile instrument panels is a problem involving human engineering. (Courtesy of Chrysler Corp.)

6.6
Market and product analysis

Areas of product analysis are

1. potential market evaluation.
2. market outlets.
3. sales features.
4. advertising methods.

POTENTIAL MARKET ANALYSIS General information about the market should be determined by including predicted age groups, income brackets, and geographical locations of prospective purchasers of the product. This kind of market information will also be helpful in planning advertising campaigns to reach the customer.

MARKET OUTLETS A product may be introduced to the market through existing distribution channels, such as retail outlets, or through newly established dealerships. Large mainframe computers are not suitable for distribution through retail outlets, so technical representatives work with clients individually. On the other hand, a sportsman's camping seat can be sold effectively through department stores and sporting goods outlets.

SALES FEATURES A list of unique sales features used in a new design should be kept to support the feasibility of the proposed design. Unique features are likely to stimulate interest in a product and attract consumers.

ADVERTISING METHODS Advertising a product can be done through several media including personal contact, direct mail, radio, television, newspapers, and periodicals. Advertising costs vary widely and should be considered before selecting a medium. For products used seasonally, timing is also important.

6.7
Physical specifications analysis

To complete the design, the product's physical specifications must be analyzed.

SIZES A product's overall size and dimensions must be evaluated to ensure it meets the standards of size (if any), such as permissible widths in an automobile design. Is its weight within the prescribed range? And what is the size of the product when extended, contracted, or positioned differently?

RANGES Many products have ranges of operation, capacities, and speeds that need to be analyzed before a design can be made final. The designer must calculate or estimate ranges such as seating capacity, miles per gallon, pounds of laundry per cycle, flow in gallons per minute, or electrical power required for operation.

SHIPPING SPECIFICATIONS Finally, the designer must be concerned with the packaging and shipping of the product. Will it be shipped by air, mail, rail, or truck? Will it be shipped one at a time or in quantities? Will it be shipped assembled, partially assembled, or disassembled? How much will the shipping crate cost, and how much will shipping cost? What will be the size of the product when crated for shipment?

6.8
Strength analysis

Much of engineering is devoted to analyzing the design's strength to support dead loads, withstand shocks, and endure repetitive usage at motions ranging from slow to fast. Data are also gathered from experiments and plotted graphically to establish strength characteristics; for example, the strength of clay tile as related to its absorption characteristics (Fig. 6.20).

In Fig. 6.21 the shear for a beam has been determined by graphical calculus. The resulting load calculations are necessary to select the materials and their sizes required by the design.

6.9
Economic analysis

Before a product is released for production, a cost analysis must be done to determine the item's production cost and the margin of profit that can be realized from it. For the average student assignment, two methods of pricing a product or project are (1) itemizing and (2) comparative pricing.

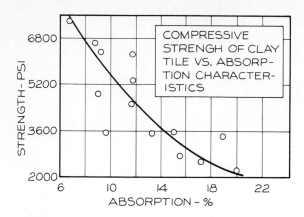

FIG. 6.20 An example of an approximate curve that represents the strength of a structural clay material related to its percent of absorption.

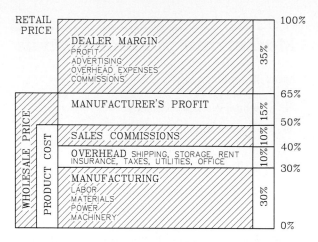

FIG. 6.22 A price structure model that can be used as a guide in the economic analysis of a product.

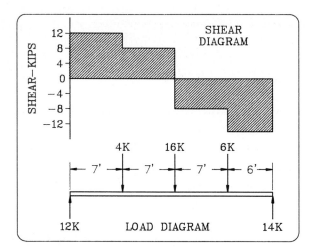

FIG. 6.21 Graphical calculus can be used as an analysis technique to determine shear in a structural beam.

ITEMIZING Industrial firms staffed with experienced estimators can determine a product's production cost by itemizing each expense. A number of costs that must be itemized are shown in Fig. 6.22; obviously, the percentages vary for different areas of manufacturing and retailing.

By studying the project's working drawings, the cost analyst can estimate the costs of materials, manufacturing, labor, overhead, and so forth. Finally, the production cost for the product, or the construction

cost for the project, is determined. The profit is then added to arrive at the wholesale price.

COMPARATIVE PRICING By comparing the proposed product with the cost of similar products on the market, the product's cost can be approximated.

An example of comparative pricing is shown in Fig. 6.23; each power tool is priced at $33.

> These tools are very similar; all have the same power sources, the same materials, and the same styling. More important, each has identical market potential.

Approximately the same numbers of drills are sold as sanders and saws; consequently, production costs and retail price will be similar for each.

Another example of comparative pricing is the price of the hunting seat (Fig. 6.24) compared with that of the exercise bench sold by Sears Company (Fig. 6.25). Both products' manufacturing requirements and market volumes are similar, both sell for about $100. On the other hand, the baby stroller sells in the $50 range because it has a larger market volume even though it is similar to the previous products (Fig. 6.26).

Comparative pricing is used by manufacturers by determining factors like cost per square foot, cost per veloping a new product. Shipping and packaging costs must be considered in pricing a product, and this may be a sizable expense.

Warehousing and storage of finished products must be absorbed in the price of a product. Coupled with warehousing is the cost of insurance, temperature control, shelving, forklifts, and of course employees. Failing to recognize expenses like these can affect a product's profitability.

FIG. 6.23 These three products retail for $99 each. You can see that each is similar in design and will have essentially identical market potential. This is an example of comparative pricing. (Courtesy of Sears, Roebuck and Co.)

mile, cost per cubic foot, or cost per day. These factors enable them to formulate rough cost estimates before doing more detailed studies.

MISCELLANEOUS EXPENSES It is easy to overlook some expenses that are usually incurred in de-

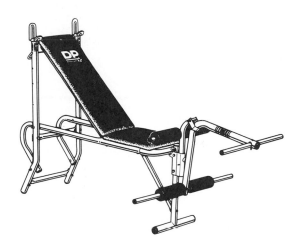

FIG. 6.25 This exercise apparatus is priced at $100, which is similar in manufacture and market appeal to the hunting seat shown in the previous figure. (Courtesy of Diversified Products Corp.)

FIG. 6.24 This hunting seat can be compared with the exercise apparatus shown in Fig. 6.25. Each is comparatively priced from $90 to $100 since both have similar manufacturing problems and market potentials. (Courtesy of Baker Manufacturing Co., Valdosta, Ga.)

FIG. 6.26 These baby strollers are priced from $55 to $75, less than the hunting seat and the exercise apparatus. The market potential for the strollers is greater than the other two products, which makes them cheaper because they can be mass marketed. (Courtesy of Strolee of California.)

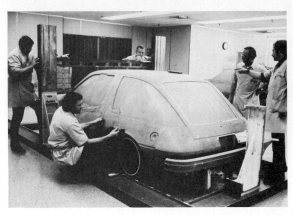

FIG. 6.27 A mock-up is a full-size dummy of a product made to represent the final appearance but not necessarily made to operate. (Courtesy of American Motors Corp.)

6.10
Model analysis

Models are effective aids in analyzing a design in the final stages of its development. A three-dimensional model can be used to study the design's proportion, operation, size, function, and efficiency. The types of models are

1. Conceptual models.
2. Mock-ups.
3. Prototypes.
4. System layout models.

CONCEPTUAL MODELS   A conceptual model is a rough model made for the designers' own use to help them analyze a design or feature.

MOCK-UPS   Mock-ups are full-size dummies of the finished design that give the proper appearance of the product (Fig. 6.27). Mock-ups are constructed more for the presentation of size, shape, appearance, and component relationships than for operational movements.

PROTOTYPES   A prototype is a full-size working model that follows the final specifications in all respects. The only exceptions may be in the use of materials. Since a prototype is made mostly by hand, materials that are easier to fabricate by hand may be used instead of those used in final production.

SYSTEM LAYOUT MODELS   System layout models are used to show the relationships between large layouts such as manufacturing systems, architectural developments, and traffic systems. System layout models of refineries are often constructed to supplement working drawings during the construction of the facility (Fig. 6.28).

Model construction

MODEL MATERIALS   Balsawood is commonly used in model construction because it is easy to shape with few tools. Standard parts such as wheels, tubing, figures, dowels, and other structural shapes can be purchased, rather than made, to save time and effort.

FIG. 6.28 A system layout model is used to analyze the details of construction of refinery design. (Courtesy of E. I. du Pont de Nemours and Co.)

FIG. 6.29 This plaster-and-clay working model of the Hoover Dam was built to show the final appearance of the dam and the surrounding terrain. (Courtesy of the U.S. Dept. of the Interior.)

For forming a shape with contoured surfaces, clay and plaster are effective materials (Fig. 6.29). Plexiglas can be used to construct models that illustrate inside and outside design features.

Models should be finished to give a realistic impression of the completed design, especially when they are used for presentation to others.

MODEL SCALE A model that is used to analyze moving parts of a functional product should be scaled so the smallest moving part will operate. For example,

a student model of a portable home caddy is shown in Fig. 6.30. This balsawood model is constructed to demonstrate the function of a linkage system that permits the wheels to be collapsed for convenience of storage. Although the model is small, the linkage system operates the same way as in the completed product.

MODEL TESTING The analysis process can be aided by testing models built to the specifications of

FIG. 6.30 A student model of a portable home caddy was designed to demonstrate how the device folds flat for ease of storage.

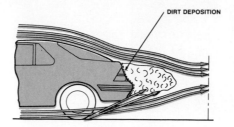
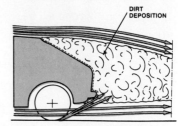

FIG. 6.31 The aerodynomic characteristics of an automobile can be determined by testing a scale model in a wind tunnel. (Courtesy of Ford Motor Co.)

FIG. 6.32 The physical relationships of design components can be observed and tested with the use of full-size models. (Courtesy of Chrysler Corp.)

the final design. The aerodynamic characteristics of the rear styling of an automobile can be evaluated by wind tunnel tests, as shown in Fig. 6.31.

Physical relationships and the functional workings of movable components, like the hatch and storage area of a car, can be tested in a prototype (Fig. 6.32). Models also are used to test consumer reactions to new products before proceeding with the production.

6.11
Hunting seat analysis

To illustrate a method of analyzing a product design, we return to the hunting seat problem.

HUNTING SEAT Many hunters hunt from trees to obtain a better vantage point. Design a seat that provides comfort and safety and that meets the requirements of economy and the limitations of hunting.

Analysis

The major areas of analysis are listed on the worksheets in Fig. 6.33. Additional worksheets should be used to elaborate on each of these areas as required.

The strength of the hunting seat's support system of one design is determined by using graphical vectors

ANALYSIS

1. Function

    A. PROVIDES A METHOD CLIMBING TREES

    B. PROVIDES COMFORTABLE SEATING

    C. PROVIDES DECK FOR STANDING

2. Human engineering

    A. SAFETY BELT
    B. 360° VISION
    C. FOOTREST FOR COMFORT
    D. EASE OF CLIMBING TREE
    E. PORTABLE: 10-15 LBS
    F. SHOULDER STRAPS FOR CARRYING

3. Market & consumer acceptance

    A. POTENTIAL MARKET
      1. STATE: 40,000
      2. NATION: 1,800,000

    B. CHEAPER THAN DEER STAND BY 100-500%

    C. AFFORDABLE AT $100

    D. ADVERTISE IN SPORTS MAGAZINES

    E. RETAIL THROUGH SPORTING GOODS STORES.

4. Physical description

    A. PLATFORM 19"X24"- STAINED

    B. FITS TREES 5"DIA-18"DIA

    C. STRAPS FOR CARRYING ON BACK

    D. HAND CLIMBER DEVICE INCLUDED

    E. SEAT FOLDS FLAT TO 20"X36"

    F. WEIGHT - 10 LBS

    G. SAFETY BELT

5. Strength

    A. SUPPORTS 300 LBS

    B. SAFETY BELT SUPPORTS 300 LBS

    C. SHOULDER STRAPS SUPPORT 300 LBS

    D. HAND CLIMBER SUPPORTS 400 LBS

6. Production procedures

    A. STRUCTURAL MEMBERS CUT FROM STANDARD ALUMINUM CHANNELS

    B. JOINTS CONNECTED WITH NUTS & BOLTS

    C. METAL EDGES SMOOTHED BY GRINDING

    D. SEAT- 3/4" PLYWOOD - STAINED GREEN

    E. SEAT COVERED WITH BLACK VINYL - BUCKLED ON

7. Economic analysis

MATERIALS	$10
LABOR	21
SHIPPING	5
WAREHOUSING	4
TOTAL	$40
SALES COMMISSION	5
PROFIT	20
WHOLESALE PRICE	65
RETAIL PRICE	$90

FIG. 6.33 Worksheet used to analyze the various features of a hunting seat design.

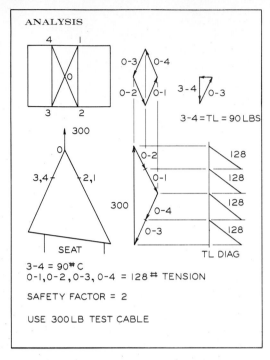

ANALYSIS

3-4 = TL = 90 LBS

3-4 = 90# C
0-1, 0-2, 0-3, 0-4 = 128# TENSION

SAFETY FACTOR = 2

USE 300 LB TEST CABLE

FIG. 6.34 On this worksheet, the support system of a proposed hunting seat is analyzed with graphical vectors.

(Fig. 6.34). Once the loads in each support cable are found, the proper size cable can be selected. For further analysis, models are constructed at a reduced scale and at full size (Fig. 6.35).

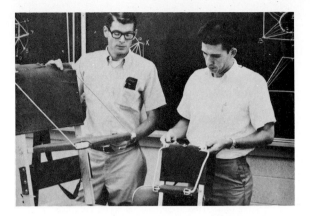

FIG. 6.35 Full-size and half-size scale models were built by Keith Sherman and Larry Oakes to aid them in analyzing their design.

FIG. 6.36 A. The hunter hugs the tree and lifts the hunting seat with his feet. B. The hunting seat used as a platform for standing. C. The hunting seat used for sitting.

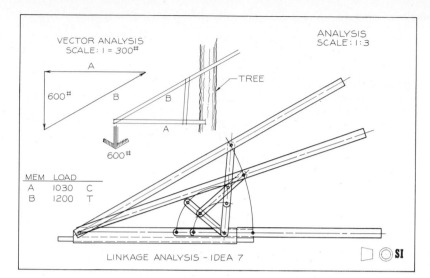

VECTOR ANALYSIS
SCALE: 1 = 300#

ANALYSIS
SCALE: 1:3

A

600#

B

B

TREE

A

600#

MEM	LOAD	
A	1030	C
B	1200	T

LINKAGE ANALYSIS - IDEA 7

SI

FIG. 6.37 The drawing is used to analyze the linkage system of the hunting seat, and a vector diagram is used to determine the forces in the members when the seat is loaded to maximum.

FIG. 6.38 A summary of the features and physical properties of the Baker Tree Stand. (Courtesy of Baker Manufacturing Co.)

This is the Original Patented Baker Tree Stand

BAKER TREE STAND FEATURES:
1. Platform 19″ x 24″ (456 sq. in.) stained.
2. Back pack wt. 10 lbs.
3. Tested to hold 560 lbs.
4. Fits trees 5″ to 18″ diameter.
5. Riveted assembly folds flat for carrying.
6. Hand Climber fits inside frame.
7. Back packs with Strap Assembly.
8. Safety Belt with Extension and a Tie Down (for safety strap) included.

A commercial version of the seat is illustrated in Fig. 6.36, where it is tested to measure its functional features including the method of using it to climb a tree. An analysis drawing of the hunting seat (Fig. 6.37) illustrates the operation of the linkage system that permits the seat to collapse into a single plane for carrying ease. The forces in the members are also found graphically by using vector analysis. The overall features and the physical properties of the Baker Tree Stand are shown in Fig. 6.38. These features are helpful to a consumer in making a purchase.

Problems

The following problems should be solved on 8½-by-11-inch paper, accompanied by the necessary drawings, notes, and text. Answers to essay problems can be typed or lettered. All sheets should be stapled together or included in a binder or folder.

General

1. Make a list of human factors that must be considered in the design of the following items: a canoe, a hairbrush, a water cooler, an automobile, a wheelbarrow, a drawing table, a study desk, a pair of binoculars, a baby stroller, a golf course, the seating in a stadium.

2. What physical quantities would have to be determined in the designs listed in Problem 1?

3. Select one of the items given in Problem 1 and make an outline of the various steps that should be taken to satisfy the following areas of analysis: (1) human engineering, (2) market analysis, (3) prototype analysis, (4) physical quantities, (5) strength, (6) function, and (7) economy.

Human engineering

4. Using your body as the average, make a drawing to indicate the optimum working areas for you when in a sitting position at a drawing table. Assume you are to use your measurements as a basis for designing a drawing table to satisfy the needs of your classmates. Your reach, posture, and vision will have considerable effect on its dimensions. Your finished drawing should give three views of the ideal working area for you while drawing; the drawing should also show the most efficient positioning of instruments for working. Experiment with the angle of tilt of the table top to determine the most comfortable position for working.

5. Using the dimensions for the average person given in this chapter, design a stadium seating arrangement that will serve the optimum needs of the spectators. Determine the dimensions shown in Fig. 6.39. A primary consideration will be the slope of the stadium seating, which should be designed to provide an adequate view of the playing

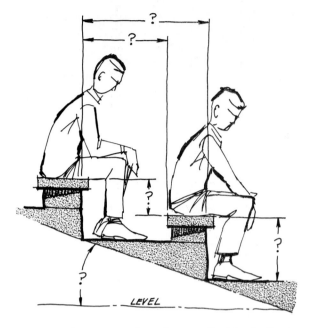

FIG. 6.39 Human engineering applied to stadium seating.

field. The comfort of the average spectator and provision for traffic between seats must also be considered.

6. Compare the dimensions of the students in your class with the standards given in Section 6.5. For example, compare the average height of your class with the national standard height.

7. Design a backpack that will be used on a camping trip. Decide on the minimum belongings a camper should carry, and use their weights and volumes in establishing the design criteria. Make sketches of the pack and the method of attaching it to the body to provide the optimum mobility, comfort, and capacity.

8. Establish the dimensions, facilities, and other provisions that would be needed in a one-person bomb shelter to provide protection for forty-eight hours. Make sketches of the interior in relation-

ship to a person and supplies. How will ventilation, water, food, and other vital resources be provided? Explain your design as it relates to human engineering needs.

9. Design a manhole access to an underground facility (Fig. 6.40). What must the diameter of the manhole be to permit a person to climb a ladder for a distance of ten feet with freedom of movement? Make a sketch of your design and explain your method of solving the problem.

10. Design a one-person facility for temporary observation service in the Arctic. This facility is to be as compact as possible to provide for the needs of a single person during the seventy-two-hour periods that the person is on duty. Determine the facilities and provisions that would be needed, including heat, ventilation, and insulation. Make sketches of your design, and explain items that you consider essential to the human engineering aspects of the problem.

11. Design a configuration for an automobile steering wheel that is different from present-day designs but is just as functional. Base your design on human factors such as arm position, grip, and vision. Make sketches of your design and list items that you considered.

12. Make sketches to indicate safety features that could be built into your automobile to reduce the seriousness of injury caused by accidents. Explain your ideas and the advantages of your designs. Primary consideration should be given to the human aspects of the designs.

13. Assume you prefer to alternate between a sitting and a standing position when working at a drawing table. Determine the ideal height of the table top for working in each position. Indicate how a table could be devised to permit instant conversion from the height for standing to the height for sitting.

14. Identify some human engineering problems that you believe in need of solving. Present several of these to your instructor for approval. Solve the approved problems. Make a series of sketches and notes to explain your approach.

Market analysis

15. Assume you are responsible for conducting a market analysis of the drill shown in Fig. 6.22. Include in your analysis all the areas covered in Section 6.6. Assume this product is new and has never been introduced in an electrically powered form before. Outline the steps you would take in conducting a product and market analysis.

16. Make a product and market analysis of the car seat shown in Fig. 6.17, following the steps suggested in Section 6.9. Arrive at a market value that you feel would be satisfactory, and determine the outlets and other information of this type that would be important to your analysis.

17. Assume the cost estimates of producing hunting seats were as follows: 100 seats, $35 each; 200 seats, $20 each; 400 seats, ten dollars each; 1000 seats, $8.50 each. Using these figures, determine the price at which you could introduce the seats to the market on a trial basis and still have some financial protection. Explain your plan.

18. List as many unique features of the hunting seat as you possibly can that would be important to a sales campaign and to advertising. Make sketches and notes to explain these features.

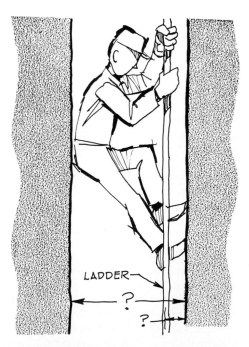

LADDER

FIG. 6.40 Optimum size for access to a manhole.

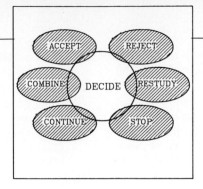

CHAPTER 7

Decision

7.1
Introduction

Once the design has been conceived, developed, refined, and analyzed, a decision must be made to determine which design is the most worthy of implementation. The decision step is based on facts and data; still, at best it is subjective and must be made by experienced persons.

7.2
Types of presentations

Presentations can be made to a few people or to a large group, and the groups can vary from project associates to laymen unfamiliar with the project and its objectives.

INFORMAL PRESENTATION Informal presentations are given to several immediate associates or to a single supervisor. Although specially prepared visual aids rarely are used in these presentations, pictorials, schematics, sketches, and models frequently are.

When only a few people are involved, ideas may be sketched. (Fig. 7.1). In a discussion involving several people, ideas, schematics, and sketches should be drawn on a blackboard (Fig. 7.2).

FORMAL PRESENTATION Formal presentations receive more emphasis in this chapter because they are usually more crucial than informal presentations.

The group to whom the presentation is made may comprise professional associates, or administrators, or laymen, or it may be a mixed group. Function and design acceptability are the primary considerations of the engineering associates. Administrators will be concerned with its economic feasibility and estimated

FIG. 7.1 A decision may be the outcome of a presentation to a single individual, where ideas and designs are discussed and sketched informally. (Courtesy of Ford Motor Co.)

profit return. Laymen could be the clients for whom the project is designed, stockholders, or members of the general public who may vote on the approval of a design.

FIG. 7.2 An informal presentation may be given to professional associates by using preliminary notes and blackboard sketches. (Courtesy of Research of New Jersey.)

Organizing a presentation

One method of planning an oral or written report is by using 3-by-5-inch index cards. Separate ideas that are to be illustrated are first written on separate cards. A rough sketch is made indicating the type of illustration required and the method of reproduction to be used, and brief notes are added outlining the discussion that will accompany the visuals. The sequence of the presentation can be easily reviewed by displaying the cards on a table, bulletin board, or planning board (Fig. 7.3). And the sequence or content of the cards can be easily changed.

FIG. 7.3 Planning cards can be used to prepare the sequence of the presentation by using a planning board (as shown here) or by merely arranging the cards on a table top.

The completed cards (Fig. 7.4) should contain the following information:

1. *Number.* The card's position in the sequence of visual aids.
2. *Illustration.* A sketch of the illustration that must be prepared.
3. *Text.* A brief outline of the oral presentation that will accompany the visual aids.

If a variety of visuals (flip charts, slides, transparencies, or combinations of these) will be used during the presentation, the material should be arranged to allow smooth transitions from one type of presentation to another. After the cards have been put in the

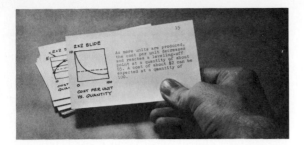

FIG. 7.4 An example of the layout of a 3" × 5" card that shows a sketch of the visual and its accompanying text.

FIG. 7.5 The flip chart can be used to communicate with small groups. (Photo by Hazel Hankin/Stock, Boston.)

FIG. 7.6 Felt-tip markers are well-suited to the preparation of flip charts and other types of visual aids.

proper sequence, they should be arranged according to similarities in illustration or production.

7.4
Visual aids for presentation

In a presentation, the visual aids most commonly used are flip charts, photographic slides, overhead projector transparencies, and models.

The following general rules will improve the effectiveness of any type of visual aid.:

1. Convey only one thought on each slide or chart.
2. Reduce lengthy statements to key phrases or words that will communicate the thought.
3. Present tabular data graphically.
4. Use slides and charts that are clearly readable.
5. Use illustrations, color, and attention-getting devices.
6. Prepare a sufficient number of slides or graphs so notes are unneeded.

Flip chart

Flip charts consist of a series of illustrations prepared on medium-weight paper and mounted on a board or an easel for presentation before a group (Fig. 7.5). Flip charts are used for small conferences or presentations in an area no larger than an average-sized classroom. A 30-by-36-inch flip chart is a common size; anything smaller may be difficult for many people to see.

PAPER Brown or white wrapping paper is suitable for a series of flip charts that will be used only once. Corrugated cardboard can be sized to serve as a backing board.

LETTERING MATERIALS The final lettering can be done with felt-tip markers, ink, tempera or sign paints, or overlay film. Felt-tip markers, used extensively for fast bold lines in a variety of colors, give sophisticated effects if correctly used (Fig. 7.6). India ink is an effective medium for lettering and for adding emphasis to a chart (Fig. 7.7). Tempera or sign paints come in many colors and can be applied with a brush for an attractive layout (Fig. 7.8).

FIG. 7.7 This flip chart was drawn on brown wrapping paper using India ink and a brush. A title is being applied with rubber cement. Colored construction paper can be used to highlight flip charts.

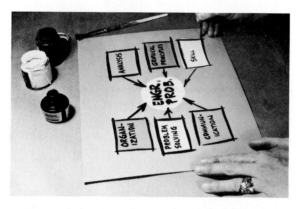

FIG. 7.8 Tempera and other water-base paints are excellent for the preparation of colorful visual aids.

COLOR Color can be added to a chart very easily through the use of construction paper and rubber cement. In general, a chart of this type should be bold and have no tedious or time-consuming details. This technique is especially useful for bar graphs (Fig. 7.9).

ASSEMBLY The finished charts are assembled in order, with a title chart covered by a blank sheet of paper. All sheets are stapled or otherwise attached to the backing board. The blank cover sheet conceals the theme of the flip charts and eliminates any anticipation of the presentation.

PRESENTATION Flipped in sequence, the charts are referred to with a pointer to ensure the speaker does not block the audience's view. Since the flip charts serve as notes, written notes can be kept to a minimum, although it may be helpful to write key points lightly (visible only to the presenter) on the charts.

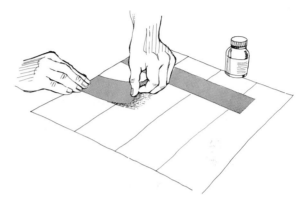

FIG. 7.9 Colored construction paper can be cemented onto flip charts and other visuals to give an attractive and colorful appearance with the minimum of effort and expense.

Photographic slides

Photographic slides are effective when flip charts are too small to be clearly seen, as in large group presentations. They may be necessary to depict actual scenes or examples.

LAYOUT OF ARTWORK FOR PHOTOGRAPHIC SLIDES Figure 7.10 illustrates a method for sizing the layout in correct proportion for a 35-mm slide. An 8-by-12-inch format is appropriate for most information presented by slides. Combining different colors on each slide will add variety to the sequence and maintain a higher level of interest. Colored construction papers, mat board, and other poster materials can be used in laying out the artwork for a slide.

Allow at least an inch margin on all sides of a layout to ensure no edges show when the art is photographed. Selecting the proper size of lettering is critical to ensure readability from any point in the audience.

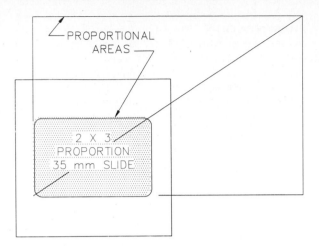

FIG. 7.10 This method of proportionately sizing artwork can be used to ensure the artwork will properly fill a photographic slide.

Only capitals should be used, with space between lines equivalent to the height of the letters.

Special effects can be achieved with three-dimensional letters (Fig. 7.11) or unusual backgrounds.

COPYING THE LAYOUTS To produce slides a camera, copy stand, light meter, and lights are needed. A 35-mm reflex camera is recommended because it has a through-the-lens viewfinder, so the photographer can accurately focus and align each slide before

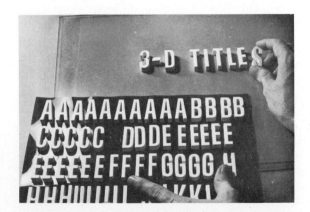

FIG. 7.11 Three-dimensional letters are available for the preparation of photographic charts.

photographing it. The copy stand holds the camera in the proper position during the photographing.

If all layouts are drawn the same size, the camera can be left in the same position. When weather permits, copy work can be photographed in natural light. Book illustrations too small for regular copying can be photographed with a close-up lens that will fill the slide with the area being photographed.

FIG. 7.12 Artwork and charts can be reproduced with a 35-mm reflex camera mounted onto a copy stand, which facilitates the positioning of the camera.

Before the presentation, the mounted slides should be reviewed and sorted to determine their quality (Fig. 7.13). The mounts of all slides should be numbered in their final sequence.

THE SLIDE SCRIPT A slide presentation that will be used repeatedly needs a script not only for presentations but also for review (Fig. 7.14).

On the left-hand side of the script are black and white photographs of the slides used and on the right-hand side is the narration. The narration is intended as a guide for the presenter.

Overhead transparencies

Overhead transparencies are reproduced on 8½-by-11-inch acetate by the heat-transfer or diazo process

FIG. 7.13 Photographic slides should be sorted and arranged in proper sequence prior to loading in the slide tray.

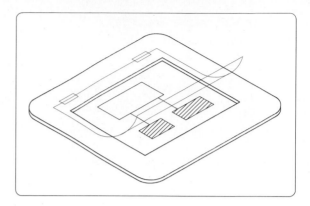

FIG. 7.15 A transparency used on an overhead projector is comprised of an $8\frac{1}{2}'' \times 9\frac{1}{2}''$ transparency mounted on a $10'' \times 12''$ frame. The projection area within the frame is about $7\frac{1}{2}'' \times 9\frac{1}{2}''$. To add interest and permit the user to show information in steps, different-colored overlay flips can be used.

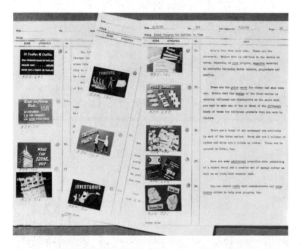

FIG. 7.14 A slide manuscript should be prepared for an important slide presentation or for one that will be given on a repetitive basis. (Courtesy of Eastman Kodak.)

COLOR OVERLAYS   Several color overlays can be hinged to the transparency mount (Fig. 7.15) for a sequential presentation showing the development of an idea or problem. The artwork for overlays is prepared on tracing paper that has been positioned over the basic layout.

PRESENTATION   WITH   TRANSPARENCIES
The presenter can stand or sit near the projector in a semilighted room and refer to the transparencies while facing the audience (Fig. 7.16). With a small pointer, the presenter can indicate important points on the stage of the projector, and the image of the pointer will project on the screen. Just as color overlays can be hinged to the mount, so can opaque paper over-

(Fig. 7.15). Tracing paper is most commonly used in the preparation of transparencies because it can be used without special equipment. For best reproduction, line work should be prepared in black India ink. Overlay materials and graphing tapes can be used to give a professional appearance. Lettering should be legible (at least one-quarter inch high). The finished tracing-paper drawing is transferred to the acetate transparencies in much the same way a diazo print is made.

FIG. 7.16 The overhead projector can be used in a semilighted room while the presenter faces the audience.

FIG. 7.17 A full-size prototype model of a design can be used as an effective presentation to a small group. (Courtesy of the Chrysler Corp.)

lays, which are used to conceal parts of the transparency and focus audience attention on a single topic at a time.

Models

A model is an excellent means of communicating the final design concept most realistically. An actual-size prototype of the completed design (Fig. 7.17) gives the most accurate impression of the finished design. Models should be at least twelve inches in size to be visible from more than twelve feet. Photographic slides can effectively supplement the model during a group presentation. A series of close-ups taken from different angles can give each member of the audience a clearer view of a relatively small model.

7.5
The group presentation

Conference rooms, classrooms, and auditoriums are usually arranged to provide good viewing of visual aids; nevertheless, conditions should be checked by viewing the materials from extreme locations in the audience area before the meeting. All projectors and other visual-aid equipment should be positioned and focused before the audience arrives. The screen or flip charts should be placed where they afford the best view from all locations. Remote controls for slide projectors should be near the speaker's position and ready for use. The speaker should be careful not to block anyone's view. Rehearsing the presentation, with an

assistant in the seating area, will help call attention to any blocking movements.

The presentation

The speaker should give the presentation at a moderate pace, and visual aids should be used as a narration aid (Fig. 7.18). A positive approach in selling ideas should not be confused with possibly deceptive high-pressure salesmanship. The presenter should be the first to point out any weaknesses in the design, but these weaknesses should be offset by alternatives that compensate for them.

Conclusions and recommendations should be given in light of available research and analysis. If after thorough analysis a designer cannot recommend a design, the reasons for that conclusion should be stated. A period for questions and answers is usually provided at the end of the presentation to clarify technical points that may not be fully understood. If possible, the technical report (discussed in the next section) should be available to the audience.

FIG. 7.18 A presentation should make full use of graphical aids and models to assist the speaker in communicating ideas. (Courtesy of Bendix Corp.)

Presentation critique

The evaluation form in Fig. 7.19 aids the student in planning as well as evaluating a presentation. The names of each team member are given at the top of the sheet and their percent contribution is decided by the team as a group. This column must add up to 100 percent. The F-factor is found for each member by multiplying the number of members on the team by the percent contribution of each. (Use the chart in Appendix 50 to find the grade of each individual.) An-

oral report

TEAM NO: _5_   PROJECT: _TOY MANUFACTURING_

NAMES NO. (N= 7) **a**	% CONTRIBUTION(C) **b**	F=NC **c**	GRADE (G)
1. _BROWN, G._	17	119	91
2. _PORTER, G._	14.3	100	87
3. _SMITH, G._	20.0	140	95
4. _REED, T._	5.7	40	63
5. _POTTER, M._	14.3	100	87
6. _FLYNN, O._	14.3	100	87
7. _ROSS, J._	14.4	100	87
8. _____			

EVALUATION BY INSTRUCTOR.
NOTE: Use this as a checklist for preparing for the oral presentation.)   100%

		MAX. VALUE	POINTS EARNED
1.	Introduction of team members	2	2
2.	Statement of purpose of the presentation	5	4
3.	Use of adequate number of visuals	10	8
4.	Continuity of the presentation	3	3
5.	Quality of visuals (including model)	15	12
6.	Use of visuals--point to important points, don't block screen, don't fumble, etc.	10	8
7.	Participation of team members (perfect score if all participate)	10	10
8.	Clear presentation of recommended design	10	8.5
9.	Presentation of alternative solution considered	2	2
10.	Coverage of economics - manufacturing, shipping, packing, overhead, mark-up, etc.	10	8.5
11.	Consideration of human factors	5	4
12.	Presentation of an effective conclusion	5	5
13.	Poise and professionalism	3	3
14.	Use of allotted time	10 / 100	9

ADDITIONAL COMMENTS BY INSTRUCTOR ON BACK OF SHEET.

grade 87

FIG. 7.19 An evaluation form for grading a team's oral presentation. Individual grades are found by using the chart in Appendix 50.

project uniqueness

TEAM NO: 5   PROJECT: _TOY MANUFACTURING_

NAMES NO. (N= 7) **a**	% CONTRIBUTION(C) **b**	F=NC **c**	GRADE (G)
1. _BROWN, G._	14.3	100	83
2. _PORTER, G._	14.3	100	83
3. _SMITH, G._	18.0	126	90
4. _REED, T._	14.3	100	83
5. _POTTER, M._	14.3	100	83
6. _FLYNN, O._	14.3	100	83
7. _ROSS, J._	10.5	74	75
8. _____			

100%

EVALUATION BY INSTRUCTOR

	UNIQUENESS	MAX. VALUE	POINTS EARNED
1.	Project serves needed function	10	8
2.	Functions effectively	10	7.5
3.	Would attract consumers	10	9
4.	Simplistic solution--minimum complication	10	8.5
5.	Reasonable cost	10	8
6.	Attractive for investment and marketing	10	7
7.	Application of imagination and ingenuity	10	9
8.	Aesthetically pleasing in appearance	10	8
9.	No existing solutions by others at present	10	9
10.	Fulfills problem statement	10 / 100	9

grade 83

ADDITIONAL COMMENTS BY IN INSTRUCTOR ON BACK OF SHEET.

FIG. 7.20 An evaluation for grading the uniqueness of a team's project. Individual grades are found by using the chart in Appendix 50.

other form (Fig. 7.20) is used to evaluate the creativity of the project being presented. This form also allows for the division of the team's grade among the members of the team based on the contribution of each.

7.6
The technical report

Engineers, technologists, and technicians at all levels must know how to prepare a written report. The report can be written for a project proposal, a progress report, or a final report (Fig. 7.21).

The project proposal

The proposal is written to substantiate the need for a given project that will require the authorization of funds and the use of manpower within the organiza-

tion. Often, the proposal is a report submitted to a client, outlining a recommendation that will require an expenditure of funds. The proposal includes data, costs, specifications, time schedules, personnel requirements, completion dates, and any other information that will aid the reader in understanding the project.

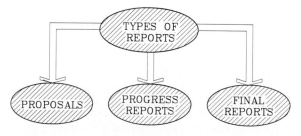

FIG. 7.21 The three basic types of technical reports.

Above all, the purpose and importance of the proposed project must be clearly stated with emphasis on the benefit of the project to the client or organization.

The proposal must also be written in the language of the reader. The businessperson will be interested in profits and benefits that a project promises, whereas the chief engineer will be concerned with its feasibility. Therefore, a typical proposal should contain the following major elements:

- *Statement of the problem.* The problem is clearly identified to present the purpose of the project.
- *Method of approach.* The procedures for attacking the problem are outlined and explained.
- *Personnel needs and facilities.* Requirements for equipment, space, and personnel are itemized.
- *Time schedule.* A time schedule gives an estimate of the completion date for each phase of the project.
- *Budget.* The funds required should be sufficiently detailed.
- *Summary.* The proposal is summarized to emphasize its most important points.

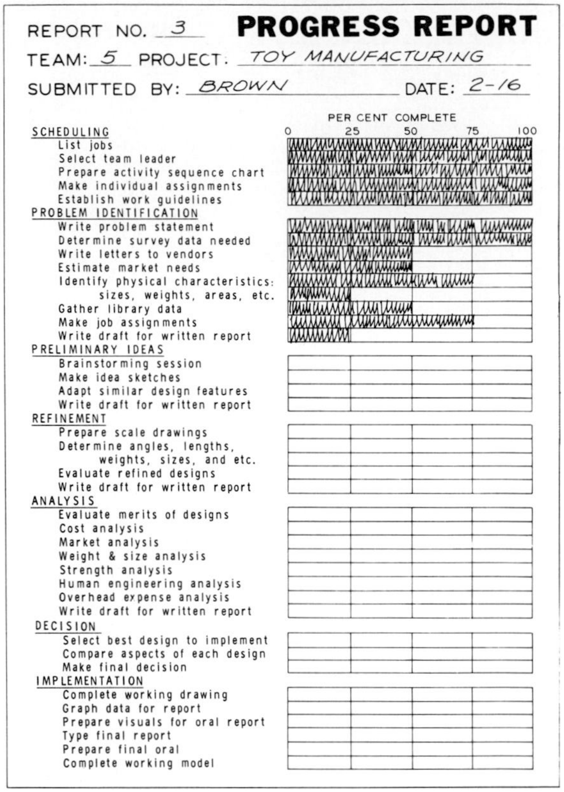
REPORT NO. 3 **PROGRESS REPORT**
TEAM: 5 PROJECT: TOY MANUFACTURING
SUBMITTED BY: BROWN DATE: 2-16

PER CENT COMPLETE
0 25 50 75 100

SCHEDULING
List jobs
Select team leader
Prepare activity sequence chart
Make individual assignments
Establish work guidelines
PROBLEM IDENTIFICATION
Write problem statement
Determine survey data needed
Write letters to vendors
Estimate market needs
Identify physical characteristics: sizes, weights, areas, etc.
Gather library data
Make job assignments
Write draft for written report
PRELIMINARY IDEAS
Brainstorming session
Make idea sketches
Adapt similar design features
Write draft for written report
REFINEMENT
Prepare scale drawings
Determine angles, lengths, weights, sizes, and etc.
Evaluate refined designs
Write draft for written report
ANALYSIS
Evaluate merits of designs
Cost analysis
Market analysis
Weight & size analysis
Strength analysis
Human engineering analysis
Overhead expense analysis
Write draft for written report
DECISION
Select best design to implement
Compare aspects of each design
Make final decision
IMPLEMENTATION
Complete working drawing
Graph data for report
Prepare visuals for oral report
Type final report
Prepare final oral
Complete working model

FIG. 7.22 A progress report form for reporting the weekly progress by a student team on a design project.

The progress report

The progress report is used to periodically review the status of a project. Progress reports which may be in the form of a letter or memo, usually give a projection of an increase or decrease in expenditures or time schedules to permit a revision of project plans. Reference to Project Evaluation and Review Techniques (PERT) introduced in Chapter 3 suggests effective methods of reporting the status of a design project. Figure 7.22 shows a weekly progress report for student projects.

The final report

The final report is written at the conclusion of a project. Recommendations in a final report are more conclusive than those in the proposal and progress report, since they are based on the results of the total project rather than on predictions and speculations.

7.7 Organization of a technical report

A good technical report comprises the following broad areas:

1. problem identification.
2. method.
3. body.
4. findings.
5. conclusions and recommendations.

The order of presenting these areas will vary with the requirements of the governing organization.

For instance, some reports are written with the conclusions and recommendations preceding the body so that the reader immediately sees the results of the report.

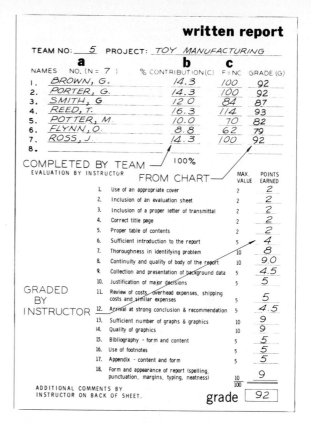

FIG. 7.24 A typical evaluation sheet for a student report. Individual grades are taken from the grading chart in Appendix 50.

Format of the report

A general sequence of the contents of a technical report is shown in Fig. 7.23.

1. *Cover.* Your finished report should always be bound in an appropriate cover. The title of the report and the name of the person or team that prepared it should be on the cover.

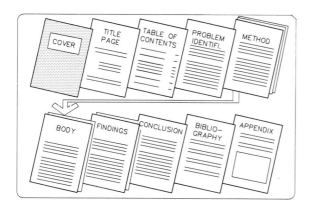

FIG. 7.23 The elements of a technical report.

2. *Evaluation sheet.* The first page of the report should be a standard evaluation sheet of the type shown in Fig. 7.24.

3. *Letter of transmittal.* The second page should be a letter of transmittal that briefly describes the contents of the report and the reasons for initiating the project (Fig. 7.25).

4. *Title page.* The title page should contain the title of the report, the name of the person or team that prepared it, and other elements shown in Fig. 7.26.

5. *Table of contents.* The major headings of the report should be shown (Fig. 7.27).

6. *Table of illustrations.* This listing of illustrations may be omitted in less formal reports.

7. *Problem identification.* (Use a heading appropriate for your report rather than this general term.) This section should establish the importance of and the need for a solution to the problem by reviewing factual information on the problem.

A typical page of a report is shown in Fig. 7.28, and a page containing a footnote is shown in Fig. 7.29. You should adhere to the margins shown here for all pages of a report, including those with illustrations. Use the third person, but do not say, "We investigated the problem and I suggested . . . ."

Drawings and illustrations should be included as numbered figures, with legends to describe them and

LETTER OF TRANSMITTAL – SECOND PAGE OF YOUR REPORT. ADDRESS TO YOUR TEACHER.

May 8, 1987

Professor J. T. Coppinger
Engineering Design Graphics Dept.
Texas A&M University
College Station, Texas 77843

Dear Professor Coppinger:

Attached is our report which outlines our recommendations for the establishment of a manufacturing operation to produce an educational toy at a rate of 5,000 per month.

We have researched and studied this problem very closely, and we feel that our conclusions and recommendations are based on sound judgement. We are hopeful that our findings meet with your approval.

Sincerely,

Gerald Brown
Gerald Brown, Chief Designer
Team 5, EDG 105, Section 145

FIG. 7.25 A letter of transmittal for a technical report.

LIST HOME ADDRESSES ON TITLE PAGE.

ORGANIZATION AND DESIGN
OF A TOY MANUFACTURING OPERATION

BY

TEAM 5

Instructor: J. T. Coppinger
Engineering Design Graphics 105
Texas A&M University
May 8, 1987

TEAM MEMBERS

Gerald Brown, 3305 NW 60 Street, Oklahoma City, OK
Greg Porter, 330 Veda Mae, San Antonio, TX
Gary Smith, 918 Sutton, San Antonio, TX
Tom Reed, 386 Highway 36, Caldwell, TX
Clifton Potter, 212 East Avenue F, Robstown, TX
Olivaree Flynn, 1002 North Flores, Rio Grande, TX
James R. Ross, 108 Ridgeway, San Marcos, TX

A

FIG. 7.26 The title page of a technical report prepared by a student team.

DOUBLE SPACE

TABLE OF CONTENTS

Introduction to the Problems 1
The Problem Statement 1
    Method of Approach 3
    Dado Connection 6
    Hammer 7
Assembly 9
Production Procedures 10
    End Sections 10
    Pegs 14
    Hammer Head 16
    Horizontal Section 17
Coloration 24
Construction Materials 27
Personnel 28
Machinery 29
Plant Layout 30
Wages and Time 33
Insurance 36
Operational Costs 37
Conclusions 39
Bibliography 41
Appendix 42

FIG. 7.27 The table of contents for a technical report.

$1\frac{1}{2}''$ USE OPAQUE PAPER. DOUBLE SPACE TEXT.

$1\frac{1}{2}''$ — INTRODUCTION TO THE PROBLEM — $1''$

Many engineering projects require that facilities be either developed or modified for the manufacture or processing of a product. Engineers must be concerned with methods of economically establishing a production system, routing of the product and management of the people and machines necessary for an efficient operation. Consideration must be given to a wide range of areas, with alternate systems for augmentation in case of emergencies. This project requires the application of these principles.

THE PROBLEM STATEMENT

An illustration of an educational toy is shown in Fig. 1 that is to be manufactured. This toy is made entirely of wood and is held together with screws. Retailers have determined that this toy has a market potential of 5,000 sales per month with expectations of increasing sales after the first six months.

A manufacturing plant must be outfitted and established for producing these toys in order to supply this market. Consideration was required of the following: square footage of the plant, types of machines, space for raw materials, office facilities and warehousing space for the finished products. Additional

$1''$

USE THESE MARGINS ON ALL PAGES.

FIG. 7.28 A typical page of text of a technical report. Notice the margins that must be used.

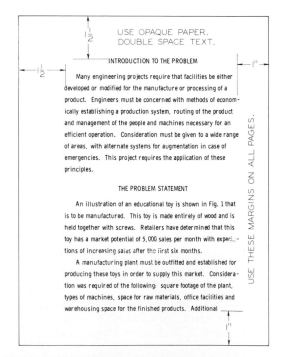

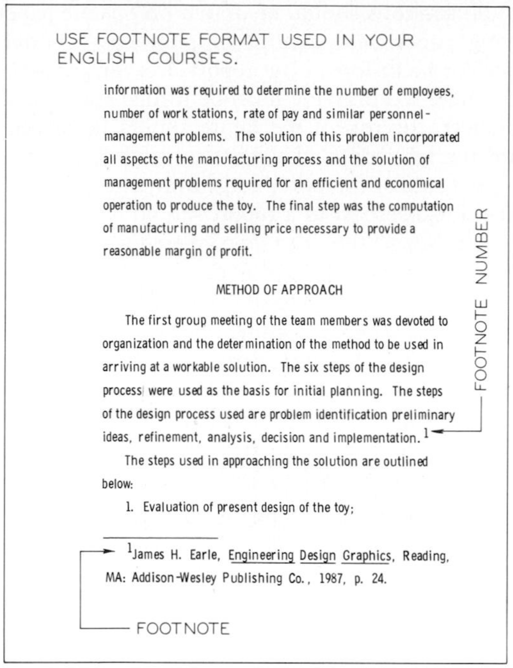

relate them to the text. The text should refer to a figure that follows; for example, "The number of boats sold between 1950 and 1986 is shown in Figure 6."

The figure should be placed immediately after the text that refers to it or on the following page. Figure 7.30 shows a figure that is referred to in Fig. 7.28.

USE FOOTNOTE FORMAT USED IN YOUR ENGLISH COURSES.

information was required to determine the number of employees, number of work stations, rate of pay and similar personnel-management problems. The solution of this problem incorporated all aspects of the manufacturing process and the solution of management problems required for an efficient and economical operation to produce the toy. The final step was the computation of manufacturing and selling price necessary to provide a reasonable margin of profit.

METHOD OF APPROACH

The first group meeting of the team members was devoted to organization and the determination of the method to be used in arriving at a workable solution. The six steps of the design process were used as the basis for initial planning. The steps of the design process used are problem identification preliminary ideas, refinement, analysis, decision and implementation.[1]

The steps used in approaching the solution are outlined below:

    1. Evaluation of present design of the toy;

[1]James H. Earle, _Engineering Design Graphics_, Reading, MA: Addison-Wesley Publishing Co., 1987, p. 24.

FOOTNOTE NUMBER

FOOTNOTE

FIG. 7.29 A page of text with a footnote.

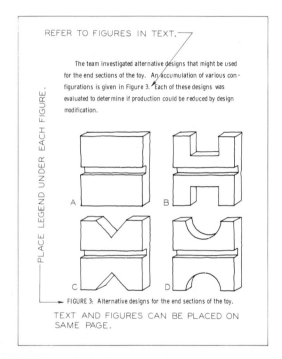

REFER TO FIGURES IN TEXT.

The team investigated alternative designs that might be used for the end sections of the toy. An accumulation of various configurations is given in Figure 3. Each of these designs was evaluated to determine if production could be reduced by design modification.

PLACE LEGEND UNDER EACH FIGURE.

A   B

C   D

FIGURE 3: Alternative designs for the end sections of the toy.

TEXT AND FIGURES CAN BE PLACED ON SAME PAGE.

FIG. 7.31 A page that contains both text and figures.

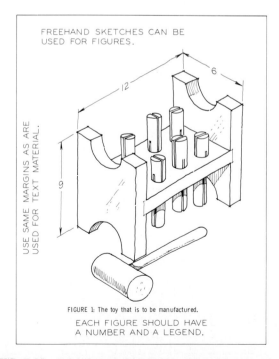

FREEHAND SKETCHES CAN BE USED FOR FIGURES.

6

12

USE SAME MARGINS AS ARE USED FOR TEXT MATERIAL.

9

FIGURE 1: The toy that is to be manufactured.

EACH FIGURE SHOULD HAVE A NUMBER AND A LEGEND.

FIG. 7.30 A freehand sketch can be used as a figure. Figures must adhere to the same margins as text.

8. _Method._ (Use a heading appropriate for your report rather than this general term.) This section should cover the general method used in solving the problem.

9. _Body._ (Use a heading appropriate for your report rather than this general term.) This part of the report elaborates on the solution of the problem. Subheadings should be used to make the parts of the report stand out. Where possible, illustrate your report with sketches, instrument drawings, graphs, or photographs (Fig. 7.30–7.33.)

TABLES OF DATA ARE EASIER TO INTERPRET IF GRAPHED.

OPERATIONAL COSTS

The monthly cost of materials required for the toy totals $1,423.42. The equipment, building, wages, insurance, materials, utilities and taxes total $57,860 for one year. For all practical purposes, this is considered as $58,000 for the first year. Although materials and insurance will cost $23,172 for one year, this is listed below as $23,000.

If the product is sold for 50¢ per toy, gross income will be $30,000 each year excluding any expansion. On a yearly basis, this process will pay for the equipment and building space after 5 years (Fig. 23). In actuality, the payments would be made on a monthly basis and over a number of years everything would be paid for. The yearly expenditures and income are shown below.

	Gross Income	Payments Necessary	Balance
First Year	$30,000	$58,000	-$28,000
Second Year	$30,000	$51,000	-$21,000
Third Year	$30,000	$44,000	-$14,000
Fourth Year	$30,000	$37,000	-$7,000
Fifth Year	$30,000	$30,000	$0,000
Sixth Year	$30,000	$23,000	+$7,000
Seventh Year	$30,000	$23,000	+$14,000
Eighth Year	$30,000	$23,000	+$21,000

FIG. 7.32 Tabular data can be given in a report, but data are easier to interpret if graphed.

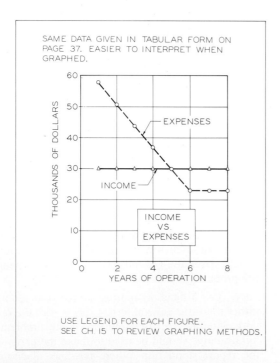

SAME DATA GIVEN IN TABULAR FORM ON PAGE 37. EASIER TO INTERPRET WHEN GRAPHED.

USE LEGEND FOR EACH FIGURE.
SEE CH 15 TO REVIEW GRAPHING METHODS.

FIG. 7.33 The data shown in Fig. 7.32 are graphed in this example for easier interpretation.

Illustrations should be drawn on opaque paper or tracing paper, and a print or photocopy should be made for inclusion in the report. Ink illustrations are preferable. If a drawing must be turned lengthwise on the sheet, the top of the figure should be placed toward the left side of the sheet so it can be read from the right side of the page. Large drawings that must be folded for insertion in a report should be folded to $8\frac{1}{2}$ by 11 inches so they can be conveniently unfolded by the reader.

10. *Findings.* (Use a heading appropriate for your report rather than this general term.) The results of the report should be tabulated and presented graphically with explanatory text. The data presented in Fig. 7.32 are easier to interpret when graphed as shown in Fig. 7.33.

11. *Conclusions.* The conclusions should summarize the entire report and include specific recommendations.

BIBLIOGRAPHY

Adams, Paul L., Manufacturing Techniques, New York, The Ajax Press, 1982.

Alexander, E. A., Production Methods, Chicago: The New Publishing Company, 1985.

Bardell, N. B., Plant Processes, Denver: Richmond Publishing Company, 1983.

Earle, James H., Engineering Design Graphics, Reading, MA: Addison-Wesley Publishing Company, 1987.

FIG. 7.34 One form of giving the bibliography of references used in the report.

FIG. 7.35 Data and survey information can be placed in the appendix.

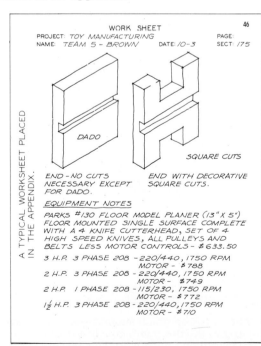

FIG. 7.36 A worksheet worthy of inclusion in the appendix. Sloppy, hard-to-read sketches should be omitted from the report.

12. *Bibliography*. This is a list of reference material—books, magazines, brochures, conversations—used in the preparation of the report. Bibliographies are usually listed alphabetically by author (Fig. 7.34).

13. *Appendix*. The appendix should include less important information (drawings, sketches, raw data, brochures, letters) not appropriate for inclusion in the main part of the report (Figs. 7.35 and 7.36).

Common omissions

Topics often omitted from technical reports that should not be omitted include cost estimates, overhead expenses, human and psychological factors, shipping costs, packing specifications, advertising methods, sales considerations, and summarizing recommendations.

7.8
Decision

The purpose of oral and technical reports is to present the basis for a decision on whether or not a recommended design should be implemented.

ACCEPTANCE   A design may be accepted in its entirety, which is a compliment to the designer's research and problem solving abilities.

REJECTION   A design recommendation may be rejected in its entirety. Changes in the economic climate or moves by competitors may make the design unprofitable.

COMPROMISE   If a design is not approved in its entirety, a compromise might be suggested in one or more areas; for example, the initial production run might be increased or decreased, or several features might be modified to make a design more attractive.

Decision—hunting seat

The design of a hunting seat is used to illustrate the decision step of the design process. The problem is restated below.

DECISION
1. Decision table for evaluation

DESIGN 1: FOLDING SEAT

DESIGN 2: CANVAS SEAT

DESIGN 3: PLATFOM SEAT

DESIGN 4:

DESIGN 5:

DESIGN 6:

Maximum value	Factors for analysis	1	2	3	4	5	6
3.0	Function	2.0	2.3	2.5			
2.0	Human factors	1.6	1.4	1.7			
.5	Market analysis	.4	.4	.4			
1.0	Strength	1.0	1.0	1.0			
.5	Production proced.	.3	.2	.4			
1.0	Cost	.7	.6	.8			
1.5	Profitability	1.1	1.0	1.3			
.5	Appearance	.3	.4	.4			
10	TOTALS	7.4	7.3	8.5			

FIG. 7.37 A worksheet with a decision table used to evaluate the developed design alternatives.

HUNTING SEAT Many hunters hunt from trees to obtain a better vantage point. Design a seat that provides the hunter with comfort and safety while hunting from a tree and that meets the requirements of economy and the limitations of hunting.

DECISION CHART The chart shown in Fig. 7.37 can be used to compare the available designs. Each idea is listed and given a number for identification.

Next, maximum values for the various factors of analysis are assigned that the total of all factors is ten points. These values will vary from product to product, so use your best judgment to determine them. You can now evaluate each factor of the competing designs.

The vertical columns of numbers are summed to determine the design with the highest total, the best design, perhaps. However, your instincts may disagree with your numerical analysis. If so, you should have enough faith in your judgment not to be restricted by your numerical analysis. Decision will always remain the most subjective part of the design process.

CONCLUSION Once a decision has been made, it should be clearly stated, along with the reasons for the design's acceptance (Fig. 7.38). Additional information may be given such as number to be produced initially, selling price per unit, expected profit per unit, expected sales during the first year, number that must be sold to break even, and its most marketable features.

It is possible that you would recommend a design not be implemented. The design process should not then be considered a failure; a negative decision could save an investor from large losses.

CONCLUSIONS   DESIGN

IMPLEMENT AND PRODUCE THE FLAT
FOLDING SEAT. THIS DESIGN IS BELIEVED
TO BE THE BEST SOLUTION AND ONE
THAT CAN BE PROFITABLY MARKETED
THROUGH SPORTING GOODS DEALERS.

SALES PRICE $ 90

SHIPPING EXPENSES $ 5

NUMBER TO SELL TO   1500
    BREAK EVEN

MANUFACTURED BY CON-
    TRACTORS AT OUTSET

ESTIMATED PROFIT $ 20
    PER SEAT

Graphics for Engineers © | NAME DOE, JOHN L. | FILE 2 SEC 41 DATE 12-1 | MIN. 5 | GRADE 10

FIG. 7.38 The decision is summarized on this worksheet to give the designer's conclusion and recommendation concerning the next step, implementation.

Problems

1. Prepare a checklist that could be used to evaluate an oral presentation of one of your classmates. List important items that should be considered, and determine a point system that could be assigned to each. Keep the form simple, yet thorough enough to be of value to the presenter. Devise a means of tabulating the evaluation to arrive at an overall rating.

2. Prepare a series of 3-by-5-inch cards to plan a flip-chart presentation that will last no more than five minutes. The subject of your flip-chart presentation may be of your choosing or assigned by your teacher. Some example topics are your career plans for the first two years after graduation, the role of this course in your total educational program, the importance of effective communications, the identification of a need for a design project you are proposing, a comparison of the engineering profession with another profession of your choice.

3. Prepare graphical aids for an oral presentation using the methods and materials indicated in this chapter.

4. Using the planning cards developed in Problem 2, prepare a five-minute briefing for a technique that you choose or your instructor assigns. Give this briefing to your class as assigned.

5. Assume you are an engineer responsible for representing your firm in the presentation of a proposal for a sizable contract. Make a list of instructions that you could give to your assistants to coordinate the preparation of your presentation. The presentation will be for a group of 20 persons ranging in background from bankers to engineers. The topic is not so important as the method of presentation. Use a topic of your choice, one assigned by your teacher, or one suggested in Problem 2. Your instructions should outline the materials you need, method of preparation of graphical aids and number required, method of projection or presentation, assistance needed in presenting materials, room seating arrangements, and other factors. Your outline should be complete enough to cover the entire program of an ideal presentation within the time you think most desirable.

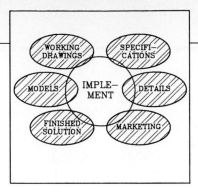

CHAPTER 8

Implementation

8.1
Introduction

The final step of the design process is implementation, during which the design becomes a reality. Graphical methods are particularly important during the initial steps of implementation, since all products are constructed from working drawing with specifications.

8.2
Working drawings

Working drawings are drawn using orthographic views that have been dimensioned and noted to show how the individual parts are to be made. An example of a working drawing generated with computer graphics is shown in Fig. 8.1, where a base plate mount is orthographically described. Notes specify the materials and the sizes of the holes. A properly drawn working drawing will result in the same product time and again, regardless of the shop in which it is made. When making working drawings, several parts can be drawn on the same sheet without attempting to arrange them in relation to one another. The names of the parts, their identifying numbers, and their materials are given near the views.

8.3
Specifications

Specifications are notes and instructions that supplement working drawings. Specifications may be given in typed documents when graphical representations are unnecessary; for example, instructions like those shown below can be given effectively as written spec-

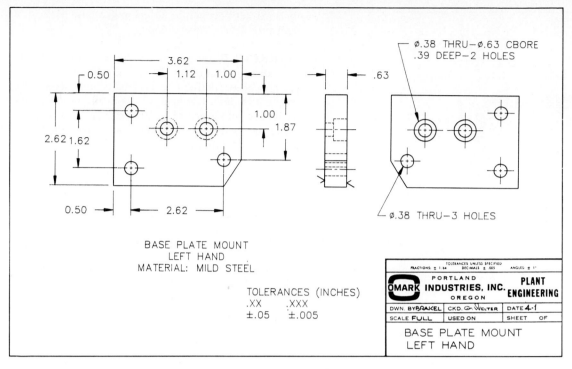

FIG. 8.1 A computer-drawn working drawing of a single part that was made as the implementation step of the design process. (Courtesy of Omark Industries, Inc.)

ifications:

> Metallurgical inspection is required before machining.

or

> Paint with two coats of flat black paint (NO. 780) after finishing.

However, when space permits, it is more convenient that specifications be given on the working drawing by the drafter.

8.4
Assembly drawings

Assembly drawings illustrate how parts are put together after they have been made. They can be drawn as pictorial or orthographic views that are fully assembled, fully exploded, or partially exploded. Figure 8.2 is a partially exploded orthographic assembly where sections have been used to clarify internal details. A parts list is usually given along with the assembly drawing.

8.5
Miscellaneous considerations

In addition to the preparation of drawings and specifications, several other aspects of the implementation step must be considered: packaging, storage, shipping, and marketing of the product.

PACKAGING In some cases, such as the toy industry, the packaging is very elaborate and may be as expensive as the product. Designers must be aware of the packaging needs as they develop a design, since a product that is difficult to package will cost more. Many products, to make them easier to package and more economical, are shipped partially disassembled.

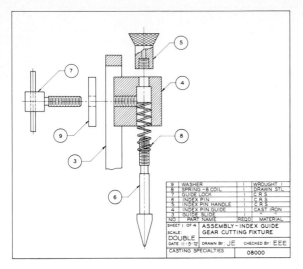

FIG. 8.2 A partially exploded orthographic assembly drawing of an index guide to a cutting fixture, which shows how the individual parts of an assembly are put together.

STORAGE Most manufacturers maintain a backlog of products for shipment. The cost of keeping an inventory must be figured as part of the product's final selling price.

SHIPPING Some industries locate their shipping facilities in the middle of their market areas to reduce shipping costs. But companies at the extreme ends of

their market's geographical region just raise the cost of their products to cover shipping.

MARKETING Designers are concerned with all aspects of a product after it enters the marketplace. They are concerned with its marketability and acceptance by the consumer. Complaints about the product's effectiveness and function are important to designers because they will want to modify any defect in future versions of the product.

8.6
Implementation—hunting seat

The design of a hunting seat illustrates the implementation step of the design process. The problem is restated below.

HUNTING SEAT Many hunters hunt from trees to obtain a better vantage point. Design a seat that provides the hunter with comfort and safety while hunting from a tree and that meets the requirements of economy and the limitations of hunting.

Four working drawing sheets (Figs. 8.3–8.6) have been prepared to present the details of the hunting seat design. The fifth sheet, Fig. 8.7, is an assembly drawing that illustrates how the parts are assembled once they have been made. (This particular design was developed, patented, and is marketed by Baker Man-

FIG. 8.3 A working drawing sheet of parts of a hunting seat design. Sheet 1 of 5.

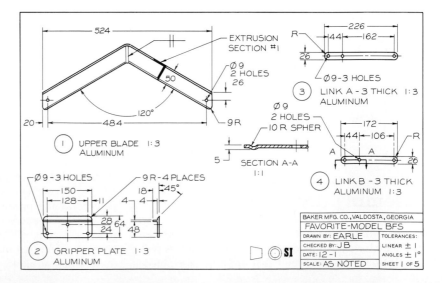

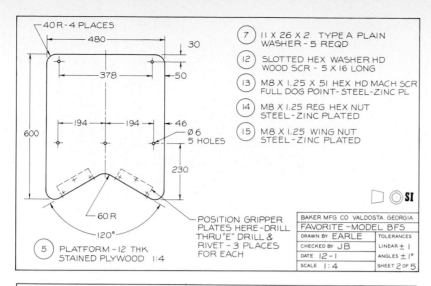

7   11 X 26 X 2   TYPE A PLAIN
WASHER - 5 REQD

12   SLOTTED HEX WASHER HD
WOOD SCR - 5 X 16 LONG

13   M8 X 1.25 X 51 HEX HD MACH SCR
FULL DOG POINT- STEEL-ZINC PL

14   M8 X 1.25 REG HEX NUT
STEEL-ZINC PLATED

15   M8 X 1.25 WING NUT
STEEL-ZINC PLATED

40 R - 4 PLACES
480
30
378
50
194   194   46
600
Ø 6
5 HOLES
230
60 R
120°
5   PLATFORM - 12 THK
STAINED PLYWOOD   1:4

POSITION GRIPPER
PLATES HERE - DRILL
THRU "E" DRILL &
RIVET - 3 PLACES
FOR EACH

BAKER MFG CO   VALDOSTA   GEORGIA
FAVORITE - MODEL BFS

DRAWN BY EARLE	TOLERANCES
CHECKED BY JB	LINEAR ± 1
DATE 12 - 1	ANGLES ± 1°
SCALE 1 : 4	SHEET 2 OF 5

FIG. 8.4 A working drawing sheet of parts of a hunting seat design. Sheet 2 of 5.

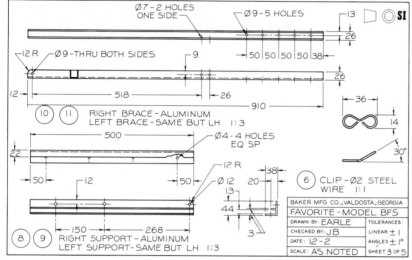

Ø 7 - 2 HOLES
ONE SIDE
Ø 9 - 5 HOLES
13
26
12 R
Ø 9 - THRU BOTH SIDES
9
50 50 50 50 38
26
12
518
26
910
36
14
30°

10   11   RIGHT BRACE - ALUMINUM
LEFT BRACE - SAME BUT LH   1:3

500
Ø 4 - 4 HOLES
EQ SP
22
50
12
12 R
50
Ø 12   20
38
13
44
3
6   CLIP - Ø2 STEEL
WIRE   1:1

8   9   RIGHT SUPPORT - ALUMINUM
LEFT SUPPORT - SAME BUT LH   1:3
150   268

BAKER MFG CO., VALDOSTA, GEORGIA
FAVORITE - MODEL BFS

DRAWN BY EARLE	TOLERANCES:
CHECKED BY JB	LINEAR ± 1
DATE 12 - 2	ANGLES ± 1°
SCALE AS NOTED	SHEET 3 OF 5

FIG. 8.5 A working drawing sheet of parts of a hunting seat design. Sheet 3 of 5.

FIG. 8.6 A working drawing sheet of parts of a hunting seat design. Sheet 4 of 5.

16   STRAP - 25 WIDE #7166
BLACK NYLON - 900 LONG

17   SHOCK CORD - Ø8 X 700 LONG
ENDS TIED TOGETHER

18   RIVET - Ø10 X 12 - STEEL
ZINC PLATED

19   RIVET - Ø10 X 22 - STEEL
ZINC PLATED

1 R X 1 DEEP GROOVES
@ 3 OC
51
4
1 R X 1 DEEP
GROOVES @ 3 OC
4
38
45°
4
EXTRUSION SECT #1
UPPER BLADE   1:1

BAKER MFG CO., VALDOSTA, GEORGIA
FAVORITE - MODEL BFS

DRAWN BY: EARLE	TOLERANCES:
CHECKED BY JB	LINEAR ± 0.5
DATE: 12 - 5	ANGLES ± 1°
SCALE:	SHEET 4 OF 5

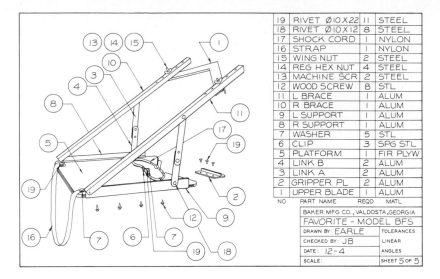

19	RIVET Ø10 X 22	11	STEEL
18	RIVET Ø10 X 12	8	STEEL
17	SHOCK CORD	1	NYLON
16	STRAP	1	NYLON
15	WING NUT	2	STEEL
14	REG HEX NUT	4	STEEL
13	MACHINE SCR	2	STEEL
12	WOOD SCREW	8	STL
11	L BRACE	1	ALUM
10	R BRACE	1	ALUM
9	L SUPPORT	1	ALUM
8	R SUPPORT	1	ALUM
7	WASHER	5	STL
6	CLIP	3	SPG STL
5	PLATFORM	1	FIR PLYW
4	LINK B	2	ALUM
3	LINK A	2	ALUM
2	GRIPPER PL	2	ALUM
1	UPPER BLADE	1	ALUM
NO	PART NAME	REQD	MATL

BAKER MFG CO., VALDOSTA, GEORGIA
FAVORITE - MODEL BFS

DRAWN BY: EARLE	TOLERANCES
CHECKED BY: JB	LINEAR
DATE: 12-4	ANGLES
SCALE:	SHEET 5 OF 5

FIG. 8.7 An assembly drawing that demonstrates how the parts of the hunting seat design are assembled. Sheet 5 of 5.

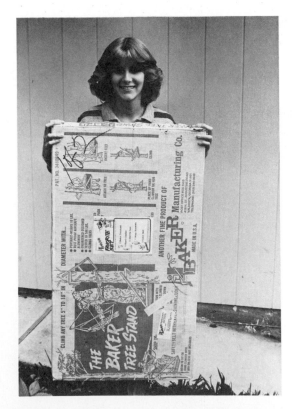

FIG. 8.8 The Baker Favorite Seat is shipped in a corrugated cardboard box to the retailer or the consumer.

ufacturing Company, Valdosta, Georgia. It is the Baker Favorite Seat, Patent No. 3460649.)

All parts have been dimensioned in millimeters, which are metric units. Standard parts that are purchased from suppliers are not drawn, but are itemized on the drawing, given parts numbers, and listed in the parts list on the assembly drawing. Figure 8.7 is an assembly drawing, which is a pictorial with the different parts identified by balloons attached to leaders. Each part is listed in the parts list by number with general information to describe it.

PACKAGING The Baker Favorite Seat is packaged in a corrugated cardboard box and weighs approximately ten pounds (Fig. 8.8). The box has been customized to advertise the product and for easy recognition of each model by shipping personnel (Fig. 8.9).

STORAGE The periods before hunting seasons will require more inventory than other times of the year. An inventory of seats waiting to be sold adds to the cost of overhead in the form of interest payments, warehouse rent, warehouse personnel, and loading equipment.

SHIPPING Shipping costs for all types of carriers (rail, motor freight, air delivery, mail services) must be evaluated. The shipping cost for a Baker Favorite Seat with its accessories is $5–$10 depending on distance when shipped one at a time by United Parcel Service. The cost per unit is reduced by about fifty percent when they are shipped in bundles of ten to the same destination.

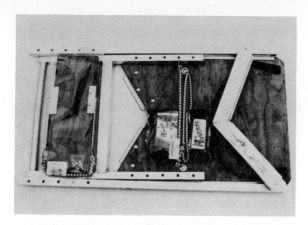

FIG. 8.9 The hunting seat is folded into a flat position for ease of packaging before shipment.

ACCESSORIES Examples of accessories are fold-down seats, hand climbers, and add-on seats (Fig. 8.10). Accessories provide for the special needs of the hunters, increase the marketability of the product, and increase sales volume.

The retail price of the Baker Seat is about ninety dollars, five or six times more than the cost of the materials and labor to manufacture it. Retailers are given approximately a forty percent margin, and distributors earn about ten percent. The remainder of the overhead includes advertising costs and other expenses already mentioned. All expenses must be absorbed by the consumer who ultimately purchases the product.

8.7
Patents

A designer who has developed an original and novel solution to a design problem should investigate the possiblities of obtaining a patent from the U.S. Patent and Trademark Office (PTO) (Fig. 8.11). This must be done before disclosing the invention, since premature disclosure may forfeit the patent right.

A good source for the specific details of the patent procedure is *General Information Concerning Patents*. (This publication, available at no cost from the PTO, was the primary reference for the following text.)

General patent requirements

Designers must have a general understanding of patent requirements, that is, what can be patented and who is eligible to apply for a patent.

FIG. 8.10 Examples of accessories designed to accompany the hunting seat. (Courtesy of Baker Manufacturing Co., Valdosta, Ga.)

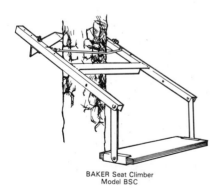

BAKER Seat Climber
Model BSC

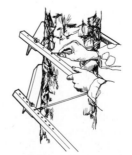

A Conversion Kit, Model CK, will convert Hand Climbers to Seat Climbers.

An accessory to the Seat Climber is the Padded Pouch — Model PP.

Secure Seat Climber and Tree Stand with tie down for added safty while hunting.

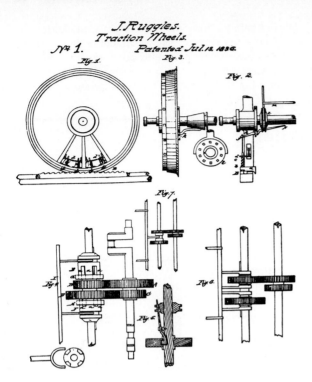

FIG. 8.11 The first patent, issued by the U.S. Patent Office in 1836.

however, be made by the executor of a deceased inventor's estate. Two or more persons may apply for a patent as joint inventors.

PATENT RIGHTS An inventor granted a patent has the right to exclude others from making, using, or selling the invention throughout the United States for seventeen years.

After expiration of the seventeen-year patent term, the invention may be made, used, or sold by anyone without authorization from the patent holder.

Patented articles must be marked with the word "Patent" and the number of the patent. Markings using the terms "Patent Pending" have no legal effect, since protection does not begin until the actual grant is made.

WHAT CAN BE PATENTED Any person who "invents or discovers any new and useful process, machine, manufacture, or composition of matter, or any new and useful improvement thereof, may obtain a patent," subject to the conditions and requirements of law. Essentially, these categories include everything made by humans and the processes for making them (Fig. 8.12).

Inventions used solely for the development of nuclear and atomic weapons for warfare are not patentable since they are not considered "useful." Also, a design for a functional mechanism that will not operate in keeping with its intended purpose is not patentable. An *idea* for a new invention or machine is not patentable; the specific design and description of the machine must be available before it can be considered for patent registration.

WHO CAN APPLY FOR A PATENT Only the inventor of a device may apply for a patent. A patent given to a person who was not the inventor would be void, and the person would be subject to prosecution for committing perjury. Application for a patent can,

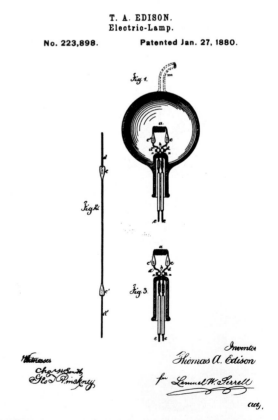

FIG. 8.12 Thomas Edison's patent drawing for the electric lamp, 1880.

Application for a patent

An inventor applying for a patent must include the following:

1. A written document that comprises a petition, a specification (description and claims), and an oath or declaration.

2. A drawing if a drawing is possible.

3. The filing fee.

PETITION AND OATH The petition and oath are usually on one form. On this form, the inventor petitions or requests to be given a patent on the invention and declares that he or she is the original and first inventor of the device described in the application.

SPECIFICATION Written specifications of a patent must be attached to the application describing the in-

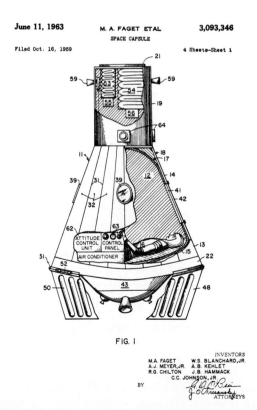

FIG. I

INVENTORS
M.A. FAGET   W.S. BLANCHARD, JR.
A.J. MEYER, JR.   A.B. KEHLET
R.G. CHILTON   J.B. HAMMACK
C.C. JOHNSON, JR.
BY

FIG. 8.13 The patent drawing of a space capsule developed by the National Aeronautics and Space Administration (NASA). (Courtesy of the U.S. Patent and Trademark Office.)

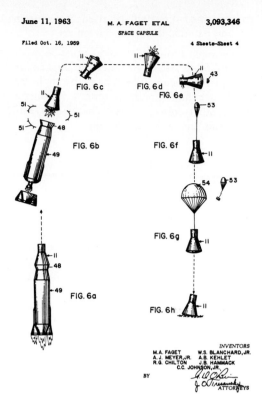

FIG. 8.14 The sequence of events showing the space capsule from launch to landing. (Courtesy of the U.S. Patent and Trademark Office.)

vention in detail so that a person skilled in the field to which the invention pertains can produce the item. Drawings should be referred to in the text by figures and part numbers (Figs. 8.13 and 8.14).

The following format is suggested for specifications:

1. Title of the invention, or a preamble stating the name, citizenship, and residence of the applicant and the title of the invention.

2. Summary of the invention.

3. If there are drawings, a brief description of several views given..

4. Detailed description.

5. Claims.

Claims are brief descriptions of the features of the invention that distinguish it from already patented material. The claims are the most significant part of the patent, since they will be used to ascertain the novelty and patentability of an invention.

FEE The application for a patent must be accompanied by the filing fee, $170; after the application has been accepted, notice will be sent to the applicant giving him or her three months to remit a $280 issue fee. There are other miscellaneous charges for multiple claims within the patent application.

8.8
The preparation
of patent drawings

When drawings are necessary to describe an invention, the applicant must submit them as well. A booklet, *Guide for Patent Draftsmen* (available from the U.S. Government Printing Office), outlines the procedures for preparing patent drawings. Most of these guidelines are presented here. If the inventor cannot furnish drawings, the Patent Office will recommend a drafter who can prepare them at the inventor's expense.

Patent drawing standards

When the patent is issued, the completed drawings are printed and published. R.H. Goddard's rocket patent was issued, and his drawings published, in 1914 (Fig. 8.15). To prevent rejection of the application, follow these drawing standards as closely as possible.

PAPER AND INK Drawings must be made on pure white paper of thickness corresponding to a two- or three-ply Bristol board. The surface must be calendered and smooth to permit erasure and correction. Only India ink will secure perfectly black solid lines. The use of white pigment to cover lines is not acceptable.

SHEET SIZE AND MARGINS The sheet size must be exactly 8½ by 14 inches (21.6 by 35.6 cm) or exactly 21.0 by 29.7 cm. All sheets in a particular application must be the same size. One of the shorter sides is regarded as the top of the sheet. On 8½-by-14-inch sheets, the top margin must be 2 in. and the side and bottom margins ¼ in. Margin border lines cannot be drawn on the sheets, but all work must be included within the margins. The sheets may be punched with two ¼-in. holes with their centerlines $\frac{11}{16}$ in. below the

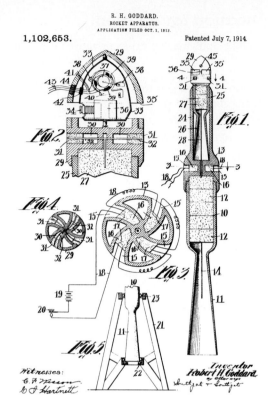

FIG. 8.15 R.H. Goddard's patent on a rocket, issued in 1914. (Courtesy of the U.S. Patent and Trademark Office.)

top edge and 2¾ in. apart and centered from the sides of the sheet. The margins for 21.0-by-29.7-cm sheets are 2.5 cm from the top, 2.5 cm from the left, 1.5 cm from the right, and 1 cm from the bottom.

CHARACTER OF LINES All lines and lettering must be absolutely black regardless of how fine the lines may be. Freehand work should be avoided.

HATCHING AND SHADING Hatching lines, used to shade the surface of an object, should be parallel and not less than $\frac{1}{20}$ in. apart (Fig. 8.16). Heavy lines are used on the shade side of the drawing; however, they should not be used if they are likely to confuse the drawing. The light is assumed to come from the upper left-hand corner at an angle of forty-five degrees (Fig. 8.17). Types of surface delineation are given in Figs. 8.18 and 8.19.

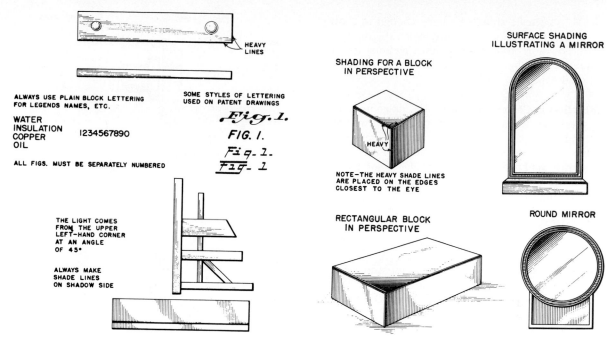

HEAVY
LINES

ALWAYS USE PLAIN BLOCK LETTERING
FOR LEGENDS NAMES, ETC.

SOME STYLES OF LETTERING
USED ON PATENT DRAWINGS

WATER
INSULATION
COPPER   1234567890
OIL

Fig. 1.

FIG. I.

Fig. 2.

FIG. 2.

ALL FIGS. MUST BE SEPARATELY NUMBERED

THE LIGHT COMES
FROM THE UPPER
LEFT-HAND CORNER
AT AN ANGLE
OF 45°

ALWAYS MAKE
SHADE LINES
ON SHADOW SIDE

SHADING FOR A BLOCK
IN PERSPECTIVE

HEAVY

NOTE—THE HEAVY SHADE LINES
ARE PLACED ON THE EDGES
CLOSEST TO THE EYE

SURFACE SHADING
ILLUSTRATING A MIRROR

RECTANGULAR BLOCK
IN PERSPECTIVE

ROUND MIRROR

FIG. 8.16 Typical examples of lines and lettering recommended for patent drawings.

FIG. 8.17 Techniques of shading patent drawings.

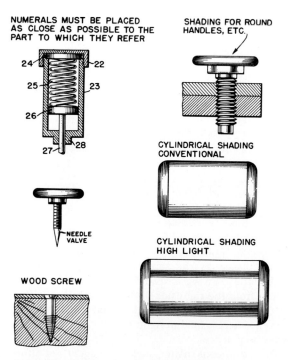

NUMERALS MUST BE PLACED
AS CLOSE AS POSSIBLE TO THE
PART TO WHICH THEY REFER

24   22
25   23
26
27   28

NEEDLE
VALVE

WOOD SCREW

SHADING FOR ROUND
HANDLES, ETC.

CYLINDRICAL SHADING
CONVENTIONAL

CYLINDRICAL SHADING
HIGH LIGHT

FIG. 8.18 Methods of numbering parts and rendering details for patent drawings.

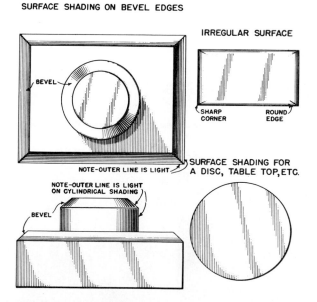

SURFACE SHADING ON BEVEL EDGES

IRREGULAR SURFACE

BEVEL

SHARP
CORNER

ROUND
EDGE

NOTE—OUTER LINE IS LIGHT

SURFACE SHADING FOR
A DISC, TABLE TOP, ETC.

NOTE—OUTER LINE IS LIGHT
ON CYLINDRICAL SHADING

BEVEL

FIG. 8.19 Techniques of representing sufaces and beveled planes on patent drawings.

SCALE The scale should be large enough to show the mechanism without crowding when the drawing is reduced for reproduction. Certain portions of the mechanism may be drawn at a larger scale to show additional details.

REFERENCE CHARACTERS The different views of a mechanism should be identified by consecutive figure numbers. Use plain, legible numerals at least ⅛ in. high, not encircled, and placed close to the parts to which they apply (Fig. 8.20). A leader is sometimes used to indicate these parts. Numbers should not be placed on hatched surfaces unless a blank space is provided for them. The same part appearing in more than one view of the drawing should be designated by the same character.

SYMBOLS Symbols used to represent various materials in sections, electrical components, and mechanical devices are suggested by the Patent Office, but these conform to the engineering drawing standards covered in Chapter 17.

SIGNATURE AND NAMES The signature of the applicant, or the name of the applicant and the signature of the attorney or agent, may be placed in the

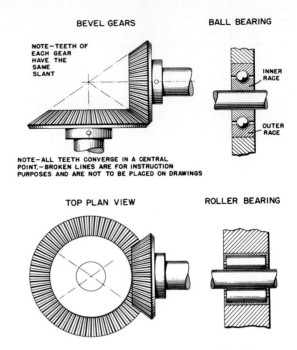

FIG. 8.21 Representation of gears and ball bearings on patent drawings.

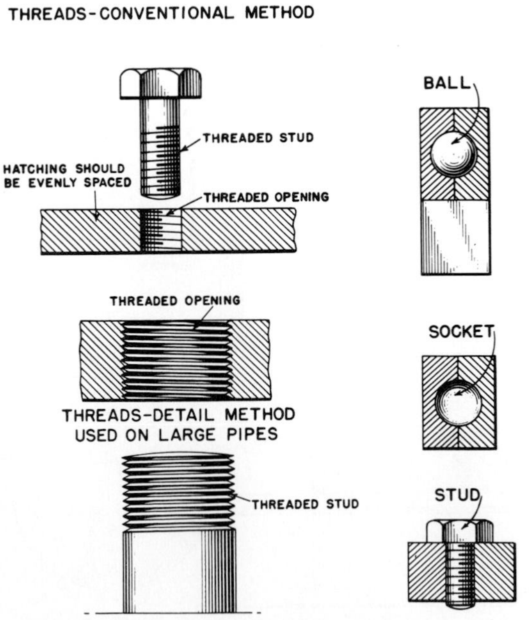

THREADS-CONVENTIONAL METHOD

FIG. 8.20 Techinques of representing threads and small components on patent drawings.

lower right-hand corner of each sheet within the marginal lines or below the lower marginal lines.

VIEWS Figures should be numbered consecutively in order of their appearance. Figures may be plan, elevation, section, perspective, or detail views (Figs. 8.20 and 8.21). Shading is used to show the shape of the components and the details of each part. Exploded views can be used to advantage to describe the assembly of a number of parts. Large parts may be broken into sections and drawn on several sheets if this does not confuse the matter. Removed sections can be used, provided that the cutting plane is labeled to indicate the section by number. Views should not be connected by projection lines. All sheet headings and signatures will be placed in the same position on the sheet whether the drawing is read from the bottom of the sheet or from the right side of the sheet. It is desirable that the drawing be positioned so that it can be read when the sheet is held upright; however, this may not always be practical.

No extraneous matter, such as an agent's or attorney's stamp or address, is permitted on the face of the drawing. The completed drawings should be sent flat, protected by heavy board or rolled in a suitable mailing tube. Drawings used for an accepted patent will not be returned to the applicant.

8.9
Patent searches

A patent can be granted only after the PTO examiners have searched existing patents to determine whether the invention has been previously patented. With more than 4,500,000 patents on record, a search is the most time-consuming part of obtaining a patent. Many inventors employ patent attorneys or agents to conduct a preliminary search of existing patents to discover whether their invention infringes on another. Patents are filed in the Search Room of the Patent Office by classes and subclasses according to subject matter. Searchers review the subclass that covers patents in the field of the application they are working on. However, other seemingly unrelated patents may cover the invention being submitted, thereby disallowing the patent.

8.10
Questions and answers about patents

For best results, patent applications must be handled with the assistance of a patent attorney or patent agent. The Patent Office has published a pamphlet, *Questions and Answers About Patents*. Most of these questions and answers are listed here.

Nature and duration of patents

1. Q. *What is a patent?*
A. A patent is a grant issued by the U.S. Government giving an inventor the right to exclude all others from making, using, or selling his or her invention within the United States, its territories and possessions.

2. Q. *For how long a term of years is a patent granted?*
A. Seventeen years from the date on which it is issued; except for patents on ornamental designs, which are granted for terms of $3\frac{1}{2}$, 7, or 14 years.

3. Q. *May the term of a patent be extended?*
A. Only by special act of Congress, and this oc-

curs very rarely and only in most exceptional circumstances.

4. Q. *Does the patentee continue to have any control over the use of the invention after the patent expires?*
A. No. Anyone has the free right to use an invention covered in an expired patent, so long as he or she does not use features covered in other unexpired patents in doing so.

5. Q. *On what subject matter may a patent be granted?*
A. A patent may be granted to the inventor or discoverer of any new and useful process, machine, manufacture, or composition of matter, or any new and useful improvement thereof, or on any distinct and new variety of plant, other than a tuber-propagated plant, which is asexually reproduced, or on any new, original, and ornamental design for an article of manufacture.

6. Q. *On what subject matter may a patent not be granted?*
A. A patent may not be granted on a useless device, on printed matter, on a method of doing business, on an improvement in a device which would be obvious to a person skilled in the art, or on a machine which will not operate, particularly on an alleged perpetual motion machine.

Meaning of words "patent pending"

7. Q. *What do the terms ''patent pending'' and ''patent applied for'' mean?*
A. They are used by a manufacturer or seller of an article to inform the public that an application for patent on that article is on file in the Patent Office. The law imposes a fine on those who use these terms falsely to deceive the public.

Patent applications

8. Q. *I have made some changes and improvements in my invention after my patent application was filed in the Patent Office. May I amend my patent application by adding a description or illustration of these features?*

A. No. The law specifically provides that new matter shall not be introduced into the disclosure of a patent application. However, you should call the attention of your attorney or patent agent promptly to any such changes you may make or plan to make, so that he or she may take or recommend any steps that may be necessary for your protection.

9. Q. *How does one apply for a patent?*
A. By making the proper application to the Commissioner of Patents, Patent and Trademark Office, Washington, D.C., 20231.

10. Q. *What is the best way to prepare an application?*
A. As the preparation and prosecution of an application are highly complex proceedings, they should preferably be conducted by an attorney trained in this specialized practice. The Patent Office therefore advises inventors to employ a patent attorney or agent who is registered in the Patent Office.

11. Q. *Of what does a patent application consist?*
A. An application fee, a petition, a specification, claims describing and defining the invention, an oath or declaration, and a drawing if the invention can be illustrated.

12. Q. *What are the Patent Office fees in connection with filing of an application for patent and issuance of the patent?*
A. A filing fee of $170 plus certain additional charges for claims, depending on their number and the manner of their presentation, are required when the application is filed. A final or issue fee of $280 plus certain printing charges are also required if the patent is to be granted. The final fee is not required until your application is allowed by the Patent Office.

13. Q. *Are models required as a part of the application?*
A. Only in the most exceptional cases. The Patent Office has the power to require that a model be furnished, but rarely exercises it.

14. Q. *Is it necessary to go to the Patent Office in Washington to transact business concerning patent matters?*
A. No; most business with the Patent Office is conducted by correspondence. Interviews regarding pending applications can be arranged with examiners if necessary, however, and are often helpful.

15. Q. *Can the Patent Office give advice as to whether an inventor should apply for a patent?*
A. No. It can only consider the patentability of an invention when this question comes regularly before it in the form of a patent application.

16. Q. *Is there any danger that the Patent Office will give others information contained in my application while it is pending?*
A. No. All patent applications are maintained in the strictest secrecy until the patent is issued. After the patent is issued, however, the Patent Office file containing the application and all correspondence leading up to issuance of the patent is made available in the Patent Office Search Room for inspection by anyone, and copies of these files may be purchased from the Patent Office.

17. Q. *May I write to the Patent Office directly about my application after it is filed?*
A. The Patent Office will answer an applicant's inquiries as to the status of the application and inform him or her whether the application has been rejected, allowed, or is awaiting action by the Patent Office. However, if you have a patent attorney or agent, the Patent Office cannot correspond with both you and the attorney concerning the merits of your application. All comments concerning your invention should be forwarded through your patent attorney or agent.

18. Q. *What happens when two inventors apply separately for a patent on the same invention?*
A. An "interference" is declared and testimony may be submitted to the Patent Office to determine which inventor is entitled to the patent. Your attorney or agent can give you further information about this if it becomes necessary.

19. Q. *Can the six-month period allowed by the Patent Office for response to an office action in a pending application be extended?*
A. No. This time is fixed by law and cannot be extended by the Patent Office, but it may be reduced to not less than thirty days. The application will be abandoned unless proper response is received in the Patent Office within the time allowed.

20. Q. *May applications be examined out of their regular order?*
A. No. All applications are examined in the order in which they are filed, except under certain very special conditions.

When to apply for a patent

21. Q. *I have been making and selling my invention for the past 13 months and have not filed any patent application. Is it too late for me to apply for a patent?*
A. Yes. A valid patent may not be obtained if the invention was in public use or sale in this country for more than one year prior to the filing of your patent application. Your own use and sale of the

invention for more than a year before your application is filed will bar your right to a patent just as effectively as though this use and sale had been done by someone else.

22. Q. *I published an article describing my invention in a magazine 13 months ago. Is it too late to apply for a patent?*

A. Yes. The fact that you are the author of the article will not save your patent application. The law provides that the inventor is not entitled to a patent if the invention has been described in a printed publication anywhere in the world more than a year before his patent application is filed.

Who may obtain a patent

23. Q. *Is there any restriction as to persons who may obtain a United States patent?*

A. No. Any inventor may obtain a patent regardless of age or sex, by complying with provisions of the law. A foreign citizen may obtain a patent under exactly the same conditions as a United States citizen.

24. Q. *If two or more persons work together to make an invention, to whom will the patent be granted?*

A. If each had a share in the ideas forming the invention, they are joint inventors and a patent will be issued to them jointly on the basis of a proper patent application filed by them jointly. If, on the other hand, one of these persons has provided all of the ideas of the invention, and the other has only followed instruction in making it, the person who contributed the ideas is the sole inventor and the patent application and patent should be in his or her name only.

25. Q. *If one person furnishes all of the ideas to make an invention and another employs him or her or furnishes the money for building and testing the invention, should the patent application be filed by them jointly?*

A. No. The application must be signed, executed, sworn to, and filed in the Patent Office in the name of the inventor. This is the person who furnishes the ideas, not the employer or the person who furnishes the money.

26. Q. *May a patent be granted if an inventor dies before filing his application?*

A. Yes; the application may be filed by the inventor's executor or administrator.

27. Q. *While in England this summer, I found an article on sale which was very ingenious and has not been introduced into the United States or patented or described. May I obtain a United States patent on this invention?*

A. No. A United States patent may be obtained only by the true inventor, not by someone who learns of an invention of another.

Ownership and sale of patent rights

28. Q. *May the inventor sell or otherwise transfer his or her right to his patent or patent application to someone else?*

A. Yes. He or she may sell all or part of the interest in the patent application or patent to anyone by a properly worded assignment. The application must be filed in the Patent Office as the invention of the true inventor, however, and not as the invention of the person who has purchased the invention from him.

29. Q. *Is it advisable to conduct a search of patents and other records before applying for a patent?*

A. Yes. If it is found that the device is shown in some prior patent it is useless to make application. By making a search beforehand the expense involved in filing a needless application is often saved.

Patent searching

30. Q. *Where can a search be conducted?*

A. In the Search Room of the Patent Office in the Department of Commerce Building, 14th and E Street, NW, Washington, D.C. Classified and numerically arranged sets of United States and foreign patents are kept there for public use.

31. Q. *Will the Patent Office make searches for individuals to help them decide whether to file patent applications?*

A. No. But it will assist inventors who come to Washington by helping them to find the proper patent classes in which to make their searches. For a reasonable fee it will furnish lists of patents in any class and subclass, and copies of the patents may be purchased for $1.50 each.

Technical knowledge available from patents

32. Q. *I have not made an invention but have encountered a problem. Can I obtain knowledge through patents of what has been done by others to solve the problem?*

A. The patents of the Patent Office Search Room in Washington contain a vast wealth of technical information and suggestions, organized in a manner that will enable you to review those most closely related to your field of interest. You may come to Washington and review these patents, or engage a patent practitioner to do this for you and to send you copies of the patents most closely related to your problem.

33. Q. *Can I make a search or obtain technical information from patents at locations other than the Patent Office Search Room in Washington?*

A. Yes. Libraries have sets of patent copies numerically arranged in bound volumes, and these patents may be used for search or other information purposes as discussed in the answer to Question 34.

34. Q. *How can technical information be found in a library collection of patents arranged in bound volumes in numerical order?*

A. You must first find out from the *Manual of Classification* in the library the Patent Office classes and subclasses which cover the field of your invention or interest. You can then, by referring to microfilm reels or volumes of the *Index of Patents* in the library, identify the patents in these subclasses and, thence, look at them in the bound volumes. Further information on this subject may be found in the leaflet *Obtaining Information from Patents,* a copy of which may be requested from the Patent Office.

Infringement of others' patents

35. Q. *If I obtain a patent on my invention, will that protect me against the claims of others who assert that I am infringing their patents when I make, use, or sell my own invention?*

A. No. There may be a patent of a more basic nature on which your invention is an improvement. If your invention is a detailed refinement or feature of such a basically protected invention, you may not use it without the consent of the patentee, just as no one will have the right to use your patented improvement without your consent.

Enforcement of patent rights

36. Q. *Will the Patent Office help me to prosecute others if they infringe the rights granted to me by my patent?*

A. No. The Patent Office has no jurisdiction over questions relating to the infringement of patent rights. If your patent is infringed, you may sue the infringer in the appropriate United States court at your own expense.

Patent protection in foreign countries

37. Q. *Does a United States patent give protection in foreign countries?*

A. No. The United States patent protects your invention only in this country. If you wish to protect your invention in foreign countries, you must file an application in the Patent Office of each such country within the time permitted by law.

Problems

Working drawings

1. Prepare working drawings as the implementation step of the design process of one of the problems that was refined at the end of Chapter 5. Draw the working drawings and assembly on 11-by-17-inch sheets of tracing vellum or film.

2. Prepare working drawings of one of the problems assigned or selected from those at the end of Chapter 24. Draw the working drawings and assembly on 11-by-17-inch sheets of tracing vellum or film.

Patents

3. Write for a copy of a patent that would be of interest to you. Make a list of the features that were used as a basis for obtaining the patent.

4. Suggest modifications to the patent mentioned in the previous problem. Make sketches of innovations that would be possible improvements of the patented mechanism.

5. Write the U.S. Patent and Trademark Office for patent application forms. Prepare a patent application for a simple invention that has been previously patented, such as a fountain pen, drafting instrument, or similar item. Determine what drawings and materials are needed to complete your application.

6. Prepare patent drawings in accordance with the standards established in Section 8.8 to depict a simple patented object, such as those mentioned in Problem 5. Strive for a finished technique that would make your drawing acceptable as a patent drawing.

7. Make a list of ideas for products that you believe to be patentable. These may be ideas that you have developed during work on design problems assigned in a class.

8. Write a technical report investigating the history and significance of the patent system and its role in our industrial society. Consult your library and available government publications on patents. Give information and data that will improve your understanding of patents.

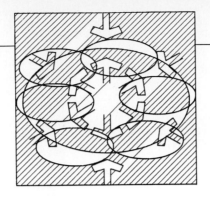

CHAPTER 9

Design Problems

9.1
Introduction

Engineering problems are seldom compartmentalized by subject area in a clearly defined structure suggesting a single solution. More common is it for problems to contain a blend of several areas. This chapter will introduce problems that can be used for class assignments, team projects, and other combinations of approaches to provide experience in applying the principles and techniques covered in this text.

9.2
The individual approach

The short problem—requiring one or two hours of work—will probably be assigned for solution by students working individually. Though simple design problems may involve fewer details and less depth, all design steps are applied as in more sophisticated, comprehensive problems.

9.3
The team approach

Obviously, an effectively organized team working on a single problem has access to more talent than does the typical individual who devotes the same number of hours to the project. The application of this talent will be a problem if the team is not properly organized. Team management and working effectively with others are invaluable skills that should be developed.

TEAM SIZE A student design team should have from three to seven members. Three is considered the minimum number for a valid team experience. The optimum size, four, lessens the possibility of domination by one or more members.

TEAM COMPOSITION In practice, the engineering team may be composed of professionals from different firms who may be total strangers. This arrangement can be advantageous in that it reduces the impact of personalities.

TEAM LEADER A leader is necessary for most teams to function effectively. This person takes the responsibility for making assignments, seeing that deadlines are met, and acting as arbitrator during any disagreements.

9.4
The selection of a problem

The best problem for a student design project is one that involves familiar and accessible conditions. A design for a water-ski rack for an automobile is more feasible than one for a support bracket for an airplane.

STUDENT-PROPOSED PROBLEM The best design problem is one that has been recognized and proposed by the students who will be working toward its solution. When the members of a design team decide on a design project, they should prepare a written proposal identifying the problem and outlining its limits.

INSTRUCTOR-ASSIGNED PROBLEMS Assigned problems are appropriate for classroom situations, and they are analogous to the assignment method used in industry.

9.5
Problem specifications

An individual or design team must consider the following specifications when preparing a design proposal or outlining an assignment.

THE SHORT PROBLEM The specifications for a short problem could include all or part of the following.
1. Completed worksheets illustrating the development of the design process (Chapter 2).

2. Freehand sketches of the design for implementation (Chapter 4).
3. An instrument drawing of the proposed design (Chapters 8 and 24).
4. A dimensioned instrument drawing of the proposed design (Chapter 24).
5. A pictorial sketch (or one made with instruments) illustrating the design (Chapters 3 and 26).
6. Visual aids, flip charts, or other media to present the design to a group (Chapter 7).

THE COMPREHENSIVE PROBLEM Comprehensive problems vary in time; typically, one takes an average of 80 to 120 work-hours.

The following specifications apply to a comprehensive problem of the systems- or product-development type.
1. A proposal outlining the problem, the method of approach, and the specifications used in solving the problem (Chapter 7).
2. Completed worksheets illustrating the development of the design process (Chapter 2).
3. Schematic diagrams, flowcharts, or other symbolic methods of illustrating the design and its function (Chapter 33).
4. An opinion survey determining interest for the proposed design (Chapter 6).
5. A market survey evaluating the product's possible acceptance and estimated profit (Chapter 3).
6. A model or prototype for analysis or presentation (Chapter 6).
7. Pictorials explaining features of the final design solution that are not clearly shown in other drawings (Chapter 26).
8. Dimensioned engineering drawings giving details and specifications of the design (Chapter 24).
9. A technical report, fully illustrated with charts and graphs, explaining the activities leading to the solution and ending with conclusions and recommendations (Chapter 7).

9.6
Scheduling team activities

The following schedule can be used as a guideline to assist team members in working toward the completion of their project.

WEEK 1 Team organization: Each team should discuss each part of the project as a group. Each team should also appoint a project chief, an assignment that can be for the duration of the project or that can be rotated among other team members.

WEEK 2 First meeting: The following activities should be discussed and assigned to specific team members.
1. Team members should review Chapters 2, 3, 4, and 9 before the first group meeting.
2. Do not attempt to solve the problem, but try to identify the factors affecting it. Use worksheets (Section 3.2) to list all facts and information as a permanent record.
3. Determine the data that must be gathered to identify your problem—dimensions, trends, surveys, and similar information (Chapter 3).

WEEK 3 Second meeting: Problem identification.
1. List jobs that must be performed on the DS&PR form (Chapter 3).
2. Prepare an Activities Network (Chapter 3).
3. List the activities on an Activity Sequence Chart (ASC) (Chapter 3), and assign each job to a team member.
4. Each team member should have a copy of the ASC for reference.
5. Prepare a progress report to cover this period's effort (Chapter 7).

WEEK 4 Third meeting: Preliminary ideas.
1. Notes and sketches should be prepared legibly and kept for group discussion (Chapter 4).
2. Review individual ideas as a group (Chapter 4).
3. Conduct a brainstorming session (Chapter 4).
4. Make assignments to team members.
5. Discuss the need for a survey (Chapter 4).
6. Prepare a progress report for your instructor.

WEEK 5 Visiting engineers.
1. Invite engineers from industry or representatives from other departments on your campus to serve as consultants to your class during one class period.
2. The consultants cannot be expected to solve your problems; instead, they will serve as counselors and coaches guiding each team.

3. Each team should make a brief progress report to each consultant at the outset of the meeting.

WEEK 6 Fourth meeting: Preliminary ideas.
1. Continue collecting ideas. Keep all worksheets and notes for inclusion in report (Chapter 4).
2. Prepare a progress report for your instructor.

WEEK 7 Fifth meeting: Design refinement.
1. Begin making scale drawings and determining specific characteristics of your project (Chapter 5).
2. Prepare graphs, diagrams, and schematics that can be used in your report and as visual aids in the oral presentation (Chapter 33).

WEEK 8 Sixth meeting: Analysis.
1. Review all collected data, preliminary ideas, and refined designs. Consider weights of materials, strength, economy, market prospects, and human factors as they might affect your design (Chapter 6).
2. Apply engineering fundamentals learned in other courses.
3. Prepare a progress report for your instructor.

WEEK 9 Seventh meeting: Decision and implementation.
1. Select or reject the best solution. If a solution is not feasible, your report should clearly state why.
2. Present all findings in a technical report; a model may be required for product designs (Chapter 8).
3. Assign responsibilities for the preparation of the final drawings and the text of the report.
4. Prepare a progress report.

WEEK 10 Eighth meeting: Implementation and report.
1. Continue with the various parts of the written report (Chapter 8).
2. Prepare a progress report.

WEEK 11 Ninth meeting: Complete report.
1. Prepare an evaluation form that reports each member's contribution (Chapter 7).
2. Complete the first draft of final report.

WEEK 12 Tenth meeting: Preparation for presentation.
1. Organize your presentation. Determine the types of visuals that are needed (Chapter 7).

2. Prepare 3-by-5-inch cards for each visual (Chapter 7).

3. Assign responsibilities for the preparation of visuals.

4. Refer to Chapter 33 for good practices of preparing graphs and visuals.

WEEK 13 Eleventh meeting: Preparation for presentation.
1. Conduct a practice presentation (Chapter 7).
2. Refer to oral report evaluation forms.
3. Have as many team members share in the presentation as possible.

WEEK 14 Twelfth meeting: Final presentation.
1. Give final presentation to the returning consultants.
2. Complete evaluation forms for oral presentation.
3. Classes will evaluate each team's presentation using the forms provided. Teams will not grade their own presentation (Fig. 7.26).

9.7
Short design problems

The following are short problems that can be completed in less than two periods.

1. Lamp bracket. Design a simple bracket to attach a desk lamp to a vertical wall for reading in bed. The lamp should be easily removable so that it can be used as a conventional desk lamp.

2. Towel bar. Design a towel bar for a kitchen or bathroom. Determine optimum size, and consider styling, ease of use, and method of attachment.

3. Pipe aligner for welding. The initial problem of joining pipes with a butt weld is the alignment of the pipes in the desired position. Design a device with which to align pipes for on-the-job welding. For ease of operation, a hand-held device would be desirable. Assume the pipes will vary in diameter from two to four inches.

4. (Fig. 9.1) **Film reel design.** The film reel used on projectors is often difficult to thread because of the limited working space. A typical twelve-inch diameter movie reel is shown in the figure. Redesign this type of reel to allow more space for threading.

FIG. 9.1 Problem 4. A typical movie projector reel.

5. Sideview mirror. In most cars, rearview mirrors are attached to the side of the automobile to improve the driver's view of the road. Design a sideview mirror that is an improvement over those you are familiar with. Consider the aerodynamics of your design, protection from inclement weather, and other factors that would affect the function of the mirror.

6. Railing-post mount. An ornamental iron railing is to be attached to a wooden porch surface supported by several one-inch-square tubular posts. Design the mounting piece that will attach the posts to the surface. Figure 9.2 shows a typical railing.

A second attachment is needed to assemble the railing with the support post at each end. If two screw holes are to be used, design the part that can secure the two perpendicular members.

7. Nail feeder. Workers lose time in covering a roof with shingles if they have to fumble for nails. Design a device that can be attached to a worker's chest and will hold nails in such a way that these

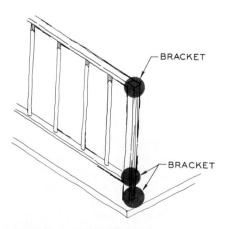

FIG. 9.2 Problem 6. A porch railing system.

will be fed in a lined-up position ready for driving.

8. **Cupboard door closer.** Kitchen cupboard and cabinet doors are usually not self-closing and thus are safety hazards and unsightly. Design a device that will close doors left partly open. It would be advantageous to provide a means for disengaging the closer when desired.

9. **Paint-can holder.** Paint cans are designed with a simple wire bail that, when held, makes it difficult to get a paint brush in the can. Wire bails are also painful to hold for any length of time. Design a holding device that can be easily attached and removed from a gallon size paint can ($6\frac{1}{2}$ by $7\frac{1}{2}$ inches). Consider human factors such as comfort, grip, balance, and function.

10. **Self-closing, self-opening hinge.** Interior doors of residences tend to remain partly open. The edge of the door that is ajar in the middle of the hall can be a hazard. Design a hinge that will hold the door in a completely open position.

11. **Tape cartridge storage unit.** Tape players are used frequently in automobiles. Design a storage unit that will hold a number of these cartridges in an orderly fashion, so that the driver can select and insert them with a minimum of motion and distraction. Determine the best location for this unit and the method of attachment to an automobile.

12. (Fig. 9.3) **Slide projector elevator.** Most commercial slide and movie projectors have adjustment feet used to raise the projector to the proper position for casting an image on a screen. Study a slide projector available to you to determine the specific needs and limitations of an adjustment. Design a device to serve this purpose as part of the original design or as an accessory that could be used on existing projectors.

13. **Book holder for reading in bed.** As a student, you may often desire to read while lying in bed. Design a holder that can be used for supporting a book in the desired position.

14. **Table leg design.** Do-it-yourselfers build a variety of tables using slab doors, plywood, and commercially available legs. Table tops come in a number of sizes, but table heights are fairly standard. Determine what the standard table heights are, and design a family of legs that can be at-

FIG. 9.3 Problem 12. A movie projector. (Courtesy of Eastman Kodak.)

tached to table tops with screws. Indicate the method of manufacture, size, cost and method of attachment.

15. **Canoe mounting system.** Canoes and light boats are often transported on the top of automobiles on a luggage rack or similar attachment. Design an accessory that will enable a single person to remove and load a boat on top of an automobile. This attachment should accommodate aluminum boats from fourteen to seventeen feet long and weighing from one hundred to two hundred pounds. Give specifications for a method of securing the boat after it is on top of the automobile.

16. **Toothbrush holder.** Design a toothbrush holder that can be attached to a bathroom wall and that can hold a drinking cup and two toothbrushes.

17. **Napkin holder.** Design a device that will hold twenty-five paper napkins on a dining room table for easy access.

18. **Book holder.** Design a holder that will support your textbook on your drawing table in a position that will make it more readable and accessible.

19. **Clothes hook.** Design a clothes hook that can be attached to a closet door for hanging clothes. It should be easy to manufacture and simple to use.

20. **Automobile ashtray.** Design an ashtray that can be attached to the dashboard of any automobile. It should be easy to attach and to remove for emptying.

21. Pencil holder. Design a holder that can be attached to the interior of the car for holding pencils or pens.

22. Door stop. Design a door stop that can be attached to a vertical wall or floor to prevent the door knob from bumping the wall.

23. Teaching aid. Design an apparatus that can be used by a teacher to illustrate the basic principles of orthographic projection. Investigate the market potential of such a product.

24. Cup dispenser. Design a paper-cup dispenser that can be attached to a vertical wall. This dispenser should hold a series of cups two inches in diameter that measure six inches tall when stacked together.

25. Drawer handle. Design a handle that would be satisfactory for a standard file cabinet drawer.

26. Paper dispenser. Design a dispenser that will hold a 6-by-24-inch roll of wrapping paper. The paper will be used on a table top for wrapping packages.

27. (Fig. 9.4) **Handrail bracket.** Design a bracket that will support a tubular handrail to be used on a staircase. Consider the weight that the handrail must support.

28. (Fig. 9.5) **Latchpole hanger.** Design a hanger that can be used to support a latchpole from a vertical wall. It should be easy to install and use.

29. (Fig. 9.6) **Pipe clamp.** A pipe with a four-inch diameter must be supported by angles that are spaced eight feet apart. Design a clamp that will

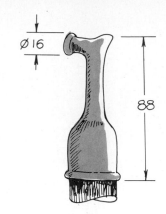

FIG. 9.5 Problem 28. A latchpole hanger.

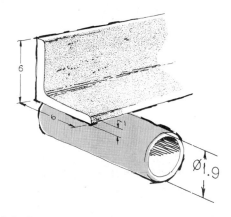

FIG. 9.6 Problem 29. A pipe clamp.

support the pipe without drilling holes in the angles.

30. TV yoke. Design a yoke that will support a TV set from the ceiling of a classroom and that will permit it to be adjusted at the best position for viewing.

31. Flagpole socket. Design a socket for flags that is to be attached to a vertical wall. Determine the best angle of inclination for the flagpole.

32. Crutches. Design a portable crutch that could be used by a person with a temporary leg injury.

33. Cup holder. Design a holder that will support a soft-drink can or bottle on the interior of an automobile.

34. Gate hinge. Design a hinge that could be attached to a three-inch-diameter tubular post to support a three-foot-wide gate.

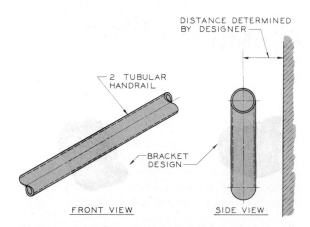

FIG. 9.4 Problem 27. A handrail bracket.

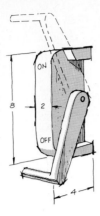

FIG. 9.7 Problem 35. A safety lock.

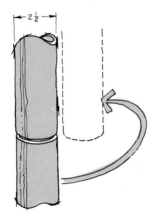

FIG. 9.8 Problem 36. A tubular hinge.

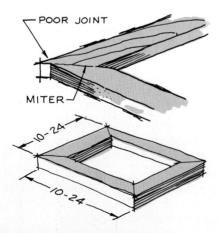

FIG. 9.9 Problem 37. A miter jig.

35. (Fig. 9.7) **Safety lock.** Design a safety lock that will hold a high-voltage power switch in either the "off" or "on" positions to prevent an accident.

36. (Fig. 9.8) **Tubular hinge.** Design a hinge that can be used to hinge 2.5-inch OD high-strength aluminum pipe in the manner shown in this figure. A hinge of this type is needed for portable scaffolding.

37. (Fig. 9.9) **Miter jig.** Design a jig that can be used for assembling wooden frames at 90° angles. The stock for the frames is to be rectangular in cross sections that vary from 0.75 by 1.5 inches to 1.60 by 3.60 inches. Outside dimensions vary from ten to twenty-four inches.

38. (Fig. 9.10) **Base hardware.** Design the hardware needed at the points indicated for a standard volleyball net. The seven-foot pipes are supported by crossing 2-by-4-inch boards. Design the hardware needed at points *a*, *b*, and *c*.

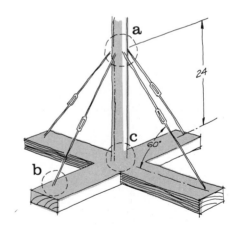

FIG. 9.10 Problem 38. Base hardware for a volleyball net.

39. (Fig. 9.11) **Conduit connector hanger.** Design a support that will attach to a three-quarter-inch conduit that will support a channel used as an adjustable raceway for electrical wiring. Your design should permit ease of adjustment.

40. (Fig. 9.12) **Fixture design.** Design a fixture that will permit a small-scale manufacturer to saw the corner of the block as shown in this figure.

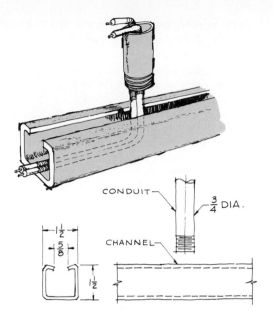

FIG. 9.11 Problem 39. A conduit connector hanger.

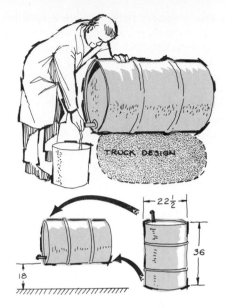

FIG. 9.13 Problem 41. A drum truck.

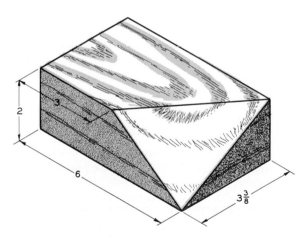

FIG. 9.12 Problem 40. A fixture design.

41. (Fig. 9.13) **Drum truck.** Design a truck that can be used for moving a fifty-five-gallon drum of turpentine (7.28 pounds per gallon). The drum will be kept in a horizontal position, but it would be advantageous to incorporate a feature into the truck permitting the drum to be set in an upright position.

9.8
Systems design problems

The systems design problem is a broad engineering problem that requires studying the interrelationships of various components—social, economic, physical, and management considerations. The end result of a systems problem may be a plan, a conclusion, or a recommendation.

42. **Multipurpose utility meter.** Today's residential units have separate meters for electricity, water, and gas that are checked each month by separate utility companies. Consider the feasibility of combining all meters into a single unit that could be read by a meter-reading service. A unit of this type would reduce the number of meter readers to one third. Outline how such a system could be organized and implemented.

43. **Portable bleachers.** Design a portable bleacher system that can be assembled and disassembled with a minimum of effort and that can be stored. Consider the structure, size, materials and method of construction. Identify a variety of uses for these bleachers that would justify their production for the general market.

44. **Archery range.** Determine the feasibility of providing an archery range for beginning and experienced archers that can be operated at a profit. Your problem is to investigate the need and potential market for such a facility, as well as other factors that would bear upon your decision, including location, equipment needed, method of operation, and costs and fees. Consider site preparation, utilities, concessions, and parking.

45. **Car rental system.** Investigate the possibilities and feasibility of a student-operated rental car system. Data must be collected to determine student interest, cost factors, number of cars needed, rates, personal needs, storage, maintenance, and so forth. Specify the details of your system, including location, cost of operation, and expected income.

46. **Model airplane field.** Model airplane hobbyists who build and fly gasoline engine models often have inadequate facilities for pursuing their hobby. Analyze their needs, including space requirements, type of surfaces, control of sound, safety factors, and method of operation. Come up with a total evaluation of the system. Select a site on or near your campus that you feel is adequate for this facility. Evaluate the equipment, utilities, and site preparation that would be required.

47. **Overnight campsite.** Analyze the feasibility of converting a local lot near a major highway into a complex of campsites for overnight campers. Determine the facilities that would be desired by the campers and the expenditures that would be required for operation. Determine the profit margins for your particular design.

48. **Swimming pool study.** A group of students wishes to construct a swimming pool on your campus that would be self-supporting. Determine the cost of the pool, the area in which it would be located, the students it would serve, and the equipment and labor necessary to operate it. As the conclusion of your problem, determine whether or not it would be feasible.

49. **Outdoor shower.** You are the owner of a weekend cottage, and you do not have a hot and cold water system—you have only cold water. Design a system that would take advantage of the sun's heat in the summer to heat the water used for bathing and kitchen chores. Devise a means of using the same system in the winter with heat provided from some other source (a wood fire, kerosene, or other method). Can you design a totally portable shower that can be used on camping trips?

50. **Information center.** Visitors to a typical college campus often find it difficult to locate buildings, parking lots, and campus facilities.

Analyze your college campus to determine the most logical location for a drive-in information center. Determine the informational material that should be included to assist a visitor. Consider using slides, photographs, maps, sound, and other audio-visual aids to accomplish your goal.

Your design should include the location of the system, its plot plan, equipment, housing and a detailed description of its operation.

51. **Golf driving-range ball return system.** Golf driving ranges are usually designed in such a way that the balls must be retrieved by hand or by means of a specially designed vehicle that collects them from the range. Design a system capable of automatically returning the balls to the driving area.

52. **Instructional system.** Classroom instruction could be improved by providing two-way communication between the teacher and the students. For example, the teacher could proceed at a more efficient rate if he or she had some idea of whether the class understood the points being made. Determine whether a system could be developed that would give the student some means of signaling understanding, or lack of understanding, of the lecture without having to ask questions.

53. (Fig. 9.14) **Instructional console system.** Determine the optimum class size, room layout, lighting requirements, choice of furniture, and other factors of this type that affect the ideal classroom. Study the possibility of developing a visual-aid console that incorporates the latest projectors, recorders, and other devices that would improve instruction and learning.

54. **Pedestrian transport system.** Your campus has considerably more pedestrian traffic in some areas than in others, causing traffic congestion. Consider the possiblity of developing a system that would provide a more even flow of pedestrian traffic.

55. **Instant motel.** Many communities have periodic needs for more housing than is available on a regular basis for events like ball games and celebra-

FIG. 9.14 Problem 53. An instructional console system.

tions. Investigate the different methods of providing an "instant motel" that could fulfill this need for a few days. Consider using tents, vans, trailers, train cars, and so on. Determine the profit margins for your proposed solution.

56. **Helicopter service.** You have been assigned the responsibility of planning a helicopter passenger service that will connect with the local airlines in your community. Analyze the needs of your campus to determine whether such a system could be feasible and self-supporting. Determine the helicopter landing area on your campus and the flight schedule.

57. **Mountain lodge.** Determine what provision should be made for a mountain lodge that was designed for six people to use during the winter months. Assume they may be unexpectedly snowed-in for up to two weeks at a time. What features should be included in the design of the lodge. What food supplies and other provisions should be stored for their use?

58. **Campus planning.** Assume your college campus is to begin planning for full-capacity use, that is the total use of your present facilities twelve months a year, twenty-four hours a day. Determine how many students could be accommodated in your classrooms, dormitories, and in other facilities without adding new buildings.

Teaching schedules, faculty, and service personnel should all be evaluated to identify possible problems. Also evaluate the changes that will occur in parking systems, pedestrian traffic, and the management of the total educational system.

59. **Diazo machine operation.** A company's reproduction department will have to reproduce many prints, so a full-time diazo machine operator is required. Establish a system to provide the most efficient use of the equipment and the operator's time to meet the specifications below.

The machine will accept individual drawings or groups of drawings along its forty-inch belt at a rate of ten feet per minute. The drawing size used most frequently by your company is 11 by 17 inches. The diazo paper must be run through the developing chamber of the machine directly above the intake at a speed of ten feet per minute. The machine is sixty-inches wide, twenty-four-inches deep, and forty-eight-inches high. Determine the equipment and tables and their optimum arrangement for most efficient operation.

What is the cost per drawing of the operator working at peak efficiency? Consider the other duties of the operator, such as stapling sets of drawings together and gathering original drawings to be returned with their prints.

60. **Drive-in theater.** Develop an area for a drive-in motion picture theater. As the chief designer you must determine the optimum size of the drive-in to provide adequate parking and viewing from the audience. Consideration must be given to traffic flow, screen size and position, electrical problems, drainage, utilities, concessions, and other facilities commonly found in a theater of this type. Determine the overall layout of the drive-in, its traffic system, and its major components. Detail a typical parking space for a single car, indicating the contour of the surface to provide the proper viewing angle for the car. List the areas that would require specialists such as electrical engineers and civil engineers.

61. **Boat-launching facility.** Design a boat-launching area at a lake where boat trailers could be positioned. Analyze the requirements of a workable system that will control traffic with a minimum of confusion. Thought must be given to the unloading area, space required, parking area, and type of terrain required. Also determine the charges required to maintain this facility and to pay for any needed help.

62. Football stadium expansion. Study the attendance figures of your football stadium to determine what the future holds for it. Will a larger stadium be required? If so, how much extra seating will be required and when? Search historical data for information that would help you make your recommendation.

63. Shopping checkout system. An acute problem for shopping centers and grocery markets is checking out goods, payment, and delivering purchases to the customers' automobiles. Develop a system for expediting the transfer of goods selected by the customer, from the store's shelf to the customer's car. Also consider how to expedite packaging and payment.

64. Injury-proof playground. Most playgrounds are unsafe for children. Study the activities of children at play. Design a playground system that permits the greatest degree of participation by the children with the least risk of injury.

65. (Fig. 9.15) Loading dock system. The trucking industry provides a sizable portion of the transportation of goods and supplies. Unloading is usually performed by a fork lift at a loading dock that is level with the truck's floor. Because the height of trucks may vary, design a method ensuring that the dock can be adjusted to the level of the truck's floor.

Design a system that will allow the truck to be enclosed or protected from the cold weather during its loading or unloading while taking advantage of the warehouse's heating.

66. Modification of a drive-in theater. Assume you are the owner of a drive-in movie theater that has fallen on bad times, and you wish to convert this facility into a different operation that would be profitable. Consider the options available to you having the least expense and the greatest possibility for profit.

67. Modification of an existing facility. Isolate an area on your campus or in your community that is inadequate for the demands made on it. This could be a traffic intersection, parking lot, recreational area, or a classroom. Identify the problem and the deficiencies that should be corrected. Propose modifications that would improve the existing facility.

68. Tape-recording system. In most classes, students spend much of their time taking notes, which interferes with concentrating on the concepts being presented. Design a system that would provide students with a tape recording of class lectures, which they could take to their rooms and review. Also consider the possibility of providing an automatic system that would record illustrations and diagrams presented by the instructor. Estimate the cost of such a system and determine its value to the educational program.

69. Service station modification. Service stations today differ little in arrangement and function from the first filling stations. One major change is the conversion to the credit card system. But the method of servicing automobiles—cleaning windshields, checking oil, and other routine chores—is done in the same manner, with little improvement in technique. Consider the usual operations performed in servicing an automobile to develop a more efficient system.

70. Recreational facility. Analyze the various recreational activities on your campus that could be improved with the construction of a multipurpose facility to accommodate them. These activities might include outdoor movies, plays, sports, meetings, and dances. The facility should be analyzed as an outdoor installation with the minimum of conventional structures.

71. (Fig. 9.16) Educational toy. Assume you are assigned the responsibility of establishing the production system for manufacturing the educational toy shown in the figure at a rate of five thousand

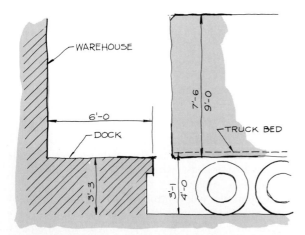

FIG. 9.15 Problem 65. The dimensions of an average truck.

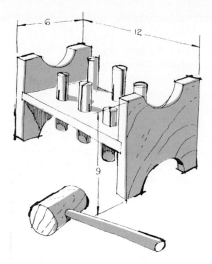

FIG. 9.16 Problem 71. An educational toy.

per month. You must determine the square footage needed, the types of machines required, and the space for raw materials, office facilities, and warehousing necessary to sustain this operation. Also determine the number of employees needed, their rate of pay, and the number of work stations required. Compute the expense to produce the item and the selling price necessary to provide the required profit margin.

72. (Fig. 9.17) **Child's furniture.** Proceed as in Problem 71 but assume the child's furniture

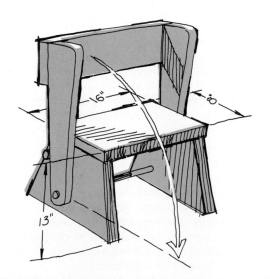

FIG. 9.17 Problem 72. Child's furniture.

shown in the figure is the product to be produced. Determine the number to be manufactured per month to break even and establish the selling price. Establish scales for selling prices in quantities in excess of the break-even level.

73. Simple product. Select a simple product and establish the production system and requirements for its manufacture proceeding as in Problem 71. Have the product approved by your instructor before going ahead.

74. (Fig. 9.18) **Conveyor system.** You have been assigned the responsibility of designing a conveyor system for delivery of green tire carcasses to the vulcanizing press. The carcasses from the overhead chain conveyor must be transferred to the belt in an upright position as shown in the figure. Each carcass weighs approximately thirty pounds and measures 12 by 18 inches. Your solution will probably involve using standard conveyor components.

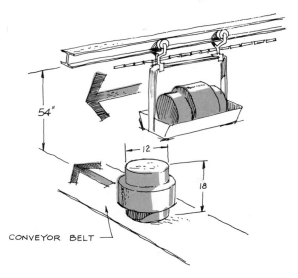

FIG. 9.18 Problem 74. A conveyor system.

9.9

Product design

A product design involves the development of a device that will perform a specific function and that will be mass-produced and sold to a broad market.

75. Hunting blind. Hunters of geese and ducks must remain concealed while hunting. Design a portable hunting blind to house two hunters. This

FIG. 9.19   Problem 77. Folding chair with a writing tablet attached.

blind should be completely portable so that it can be carried in separate sections by each of the hunters. Specify its details and how it is to be assembled and used.

76. **Convertible drafting table.** Design a drafting table that can enclose a drafting machine, thereby concealing it and protecting it when it is not needed.

77. (Fig. 9.19) **Writing table for a folding chair.** Design a writing-tablet arm for a folding chair that could be used in an emergency or when a class needs more seating. To allow easy storage, the arm must fold with the chair.

78. (Fig. 9.20) **Portable truck ramp.** Delivery trucks need ramps to load and unload supplies and materials. Assume that the bed of the truck is twenty inches from ground level. Design a portable ramp that would reduce manual lifting and

FIG. 9.20   Problem 78. A delivery ramp for unloading goods.

permit the unloading of goods with the use of a hand dolly.

79. (Fig. 9.21) **Sensor-retaining device.** The Instrumentation Department of the Naval Oceanographic Office uses underwater sensors to learn more about the ocean. These sensors, which each weigh seventy-five pounds, are submerged on cables from a boat on the surface. The winch used to retrieve the sensor frequently overruns (continues pulling when it has been retrieved), causing the cable to break and the sensor to be lost. Design a safety device that will retain the sensor if the cable is broken when the sensor reaches a pulley.

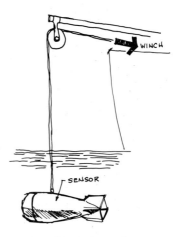

FIG. 9.21   Problem 79. Underwater sensor.

80. **Flexible trailer hitch.** In combat zones, vehicles must tow trailers where terrain may be uneven and hazardous. Design a trailer hitch that will provide for the most extreme conditions possible. Study the problem requirements and limitations to identify the parameters within which your design must function.

81. **Workers' stilts—human engineering.** Workers who apply gypsum board and other types of wallboards to the interiors of buildings and homes must work on scaffolds or wear some type of stilts to be able to reach the ceiling to nail the 4-by-8-foot boards into position. Design stilts that will provide workers with access to an eight-foot ceiling while permitting them to perform the job of nailing ceiling panels with comfort.

82. **Pole-vault uprights.** Many pole-vaulters are exceeding the eighteen-foot height in track meets,

which introduces a problem for the officials of this event. The pole-vault uprights must be adjusted for each pole-vaulter by moving them forward or backward plus or minus eighteen inches. Also, the crossbar must be replaced with great difficulty at these heights by using forked sticks and ladders. Develop a more efficient set of uprights that can be easily repositioned and that will allow the crossbar to be replaced with greater ease.

83. **Sportsman's chair.** Analyze the need for a sportsman's chair that could be used for camping, for fishing from a bank or boat, at sporting events, and for as many other purposes as you can think of. The need is not for a special-purpose chair but for a chair suitable for a variety of uses to fully justify it as a marketable item.

84. **Portable toilet.** Design a portable toilet unit for the camper and outdoorsperson. The unit should be highly portable, with consideration given to the method of waste disposal. Evaluate the market potential for this product.

85. **Child carrier for a bicycle.** Design a seat that can be used to carry a small child as a passenger on a bicycle. Assume the bicycle will be ridden by an older youth or an adult. Determine the age of the child who would probably be carried as a passenger.

86. **Lawn-sprinkler control.** Design a sprinkler that can be used to water irregularly shaped yards while giving a uniform coverage. This sprinkler should be adjustable so that it can be adapted, within its range, to yards of any shape. Also consider a method of cutting the water off at certain sprinkler positions to prevent the watering of patios or other areas that are to remain dry.

87. **Power lawn-fertilizer attachment.** The rotary-power lawn mower emits a force through its outlet caused by the air pressure from the rotating blades. This force might be used to distribute fertilizer while the lawn is being mowed. Design an attachment for a power mower that could spread fertilizer while the mower is performing its usual cutting operation.

88. **Car and window washer.** Design an attachment for the typical garden hose that would apply water and agitation (for optimum action) to the surface being cleaned. Consider other applications of the force exerted by water pressure in the performance of yard and household chores.

89. **Projector cabinet.** Design a cabinet that could serve as an end table or some other function while also housing a slide projector ready for use at any time. The cabinet might also serve as storage for slide trays. It should have electrical power for the projector. Evaluate the market potential for a multipurpose cabinet of this type.

90. **Heavy-appliance mover.** Design a device that can be used for moving large appliances, such as stoves, refrigerators, and washers, about the house. This product would not be used often—only for rearranging, cleaning, and servicing the appliances.

91. **Car jack.** The average car jack does not attach itself adequately to the automobile's frame or bumper, introducing a severe safety problem. Design a jack that would be an improvement over existing jacks and possibly employ a different method of applying a lifting force to a car. Consider the various types of terrain on which the device must serve.

92. **Map holder.** The driver of an automobile traveling alone in an unfamiliar part of the country must frequently refer to a map. Design a system that will give the driver a ready view of the map in a convenient location in the car. Also provide a means of lighting the map during night driving that will not distract the driver.

93. **Bicycle-for-two adapter.** Design the parts and assembly required to convert two bicycles into one bicycle-built-for-two (tandem). Work from an existing bicycle, and consider how each rider can equally share in the pedaling. Determine the cost of your assembly and its method of attachment to the average bicycle.

94. **Automobile unsticker.** Design a kit to be carried in the car trunk in a minimum of space containing the items required to "unstick" a car when no other help is available. This kit can contain one or several items. Investigate the need for such a kit and the main factors that lead to the loss of traction.

95. **Stump remover.** Assume a number of tree stumps must be removed from the ground to clear land for construction. The stumps are dead with partially deteriorated root systems and require a force of approximately two thousand pounds to remove them. Design an apparatus that could be attached to a car bumper that could be used to remove the stumps by either pushing or pulling.

96. **Gate opener.** An annoyance to farmers and

ranchers is the necessity of opening and closing gates when driving from one fenced area to the next. A gate that could be opened and closed without getting out of the vehicle would be desirable. Design a manually operated gate that would appeal to this market.

97. Paint mixer. Design a product that could be used by the paint store or the paint contractor to quickly mix paint in the store and on the job. Determine the standard-size paint cans for which your mixer will be designed.

98. Mounting for an outboard motor on a canoe. Unlike a square-end boat, the pointed-end canoe does not provide a suitable surface for attaching an outboard motor. Design an attachment that will adapt an outboard motor to a canoe.

99. Automobile coffee maker. Adequate heat is available in the automobile's power system to prepare coffee in minimum time. Design an attachment as an integral system of an automobile that will serve coffee from the dashboard area. Consider the type of coffee to be used (instant or regular), the method of changing or adding water, the spigot system, and similar details.

100. Baby seat (cantilever). Design a child's chair that can be attached to a standard table top and that will support the child at the required height. The chair should be designed to ensure that the child cannot crawl out or detach it from the table top. A possible solution could be a design that would cantilever from the table top, using the child's body as a means of applying the force necessary to grip the table top. The design would be further improved if the chair were collapsible or suitable for other purposes.

101. Miniature-TV support. Miniature television sets for close viewing are available with a 5-by-5-inch screen. An attachment is needed that would support sets ranging in size from 6 by 6 inches to 7 by 7 inches for viewing from a bed. Determine the placement of the set for best viewing results. Provide adjustments on the support that will be used to position the set properly.

102. Panel applicator. A worker who applies 4-by-8-foot gypsum board or paneling must be assisted by a helper who holds the panel in position while it is being nailed to the ceiling. Design a device that could be used in this capacity; it should be collapsible, economical, and versatile. The average ceiling height is eight feet, but provide adjustments that would adapt the device to lower or higher ceilings.

103. Backpack. Design a backpack that can be used by the outdoorsperson who must carry supplies while hiking. Most of your design effort should be devoted to adapting the backpack to the human body for maximum comfort and the leverage for carrying a load over a long time. Can other uses be made of your design?

104. Automobile controls. Design driving controls that can be easily attached to the standard automobile that will permit an injured person to drive a car without the use of his or her legs. This device should be easy to operate with the maximum of safety.

105. Bathing apparatus. Design an apparatus that would help a wheelchair-bound person, who does not have use of his or her legs, to get in and out of a bathtub without assistance from others.

106. Adjustable TV base. Design a TV base to support full-size TV sets that would allow the maximum of adjustment: up and down, rotation about a vertical axis and about a horizontal axis. Design the base to be as versatile as possible.

107. Trailer jack. Design a trailer jack that can be used to repair flat tires that may occur on a boat trailer. It may be possible to build in jack devices as permanent features of the trailer.

108. Projector cabinet. Design a portable cabinet, which could be left permanently in a classroom, that would house a slide projector and a movie projector. The cabinet should provide both convenience and security from vandalism and theft.

109. (Fig. 9.22) **Boat loader.** Design a rack and a system whereby a single person could load a boat weighing 110 pounds on top of a car for transporting from site to site. Use the boat specifications shown in the figure.

110. (Fig. 9.23) **Door opener.** Design a method whereby a trucker at a loading dock could open the warehouse door without having to get out of the truck. The doors are dimensioned in the figure, and the dock extends eight feet from the doors.

111. Projector eraser. Design a device that would erase grease pencil markings from the acetate roll of an overhead projector as the acetate is cranked past the stage of the projector.

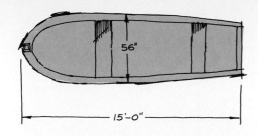

FIG. 9.22 Problem 109. Boat specifications.

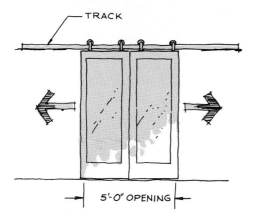

FIG. 9.23 Problem 110. Door dimensions.

112. Cement mixer. Design a portable and simple cement mixer that can be operated manually by a home owner. The mixer is seldom used so it should be sufficiently simple and economical to justify its being built.

113. Boat trailer. Design a trailer that supports the boat under the trailer rather than on top of it. Develop this design so that a boat can be launched in water more shallow than is required now.

114. Washing machine. Design a manually operated washing machine that can be used by less-developed countries without power. This could be considered an "undesign" of a powered washing machine.

115. (Fig. 9.24) Pickup truck hoist. Design a tailgate that can be attached to the tailgate of a

FIG. 9.24 Problem 115. A pickup truck hoist.

pickup truck and that can be used for raising and lowering loads from the ground to the floor of the truck. Design it to be operated without a motor.

116. (Fig. 9.25) Writing tablet for the handicapped. Design a writing tablet that will permit a person to write from a prone position using a series of mirrors. This problem involves application of human engineering considerations.

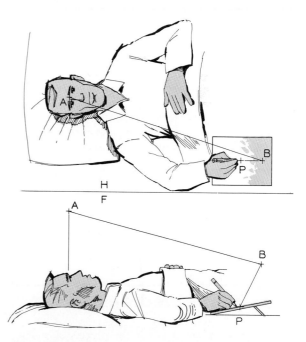

FIG. 9.25 Problem 116. A writing tablet for the handicapped.

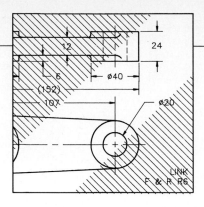

CHAPTER 10

The Computer in Design and Drafting

10.1
Introduction

The use of computers in engineering and related fields is widespread, and great growth is still expected. It is increasingly important for students of engineering and technology to become familiar with the nature and prospects of computer technologies. This chapter is devoted to computer use in drafting and design. Emphasis is placed on techniques of computer-aided design drafting (CADD), which will give you familiarity with how computer graphics can help apply the principles learned in this text. These techniques are introduced in this chapter and then integrated throughout the remainder of the text.

10.2
Computer-aided design

Computer-aided design (CAD) is the process of solving design problems with the aid of computers. This includes computer generation and modification of graphic images on a video display, printing these images as hard copy on a plotter or printer, analysis of design data, and electronic storage and retrieval of design information. Many CAD systems perform these functions in an integrated fashion, which can increase the designer's productivity manyfold.

Computer-aided design drafting (CADD), a subset of CAD, is the computer-assisted generation of working drawings and other engineering documents. The CADD user generates graphics by interactive communication with the computer. Graphics are displayed on a video display and can be converted into hard copy with a plotter or printer.

Most engineers agree that the computer does not change the nature of the design process and that it is simply a tool to improve efficiency and productivity. It is appropriate to view the designer and the CAD system as a design team: The designer provides knowledge, creativity and control, and the computer assists with accurate, easily modifiable graphics and the capacity to perform complex design analysis at great speeds and to store and recall design information. Occasionally, the computer can augment or replace many of the engineer's other tools, but it is important to remember that it does not change the fundamental role of the designer.

Advantages of using the computer in design and drafting

Depending on the nature of the problem and the sophistication of the computer system, there are several potential advantages that computer use can afford the designer or drafter.

1. *Easier creation and correction of working drawings.* In a CAD system working drawings can be created using function commands and digitizing (assigning numerical coordinates). Complicated changes and corrections are made using a few keystrokes.

2. *Easier visualization of drawings.* In many systems different views of the same part can be displayed quickly and easily. In systems with three-dimensional capabilities, a part can even be rotated on the CRT screen.

3. *Drawings can be stored and easily referenced for modification.* Modified designs can be made from one original in far less time than it would take using a manual approach. Design databases (libraries of designs) can be created in some systems. These databases can also store standard parts and symbols for easy recall. Many systems are configured so that information in the databases can be easily accessed by others in an organization, such as management or production personnel.

4. *Quick and convenient solution of computational design analysis problems.* Because the computer offers a tremendous advantage in ease of design analysis, the designer can rigorously analyze each design, thus speeding up the design refinement stage.

5. *Simulation and testing of designs.* Some systems enable the engineer to simulate the operation of a design and to perform tests and analyses in which the part is subjected to a variety of conditions or stresses. This capacity may improve or replace the process of building models and prototypes.

6. *Increased accuracy.* The accuracy of the computer lessens the chance for error. Many CAD systems are capable of detecting errors and will inform the user when data or designs are incorrect.

10.3
Some applications of CAD

A natural application of computer-aided design is the design and drawing of printed circuits of the type shown in Fig. 10.1. Printed circuits are drawn as much as five times their true size and are then photographically reduced. A computer-driven plotter will draw the circuit within an accuracy of approximately 0.001 in.

Piping systems can be designed and represented in both orthographic and pictorial views. Once the system has been completed, the design and its specifications can be stored for rapid recall when the system

Fig. 10.1 An often-used application of computer graphics is the drawing of printed circuits for electronic systems. (Courtesy of Prime Computer, Inc.)

Fig. 10.2 An example of a computer-graphics workstation in the BruningCAD Spectra II. (Courtesy of Bruning.)

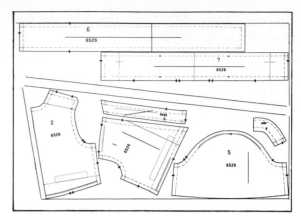

Fig. 10.4 This clothing pattern was designed, plotted, and cut using computer graphics on a Versatec plotter. Programs are available for cutting patterns for graduated sizes of clothing. (Courtesy of Versatec.)

needs to be reexamined (Fig. 10.2).

CAD is used as an aid in finite element analysis, where a series of elements is used to represent an irregular three-dimensional shape. In Fig. 10.3, the designer is digitizing the elements of a gun mount by working from a multiview drawing at his right. The mount is then displayed by the computer system as an isometric on the screen of the CRT.

Clothing patterns for a wide selection of graduated sizes (Fig. 10.4) can be drawn and cut by using CAD. The automatic cutter follows the computer-generated path to cut the most economical patterns from a section of material.

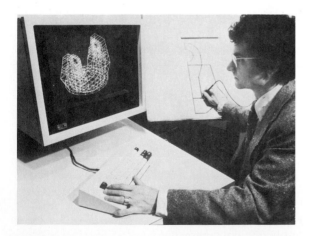

Fig. 10.3 This engineer has constructed a 3-D finite element model of a gun mount on the Applicon display by digitizing a multiview engineering drawing. (Courtesy of Applicon.)

10.4
CAD/CAM

An important application of CAD lies in the field of manufacturing. *Computer-aided design/computer-aided manufacturing—(CAD/CAM)*—describes a system that can design a part or product, devise the essential production steps, and electronically communicate this information to manufacturing equipment like robots (Fig. 10.5). A CAD/CAM system offers many potential advantages over traditional manufacturing systems, including less design effort through use of CAD and CAD databases, more efficient material use, reduced lead time, greater accuracy, and improved inventory functioning.

In *computer-integrated manufacturing (CIM)* a computer or system of computers coordinates all stages of manufacturing, which will enable manufacturers to custom design products efficiently and economically.

Fig. 10.5 Automatic welding system for Chrysler LeBaron GTS and Doge Lancer car bodies uses computer-controlled robots for consistent welds of all components in the unitized body structure. This assembly plant also features energy-efficient electric robot welders that require less maintenance and provide a high degree of accuracy. (Courtesy of Chrysler Corp.)

Fig. 10.6 Designer center ME Series 102–D design and drafting system is a configurable family of mechanical-engineering design solutions with direct links to finite-element analysis and numerical control. The system is based on the HP 9000 Series 300 engineering workstation with HP-UX 5.1 operating system. (Courtesy of Hewlett-Packard.)

10.5
Hardware systems

Computer-aided design systems have three major components: the designer, hardware, and software. *Hardware* is the physical components of a computer system, whereas *software* is the programmer's instructions to the computer. The hardware of a computer graphics system includes the computer, terminal, input devices (digitizers, light pens), and output devices (plotters, printers) (Fig. 10.6).

Computer

Computers receive input from the user, execute the instructions contained in the input, and then produce some form of output. A sequence of instructions called a *program* controls the computer's activities. The part of the computer that follows the program's instructions is the *central processing unit* or *CPU*.

Computers are often classified by size. *Mainframes,* the largest computers, are big, fast, powerful, and expensive. *Minicomputers,* smaller and less costly than mainframes, are used by many small businesses. *Microcomputers,* the smallest computers, are widely used for both personal and business applications. Rapid advances in hardware technology and software capability, along with continually falling prices, have brought microcomputers into wide use in engineering graphics applications in industry, government, and education.

Terminal

The *terminal* allows the user to communicate with the computer. It typically consists of a keyboard, a cathode-ray tube, and the interconnections between these devices and the computer (Fig. 10.7).

KEYBOARD The keyboard allows the user to communicate with the computer through a set of alphanumeric and function keys. Keyboards generally resemble typewriters but usually include many other function keys, some of which may be user-defined.

CATHODE-RAY TUBE (CRT) A *CRT* is a video display device consisting of a tube with a phosphor-coated screen. An electron gun throws a beam that sweeps out rows, called *raster lines,* on the screen. Each raster line consists of a number of dots called

Fig. 10.7 These students are using IBM PCs with CADD-23 software to solve their engineering graphics problems. (Courtesy of Brodhead-Garrett Co., CADD-23.)

Fig. 10.8 A close-up view of a flatbed plotter that uses four pens of different colors to plot the computer's output. (Courtesy of ComputerVision.)

pixels. Images are generated on the screen by turning pixels on and off. Raster-scanned CRTs refresh the picture display many times per second. One measure of the quality of the pictures that can be produced on a screen is *resolution,* the number of pixels per inch that can be drawn on the screen. Higher quality graphics can be drawn on higher resolution screens.

Raster-scan technology has largely replaced the *vector-refreshed tube* used in early video displays. In this type of display each line in the picture is continuously redrawn by the computer. Because the display produced a collection of lines instead of a set of individual pixels, vector-display pictures are less realistic than raster-scanned displays.

Plotters

The plotter is a machine directed by the computer to make a drawing. Two types of plotters are the *flatbed* plotter and the *drum,* or *roll-feed,* plotter. In the flatbed plotter, paper is attached to the bed, and the pen is moved about the paper in raised or lowered position to complete the drawing (Fig. 10.8). In the drum plotter, a special type of paper is held on a spool and rolled over a rotating drum (Fig. 10.9). As the drum rotates, the pen suspended above the paper moves left or right along the drum.

Fig. 10.9 A roll-feed plotter used for plotting the output. (Courtesy of Hewlett-Packard.)

126

Printers

The printer is a device operated by the computer that makes images on paper. An *impact printer,* which works like a typewriter, forms characters by forcing typefaces to impact with an inked ribbon and paper. *Dot-matrix printers,* popular, inexpensive type of impact printer, have print heads composed of a rectangle of pins, each of which can be raised or lowered to form a character. These patterns of pins are forced against a ribbon to make dotted characters on paper. Dot-matrix printers are useful for graphics because the patterns of dots can be made to correspond to lighted pixels on a CRT screen. However, the output of dot-matrix printers is of lower quality than some other printers, so they have limited usefulness in graphics applications.

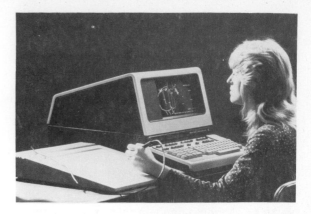

Fig. 10.11 A digitizer board is made up of a set of coordinates that correspond to points on the CRT screen. (Courtesy of Hewlett-Packard.)

Fig. 10.10 This compact ink-jet printer was designed to be used with personal computers. (Courtesy of Hewlett-Packard.)

Digitizers

The *digitizer* is a graphics input device that can communicate information in a picture to the computer for display, storage, or modification (Fig. 10.11). The user lays a drawing or sketch on a digitizer board and scans it, thereby converting the picture to a digital or computerized form based on the *xy*-coordinates of individual points. Since a digitizer can also input symbols, a fully labeled drawing can be represented on a CRT screen and stored for later use.

Digitized pictures on a screen can be created or modified point by point using a *light pen.* The user places this device in contact with the CRT screen, thereby telling the computer the position of the pen on the screen (Fig. 10.12).

Nonimpact printers form characters from a distance by using ink sprays, laser beams, photography, or heat. One example, the *ink-jet printer,* uses electrical fields to direct jets of ink to appropriate spots on the paper (Fig. 10.10). Like the dot-matrix printer, the ink-jet printer forms a pattern of dots, thus limiting its accuracy or resolution. One feature of this type of printer is that multicolor graphics can be generated by using multiple jet nozzles.

As printer technology makes rapid strides in cost reduction and output quality, it will be found increasingly useful for graphics applications. Still, the need for strict accuracy makes the use of plotters more attractive for many applications.

Fig. 10.12 A light pen enables the user to modify drawings by making contact with the CRT screen. (Courtesy of Hewlett-Packard Co.)

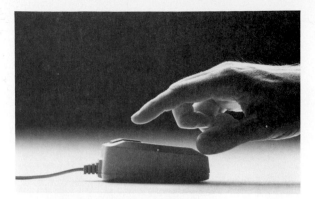

Fig. 10.13 A mouse can move the cursor around a CRT screen. The user activates commands or makes changes by pressing the button(s). (Courtesy of Apollo Computer, Inc.)

Fig. 10.14 A joy stick can be used to "steer" a cursor around a CRT screen. (Courtesy of Apple Computer, Inc.)

A *mouse* is a device that can be rolled on a table top to move a cursor (a mark on the screen that indicates location) around a CRT screen. The mouse can be used to activate commands or change information on the screen (Fig. 10.13).

A *joy stick* allows the user to "steer" a cursor around the screen by tilting a lever in appropriate directions (Fig. 10.14).

10.6
CAD software

There are hundreds of software packages commercially available for computer-aided design. Though they range widely in price and sophistication, they all give the computer the appropriate instructions to generate computer graphics, store data, perform analysis, and so forth.

Writing computer graphics programs

Even the most basic microcomputers can be programmed to perform certain graphics functions. Using a high-level language like BASIC or FORTRAN, the user can write programs that draw graphics on a CRT screen or produce hard copy on a plotter or printer.

A simple graphics program, written in BASIC, is shown in Fig. 10.15. This program draws a line from *A* to *B* when the coordinates of each point are known. The coordinates are in the statements in lines 100, 110, and 120. (These REM or remark statements communicate background information to the user and do not send instructions to the computer.)

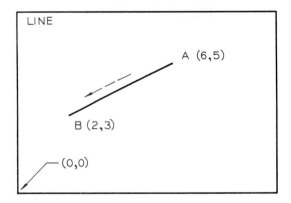

```
100 REM PORGRAM TO DRAW A LINE FROM A TO B
110 REM POINT A ---> (X=6,Y=5)
120 REM POINT B ---> (X=2,Y=3)
130 IP%=0 : GOSUB 9000
140 XP=0 : YP=0 : IP%=3 : GOSUB 9000
150 XP=6 : YP=5 : IP%=3 : GOSUB 9000
160 XP=2 : YP=3 : IP%=2 : GOSUB 9000
170 XP=0 : YP=0 : IP%=3 : GOSUB 9000
180 STOP
190 END
```

Fig. 10.15 A program and the resulting plot on the screen for drawing a line from point *A* to point *B*.

The pen of a plotter is initialized at 0,0 by statements 130 and 140. Statement 150 directs the pen to point *A*(6,5) with the pen up *(IP% is pen position, so IP% = 3)*. Statement 160 lowers the pen and moves it to point *B* at coordinates (2,3), and statement 170 moves the pen back to the point of origin.

Experienced programmers are capable of writing graphics programs for important professional applica-

Fig. 10.16 A general-purpose drafting program, VERSACAD ADVANCED is one example of CAD software designed for the new generation of personal computers. (Courtesy of T&W Systems.)

tions. Most technical professionals, however, rely on commercial CAD software packages.

Commercial CAD packages and the microcomputer

More and more commercial CAD software packages are becoming available for use with microcomputers. Micro-based systems offer strong competition for the mainframe CAD workstations, such as CADAM and Computervision, which dominated the market until the early 1980s (Fig. 10.16). The two clearest advantages of CAD microcomputer software packages are cost and ease of upgrading when improvements are required or the manufacturer makes updates available.

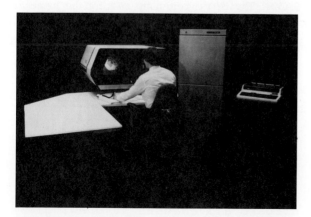

Fig. 10.17 A mainframe-based CAD Workstation. (Photo courtesy of GE Calma.)

Several of the most popular computer graphics software packages for the microcomputer are: AutoCAD, CADD-23, CADplan, Personal Designer, and VersaCAD. Each of these programs works best if used with a computer with a hard disk and 512Kb of RAM. A typical IBM microcomputer graphics system with a tablet digitizer is shown in Fig. 10.17.

10.7
AutoCAD computer graphics

The remainder of this chapter is devoted to the use of AutoCAD 2.15 software, (AutoCAD is a trademark of Autodesk, Inc.) with an IBM PC XT (512Kb RAM, 10Mb hard disk), a Mouse Systems mouse, and a Hewlett-Packard 7475A plotter. Although CAD software differs, it is possible to transfer knowledge from one system to the next. A general coverage of AutoCAD commands and operations are given in this chapter, and specialized applications will be covered throughout the remaining chapters.

To apply the principles and techniques covered in this book, emphasis will be placed on the use of AutoCAD as a two-dimensional drawing tool. As such, the reader should regard AutoCAD as a sophisticated drawing tool and not become confused by trying to think of it in comparison with sophisticated CAD packages with massive analysis and storage capabilities.

All commands and data can be input through the keyboard when using AutoCAD, but the drafter will find it advantageous to use a tablet or mouse to provide additional speed while working at the terminal. Some Mice work on a mechanical principle, much like a large ball bearing, whereas others use an optical light that is transmitted to a light-sensitive grid. The mouse shown in Fig. 10.18 is equipped with 2 buttons that can be used to select points, and give a carriage return.

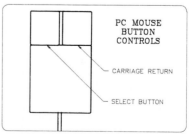

Fig. 10.18 A three-button mouse that is used with a pad to provide two functions.

10.8
Starting up

Begin AutoCAD by booting up the system and typing *ACAD*. The Main Menu will appear on the screen:

```
    0. Exit AutoCAD
    1. Begin a NEW drawing
    2. Edit an EXISTING drawing
    3. Plot a drawing
    4. Printer plot a drawing
    5. Configure AutoCAD
    6. File Utilities
    7. Compile shape/font descrip-
       tion file
    8. Convert old drawing file

    Enter Selection: 1 (CR)
    Enter NAME of drawing: DRAW1
    (CR)
```

The number 1 was entered to specify that a new drawing, called DRAW1, is to be drawn. Names of drawings can be no longer than eight characters with no blank spaces, exclusive of drive prefix (A:, for example). AutoCAD automatically adds a file type of ".DWG" following the assigned name, which will be filed as A:DRAW1.DWG. The Main Menu leaves the screen, and the drawing area with the Root Menu ready for making a drawing is displayed.

Had you wished to edit a previous drawing, you could have responded:

```
    Enter selection: 2 (Edit an existing
    drawing)
    Enter NAME of drawing: A:DRAW2
    (CR)
```

The drawing, DRAW2, would then be displayed on the screen as it appeared when last exited from. You may now edit (change) the drawing as a continuation of the last drawing session.

If you have forgotten the names of the drawing files in memory, enter selection 6 from the Main Menu for a listing of them. Type 0 to return to the Main Menu, and enter 2 to name the drawing to edit.

10.9
Experimenting

The beginner who has not used any version of computer graphics, or AutoCAD in particular, can benefit from turning the machine on, loading the system by typing ACAD, and responding to the screen prompts to see how much could be understood from the prompts alone. For example, to draw a line, you would select DRAW from the Root Menu, and LINE from the submenu. By experimentation, you will become familiar with the system, its prompts and menus. But you will soon find the need for specific instructions.

10.10
Shutting down

Turn the machine off by leaving the Drawing Editor, going to the Main Menu, exiting from AutoCAD, and returning to the operating system. The first step is to select the UTILITY command from the Root Menu, which gives the following options: QUIT, SAVE, END, and ENDSV.

END By selecting END, your drawing is saved under the name it was given at the start of the session; your drawing leaves the screen, and the Main Menu reappears.

SAVE The SAVE command asks for the name to save the drawing under. You may give it a new name and keep the current drawing on the screen in the Drawing Editor. SAVE can also be used to protect against a power failure by periodically saving the drawing under its original name. For example, place an "A:" in front of the name (A:DRAW3) if you want to save the drawing to drive A.

QUIT If you are changing a drawing or looking at a previously made drawing and wish to return to the Main Menu and discard all changes made during the current session, use the QUIT command. You will receive the prompt Do you really want to discard all changes to drawing? By responding *Yes* or *Y*, you will be returned to the Main Menu, and the drawing will be left unchanged.

ENDSV The ENDSV command can be used to save your drawing as a vector file that will display the drawing more rapidly than when saved with the END command. However, this command uses more memory than the END command, and ENDSV files should be erased when they have served their purpose to free up disk space.

10.11
Drawing layers

AutoCAD provides an infinite number of layers on which to make a drawing. Each layer is assigned its own name, color, and line type. For example, you may have a layer name HIDDEN that appears on the screen in yellow, and the line types drawn on it will be dashed lines to represent hidden lines. No more than one name, color, and line type can be assigned to a single layer.

Layers can be used to reduce the duplication of effort. For example, architects commonly use the same floor plan for a number of different applications: one for dimensions, one for furniture arrangement, one for floor finishes, one for electrical details, and so forth. The same basic plan is used for all these applications by turning on the needed layers and turning off others.

SETTING LAYERS The following layers are sufficient for most working drawings:

Layer Name	State	Color	LineType
0	ON/OFF	7 (White)	CONTINUOUS
VISIBLE	ON/OFF	1 (Red)	CONTINUOUS
HIDDEN	ON/OFF	2 (Yellow)	HIDDEN
CENTER	ON/OFF	3 (Green)	CENTER
HATCH	ON/OFF	4 (Cyan)	CONTINUOUS
DIMEN	ON/OFF	5 (Blue)	CONTINUOUS
CUT	ON/OFF	6 (Magenta)	PHANTOM

Layers are given names that correspond with their line types, and each is assigned a different color to distinguish them on a color monitor. The 0 (zero) layer is the system default layer, which can be turned off but not deleted.

By selecting LAYERS from the Root Menu, you will receive a subcommand of LAYER, which is selected from the keyboard or mouse. The following is AutoCAD's dialogue with you while you are setting layers:

```
Command: LAYER (CR)
?/Set/New/On/Off/Color/Ltype/
Freeze/Thaw: New or N (CR)*
Layer name(s): VISIBLE,
HIDDEN, CENTER, HATCH, DIMEN,
CUT (CR)
```

This series of responses has created six new layers by name, and each will have a white default color and a continuous line type. To set the **COLOR** of each layer, we must respond in the following manner:

```
?/Set/New/On/Off/Color/Ltype/
Freeze/Thaw: COLOR (Or C) (CR)
Color: RED or 1 (CR)
Layer name(s) for color 1
(red) <VISIBLE>: VISIBLE (CR)
```

The Visible layer has been assigned the color red, which could have been assigned with the number 1 instead of the word, RED. The colors of the other layers must be assigned in the same manner, one at a time. **LINE TYPES** are assigned to layers in the following manner:

```
?/Set/New/On/Off/Color/Ltype/
Freeze/Thaw: Ltype (Or L) (CR)
Linetype (or ?) <CONTINUOUS>:
HIDDEN (CR)
Layer name(s) for linetype
HIDDEN <0>: HIDDEN (CR)
```

The lines on the Hidden layer will now be drawn with dashed (Hidden) lines.

SET We must SET a layer to make it the current layer in order to draw on it by responding in the following manner:

```
?/Set/New/On/Off/Ltype/Freeze/
Thaw: SET (Or S) (CR)
New current layer <0>:
VISIBLE (CR) (CR) (Two returns)
```

Any line drawn will appear on the Visible layer in red with continous lines. Once we have defined a number of layers and have drawn on each, they can be turned on or off by the `On/Off` command in the following manner:

```
?/Set/New/On/Off/Ltype/Freeze/
Thaw: ON or OFF
Layer name(s) to turn on:
VISIBLE or * (For all layers)
```

(Or, <u>VISIBLE</u>, <u>HIDDEN</u>, <u>CENTER</u>, to turn these layers <u>ON</u> or <u>OFF</u>) (CR)

Even though a number of layers are on, you can draw on only one layer, the *current layer*. By selecting the question-mark option (?), the screen will display the current listing of the layers, their line types, colors, and on/off status. Press function key, *F1*, to change the screen back to the graphics editor.

FREEZE AND THAW `FREEZE` and `THAW` are options under the `LAYER` command. By `FREEZE`-ing a specified layer, it will not be redrawn or plotted until it has been `THAW`ed; consequently, a drawing can be regenerated much faster on the screen. This option allows unneeded layers to be turned off much like the layer option, `OFF`. To save your file of layers and their specifications, use the utility command `END`, which will return you to the Main Menu.

But before we `END` let's set additional parameters as shown in the next section.

10.12
Setting screen parameters

To set screen parameters, you must become familiar with `LIMITS`, `GRID`, `SNAP`, `UNITS`, `ORTHO`, and the function keys.

LIMITS The `LIMITS` command is used to establish the size of the screen area which represents the size of the paper on which a drawing will be plotted. A full-size drawing that will fit on an A-size sheet (11 by 8.5 inches) will have a drawing area of about 10.5 by 8 inches for most plotters. If millimeters are used,

the limits will be approximately 254, 198. `LIMITS` are set as follows:

```
Command: LIMITS (CR)
ON/OFF Lower Left corner:
<0.00,0.00>: (CR) (Accept default value.)
Upper right corner:
<24.00,36.00>: 10,7.8 (CR)
```

GRID A grid on the screen can be set in the following manner:

```
Command: GRID (CR)
On/Off/Value (X)/Aspect
<0.00>: .2 (CR)
```

This command paints the screen with a pattern of square dots that are each 0.2 in. apart (Fig. 10.19). To make the newly assigned limits fill the working area of the screen, use the `ZOOM ALL` command from the `DISPLAY` command on the Root Menu.

UNITS The `UNITS` used in a drawing must be assigned before the limits can be set. By typing or selecting the command, `UNITS`, AutoCAD will give the following listing to select from:

```
1. Scientific (2.67E+02)
2. Decimal (267.00 inches or millimeters)
3. Engineering (12'-4.50")
4. Architectural (16'-3 1/2")

Enterchoice, 1 to 4 <2>: 3 (CR)
```

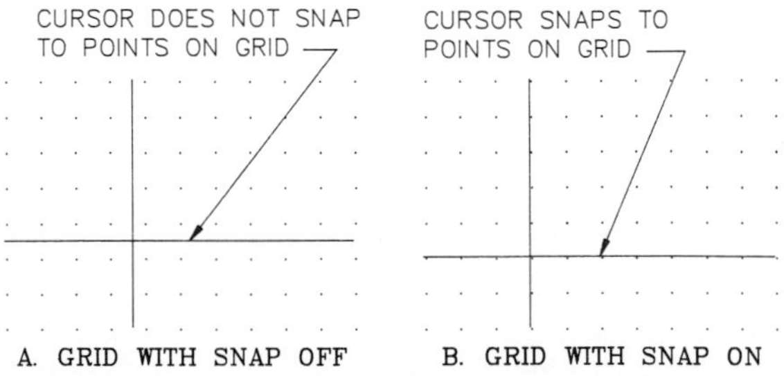

A. GRID WITH SNAP OFF   B. GRID WITH SNAP ON

Fig. 10.19 The SNAP command can be used to make the cursor stop on a visible (or invisible) grid on the screen.

---

*(CR) will be used to indicate a carriage return.

Select the type of units and respond to the following prompts to specify the details desired, such as the number of decimal places. Decimal units should be used for the metric system (millimeters) and the English system (inches). To return to the Drawing Editor, press the *F1* function key.

SNAP The SNAP command can be used to make the cursor on the screen snap to a visible or invisible grid. If your grid is set at spacings of 0.2 in., the cursor can be made to snap to the grid or to an invisible 0.1-in. grid between the visible grid points (Fig. 10.19).

```
Command: SNAP (CR)
On/Off/Value/Aspect/
Rotate/Style: 0.1 (CR)
```

AXIS The AXIS command is used to display ruler lines of a specified spacing on the screen across the bottom and right sides, in the following manner:

```
Command: AXIS (CR)
On/Off/Tick spacing
(X)/Aspect:    0.20 (CR)
```

This response sets the ruler with tick marks 0.20 units apart. By responding with *5X*, the tick marks are spaced at a multiple of five times the snap resolution, which places tick marks at every fifth snap point. The ASPECT option allows you to set different values on each scale of the axis.

ORTHO The ORTHO command forces all lines to be either vertical or horizontal, which aids in making orthographic views where most lines meet at right angles.

```
        Command: ORTHO (Select On or Off from the
screen menu.) (CR)
```

ORTHO can be turned off by using *CTRL 0* or by pressing function key *F8* (Fig. 10.20). Function keys can be used to turn screen parameters on and off once they have been set. *Running coordinates* can be obtained at the top of the screen to give the coordinates of the cursor's position by pressing *F6*. The *status line* (Fig. 10.21) at the top of the screen gives the current layer by name; shows that SNAP is on when an S appears, ORTHO is on when an O appears, and

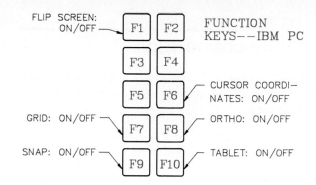

Fig. 10.20 The function keys on the IBM, can be used to turn six functions on or off.

FILL is on when an F appears; and displays coordinates when this feature is on.

If these parameters were specified while you were in the process of setting layers, they would be saved when you ENDed the file. This file should be saved with its parameters as an "empty" file with no drawing, and named FORMAT so it can be called up as a drawing to be edited on which to begin a new drawing. Once the drawing is completed in this FORMAT file, SAVE the drawing and give it a name different from FORMAT, then QUIT, and respond YES when asked if you wish to discard all changes in the drawing. In this way you have returned to the original file, FORMAT, without disturbing its parameters, allowing it to be used again, and you have saved a drawing that was made from it.

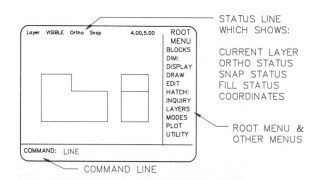

Fig. 10.21 The status line at the top of the screen gives the status of the modes in which you are operating. The command line appears at the bottom of the screen.

10.13
Utility commands

Utility commands can be used to control the operation of AutoCAD and make changes in the files that are developed.

HELP The HELP (?) command gives a list of commands on the screen. Press *CTRL C* to abort the listing process. Press *F1* to return to the graphics mode on the screen. To obtain additional information on a specific command, respond to the prompts as follows:

```
Command: HELP (Or ?) (CR)
Command name (Return for list):
LINE (CR)
```

The screen will give information about the LINE command.

FILES The FILES command permits you to list, delete, rename, and copy files from the Drawing Editor. By typing FILES, the following listing will appear on the screen:

```
File Utility Menu
0. Exit file utility menu
1. List drawing files
2. List user specified files
3. Delete files
4. Rename files
5. Copy files

Enter selection: 1 (CR)
```

By typing *1*, you will be prompted for the disk drive to search, Enter prefix. Enter *C:* if that is the file that you wish to have searched.

Option 2, List user-specified files, can be used to display the files that you specify. For example, by responding to ENTER FILE SEARCH SPECIFICATIONS: with *C:*.BAK*, you will get a listing of all drawing files with a suffix .BAK that are stored on disk drive C.

Option 3, Deleting files, allows you to eliminate unneeded files on your disks. You can specify a file such as DRAW1.DWG, or you can use wild cards such as *.DWG to delete all files with a DWG suffix. When using wild cards, each file that meets the specifications will be listed one at a time, and you are given the option to delete them by entering *Y* or *N* as they are listed.

Option 4, Rename files, is used to change the name of a file by responding to the prompts as follows:

```
Enter current file name:
A:DRAW1.DWG (CR)
Enter new file name:
A:DRAW2.DWG (CR)
```

Option 5, Copy file, lets you copy a file with the following prompts:

```
Enter name of Source file
name: B:DRAW1.DWG (CR)
Enter name of Destination file
name:
C:DRAW1.DWG (CR)
```

SHELL The SHELL command gives access to the DOS operating system while remaining in the Drawing Editor by responding:

```
Command: SHELL (CR)
DOS command: DIR A: (Or similar com-
mand) (CR)
Command: (Reappears on screen)
```

PURGE The PURGE command can be used when you first begin editing an existing drawing in the following manner:

```
Command: PURGE (CR)
Purge unused Blocks/Layers/
LTypes/SHapes/STyles?
All: ALL (CR)
```

The ALL response is used to eliminate any unused objects that are in the drawing file. The other options can be used to purge specific features of a drawing.

10.14
Basics of drawing lines

From the Main Menu, enter *2* to edit a drawing, and respond with FORMAT when prompted for a drawing to edit. The empty drawing file with its assigned parameters will appear on the screen ready for a new drawing. Select DRAW from the Root Menu and

LINE from the submenu that follows to draw lines (Fig. 10.22).

The command line on the screen prompts you as follows:

Command: <u>LINE</u> (CR)
From point: (Select a point or give coordinates.)
To point: (Select second point or give coordinates.)
To point: (You can continue in this manner with a series of connected lines.)

As the lines are drawn by moving the cursor about the screen, the current line will *rubber band* from its last point (Fig. 10.23). Points can be entered at the keyboard as absolute or relative coordinates (Fig. 10.24). *Absolute coordinates* are located with respect to 0,0, the lower left corner of the drawing area, whereas *relative coordinates* are measured from the last

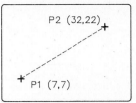

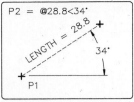

A. ABSOLUTE COORDINATES   **B. RELATIVE COORDINATES**

Fig. 10.24 **A.** Absolute coordinates can be entered at the keyboard to establishe the ends of a line. **B.** Relative coordinates are relative to the previous point on the screen and can be entered at the keyboard.

point. Relative points are located by typing the symbol @ followed by the distance from the last point, for example, @2,4.0. By inserting @, followed by the distance from the previous point, and <, followed by the angle the line makes the horizontal direction measured in a counterclockwise direction, the coordinates are given in polar form. For example, @ 28.8<34 locates a point 28.8 units from the previous point at a 34° angle with the horizontal.

You can correct errors when typing commands by one of the following methods:

CTRL X (Deletes the line)
CTRL C (Cancels the current command and returns the "Command:" prompt)
Backspace (Deletes one character at a time)

The status line gives the length and angle from the last point when you are rubber banding from point to point. When drawing polygons, the last side can be made to close by using the CLOSE command as shown in Fig. 10.25.

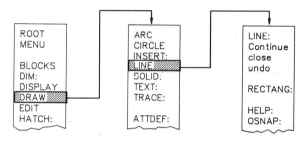

Fig. 10.22 You may progress from the Root Menu to levels of submenus by selecting commands with the keyboard or mouse.

10.15
TRACE command

Wide lines, thicker than the point of a pen, can be drawn using the TRACE command in the following manner:

Command: <u>TRACE</u> (CR)
Width: <u>0.4</u> (CR)
From point: 2,3 (CR)
To point: 4,6 (CR)
To point: 6,2 (CR)

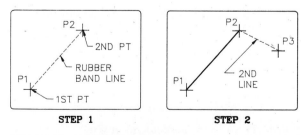

STEP 1   **STEP 2**

Fig. 10.23 Drawing a line.

Step 1 Command: LINE
From point: **P1** (Select point.)
To point: **P2** (The line is drawn.)

Step 2 To point: **P3** (The line is drawn). (CR) (Return to disengage the rubber band.)

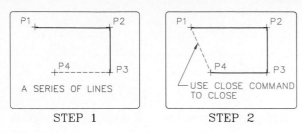

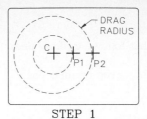

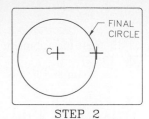

Fig. 10.27 Drawing a circle.
Step 1   Command: *CIRCLE*
3P/2P/<Center point>: Locate center **C** with cursor.
Diameter/<Radius>:DRAG *(Select center **P1**, move to **P2**, and the circle will dynamically change if* DRAGMOD: *is* ON.*)*

Step 2   The final radius is selected with the select button and the circle is drawn.

Fig. 10.25   CLOSE command.

Step 1   A series of lines are drawn using the LINE Command from **P1** through **P4**. (The last line is a rubber-band line until the next point is selected.)

Step 2   After **P4** has been selected, select CLOSE, and the line will be drawn to the first point of the series, **P1**.

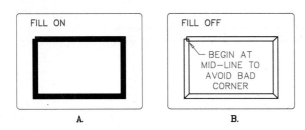

Fig. 10.26   TRACE command.

A. When the TRACE command is used with FILL ON, lines are drawn solid to the width specified.

B. When FILL is turned off, parallel lines are drawn.

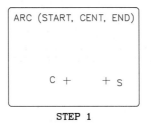

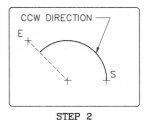

Fig. 10.28   ARC command.

Step 1   Command: ARC (Select S,C,E)
Arc Center /<Start>: (Select start pt.)
Center/End/<Second point>: C CENTER (Select center.)

Step 2   Angle/Length of chord/<Endpoint>: DRAG *Select Pt.* **E** (The arc is drawn to an imaginary line from **C** to **E** in a counterclockwise direction.)

With FILL on, the line will be drawn as shown in Fig. 10.26A. When FILL is off, the line will be drawn as parallel lines (Fig. 10.26B). The plotting of the line being entered will not be plotted until the endpoint of the next line is indicated in order for the program to compute the "miter" angles at each corner.

10.16
Drawing circles

The command for drawing circles is found under the DRAW submenu, where you may give the center and radius, the center and diameter, or three points. Used with DRAG, you can move the cursor and see the cir-

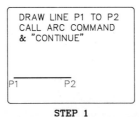

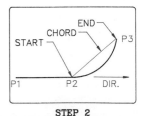

Fig. 10.29   Arc tangent to end of line.

Step 1   Command: *Line* From point: *P1*
To point: *P2* (CR)

Step 2   Command: ARC CONTINUE
Endpoint: DRAG (Select *P3*)
(Arc is drawn tangent to the line.)

cle change until it is the size you want it to be (Fig. 10.27). DRAG can be turned On or Off by inserting the command, DRAGMODE, followed by On or Off.

10.17
Drawing arcs

The ARC command is found under the DRAW command, where arcs can be drawn using nine combinations of variables, including starting point, center, angle, ending point, length of arc, and radius. For example, the *S,C,E* version requires that you locate the starting point *(S)*, the center *(C)*, and the ending point *(E)* (Fig. 10.28). The arc begins at point *S*, but point *E* need not lie on the arc. Arcs are drawn in a counterclockwise direction by default.

To continue a line with an ARC drawn from the last point of the line and tangent to it, respond as follows:

```
Command: ARC (CR)
Center/Start point>: CONTINUE
```

An arc may now be drawn tangent to the last point of the line, which is useful for drawing runouts on a part (Fig. 10.29).

With the only difference being the order of the commands, the same technique can be used for drawing a line from a previously drawn arc.

The DRAGMODE command can be turned on to allow the arcs to be seen before they are selected for their final positions. An example of drawing straight lines and arcs in combination is shown in Fig. 10.30.

An arc that is to be tangent of a line and an arc can be constructed by drawing a circle using the radius of the tangent arc. The circle is DRAGged into position to be tangent (Step 1). In Step 2, the line, arc, and circle are broken to give a smooth transition.

10.18
Enlarging and reducing drawings

Parts of a drawing or the entire drawing can be enlarged by the ZOOM command, a submenu under the DISPLAY Command. A part of the drawing in Fig. 10.31 is too small to read at its present size, but by

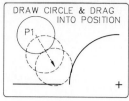

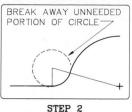

Fig. 10.30 ARC tangent to line and arc.

Step 1 Command: *CIRCLE*
3P/2P/<Center point>:*P1*
Diameter/<Radius>: *Select R*
Command: MOVE, Select Objects or Window or Last: *Window circle*
Base point of displacement: *P1* (Use center.)
Second point or displacement: *DRAG* (Move circle to be tangent to the line and arc.)

Step 2 BREAK all three entities at the intersection points. Draw lines that locate points of tangency. (See Section 10.17 BREAKing entities.)

using a ZOOM WINDOW, you can select the part that you want enlarged to fill the screen.

Instead of responding with WINDOW, you could have responded with ALL, CENTER, EXTENTS, LEFT, PREVIOUS, or a number to indicate the factor by which you want the present drawing changed in size.

ALL The ALL response makes the drawing's limits as large as possible while fitting it within the display screen.

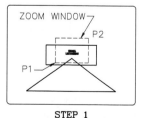

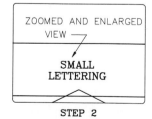

Fig. 10.31 ZOOM command.

Step 1 Command: *ZOOM*
Magnification or type (ACELPW): *W* (CR)
First point: *P1* (Select pt.)

Step 2 Second point: *P2* (Select pt.)
(The window will be enlarged to fill the screen.)

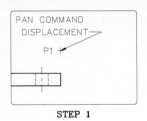

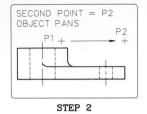

STEP 1   STEP 2

Fig. 10.32 *PAN* command.

Step 1 Command: *PAN*
Displacement: *P1* (Select pt.)

Step 2 Second point: *P2* (Select new pt.)
(The object will be moved to new position.)

CENTER The CENTER reply allows you to pick the center of the drawing and degree of magnification or reduction desired.

EXTENTS The EXTENTS response enlarges the drawing as much as possible while disregarding its limits if it does not fill them.

LEFT By responding with L (left), you can pick the lower left corner and the height of the drawing that you wish to have enlarged.

PREVIOUS The PREVIOUS reply displays the last view that was used. Views can be ZOOMed an infinite number of times.

PAN The PAN command moves a part of the drawing about the display screen, which is useful when the drawing extends beyond the screen's limits. The drawing in Fig. 10.32 is PANned by entering two points, the displacement point (a point on the screen to be moved) and the second point (the new position of the first point).

10.19
Erasing lines

The ERASE command, a subcommand under EDIT, is used to remove parts of a drawing. The LAST command calls for the erasure of single entities (lines, text, circles, arcs, and blocks) one at a time working backward from the one most recently drawn. The use of a WINDOW is the second option for erasing, where en-

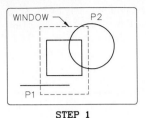

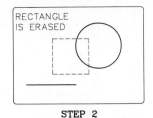

STEP 1   STEP 2

Fig. 10.33 ERASE command.

Step 1 Command: ERASE
Select Objects or Window or Last: *W*
(CR)
First point: *P1* (Select pt.)
Second point: *P2* (Select pt.)

Step 2 Select Objects or Window or Last: (CR)
(The rectangle within the window is erased. Entities partially within the window are not erased.)

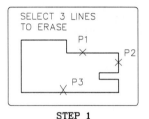

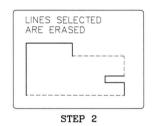

STEP 1   STEP 2

Fig. 10.34 ERASE command—entities.

Step 1 Command: *ERASE*
Select Objects or Window or Last:(O is default) (Select entities (lines) with *P1, P2,* and *P3*.)

Step 2 Press (CR) again and the lines are erased. Use Command: OOPS to recall erased entities.

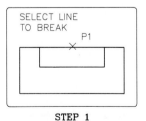

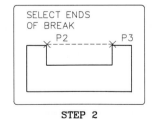

STEP 1   STEP 2

Fig. 10.35 BREAK command.

Step 1 Command: *BREAK* (CR)
Select Object: *P1* (On line to be broken)
Enter second point or F: *F* (CR)

Step 2 Enter first point: *P2*
Enter second point: *P3* (The line is broken from *P2* to *P3*.)

tities that lie completely within the window will be erased as shown in Fig. 10.33.

```
Command: ERASE (CR)
Select objects or Window or
Last: WINDOW (Or W) (CR)
First point: Select P1
Second point: Select P2, 4 found
Select objects or Window or
last: (CR)
```
(The entities are erased.)

The default of the ERASE command is Select Objects which requires that you point to entities on the screen with the cursor to indicate those to erase (Fig. 10.34). By pressing RETURN, the entities are erased.

Entities lying within the window that should not be erased can be removed by using the REMOVE command and selecting the entities with the cursor. When selected, the entities become dashed lines on the screen indicating they have been marked for erasure. Additional lines can be marked for erasure by the ADD command in the same manner as the REMOVE command but with an opposite effect. Should you mistakenly erase something, the OOPS command can be used to restore the last erasure, but only the last erasure.

The BREAK command is used to remove a part of an entity, such as a line (Fig. 10.35).

```
Command: BREAK (CR)
Select object: Select Point P1 on the line
to be broken
Enter second point or F:F (CR)
Enter first point: Select P2
Enter second point: Select P3
```
(The line is broken from P2 to P3.)

Had the break not begun and ended at intersections of lines, you could have omitted the step in which the *F* response was given. But this extra step ensures that the computer understands which line is to be broken.

10.20
CHANGE command

The CHANGE command permits changes in LINES, LAYERS, BLOCKS, CIRCLES, and TEXT. For example, to change the line in Fig. 10.36 to be tangent

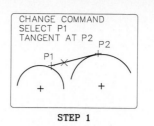 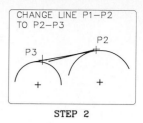

Fig. 10.36 CHANGE command.

Step 1 Command: *CHANGE*
Select Objects or Window or Last: *PX*
(Select pt on line.) (CR)

Step 2 Change point (or Layer or Elevation): *P3* (Select new location, *P3*, and *P1* will be moved to *P3*.)

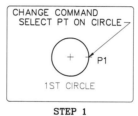 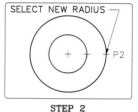

Fig. 10.37 CHANGE circle command.

Step 1 Command: *CHANGE* (CR)
Select Objects or Window or Last: *P1*
(Select pt on circle.) (CR)

Step 2 Change point (or Layer or Elevation): *P2* (Select new point to change radius.)

to a circle, you may use the CHANGE command, select a point on the line, and move it to be tangent to the circle (by using OSNAP), and the end of the line is moved to be tangent to the circle. The size of a circle can be changed by pointing to the circle and selecting a new radius, and the circle will be enlarged or reduced on the screen (Fig. 10.37). Blocks can be moved or rotated with the CHANGE command, as shown in Fig. 10.38. Text can be moved or changed.

If you wish to change part of a drawing to a different layer, execute CHANGE and select the part to move by pointing to entities, windowing, or using the "last" option. When prompted for Change point, respond with L (layer), and give the name of the layer to which the drawing will be moved (Fig. 10.39).

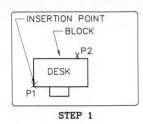
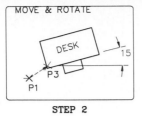

STEP 1   STEP 2

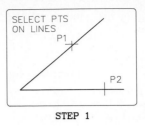
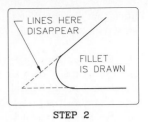

STEP 1   STEP 2

Fig. 10.38 CHANGE blocks command.

Step 1 Command: <u>CHANGE</u> (CR)
Select Objects or Window or Last: <u>P1</u>
(Point to block) (CR)
Change point (or Layer or Elevation):
<u>P2</u> (Select new position.)

Step 2 New rotation angle <0>: <u>45</u> (CR)
(The block is rotated 45 degrees.)

Fig. 10.40 FILLET command.

Step 1 Command: <u>FILLET</u> (CR)
Polyline Radius/<Select two lines>: <u>R</u> (CR)
Enter fillet radius <0.0000>: <u>1.5</u> (CR)
Command: (CR)

Step 2 FILLET Polyline Radius/<Select
two lines>: <u>P1</u> and <u>P2</u> (The fillet is drawn and
the lines are trimmed.)

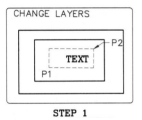
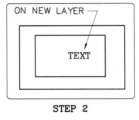

STEP 1   STEP 2

Fig. 10.39 CHANGE layers command.

Step 1 Command: <u>*CHANGE*</u> (CR)
Select Objects or Window or Last: <u>W</u>
(CR)
First point: *P1* Second point: <u>*P2*</u> (CR)

Step 2 Change point (or Layer or
Elevation): <u>L</u>
New layer name: <u>*VISIBLE*</u> (Name of existing
layer) (CR)

10.21
FILLET command

FILLETS can be drawn to any desired radius between two nonparallel lines, whether or not they intersect. The fillet is drawn and the lines are trimmed
as shown in Fig. 10.40.

By entering a fillet radius of 0, nonintersecting
lines will be automatically extended to form a perfect
intersection. Once the radius is assigned, it remains in
memory as the default radius for drawing additional
fillets.

10.22
POLYLINE (PLINE) command

The PLINE command is used to connect a series
of lines and arcs of varying widths to form a
POLYLINE. The prompt after executing PLINE
isArc/Close/Halfwidth/Length/Undo/
Width/<End point of line>:. The default
response, End point of line, shown in
brackets, will result in a line drawn from the last point
on the screen at the time you executed PLINE. You
will be prompted for the width of the line at its beginning and end. This command requires experimentation to become familiar with its many options.

STRAIGHT LINES PLINE command defaults to
the straight-line mode and prompts:

From point: *Select point*
Current line width is 0.3:
Arc/Close/Halfwidth/Length/
Undo/Width/<End point of
line>: <u>W</u> (For Width)
Starting width <0.000>: <u>0.4</u>
(CR)
Ending width <1.000>: <u>0.6</u> (CR)
(Select endpoint.)

The line is drawn. The longer the line, the thinner and
more tapered it becomes.

A value of zero can be entered for the thinnest
lines that can be drawn by the system. CLOSE will

automatically close to the beginning point of the PLINE and terminate the command. LENGTH lets you continue a PLINE at its last angle by specifying the length of the segment. If the first line was an arc, this command will produce a line that is tangent to the arc.

The UNDO option erases the last segment of the polyline, and it can be repeated to continue erasing segments of the PLINE. The HALFWIDTH option lets you specify the width of the line from the center of a wide line, as shown in Fig. 10.41.

ARCS When you respond to the PLINE option line with an *A* (ARC) you can specify arc segments of a PLINE. AutoCAD will give the following command line:

```
Angle/Center/Close/Direction/
Halfwidth/Line/Radius/Second
pt/Undo/Width <End point of
arc>:
```

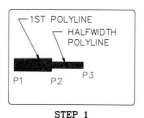
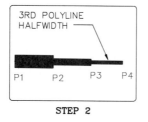

Fig. 10.41 PLINE (CR) command.

Step 1 Command: *PLINE*
From point: *P1* (Select pt.)
Current line width is 0.0000
Arc/Close/Halfwidth/Length/
Undo/Width/<Endpoint of line>: WIDTH
(CR)
Starting width <0.0000>: .6 (CR)
Ending width <0.6000>: (CR)
Arc/Close/ . . . /<Endpoint of line>:
P2
Arc/Close/ . . . /<Endpoint of line>:
Halfwidth
Starting half-width <0.3000>: .3 (CR)
Ending half-width <0.3000>: (CR)
Arc/Close/ . . . /<Endpoint of line>:
P3

Step 2 Arc/Close/ . . . /<Endpoint of line>: *Halfwidth*
Starting half-width <0.2>: .1 (CR)
Ending half-width <0.1000>: (CR)
Arc/Close/ . . . ?<Endpoint of line>:
P4

The default assumes that an arc will be drawn tangent to the last line drawn and will pass through the next point selected.

By using the ANGLE option you may give the Included angle: as a positive or negative value, and the next prompt will ask for Center/Radius/<End point>:. AutoCAD then draws an arc that is tangent to the previous line segment. If you select Center, you will be asked to give the center of the next arc segment. The next prompt asks for Angle/Length/<End point>:, where Angle refers to the included angle, and Length is length of the arc's chord.

The CLOSE option causes the PLINE to be closed with an arc segment to the beginning point. DIRECTION allows you to override the default, which draws the next arc tangent to the last PLINE segment. AutoCAD prompts direction from starting point:, and you can point to the desired beginning point and respond to the next prompt, Endpoint, to give the direction of the arc.

The LINE option switches the PLINE command back to the straight-line mode. The RADIUS option gives a prompt, Radius:, that allows you to specify the radius of the next arc. The following prompt, Angle/Length/<End point>:, lets you specify the included angle or the length of the arc's chord.

SECOND PT is used to select the second point and the endpoint of a three point arc. The two prompts are Second point: and Endpoint:.

10.23
PEDIT command

The PEDIT command is used to edit polylines drawn with the PLINE command. The prompts for this command are:

```
Select polyline: Select line with cursor
Entity selected is not a
polyline
Do you want it turned into
one? Y (YES)
Close/Join/Width/Edit ver-
tex/Fit curve/Uncurve/exit
<X>: CLOSE (CL)
(Used to close PLINE)
```

If the PLINE is already closed, the CLOSE com-

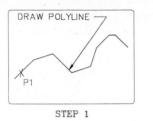 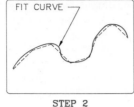

STEP 1   STEP 2

Fig. 10.42 PEDIT command.

Step 1 Command: *PEDIT* (CR)
Select Polyline: *P1* (Select line.)

Step 2 Close/Join/Width/Edit vertex/Fit curve/Uncurve/eXit <X>: *FIT CURVE* (CR) (The curve is smoothed.)

mand will OPEN it by removing the closing segment that was drawn with the CLOSE option.

JOIN The JOIN option lets you respond to the prompt Select objects Window or Last: by selecting segments that are to be joined to the polyline. Once chosen, these segments become part of the polyline. Segments must have exact meeting points and must not meet with an overlapping intersection for joining to take place.

WIDTH The WIDTH option gives the prompt Enter new width for all segments:. Enter the new width from the keyboard, and the Pline is redrawn to this new width.

FIT CURVE The FIT CURVE(F) option constructs a smooth curve that passes through all vertices of the PLINE with pairs of arcs that join sequential vertices (Fig. 10.42). To change the resulting curve to better suit your needs, use the EDIT VERTEX command discussed below.

UNCURVE UNCURVE (U) removes extra vertices inserted by the FIT CURVE option above. EXIT (X) is used to exit from the PEDIT command and return to the command prompt.

EDIT VERTEX EDIT VERTEX (E) allows you to select a single vertex of the Pline and edit it. When this option is used, the first vertex of the Pline will be marked with an X on the screen. An arrow will be shown if you have specified a tangent direction for

the vertex, and you will receive the following prompt:

Next/Previous/Break/Insert/
Move/Regen/Straighten/Tangent/
Width/eXit/<N>:

NEXT and PREVIOUS The NEXT (N) and PREVIOUS (P) options move the X-marker to the next or previous vertex. To move to a vertex several vertices away, select NEXT or PREVIOUS and press *Return* until the vertex is reached.

BREAK When the BREAK (B) option is selected, the location of the position of the X is shown and the following prompt appears:

Next/Previous/Go/eXit <N>:

By using NEXT or PREVIOUS you can select a second point and enter GO, and the line between the two points will be erased (Fig. 10.43). Enter EXIT and the break will be canceled and you will return to EDIT VERTEX.

INSERT The INSERT option gives the prompt:

Enter location of new vertex:

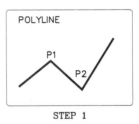 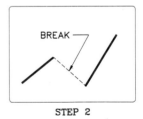

STEP 1   STEP 2

Fig. 10.43 PEDIT—break polyline.

Step 1 Command: *PEDIT* (CR) Select polyline: (Selectline)
Close/Join/Width/Edit vertex/Fit curve/Uncurve/eXit <X>: *E* (Edit vertex)
Next/Previous/Break/Insert/Move/Regen/
Straighten/Tangent/Width/eXit <N>:
(Move cursor to where break is to begin with Previous or Next) *P1*. BREAK
Next/Previous/Go/eXit <N>: Move to *P2* with Next

Step 2 Next/Previous/Go/eXit<N>: *Go* (CR) (The line will break between *P1* and *P2*.)

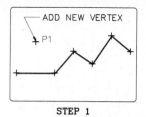
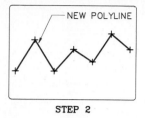

STEP 1 — STEP 2

Fig. 10.44 PEDIT—add vertex.

Step 1 Use EDIT VERTEX command and place "X" on line before the new vertex.
```
Next/Previous/Break/Insert/Move/Regen/
Straighten/Tangent/Width/eXit <N>:
Insert
Enter location of new vertex: P1 (New pt.)
```

Step 2 Press (CR), and the new vertex will be inserted and the polyline will pass through it.

This lets you add a new vertex to a polyline (Fig. 10.44). Also, a vertex can be moved by the MOVE (M) option (Fig. 10.45).

STRAIGHTEN A polyline can be straightened by the STRAIGHTEN (S) option, which saves the current location of the vertex specified by an "X" and prompts:

```
Next/Previous/Go/eXit/<N>:
```

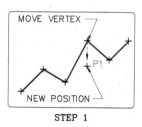
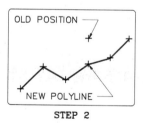

STEP 1 — STEP 2

Fig. 10.45 PEDIT—move vertex.

Step 1 Command: PEDIT (CR)
```
Close/Join/Width/Edit vertex/Fit
curve/Uncurve/eXit <X>: E (Edit vertex)
(Move X to vertext to be moved.)
Next/Previous/Break/Insert/Move/Regen/
Straighten/Tangent/Width/eXit <N>:
Move
Enter location of new vertex: P1
```

Step 2 Press (CR), and the polyline will be changed to pass through the moved vertex.

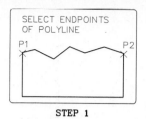
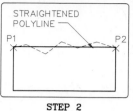

STEP 1 — STEP 2

Fig. 10.46 PEDIT—straighten line.

Step 1 Use EDIT VERTEX option of PEDIT, and place the "X" at the vertex at the beginning of the line to be straightened, P1.
```
Next/Previous/Break/Insert/Move/
Regen/Straighten/Tangent/Width/eXit
<N>: Straighten
```

Step 2 Next/Previous/Go/eXit <N>: Next (Move to P2.)
```
Next/Previous/Go/eXit <N>: Go (Line P1–P2
is straightened.)
```

By moving the "X" to a new vertex on the line and specifying GO, the line will be straightened between the two vertices (Fig. 10.46). Enter X for eXit, if you change your mind, and you will be returned to the EDIT VERTEX prompt.

The TANGENT (T) suboption lets you indicate a tangent direction at the vertex marked by the "X" for use in curve fitting when it is used next. The prompt is:

```
Direction of tangent:
```

Enter the angle from the keyboard or select a point with the cursor on the screen from the current point.

The WIDTH (W) suboption lets you change the beginning and ending widths of an existing line segment from the X-marked vertex. Use the NEXT and PREVIOUS options to confirm which direction the line will be drawn. To draw the changed polyline on the screen use the REGEN (R) option.

10.24
CHAMFER command

The CHAMFER command is used to construct angular bevels at the intersections of lines or polylines. Select two lines and the lines are trimmed or extended, and the CHAMFER is drawn. Press *Return* and you are

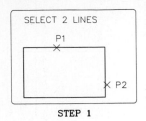
STEP 1

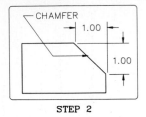
STEP 2

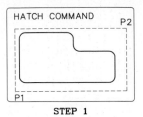
STEP 1

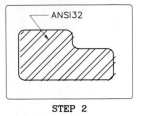
STEP 2

Fig. 10.47 CHAMFER command.

Step 1 Command: *CHAMFER* (CR)
Polyline/Distance/<Select first line>: *D* (Used for nonpolylines)
Enter first chamfer distance <0.4>: *1.00* (CR)
Enter second chamfer distance (.4): *1.00* (CR)

Step 2 Command: (CR)
CHAMFER Polyline/Distance/<Select first line>: *P1*
Select second line: *P2*

ready to repeat this command using the last values if other corners are to be chamfered (Fig. 10.47). A polyline is chamfered in the same manner.

10.25
HATCH command

The HATCH command is used to crosshatch an area that has been sectioned (Fig. 10.48). The prompts are:

    Command: HATCH (CR)
    Pattern (? or name/U,style):
    <default>: ANSI31 (CR)

By responding with ?, you will be given a list of the standard patterns in ACAD.PAT. A response of U is a user-defined pattern with the following prompts:

    Angle for crosshatch lines
    <0>: 45 (Or show with the pointer) (CR)
    Spacing between lines <0.1>:
    0.2 (Or show with the pointer) (CR)
    Double hatch area? <N> (CR)
    Select Objects or Window or
    Last:

By responding as shown above, equally spaced hatch lines will be drawn 0.2 units apart and at 45° with the horizontal.

Fig. 10.48 HATCH command.

Step 1 Command: *HATCH* (CR)
Pattern (? or name/U,style)
<default>: *ANSI32*
Scale for pattern <default>: *1.00*
Angle for pattern <default>: *0*
Select Objects or Window or Last: *W*
(CR)(Window object)

Step 2 Press (CR) and the plotting will be completed. Hit CTRL C to terminate hatching, if desired.

The letters N, O, or I can be added to your response to the PATTERN prompt. For example, PATTERN: ANSI32,N. These letters will cause the crosshatching to be given in the style shown in Fig. 10.49 and as defined below:

> *N-Normal* (Hatches alternate areas beginning with the outermost)
>
> *O-Outermost* (Hatches only the outermost areas)
>
> *I-Ignore* (Hatches all inside areas ignoring contents)

If a pattern is selected from the *ACAD.PAT* file, you will receive the following prompt:

    Scale for pattern <1.00>: .75
    Angle for pattern <0>: 0
    Select Objects or Window or
    Last:

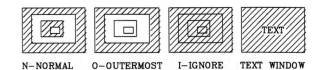

N–NORMAL   O–OUTERMOST   I–IGNORE   TEXT WINDOW

Fig. 10.49 Whenever you specify a hatching pattern, enter a comma and the letter N, O, or I after the pattern, and the hatching will be applied as shown above, for example, "Pattern: ANSI31,N," to hatch the outside and alternate layers inside the figure. Hatching automatically leaves a window around text within the area.

Both responses can be made at the keyboard or by selecting two points on the screen for each to indicate the angle and the scale by *CTRL C*. All hatch lines can be removed as a group, since they are entered as a block. If entered with an ''*'' preceding the pattern name, they can be edited one line at a time. The area to be hatch is selected one line at a time or windowed as shown in Fig. 10.49.

10.26
Text and numerals

The TEXT command, a subcommand under DRAW can be used for inserting plotted text in a drawing from the keyboard at any size and angle (Fig. 10.50).

```
Command: TEXT (CR)
Starting point or (ACRS): Select
point with cursor
Height <0.18>: 0.5 (CR)
Rotation angle <0>: (CR)
Text: TEXT (CR) (The word, TEXT, is
```
written beginning with the point selected and is left-justified.)

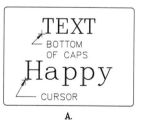
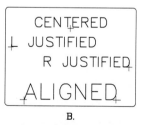

Fig. 10.50 A. The TEXT Command defaults to left-justified text, and the point indicated represents the lower left corner of capital letters. **B.** Text can also be centered, right justified, or aligned so that the first and last letters of a word or line span the distance between two selected points.

Other responses are A (aligned), which means the text is changed to fit exactly between two endpoints (Fig. 10.50A); C centers the text about a point; R (right) justifies the text to point located at the right; and S (style) selects a text font from those available.

10.27
Text style

Five text fonts are available with AutoCAD: TXT, SIMPLEX, COMPLEX, ITALIC, and VERTICAL (Fig. 10.51). The default style is STANDARD, which uses TXT. The STYLE command is used to create and name variations of the four standard fonts in the following manner:

```
Command: STYLE (CR)
Text style name (or?): PRETTY
Font file <TXT>: COMPLEX (CR)
Height (0.20): 0 (CR)
Width factor <default>: 1.00
(CR)
Obliquing angle <45>: 0 (CR)
Backwards? (Y/N): N (CR)
Upside-down? <Y/N>: N (CR)
```

STYLE NAME	FONT	VERTICAL STYLE
STANDARD	TXT	V E R T I C A L
SIMPLEX	SIMPLEX	
COMPLEX	COMPLEX	
ITALIC	*ITALIC*	

Fig. 10.51 Five fonts of text are available from AutoCAD: Txt, Simplex, Complex, Italic, and Vertical.

Responding to the first prompt with ? will give a list of the defined text styles. The text style created above is named PRETTY. Each time PRETTY is specified, it will be unnecessary to reassign the values given above unless you wish to change them. However, if you select a new font file such as SIMPLEX for PRETTY, then previously drawn text will be redrawn when regenerated. By giving a height of 0, you will be prompted for the height of the lettering desired whenever PRETTY is specified as the style of text.

The fifth font, VERTICAL, can be used for drawing vertical lines of text with letters stacked one under the other. Text must be entered with a rotation angle of 270° when using this font.

The QTEXT command can be used to reduce the time required to display a drawing on the screen when

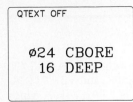

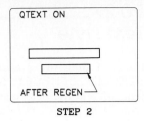

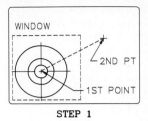

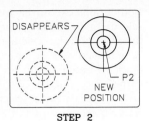

STEP 1   STEP 2   STEP 1   STEP 2

Fig. 10.52 QTEXT command.

Step 1   Command: *QTEXT* (CR) On/Off <current>: *ON*

Step 2   Command: *REGEN* (Text is shown as boxes.)

it is redrawn. QTEXT (quick text) can be set to ON, and the text will be drawn on the screen as a series of boxes representing the space required for the text (Fig. 10.52). When QTEXT is turned OFF, the full text will be plotted on the screen.

Special characters must be entered from the keyboard when using the TEXT command. By preceding the codes with a double percent sign, %%, the following characters will be entered:

%%O	Toggle overscore mode on/off
%%U	Toggle underscore mode on/off
%%d	Draw "degrees" symbol
%%p	Draw "plus/minus" tolerance symbol
%%c	Draw "circle diameter" dimensioning symbol
%%%	Force a single percent sign
%%nnn	Draw special character number "nnn" (ASCII)

10.28
Moving and copying drawings

A drawing can be moved to a new position by the MOVE command (Fig. 10.53), or it can be duplicated by the COPY command. When copied, the original drawing is left in its original position, and a copy of it is located where specified.

The COPY procedure is the same except the beginning command is COPY, instead of MOVE.

A drawing can be moved or copied by dragging it into position by the DRAG command (Fig. 10.54). Drag is activated by responding to Second point of displacement: with DRAG. When the cur-

Fig. 10.53 MOVE command.

Step 1   Command: *MOVE* (CR)
Select Objects or Window or Last: *W* (CR)
First point: (Select pt)
Second point: (Select pt): 5 found.
Select Objects or Window or Last: (CR)
Base point or displacement: (Select 1st pt)

Step 2 Second point of displacement: (Select 2nd pt, *P2*) (The object is drawn at its new position, *P2*, and the original drawing disappears.)

sor is moved about the screen, the drawing is dynamically moved until it is set by pressing the select button of the mouse.

In addition to a window, you may use the default, which is set to select the entities to be moved with the cursor, one at a time. Also, you may select LAST, which will move the last entity, such as a LINE, ARC, CIRCLE, TEXT, or BLOCK.

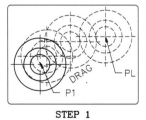

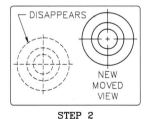

STEP 1   STEP 2

Fig. 10.54 DRAG command.

Step 1   Command: *MOVE* (CR)
Select Objects or Window or Last: *W* (Window object)
Base point or displacement: *P1* (or X,Y distance)
Second point of displacement: *DRAG* (As the cursor is moved, the drawing is DRAGged about the screen.)

Step 2 When moved to the desired location, press the Select button to draw the object in its final position. (The original drawing disappears.)

10.29
Mirroring drawings

Symmetrical objects can be drawn by drawing a portion of the figure and then mirroring the drawing about one or more axes. The schematic threads in Fig. 10.55 are drawn using the MIRROR command. If a line coincides with the MIRROR line, such as *P1–P2* in Fig. 10.55, the line will be drawn twice when mirrored. Therefore a line of this type should be drawn after the view has been mirrored.

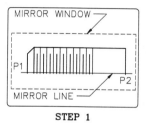

Fig. 10.55 MIRROR command.

Step 1 Command: *MIRROR*
Select Objects or Window or Last:
(Window drawing to be mirrored)(CR)
First point or mirror line: *P1*
Second point: *P2*

Step 2 Delete old objects? <N>: *N* (CR)
(The object is mirrored about the mirror line. Draw the centerline last.)

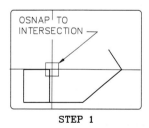

Fig. 10.56 OSNAP option.

Step 1 Command: *LINE* (CR)
From point: Select OSNAP on the screen menu
Select INTERSEC (Intersection)
Interce of (Select intersection.)

Step 2 To point: (Select ENDPOINT)
To point: endpoint of (Move cursor to endpoint of line, and press Select button. The line is drawn.)

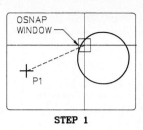 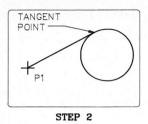

Fig. 10.57 OSNAP tangent option.

Step 1 Command: *LINE* (CR)
Line from point: Select *P1*
To point: Select OSNAP from the menu and select tangent

Step 2 To point: tangent to select point on circle. (The line is drawn from *P1* tangent to the circle on the side where the tangent point was selected.)

10.30
Snapping to objects (OSNAP)

On some occasions, you will want to draw lines that SNAP to features of other objects within drawings rather than snap to a grid. This type of snap is called an OSNAP, which is short for object snap. For example, the steps in drawing a line from an intersection of lines to the endpoint of a line is illustrated in Fig. 10.56.

From *P1* in Fig. 10.57 a line is drawn tangent to the circle. The same procedure as shown above is used, but instead of ENDPOINT for the second point, TANGENT is selected on the side of the circle where the tangent point is to be located. The tangent point is found automatically, and the line is drawn.

Other OSNAP options permit you to snap to the following: *centers of arcs and circles, insertion points of blocks, midpoints of lines, nearest points of an object, points (nodes), perpendicular to lines, tangent to arcs and circles, and quadrant points of a circle.*

10.31
ARRAY command

The ARRAY command is used to draw repetitive shapes in circular and rectangular patterns. For example, a series of holes can be located on a bolt circle by drawing the first hole in its desired position and

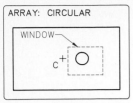

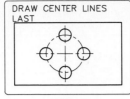

STEP 1   STEP 2

Fig. 10.58 ARRAY—circular.

Step 1 Begin by drawing the figure to array
Command: *ARRAY* (CR)
Select Objects or Window or Last: *W*
(Window the hole) (CR)
Rectangular or Circular array (R/C):
C
Center point of array: *C* (Select point.)

Step 2 Angle between items (+=CCW,
−=CW): *30* (CR)
Number of items or −(degrees to
fill): *−360* (CR)
Rotate objects as they are copied?
<N>: *Y* (CR)
(The array is drawn.)

then activating the ARRAY command, as shown in
Fig. 10.58.

A rectangular ARRAY is begun by drawing the
first part in the lower left corner of the array. Once
drawn, follow the commands of the rectangular AR-
RAY shown in Fig. 10.59. Termination of the array
can be caused by *CRTL C.*

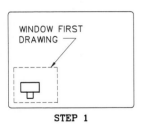

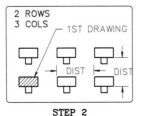

STEP 1   STEP 2

Fig. 10.59 ARRAY—rectangular.

Step 1 (Draw desk in lower left of ARRAY.)
Command: *ARRAY* (CR)
Select Objects or Window or Last: *W*
(Window desk) (CR)
Rectangular of Circular array (R/C): *R*

Step 2 Number of rows: *2* (CR)
Number of columns: *3* (CR)
Unit cell distance between rows: *4* (CR)
Distance between columns: *3.5* (CR)
(CR) (The array is drawn.)

10.32
BLOCKS

One of the more powerful features of computer graphics is the option of building a file of drawings or symbols to be used repetitively on drawings. In AutoCAD,
these drawing files are called BLOCKS.

BLOCKS, such as the SI symbol in Fig. 10.60,
are drawn in the conventional manner and are
blocked as shown below:

Command: BLOCK (CR)
BLOCK name (or ?): SI (CR)
Insertion base point: *Select insert
point*
Select objects or Window or
Last: WINDOW (CR)
First point: *Select pt*
Second point: *Select pt*
Select objects or Window or
Last: (CR)
(The BLOCK is filed into memory and disappears
from screen.)

The BLOCK is inserted into the drawing by the
steps shown in Fig. 10.60. BLOCKS are inserted as

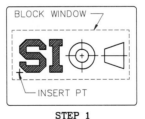

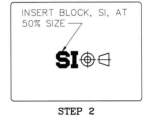

STEP 1   STEP 2

Fig. 10.60 BLOCK command.

Step 1 Make a drawing that you wish to BLOCK,
and respond to the BLOCK Command as follows:
Block name (or ?): *SI*
Insertion point: (Select insert point)
Select Objects or Window or Last: *W*
(Window the drawing and the object disappears
into memory.)

Step 2 To Insert, Command: *INSERT* (CR)
Block name (or ?): *SI* (CR)
Insertion point: (Select point with cursor)
X-scale factor <1>/ Corner/XYZ: *0.5* (CR)
Y-scale factor <default=X>: (CR)
Rotation angle <0>: (CR)
(Block SI is inserted at 50% size.)

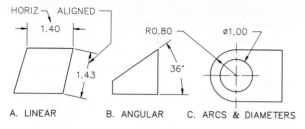

Fig. 10.61 The types of dimensions that appear on a drawing.

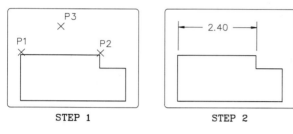

Fig. 10.62 Dimensioning a line.

Step 1 Command: *DIM*
DIM: LINEAR/ HORIZ:
First extension line origin or RETURN to select: *P1*
Second extension line origin: *P2*
Dimension line location: *P3*

Step 2 Press (CR), and the measurement appears at the command line, 2.40. Press (CR) to accept this value (or give a different value), and the dimension line is drawn.

DIM VARS	DEFAULT	
DIMSCALE	1.0000	Overall scale factor
DIMASZ	0.1800	Arrow size
DIMCEN	0.0900	Center mark size
DIMEXO	0.0645	Extension line offset
DIMDLI	0.3800	Dimension line increment
DIMEXE	0.1800	Extension beyond dim. line
DIMTP	0.0000	Plus tolerance
DIMTM	0.0000	Minus tolerance
DIMTXT	0.1800	Text height
DIMTSZ	0.0000	Tick size
DIMTOL	*OFF	Add +/− to dimension text
DIMLIM	*OFF	Generate dimension limits
DIMTIH	ON	Text inside extension is horizontal
DIMTOH	ON	Text outside extension is horizontal
DIMSE1	OFF	Suppress first extension line
DIMSE2	OFF	Suppress second extension line
DIMTAD	OFF	Place text above dimension line

Fig. 10.63 A listing of the DIM VARS that can be selected and changed when dimensioning.

entities, which means they cannot be edited by erasing parts of them or breaking lines within them.

When an attempt is made to erase a portion of a BLOCK, the whole BLOCK is erased. However, BLOCKS can be edited if a star is inserted in front of the block name, for example, Block name (or?): *SI.

BLOCKS can be used only on the current drawing file unless they are converted to WBLOCKS, Write Blocks, which become permanent. This conversion is performed as follows:

Command: WBLOCK (CR)
File Name: SI (CR) (This assigns the name of the WBLOCK.)
Block Name: SI (CR) (This is the name of the BLOCK that is being changed to a WBLOCK.)

A library of WBLOCKS that can be inserted in different files can relieve drafters of making drawings and will greatly improve their productivity.

10.33
Dimensioning

The types of dimensions that can be used with Auto-CAD are shown in Fig. 10.61; these options can be found in the submenu of the DIM: command. When a line is dimensioned, you are prompted to specify if the dimension line is to be horizontal, aligned, or rotated. In Fig. 10.62 a horizontal line is dimensioned by selecting its two endpoints (**P**1 and **P**2) and locating the dimension line with **P**3. The dimension of 2.40 in., which is measured by the program, is accepted by a <u>Return</u>, and the dimension is shown on the drawing (Step 2).

> All drawings should be drawn full size when using computer graphics. The scaling process will take place at the time of plotting the drawings.

Before any dimensioning attempts, you should check the dimensioning variables, Dim Vars, which can be called up on the screen and displayed one at a time (Fig. 10.63). All ratios of these variables are based on the letter height of the dimension numerals and letters, as shown in Fig. 10.64. Once the variables are inserted, they become default variables.

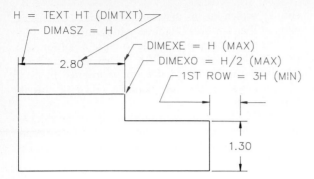

Fig. 10.64 Dimensioning variables are based on the height of the lettering, usually about one-eighth inch.

The UNIT command must be used to assign the number of decimals that the dimension units are to have. Two decimal places are used for inches, and none for millimeters. Architectural units are feet and inches.

Dimensions can be placed over or within the dimension lines by turning the DIMTAD command on or off (Fig. 10.65).

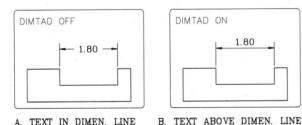

A. TEXT IN DIMEN. LINE   B. TEXT ABOVE DIMEN. LINE

Fig. 10.65 **A.** When DIMTAD is OFF, the text is inserted within the dimension line. **B.** When DIMTAD is ON, the text is placed above the dimension line.

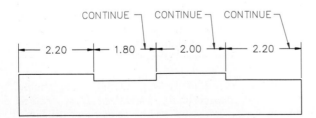

Fig. 10.66 When dimensions are placed end to end, the command, CONTINUE, can be used to specify the "First extension line origin" after the first dimension line has been drawn.

When a series of horizontal dimensions link together end-to-end, the CONTINUE command can be used to join successive dimension lines in a line, as shown in Fig. 10.66. This command suppresses the first extension line, since one is left from the previous dimension.

BASELINE dimensioning can be used to give a series of dimensions that originate from the same baseline. The DIMDLI variable (dimension line increment for continuation) automatically separates the parallel dimension lines (Fig. 10.67).

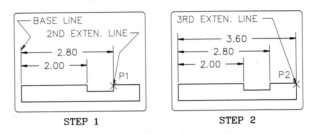

STEP 1   STEP 2

Fig. 10.67 BASELINE command.

Step 1 Command: *BASELIN*
Dimension the first line as shown in Fig. 10.66. When prompted for the "First extension line" for the next dimension, respond with BASELIN, and you will be asked for the "Second extension line." (Select *P1*.)

Step 2 The second dimension line will be drawn. Respond to "DIM:" with BASELIN again, and when asked for "Second extension line," select *P2*. Continue in this manner for any number of dimensions using the same baseline.

10.34

Dimensioning arcs and circles

Dimensions of circles will be given, as shown in Fig. 10.68, based on the size of the circle unless you override the defaults in the system. The computer follows the same decision process in selecting the style of dimensions as you would when using a pencil. The process of dimensioning a circle is shown in Fig. 10.69. A point on the circumference is selected with the cursor, and the diameter is automatically computed and placed across the circle.

Arcs are dimensioned with an **R** placed in front of the dimension of the radius (Fig. 10.70). LEAD-

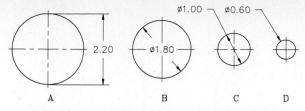

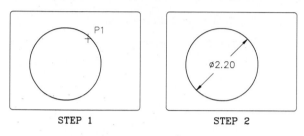

Fig. 10.68 Types of dimensions available for dimensioning circles with AutoCAD.

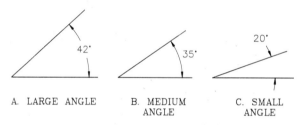

Fig. 10.69 Dimensioning a circle.

Step 1 Command: *DIAMETER* (CR)
Select arc or circle: (Select *P1*)

Step 2 Dimension text <2.20>: (CR) to accept this dimension. The diametric dimension is drawn from *P1* through the center.)

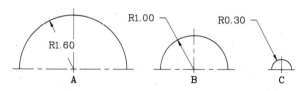

Fig. 10.70 Arcs will be dimensioned by one of the formats given here, depending on the size of the radius.

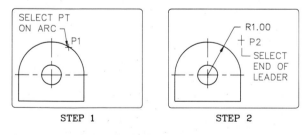

Fig. 10.71 Dimensioning with a LEADER.

Step 1 Command: *LEADER*
Leader start: *(Select P1)*

Step 2 To point: *(Select P2)*
Dimension text <1.00>: (CR) to accept this value. (The leader and dimension are drawn.)

ERS can be used to position diametric and radial dimensions at locations of your choosing while still giving the **R** and diameter symbols in front of the measurements. The steps of dimensioning an arc are shown in Fig. 10.71.

10.35
Dimensioning angles

Variations in dimensioned angles, which occur because of inadequate space, are shown in Fig. 10.72. Begin by selecting two lines of the angle, and then select the location for the dimension line arc. Where room permits, the dimension value will be centered in the arc between the arrows. The commands for dimensioning an angle are shown in Fig. 10.73.

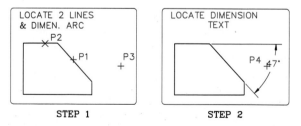

A. LARGE ANGLE   B. MEDIUM ANGLE   C. SMALL ANGLE

Fig. 10.72 Angles will be dimensioned in one of the following formats using AutoCAD.

Fig. 10.73 ANGULAR command.

Step 1 Command: DIM: *ANGULAR*
Select first line: *P1*
Select second line: *P2*
Enter dimension line arc location: *P3*
(CR)
Dimension text <47>: (CR)

Step 2 Enter text location: *P4* or (CR) (The angular dimension is drawn.)

10.36
Toleranced dimensions

Dimensions can be toleranced automatically, using any of the forms shown in Fig. 10.74, by setting the Dim Vars—DIMTP (plus tolerance), DIMTM (minus tolerance)—and by setting the DIMTOL to *ON*.

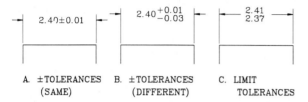

A. ±TOLERANCES   B. ±TOLERANCES   C. LIMIT
   (SAME)   (DIFFERENT)   TOLERANCES

Fig. 10.74 Dimensions can be toleranced in any of these three formats.

As long as DIMTOL is on, all dimensions will have tolerance applied to them. By turning ON the Dim Vars, DIMLIM, AutoCAD will compute the upper and lower limts of the dimension and apply them to the dimension line. The steps of giving tolerance limits on a dimension are shown in Fig. 10.75.

10.37
Oblique pictorials

An oblique pictorial can be constructed as shown in Fig. 10.76. The front, orthographic view is constructed and then COPYed behind the first view at the angle desired for the receding axis. Using OSNAP, the visible endpoints are connected and invisible lines are erased. Circles can be drawn as true circles on the true-size front surface, but circular features should be avoided on the receding planes, since their construction is complex.

10.38
Isometric pictorials

The STYLE subcommand of the SNAP command can be used to change the rectangular GRID (called STANDARDS) to ISOMETRIC (I) where the dots are plotted vertically and at 30 degrees with the

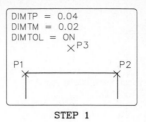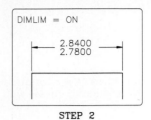

STEP 1   STEP 2

Fig. 10.75 Toleranced dimensions.

Step 1 To tolerance a dimension, set Dim Vars, DIMTOL to ON. Set DIMTP (plus tolerance) and DIMTM (minus tolerance) to the desired values. Set DIMLIM to ON to convert the tolerances into the limit form.

Step 2 Dimension the line by using *P1*, *P2*, and *P3* to specify the first extension line, second extension line, and the dimension location. (The toleranced dimension is drawn.)

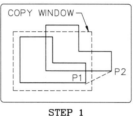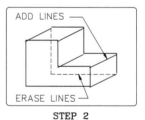

STEP 1   STEP 2

Fig. 10.76 An oblique pictorial.

Step 1 Draw the frontal surface of the oblique, and COPY this view at *P2*, which is the desired angle and distance from *P1*.

Step 2 Connect the corner points, and erase the invisible lines to complete the oblique.

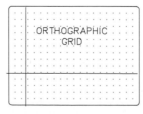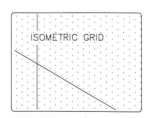

A. ORTHOGRAPHIC GRID   B. ISOMETRIC GRID

Fig. 10.77 **A.** The orthographic grid is called the STANDARD style of the SNAP mode.
B. The ISOMETRIC style of the SNAP mode.

horizontal (Fig. 10.77). The lines of the cursor on the screen align with two of the isometric axes. The axes can be rotated by the use of *Control-E* or by activation of the ISOPLANE command from the screen menu. When ORTHO is ON, lines are forced to be drawn parallel to the isometric axes. The steps for constructing an isometric using the grid are shown in Fig. 10.78.

To depict circular features in isometric, a four-center ellipse one unit across its diameter is constructed (Fig. 10.79). This ellipse should be made into a BLOCK and inserted where needed. It can be scaled to fit different drawings by giving it an X-VALUE

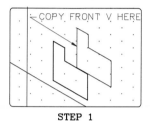
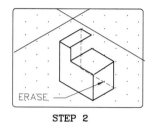

STEP 1   STEP 2

Fig. 10.78 The isometric pictorial.

Step 1 Draw the front view of the isometric pictorial. COPY this front view at its proper location.

Step 2 Connect the corner points and erase the invisible lines. The cursor lines can be moved into three positions using Control E or ISOPLANE.

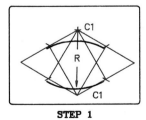
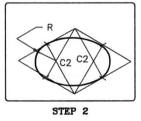

STEP 1   STEP 2

Fig. 10.79 The four-center ellipse block.

Step 1 Construct a rhombus that is one unit along both axes. Using the four-center ellipse method, construct an ellipse using four centers and four arcs. (The rhombus should be drawn on a construction layer that can be turned off so the lines will now show.)

Step 2 Make a WBLOCK of this ellipse using the center as the insertion point. This block can be INSERTed as a *BLOCK so it can be broken when needed.

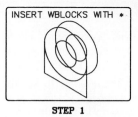
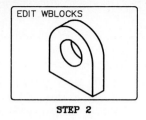

STEP 1   STEP 2

Fig. 10.80 An isometric drawing with elliptical features

Step 1 The UNIT BLOCK of the ellipse developed in Fig. 10.79 is inserted and sized (X-scale factor) to match the drawing's dimensions. The ellipse can be rotated as a BLOCK to fit any of the three isometric surfaces.

Step 2 The back side of the isometric is drawn. The hidden lines are removed with the BREAK command.

factor at the time of insertion, and it can be rotated to fit each isometric plane.

An isometric pictorial with elliptical features is shown in Fig. 10.80. The ellipse was inserted using the four-center ellipse BLOCK developed in Fig. 10.79. The size of the block is changed by entering the size factor when prompted for the X-value. It is advantageous if the block is designed to be one unit across (a UNIT block), so its size can be scaled with ease. For example, a UNIT BLOCK will be multiplied by an X-value of 2.48 to give a BLOCK that is 2.48 in.

10.39
Plotting

A hard copy of a drawing can be plotted with either a pen plotter or a printer plotter (a typewriter printer with graphics capability). You may initiate the commands for plotting from the Main Menu or from the Root Menu, when the drawing to be plotted appears on the display. The command for activating the pen plotter is PLOT, and PRPLOT is used for the printer plotter.

You must tell AutoCAD which part of the drawing to plot by responding to the following prompt:

```
What to plot Display, Extents,
Limits, View, or Window <D>: L
```

The meanings of these options are:

DISPLAY *(D)* will plot the view currently visible on the display screen, or the last view that was displayed before SAVE, END, or ENDSV.

EXTENTS *(E)* will plot the entities to fill the drawing area and be as large as possible, even though they may fill only part of the limits.

LIMITS *(L)* will plot the drawing area defined by the limits.

VIEW *(V)* will plot a named view that was saved by the VIEW command. You will be prompted for the VIEW name.

WINDOW *(W)* will plot a selected part of the drawing that is enclosed in a window specified by two diagonal points. When PLOT is begun from the Drawing Editor, you can point to the corners of the window, but you must use coordinates of the window entered at the keyboard when beginning at the Main Menu.

The plotter origin for a Hewlett-Packard 7475 is located at 0.42 in., 0.35 in. from the lower left corner of an 11 × 8½-inch sheet, which gives an effective plotting area of 10.15 by 7.8 inches when you instruct the plotter to place the origin at 0,0.

After selecting the part of the drawing to be plotted, you will be given the following specifications on the display screen:

```
Sizes are in Inches
Plot origin is at (0.00, 0.00)
Plotting area is 10.50 wide by
8.00 high (A size)
Plot is NOT rotated 90 degrees
Pen width is 0.010
Area fill will NOT be adjusted
for pen width
Hidden lines will NOT be re-
moved
Scale is 1 = 1

Do you want to change anything?
<N>:
Yes or No, please.
```

You can use these values by responding with *N* or *NO*, or you can change them by entering *Y* or *YES*. When YES is entered you will be given the option of changing as shown in the Computer Example 10.1.

Computer Example 10.1

Layer Color	Pen No.	Line Type	Pen Speed	Layer Color	Pen No.	Line Type	Pen Speed
1 (Red)	1	0	9	9	3	0	2
2 (Yellow)	2	0	9	10	4	0	38
3 (Green)	2	0	9	11	5	0	38
4 (Cyan)	2	0	9	12	6	0	38
5 (Blue)	2	0	9	13	1	0	38
6 (Magenta)	1	0	9	14	2	0	38
7 (White)	2	0	9	15	3	0	38
8	2	0	38				

```
Line types   0 = continuous line
             1 = ................................
             2 = ----     ----     ----     ----
             3 = -----     -----     -----     -----
             4 = ------.  ------.  ------.  ------.
             5 = ---- -   ---- -   ---- -   ---- -
             6 = --- - -  --- - -  --- - -  --- - -
Do you want to change any of these parameters? <N>
```

By pressing RETURN for *NO,* you accept these specifications. If you enter *Y* for *YES,* you will be given a
chance to change any of these values. Do *not* change
the line types if your plotter supports multiple pen
types; leave the line type set at 0 (continuous line).
Pen speed can be set to yield the best line for the type
of pen you use. PEN NO. gives the location of the
pen in a multiple-pen plotter.

After *Y* has been entered to indicate that you wish
to change parameters, AutoCAD will list Layer
Color, Pen No., Line type, and Pen
Speed one at a time. You may accept the current
value for each by pressing RETURN, or change the
values by typing the new value after each default is
displayed. When completed, type **S** to obtain an updated display of your changes. Type **X** to exit from this
portion of the program when the changes are correct.

AutoCAD will now prompt you for additional
plot specifications:

Size units (Inches or millimeters)
<1>: I (I for inches and M for millimeters)
Plot origin in units <0.00,0.00>:
2,1 (Enter coordinates of the orgin from the
"home" position of your plotter, using millimeters or inches, depending on the units you have
specified.)

AutoCAD will list the standard plotting sizes:

Standard values for plotting
size

Size	Width	Height
A	10.50	8.00
B	16.00	10.00

Enter the Size or Width,
Height (in units) <B>: A
(A-size sheet 11 by 8.5 inches or MAX for the
largest size the plotter will accept, perhaps a C-
size sheet.)

Rotate 2D plots 90 degrees
clockwise? <N>: N
Pen width <0.10>: (CR to accept or enter a new value)
Adjust area fill boundaries
for pen width? <N>: N
(By responding "Y" the boundaries of filled areas
will be moved inward one-half pen width for a
higher degree of accuracy.)

Remove hidden lines? <N> *(CR)*
Specify scale by entering:
Plotted units=Drawing units or
Fit or? <Fit>:

If your drawing was to be plotted in inches, enter
1 = 1 for a full-size drawing where a plotted inch was
equal to 1 inch on the screen. For an architectural
scale such as $\frac{1}{2}'' = 1'-0''$, the unit at the left of the
equal sign should be converted to 1. This conversion
results in an equality of $\frac{1}{2}$ in. = 12 in. or 1 in. = 24
in. If your drawing was drawn in millimeters, when
prompted, "Size units (Inches or Millimeters) ⟨I⟩:",
enter M, and the drawing will be plotted in metric
units. A scale of $1'' = 1''$ is equivalent to $1'' = 25.4$
mm. Respond to the plotting prompts as follows:

Specify scale by entering:
Plotted   Millimeters=Drawing
units
or Fit or ? <F>: 1=25.4
(To make 1 inch represent 25.4 millimeters)

By responding with **F,** the drawing will be scaled
to fill the available space, but the scale will be a nonstandard scale. Responding with **?** will give a list of
the various scaling options and their descriptions.

Plot specifications are saved by AutoCAD so you
will not have to change plot parameters until you
wish. AutoCAD will display the message:

Effective plotting area: 10.50
wide by 8.00 high
Position the paper in plotter (Pause
for this step.)
Press RETURN to continue or S
to Stop for hardware setup
Press RETURN to begin plotting

Some plotters have other features that can be adjusted in accordance with manufacturer's specifications. However, the preceding steps are typical of the
ones used by plotters and printers.

10.40
Grid rotation

To draw lines parallel or perpendicular to a given line,
the grid on the screen can be rotated to align with
existing lines. The ROTATE command is an option

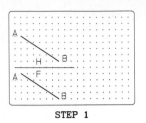 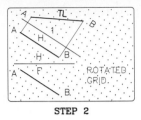

STEP 1   STEP 2

Fig. 10.81 Descriptive geometry problem.

Step 1 Find the true-length view of AB by the following steps: Command: *SNAP*
`On/Off/Value(X)/Aspect/Rotate/Style: R`
`Base point <0,0>:` (Select Pt A)
`Rotation angle <0>:` (Select Pt B)(The grid will rotate to be parallel to AB.)

Step 2 DRAW H–1 reference line. OSNAP from pts A and B perpendicular to H–1. Extend the projectors to locate the true-length view.

under the SNAP command. The prompts are as follows:

> `Command: SNAP` (CR)
> `On/Off/Value/Aspect/Rotate/`
> `Style:R` (Rotate)
> `Base point <0,0>:` *Select a point*
> `Rotation angle <0>:` *Select a second point at the screen or from the keyboard*

The grid will be plotted on the screen in alignment with the selected points. An example of a descriptive geometry problem with a true-length auxiliary view of line **AB** is shown in Fig. 10.81.

The grid is returned to its original position by selecting the SNAP command and the ROTATE option prompts:

> `Base point <2,3>:` (CR)
> `Rotation angle <37>:` 0 (Realigns grid to its original position)

10.41
Applications

Practically any type of drawing can be made using computer graphics.

GRAPHS The circle graph in Fig. 10.82 was generated from drawing commands discussed in this chapter.

TWO-DIMENSIONAL DRAWINGS The Lufkin oil well pump in Fig. 10.83 was drawn using VersaCAD™ software. Notice that this drawing is a combination of points, lines, arcs, and circles.

THREE-DIMENSIONAL DRAWINGS Figure 10.84 shows a three-dimensional drawing of a pulley drawn by using elements of the part's geometry.

Fig. 10.82 A circle graph is an example of drawing plotted on a flatbed plotter. (Courtesy of Houston Instrument, Inc.)

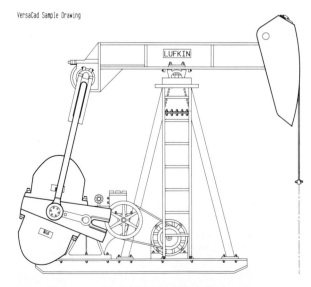

VersaCad Sample Drawing

Fig. 10.83 A two-dimensional drawing of an oil pump. (Courtesy of T&W Systems, Inc.)

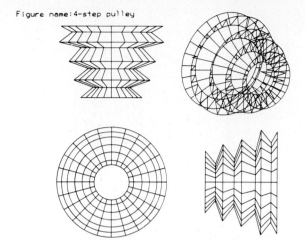

Figure name:4-step pulley

File Name:B:4-step pulley.gra Date:01-01-1985 Time:00:01:48

Fig. 10.84 A two-dimensional and a three-dimensional plot of a pulley. (Courtesy of Brodhead-Garrett Co., (ADD-23.)

10.42
Summary of AutoCAD

The previous examples have given the basic applications of AutoCAD. AutoCAD applications to sample problems will be given in the following chapters. For best results, use the AutoCAD manual along with this text.

Suppliers of microcomputer software

Some sources of microcomputer software programs that can be used for the production of graphics are as follows:

AutoCAD
Autodesk, Inc.
2320 Marinship Way
Sausalito, CA 94965
(415) 331-0356

CADD-23
Brodhead Garrett Co.
4560 East 71st Street
Cleveland, OH 44105
(800) 321-6730

CADplan
Personal CAD systems, Inc.
981 University Avenue
Los Gatos, CA 95030
(408)-354-7193

MEGA CADD
MEGA CADD, Inc.
The Court in the Square
401 Second Avenue South
Seattle, Washington 98104
(206) 623-6245

Personal Designer
Computervision, Inc.
50 Mall Road
Burlington, MA 01803
(617) 275-1800

VersaCAD
T&W Systems
7372 Prince Drive, Suite 106
Huntington Beach, CA 92647
(714) 847-9960

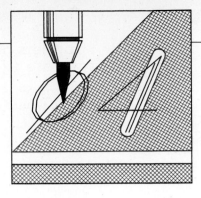

CHAPTER 11

Drawing Instruments

11.1
Introduction

The preparation of technical drawings is possible only by being knowledgeable of and skillful in the use of drafting instruments. You will find that your skill and productivity will increase as you become more familiar with using the available tools. Persons with little artistic ability can produce professional technical drawings when they learn to use drawing instruments properly.

11.2
Pencil

Pencils may be the conventional wood pencil or the lead holder, which is a mechanical pencil (Fig. 11.1). Both types are identified by numbers and letters at their ends. Sharpen the end of the pencil opposite

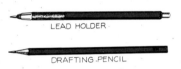

FIG. 11.1 The mechanical pencil (lead holder) or the wood pencil can be used for mechanical drawing. The ends of these pencils are labeled to indicate the grade of the pencil lead.

these markings so you will not sharpen away the identity of the grade of lead.

Pencil grades range from the hardest, 9H, to the softest, 7B (Fig. 11.2). The pencils in the medium-grade range, 3H–3B, are most often used for drafting work.

It is important that your pencil be properly sharpened. This can be done with a small knife or a drafter's pencil sharpener, which removes the wood and leaves approximately three-eighths inch of lead exposed (Fig. 11.3). The point can then be sharpened to a conical

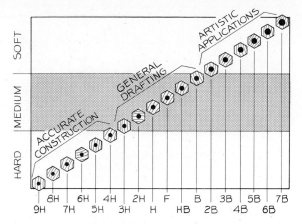

FIG. 11.2 The hardest pencil lead is 9H, and the softest is 7B. Note the diameter of the hard leads is smaller than the soft leads.

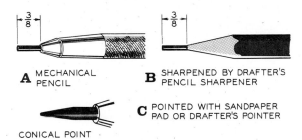

A MECHANICAL PENCIL

B SHARPENED BY DRAFTER'S PENCIL SHARPENER

C POINTED WITH SANDPAPER PAD OR DRAFTER'S POINTER

CONICAL POINT

FIG. 11.3 The drafting pencil should be sharpened to a tapered conical point (not a needle point) with a sandpaper pad or other type of sharpener.

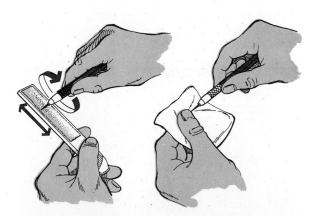

FIG. 11.4 The drafting pencil is revolved about its axis as you stroke the sandpaper pad to form a conical point. The graphite is wiped from the sharpened point with a tissue or cloth.

point with a sandpaper pad by stroking the sandpaper with the pencil point while revolving the pencil (Fig. 11.4). Excess graphite is wiped from the point with a cloth or tissue.

A pencil pointer used by professional drafters (Fig. 11.5) can be used to sharpen wood and mechanical pencils. Simply insert the pencil in the hole and revolve it to sharpen the lead. Other types of small hand-held point sharpeners are also available.

FIG. 11.5 The professional drafter often uses a pencil pointer of this type to sharpen pencils.

11.3
Papers and drafting media

SIZES The surface on which a drawing is made must be carefully selected to yield the best results for a given application. Sheet sizes are specified by letters such as Size A, Size B, and so forth. These sizes, which are listed below, are multiples of either the standard $8\frac{1}{2}$- × -11-inch sheet or the 9- × -12-inch sheet.

Size A	$8\frac{1}{2}'' \times 11''$	$9'' \times 12''$
Size B	$11'' \times 17''$	$12'' \times 18''$
Size C	$17'' \times 22''$	$18'' \times 24''$
Size D	$22'' \times 34''$	$24'' \times 36''$
Size E	$34'' \times 44''$	$36'' \times 48''$

DETAIL PAPER When drawings are not to be reproduced by the diazo process, (which is a blue-line print) an opaque paper, called *detail paper*, can be used as the drawing surface. The higher the rag content (cotton additive) of the paper, the better its quality and durability. Preliminary layouts can be drawn on detail paper and then traced onto the final tracing surface.

TRACING PAPER A thin, translucent paper used for making detail drawings is *tracing paper* or *tracing vellum*. These papers permit the passage of light through them so drawings can be reproduced by the diazo process (blue-line process). The tracing papers that yield the best reproductions are most translucent. Vellum is tracing paper that has been chemically treated to improve its translucency; unfortunately, it does not retain its original quality as long as does high-quality, untreated tracing paper.

TRACING CLOTH *Tracing cloth* is a permanent drafting medium for both ink and pencil drawings. It is made of cotton fabric that has been covered with a compound of starch to provide a tough, erasable drafting surface that yields excellent blue-line reproductions. More stable than paper, tracing cloth does not change its shape with variations in temperature and humidity as much as does tracing paper. Erasures can be made on tracing cloth repeatedly without damaging the surface, which is especially important when drawing with ink.

POLYESTER FILM An excellent drafting surface is polyester film, which is available under several trade names such as Mylar. It is more transparent, stable, and tough than paper or cloth. It is also waterproof and difficult to tear.

Mylar film is used for both pencil and ink drawings. Some films specify that a plastic-lead pencil be used, whereas others adapt well to standard lead pencils. Ink will not wash off with water and will not erase with a dry eraser; erasures can be made with a dampened hand-held eraser. Unless recommended by the manufacturer of the polyester film, an electric eraser is not recommended for use with this medium.

11.4
T-square and board

The T-square and drafting board are the basic equipment used by the beginning drafter (Fig. 11.6). With its head in contact with the edge of the drawing board, the T-square can be moved for drawing parallel hori-

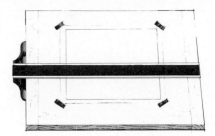

FIG. 11.6 The T-square and drafting board are the basic tools used by the student drafter. The drawing paper is taped to the board with drafting tape.

FIG. 11.7 The drafting machine is often used instead of the T-square and drafting board. (Courtesy of Keuffel & Esser Co., Morristown, N.J.)

FIG. 11.8 The professional who uses a drafting station may work in an environment similar to the one shown here. (Courtesy of Martin Instrument Co.)

zontal lines. Drawing paper should be attached to the board parallel to the blade of the T-square (Fig. 11.7). The drafting board is made of basswood, which is lightweight but strong. Standard sizes of boards are 12 × 14 inches, 15 × 20 inches, and 21 × 26 inches.

11.5
Drafting machines

Although the T-square is used in industry and in the classroom, most professional drafters prefer the *mechanical drafting machine* (Fig. 11.7). This machine which is attached to the table top, has fingertip controls for drawing lines at any angle.

The professional who must have access to a drafting workstation will likely use equipment more sophisticated than that used by the student. A modern, fully equipped drafting station is shown in Fig. 11.8. Today, many offices are equipped with computer graphics stations to supplement manual equipment and techniques (Fig. 11.9).

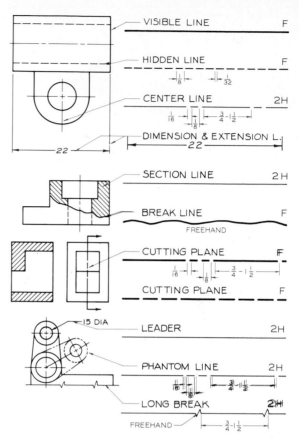

FIG. 11.10 The alphabet of lines varies in width. The full-size lines are shown in the right column along with the recommended pencil grades for drawing them.

FIG. 11.9 More and more professional work stations are being equipped with computer graphics equipment. (Courtesy of Bausch & Lomb.)

11.6
Alphabet of lines

The type of line produced by a pencil depends on the hardness of the lead, the drawing surface, and the technique of the drafter. Examples of the standard

lines, or the *alphabet of lines*, is shown in Fig. 11.10, along with the recommended pencils for drawing the lines. These pencil grades may vary greatly with the drawing surface being used. Guidelines are very light lines (just dark enough to be seen) that are used to aid in lettering and in laying out a drawing; a 4H pencil is recommended for drawing most guidelines.

11.7
Horizontal lines

A horizontal line is drawn using the upper edge of your horizontal straightedge and, for the right-handed person, drawing the line from left to right (Fig. 11.11). Your pencil should be held in a vertical plane to make

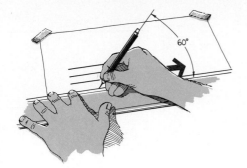

FIG. 11.11 Horizontal lines are drawn with a pencil held in a plane perpendicular to the paper and at 60° to the surface. These lines are drawn left to right along the upper edge of the T-square.

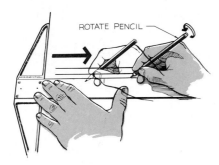

FIG. 11.12 As the horizontal lines are drawn, the pencil should be rotated about its axis so that the point will wear down evenly.

a 60° angle with the drawing surface. As horizontal lines are drawn, the pencil should be rotated about its axis to allow its point to wear evenly (Fig. 11.12). If necessary, lines can be darkened by drawing over them one or more times. For the best line, a small space should be left between the straightedge and the pencil point (Fig. 11.13).

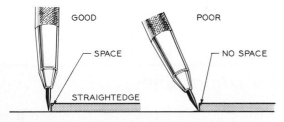

FIG. 11.13 The pencil point should be held in a vertical plane and inclined 60° to leave a space between the point and the straightedge.

11.8
Vertical lines

A triangle is used with a straightedge for drawing vertical lines. While the straightedge is held firmly with one hand, the triangle can be positioned where needed and the vertical lines drawn (Fig. 11.14). Vertical lines are drawn upward along the left side of the triangle while holding the pencil in a vertical plane at 60° to the drawing surface.

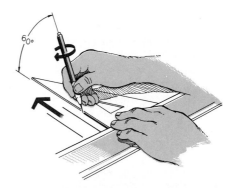

FIG. 11.14 Vertical lines are drawn along the left side of a triangle in an upward direction with the pencil held in a vertical plane at 60° to the surface.

11.9
Drafting triangles

The two most often used triangles are the 45° triangle and the 30°–60° triangle. The 30°–60° triangle is specified by the longer of the two sides adjacent to the 90° angle (Fig. 11.15). Standard sizes of 30°–60° triangles range, in two-inch intervals, from four to twenty-four inches.

The 45° triangle is specified by the length of the sides adjacent to the 90° angle. These range in size from 4 to 24 inches at two-inch intervals, but the six- and ten-inch sizes are adequate for most classroom applications. The various angles that can be drawn with this triangle are shown in Fig. 11.16. By using the 45° and 30°–60° triangles in combination, angles can be drawn at 15° intervals throughout 360° (Fig. 11.17).

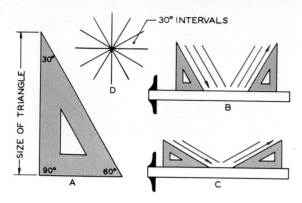

FIG. 11.15 The 30°–60° triangle can be used to construct lines spaced at 30° intervals throughout 360°.

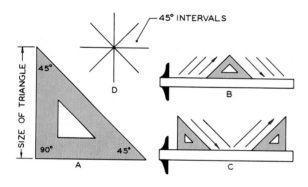

FIG. 11.16 The 45° triangle can be used to draw lines at 45° intervals throughout 360°.

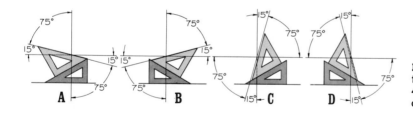

FIG. 11.17 By using the 30°–60° triangle in combination with the 45° triangle, angles can be drawn at 15° intervals.

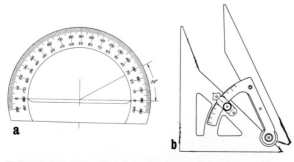

FIG. 11.18 The semicircular protractor can be used to measure angles. The adjustable triangle can be used as a drawing edge and to measure angles.

11.10
Protractor

When lines must be drawn or measured at multiples of other than 15°, a *protractor* is used (Fig. 11.18). Protractors are available as semicircles (180 degrees) or circles (360°). An adjustable triangle serves as a protractor and a drawing edge at the same time (Fig. 11.18B).

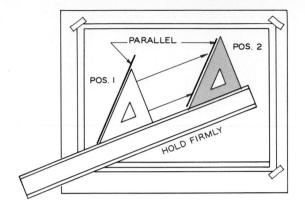

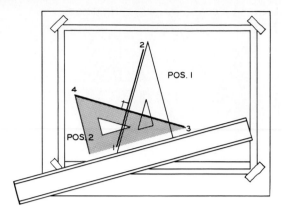

FIG. 11.19 A straightedge and a 45° triangle can be used to draw a series of parallel lines. The straightedge is held firmly in position and the triangle is moved from position 1 to position 2.

FIG. 11.20 A 30°–60° triangle and a straightedge can be used to construct a line perpendicular to line 1–2. The triangle is aligned with 1–2 and is then rotated to position 2 to construct line 3–4.

11.11 _____
Parallel lines

A series of lines can be drawn parallel to a given line by using a triangle and a straightedge (Fig. 11.19). The 45° triangle is placed parallel to a given line and is held in contact with the straightedge (which may be another triangle). By holding the straightedge in one position, the triangle can be moved to various positions for drawing series of parallel lines.

11.12 _____
Perpendicular lines

Perpendicular lines can be constructed by using either of the standard triangles. A 30°–60° triangle is used with a straightedge or another triangle to draw line 3–4 perpendicular to line 1–2 (Fig. 11.20). One edge of the triangle is placed parallel to line 1–2 in position 1 with the straightedge in contact with the triangle. By holding the straightedge in place, the triangle is ro-

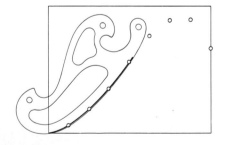

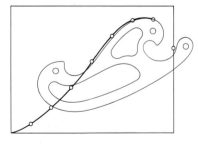

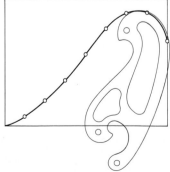

FIG. 11.21 Use of the irregular curve.

Step 1 The irregular curve is positioned to pass through as many points as possible, and a portion of the curve is drawn.

Step 2 The irregular curve is positioned for drawing another portion of the connecting curve.

Step 3 The last portion is drawn to complete the smooth curve. Most irregular curves must be drawn in separate steps.

tated and moved to position 2 to draw the perpendicular line.

11.13
Irregular curves

Curves that are not arcs must be drawn with an instrument called an *irregular curve*. These plastic curves come in a variety of sizes and shapes, but the one shown in Fig. 11.21 is typical of those that are used. The use of the irregular curve is shown in this figure where a series of points is connected to form a smooth curve. The *flexible spline* is an instrument used for drawing long, irregular curves (Fig. 11.22). The spline is held in position by weights while the curve is drawn.

FIG. 11.22   A flexible spline can be used for drawing large irregular curves. The spline is held in position by weights.

11.14
Erasing

Erasing should be done with the softest eraser that will serve the purpose. For example, ink erasers should not be used to erase pencil lines, because ink erasers are coarse and may damage the surface of the paper. When working in small areas, an *erasing shield* can

FIG. 11.23   The erasing shield is used for erasing in tight spots. The dusting brush is used to remove the erased material. Brushing with the palm of your hand will smear the drawing.

prevent accidentally erasing adjacent lines (Fig. 11.23). All erasing should be followed by removing the "crumbs" with a dusting brush; do not use your hands for this because you may smudge your drawing. A cordless model electric eraser (Fig. 11.24) uses erasers available in several grades for erasing ink and pencil lines.

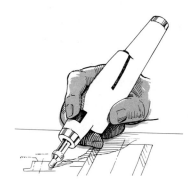

FIG. 11.24   This cordless electric eraser is typical of those used by professional drafters.

11.15
Scales

All engineering drawings require the use of scales to measure lengths, sizes, and so forth. Scales may be flat or triangular and are made of wood, plastic, or metal. Triangular engineers' and architects' scales are shown in Fig. 11.25. Most scales are either six- or twelve-inches long. In this section, we cover the architects', engineers', mechanical engineers', and metric scale.

Architects' scale

The architects' scale is used to dimension and scale features encountered by the architect such as cabinets, plumbing, and electrical layouts. Most indoor measurements are made in feet and inches with an architects' scale. Figure 11.26 shows the basic form for indicating the scale being used. This form should be used in the title block or in some other prominent location on the drawing. Since the dimensions made with the architects' scale are in feet and inches, it is necessary to convert all dimensions to decimal equivalents (all feet or all inches) before the simplest arithmetic can be performed.

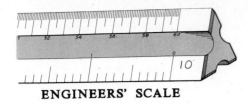

ENGINEERS' SCALE

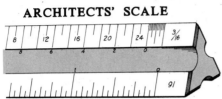

ARCHITECTS' SCALE

FIG. 11.25 The architects' scale measures in feet and inches, whereas the engineers' scale measures in decimal units.

ARCHITECTS' SCALE

BASIC FORM $SCALE: \dfrac{X}{X} = 1'-0$ ┌FROM END OF SCALE

TYPICAL SCALES

SCALE: FULL SIZE (USE 16-SCALE)

SCALE: HALF SIZE (USE 16-SCALE)

SCALE: 3 = 1'-0   *SCALE: $1\frac{1}{2}$ = 1'-0*

SCALE: $1\frac{1}{2}$ = 1'-0   *SCALE: $\frac{3}{4}$ = 1'-0*

SCALE: $\frac{1}{2}$ = 1'-0   *SCALE: $\frac{3}{8}$ = 1'-0*

SCALE: $\frac{3}{16}$ = 1'-0   *SCALE: $\frac{1}{8}$ = 1'-0*

SCALE: $\frac{3}{32}$ = 1'-0

FIG. 11.26 The basic form for indicating the scale on the architects' scale and the variety of scales available.

SCALE: FULL SIZE The 16 scale is used for measuring full-size lines (Fig. 11.27A). An inch on the 16 scale is divided into sixteenths to match the ruler used by the carpenter. This example is measured to be $3\frac{1}{8}''$. Note that when the measurement is less than one foot, a zero may precede the inch measurements; note also that the inch marks are omitted.

SCALE: 1 = 1'-0 In Fig. 11.27B, a line is measured to its nearest whole foot (2 ft in this case), and the remainder is measured in inches at the end of the scale ($3\frac{1}{2}$ in.) for a total of 2'-$3\frac{1}{2}$. At the end of each architects' scale, a foot is divided into inches for measuring dimensions less than a foot. The scale 1" = 1'-0 is the same as saying 1 in. is equal to 12 in., or a $\frac{1}{12}$ size.

SCALE: $\frac{3}{8}$ = 1'-0 When this scale is used, $\frac{3}{8}$ in. represents 12 in. on a drawing. Figure 11.27C is measured to be 7'-5.

SCALE: $\frac{1}{2}$ = 1'-0 A line is measured to be 5'-$8\frac{1}{2}$ in Fig. 11.27D.

SCALE: HALF SIZE The 16 scale is used to measure or draw a line that is half size. This is sometimes specified as Scale: 6 = 12 (inch marks omitted). The line in Fig. 11.27E is measured to be 0'-$6\frac{3}{8}$.

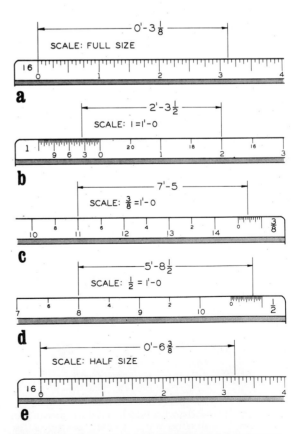

FIG. 11.27 Examples of lines measured using an architects' scale.

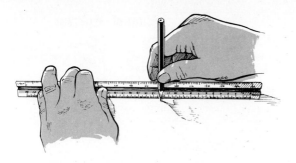

FIG. 11.28 When marking off measurements along a scale, hold your pencil vertically for the most accurate measurement.

OMIT INCH MARKS

ZERO HERE

ZERO OPTIONAL

FIG. 11.29 Inch marks are omitted according to current standards, but foot marks are shown. When the inch measurement is less than a whole inch, a leading zero is used. When representing feet, a zero is optional if the measurement is less than a foot.

A couple of pointers: When marking measurements, hold your pencil vertically for the greatest accuracy (Fig. 11.28). And when indicating dimensions in feet and inches, they should be in the form shown in Fig. 11.29. (Notice the fractions are twice as tall as the whole numerals.)

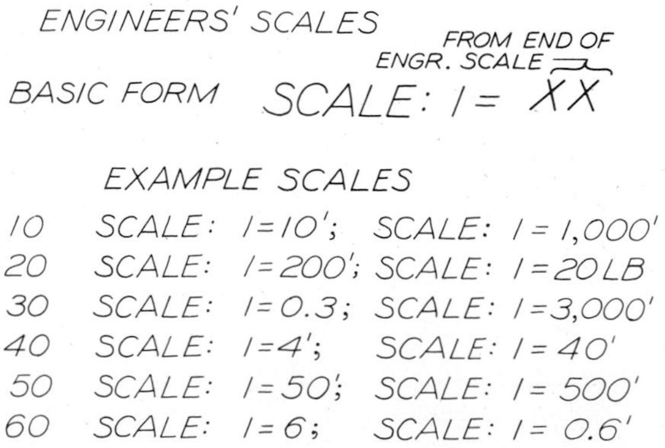

ENGINEERS' SCALES

FROM END OF ENGR. SCALE

BASIC FORM   SCALE: 1 = XX

EXAMPLE SCALES

10	SCALE: 1=10';	SCALE: 1 = 1,000'
20	SCALE: 1=200';	SCALE: 1=20 LB
30	SCALE: 1=0.3;	SCALE: 1=3,000'
40	SCALE: 1=4';	SCALE: 1 = 40'
50	SCALE: 1=50';	SCALE: 1 = 500'
60	SCALE: 1=6';	SCALE: 1 = 0.6'

FIG. 11.30 The basic form for indicating the scale on the engineers' scale and the variety of scales available.

Engineers' scale

The engineers' scale is a decimal scale on which each division is a multiple of ten units. Because it is used for making drawings of engineering projects that are located outdoors—streets, structures, land measurements, and other large topographical dimensions—it is sometimes called the civil engineers' scale.

Since the measurements are in decimal form, it is easy to perform arithmetic operations; there is no need to convert from one unit to another, as there is when using the architects' scale. The form for specifying scales on the engineers' scale is shown in Fig. 11.30; for example, Scale: 1 = 10'. Each end of the scale is labeled 10, 20, 30 (and so on), which indicates the number of units per inch on the scale. Many combinations may be obtained by moving the decimal places of a given scale, as indicated in Fig. 11.30.

10 SCALE In Fig. 11.31A, the 10 scale is used to measure a line at the scale of 1 = 10'. The line is 32.0 feet long.

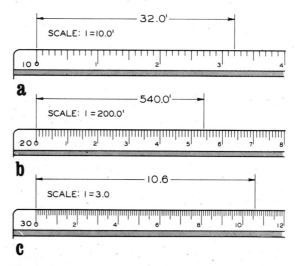

FIG. 11.31 Examples of lines measured with the engineers' scale.

20 SCALE In Fig. 11.31B, the 20 scale is used to measure a line drawn at a scale of 1 = 200.0'. The line is 540.0 feet long.

FIG. 11.32 When using English units (inches), decimal fractions do not have leading zeros and inch marks are omitted. Be sure to provide adequate space for decimal points between the numbers. Foot marks are shown.

30 SCALE A line of 10.6 (inch marks omitted) is measured using the scale of 1 = 3.0 in Fig. 11.31C.

The format for indicating measurements in feet and inches is shown in Fig. 11.32.

Mechanical engineers' scale

The mechanical engineers' scale is used to draw small parts (Fig. 11.33) in inches using common fractions. These scales are available in ratios of half size, one-quarter size, and one-eighth size. For example, on the half-size scale, 1 inch is used to represent 2 inches. On the quarter-size scale, 1 inch would represent 4 inches.

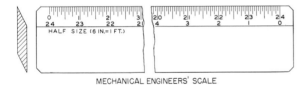

MECHANICAL ENGINEERS' SCALE

FIG. 11.33 The mechanical engineers' scales are used for measuring small parts at scales of half size, quarter size, and one-eighth size. These units are in inches with common fractions.

English system

The English system (Imperial system) of measurement has been used in the United States, Britain, and Canada since these countries were established. The English system was based on arbitrary units of the inch, foot, cubit, yard, and mile (Fig. 11.34). Because there is no common relationship between these units, the system is cumbersome to use when simple arithmetic is performed; for example, finding the area of a rectangle that is 25 inches by $6\frac{3}{4}$ yards is a complex problem.

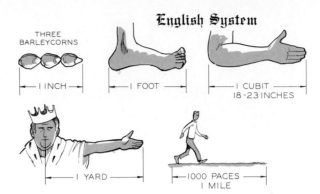

FIG. 11.34 The units of the English system were based on arbitrary dimensions.

Metric system—SI units

The metric system was proposed by France in the fifteenth century. In 1793, the French National Assembly agreed that the meter (m) would be one ten-millionth of the meridian quadrant of the earth (Fig. 11.35). Fractions of the meter were expressed as decimal fractions. Debate continued until an international commission officially adopted the metric system in 1875. Since a slight error in the first measurement of the meter was found, the meter was later established as equal to 1,650,763.73 wavelengths of the orange-red light given off by krypton-86 (Fig. 11.35).

The international organization charged with the establishment and promotion of the metric system is

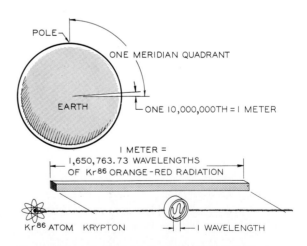

FIG. 11.35 Originally based on the dimensions of the earth, the meter was later based on the wavelength of krypton-86. A meter is 39.37 inches.

SI UNITS			DERIVED UNITS		
LENGTH	METER	m	AREA	SQ METER	m²
MASS	KILOGRAM	kg	VOLUME	CU METER	m³
TIME	SECOND	s	DENSITY	KILOGRAM/CU MET	kg/m³
ELECTRICAL			PRESSURE	NEWTON/SQ MET'	N/m²
CURRENT	AMPERE	A			
TEMPERATURE	KELVIN	K			
LUMINOUS					
INTENSITY	CANDELA	cd			

FIG. 11.36 The basic SI units and their abbreviations. The derived units have come into common usage.

the *International Standards Organization* (ISO). The system they have endorsed is called *Système International d'Unités* (International System of Units) and is abbreviated SI. The basic SI units with their abbreviations are shown in Fig. 11.36. It is important that lowercase and uppercase abbreviations be used properly as shown.

To make them easier to use, several practical units of measurement have been derived from basic SI units (Fig. 11.37). Note that degrees Celsius (centigrade) is recommended over the official temperature measurement, Kelvin. When using Kelvin, the freezing and boiling temperatures are 273.15° K and 373.15 K, respectively.

PARAMETER	PRACTICAL UNITS		SI EQUIVALENT
TEMPERATURE	DEGREES CELSIUS	°C	0°C = 273.15 K
LIQUID VOLUME	LITER	l	l = dm³
PRESSURE	BAR	BAR	BAR = 0.1 MPa
MASS WEIGHT	METRIC TON	t	t = 10³ kg
LAND MEASURE	HECTARE	ha	ha = 10⁴ m²
PLANE ANGLE	DEGREE	°	1° = π/180 RAD

FIG. 11.37 These practical metric units are a few of those that are widely used because they are easier to deal with than the official SI units.

VALUE			PREFIX	SYMBOL
1 000 000	=	10^6	= MEGA	M
1 000	=	10^3	= KILO	k
100	=	10^2	= HECTO	h
10	=	10^1	= DEKA	da
1	=	10^0		
0.1	=	10^{-1}	= DECI	d
0.01	=	10^{-2}	= CENTI	c
0.001	=	10^{-3}	= MILLI	m
0.000 001	=	10^{-6}	= MICRO	u

FIG. 11.38 The prefixes and abbreviations used to indicate the decimal placement for SI measurements.

Many SI units have prefixes to indicate placement of the decimal, the more common of these are shown in Fig. 11.38. And several comparisons of English and SI units are given in Fig. 11.39.

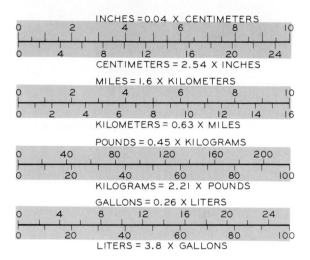

FIG. 11.39 A comparison of metric units with those used in the English system of measurement.

11.16
Metric scales

The basic unit of measurement on an engineering drawing is the millimeter (mm), which is one-thousandth of a meter, or one-tenth of a centimeter.

These units are understood unless otherwise specified on a drawing. The width of the fingernail of your index finger can serve as a convenient gage to approximate the dimension of one centimeter, or ten millimeters (Fig. 11.40). The form for indicating metric scales is shown in Fig. 11.41. The millimeter is understood to be the unit of measurement.

Decimal fractions are unnecessary on drawings dimensioned in millimeters; consequently, the dimensions are usually rounded off to whole numbers except for those measurements that are dimensioned with specified tolerances.

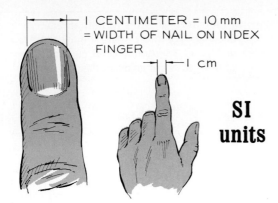

I CENTIMETER = IO mm
= WIDTH OF NAIL ON INDEX
FINGER

I cm

SI units

FIG. 11.40 The width of the nail on your index finger is approximately equal to a centimeter or ten millimeters.

METRIC SCALES   *FROM END OF SCALE*

BASIC FORM   *SCALE: 1:2*

TYPICAL SCALES

SCALE: 1:1 (1mm=1mm; 1cm=1cm: ETC)

SCALE: 1:20 (1mm=20mm; 1mm=2cm)

SCALE: 1:300 (1mm=300mm; 1mm=0.3m)

OTHERS: 1:125; 1:250; 1:500

FIG. 11.41 The basic form for indicating the scale on the metric scale and the variety of scales available.

For metric units less than 1, a zero is placed in front of the decimal. In the English system, the zero is omitted from inch measurements (Fig. 11.42).

SCALE 1: 1 The full-size metric scale (Fig. 11.43) shows the relationship between the metric units of the

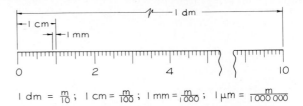

$$1\ dm = \tfrac{m}{10}\ ;\quad 1\ cm = \tfrac{m}{100}\ ;\quad 1\ mm = \tfrac{m}{1000}\ ;\quad 1\ \mu m = \tfrac{m}{1\,000\,000}$$

FIG. 11.43 The dekameter is one tenth of a meter; the centimeter is one hundredth of a meter; a millimeter is one thousandth of a meter; and a micrometer is one millionth of a meter.

dekameter, centimeter, millimeter, and micrometer. There are 10 dekameters in a meter; 100 centimeters in a meter; 1000 millimeters in a meter; and 1,000,000 micrometers in a meter. A line of 59 mm is measured in Fig. 11.44A.

SCALE 1: 2 This scale is used when 1 mm is equal to 2 mm, 20 mm, 200 mm, and so forth. The line in Fig. 11.44B is 106 mm long.

SCALE 1: 3 A line of 165 mm is measured in Fig. 11.44C, where 1 mm is used to represent 30 mm.

Other scales

Many other metric (SI) scales are used: 1:250, 1:400, 1:500, and so on. The scale ratios mean that one unit represents the number of units on the right of the colon. For example, 1:20 means that one millimeter

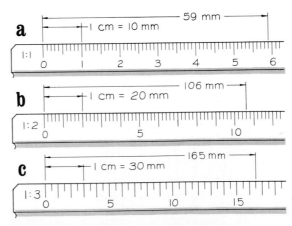

FIG. 11.44 Examples of lines measured with metric scales.

2|2|.|0   4|13|.5   0|.|14

ZERO HERE

|47 |47 |4.7

POOR-CROWDED   GOOD-SPACE

FIG. 11.42 When decimal fractions are shown in metric units, a zero precedes the decimal. Be sure to allow adequate space for the decimal point when numbers with decimals are lettered.

a METRIC UNITS—
3ʳᵈ ANGLE PROJ.

b METRIC UNITS—
1ˢᵀ ANGLE PROJ.

FIG. 11.45 The large SI indicates that the measurements are in metric units. The partial cones indicate that the views are arranged using the third-angle projection (the U.S. system) or the first-angle projection (the European system).

equals 20 mm, or one centimeter equals 20 cm, or one meter represents 20 m.

Metric symbols

When drawings are made in metric units, this can be noted in the title block or elsewhere using the SI symbol (Fig. 11.45), which indicates Système International. The two views of the partial cone are used to denote whether the orthographic views were drawn in the U.S. system (third-angle projection) or the European system (first-angle projection).

Scale conversion

Tables for converting inches to millimeters are given in Appendix 2; however, this conversion can be performed by multiplying decimal inches by 25.4 to obtain millimeters.

An architect's scale must be multiplied by 12 to convert it to an approximate metric scale. For example, Scale: $\frac{1}{8}$ = 1′–0 is the same as $\frac{1}{8}$ in. = 12 in. or 1 in. = 96 in. This scale closely approximates the metric scale of 1:100. Many of the scales used in the metric system cannot be converted to exact English scales, but the metric scale of 1:60 converts exactly to the scale of 1-5′.

Expression of metric units

The general rules for expressing SI units are given in Fig. 11.46. Commas are not used between sets of zeros; instead, a space is left between them.

11.17
The instrument set

A basic set of drawing instruments is shown in Fig. 11.47. Although these can be purchased separately, they are available as a set in a case similar to the one shown in Fig. 11.48.

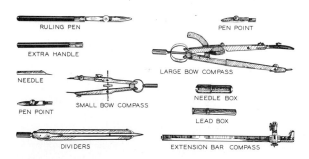

RULING PEN

PEN POINT

EXTRA HANDLE

NEEDLE

LARGE BOW COMPASS

SMALL BOW COMPASS

NEEDLE BOX

PEN POINT

LEAD BOX

DIVIDERS

EXTENSION BAR COMPASS

FIG. 11.47 The parts of a set of drafting instruments.

OMIT COMMAS AND GROUP INTO THREES

1 000 000 000   NOT   1,000,000,000

USE RAISED DOT FOR MULTIPLICATION

N•M   OR   NM

INDICATE DIVISION BY EITHER

kg/m   OR   kg m⁻¹

USE ZERO PRECEDING DECIMALS

0.72 mm   NOT   .72 mm

INDICATE SI SCALES AS

SCALE: 1:2 SI

FIG. 11.46 General rules to be used with the SI system.

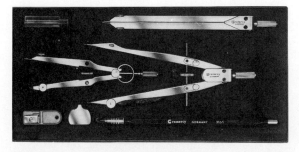

FIG. 11.48 Instruments usually come as a cased set. (Courtesy of Gramercy Guild.)

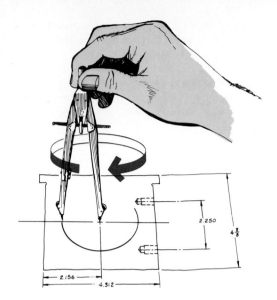

FIG. 11.49 The compass is used for drawing circles.

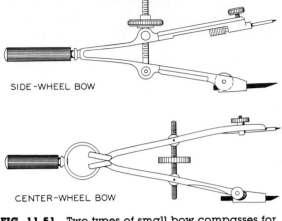

SIDE-WHEEL BOW

CENTER-WHEEL BOW

FIG. 11.51 Two types of small bow compasses for drawing circles of about one-inch radius.

Compass

The *compass* is used to draw circles and arcs in ink and in pencil (Fig. 11.49). To obtain good results with the compass, its pencil point must be sharpened on its outside with a sandpaper board (Fig. 11.50). A bevel cut of this type gives the best all-around point for drawing a circle. When the compass point is set in the drawing surface, it should be inserted just enough for a firm set, not to the shoulder of the point. When the

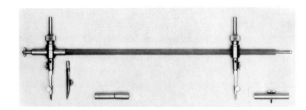

FIG. 11.52 Large circles can be drawn with the beam compass. Note that ink attachments are available also.

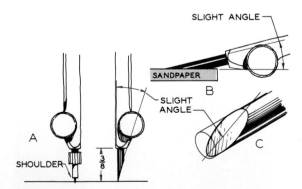

FIG. 11.50 A. The pencil point should be about the same length as the compass point. B. The compass lead should be sharpened from the outside on a sandpaper pad. C. The lead should be sharpened to a wedge point.

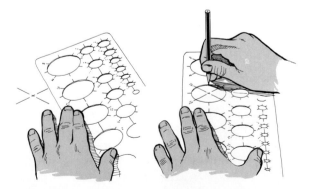

FIG. 11.53 Circle templates can be used for drawing circles without the use of a compass. The circle or ellipse is aligned with the centerlines.

table top has a hard covering, several sheets of paper should be placed under the drawing to provide a seat for the compass point.

Bow compasses are provided in some sets (Fig. 11.51) for drawing small circles. For large circles, bars are provided to extend the range of the large bow compass. A beam compass can be used (Fig. 11.52) for still larger circles. Small circles and ellipses can be effectively drawn with a circle template that is aligned with the perpendicular centerlines of the circle. The circle or ellipse is drawn with a pencil to match the other lines of the drawing (Fig. 11.53).

Dividers

The *dividers* look much like a compass but are used for laying off and transferring dimensions onto a drawing. For example, equal divisions can be stepped off rapidly along a line (Fig. 11.54). As each measurement is made, a slight impression is made in the drawing surface with the dividers' points.

Dividers can be used to transfer dimensions from a scale to a drawing (Fig. 11.55) or to divide a line into a number of equal parts. Small bow dividers can be used for transferring smaller dimensions, such as the spacing between the guidelines for lettering (Fig. 11.56).

Proportional dividers

Dimensions can be transferred from one scale to another by using a special type of dividers, the *proportional dividers*. The central pivot point can be moved to vary the ratio of the spacing at one end of the dividers to the ratio at the other end (Fig. 11.57).

11.18
Ink drawing

Unlike pencil drawings, ink drawings remain dark and distinct. Ink lines also reproduce better than pencil.

Materials for ink drawing

A good grade of tracing paper can be used for ink drawings, but erasing errors may result in holes in the paper and the loss of your time. Therefore tracing film

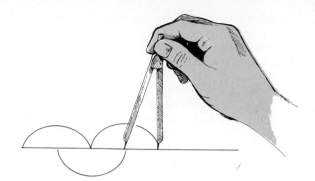

FIG. 11.54 Dividers are used to step off measurements.

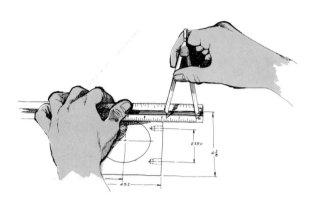

FIG. 11.55 Dividers are also used to transfer dimensions from a scale to a drawing.

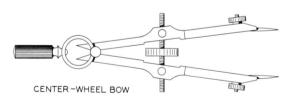

CENTER–WHEEL BOW

FIG. 11.56 A type of a bow divider for transferring small dimensions, such as the spacing between guidelines for lettering.

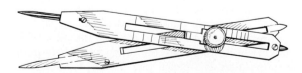

FIG. 11.57 A proportional divider can be used for making measurements that are proportional to other dimensions.

or tracing cloth should be used, which will withstand many erasings and corrections.

When using drafting film, the drawing should be made on the matte surface according to the manufacturer's directions. A cleaning solution is available that can be used to prepare the surface for ink and to remove spots that might not take the ink properly. Tracing cloths need to be prepared for inking by applying a coating of powder or pounce to absorb oily spots that will otherwise repel an ink line. India ink—a dense, black carbon ink that is much thicker and faster-drying than regular fountain pen ink—is used for engineering drawings.

To prevent clogging, ink should be removed from instruments before drying.

FIG. 11.59 The pen is inked between the nibs with the spout on the ink bottle cap.

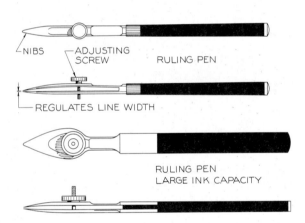

FIG. 11.58 The ruling pen has two nibs that are adjusted by a set screw to vary the width of the ink lines.

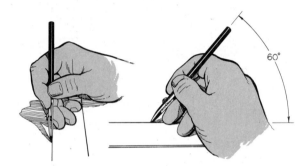

FIG. 11.60 The ruling pen is held in a vertical plane at 60° to the drawing surface.

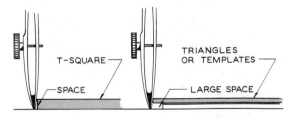

FIG. 11.61 The ruling pen should be held in a vertical plane so that there will be a space between the T-square and the nibs. A triangle or template can be placed under the straightedge for a greater margin of safety.

Ruling pen

Two types of ruling pens are shown in Fig. 11.58. Both have set screws for varying the widths of the lines that are drawn. The ruling pen should be inked with the spout on the cap of the ink bottle (Fig. 11.59). Experiment with your particular pen to learn the proper amount of ink to apply to the nibs. When drawing horizontal lines, the ruling pen is held in the same position as a pencil (Fig. 11.60), maintaining a space between the nibs and the straightedge. An extra margin of safety can be obtained by placing a triangle or template under the straightedge, as shown in Fig.

11.61. Examples of poorly drawn ink lines are shown in Fig. 11.62.

An alternative pen that can be used is the technical ink fountain pen (Fig. 11.63). These pens come

A. GOOD-EVEN LINE

B. POOR-INK RAN UNDER STRAIGHTEDGE

C. POOR-TOO MUCH INK; SLOW AT ENDS

D. POOR-NIBS TOUCH IMPROPERLY

FIG. 11.62 Examples of poor ink lines are shown at B, C, and D.

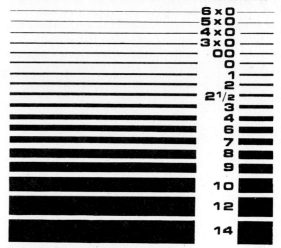

Available in 18 "Kolor-Koded" line widths *

6x0
5x0
4x0
3x0
00
0
1
2
2½
3
4
6
7
8
9
10
12
14

*Approximate only. (Line widths will vary, depending on type of surface, type of ink, speed at which line is drawn, etc.)

FIG. 11.65 This chart shows the variety of technical pens available for drawing lines of graduated widths. (Courtesy of Koh-I-Noor Rapidograph, Inc.)

FIG. 11.63 An India ink technical pen that can be used for making ink drawings.

in sets (Fig. 11.64) with pen points of various sizes that are used for the alphabet of lines (Fig. 11.65). Lines drawn with this type of pen dry faster than those drawn with ruling pens because the ink is applied in a thinner layer.

Inking compass

The inking compass is usually the same compass used for the circles drawn by pencil, with the inking attachment inserted in place of the pencil attachment. The circle can be drawn with one continuous line, as shown in Fig. 11.66. A compass for drawing smaller

FIG. 11.64 A set of inking pens available in various line weights.

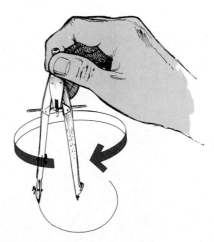

FIG. 11.66 An attachment can be used with a large bow compass for drawing circles and arcs in ink.

circles is shown in Fig. 11.67. The spring bow compass (Fig. 11.68) can be used to draw pencil and ink circles that are about one-eighth inch in radius. Larger circles can be drawn using the extension bar with the large compass (Fig. 11.69). Special compasses are available for drawing circles with a technical fountain pen (Fig. 11.70). These pens screw into an adapter that fits the compass.

Order of inking

When a drawing is to be inked, begin by locating the centerlines and tangent points of the arcs, as shown in Step 1 of Fig. 11.71. This construction should be laid out in pencil before inking. When a drawing is composed mostly of straight lines, begin at the top and draw all the horizontal lines as you progress from the

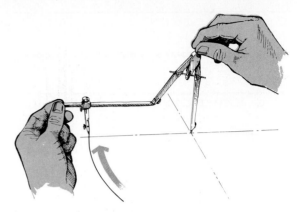

FIG. 11.69 An extension bar can be used with a bow compass for drawing large arcs.

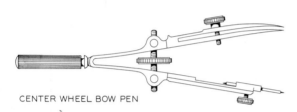

CENTER WHEEL BOW PEN

FIG. 11.67 A small bow compass for drawing arcs in ink up to a radius of about one inch.

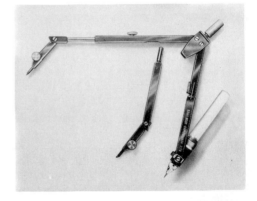

FIG 11.70 Special compasses with adapters are available for using technical ink pens for drawing arcs. (Courtesy of Koh-I-Noor Rapidograph, Inc.)

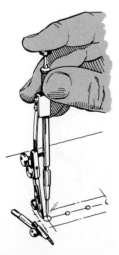

FIG. 11.68 The spring bow compass is used for drawing small circles of about one-quarter inch. A pencil attachment is available.

top of the sheet to the bottom. After allowing the last horizontal line to dry, ink the vertical lines by beginning with the far left (if you are right-handed) and moving across the drawing to the right, away from the wet lines.

Templates

A wide variety of templates (Fig. 11.72) are available for drawing nuts and bolts, circles and ellipses, architectural symbols, and many other applications. Templates work best when used with technical fountain pens rather than with the traditional ruling pen.

CONSTRUCTION LAYOUT   INK ARCS   INK STRAIGHT LINES

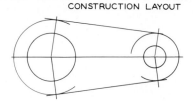 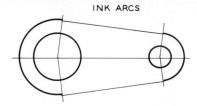 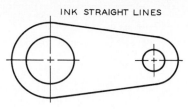

FIG. 11.71 Order of inking.

Step 1 The drawing is laid out with light pencil construction lines. All centers and tangent points are accurately located.

Step 2 The arcs and circles are always inked first. Arcs should stop at their points of tangency.

Step 3 Straight lines are drawn to match the ends of the arcs. Centerlines are shown to complete the drawing.

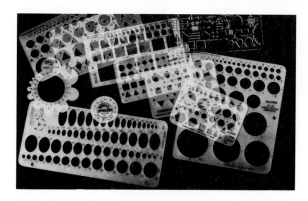

FIG. 11.72 Many types of templates are available to aid drafters in their work. (Courtesy of Rapidesign.)

11.19
Solutions of problems

The following formats are suggested for the layout of problem sheets. Most problems will be drawn on 8½-×-11-inch sheets, as shown in Fig. 11.73. A title strip is suggested in this figure, with a border as shown. Guidelines should be drawn very lightly to be only faintly visible. The 8½-×-11-inch sheet in the vertical format is called Size AV throughout the remainder of this textbook. When this sheet is in the horizontal format, as shown in Fig. 11.74, it will be called Size AH.

The standard sizes of sheets from Size A through Size E are shown in Fig. 11.74. An alternative title strip for Sizes B, C, D, and E is shown in Fig. 11.75. Guidelines should always be used for lettering title strips.

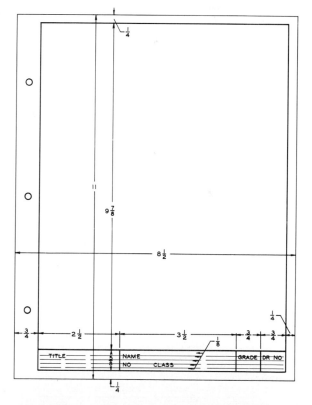

FIG. 11.73 The format and title strip for a Size A sheet (8½" × 11") suggested for solving the problems at the end of each chapter. When the sheet is in the vertical format it will be called a Size AV.

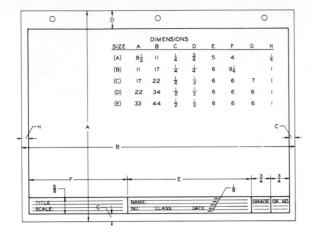

DIMENSIONS								
SIZE	A	B	C	D	E	F	G	H
(A)	$8\frac{1}{2}$	11	$\frac{1}{4}$	$\frac{3}{4}$	5	4		$\frac{1}{4}$
(B)	11	17	$\frac{1}{4}$	$\frac{1}{4}$	6	$8\frac{1}{4}$		1
(C)	17	22	$\frac{1}{2}$	$\frac{1}{2}$	6	6	7	1
(D)	22	34	$\frac{1}{2}$	$\frac{1}{2}$	6	6	6	1
(E)	33	44	$\frac{1}{2}$	$\frac{1}{2}$	6	6	6	1

FIG. 11.74 The format for Size AH (an 8½″ × 11″ sheet in a horizontal position) and the sizes of other sheets. The dimensions under columns A through H give the various layouts.

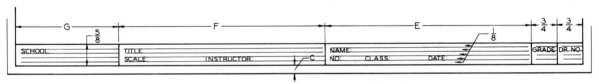

FIG. 11.75 A title strip that can be used on sheet Sizes B, C, D, and E instead of the one given in Fig. 11.74.

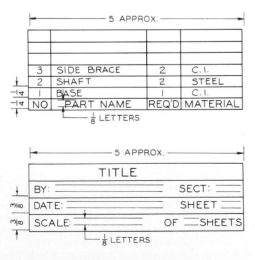

FIG. 11.76 A title block and parts list that can be used on some problem sheets if needed.

Another title block and parts list is given in Fig. 11.76, which are placed in the lower right-hand corner of the sheet against the borders. When both are used on the same drawing, the parts list is placed directly above and in contact with the title block or title strip.

Problems

Construct the problems (Figs. 11.77–11.79) on Size AH (8½- × 11-inch) paper, plain or with a printed grid, using the format shown in Fig. 11.73. Use pencil or ink as assigned by your instructor. Two problems can be drawn per sheet using the scale of each square equals 1.2 in. or 5 mm. One double-size problem can be drawn per sheet.

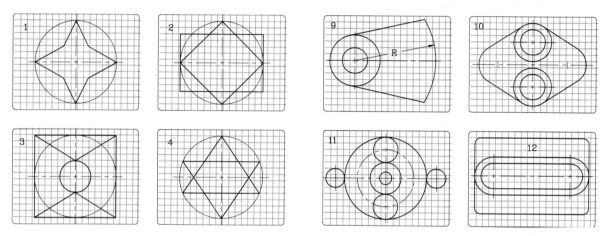

FIG. 11.77 Problems 1–4.

FIG. 11.79 Problems 9–12.

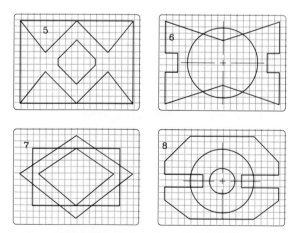

FIG. 11.78 Problems 5–8.

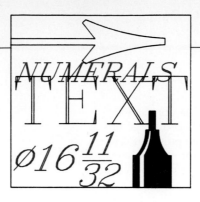

CHAPTER 12

Lettering

12.1
Lettering

All drawings are supplemented with notes, dimensions, and specifications that must be lettered. The ability to construct legible freehand letters is an important skill to develop since it affects the usage and interpretation of a drawing.

12.2
Tools of lettering

The best pencils for lettering on most surfaces are in the H–HB grade range, with an F pencil being the most commonly used grade. Some papers and films are coarser than others and may require a harder pencil lead. To give the desired line width, the point of the pencil should be slightly rounded (Fig. 12.1); a needle point will break off when pressure is applied.

When lettering, the pencil should be revolved slightly between your fingers as the strokes are being made so the lead will wear down gradually and evenly. Bear down firmly to make letters black and bright for good reproduction. To prevent smudging

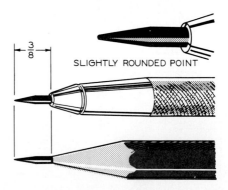

FIG. 12.1 Good lettering begins with a properly sharpened pencil point. The point should be slightly rounded, not a needle point. The F pencil is usually the best grade for lettering.

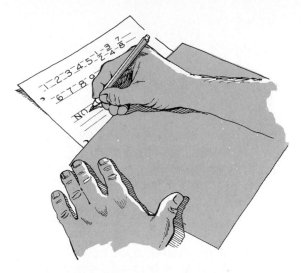

FIG. 12.2 When lettering a drawing, use a protective sheet under your hand to prevent smudges. Your lettering will be best when you are working from a comfortable position; you may wish to turn your paper for the most natural strokes.

your drawing while lettering, place a sheet of paper under your hand to protect the drawing (Fig. 12.2).

A type of inking pen that is widely used is the India ink technical pen (Fig. 12.3). These pens have tubular points that are kept clear by gently shaking the

FIG. 12.3 Ink fountain pens are widely used by professional drafters. Each pen has a number that indicates the line width of the pen. (Courtesy of J.S. Staedtler, Inc.)

pen up and down to activate the plunger inside the tubes. Technical pens can hold enough ink for hours of use without refilling.

12.3
Gothic lettering

The type of lettering recommended for engineering drawings is *single-stroke Gothic lettering,* so called because the letters are made with a series of single strokes and the letter form is a variation of Gothic lettering.

> Two general categories of Gothic lettering are *vertical* and *inclined* lettering (Fig. 12.4). Although each is equally acceptable, they both should not be used on the same drawing.

VERTICAL GOTHIC

INCLINED GOTHIC

FIG. 12.4 Two types of Gothic lettering recommended by engineering standards are vertical and inclined lettering.

12.4
Guidelines

> The most important rule of lettering is: *Use guidelines at all times.*

This applies whether you are lettering a paragraph or a single letter or numeral. The method of constructing and using guidelines can be seen in Fig. 12.5. Use a sharp pencil in the 3H–5H grade range, and draw these lines lightly, just dark enough for them to be seen.

Most lettering is done with the capital letters $\frac{1}{8}$ in. (3 mm) high. The spacing between the lines of lettering should be no closer than half the height of the capital letters, $\frac{1}{16}$ in. in this case.

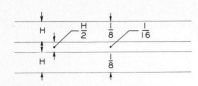

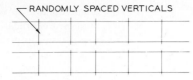

FIG. 12.5 Lettering guidelines.

Step 1 Letter heights, *H*, are laid off, and light guidelines are drawn with a 4H pencil. The spacing between the lines should be no closer than *H*/2.

Step 2 Vertical guidelines are drawn as light, thin lines. These are randomly spaced to serve as visual guides for lettering.

Step 3 The letters are drawn with single strokes using a medium-grade pencil, H–HB. The guidelines need not be erased since they are drawn lightly.

Lettering guides

The two instruments used most often for drawing guidelines are the *Braddock-Rowe lettering triangle* and the *Ames lettering instrument*.

The Braddock-Rowe triangle is pierced with sets of holes for spacing guidelines (Fig. 12.6A). The numbers under each set of holes represent thirty-seconds of an inch. For example, the numeral 4 represents $\frac{4}{32}$ in., or holes that are $\frac{1}{8}$-in. apart for making uppercase (capital) letters. Some triangles are marked for metric lettering in millimeters. Note in Fig. 12.6 that intermediate holes are provided for guidelines for lowercase letters, which are not as tall as the capital letters.

With a horizontal straightedge held firmly in position, place the Braddock-Rowe triangle against its edge. A sharp 4H pencil is placed in one hole of the desired set of holes to contact the drawing surface, and the pencil point is guided across the paper to draw the guideline while the triangle slides against the straightedge. This is repeated as the pencil point is moved successively to each hole until the desired number of guidelines are drawn. An oblique slot for drawing guidelines for inclined lettering is cut in the triangle. These slanting guidelines are spaced randomly by eye.

The Ames lettering guide (Fig. 12.6B) is a similar device with a circular dial for selecting the proper spacing of guidelines. Again, the numbers around the dial represent thirty-seconds of an inch. The number 8 represents $\frac{8}{32}$ in., or guidelines for drawing capital letters that are $\frac{1}{4}$ in. tall.

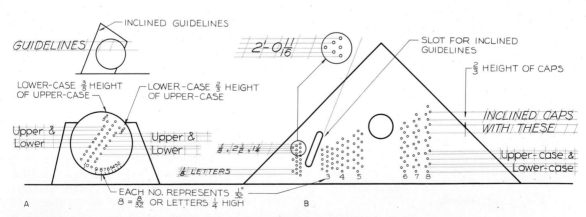

FIG. 12.6 A. The Braddock-Rowe triangle can be used as a 45° triangle and an instrument for constructing guidelines. The numbers designating the guidelines represent thirty-seconds of an inch. **B.** The Ames letter guide can be used for drawing guidelines for vertical or inclined uppercase and lowercase letters. The dial is set to the desired number of thirty-seconds of an inch for the height of uppercase letters.

12.5
Vertical letters

Vertical capital letters

The capital letters for the single-stroke Gothic alphabet are shown in Fig. 12.7. Each letter is drawn inside a square box of guidelines to help you learn their correct proportions. Some letters require the full area of the box, some require less space, and a few require more space. Each straight-line stroke should be drawn as a single stroke; for example, the letter A is drawn with three single strokes. Letters composed of curves can best be drawn in segments; the letter O can be drawn by joining two semicircles to form the full circle.

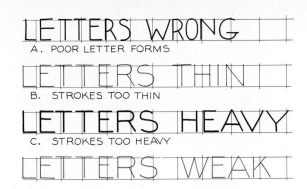

FIG. 12.8 There are many ways to letter poorly. A few of them, and the reasons why the lettering is inferior, are shown here.

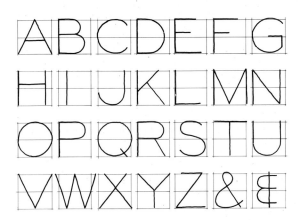

FIG. 12.7 The uppercase letters used in single-stroke Gothic lettering. Each letter is drawn inside a square to help you learn their correct proportions.

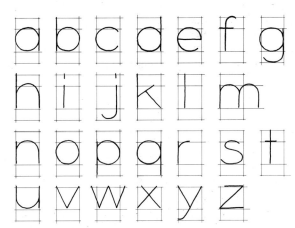

FIG. 12.9 The lowercase alphabet used in single-stroke Gothic lettering. The body of each letter is drawn inside a square to help you learn the proportions.

Memorize the shape of each letter given in this alphabet. Small wiggles in your strokes will not detract from your lettering if the letter forms are correct. Examples of poor lettering are shown in Fig. 12.8.

Vertical lowercase letters

An alphabet of lowercase letters is shown in Fig. 12.9. Lowercase letters are either two thirds or three fifths as tall as the uppercase letters they are used with. Both of these ratios are labeled on the Ames guide, but only the two-thirds ratio is available on the Braddock-Rowe triangle.

Some lowercase letters have *ascenders* that extend above the body of the letter, such as the letter b; some have *descenders* that extend below the body, such as the letter y. The ascenders are the same length as the descenders.

The guidelines in Fig. 12.9 form perfect squares about the body of each letter to illustrate the proportions. A number of these letters have bodies that are perfect circles that touch all sides of the squares. In Fig. 12.10 capital and lowercase letters are used together.

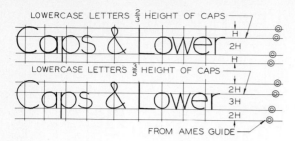

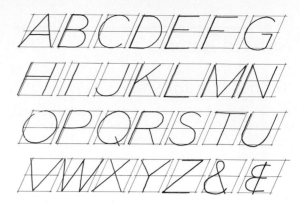

FIG. 12.10 Uppercase and lowercase letters are sometimes used together. The ratio of the lowercase letters to the uppercase letters will be either two thirds or three fifths. The Ames guide has both, and the Braddock-Rowe triangle has only the three-fifths ratio.

FIG. 12.12 The inclined uppercase alphabet for single-stroke Gothic lettering.

Vertical numerals

Vertical numerals are shown in Fig. 12.11, where each number is enclosed in a square box of guidelines. Each number is made the same height as the capital letters being used, usually $\frac{1}{8}$ in. high. The numeral, zero, is an oval, whereas the letter, *O,* is a perfect circle in vertical lettering.

Inclined lowercase letters

Inclined lowercase letters are drawn in the same manner as vertical lowercase letters (Fig. 12.13). Ovals (ellipses) are used instead of the circles used in vertical lettering. The angle of inclination is 68°, the same as is used for uppercase letters.

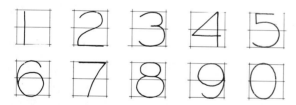

FIG. 12.11 The numerals for single-stroke Gothic lettering. Each is drawn inside a square to help you learn the proportions.

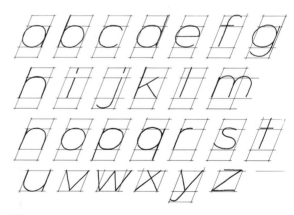

FIG. 12.13 The inclined lowercase alphabet for single-stroke Gothic lettering. The body of each letter is drawn inside a rhombus to help you learn the proportions.

12.6

Inclined letters

Inclined uppercase letters

Inclined uppercase letters (capitals) have the same heights and proportions as vertical letters; the only difference is their 68° inclination (Fig. 12.12). Inclined guidelines should be drawn using the Braddock-Rowe triangle or the Ames guide.

Inclined numerals

The inclined numerals that should be used with inclined lettering are shown in Fig. 12.14. The use of inclined letters and numbers in combination is seen in Fig. 12.15. The guidelines in this example were constructed using the Braddock-Rowe triangle.

FIG. 12.14 The inclined numerals for single-stroke Gothic lettering. Each number is drawn inside a rhombus to help you learn the proportions.

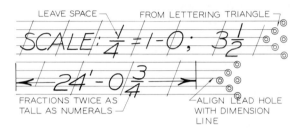

FIG. 12.15 Inclined common fractions are twice as tall as single numerals. Inch marks are omitted when numerals are used to show dimensions.

12.7
Spacing numerals and letters

Common fractions are twice as tall as single numerals (Fig. 12.16). A separate set of holes for common fractions is given on the Braddock-Rowe triangle and on the Ames guide. These are equally spaced $\frac{1}{16}$ in. apart with the centerline being used for the fraction's crossbar.

When numbers are used with decimals, space should be provided for the decimal point (Fig. 12.17).

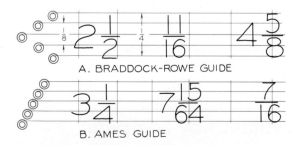

FIG. 12.16 Common fractions are twice as tall as single numerals. Guidelines for these can be drawn by using the Ames guide or the Braddock-Rowe triangle.

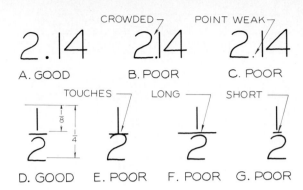

FIG. 12.17 Examples of poorly spaced numerals that result in inferior lettering.

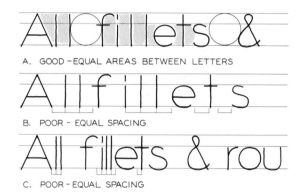

FIG. 12.18 Proper spacing of letters is necessary for good lettering and appearance. The areas between letters should be approximately equal.

The correct method of drawing common fractions is illustrated in Fig. 12.17D, and several of the often encountered errors are shown in Figs. 12.17E–G.

When letters are grouped together to spell words, the area between the letters should be approximately equal for the most pleasing result (Fig. 12.18). The incorrect use of guidelines and other violations of good lettering practice are shown in Fig. 12.19.

12.8
Mechanical lettering

Drawings and illustrations to be reproduced by a printing process are usually drawn in India ink. Several mechanical aides for ink lettering are available.

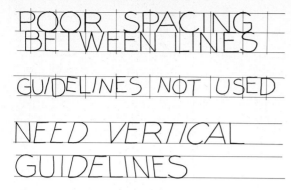

FIG. 12.19 Always leave space between lines of lettering. After constructing guidelines, use them. Use vertical guidelines to improve the angle of your vertical strokes.

FIG. 12.20 A Wrico lettering template can be used for mechanical lettering. (Courtesy of Wood-Reagan Instrument Co.)

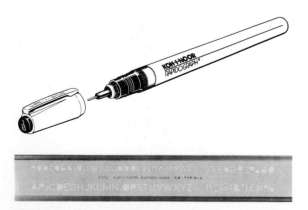

FIG. 12.21 A typical India ink fountain pen and template that can be used for mechanical lettering. (Courtesy of Koh-I-Noor Rapidograph, Inc.)

The Wrico lettering template (Fig. 12.20) can be placed against a fixed straightedge for aligning the letters. You move the template from position to position while drawing each letter (with a lettering pen) through the raised portion of the template where the holes form the letters. A slightly different template and pen are used by the Rapidograph system (Fig. 12.21). A variety of pen sizes are available from each manufacturer for drawing different-sized letters and numbers with thin or bold lines.

Another system of mechanical lettering uses a grooved template along with a scriber that follows the grooves and inks the letters on the drawing surface (Fig. 12.22). Note that a standard India ink technical pen can be unscrewed from its barrel and attached to the scriber (Fig. 12.22). Many templates of varying styles of lettering and symbols are available for this system of mechanical lettering.

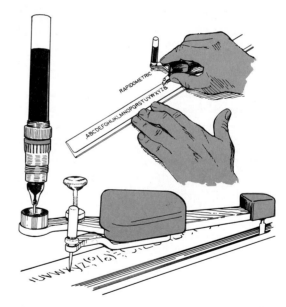

FIG. 12.22 Templates and scribers of this type are available for mechanical lettering.

12.9
Lettering by computer

Lettering of all types can be done by using AutoCAD, as mentioned in Chapter 10. Letters and numbers made with the Simplex font are shown in Fig. 12.23.

FIG. 12.23 Characters drawn by the S i m p l e x font provided by AutoCAD.

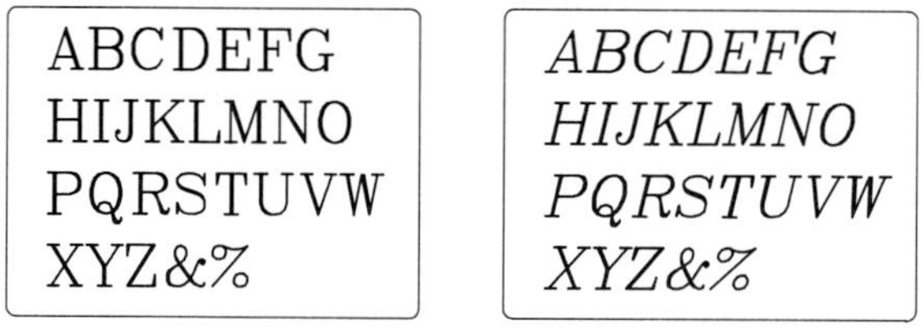

A   *B*

FIG. 12.24 A. Vertical uppercase letters drawn with the COMPLEX font. B. Inclined uppercase letters drawn with the COMPLEX font. The angle of inclination is the angle with the vertical in a clockwise direction.

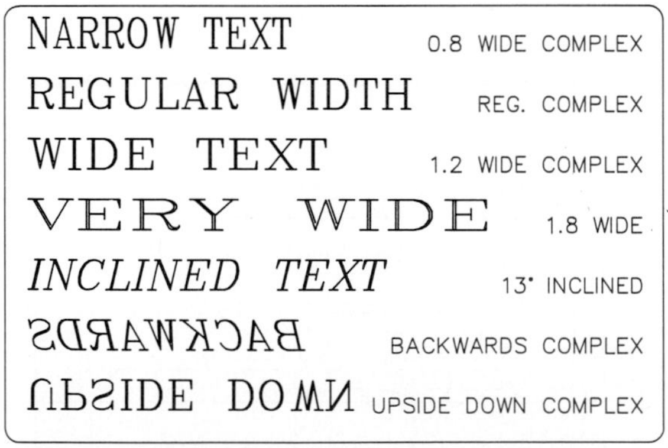

FIG. 12.25 STYLES of lettering can be created to your specifications by changing the variables of the STYLE command.

The Simplex font has smoother curves than does the Txt font, which is the default font.

The Complex font available from AutoCAD can be inclined at any angle, as illustrated in Fig. 12.24. This font is drawn using multiple strokes to form letters of varying widths to generate a Roman style.

The Complex font is used in Fig. 12.25 to illustrate some of the many styles of text that can be created using AutoCAD options. Text can be inclined, backwards or upside-down, and it can be varied in width to any degree.

When you begin a new drawing, use the following text STYLE Command, unless you have previously SAVd a different STYLE in a Default Drawing:

```
Command: Style Name (or ?):
Standard (CR)
Font file <TXT>: TXT (CR)
Height: <0.00>: 0 (Not fixed)
(CR)
Width factor: <1.00>: 1 (CR)
Obliquing angle: <0>: 0 (CR)
Backwards: <N>: No (CR)
Upside-down: <N>: No (CR)
```

You can change the the Text Style definitions with the STYLE command in the following manner:

```
Command: STYLE Text style name
(or ?): Pretty (CR)
Font file <default>: Simplex
(CR)
Height <default>: 0 (CR)
Width factor (default): 1 or
(CR)
Obliquing angle (default): 15
(CCW angle from vertical) (CR)
Backwards? <Y/N> N (CR)
Upside-down? <Y/N> N (CR)
```

The style, Pretty, has been created, and it can be called by using the S ' ' (STYLE) option under the TEXT command. When prompted for the style name, enter Pretty, and the font can be used with all the variables that were defined above. When asked for height, you must specify the lettering height when it is first used, since it was specified to have a 0 height. The value that you enter will remain the default height by pressing the space bar or (CR) when asked for the desired height. You can see that numerous styles can be created and named for ready use.

Problems

Lettering problems are to be presented on Size AV (8½-by-11-inch) paper, plain or grid, using the format shown in Fig. 12.26.

1. Practice lettering the vertical uppercase alphabet shown in Fig. 12.7. Construct each letter four times: four As, four Bs, and so on. Use a medium-weight pencil—H, F, or HB.

2. Practice lettering vertical numerals and the lowercase alphabet as shown in Fig. 12.27. Construct each letter and numeral three times: three 1s, three 2s, and so on. Use a medium-weight pencil—H, F, or HB.

3. Practice lettering the inclined uppercase alphabet shown in Fig. 11.14. Construct each letter four times. Use a medium-weight pencil—H, F, or HB.

4. Practice lettering the vertical numerals and the lowercase alphabet shown in Figs. 12.9 and 12.11. Construct each letter three times. Use a medium-weight pencil—H, F, or HB.

5. Construct guidelines for ⅛ in. capital letters starting ¼ in. from the top border. Each guideline should end ½ in. from the left and right borders. Using these guidelines, letter the first paragraph of the text of this chapter. Use all vertical capitals. Spacing between the lines should be ⅛ in.

6. Repeat Problem 5, but use all inclined capital letters. Use inclined guidelines to assist you in slanting your letters uniformly.

7. Repeat Problem 5, but use vertical capitals and lowercase letters in combination. Capitalize only those words that are capitalized in the text.

8. Repeat Problem 5, but use inclined capitals and lowercase letters in combination. Capitalize only those words that are capitalized in the text.

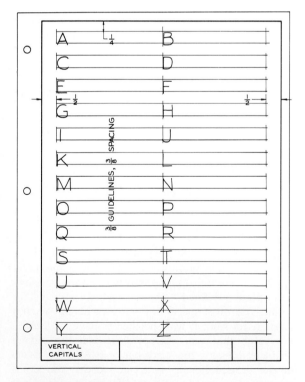

FIG. 12.26 Problem 1. Construct each vertical uppercase letter four times.

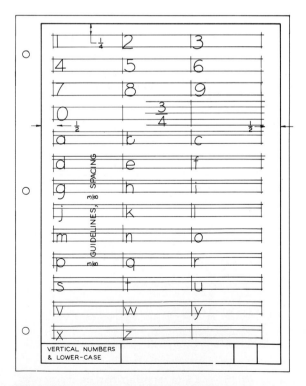

FIG. 12.27 Problem 2. Construct each vertical numeral and lowercase letter three times.

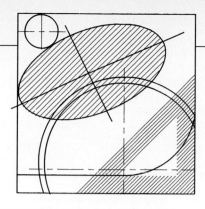

CHAPTER 13

Geometric Construction

13.1
Introduction

Many graphical problems can be solved only by using geometry and geometric construction. Mathematics was an outgrowth of graphical construction, so the two areas are closely related. The proofs of many principles of plane geometry and trigonometry may be developed by using graphics. Moreover, graphical methods can be applied to algebra and arithmetic, and virtually all problems of analytical geometry can be solved graphically.

13.2
Angles

A fundamental requirement of geometric construction is the construction of lines that join at specified angles with each other. The definitions of various angles are given in Fig. 13.1.

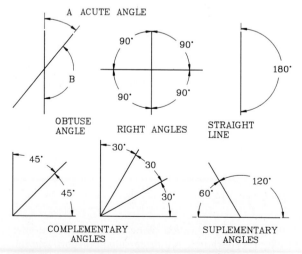

FIG. 13.1 Standard types of angles and their definitions.

The unit of angular measurement is the degree, and a circle has 360 degrees. A degree (°) can be divided into 60 parts called minutes ('), and a minute can be divided into 60 parts called seconds ("). An angle of 15°32′14″ is an angle of 15 degrees, 32 minutes, and 14 seconds.

13.3
Triangles

The *triangle* is a three-sided polygon (or figure) that is named according to its shape. The four types of triangles are the *scalene, isosceles, equilateral,* and *right triangle* (Fig. 13.2). The sum of the angles inside a triangle is always 180 degrees.

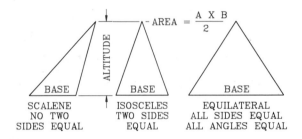

FIG. 13.2 Types of triangles and their definitions.

13.4
Quadrilaterals

A *quadrilateral* is a four-sided figure of any shape. The sum of the angles inside a quadrilateral is 360°. The various types of quadrilaterals are shown in Fig. 13.3, along with the equations for the areas of these figures.

13.5
Polygons

A *polygon* is a multisided plane figure of any number of sides. (The triangle is a three-sided polygon, and the quadrilateral is a four-sided polygon.) If the sides of the polygon are equal in length, the polygon is a *regular polygon.* Four types of regular polygons are shown

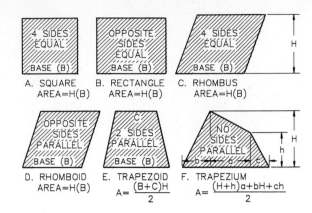

FIG. 13.3 Types of quadrilaterals (four-sided plane figures).

in Fig. 13.4. Note that a regular polygon can be inscribed in a circle and that all the corner points will lie on the circle.

Other regular polygons not pictured are the *heptagon* (seven sides), the *nonagon* (nine sides), the *decagon* (ten sides), and the *dodecagon* (twelve sides). The sums of the angles inside any polygon can be found by the equation

$$S = (n - 2) \times 180°,$$

where *n* is equal to the number of sides of the polygon.

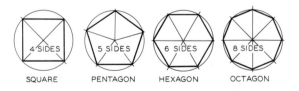

FIG. 13.4 Regular polygons inscribed in circles.

13.6
Elements of circles

A circle can be divided into a number of parts, each of which has its own special name (Fig. 13.5). The equation for finding the area of a circle is

$$A = \pi r^2,$$

where r_n is the radius and π is equal to 3.14 (pi). The

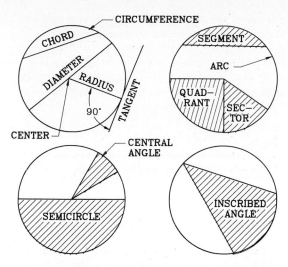

FIG. 13.5 Definitions of the elements of a circle.

equation for finding the circumference is

$$C = \pi d,$$

where d is the diameter.

13.7
Geometric solids

The various types of solid geometric shapes, along with their names and definitions, are shown in Fig. 13.6.

POLYHEDRA A multisided solid formed by intersecting planes is called a *polyhedron*. If the faces of a polyhedron are regular polygons, it is called a *regular polyhedron*. The five regular polyhedra are the *tetrahe-*

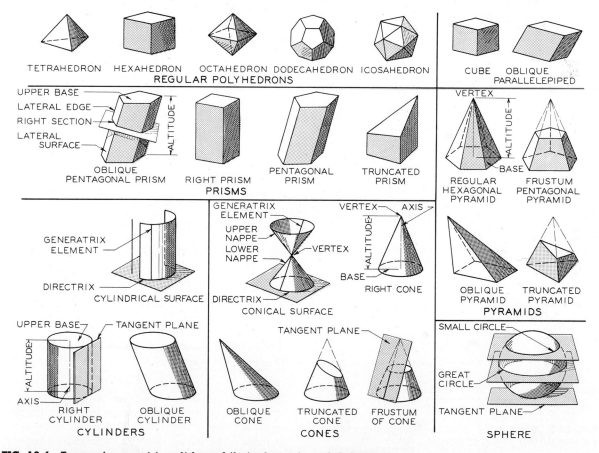

FIG. 13.6 Types of geometric solids and their elements and definitions.

dron (four sides), the *hexahedron* (six sides), the *octahedron* (eight sides), the *dodecahedron* (twelve sides), and the *icosahedron* (twenty sides).

PRISMS A *prism* is a solid that has two parallel bases that are equal in shape. The bases are connected by sides that are parallelograms. The line from the center of one base to the other is called the *axis*. An axis that is perpendicular to the bases is an *altitude*, and the prism is a *right prism*. If the axis is not perpendicular to the base, the prism is an *oblique prism*. A prism that has been cut off to form a base that is not parallel to the other is called a *truncated prism*. A *parallelepiped* is a prism with a base that is either a rectangle or a parallelogram.

PYRAMIDS A *pyramid* is a solid with a polygon as a base and triangular faces that converge at a point called the *vertex*. The line from the vertex to the center of the base is called the *axis*. If the axis is perpendicular to the base, it is the *altitude* of the pyramid, and the pyramid is a *right pyramid*. If the axis is not perpendicular to the base, the pyramid is an *oblique pyramid*. A truncated pyramid is called a *frustum* of a pyramid.

CYLINDERS A *cylinder* is formed by a line or element (called a *generatrix*) that moves about the circle while remaining parallel to its axis. The axis of a cylinder connects the centers of each end of a cylinder. If the axis is perpendicular to the bases, it is the *altitude* of a *right cylinder*. If the axis does not make a 90° angle with the base, the cylinder is an *oblique cylinder*.

CONES A *cone* is also formed by a generatrix, one end of which moves about the curved base while the other end remains at a fixed point called the *vertex*. The line from the center of the base to the vertex is called the *axis*. If the axis is perpendicular to the base,

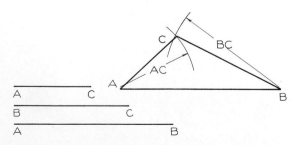

FIG. 13.7 When three sides are given, a triangle can be drawn with a compass.

it is called the *altitude*, and the cone is a *right cone*. A truncated cone is called a *frustum* of a cone.

SPHERES A *sphere* is generated by the plane of a circle that is revolved about one of its diameters to form a solid. The ends of an axis through the center of the sphere are called *poles*.

13.8
Constructing triangles

When three sides of a triangle are given, the triangle can be constructed by using a compass, as shown in Fig. 13.7. Only one triangle can be found when the sides are given by this method, called triangulation. A right triangle can be constructed by inscribing it in a semicircle, as shown in Fig. 13.8. Any triangle inscribed in a semicircle will always be a right triangle.

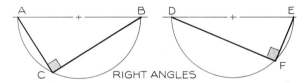

FIG. 13.8 Any angle that is inscribed in a semicircle will be a right angle.

13.9
Constructing polygons

A regular polygon (having equal sides) can be inscribed in a circle or circumscribed about a circle. When inscribed, all the corner points will lie along the circle (Fig. 13.9). For example, a twelve-sided polygon is constructed by dividing the circle into twelve sectors and connecting the points to form the polygon.

13.10
Hexagons

The *hexagon*, a six-sided regular polygon, can be inscribed and circumscribed, shown in Fig. 13.10. Hexagons are drawn with 30°–60° triangles either inside or outside the circles. Note that the circle represents

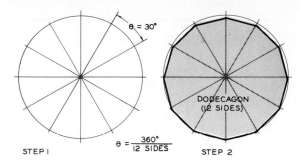

STEP I   STEP 2

FIG. 13.9 The regular polygon.

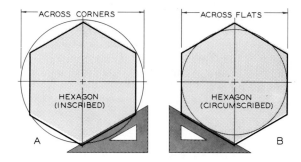

A   B

FIG. 13.10 A circle can be inscribed or circumscribed to form a hexagon by using a 30°–60° triangle.

the distance from corner to corner when inscribed, and from flat to flat when circumscribed.

13.11
Octagons

The *octagon,* an eight-sided regular polygon, can be inscribed in, or circumscribed about, a circle (Fig. 13.11) by using a 45 degree triangle. A second method inscribes the octagon inside a square (Fig. 13.12).

13.12
Pentagons

The *pentagon,* a five-sided regular polygon, can be inscribed in, or circumscribed about, a circle. Another method of constructing a pentagon is shown in Fig. 13.13. This construction is performed with a compass and straightedge.

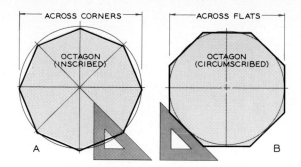

A   B

FIG. 13.11 A circle can be inscribed or circumscribed to form an octagon with a 45° triangle.

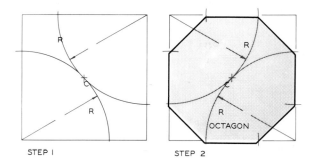

STEP I   STEP 2

FIG. 13.12 Octagon in a square.

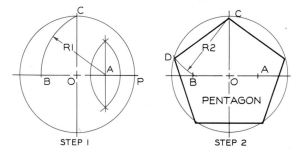

STEP I   STEP 2

FIG. 13.13 The pentagon.

Step 1 Bisect radius OP to locate point A. With A as the center and AC as the radius R_1, locate point B on the diameter.

Step 2 With point C as the center and BC as the radius R_2, locate point D on the arc. Line CD is the chord that can be used to locate the other corners of the pentagon.

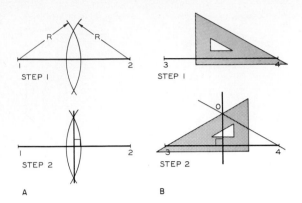

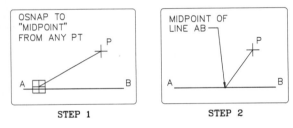

FIG. 13.14 Bisecting a line.

A line can be bisected by using a compass and any radius or a standard triangle and a straightedge.

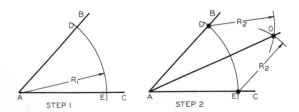

FIG. 13.15 Midpoint by computer.

Step 1 Find the midpoint of *AB* in the following manner:
`Command: `*LINE*` From point: `*P* (Locate *P* anywhere.)
`To point: `*OSNAP* (Select `MIDPOINT` mode.)
(Select any point on line *AB*.)

Step 2 The line from Pt *P* will be drawn to the midpoint of line *AB*.

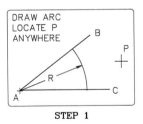

FIG. 13.16 Bisecting an angle.

Step 1 Swing an arc of any radius to locate points *D* and *E*.

Step 2 Draw two equal arcs from *D* and *E* to locate point 0. Line *A0* is the bisector of the angle.

13.13
Bisecting lines and angles

Two methods of finding the midpoint, or the perpendicular bisector, of a line are illustrated in Fig. 13.14. The first method, which can be used to find the midpoint of an arc or a straight line uses a compass to construct a perpendicular to a line. The second method uses a standard triangle.

> **COMPUTER METHODS** The midpoint of a line can be found by computer (Fig. 13.15) by using the `Midpoint` mode of `OSNAP` and drawing a line from any point, *P*, to the line. The line will automatically snap to the given line's midpoint. The angle in Fig. 13.16 can be bisected with a compass by drawing three arcs.

> **COMPUTER METHOD** A second computer method is to draw an arc at any position between the two given lines using their point of intersection as the center (Fig. 13.17). Using the `DRAW` command and the `Midpoint` mode of `OSNAP`, a line is drawn from any point to the arc, which will be the arc's midpoint. The bisector can then be drawn from Point A to this midpoint.

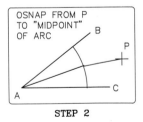

FIG. 13.17 Bisecting an angle by computer.

Step 1 Using the ARC Command, any radius, and center *A*, draw an arc that SNAPs to lines *AC* and *AB*.

Step 2 `Command: `*LINE*` From point: `(Select *P* anywhere.)
`To point: `(`OSNAP`)` Midpoint of `(Select arc.)
(The line from *P* is drawn to the midpoint of the arc, which locates the bisector.)

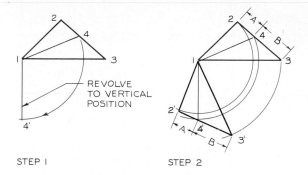

FIG. 13.18 Rotation of a figure.

Step 1 A plane figure can be rotated about any point. Line 1–4 is rotated about point 1 to its desired position with a compass.

Step 2 Points 2′ and 3′ are located by measuring distances A and B from 4.

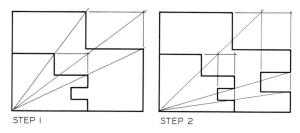

FIG. 13.19 Enlargement of a figure.

Step 1 A proportional enlargement is made by using a series of diagonals drawn through a single point, the lower left corner in this case.

Step 2 Additional diagonals are drawn to locate the other features of the object. This process can be reversed for reducing an object.

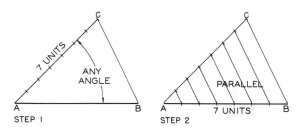

FIG. 13.20 Division of line.

Step 1 Line AB is divided into seven equal divisions by constructing a line through A and dividing it into seven known units with your dividers. Point C is connected to point B.

Step 2 A series of lines are drawn parallel to CB to locate the divisions along line AB.

13.14
Revolution of figures

Rotating a triangle about one of its points is demonstrated in Fig. 13.18. Where the triangle is rotated about point 1 of lines 1–4, Point 4 is rotated to its desired position using a compass. Points 2 and 3 are found by triangulation by drawing arcs with radii 4–2 and 4–3 from 4′ to complete the rotated view.

13.15
Enlargement and reduction of figures

In Fig. 13.19, the small figure is enlarged by using a series of radial lines from the lower left corner. The smaller figure is completed as a rectangle, and the larger rectangle is drawn proportional to the small one. The upper right notch is located using the same technique. This method can also be used to reduce a larger drawing.

13.16
Division of lines

It is often necessary to divide a line into a number of equal parts when a convenient scale is not available for this purpose. For example, suppose a six-inch line is to be divided into seven equal parts. The method shown in Fig. 13.20 is an efficient way to solve this problem.

An application of this principle is used for locating lines on a graph that are equally spaced (Fig. 13.21). A scale with the desired number of units (0 to 5) is laid across from left to right on the graph. Vertical lines are drawn through these points to divide the graph into five equal divisions.

13.17
Arcs through three points

An arc can be drawn through any three points by connecting the points with two lines (Fig. 13.22). Perpendicular bisectors are found for each line to locate the

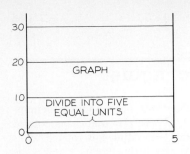 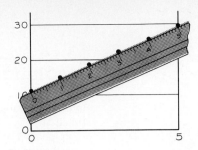 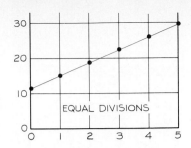

FIG. 13.21 Division of a space.

Step 1 To divide a graph into five equal divisions along the *x*-axis, a scale with five units of measurement that approximate the horizontal distance is laid across the graph. Align the 0 and 5 markings with the lines.

Step 2 Construct vertical lines through the points that were found in Step 1. This method can also be used to calibrate the divisions along the *y*-axis.

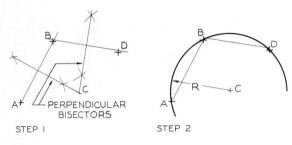

FIG. 13.22 Arc through three points.

Step 1 Connect the points with two lines and find their perpendicular bisectors. The bisectors will intersect at the center, *C*.

Step 2 Using the center, *C*, and the distance to the points as the radius, construct the arc through the points.

center at point *C*. The radius is drawn, and the lines *AB* and *BD* become chords of the circle.

This system can be reversed to find the center of a given circle or arc. Draw two chords that intersect at a point on the circumference and bisect them. The perpendicular bisectors will intersect at the center of the circle.

COMPUTER METHOD Using the ARC command and 3-point option, a counterclockwise arc can be drawn through any three points that are selected on the screen (Fig. 13.23).

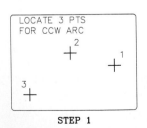 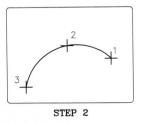

STEP 1   **STEP 2**

FIG. 13.23 Three-point arc by computer.

Step 1 Command: <u>ARC</u>
ARC Center/<Start point>: (Locate pt 1)
CenterEnd <Second point>:(Locate pt 2)

Step 2 Endpoint: <u>DRAG</u> (Locate 18 3) (The arc is drawn counterclockwise through the three points.)

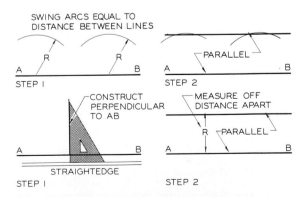

FIG. 13.24 Construction of parallel lines.

Either of the above methods can be used for constructing one line parallel to another. The first method uses a compass and a straightedge; the other method uses a triangle and a T-square.

13.18
Parallel lines

A line can be drawn parallel to another by using either of the methods shown in Fig. 13.24. The first method uses a compass to draw two arcs to locate a parallel line that is the desired distance away. The second method requires constructing a perpendicular from a given line and measuring the distance, *R,* to locate the parallel, which is drawn with a straightedge.

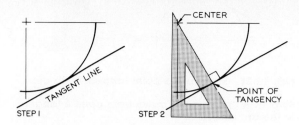

FIG. 13.25 Locating a tangent point.

Step 1 Align your triangle with the tangent line while holding it firmly against a straightedge.

Step 2 Hold the straightedge in position, rotate the triangle, and construct a line through the center that is perpendicular to the line.

13.19
Points of tangency

A point of tangency is the theoretical point where a straight line joins an arc or where two arcs join. In Fig. 13.25, a line is tangent to an arc. The point of tangency is located by constructing a thin perpendicular to the line from the center of the arc. The conventional methods of marking points of tangency are shown in Fig. 13.26.

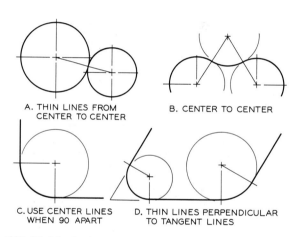

FIG. 13.26 Thin lines that extend beyond the curves from the centers are used to mark points of tangency. These lines should always be shown in this manner.

13.20
Line tangent to an arc

The method of finding the exact point of tangency between a line and an arc is shown in Fig. 13.27. Point *A* and the arc are given. Point *A* is connected to the center in Step 1, *AC* is bisected in Step 2, and *T* is

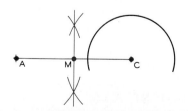
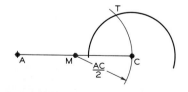
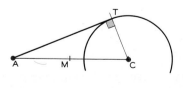

FIG. 13.27 Line from a point tangent to an arc.

Step 1 Connect point *A* with center *C.* Locate point *M* by bisecting *AC.*

Step 2 Using point *M* as the center and *MC* as the radius, locate point *T* on the arc.

Step 3 Draw the line from *A* to *T* that is tangent to the arc of point *T.*

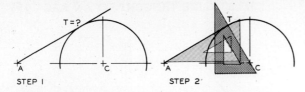

STEP I   STEP 2

FIG. 13.28   Line from a point tangent to an arc.

Step 1   A line can be drawn from point *A* tangent to the arc by eye.

Step 2   By rotating your triangle, the point of tangency can be located at the 90° angle with the line that passes through the center.

located in Step 3. The point of tangency could also have been found by using a standard triangle, as shown in Fig. 13.28.

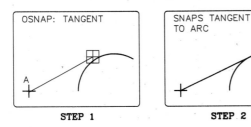

STEP 1   STEP 2

FIG. 13.29   Line tangent to an arc by computer.

Step 1   Command: <u>LINE</u> From point: <u>A</u> To point: (OSNAP—Tangent mode) Tangent to:

Step 2   Select point on the arc, and line *AB* is drawn tangent to the arc.

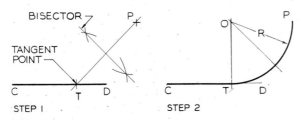

STEP I   STEP 2

FIG. 13.30   Arc through two points.

Step 1   If an arc must be tangent to a given line at *T* and pass through *P*, find the perpendicular bisector of the line *TP*.

Step 2   Construct a perpendicular to the line at *T* to intersect the bisector. The arc is drawn from center *O* with radius *OT*.

COMPUTER METHOD   A line can be drawn from a point tangent to an arc by using the Tangent *option* of the OSNAP command (Fig. 13.29). When prompted for the second point on the line, the snap target is placed over the arc, and the tangent is drawn.

13.21
Arc tangent to a line from a point

If an arc is to be constructed tangent to line *CD* at *T* (Fig. 13.30) and pass through point *P*, a perpendicular bisector of *TP* is drawn. A perpendicular to *CD* is drawn at *T* to locate the center at 0. A similar problem in Fig. 13.31 requires you to draw an arc of a given radius that will be tangent to line *AB* and pass through point *P*. In this case the point of tangency on the line is not known until the problem has been solved.

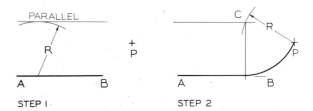

STEP I   STEP 2

FIG. 13.31   Arc tangent to a line and a point.

Step 1   When an arc of a given radius is to be drawn tangent to a line and through a point *P*, draw a line parallel to *AB* and *R* from it.

Step 2   Draw an arc from *P* with radius *R* to locate the center at *C*. The arc is drawn with radius *R* and center *C*.

COMPUTER METHOD   A line can be drawn from a point tangent to an arc by using the Tangent *option* of the OSNAP command (Fig. 13.29). When prompted for the second point on the line, the snap target is placed over the arc, and the tangent is drawn.

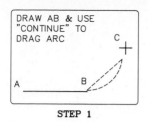
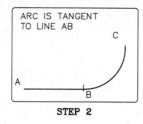

FIG. 13.32 Arc tanget to a line by computer.

Step 1 Command *LINE* From point: **A**
To point: *B*
Command: ARC Center/<Start point>:
CONTINUE
End point: *DRAG*

Step 2 Move cursor to Pt. *C* to locate the end of the curve that is tangent to *AB* at *B*.

13.22
Arc tangent to two lines

An arc of a given radius can be constructed tangent to two nonparallel lines if the radius is given. This method is shown in Fig. 13.33, where two lines form an acute angle. The same steps are used to find an arc tangent to two lines that form an obtuse angle (Fig. 13.34). In both cases the points of tangency are located with thin lines drawn from the centers through the points of tangency.

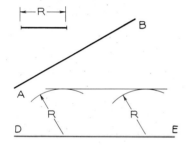
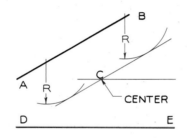
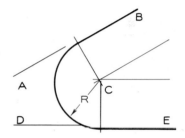

FIG. 13.33 Arc tangent to two lines—acute angle.

Step 1 Construct a line parallel to *DE* with the radius of the specified arc *R*.

Step 2 Draw a second construction line parallel to *AB* to locate the center *C*.

Step 3 Thin lines are drawn from *C* perpendicular to *AB* and *DE* to locate the points of tangency. The tangent arc is drawn using the center *C*.

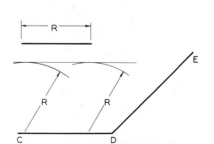
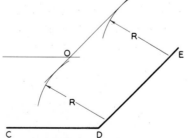
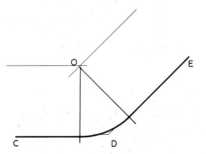

FIG. 13.34 Arc tangent to two lines—obtuse angle.

Step 1 Using the specified radius *R*, construct a line parallel to *CD*.

Step 2 Construct a line parallel to *DE* that is distance *R* from it to locate center *0*.

Step 3 Construct thin lines from center *0* perpendicular to lines *CD* and *DE* to locate the points of tangency. Draw the arc using radius *R* and center *0*.

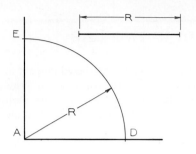
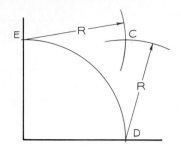
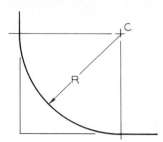

FIG. 13.35 Arc tangent to perpendicular lines.

Step 1 Using the specified radius, R, locate points D and E by using center A.

Step 2 To locate point C, swing two arcs using the radius, R, that was used in Step 1.

Step 3 Locate the tangent points with lines from C. Draw the arc with radius R.

A different technique can be used to find an arc of a given radius that is tangent to perpendicular lines (Fig. 13.35). This method will work only for perpendicular lines.

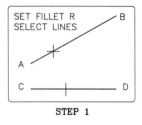
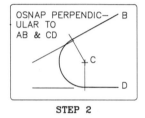

STEP 1   STEP 2

FIG. 13.36 Arc tangent to two lines by computer.

COMPUTER METHOD An arc can be drawn tangent to two nonparallel lines by using the F I L - LET command (Fig. 13.36). You may assign the radius length and select a point on each line; the arc will be drawn and the lines will be automatically trimmed. If tangent points need to be shown, locate the center of the arc by using the CENTER option of the DIM: command. Lines can be O - SNAPped perpendicular to AB and CD from the center, C.

Step 1 Command: *FILLET*
PolylineRadius<Select two lines.>: _R_ (to define radius.)
Enter fillet radius: _.75_ (CR) (CR)
Polyline/Radius/ (Select two lines.): (Pts. on *AB* and *CD*)

Step 2 Command: *LINE* From point: OSNAP-center center of (Select point on the arc.) To point: *perpend* perpend to (Select point on *AB* and a perpendicular is drawn to the point of tangency. Extend the line beyond *AB*. Locate the tangent point on *CD* in the same manner.)

13.23
Arc tangent to an arc and a line

The steps for constructing an arc tangent to an arc and a line are shown in Fig. 13.37. A variation of this principle of construction is shown in Fig. 13.38, where the arc is drawn parallel to an arc and line with the arc in a reverse position.

13.24
Arc tangent to two arcs

A third arc is drawn tangent to two given arcs in Fig. 13.39. Thin lines are drawn from the centers to locate the points of tangency. This tangent arc is concave from the top. A convex arc can be drawn tangent to the given arcs if the radius of the arc is greater than the radius of either of the given arcs (Fig. 13.40).

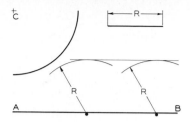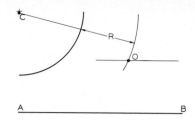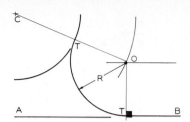

FIG. 13.37 Arc tangent to an arc and a line.

Step 1 Construct a line parallel to *AB* that is *R* from it. Use thin construction lines.

Step 2 Add radius *R* to the extended radius from point *C*. Use this lengthened radius to locate point 0.

Step 3 Lines *0C* and *0T* are drawn to locate the tangency points. The arc with radius *R* and center 0 is drawn.

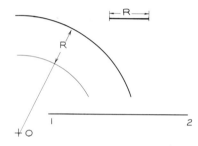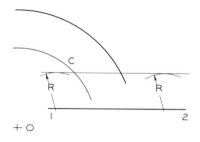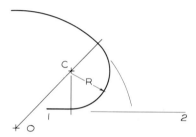

FIG. 13.38 Arc tangent to an arc and a line.

Step 1 The specified radius, *R*, is subtracted from the radius through the arc's center at 0. A concentric arc is drawn with the shortened radius.

Step 2 A line parallel to 1–2 is drawn a distance of *R* from it to locate the center, point *C*.

Step 3 The tangent points are located with lines from 0 through *C*, and through *C* perpendicular to 1–2. Draw the tangent arc with radius *R*.

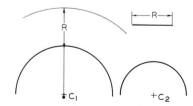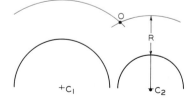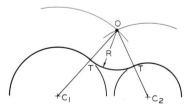

FIG. 13.39 Arc tangent to two arcs—concave arc.

Step 1 The radius of one circle is extended and the radius *R* is added to it. The extended radius is used for drawing a concentric arc.

Step 2 The radius of the other circle is extended and the radius *R* is added to it. The extended radius is used to construct an arc and to locate point 0, the center.

Step 3 The centers are connected with point 0 to locate the points of tangency. The arc is drawn tangent to the two arcs with radius *R*.

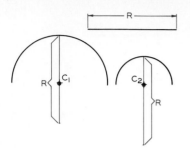
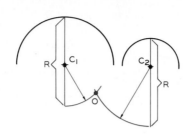
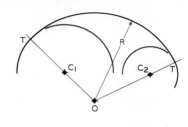

FIG. 23.40   Arc tangent to two arcs—convex arc.

Step 1   The radius of each arc is extended from the arc past its center, and the specified radius R is laid off from the arcs along these lines.

Step 2   The distance from each center to the ends of the extended radii are used for drawing two arcs to locate the center O.

Step 3   Thin lines from O through centers C_1 and C_2 locate the points of tangency. The arc is drawn using point O as the center.

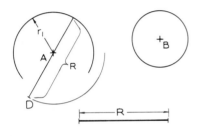
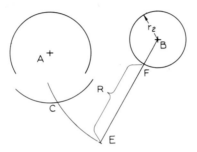
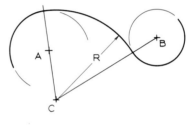

FIG. 13.41   Arc tangent to two circles.

Step 1   The specified radius R is laid off from the arc along the extended radius to locate point D. Radius AD is used to construct a concentric arc.

Step 2   The radius through center B is extended, and the radius R is added to it from point F. Radius BE is used to locate the center C.

Step 3   The tangent arc is drawn with center C and radius R. The points of tangency are located with thin lines from C through the given centers.

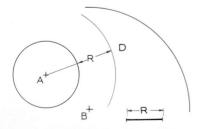
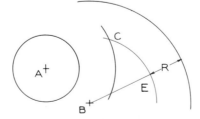
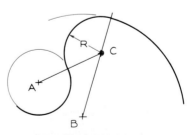

FIG. 13.42   Arc tangent to two arcs.

Step 1   Radius R is added to the radius from center A. Radius AD is used to draw a concentric arc with center A.

Step 2   Radius R is subtracted from the radius through B. Radius BE is used to construct an arc to locate point C.

Step 3   The points of tangency are located with thin lines BC and AC. The tangent arc is drawn with center C.

A variation of this problem is shown in Fig. 13.41, where an arc of a given radius is drawn tangent to the top of one arc and the bottom of the other. A similar problem is shown in Fig. 13.42, where an arc is drawn tangent to a circle and a larger arc.

COMPUTER METHOD   An arc is drawn tangent to two arcs by extending the radii from the centers of the arcs and by adding the radius of the tangent arc from the arcs (Fig. 13.43). The extended radii are used to draw arcs that intersect at C, the center of the tangent arc. Tangent points are located by drawing lines from C to the centers, C1 and C2.

Using the ARC command, an arc is drawn from center C that is tangent to the arcs. The arcs are trimmed to the proper length using the BREAK command.

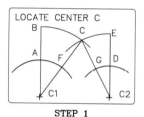 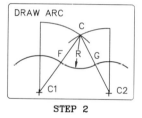

FIG. 13.43   Arc tangent to two arcs.

Step 1   Command: *LINE* From point: C1
To point: *(OSNAP Nearest)* A (CR) (CR)
LINE From point: A
To point: B (Distance AB should be the radius.)
(Repeat these steps from C2, and draw arcs using the extended radii to located center, C. Draw lines C–C1 and C–C2.)

Step 2   Command *ARC (C, S, E)*
ARC Center/<Start point>: (OSNAP to C)
Center/End/<Second point>: (OSNAP to F)
Endpoint: *DRAG* (OSNAP to G)

13.25
Ogee curves

The *ogee curve* is a double curve formed by tangent arcs. By constructing two arcs tangent to three intersecting lines, the ogee curve in Fig. 13.44 was found.

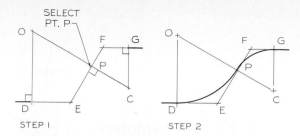

FIG. 13.44   An ogee curve.

Step 1   To draw an ogee curve between two parallel lines, draw line EF, at any angle. Locate a point of your choosing along EF, P in this case. Find the tangent points by making FG equal to FP and DE equal to EP. Draw perpendiculars at G and D to intersect the perpendicular bisector of EF.

Step 2   Using radii CP and OP, draw the two tangent arcs to complete the ogee curve.

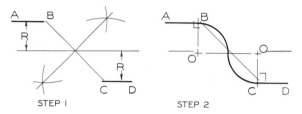

FIG 13.45   An ogee curve.

Step 1   To draw an ogee curve formed by two equal arcs passing through points B and C, draw a line between the points. Bisect the line BC, and draw a line parallel to AB and CD to find the radius, R.

Step 2   Construct perpendiculars at B and C to locate the centers at both points 0. Draw the arcs to complete the ogee curve.

An ogee curve can be drawn between two parallel lines (Fig. 13.45) from points B to C by geometric construction. An alternative method of drawing an ogee curve that passes through points B, E, and C is illustrated in Fig. 13.46.

13.26
Curve of arcs

An irregular curve formed with tangent arcs can be constructed as shown in Fig. 13.47. The radii of the arcs are selected to give the desired curve by moving from one set of points to the next.

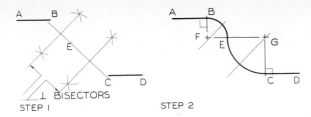

STEP I STEP 2

FIG. 13.46 The unequal ogee curve.

Step 1 Two parallel lines are to be connected by an ogee curve that passes through *B* and *C*. Draw line *BC* and select point *E* on the line. Bisect *BE* and *EC*.

Step 2 Construct perpendiculars at *B* and *C* to intersect the bisectors to locate centers *F* and *G*. Locate the points of tangency and draw the ogee curve using radii *FB* and *GC*.

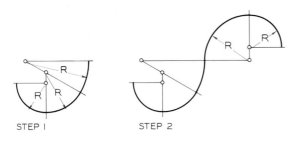

STEP I STEP 2

FIG. 13.47 A curve of arcs.

Step 1 A series of arcs can be joined to form a smooth curve. Begin with the small arc, extend the radius through its center, draw the second and then the third.

Step 2 The curve can be reversed by extending the radius in the opposite direction and repeating the same process.

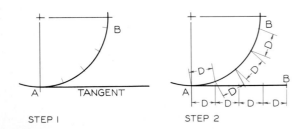

STEP I STEP 2

FIG. 13.48 To rectify an arc.

Step 1 Construct a line tangent to the arc and divide the arc into a series of equal divisions from *A* to *B*.

Step 2 The chordal distances, *D*, along the arc are laid out along the straight line until point *B* is located.

13.27
Rectifying arcs

An arc is rectified when its true length is laid out along a straight line. A method of rectifying an arc is shown in Fig. 13.48. Another method of rectifying an arc uses the mathematical equation for finding the circumference of the circle. Since a circle has 360°, the arc of a 30° sector is one twelfth of the full circumference (360 ÷ 30 = 12). Therefore if the circumference is twelve inches, the 30° arc is equal to one inch.

13.28
Conic sections

Conic sections are plane figures that can be described graphically as well as mathematically. They are formed by passing imaginary cutting planes through a right cone (Fig. 13.49).

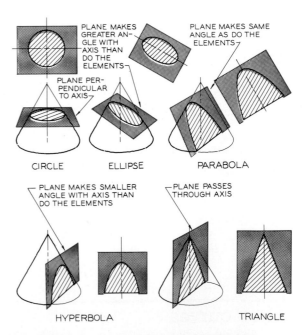

FIG. 13.49 The conic sections are formed by passing cutting planes at various angles through right cones. The conic sections are the circle, ellipse, parabola, hyperbola, and triangle.

13.29
Ellipses

The *ellipse* is a conic section formed by passing a plane through a right cone at an angle (Fig. 13.49). The ellipse is mathematically defined as the path of a point that moves in such a way that the sum of the distances from two focal points is a constant. The largest diameter of an ellipse is always the true length and is called the *major diameter*. The shortest diameter is perpendic-

ular to the major diameter and is called the *minor diameter*.

The construction of an ellipse is found by revolving the edge view of a circle, as shown in Fig. 13.50. This ellipse could have been drawn using the ellipse template shown in Fig. 13.51. The angle between the line of sight and the edge view of the circle (or the one closest to this size) is the angle of the ellipse template that should be used. Ellipse templates are available in intervals of five degrees and in variations in size of the major diameter of about ⅛ inch (Fig. 13.52).

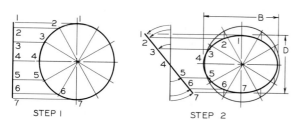

FIG. 13.50 An ellipse by revolution.

Step 1 When the edge view of a circle is perpendicular to the projectors between its adjacent view, the view will be a true circle. Mark equally spaced points along the arc and project them to the edge.

Step 2 Revolve the edge of the circle to the desired position, and project the points to the circular view, which will now appear as an ellipse. The points are projected vertically downward to new positions.

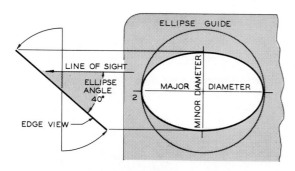

FIG. 13.51 The ellipse template.

When the edge view of a circle is revolved so the line of sight between the two views is not perpendicular to the edge view, the circle will appear as an ellipse. The major diameter remains constant, but the minor diameter will vary. The angle between the line of sight and the edge view of the circle is the angle of the ellipse template.

FIG. 13.52 Ellipse templates are calibrated at 5° intervals from 15° to 60°. (Courtesy of Timely Products, Inc.)

The ellipse can also be constructed inside a rectangular box or a parallelogram (Fig. 13.53), where a series of points is plotted to form an elliptical curve. Two circles can be used for constructing an ellipse by making the diameter of the large circle equal to the major diameter and the diameter of the small circle equal to the minor diameter (Fig. 13.54).

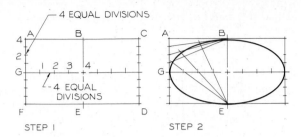

FIG. 13.53 Ellipse—parallelogram method.

Step 1 An ellipse can be drawn inside of a rectangle or a parallelogram by dividing the horizontal centerline into the same number of equal divisions as the shorter sides, *AF* and *CD*.

Step 2 The construction of the curve in one quadrant is shown by using sets of rays from *E* and *B* to plot the points.

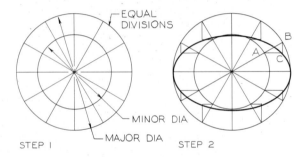

FIG. 13.54 Ellipse—circle method

Step 1 Two concentric circles are drawn with the large one equal to the major diameter and the small one equal to the minor diameter. Divide them into equal sectors.

Step 2 Plot points on the ellipse by projecting downward from the large curve to intersect horizontal construction lines drawn from the intersections on the small circle.

COMPUTER METHOD Using the `ELLIPSE` command, a box can be constructed that defines the major and minor diameters of an ellipse (Fig. 13.55). You will be prompted for the angle of inclination of the ellipse, and the ellipse will be drawn at the angle specified. The `CENTER` command cannot be used to locate the ellipse's center since it is a `BLOCK`. The center can be found by using the `INQUIRY` (Distance option) command to determine the major and minor diameter lengths, which are then halved.

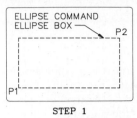
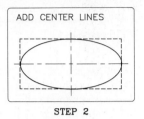

FIG. 13.55 Ellipse—computer method.

Step 1 Command *ELLIPSE*
Block name (or ?) <ELLIPSE>: *ELLIPSE*
Insertion point: *P1*
X-scale factor 1CornerXYZ: *DRAG P2*

Step 2 Rotation angle 0: (CR) (to accept 0 and the ellipse is drawn. Centerlines are added.)

The *conjugate diameters* of an ellipse are diameters parallel to the tangents at the ends of each, as illustrated in Fig. 13.56. A single ellipse has an infinite number of sets of conjugate diameters.

When the ellipse and a pair of conjugate diameters are given, the major and minor diameters of the ellipse can be found by using the method illustrated in Fig. 13.57. When the conjugate diameters are not given, they can be constructed as shown in Fig. 13.56, and the major and minor diameters found. The major and minor diameters are necessary for using the ellipse template.

The mathematical equation of an ellipse is

$$\frac{x^2}{a^2} + \frac{y^2}{b^2} = 1, \quad \text{where } a, b \neq 0.$$

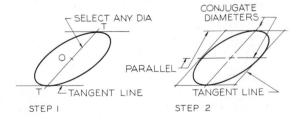

FIG. 13.56 Conjugate diameters.

Step 1 A conjugate diameter is parallel to the tangents at the ends of another conjugate diameter. A diameter is selected and parallel tangents are drawn.

Step 2 A conjugate diameter is drawn parallel to the horizontal tangents, and the inclined tangents are drawn parallel to the conjugate diameter found in Step 1.

13.30
Parabolas

The *parabola* is mathematically defined as a plane curve, each point of which is equidistant from a straight line (called a *directrix*) and its focal point. The parabola is a conic section formed when the cutting

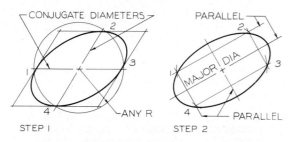

STEP I   STEP 2

FIG. 13.57 Finding the axes of an ellipse.

Step 1 When the conjugate diameters of an ellipse are given, you can find the major and minor diameters of the ellipse by drawing a circle of any radius from the intersection of the diameters.

Step 2 The circle cuts four points along the ellipse. The points are connected to form a rectangle. The major and minor diameters are parallel to the sides of the rectangle.

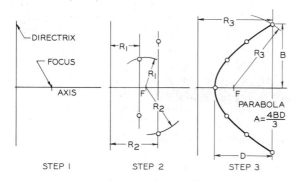

STEP I   STEP 2   STEP 3

FIG. 13.58 Parabola—mathematical method.

Step 1 Draw an axis perpendicular to a line (a directrix). Choose a point for the focus, *F*.

Step 2 Locate points by using a series of selected radii to plot points on the curve. For example, draw a line parallel to the directrix and R_2 from it. Swing R_2 from *F* to intersect the line and plot the point.

Step 3 Continue the process with a series of arcs of varying radii until an adequate number of points have been found to complete the curve.

plane makes the same angle with the base of a cone as do the elements of the cone.

The construction of a parabola using its mathematical definition is shown in Fig. 13.58. A parabolic curve can also be constructed geometrically by dividing the two perpendicular lines into the same number of divisions (Fig. 13.59). The parabola is drawn through the plotted points with an irregular curve. A third method of construction, using parallelograms, is illustrated in Fig. 13.60.

The mathematical equation of the parabola is

$$y = ax^2 + bx + c, \quad \text{where } a \neq 0.$$

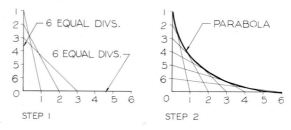

STEP I   STEP 2

FIG. 13.59 Parabola—tangent method.

Step 1 Construct two lines at a convenient angle and divide each of them into the same number of divisions. Connect the points with a series of diagonals.

Step 2 When finished, construct the parabolic curve to be tangent to the diagonals.

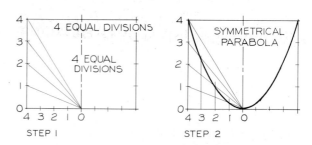

STEP I   STEP 2

FIG. 13.60 Parabola—parallelogram method.

Step 1 Construct a rectangle or a parallelogram to contain the parabola, and locate its axis parallel to the sides through *0*. Divide the sides into equal divisions. Connect the divisions with point *0*.

Step 2 Construct lines parallel to the sides (vertical in this case) to locate the points along the rays from *0*. Draw the parabola.

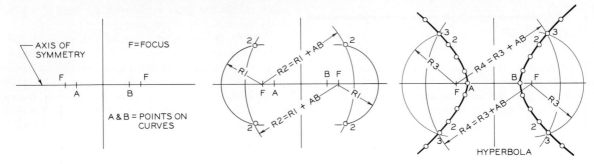

FIG. 13.61   The hyperbola.

Step 1   A perpendicular is drawn through the axis of symmetry, and focal points *F* are located equidistant from it on both sides. Points on the curve, *A* and *B*, are located equidistant from the perpendicular at a location of your choice but between the focal points.

Step 2   Radius *R1* is selected to draw arcs using focal points *F* as the centers. *R1* is added to *AB* (the distance between the nearest points on the hyperbolas) to find *R2*. Radius *R2* is used to draw arcs using the focal points as centers. The intersections of *R1* and *R2* establish points 2 on the hyperbola.

Step 3   Other radii are selected and added to distance *AB* to locate additional points in the same manner as described in Step 2. A smooth curve is drawn through the points to form the hyperbolic curves.

13.31
Hyperbolas

The *hyperbola* is a two-part conic section. Mathematically, it is defined as the path of a point that moves in such a way that the difference of its distances from two focal points is a constant. Figure 13.61 shows the construction of a hyperbola using this definition.

A second method of construction is shown in Fig. 13.62. Two perpendicular lines are drawn through point *B* as asymptotes. The hyperbolic curve becomes more nearly parallel and closer to the asymptotes as the hyperbola is extended, but the curve never merges with the asymptotes.

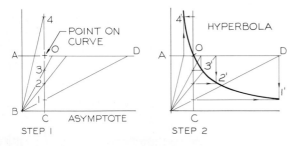

FIG. 13.62   The equilateral hyperbola.

Step 1   Two perpendiculars are drawn through *B*, and any point *0* on the curve is located. Horizontal and vertical lines are drawn through *0*. Line *C0* is divided into equal divisions, and rays from *B* are drawn through them to the horizontal line.

Step 2   Horizontal construction lines are drawn from the divisions along line *0C*, and lines from *AD* are projected vertically to locate points 1′ through 4′ on the curve.

13.32
Spirals

The *spiral* is a coil that begins at a point and becomes larger as it travels around the origin. A spiral lies in a single plane. The steps for construction of a spiral are shown in Fig. 13.63.

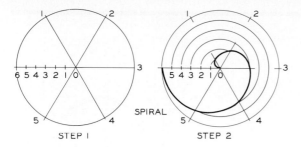

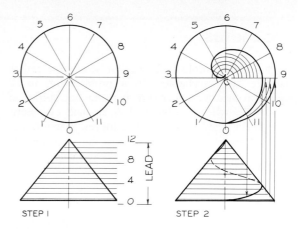

FIG. 13.63 The spiral.

Step 1 Draw a circle and divide it into equal parts. The radius is divided into the same numer of equal parts (six in this example).

Step 2 By beginning on the inside, draw arc 0–1 to intersect radius 0–1. Then swing arc 0–2 to radius 0–2 and continue until the last point is reached at 6, which lies on the original circle.

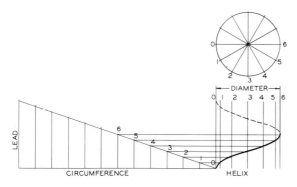

FIG. 13.64 The helix.

Divide the top view of the cylinder into equal divisions and project them to the front view. Lay out the circumference and the height of the cylinder, which is the lead. Divide the circumference into the same number of equal parts by taking the measurements from the top view. Project the points along the inclined rise to their respective elements to find the helix.

13.33
Helixes

The *helix* is a curve that coils around a cylinder or a cone at a constant angle of inclination. Examples of helixes are corkscrews or threads on a screw. A helix is constructed about a cylinder in Fig. 13.64 and about a cone in Fig. 13.65.

FIG. 13.65 A conical helix.

Step 1 Divide the cone's base into equal parts. Pass a series of horizontal cutting planes through the front view of the cone. Use the same number as the divisions on the base, twelve in this case.

Step 2 Project all the divisions along the front view of the cone to line $C\,9$, and draw a series of arcs from center C to their respective radii in the top view to plot the points. Project the points to their respective cutting planes in the front view.

13.34
Involutes

The *involute* is the path of the end of a line as it unwinds from a line or a plane figure. In Fig. 13.66 an involute is formed by unwinding a line from a rectangle. Successively different radii that are equal in length to the sides are used to develop the involute.

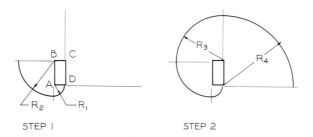

FIG. 13.66 The Involute.

Step 1 Side AD is used as a radius for drawing arc AD. AB is added to AD to form radius R_2. Draw a second arc using center B.

Step 2 The two remaining sides are used to unwind the involute back to its point of origin.

Problems

The following problems are similar to those solved as examples in this chapter. These problems are to be solved on Size AV paper similar to the one shown in Fig. 13.67, where Problems 1–5 are laid out. Each inch on the grid is equal to 0.20 inches; therefore use your engineers' 10 scale to lay out the problems. By equating each grid to 5 mm, you can use your full-size metric scale to lay out and solve the problems.

Show your construction and mark all points of tangency, as discussed in the chapter. (Refer to Figs. 13.67–13.77 for Problems 1–45.)

1. Draw triangle *ABC* using the given sides.

2–3. Inscribe an angle in the semicircles with the vertexes at point *P*.

4. Inscribe a nine-sided regular polygon inside the circle.

5. Circumscribe a ten-sided regular polygon about the circle.

6. Circumscribe a hexagon about the circle.

7. Inscribe an octagon in the circle.

8. Circumscribe an octagon about the circle.

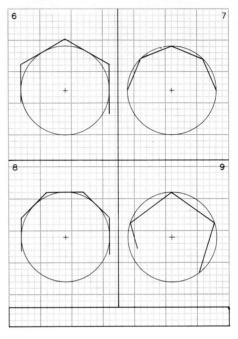

FIG. 13.68 Problems 6–9. Construction of regular polygons.

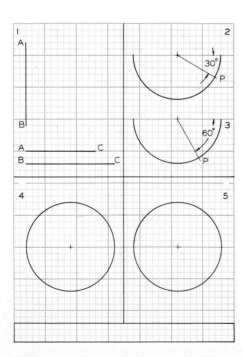

FIG. 13.67 Problems 1–5. Basic constructions.

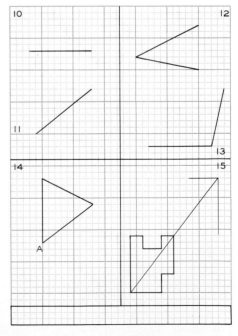

FIG. 13.69 Problems 10–15. Basic constructions.

9. Construct a pentagon inside the circle using the compass method.

10–11. Bisect the lines.

12–13. Bisect the angle.

14. Rotate the triangle 80° in a clockwise direction about point *A*.

15. Enlarge the given shape to the size indicated by the diagonal.

16. Divide *AB* into seven equal parts. Draw the construction line through *B* for your construction.

17. Divide the two vertical lines into four equal divisions. Draw three equally spaced vertical lines at the divisions.

18. Construct an arc with radius *R* that is tangent to the line at *J* and passes through point *P*.

19. Construct an arc with radius *R* that is tangent to the line and passes through *P*.

20. Construct a line from *P* that is tangent to the semicircle. Locate the points of tangency. Use the compass method.

21–23. Construct arcs with a given radii tangent to the lines.

24–27. Construct arcs that are tangent to the arcs and/or lines. The radii are given for each problem.

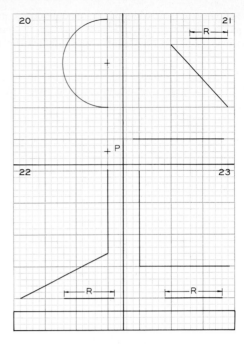

FIG. 13.71 Problems 20–23. Tangency construction.

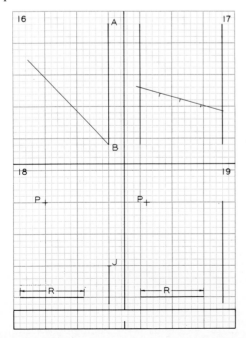

FIG. 13.70 Problems 16–19. Tangency construction.

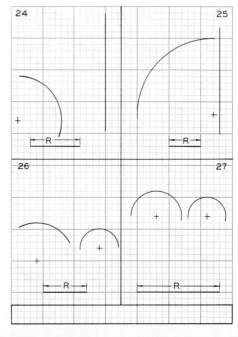

FIG. 13.72 Problems 24–27. Tangency construction.

28–31. Construct ogee curves that connect the ends of the given lines and pass through points *P* where given. In Problem 31, the radii for the arcs are given.

32–33. Using the given radii, connect the given arcs with a tangent arc as indicated in the freehand sketches.

34. Rectify the arc along the given line by dividing the circumference into equal divisions and laying them off with your dividers.

35. Rectify the arc by using the compass method shown in Fig. 13.48.

36. Construct an ellipse inside the rectangular layout.

37. Construct an ellipse inside the large circle. The small circle represents the minor diameter.

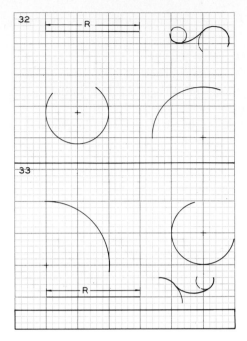

FIG. 13.74 Problems 32–33. Tangency construction.

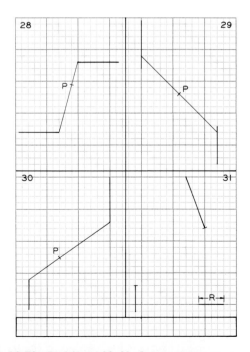

FIG. 13.73 Problems 28–31. Ogee curve construction.

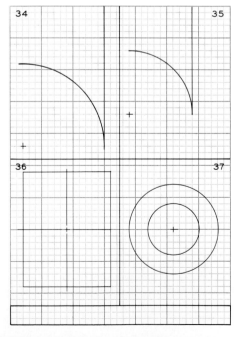

FIG. 13.75 Problems 34–37. Rectifying an arc, ellipse construction.

38. Construct an ellipse inside the circle when the edge view has been rotated 45° as shown.

39. Using the focal point *F* and the directrix, plot and draw the parabola formed by these elements.

40. Construct a parabola using perpendicular lines by either of the methods shown in Figs. 13.59 and 13.60.

41. Using the focal point *F*, points *A* and *B* on the curve, and the axis of symmetry, construct the hyperbolic curve.

42. Construct a hyperbola that passes through 0. The perpendicular lines are asymptotes.

43. Construct a spiral by using the four divisions marked along the radius.

44–45. Construct a helix that has a rise equal to the heights of the cylinder and cone. Show construction and the curve in all views.

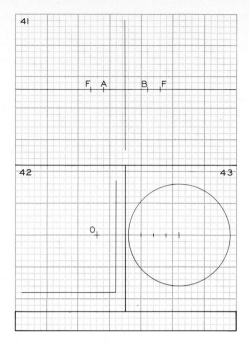

FIG. 13.77 Problems 41–43. Hyperbola and spiral construction.

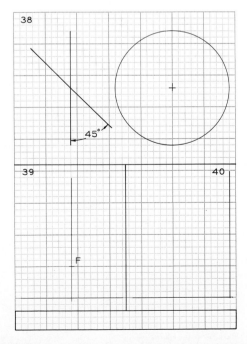

FIG. 13.76 Problems 38–40. Ellipse and parabola construction.

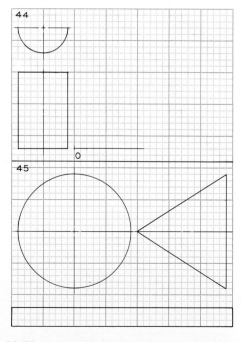

FIG. 13.78 Problems 44–45. Helix construction.

46–56. (Fig. 13.79–13.89) Construct these problems on Size A sheets, one problem per sheet. Select the proper scale that will best fit the problem to the sheet. Mark all points of tangency and strive for good line quality.

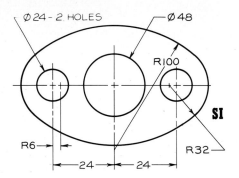

FIG. 13.79 Problem 46. Gasket.

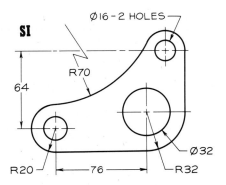

FIG. 13.80 Problem 47. Lever crank.

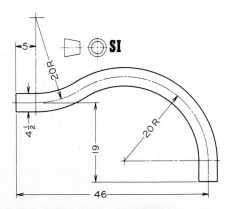

FIG. 13.81 Problem 48. Road tangency.

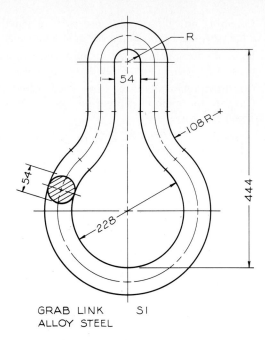

GRAB LINK   SI
ALLOY STEEL

FIG. 13.82 Problem 49. Grab link.

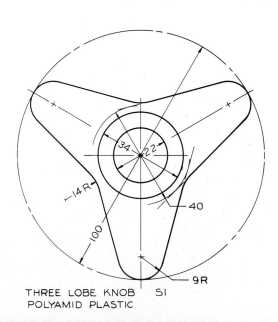

THREE LOBE KNOB   SI
POLYAMID PLASTIC

FIG. 13.83 Problem 50. Three-lobe knob.

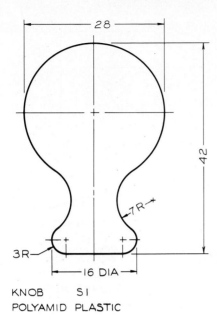

KNOB   SI
POLYAMID PLASTIC

FIG. 13.84 Problem 51. Knob.

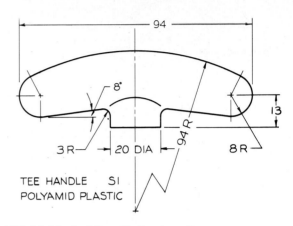

TEE HANDLE   SI
POLYAMID PLASTIC

FIG. 13.86 Problem 53. Tee handle.

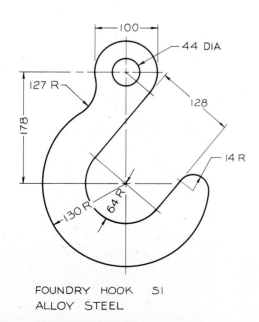

FOUNDRY HOOK   SI
ALLOY STEEL

FIG. 13.85 Problem 52. Foundry hook.

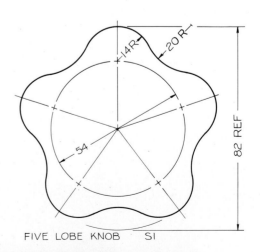

FIVE LOBE KNOB   SI

FIG. 13.87 Problem 54. Five-lobe knob.

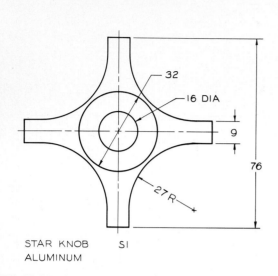

STAR KNOB   SI
ALUMINUM

FIG. 13.88 Problem 55. Star knob.

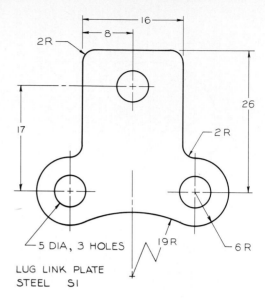

LUG LINK PLATE
STEEL   SI

FIG. 13.89 Problem 56. Lug link plate.

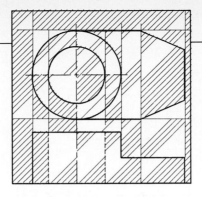

CHAPTER 14

Multiview Sketching

14.1
The purpose of sketching

Sketching is as much a thinking process as it is a communication technique. Designers develop their ideas by making many sketches and revising them before finally arriving at the desired solution.

Sketching should be a rapid method of drawing. Designers who develop their sketching skills can assign drafting work to assistants who can then prepare the finished drawings from the designers' sketches. If sketches are not sufficiently clear to communicate the ideas to someone else, then the designer has likely not thought out the solution well enough. The ability to communicate by any means is a great asset, and sketching is one of the more powerful techniques of communication.

14.2
Shape description

A pictorial of an object is shown in Fig. 14.1 with three arrows that indicate the directions of sight that will give top, front, and right-side views of the object. Each view is a two-dimensional view.

This system is called *orthographic*, or *multiview*, *projection*. In multiview projection, it is important that the views be located as shown in Fig. 14.1. The top view is placed over the front view, since both views share the dimension of width. The side view is placed to the right of the front view where these views share the dimension of height. The distance between the views can vary, but the views must be positioned so that they project from each other as shown here.

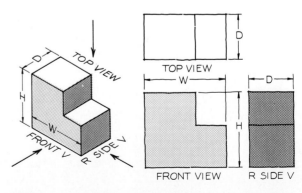

FIG 14.1 Three views of an object can be found by looking at the object in this manner. The three views—the top, front, and right side—describe the object.

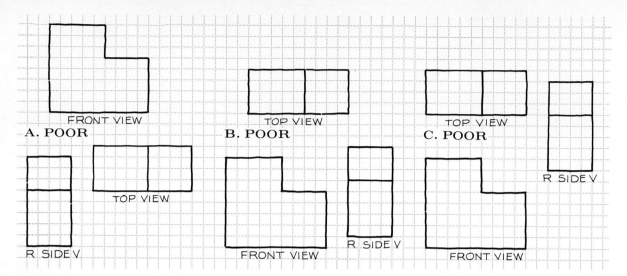

FIG. 14.2 Arrangement of views.

A. These views are sketched incorrectly; the views are scrambled.

B. These views are nearly correct, but they do not project from view to view.

C. The top and front views are correctly positioned, but the right-side view is incorrectly positioned.

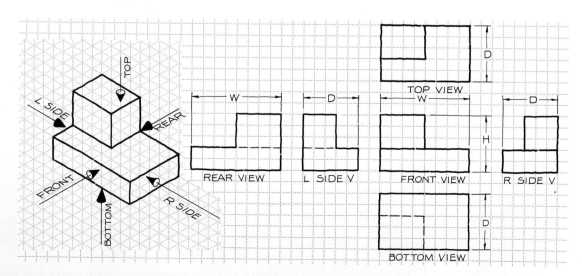

FIG. 14.3 Six principal views can be sketched by looking at the object in the directions indicated by the lines of sight. Note how the dimensions are placed on the views. Height (*H*) is shared by all four of the horizontally positioned views.

Several examples of poorly arranged views are shown in Fig. 14.2. Although these views are correct, they are hard to interpret because the views are not placed in their standard positions.

14.3
Six-view drawings

Six principal views may be found for any object by using the rules of orthographic, or multiview, projection. The directions of sight for the six orthographic views are shown in Fig. 14.3, where the views are drawn in their standard positions. The width dimension is common to the top, front, and bottom views. Height is common to the right-side, front, left-side, and rear views.

Seldom will an object be so complex as to require six orthographic views, but if six views are needed, they should be arranged as shown in this figure.

14.4
Sketching techniques

Sketching means freehand drawing without the use of instruments or straightedges. Medium-weight pencils, such as H, F, or HB grades are the best pencils for sketching. The standard lines used in multiview drawing and their respective line weights are shown in Fig. 14.4.

By sharpening the pencil point to match the desired line width, you will be able to use the same

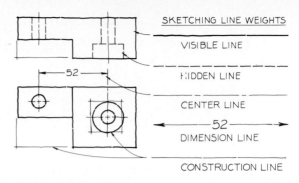

FIG. 14.4 These lines are examples of those that you should sketch with an F or an HB pencil when drawing views of an object. Some lines are thinner than others and all, except construction lines, are black.

grade of pencil for all lines. The different point sizes are shown in Fig. 14.5. A line that is drawn freehand should have a freehand appearance; no attempt should be made to give the line the appearance of one drawn by instruments. However, sketching technique can be improved by using a printed grid on sketching paper or by overlaying a printed grid with translucent tracing paper (Fig. 14.6) so that the grid can be seen through the paper.

When a freehand sketch is made, some lines will be vertical, others horizontal or angular. If you do not tape your drawing to the table top, you will be able to position the sheet for the most comfortable strokes, which are (for the right-handed drafter) from left to

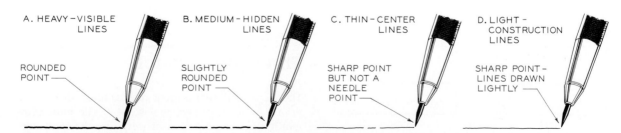

FIG. 14.5 The alphabet of lines that are sketched freehand are all made with the same pencil grade (F or HB). The variation in the lines is achieved by varying the sharpness of the pencil point.

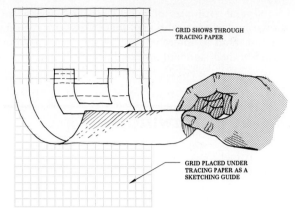

GRID SHOWS THROUGH
TRACING PAPER

GRID PLACED UNDER
TRACING PAPER AS A
SKETCHING GUIDE

FIG. 14.6 To aid you in freehand sketching, a grid can be placed under a sheet of tracing paper.

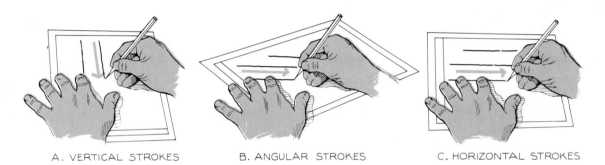

A. VERTICAL STROKES   B. ANGULAR STROKES   C. HORIZONTAL STROKES

FIG. 14.7 Freehand sketching techniques.

A. Vertical lines should be sketched in a downward direction.

B. Angular strokes can be sketched left to right, if you rotate your sheet slightly.

C. Horizontal strokes are made best in a left-to-right direction. Always sketch from a comfortable position, and turn your paper if necessary.

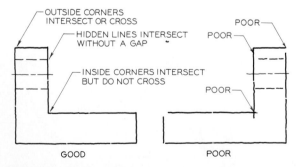

OUTSIDE CORNERS
INTERSECT OR CROSS

HIDDEN LINES INTERSECT
WITHOUT A GAP

INSIDE CORNERS INTERSECT
BUT DO NOT CROSS

POOR

POOR

POOR

GOOD   POOR

FIG. 14.8 For good sketches, follow these examples of good technique. Compare the good drawing with the poor one.

right (Fig. 14.7). Finally, for the best effect, the lines that are sketched to form the various views should intersect as indicated in Fig. 14.8.

14.5
Three-view sketch

The steps of drawing three orthographic views on a printed grid are shown in Fig. 14.9. The most commonly used combination of views are the front, top, and right-side views, as shown in this figure. First, the overall dimensions of the object are sketched. Then the slanted surface is drawn in the top view and pro-

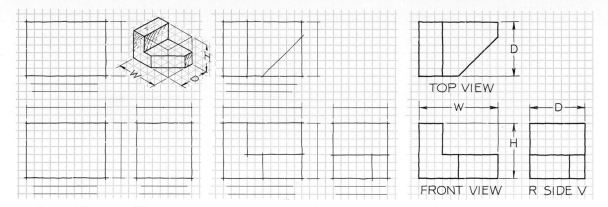

FIG. 14.9 Three-view sketching.

Step 1 Block in the views by using the overall dimensions. Allow proper spacing for labeling and dimensioning the views.

Step 2 Remove the notches and project from view to view.

Step 3 Check your layout for correctness; darken the lines, and complete the labels and dimensions.

jected to the other views. Lastly, the lines are darkened, the views labeled, and the overall dimensions of height, width, and depth applied.

When surfaces are slanted, they will not appear true shape in the principal views of orthographic projection (Fig. 14.10). Surfaces that do not appear true size either are *foreshortened* or appear as *edges*. In Fig. 14.10C, two planes of the object are slanted; thus both appear foreshortened in the right-side view.

A good exercise for analyzing the given views is to find the missing view when two views are given (Fig. 14.11). The right-side view is found in Fig. 14.12, where the top and front views are given. The right-side view has the depth dimension in common with the top view, and height in common with the front view. Knowing this enables us to block in the side view in Step 1. The side view is developed in Step 2 and completed in Step 3.

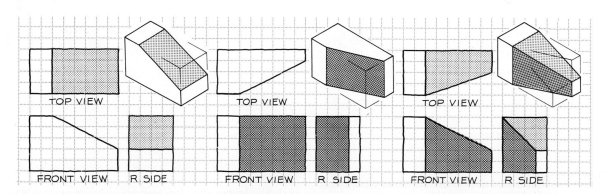

FIG. 14.10 Views of planes.

A. The plane appears as an edge in the front view, and it is foreshortened in the top and side views.

B. The plane is an edge in the top view, and foreshortened in the front and side views.

C. These two planes appear foreshortened in the right-side view. Each appears as an edge in the top and front views.

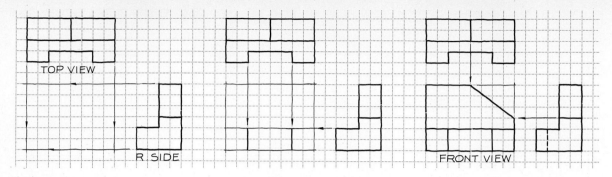

FIG. 14.11 Missing front views.

Step 1 When two views are given and the third is required, begin by projecting the overall dimensions from the top and right-side views.

Step 2 The various features of the object are sketched using construction lines.

Step 3 The features are completed, the view is checked for correctness, and the lines are darkened to the proper line quality.

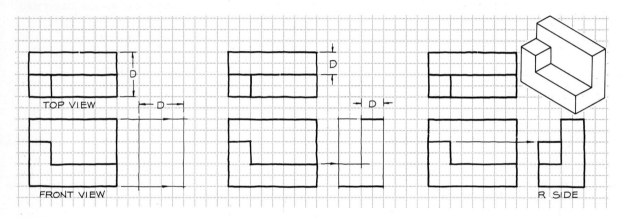

FIG. 14.12 Missing side view.

Step 1 To find the right-side view when the top and front views are given, block in the view with the overall dimensions.

Step 2 Develop the features of the view by analyzing the views together. Use light construction lines.

Step 3 Check the view for correctness, and darken the lines to their proper line weight.

Another exercise is completing the views when some or all of them have missing lines (Fig. 14.13). A pictorial is provided to help you analyze the given views.

14.6
Circular features

Centerlines indicate the features of true circles or cylinders. Examples of these are shown in Fig. 14.14. In circular views, centerlines cross to indicate the center of the circle. Centerlines consist of short dashes, which should cross in the circular views, spaced at inch intervals along the line. If a centerline coincides with an object line—visible or hidden—the centerline should be omitted since the object lines are more important (Fig. 14.14C).

The application of centerlines is shown in Fig. 14.15, where they indicate whether or not circles and arcs are concentric (share the same centers). The centerline should extend beyond the arc by about one eighth of an inch. The correct manner of applying centerlines is shown in Fig. 14.16. The circular view

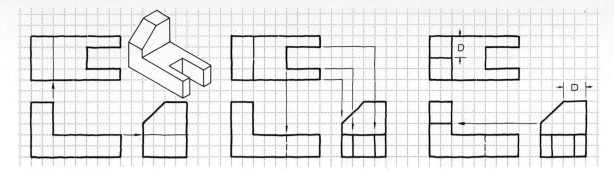

FIG. 14.13 Missing lines.

Step 1 Lines may be missing in all views in this type of problem. The first missing line is found by projecting the edges of the planes from the front to the top and side views.

Step 2 The notch in the top view is projected to the front and side views. The line in the front view is a hidden line.

Step 3 The line formed by the beveled surface is found in the front view by projecting from the side view.

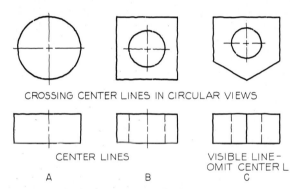

FIG. 14.14 Centerlines are used to indicate the centers of circles, the axes of cylinders, and are drawn as very thin lines. When they coincide with visible or hidden lines, centerlines are omitted.

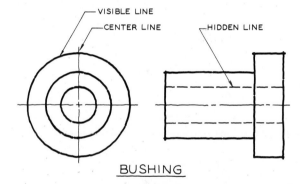

FIG. 14.16 Here you can see the application of centerlines of concentric cylinders and the relative weight of hidden lines, visible lines, and centerlines.

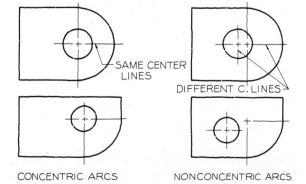

FIG. 14.15 The centerline should extend beyond the last arc that has the same center. When the arcs are not concentric, separate centerlines should be drawn.

clearly indicates that the cylinders are concentric since each shares the same centerlines.

Sketching circles

Circles can be sketched by using light guidelines along with centerlines (Fig. 14.17). Since it is difficult to draw a freehand circle in one continuous line, short arcs are drawn using the guidelines. If you fail to become reasonably skilled at sketching circles, use a circle template or compass to lightly draw the circle or arc, then darken the line freehand to match the other lines of your sketch.

The construction of three orthographic views with circular features is shown in Fig. 14.18. The circular

223

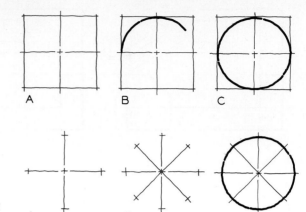

FIG. 14.17 Circles can be sketched using either of the construction methods shown here. The use of guidelines is essential to freehand sketching of circles and arcs.

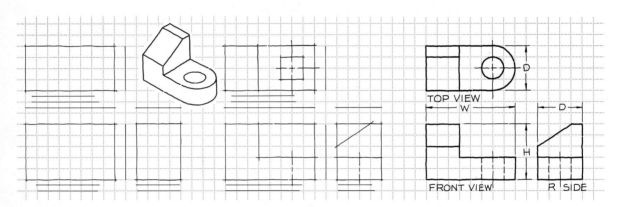

FIG. 14.18 Circular features in orthographic views.

Step 1 To draw orthographic views of the object shown, begin by blocking in the overall dimensions. Leave room for the labels and dimensions.

Step 2 Construct the centerlines and the squares about the centerlines in which the circles will be drawn. Show the slanted surface in the side view.

Step 3 Sketch the arcs, and darken the final lines of the views. Label the views, and show the dimensions of W, D and H.

features are located with centerlines and guidelines in Step 2; then they are sketched in and darkened in Step 3.

A similar example is given in Fig. 14.19, where the object consists of circular features and arcs. The circles should be drawn first so that their corresponding rectangular views (such as the hidden hole in the top view) can be found by projecting from the circular view.

14.7
Isometric sketching

Another type of three-dimensional pictorial is the *isometric drawing*, which may be drawn on a specially printed grid composed of a series of lines making sixty-degree angles with one another (Fig. 14.20). The squares in the orthographic views can be laid off along

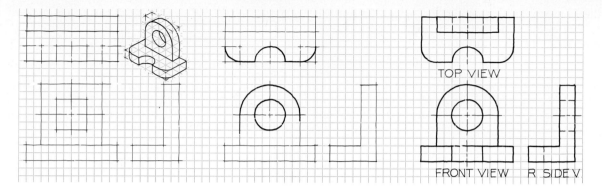

FIG. 14.19 Orthographic views with multiple-circular features.

Step 1 When sketching orthographic views with circular features, you should begin by sketching the centerlines and guidelines first.

Step 2 Using the guidelines, sketch the circular features. These can be darkened as they are drawn if they will be final lines.

Step 3 The outlines of the circular features are found by projecting from the views found in Step 2. All final lines are darkened.

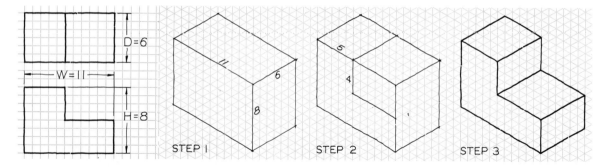

FIG. 14.20 Isometric pictorial sketching.

Step 1 When orthographic views of a part are given, an isometric pictorial can be sketched by using a printed isometric grid. Begin by constructing a box using the overall given dimensions.

Step 2 The notch can be located by measuring over five squares as shown in the orthographic views. The notch is measured four squares downward by counting the units.

Step 3 The pictorial is completed, and the lines are darkened. This is a three-dimensional pictorial, whereas the orthographic views are each two-dimensional.

the isometric grid, as shown in Step 1. The notch is located in the same manner in Steps 2 and 3 to complete the isometric pictorial.

Angles cannot be measured with a protractor in isometric pictorials; they must be drawn by measuring coordinates along the three axes of the printed grid. In Fig. 14.21, the ends of the angular slope are located by measuring the direction of width and height. When an object has two sloping planes that intersect (Fig. 14.22), it is necessary to draw the sloping planes one

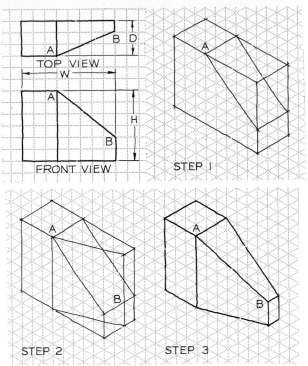

FIG. 14.22 Double angles in isometric pictorials.

Step 1   When part of an object has a double angle, begin by constructing the overall box and finding one of the angles.

Step 2   Find the second angle that locates point B. Point A will connect to point B to give the intersection line.

Step 3   The final lines are darkened. Line AB is the line of intersection between the two sloping planes.

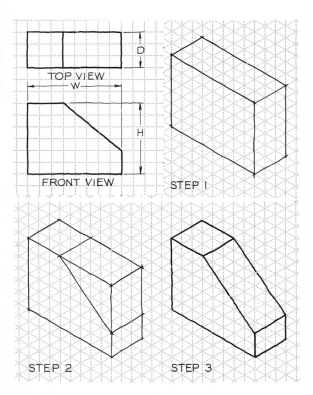

FIG. 14.21 Angles in isometric pictorials.

Step 1   Begin by drawing a box using the overall dimensions given in the orthographic views. Count the squares, and transfer them to the isometric grid.

Step 2   Angles cannot be measured with a protractor. Angles will be either larger or smaller than their true measurement. Find each end of the angle by measuring along the axes.

Step 3   The ends of the angles are connected. Dimensions can be measured only in directions parallel to the three axes.

at a time to find point B. The line from A to B is the line of intersection between the two planes.

Circles in isometric pictorials

Circles, which will appear as ellipses in isometric pictorials can be sketched by locating their centerlines as shown in Step 1 of Fig. 14.23. The center must be equidistant from the top, bottom, and end of the front view.

Figure 14.24 shows two methods of constructing elliptical views of circles, and Fig. 14.25 shows several

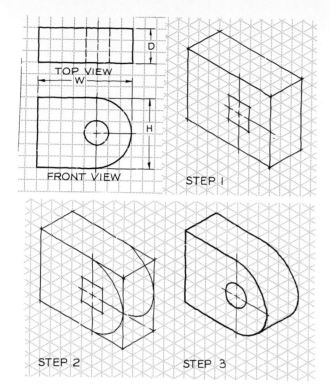

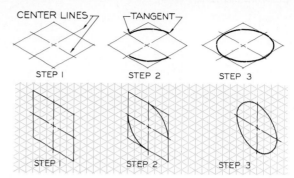

FIG. 14.24 Sketching ellipses.

Two methods of sketching ellipses are shown: one without a grid and one with a grid. When a grid is not used, the centerlines are drawn at 30° to the horizontal (for a horizontal circle). A rhombus is drawn about them, and the ellipse is sketched inside the guidelines. When there is a grid, the same technique is used except that the lines of the grid become the guidelines.

FIG. 14.23 Circles in isometric pictorials.

Step 1 Construct a box using the overall orthographic dimensions. Draw the centerlines and a square (rhombus) of guidelines around the circular hole.

Step 2 Draw the pictorial views of the arcs tangent to the boxes formed by the guidelines. These arcs will appear elliptical rather than circular.

Step 3 Construct the hole, and darken the lines. Hidden lines are normally omitted in pictorial sketches.

techniques of sketching circular and cylindrical shapes.

A part composed of a number of cylindrical forms is shown in Fig. 14.26A. These three views are used as the basis for an isometric pictorial shown in the steps of construction. When an isometric grid is not used, the axes of the isometric sketch are positioned 120 degrees apart. In other words, the height dimension is vertical, and the width and depth dimensions make 30° angles with the horizontal direction on your paper.

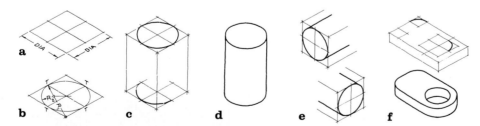

FIG. 14.25 Sketching circular features.

Several examples of sketching circles and cylinders are shown here. In all cases, guidelines are used to aid in proportional sketching.

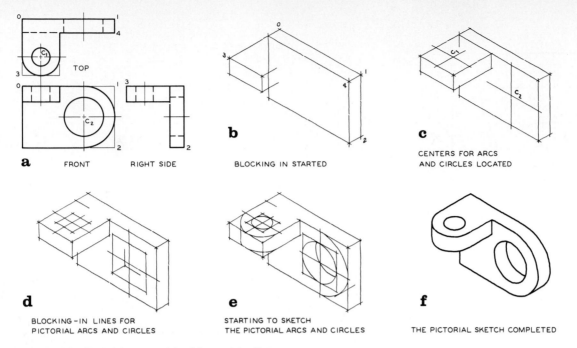

FIG. 14.26 Sketching an object isometrically.

The five steps of constructing an isometric pictorial are shown here. The guidelines are drawn vertical and at 30° to the horizontal. Centerlines are used for locating the circular features (C). The object is developed (D and E) and darkened (F).

Problems

These sketching problems should be drawn on Size A (8½-by-11-inch) paper with or without a printed grid. A typical format for this size sheet is shown in Fig. 14.27, where a 0.20″ grid is given. (This grid can be converted to an approximate metric grid by equating each square to 5 mm.) All sketches and lettering should be neatly executed by applying the principles covered in this chapter. Figures 14.28 and 14.29 contain the problems and instructions.

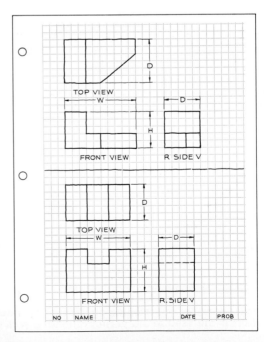

FIG. 14.27 The layout of a Size A sheet for sketching problems.

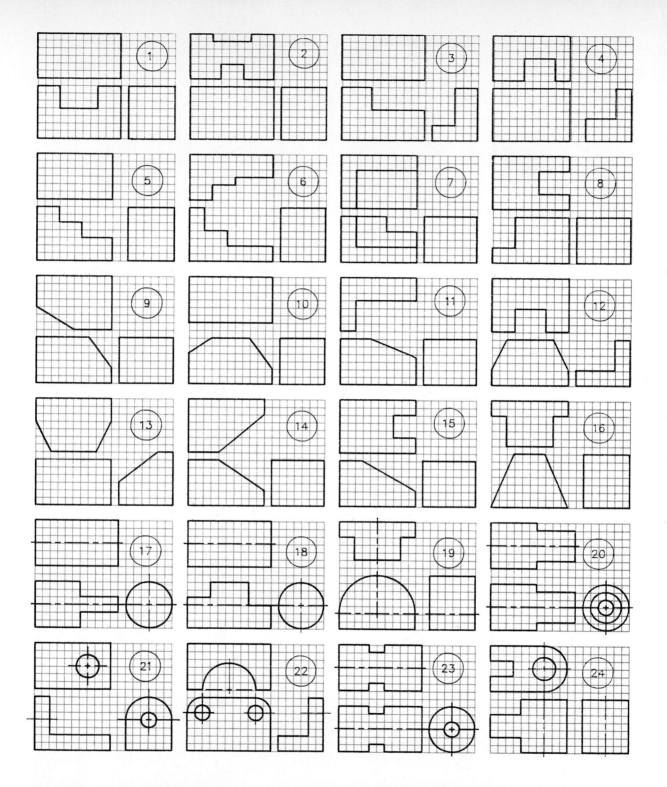

FIG. 14.28 On Size A paper sketch top, front, and right-side views of the problems assigned. Two problems can be drawn on each sheet. Give the overall dimensions of W, D, and H, and label each view.

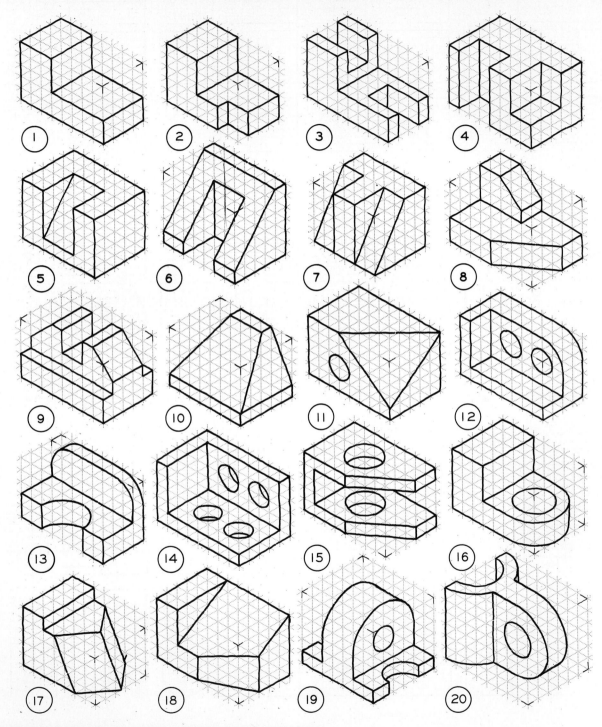

FIG. 14.29 Multiview problems and Isometric sketching.

On Size A paper, sketch the top, front, and right-side views of the problems assigned. Supply the lines that may be missing from all views. Then sketch isometric pictorials of the object assigned, two per sheet.

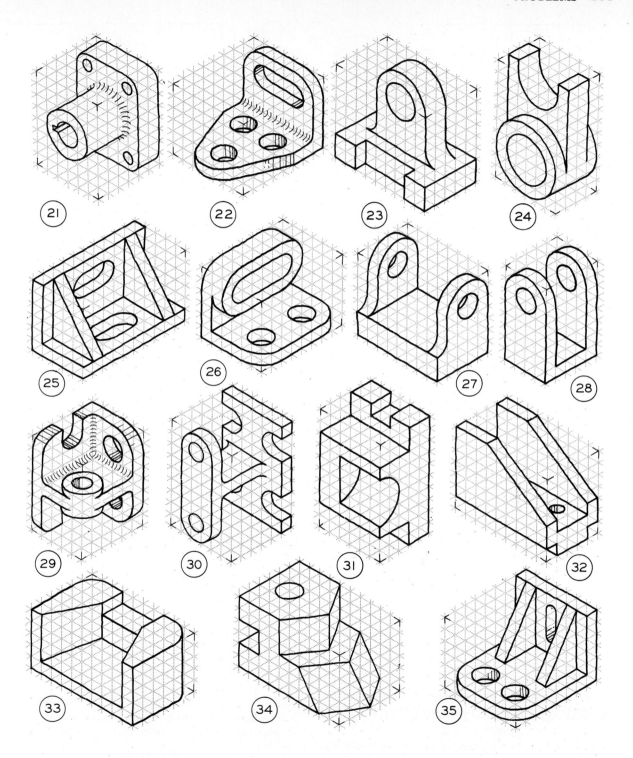

21 22 23 24
25 26 27 28
29 30 31 32
33 34 35

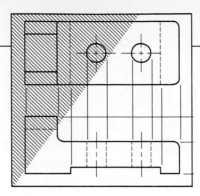

CHAPTER 15

Multiview Drawing With Instruments

15.1
Introduction

Multiview drawing, the system of representing three-dimensional objects by separate views arranged in a standard manner, is readily understood by the technical community. Because multiview drawings are usually executed with instruments and drafting aids (see Chapter 11), they are often referred to as *mechanical drawings*. They are called *working*, or *detail, drawings* when dimensions and notes are added to complete the specifications of the parts that have been drawn.

15.2
Orthographic projection

The artist is likely to represent objects impressionistically, but the drafter must represent them precisely. The method of preparing a precise, detailed, clearly understood drawing, is *orthographic projection* or *multiview drawing*.

By orthographic projection, the views of an object are projected perpendicularly onto projection planes with parallel projectors (Fig. 15.1).

The front view is projected onto a vertical frontal plane with parallel projectors. The resulting front view is two dimensional, since it has no depth and lies in a single plane described by two dimensions, width and height.

The top view of the same object is projected onto a horizontal projection plane perpendicular to the frontal projection plane in Fig. 15.2A. The right-side view is projected onto a vertical profile plane perpendicular to both the horizontal and frontal planes (Fig. 15.2B).

Imagine that the same object has been enclosed in a glass box showing the frontal, horizontal, and

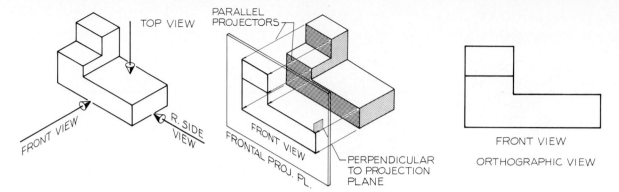

FIG. 15.1 Orthographic projection.

Step 1 Three mutually perpendicular lines of sight are drawn to obtain three views of the object.

Step 2 The frontal plane is a vertical plane on which the front view is projected with parallel projectors perpendicular to the frontal plane.

Step 3 The resulting view is the front view of the object. This is a two-dimensional orthographic view.

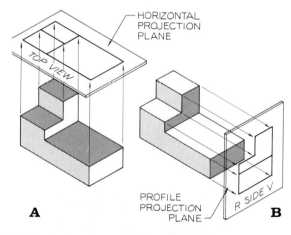

FIG. 15.2 The top view is projected onto a horizontal projection plane. The right-side view is projected onto a vertical profile plane perpendicular to the horizontal and frontal planes.

view over the front view, and the right-side view to the right of the front view.

The three principal projection planes used in orthographic projection are the *horizontal* (H), *frontal* (F), and *profile* (P) planes.

Any view projected onto one of these principal planes is called a *principal view*. The dimensions of an object used to show its three-dimensional form are *height (H)*, *width (W)*, and *depth (D)*.

15.3
Alphabet of lines

The use of proper line weights will greatly improve a drawing's readability and appearance. All lines should be drawn dark and dense as if drawn with ink. Only by their width should the lines vary—except that is, for guidelines and construction lines, which are drawn very lightly for lettering and laying out drawing.

The lines of an orthographic view are labeled, along with the suggested pencil grades for drawing them, in Fig. 15.5. The lengths of dashes in hidden lines and centerlines are drawn longer as the size of a

profile projection planes (Fig. 15.3). While in the glass box, the object's views are projected onto the projection planes. Then the box is opened into the plane of the drawing surface to give the standard positions for the three orthographic views.

A similar example of this principle is shown by the object in the projection box in Fig. 15.4. The three views are positioned in the same manner: the top

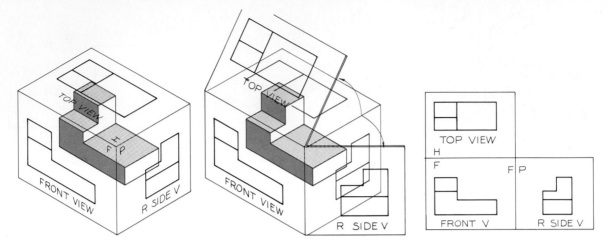

FIG. 15.3 Glass-box theory.

Step 1 Imagine that the object has been placed inside a box formed by the horizontal, frontal, and profile planes onto which the top, front, and right-side views have been projected.

Step 2 The three projection planes are then opened into the plane of the drawing surface.

Step 3 The three views are positioned with the top view over the front view and the right-side view to the right. The planes are labeled *H, F,* and *P* at the fold lines.

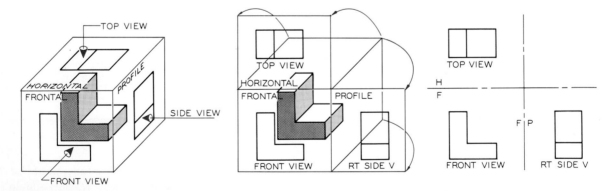

FIG. 15.4 Principal projection planes.

A. The three principal projection planes of orthographic projection can be thought of as planes of a glass box.

B. The views of an object are projected onto the projection planes, which are opened into the plane of the drawing surface.

C. The outlines of the planes are omitted. The fold lines are drawn and labeled.

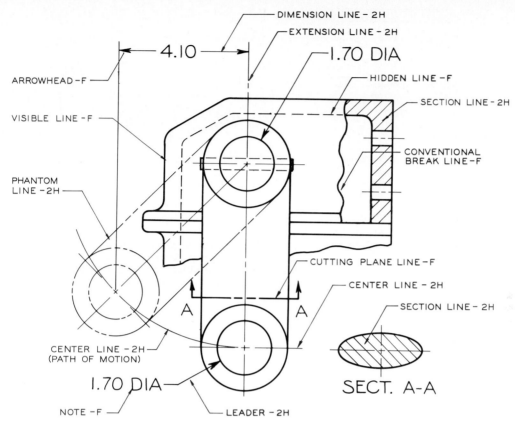

FIG. 15.5 The line weights and suggested pencil grades recommended for orthographic views.

drawing increases. Additional specifications for these lines are given in Fig. 15.6.

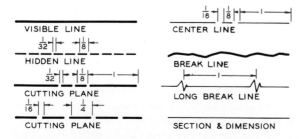

FIG. 15.6 A comparison of the line weights for orthographic views. These dimensions will vary for different sizes of drawings and should be approximated by eye.

COMPUTER LINES Smaller computer graphics plotters vary the widths of lines on a drawing by multiple-pen accessories that hold pens of different point widths, usually 0.7 mm and 0.3 mm wide. Different pen widths and/or colors are specified with the LINETYP command. AutoCAD provides a command, **LTSCALE,** that varies dash and space lengths in dashed lines (Fig. 15.7).

A comparison of lines drawn with **LTSCALE**s of 0.4 and 0.2 and pen widths of 0.7 mm and 0.3 mm is given. Once the drawing has been completed, the line scales are changed with the **LTSCALE** command by changing one number (from 0.4 to 0.2, in this example), and the views are automatically redrawn to show the revised lines (Fig. 15.8).

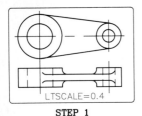

```
VISIBLE: 0.7 mm
HIDDEN: 0.3 mm
CENTER: 0.3 mm
CUTTING PLANE: 0.7 mm
        LTSCALE = 0.4
        STEP 1
```

```
VISIBLE: 0.7 mm
HIDDEN: 0.3 mm
CENTER: 0.3 mm
CUTTING PLANE: 0.7 mm
        LTSCALE: 0.2
        STEP 2
```

FIG. 15.7 Line weights by computer.

Step 1 These lines were drawn using AutoCAD's standard LINE TYPES and two pens. The LTSCALE factor that was used was 0.4, which affects the spacing between the dashes and the lengths of the dashes.

Step 2 By changing the LTSCALE factor from 0.4 to 0.2, the dashes and the spaces between them are reduced in size.

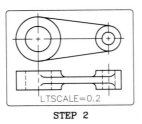

STEP 1   STEP 2

FIG. 15.8 Line weights by computer.

Step 1 This two-view drawing was made using an LTSCALE of 0.4, which resulted in the omission of dashes in centerlines and hidden lines with dashes that were too long.

Step 2 By activating the LTSCALE and giving it a factor of 0.2, the views are redrawn, and centerlines and hidden lines appear with the appropriate dashes.

15.4
Six-view drawings

If you visualize an object placed inside a glass box, you will see there are two horizontal planes, two frontal planes, and two profile planes (Fig. 15.9).

> Therefore the maximum number of principal views that can be used to represent an object is six.

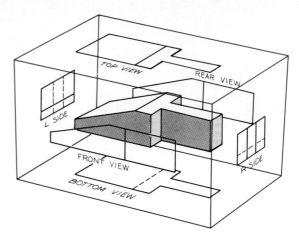

FIG. 15.9 Six principal views of an object can be drawn in orthographic projection. You can imagine that the object is in a glass box with the views projected onto the six planes.

The top and bottom views are projected onto horizontal planes, the front and rear views are projected onto frontal planes, and the right- and left-side views are projected onto profile planes.

To draw the views on a sheet of paper, imagine the glass box is opened up into the plane of the drawing paper (Fig. 15.10); the views will then appear as shown in Fig. 15.11. Note the order and arrangement of the six views, and you will see the logic behind orthographic projection. The top view is placed over, and the bottom view under, the front view; the right-side view is to the right of, and the left-side view to the left of the front view; and the rear view is to the left of the left-side view.

Height, width, and depth are three dimensions of an object necessary to give its size. The standard arrangement of the six views allows some of the views to share dimensions by projection. For example, the height dimension, which is shown only once between the front and right-side views, applies to the four horizontally arranged views. Furthermore, the width dimension is placed between the top and front views, but it also applies to the bottom view.

Projectors align the views both horizontally and vertically about the front view in Fig. 15.11. Each side of the fold lines of the glass box is labeled H, F, or P (horizontal, frontal, or profile) to identify the projection planes on a given side of the fold lines.

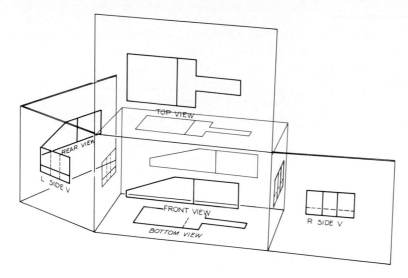

FIG. 15.10 The glass box is opened into the plane of the drawing surface, which locates the views in their standard positions.

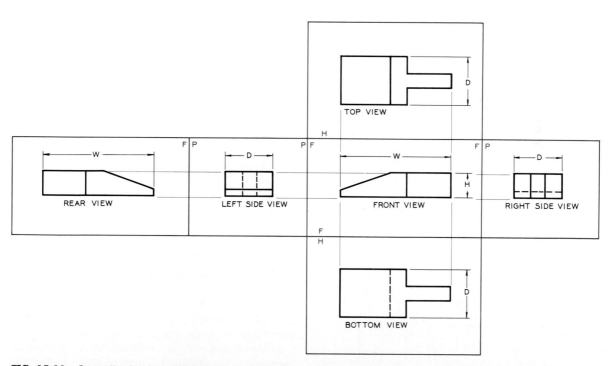

FIG. 15.11 Once the box is completely opened into a single plane, the six views are arranged to describe the object. The outlines of the planes are usually omitted. They are shown here to assist you in relating this figure to the previous one.

15.5
Three-view drawings

Because three views are usually adequate to describe an object, the most commonly used orthographic arrangement is the *three-view drawing*, consisting of front, top, and right-side views. The object used in the previous example is shown placed in a glass box in Fig. 15.12, which is opened onto the plane of the drawing surface. The resulting three-view arrangement is shown in Fig. 15.13, where the views are labeled and dimensioned.

COMPUTER VIEWS The three-view drawing of a part (Fig. 15.14) was drawn by using two pen widths, 0.7 mm and 0.3 mm, and by using the **DRAW** and **DIM**ension commands of AutoCAD, The same principles of orthographic projection were used in this computer-plotted drawing as in manually-drawn drawings.

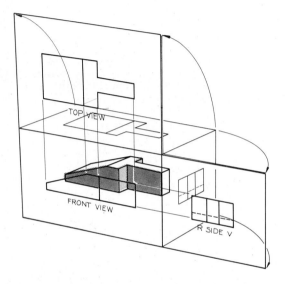

FIG. 15.12 Three-view drawings are commonly used for describing small machine parts. The glass box is used to illustrate how the views are projected onto their projection planes.

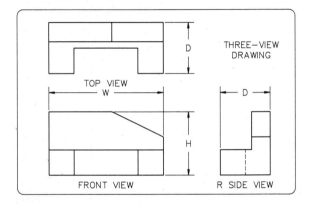

FIG. 15.14 This three-view orthographic drawing was made by AutoCAD and an A-B size plotter using two pen sizes (0.7 mm and 0.3 mm). The same principles of orthographic projection are used whether drawings are made manually or by computer.

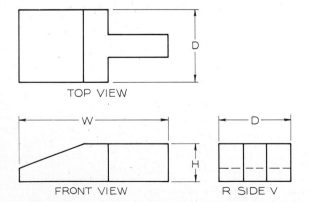

FIG. 15.13 The resulting three-view drawing of the object from the previous figure.

15.6
Arrangement of views

Figure 15.15A shows the standard positions for a three-view drawing: The top and side views are projected directly from the front view. The views are properly labeled and dimensioned. Views that are arranged in a nonstandard sequence (Fig 15.15B) and views that do not project from view to view (Fig. 15.15C) are both improperly drawn. These rules of arrangement are emphasized in Fig. 15.16.

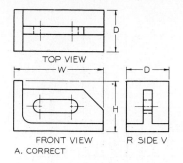 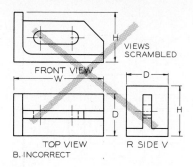 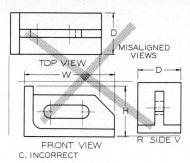

FIG. 15.15 Positioning orthographic views.

A. This is a correct arrangement of views, labels, and dimensions. The views project from each other in proper alignment.

B. These views are scrambled into unconventional positions, making it hard to interpret them. Dimensions are unnecessarily repeated.

C. These views are misaligned, so they do not project from one to the other. This is an incorrect arrangement.

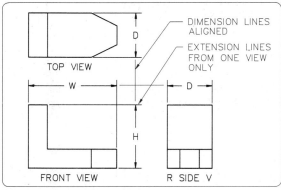

FIG. 15.16 Dimension and extension lines used in three-view orthographic projection should be aligned as shown in this computer drawing. Notice that extension lines are drawn from only one view when dimensions are placed between two views.

15.7
Selection of views

When drawing an object by orthographic projection, you should select the views with the fewest hidden lines. In Fig. 15.17A, the right-side view is preferred to the left-side view because it has fewer hidden lines.

Although the three-view arrangement of top, front, and right-side views is the most commonly used, the top, front, and left-side views is equally acceptable (Fig. 15.17B) if the left-side view has fewer hidden lines than the right-side view.

Some objects have standard views that are regarded as the front view, top view, and so forth. A chair for example, has front and top views that are recognized as such by everyone; therefore, a chair's accepted front view should be used as the orthographic front view.

15.8
Line techniques

As drawings become more complex, you will encounter more instances of lines overlapping and intersecting in ways similar to those shown in Fig. 15.18. This illustration shows the techniques of handling most types of intersecting lines. The methods of constructing hidden lines composed of straight lines and curved segments is shown in Fig. 15.19.

You should become familiar with the order of precedence (priority) of lines (Fig. 15.20): The most important line is the visible object line; it is shown regardless of any other line lying behind it. Of next importance is the hidden object line, which takes precedence over the centerline.

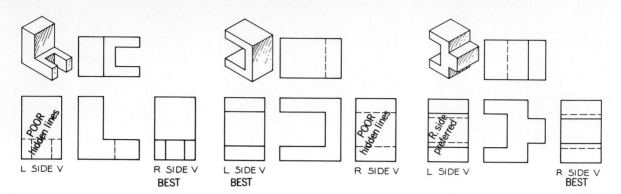

FIG. 15.17 Selection of views.

A. In orthographic projection, you should select the sequence of views with the fewest hidden lines. *

B. The left-side view has fewer hidden lines; therefore this view is selected over the right-side view.

C. When both views have an equal number of hidden lines, the right-side view is traditionally selected.

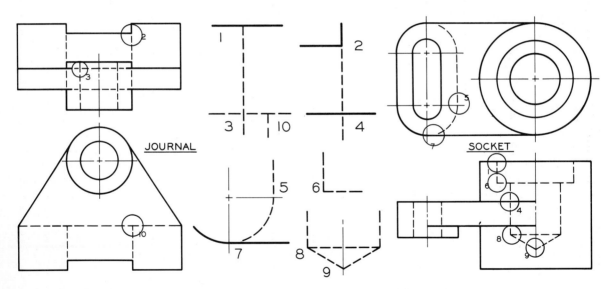

FIG. 15.18 When lines intersect in orthographic projection, they should intersect as shown here.

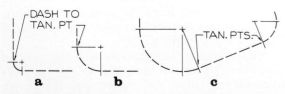

FIG. 15.19 Hidden lines in orthographic projection that are composed of curves should be drawn in this manner.

CENTERLINES BY COMPUTER A method of ensuring that centerline dashes cross properly when drawn by computer is shown in Fig. 15.21. Centerlines must be drawn in connected segments (Fig. 15.21B).

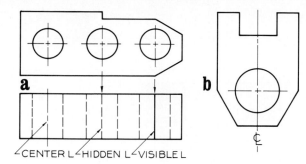

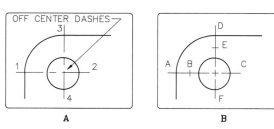

FIG. 15.20 The order of importance (precedence) of lines is visible lines, hidden lines, and centerlines. The symbol made of the letters *C* and *L* is used to label a centerline when needed on symmetrical parts.

FIG. 15.21 Centerlines by computer.

Step 1 When centerlines for a rounded corner are drawn from 1 to 2 and from 3 to 4, the center dashes do not cross at the center.

Step 2 Centerline dashes can be made to cross at the center by drawing a line from *A* to *B* to *C*, where the *BC* segment extends equally beyond the center on both sides. Line *DF* is drawn from *D* to *E* to *F* in the same manner.

15.9
Point numbering

The method of numbering points and lines of an object will be helpful to you in constructing orthographic views. For example, an object is shown in Fig. 15.22 that has been numbered to aid in the construction of the missing front view when the top and side views are given. By projecting selected points from the top and side views, the front view of the object can be found.

15.10
Line and planes

An orthographic view of a line can appear true length, foreshortened, or as a point (Fig. 15.23). When a line appears true length, it must be parallel to the reference line in the previous view. A plane in orthographic projection can appear true size, foreshortened, or as an edge (Fig. 15.23).

15.11
Alternate arrangement of views

Although the right-side view is usually placed to the right of the front view (Fig. 15.24), the side view can

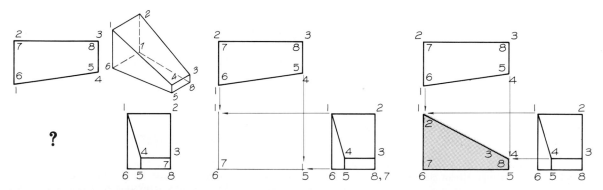

FIG. 15.22 Point numbering.

Step 1 When a missing orthographic view is to be drawn, it is helpful to number the points in the given views.

Step 2 Points 1, 5, 6, and 7 are found by projecting from the given views of these points.

Step 3 The plotted points are connected to form the missing front view.

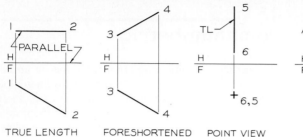

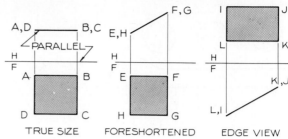

FIG. 15.23   Lines and planes.

A line will appear in orthographic projection as true length, foreshortened, or a point.

A plane in orthographic projection will appear as true size, foreshortened, or an edge.

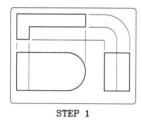

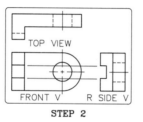

FIG. 15.24   Three views by computer.

Step 1   The three views are blocked in. Notice that the depth dimension in the top view is the same as the depth in the right-side view.

Step 2   The circular hole and its centerlines are drawn in all views. The notch is drawn in the side view and projected to the top and front views.

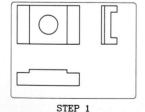

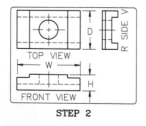

FIG. 15.25   Alternate-position side view.

Step 1   The right-side view of this part is projected from the top view to save space, since the depth dimension is large.

Step 2   The views are completed, and dimensions and labels are added.

be projected from the top view (Fig. 15.25); this is advisable when the object has a much larger depth than height.

15.12
Laying out three-view drawings

The depth dimension applies to both the top and side views, but these views are usually positioned where this dimension does not project between them (Fig. 15.26). The depth dimension can be graphically transferred by using a 45° line, an arc, or a pair of dividers. These and similar methods of transferring dimensions from view to view are illustrated in Fig. 15.27. Examples of three-view orthographic drawings of objects are shown in Figs. 15.28–15.32.

15.13
Two-view drawings

It is good economy of time and space to use only the views necessary to depict an object. Objects that require only two views are shown in Fig. 15.33. Cylindrical objects need only two views, as shown in Fig. 15.34. Because it is preferable to select the views with the fewest hidden lines, the right-side view is the better view in this example.

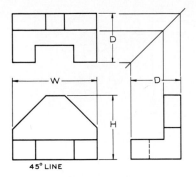
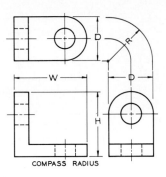
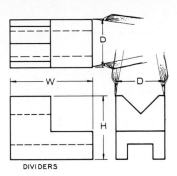

45° LINE   COMPASS RADIUS   DIVIDERS

FIG. 15.26 Transferring depth dimensions.

A. The depth dimension can be projected from the top view to the right-side view by constructing a 45° line positioned as shown.

B. The depth dimension can be projected from the top view to the side view by using a compass and a center point.

C. The depth dimension can be transferred from the top view to the side view by using dividers—the most desirable method.

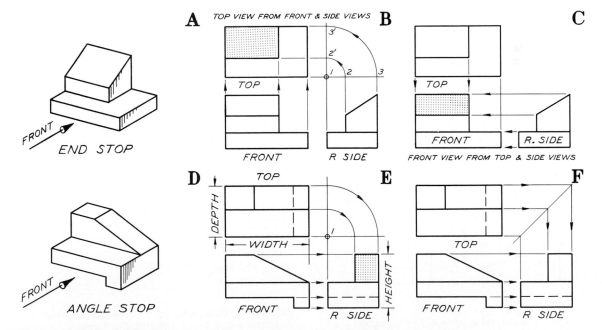

FIG. 15.27 Several methods of transferring dimensions from the views are shown here.

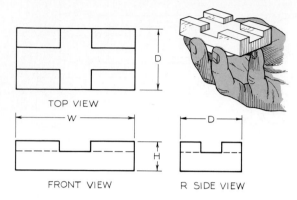

FIG. 15.28   A three-view drawing of an object.

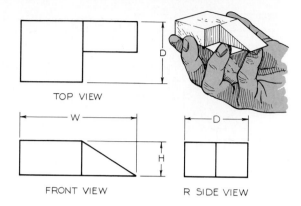

FIG. 15.29   A three-view drawing of an object.

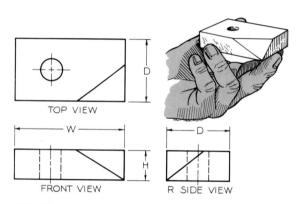

FIG. 15.30   A three-view drawing of an object.

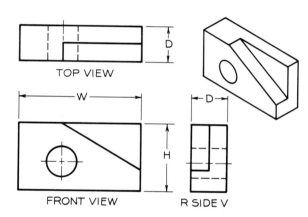

FIG. 15.31   A three-view drawing of an object.

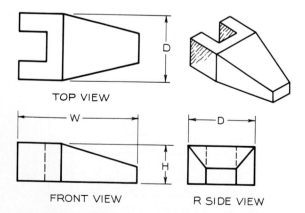

FIG. 15.32   A three-view drawing of an object.

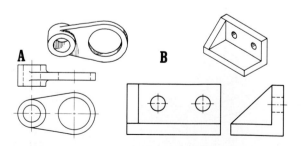

FIG. 15.33   These objects can be adequately described with two orthographic views.

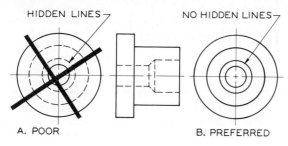

FIG. 15.34 Cylindrical objects can be depicted with two views. Always select views with the fewest hidden lines.

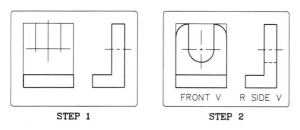

FIG. 15.35 A two-view drawing by computer.

Step 1 The front and side views are drawn using the overall dimensions.

Step 2 The FILLET command is used to draw the corners in the front view, and the ARC command is used to draw the semicircular arc.

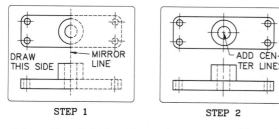

FIG. 15.36 Mirroring by computer.

Step 1 To save drawing time, the top and front views are drawn as half views, and MIRROed about the centerline.

Step 2 Centerlines that coincide with the MIRROR line should be drawn after the mirroring to prevent the centerline from being drawn twice.

TWO-VIEW DRAWINGS BY COMPUTER
The steps in making a two-view drawing of an object are shown in Fig. 15.35. The ARC and FILLET commands are used to draw the semicircular feature and the rounded corners in Step 2.

The MIRROR command can be used to reduce construction by drawing only half (or onequarter) of a view and then mirroring the drawing to give the other symmetrical half (Fig. 15.36). Centerlines should not be drawn along the mirror line before mirroring since this would cause them to be drawn twice.

The object in Fig. 15.37, which is composed of arcs and tangent lines, is constructed as a half top view, which is mirrored along line *AB*. Fillets are drawn in Step 2 with the FILLET command, and runouts are drawn in the front view with the ARC LINE, and CONTINUE commands.

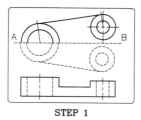 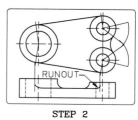

FIG. 15.37 Circular features by computer.

Step 1 The top view is drawn as a half view and MIRRORed about *AB*. Tangent points are found by using the PERPendicular option of the OSNAP command.

Step 2 Fillets are found with the FILLET command, and runouts are found with the LINE, CONTINUE, and ARC commands in the front view. The tangent arc is found in the top view using geometric construction and the ARC command.

15.14
One-view drawings

Cylindrical parts and those with a uniform thickness can be described in one view. In both cases, notes are used to explain the missing feature or dimension (Fig. 15.38).

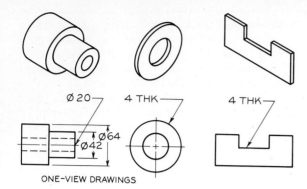

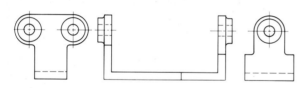

ONE-VIEW DRAWINGS

FIG. 15.38 Objects that are cylindrical or of uniform thickness can be described with only one orthographic view and supplementary notes.

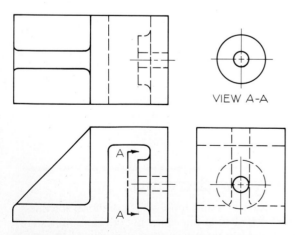

FIG. 15.39 Unnecessary and confusing hidden lines are omitted in the side views to improve their clarity.

VIEW A-A

FIG. 15.40 A removed view, indicated by the directional arrows, can be used to draw views in new hard-to-see locations.

15.15
Incomplete and removed views

The right- and left-side views of the part in Fig. 15.39 would be hard to interpret if all hidden lines were shown as specified by the rules of orthographic projection. Therefore it is best to omit lines that confuse a clear understanding of the views.

Often it is difficult to show a feature because of its location. Standard views can be confusing when lines overlap other features. The view indicated by the directional arrows in Fig. 15.40 is more clearly shown when removed to an isolated position.

15.16
Curve plotting

An irregular curve can be drawn by following the rules of orthographic projection, as shown in Fig. 15.41. Plotting begins by locating points along the curve in two given views. These points are projected to the top view where each point is located, and the points are then connected by a smooth curve. In Fig. 15.42 an ellipse is plotted in the top view by projecting from the front and side views. You will find it helpful to number points on curves that are being located by projection.

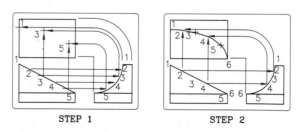

STEP 1   STEP 2

FIG. 15.41 Plotting curved lines by computer.

Step 1 A series of points is found in the front and side views by orthographic projection. Points 1, 3, and 5 are projected to the top view.

Step 2 Points 2 and 4 are projected to the top view, and a PLINE curve is drawn to FIT the points to complete the top view.

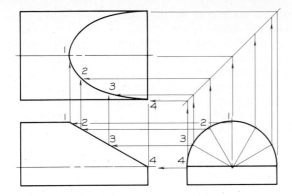

FIG. 15.42 The ellipse in the top view was found by numbering points in the front and side views and then projecting them to the top view.

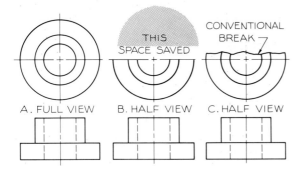

FIG. 15.43 To save space and drawing time, the top view of a cylindrical part can be drawn as a partial view using either of these methods.

15.17
Partial views

A partial view can be used to save time and space when the parts are symmetrical or cylindrical. By omitting the rear of the circular top view in Fig. 15.43, space can be saved without sacrificing clarity. Or, to make it more apparent that a portion of the view has been omitted, a break may be used.

15.18
Conventional revolutions

The readability of an orthographic view may be improved if the rules of projection are violated.

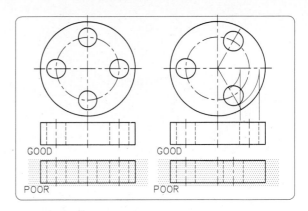

FIG. 15.44 Revolving holes.

Left: A true projection of equally spaced holes gives a misleading impression that the center hole passes through the center of the plate.

Right: A conventional view is used to show the true radial distances of the holes from the center by revolution. The third hole is omitted.

> Established violations of rules that are customarily made for the sake of clarity are called *conventional practices*.

When holes are symmetrically spaced in a circular plane, as shown in Fig. 15.44, it is conventional practice to show them at their true radial distance from the center of the plane. This requires an imagined revolution of the holes in the top view. This principle of revolution also applies to symmetrically positioned features, such as the three lugs on the outside of the part in Fig. 15.45. The conventional view is better than the true orthographic projection. The conventional and

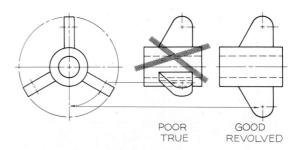

FIG. 15.45 Symmetrically positioned external features, such as these lugs, are revolved to their true-size positions for the best views.

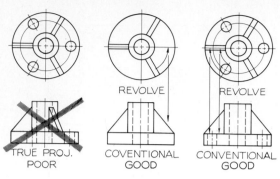

FIG. 15.46 The conventional methods of revolving holes and ribs in combination for improved clarity.

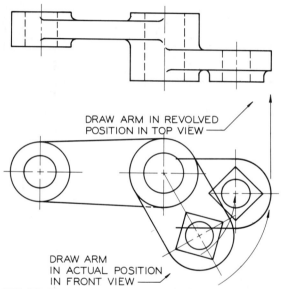

FIG. 15.47 The arm in the front view is imagined to be revolved so its true length can be drawn in the top view. This is an accepted conventional practice.

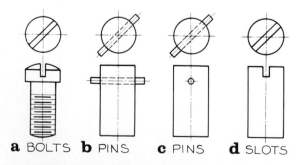

FIG. 15.48 Parts like these are drawn at 45° angles in the top views, but the front views are drawn to show the details as revolved views.

desired method of drawing holes and ribs in combination is shown in Fig. 15.46.

Another conventional practice is illustrated in Fig. 15.47, where an inclined feature is revolved to a horizontal position in the front, so it can be drawn as true size in the top view. The revolution of the part is not drawn since it is an imagined revolution.

Other parts whose views are improved by revolution are those shown in Fig. 15.48. It is desirable to show the top view features at forty five degrees, so they will not coincide with the centerlines. The front views are drawn by imagining the features have been revolved. A closely related type of conventional view is the true-size developed view where a bent piece of material is drawn as if it were flattened out (Fig. 15.49).

15.19

Intersections

In orthographic projection, the intersection between planes results in a line that describes the object. In Fig. 15.50, examples of views are shown where lines may or may not be required.

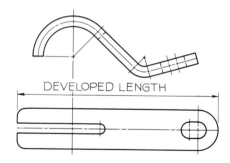

FIG. 15.49 Objects that have been shaped by bending thin stock can be shown as true-size developed views.

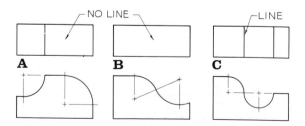

FIG. 15.50 Object lines are drawn only where there are sharp intersections or where arcs are tangent at their centerlines, as in part C.

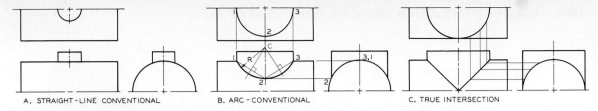

FIG. 15.51 The conventional methods of showing intersections between cylinders. These are approximations except for part C.

Figure 15.51 shows the standard types of intersections between cylinders. Figures. 15.51A and B are conventional intersections, which means they are approximations drawn for ease of construction. Figure 15.51C shows a true intersection between cylinders of equal diameters. Similar intersections are shown in Figs. 15.52 and 15.53. The types of conventional intersections formed by holes in cylinders are shown in Fig. 15.54.

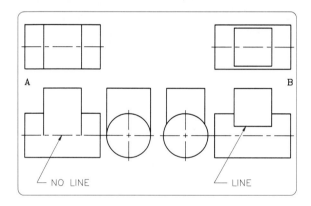

FIG. 15.52 Conventional intersections between prisms and cylindrical shapes that are drawn by computer.

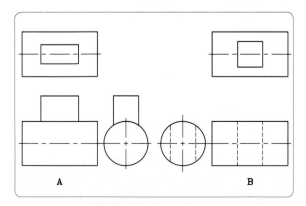

FIG. 15.53 Conventional intersections between prisms and cylindrical shapes.

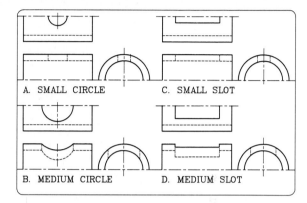

A. SMALL CIRCLE   C. SMALL SLOT
B. MEDIUM CIRCLE   D. MEDIUM SLOT

FIG. 15.54 Conventional intersections between cylinders and holes piercing them.

15.20
Fillets and rounds

Fillets and *rounds* are rounded corners used on castings, such as the body of the Collet Index Fixture shown in Fig. 15.55. A fillet is an inside rounding, and a round is an external rounding. The radii of fillets and rounds may be many sizes, but they are usually about one-quarter inch. Fillets and rounds are used on castings for added strength and improved appearance. A casting will have square corners only when its surface has been finished, which is the process of machining away a portion of the surface to a smooth finish (Fig. 15.56B).

Finished surfaces are indicated by placing a **finish mark** (V) on the edge views of the finished surfaces, whether the edges are visible or hidden. Alternative finish marks are shown in Fig. 15.57.

FIG. 15.55 The edges of this Collet Index Fixture are rounded to form fillets and rounds. Note that the surface of the casting is rough except where it has been machined. (Courtesy of Hardinge Brothers Inc.)

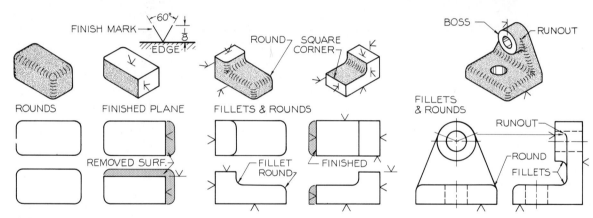

FIG. 15.56 Fillets and rounds.

A. When a surface has been finished by machining, rounds are removed and the corners are squared. A finish mark is indicated by a V placed on the edge view of the finished surfaces.

B. A fillet is a rounded inside corner. The rounds are removed when the outside surfaces are finished. The fillets can be seen only in the front view.

C. The views of an object with fillets and rounds must be drawn in a way that calls attention to them.

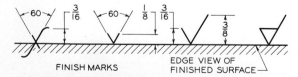

FIG. 15.57 Alternative finish marks which are applied to the edge views of the finished surface, whether hidden or visible.

Note (in Fig. 15.56C) that a *boss* is a raised cylindrical feature that is thickened to receive a shaft or to be threaded, and that the curve formed by a fillet at a point of tangency is called a *runout*. The techniques of showing fillets and rounds on orthographic views are given in Fig. 15.58.

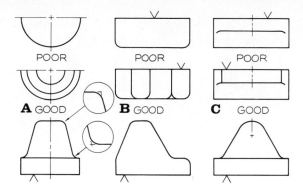

FIG. 15.58 Examples of conventionally drawn fillets and rounds.

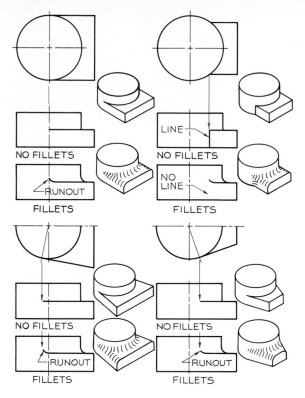

A comparison of intersections and runouts of parts with and without fillets and rounds is shown in Fig. 15.59. Large runouts are constructed as an eighth of a circle with a compass, as illustrated in Fig. 15.60. Small runouts can be drawn with a circle template. Runouts on orthographic views will reveal much about the details of an object. For example, the runout

FIG. 15.59 Intersections between features of objects. Intersections with fillets have runouts.

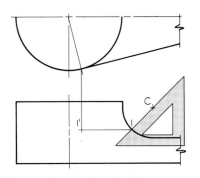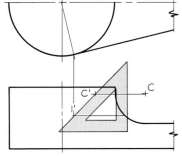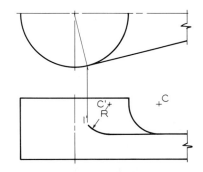

FIG. 15.60 Plotting runouts.

Step 1 Find the point of tangency in the top view and project it to the front view. A 45° triangle is used to find point 1, which is projected to point 1′.

Step 2 A 45° triangle is used to locate point C′, which is on the horizontal projector from the center of the fillet, C.

Step 3 The radius of the fillet is used to draw the runout with C′ as the center. The runout arc is equal to one eighth of a circle.

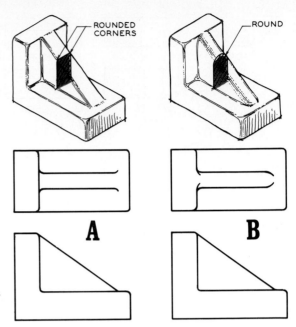

FIG. 15.61 Runouts are shown for differently shaped ribs. Part A has fillets, and part B has rounded edges.

in the top view of Fig. 15.61A tells us that the rib has rounded corners, whereas the top view of Fig. 15.61B tells us the rib is completely round. Methods of showing other types of intersections are shown in Figs. 15.62 and 15.63.

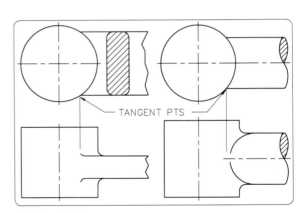

FIG. 15.63 Conventional runouts of different cross-sectional shapes.

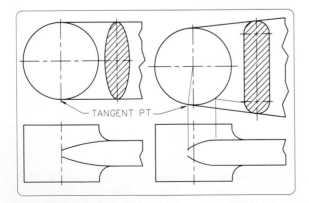

FIG. 15.62 Conventional runouts of different cross-sectional shapes.

COMPUTER METHOD A drawing of a part with runouts at its tangent points is shown in Fig. 15.64. The runouts are plotted with the **ARC** command after the tangent points have been projected from the right-side view.

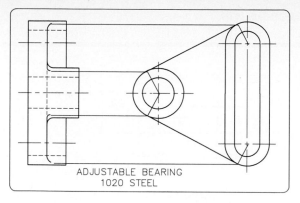

FIG. 15.64 A typical application of runouts on a part with fillets and rounds.

15.21
Left-hand and right-hand views

Two parts are often required that are "mirror images" of each other (Fig. 15.65). The drafter can reduce drawing time by drawing views of only one of the parts and labeling these views. A note can be added to indicate that the other matching part has the same dimensions.

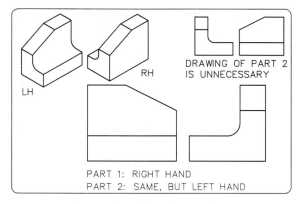

FIG. 15.65 When left- and right-hand mirror parts are needed, only one view is drawn and labeled. The other view need not be drawn but should be indicated by a note.

15.22
First-angle projection

The examples in this chapter are presented as third-angle projections, where the top view is placed over the front view, and the right-side view to the right of the front view. This method is used extensively in America, Britain, and Canada. Most of the rest of the industrial world uses *first-angle* projection.

The first-angle system is illustrated in Fig. 15.66, where an object is placed above the horizontal plane and in front of the frontal plane. When these projection planes are opened onto the surface of the drawing paper, the front view is projected over the top view, and the left-side view is placed to the right of the front view.

It is important that the angle of projection be indicated on a drawing to aid in the interpretation of the views. This is done by placing a truncated cone in or near the title block (Fig. 15.67). When metric units of measurement are used, the cone and the SI symbol are placed together on the drawing.

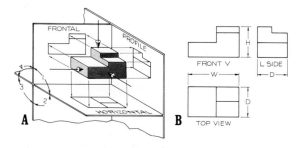

FIG. 15.66 First-angle projection.

A. The first angle of projection is used by many of the countries that use the metric system. You imagine that the object is placed above the horizontal and in front of the frontal plane.

B. The views are drawn in this location, which is different from the third-angle of projection that is used in the United States.

FIG. 15.67 The angle of projection used to prepare a set of drawings is indicated by a truncated cone, which is placed in or near the title block of a drawing.

Problems

The following problems are to be drawn as orthographic views on Size A or Size B paper, as assigned by your instructor.

1–11. (Figs. 15.68–15.78) Draw the given views using the dimensions provided, and then construct the missing top, front, or right-side views. Use Size A sheets, and draw one or two problems per sheet.

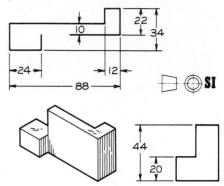

FIG. 15.70   Problem 3: Adjustable stop.

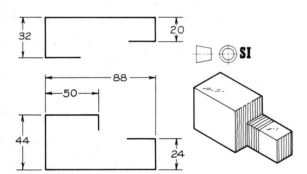

FIG. 15.68   Problem 1: Guide block.

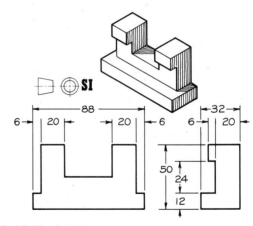

FIG. 15.71   Problem 4: Lock catch.

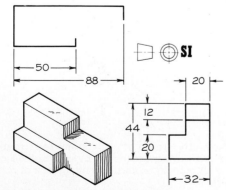

FIG. 15.69   Problem 2: Double step.

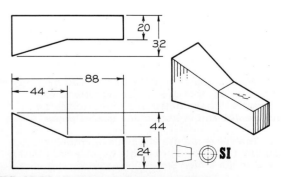

FIG. 15.72   Problem 5: Two-way adjuster.

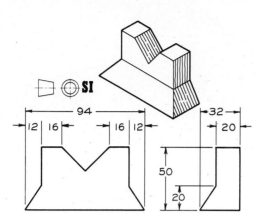

FIG. 15.73 Problem 6: Vee block.

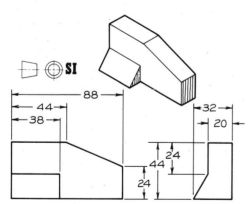

FIG. 15.74 Problem 7: Filler.

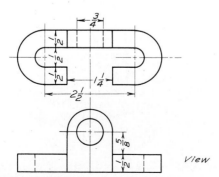

View

FIG. 15.75 Problem 8: Slide stop.

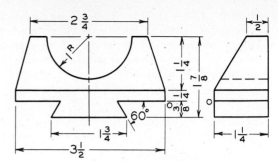

FIG. 15.76 Problem 9: Shaft support.

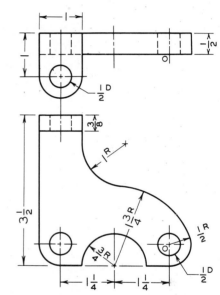

FIG. 15.77 Problem 10: Support brace.

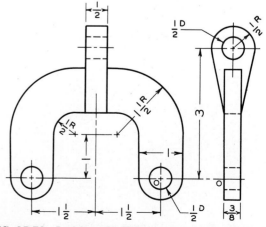

FIG. 15.78 Problem 11: Yoke.

12–48. (Figs. 15.79–15.115) Construct the necessary orthographic views to describe the objects in these figures. Draw the views on Size A or Size B sheets. Label the views and show the overall dimensions of *W*, *D*, and *H*.

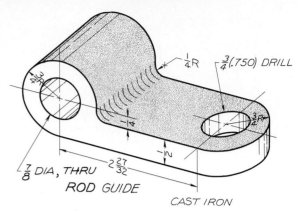

ROD GUIDE

CAST IRON

FIG. 15.81 Problem 14: Rod guide.

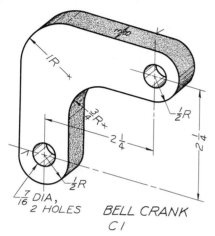

BELL CRANK
C1

FIG. 15.79 Problem 12: Bell crank.

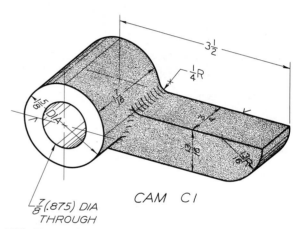

CAM C1

FIG. 15.82 Problem 15: Cam.

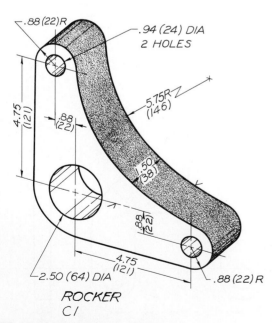

ROCKER
C1

FIG. 15.80 Problem 13: Rocker.

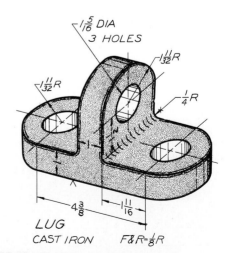

LUG
CAST IRON   F&R=⅛R

FIG. 15.83 Problem 16: Lug.

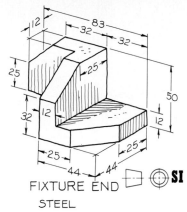

FIXTURE END ▢ ⊕ SI
STEEL

FIG. 15.84 Problem 17: Fixture end.

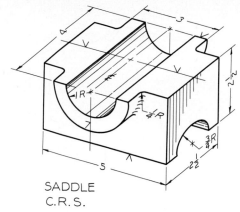

SADDLE
C.R.S.

FIG. 15.87 Problem 20: Saddle.

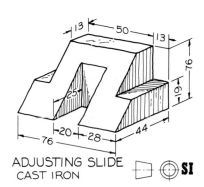

ADJUSTING SLIDE ▢ ⊕ SI
CAST IRON

FIG. 15.85 Problem 18: Adjusting slide.

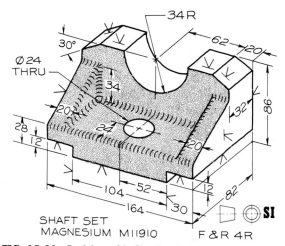

SHAFT SET
MAGNESIUM M11910   F & R 4R

FIG. 15.88 Problem 21: Shaft set.

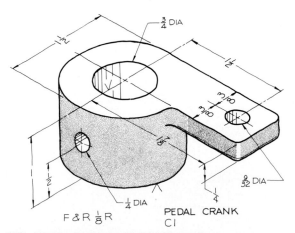

F & R ⅛ R   PEDAL CRANK
                    CI

FIG. 15.86 Problem 19: Pedal crank.

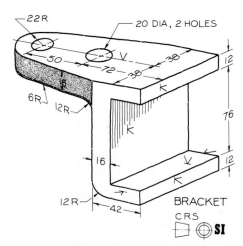

BRACKET
C R S
▢ ⊕ SI

FIG. 15.89 Problem 22: Bracket.

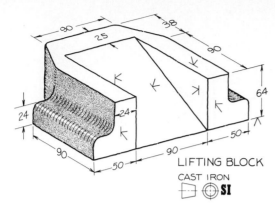

FIG. 15.90   Problem 23: Lifting block.

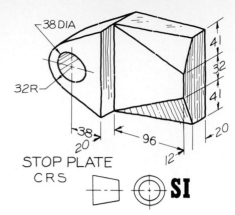

FIG. 15.93   Problem 26: Stop plate.

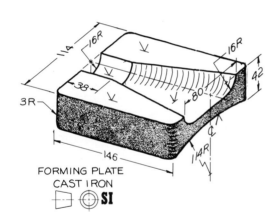

FIG. 15.91   Problem 24: Forming plate.

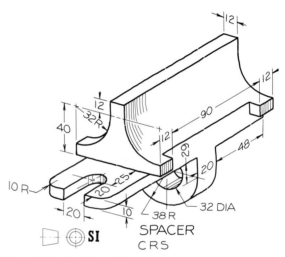

FIG. 15.94   Problem 27: Spacer.

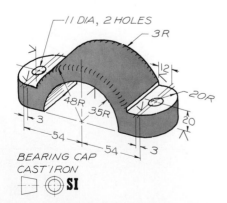

FIG. 15.92   Problem 25: Bearing cap.

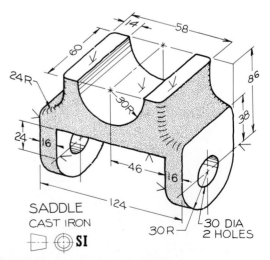

FIG. 15.95   Problem 28: Saddle.

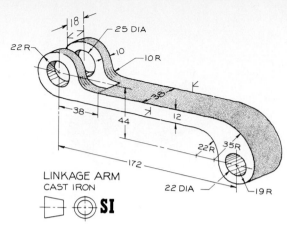

LINKAGE ARM
CAST IRON
SI

FIG. 15.96   Problem 29: Linkage arm.

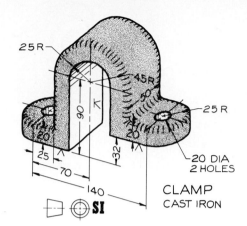

CLAMP
CAST IRON

FIG. 15.99   Problem 32: Clamp.

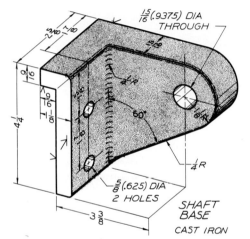

SHAFT
BASE
CAST IRON

FIG. 15.97   Problem 30: Shaft base.

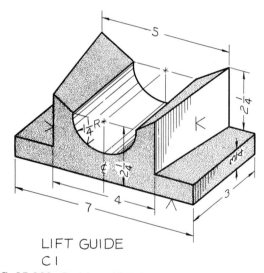

LIFT GUIDE
C I

FIG. 15.100   Problem 33: Lift guide.

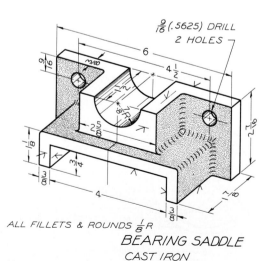

ALL FILLETS & ROUNDS $\frac{1}{8}$R

BEARING SADDLE
CAST IRON

FIG. 15.98   Problem 31: Bearing saddle.

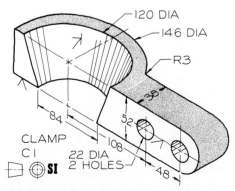

CLAMP
C I
SI

FIG. 15.101   Problem 34: Clamp.

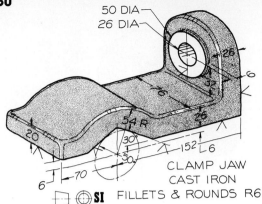

FIG. 15.102 Problem 35: Clamp jaw.

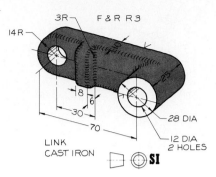

FIG. 15.105 Problem 38: Link.

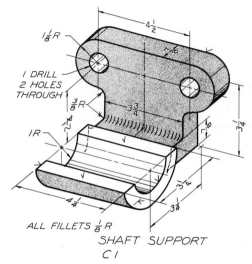

FIG. 15.103 Problem 36: Shaft support.

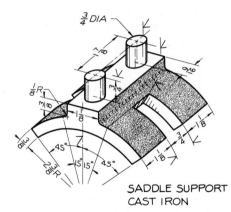

FIG. 15.106 Problem 39: Saddle support.

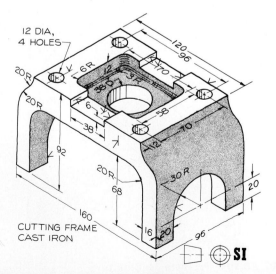

FIG. 15.104 Problem 37: Cutting frame

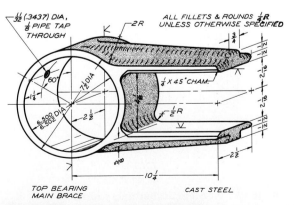

FIG. 15.107 Problem 40: Top bearing.

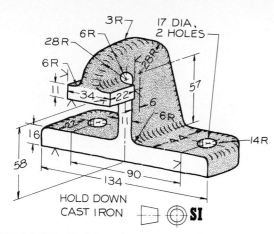

FIG. 15.108   Problem 41: Hold down.

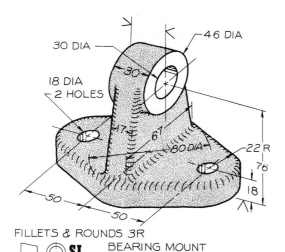

FIG. 15.109   Problem 42: Bearing mount.

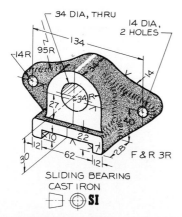

FIG. 15.110   Problem 43: Sliding bearing.

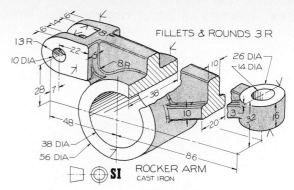

FIG. 15.111   Problem 44: Rocker arm.

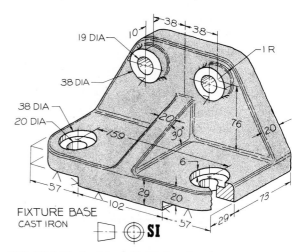

FIG. 15.112   Problem 45: Fixture base.

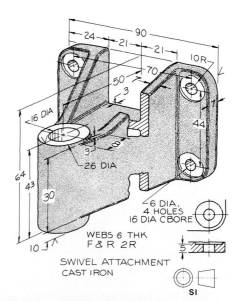

FIG. 15.113   Problem 46: Swivel attachment.

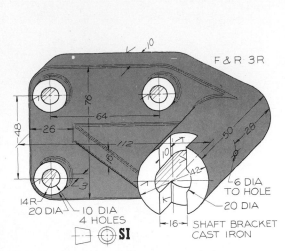

FIG. 15.114 Problem 47: Shaft bracket.

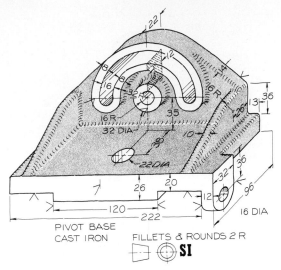

FIG. 15.115 Problem 48: Pivot base.

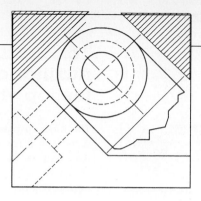

CHAPTER 16

Auxiliary Views

16.1
Introduction

A plane not parallel to one of the principal projection planes is a nonprincipal plane that will *not* appear true size in a principal view; it can be found true size only on an *auxiliary plane* parallel to the nonprincipal plane. This nonprincipal view is called an *auxiliary view*.

An inclined surface of an object (Fig. 16.1) does not appear true size in the top view because it is not parallel to the horizontal projection plane. However, the inclined surface will appear true size if an auxiliary view is projected perpendicularly from the edge view of the plane in the front view.

16.2
Folding-line approach

The three principal orthographic planes are the *frontal* (F), *horizontal* (H), and *profile* (P) planes.

A primary auxiliary plane is perpendicular to one of the principal planes but oblique to the other two, and a *primary auxiliary view* is projected from a *primary orthographic view*.

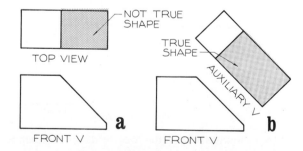

FIG. 16.1 When a surface appears as an inclined edge in a principal view, it can be found true size by an auxiliary view. In part a, the top view is foreshortened, but this plane is true size in the auxiliary view in part b.

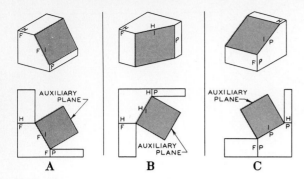

FIG. 16.2 A primary auxiliary plane can be folded from the frontal, horizontal, or profile planes. The fold lines are labeled F-1, H-1, and P-1.

Auxiliary planes can be thought of as planes that fold from principal planes, as shown in Fig. 16.2. The plane at Fig. 16.2A folds from the frontal plane to make a ninety degree angle with it. The fold line between the two planes is labeled F-1, where F is an abbreviation for frontal, and 1 represents *first*, or *primary*, auxiliary plane. Figures 16.2B and C illustrate the positions for auxiliary planes that fold from the horizontal and profile planes.

16.3
Auxiliaries projected from the top view

The inclined plane in Fig. 16.3 is an edge in the top view, and it is perpendicular to the horizontal plane. An auxiliary plane can be drawn parallel to the inclined surface of the object, and the view projected onto it will be a true-size view of the inclined plane.

> A surface must appear as an edge in a principal view before it can be found as true size in a primary auxiliary view.

Fold line H-1 is drawn parallel to the edge view of the inclined plane in Step 1. The line of sight is drawn perpendicular to the edge view. Each corner of the inclined plane is projected perpendicularly to the auxiliary plane and is located by using the dimension of height *(H)* from the side or front view.

A similar example is the object shown in the glass box in Fig. 16.4. Since this object has an inclined surface that appears as an edge in the top view, it can be found true size in a primary auxiliary view. The height *(H)* is transferred from the front view to the auxiliary view, since both views are measured from the same

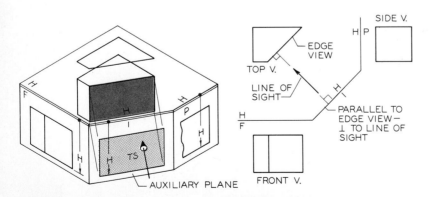

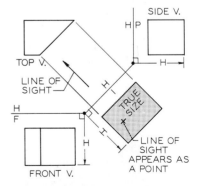

FIG. 16.3 Auxiliary from the top—folding-line method.

Given An object with an inclined surface.
Required Find the inclined surface true size by an auxiliary view.

Step 1 Construct a line of sight perpendicular to the edge view of the inclined surface. Draw the H-1 fold line parallel to the edge, and draw the H-F fold line between the top and front views.

Step 2 Project the four corners of the edge parallel to the line of sight. Locate the corners by measuring perpendicularly from the horizontal plane with height *(H)* dimensions.

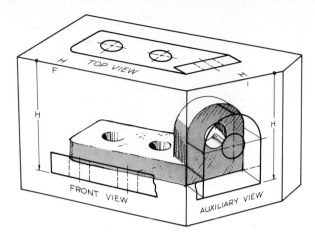

FIG. 16.4 A pictorial showing the relationship of the auxiliary projection plane is used to find the true-size view of the inclined surface.

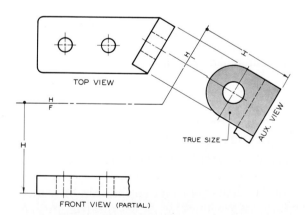

FIG. 16.5 The auxiliary plane is opened into the plane of the top view by revolving it about the *H*-1 fold line.

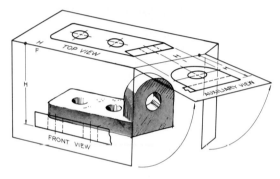

FIG. 16.6 When the object is drawn on a sheet of paper, it is laid out in this manner. The front view is drawn as a partial view since the omitted part is shown true size in the auxiliary view.

horizontal plane. The auxiliary plane is rotated about the H-1 fold line into the plane of the top view in Fig. 16.5.

When drawn on a sheet of paper, the drawing of this object would appear as shown in Fig. 16.6. The front view is shown as a partial view because the omitted portion would have been hard to draw, and it would not have been true size. The auxiliary view shows the true-size view of the inclined surface.

16.4
Auxiliaries from the top view—folding-line method

The steps of constructing an auxiliary view projected from the top view are shown in Fig. 16.7. The purpose of the auxiliary view is to find the true-size view of the inclined surface. Since the inclined surface projects as an edge in the top view, it can be found true size in a primary auxiliary view. The line of sight is drawn perpendicular to the edge view, and the fold line is drawn parallel to the edge. Height *(H)* is transferred from the front view.

The fold line, drawn thin but black, is labeled H-1. It is also helpful to number or letter points on the views. The projectors are construction lines, and they should be drawn with a hard pencil (3H–4H) just dark enough to be seen.

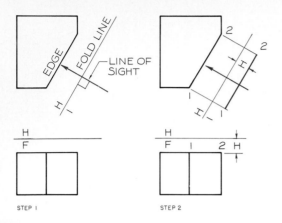

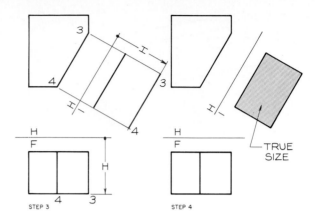

FIG. 16.7 Construction of an auxiliary view.

Step 1   The line of sight is drawn perpendicular to the edge view of the inclined surface. The H-1 fold line is drawn parallel to the edge view of the inclined surface. An H-F fold line is drawn between the given views.

Step 2   Points 1 and 2 are found by transferring the height *(H)* dimensions from the front view to the auxiliary view.

Step 3   Points 3 and 4 are found in the same manner using the dimensions of height *(H)*.

Step 4   The corner points are connected to complete the true-size view of the inclined plane.

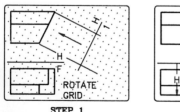

FIG. 16.8 Auxiliary view by computer.

Step 1   To perpendicularly project an auxiliary from the edge of the inclined surface in the top view, the grid is rotated, using the SNAP and ROTATE commands. The H-1 reference line is drawn parallel to the edge, and projectors are drawn perpendicular to the H-1 line.

Step 2   The auxiliary view is found by transferring height dimensions measured in the front view and transferring them perpendicularly from the H-1 line. The auxiliary view is the true-size view of the inclined surface.

> **COMPUTER METHOD**   Using computer graphics, the true-size view of a plane is found (Fig. 16.8) by projecting from the top view, where the inclined plane appears as an edge. With AutoCAD's **SNAP** command and **ROTATE** command, the grid is rotated so that it is parallel to the edge view of the plane in the top view (Step 1). The true-size auxiliary view is found by projecting perpendicularly from the top view of the plane and locating the auxiliary view with height dimensions transferred from the front view.

16.5 _____

Auxiliaries from the top view—reference-line method

A similar method of locating an auxiliary view uses a reference plane instead of the fold line (Fig. 16.9). Instead of placing a fold line between the top and front views, a reference plane is passed through the bottom of the front view. The height *(H)* dimensions are mea-

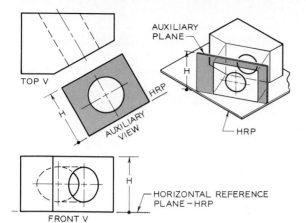

TOP V

AUXILIARY PLANE

HRP

H

AUXILIARY VIEW

HRP

HRP

H

FRONT V

H

HORIZONTAL REFERENCE PLANE—HRP

FIG. 16.9 A horizontal reference plane can be used instead of the folding-line technique to construct an auxiliary view. Instead of placing the reference plane between the top and auxiliary views, it is placed outside the auxiliary view.

sured upward from the reference plane instead of downward from a fold line.

The reference plane shown as a horizontal edge in the front view is called a *horizontal reference plane* and is labeled HRP. The HRP will appear as an edge

view of the inclined surface from which the auxiliary is projected. The auxiliary view will lie between the HRP and the top view.

16.6
Auxiliaries from the front view—folding-line method

A plane that appears as an edge in the front view (Fig. 16.10) can be found true size in a primary auxiliary view projected from the front view. Fold line F-1 is drawn parallel to the edge view of the inclined plane in the front view at a convenient location.

The line of sight is drawn perpendicular to the edge view of the inclined plane in the front view. Observed from this direction, the frontal plane appears as an edge; therefore the measurements perpendicular to the frontal plane—the depth *(D)* dimensions—will be seen true length. Depth dimensions are transferred from the top view to the auxiliary view by using dividers.

An application of this type of auxiliary view is shown in Fig. 16.11. The object is enclosed in a glass box, and an auxiliary plane is constructed parallel to

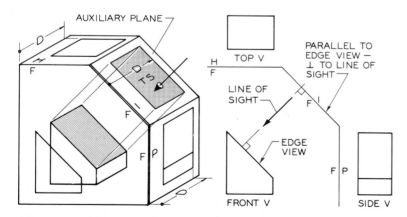

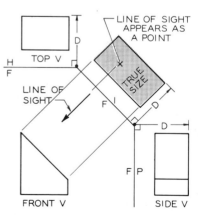

FIG. 16.10 Auxiliary from the front—folding-line method.

Given An object with an inclined surface.

Required Find the inclined surface true size by an auxiliary view.

Step 1 Draw a line of sight perpendicular to the edge view of the inclined surface. Draw the F-1 fold line parallel to the edge, and draw the H-F or F-P fold lines between the given views.

Step 2 Project the corners of the edge view parallel to the line of sight. Locate the corners of the true-size view by measuring perpendicularly from the frontal plane with depth *(D)* dimensions.

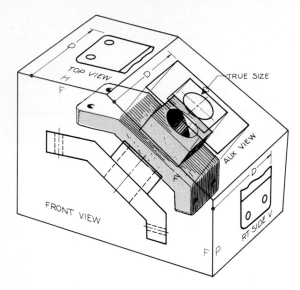

FIG. 16.11 A pictorial showing the relationship of the projection planes used to find the true-size view of the inclined surface of the object.

the inclined plane. When drawn on a sheet of paper, the object appears as shown in Fig. 16.12. The top and side views are drawn as partial views, since the auxiliary view eliminates the need for seeing their complete views. The auxiliary view, which is located by using the depth dimension measured from the edge view of the frontal projection plane, shows the surface's true size.

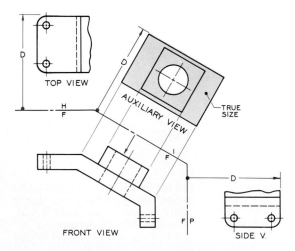

FIG. 16.12 The layout and construction of an auxiliary view of the object shown in Fig. 16.11.

COMPUTER METHOD An inclined plane that appears as an edge in the front view (Fig. 16.13) is found true size by computer graphics. **SNAP** and **ROTATE** commands are used to rotate the grid parallel to the edge view of the plane in the front view (Step 1). In Step 2, the auxiliary plane is projected from the edge view of the plane by transferring height dimensions from the top view with your dividers.

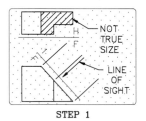 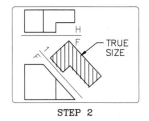

STEP 1   STEP 2

FIG. 16.13 Auxiliary view by computer.

Step 1 By using the SNAP and ROTATE commands, the grid is rotated so it is parallel to the edge view of the inclined plane in the front view. The F-1 reference line is drawn, and projectors are drawn from the edge perpendicular to the F-1 line.

Step 2 The true-size view of the inclined plane is found in the auxiliary view by transferring the depth dimension from the top view to the auxiliary view.

16.7
Auxiliaries from the front view—reference-plane method

The object in Fig. 16.14 has an inclined surface that appears as an edge in the front view; therefore this plane can be found true size in a primary auxiliary view.

Since it is symmetrical, it is advantageous to use a reference plane that passes through the center of the symmetrical top view. Because the reference plane is a frontal plane, it is called a *frontal reference plane* (FRP) in the auxiliary view. The FRP is located parallel

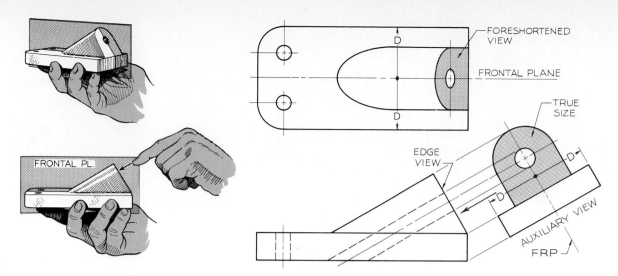

FIG. 16.14 Since the inclined surface of this part is symmetrical, it is advantageous to use a frontal reference plane (FRP) that passes through the object. The auxiliary view is projected perpendicularly from the edge view of the plane in the front view. The FRP appears as an edge in the auxiliary view, and depth (D) dimensions are made on each side of it to locate points on the true-size view of the inclined surface.

to the edge view of the inclined plane and through the center of the auxiliary view.

16.8
Auxiliaries from the profile view—folding-line method

Since the inclined surface appears as an edge in the profile plane, it can be found true size in a primary auxiliary view projected from the profile view (Fig. 16.15). The auxiliary fold line, P-1, is drawn parallel to the edge view of the inclined surface.

A line of sight perpendicular to the auxiliary plane will see the profile plane as an edge. Therefore width (W) dimensions, which are transferred from the front view to the auxiliary view, will appear true length in the auxiliary view.

16.9
Auxiliaries from the profile—reference-plane method

The object in Fig. 16.16 has two inclined surfaces that appear as edges in the right-side view, the profile view. These inclined surfaces are found true size by

using a *profile reference plane* (PRP) that is a vertical edge in the front view.

To find the inclined surface's true size, an auxiliary view is drawn. The profile reference plane is positioned at the far outside of the auxiliary view instead of between the profile and auxiliary views, as in the folding-line method. By transferring the width (W) dimensions from the edge view of the PRP in the front view to the auxiliary view, the views are found.

16.10
Auxiliaries of curved shapes

When an auxiliary view is drawn to show a curve that is not a true arc, a series of points must be plotted. The cylinder in Fig. 16.17 has a beveled surface that appears as an edge in the front view. This true-size surface is elliptical in shape.

Since the cylinder is symmetrical, it is beneficial to use an FRP through the object so that dimensions can be measured on both sides of it. Points are located about the circular right-side view, and these are projected to the edge view of the surface in the front view. The FRP is located parallel to the edge view of the plane, and the points are projected perpendicularly from the edge view of the plane. Dimensions A and B are shown as examples for plotting points in the aux-

269

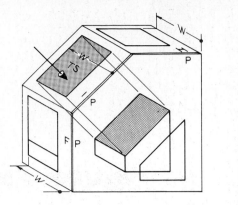

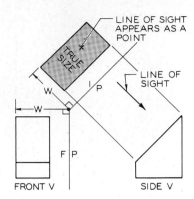

FIG. 16.15 Auxiliary from the side—folding-line method.

Given An object with an inclined surface.

Required Find the inclined surface true size by an auxiliary view.

Step 1 Draw a line of sight perpendicular to the edge view of the inclined surface. Draw the P-1 fold line parallel to the edge view, and draw the F-P fold line between the given views.

Step 2 Project the corners of the edge parallel to the line of sight. Locate the corners by measuring perpendicularly from the P-1 fold line with width (W) dimensions.

FIG. 16.16 An auxiliary view is projected from the right-side view by using a profile reference plane (PRP). The auxiliary view shows the true-size view of the inclined surface.

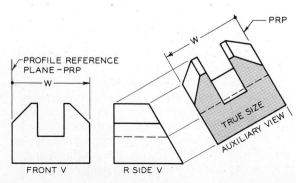

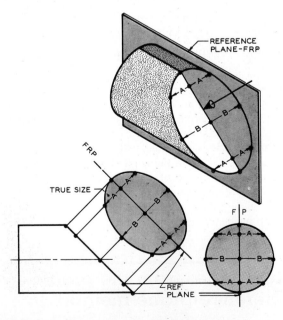

FIG. 16.17 This auxiliary view requires that a series of points be plotted. Since the object is symmetrical, the reference plane (FRP) is positioned through its center.

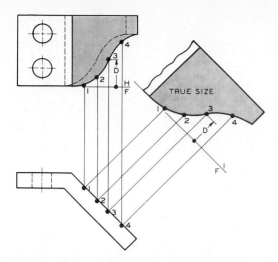

FIG. 16.18 The auxiliary view of this surface required that a series of points be located in the given view and then projected to the auxiliary view.

iliary view. To construct a smooth curve, more points are needed than are shown.

An irregular curve is found as an outline of a true-size surface in Fig. 16.18. Points are located on the curve in the top view and are projected to the front view. These points are found in the auxiliary view, by plotting each of them using the depth (D) dimensions transferred from the top view.

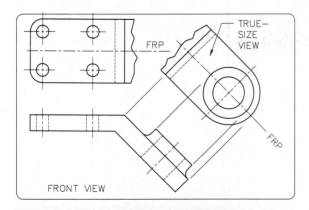

FIG. 16.19 This computer-drawn combination of views is used to represent an object, although the top and side views are partial views. The foreshortened portions of the object have been omitted. The FRP reference line passes through the center of the object in the top view, since the object is symmetrical about this line.

16.11
Partial views

An auxiliary view is a supplementary view, so some views of an orthographic arrangement can be drawn as partial views. The object in Fig. 16.19 shows a complete front view and partial auxiliary and side views. The partial views are easier to draw and are more functional without sacrificing clarity.

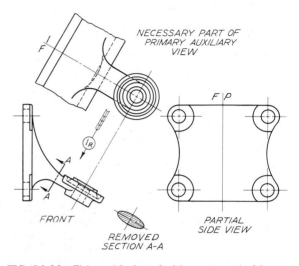

FIG. 16.20 This guide bracket is represented by partial views. The hub, which would appear elliptical, is not shown in the side view at all.

A similar example is given in Fig. 16.20, where the front view of the guide bracket is a complete view, and the other two views are partial views. An FRP is passed through the center of the side view. A photograph of the guide bracket is shown in Fig. 16.21.

16.12
Auxiliary sections

A section through a part that is projected as an auxiliary view is shown in Fig. 16.22. The section is labeled A-A, and a cutting plane passing through the object is labeled A-A. The cutting plane shows where the sectional view was projected from and where the object was cut to show the section. The auxiliary section pro-

FIG. 16.21 A photograph of the guide bracket that was drawn in Fig. 16.20.

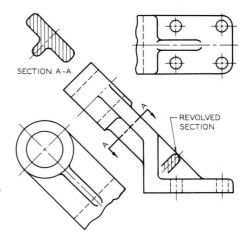

SECTION A-A

REVOLVED SECTION

FIG. 16.22 Auxiliary section A-A is projected from the cutting plane labeled A-A to show the cross section of the object.

vides a good description of the part that cannot be readily understood from the given principal views.

16.13
Secondary auxiliary views

> A *secondary auxiliary view* is projected from a primary auxiliary view

In Figure 16.23, the inclined plane is found as an edge view in the primary auxiliary view, and then a line of sight perpendicular to the edge is established. The second auxiliary view shows the inclined surface true size. Remember that it is necessary to project from an edge view of a plane before it can be found true size.

The problem in Fig. 16.24 is a secondary auxiliary projection in which the point view of a diagonal of a cube is found. When this is found, the three surfaces of the cube are equally foreshortened. The secondary auxiliary view is an *isometric projection* that is the basis for isometric pictorial drawing.

Since auxiliary views are supplementary views, they can be partial views if all features are sufficiently shown. The object in Fig. 16.25 is shown as a series of orthographic views, all of which are partial views.

16.14
Elliptical features

Occasionally, circular shapes will project as ellipses, which can be drawn using any of the techniques introduced in Chapter 13 once the necessary points have

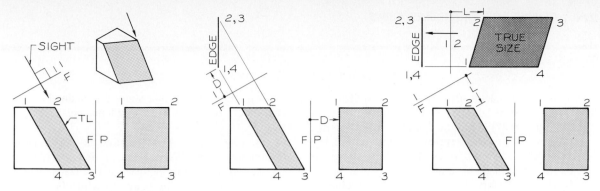

FIG. 16.23 Secondary auxiliary views.

Step 1 A line of sight is drawn parallel to the true-length view of a line on the oblique surface. The folding line F-1 is drawn perpendicular to the line of sight.

Step 2 The primary auxiliary view of the oblique surface is an edge view. Depth *(D)* is used to locate a point in the primary auxiliary view.

Step 3 A line of sight is drawn perpendicular to the edge view, and a secondary auxiliary view is projected in this direction. The dimension *L* is used to locate one of the points in the true-size view.

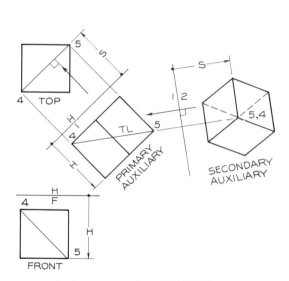

FIG. 16.24 A point view of a diagonal of a cube is found in the secondary auxiliary view. Line 4–5 is found true length in the primary auxiliary view and then as a point in the secondary auxiliary view, which is an isometric projection of the cube.

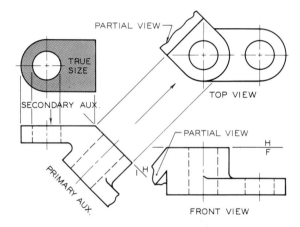

FIG. 16.25 An example of a secondary auxiliary view projected from a partial auxiliary view.

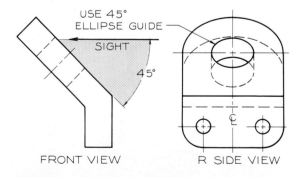

FIG. 16.26 The ellipse guide angle is the angle that the line of sight makes with the edge view of the circular feature. The ellipse guide for the right-side view is 45.

been plotted. The most convenient method of drawing ellipses is with an ellipse guide (template). The angle of the ellipse guide is the angle the line of sight makes with the edge view of the circular feature. In Fig. 16.26, the angle is found to be 45°, so the right-side view is drawn as a 45° ellipse.

Problems

The following problems are to be solved on Size A or Size B sheets, as assigned by your instructor.

1–10. (Fig. 16.27) Using the example layout, change the top and front views by substituting the top views given at the right in place of the one given in the example. The angle of inclination in the front view is forty-five degrees for all problems, and the height is 1½ inches in the front view. Con- struct auxiliary views that show the inclined sur- face true size. Draw two problems per Size A sheet.

11–32. (Figs. 16.28–16.49) Draw the necessary pri- mary and auxiliary views to describe the parts shown. Draw one per Size A or Size B sheet, as assigned. Adjust the scale of each to fit the space on the sheet.

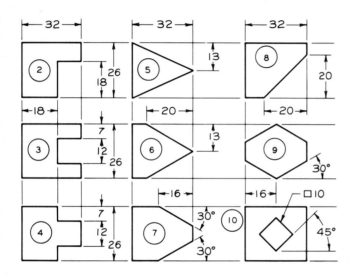

FIG. 16.27 Problems 1–10: Primary auxiliary views.

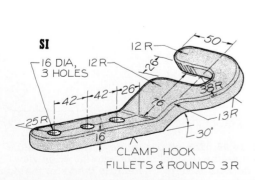

FIG. 16.28 Problem 11: Clamp hook.

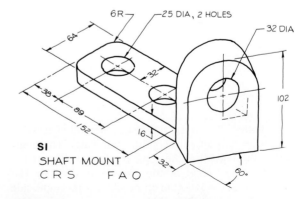

FIG. 16.29 Problem 12: Shaft mount.

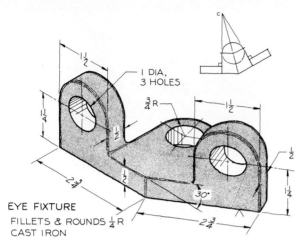

FIG. 16.30 Problem 13: Eye fixture.

EYE FIXTURE

FILLETS & ROUNDS $\frac{1}{4}$ R
CAST IRON

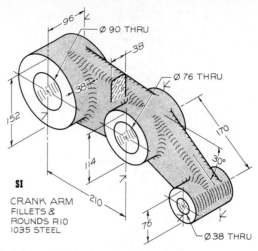

FIG. 16.33 Problem 16: Crank arm.

Ø 90 THRU

Ø 76 THRU

Ø 38 THRU

SI

CRANK ARM
FILLETS &
ROUNDS R10
1035 STEEL

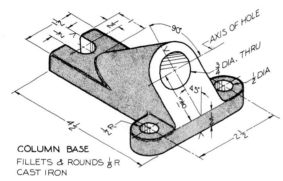

COLUMN BASE

FILLETS & ROUNDS $\frac{1}{8}$ R
CAST IRON

FIG. 16.31 Problem 14: Column base.

AXIS OF HOLE

$\frac{3}{4}$ DIA. THRU

$\frac{1}{2}$ DIA

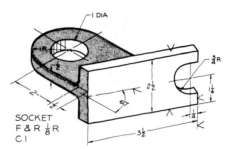

FIG. 16.34 Problem 17: Socket.

I DIA

SOCKET
F & R $\frac{1}{8}$ R
C I

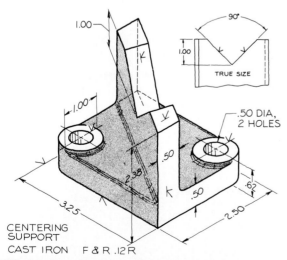

1.00

TRUE SIZE

90°

1.00

.50 DIA,
2 HOLES

CENTERING
SUPPORT
CAST IRON   F & R .12 R

FIG. 16.32 Problem 15: Centering support.

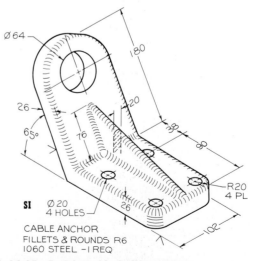

Ø 64

R20
4 PL

SI   Ø 20
4 HOLES

CABLE ANCHOR
FILLETS & ROUNDS R6
1060 STEEL –1 REQ

FIG. 16.35 Problem 18: Cable anchor.

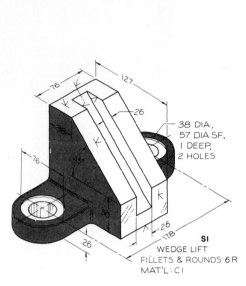

38 DIA,
57 DIA SF,
1 DEEP,
2 HOLES

SI

WEDGE LIFT
FILLETS & ROUNDS 6 R
MAT'L : CI

FIG. 16.36 Problem 19: Wedge lift.

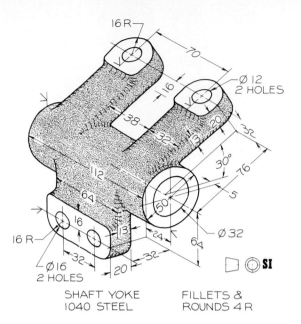

16 R

Ø 12
2 HOLES

16 R

Ø 16
2 HOLES

30°

Ø 32

SI

SHAFT YOKE
1040 STEEL

FILLETS &
ROUNDS 4 R

FIG. 16.38 Problem 21: Shaft yoke.

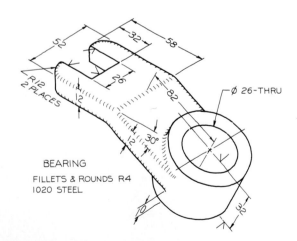

R 12
2 PLACES

Ø 26-THRU

30°

BEARING

FILLETS & ROUNDS R4
1020 STEEL

FIG. 16.37 Problem 20: Bearing.

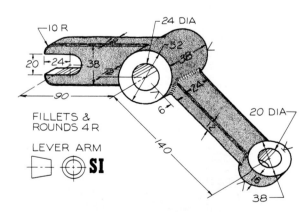

10 R

24 DIA

20 DIA

38

FILLETS &
ROUNDS 4 R

LEVER ARM

SI

FIG. 16.39 Problem 22: Lever arm.

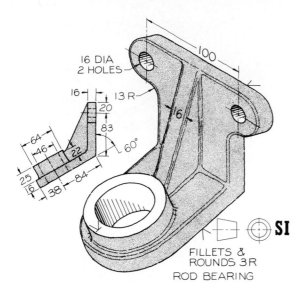

FIG. 16.40 Problem 23: Rod bearing.

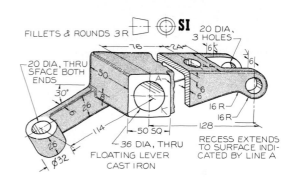

FIG. 16.42 Problem 25: Floating lever.

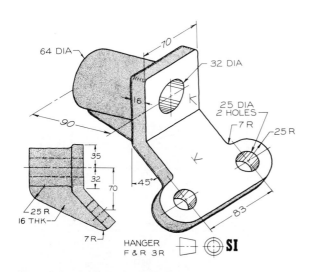

FIG. 16.41 Problem 24: Hanger.

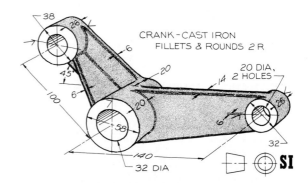

FIG. 16.43 Problem 26: Crank.

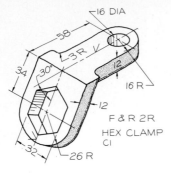

FIG. 16.44 Problem 27: Hexagon angle.

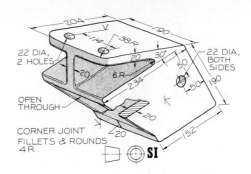

FIG. 16.47 Problem 30: Corner joint.

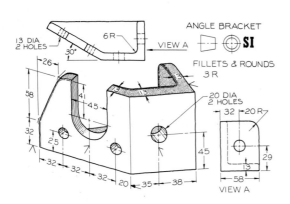

FIG. 16.45 Problem 28: Angle bracket.

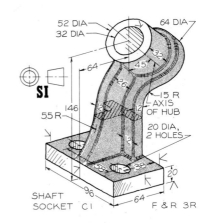

FIG. 16.48 Problem 31: Shaft socket.

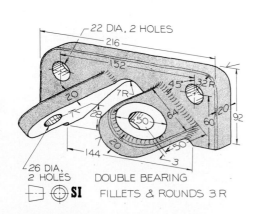

FIG. 16.46 Problem 29: Double bearing.

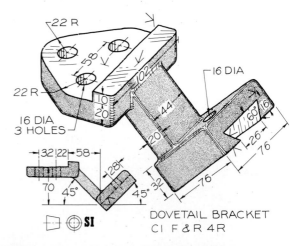

FIG. 16.49 Problem 32: Dovetail bracket.

33–34. (Figs. 16.50–16.51) Lay out these orthographic views on Size B sheets, and complete the
auxiliary and primary views.

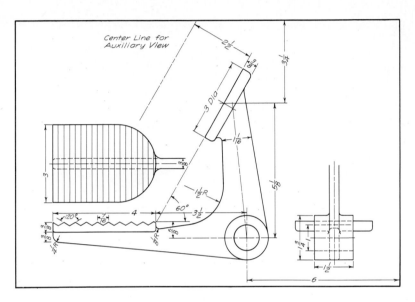

FIG. 16.50 Problem 33: Clutch pedal.

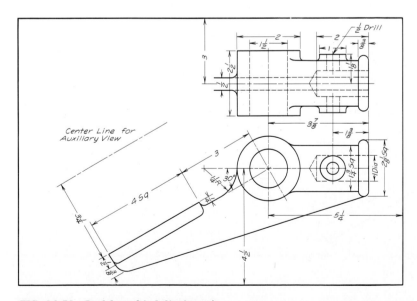

FIG. 16.51 Problem 34: Adjustment.

35–37. (Figs. 16.52–16.54) Construct orthographic views of the given objects, and using secondary auxiliary views, draw auxiliary views that give the true-size views of the inclined surfaces. Draw one per Size B sheet.

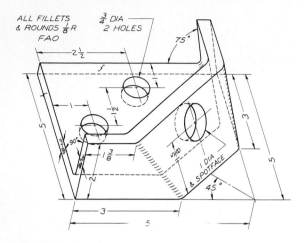

FIG. 16.52 Problem 35: Corner connector.

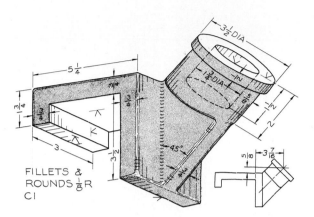

FIG. 16.53 Problem 36: Shaft bearing.

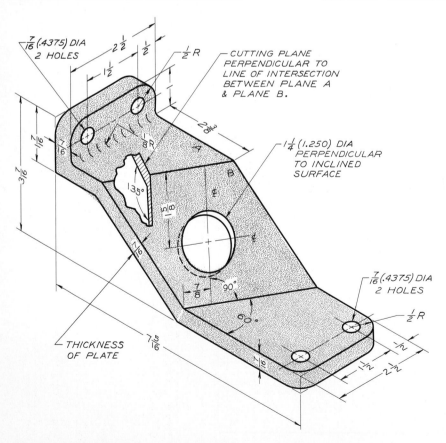

FIG. 16.54 Problem 37: Oblique support.

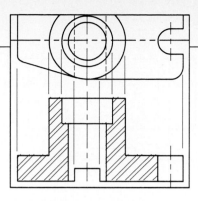

CHAPTER 17

Sections

17.1
Introduction

Standard orthographic views that show all hidden lines may not effectively reveal the true details of an object. This shortcoming can often be improved by using a technique of cutting away part of the object and looking at the cross-sectional view. Such a cutaway view is called a *section*.

A section is shown pictorially in Fig. 17.1A, where an imaginary cutting plane is passed through the object to show its internal features. The standard top and front views are shown in Fig. 17.1B, and the method of drawing a section is shown in Fig. 17.1C, where the front view has been converted to a *full section* and the cut portion is cross-hatched. Hidden lines have been omitted since they are not needed. The cutting plane is drawn as a heavy line with short dashes at intervals; this can be thought of as a knife-edge cutting through the object.

Two types of cutting planes are shown in Fig. 17.2. Either is acceptable, although the example in Fig. 17.2A is more often used. The spacing of the dashes depends on the size of the drawing. The weight of the cutting plane is the same as that of a visible object line. Letters can be placed at each end of the cutting plane to label the sectional view, such as section B-B in Fig. 17.2B.

The three basic views that appear as sections, with their respective cutting planes, are shown in Fig. 17.3. Each cutting plane has perpendicular arrows pointing in the direction of the line of sight for the section. For example, the cutting plane in Fig. 17.3A passes through the top view, the front of the top view is removed, and the line of sight is toward the remaining portion of the top view. The top view will appear as a section when the cutting plane passes through the front view and the line of sight is downward (Fig. 17.3B). When the cutting plane passes through the front view (Fig. 17.3C), the right-side view will be a section.

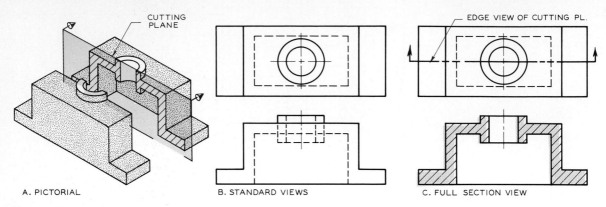

FIG. 17.1 A comparison of a regular orthographic view with a full-section view showing the internal and external features of the same object.

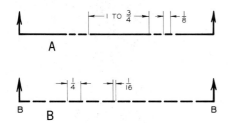

FIG. 17.2 Typical cutting-plane lines used to represent sections. The cutting plane marked B-B will produce a section labeled B-B.

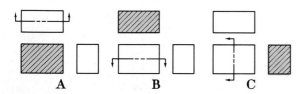

FIG. 17.3 The three standard positions of cutting planes that pass through views to result in sectional views in the front, top, and side views. The arrows point in the direction of the line of sight for each section.

17.2
Sectioning symbols

The symbols used to distinguish between different materials in sections are shown in Fig. 17.4. Although the symbols can be used to indicate the materials within a section, it is advisable to provide supplementary notes specifying the materials to avoid misinterpretation.

Cast-iron symbols are usually drawn with a 2H pencil with lines slanted at 30°, 45°, or 60° angle, and spaced about one-sixteenth inch apart.

> The cast-iron symbol of evenly spaced section lines can be used to represent any material and is the most often used sectioning symbol.

> **COMPUTER METHOD** A few of the many cross-sectional symbols available from AutoCAD are shown in Fig. 17.5. The spacing between the lines and dash lengths may be varied by changing the pattern scale factor.

The proper spacing of section lines is shown in Fig. 17.6A, where the lines are evenly spaced. Common errors of lining are shown in the other parts of the figure.

Extremely thin parts such as sheet metal, washers, or gaskets (Fig. 17.7) are sectioned by blacking in the areas completely rather than using section lines. Large parts are sectioned with an *outline section* to save time and effort. The section lines are drawn closer together in small parts than in larger parts.

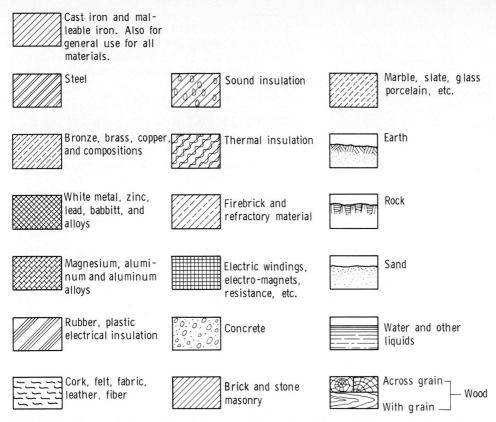

Cast iron and malleable iron. Also for general use for all materials.

Steel

Bronze, brass, copper, and compositions

White metal, zinc, lead, babbitt, and alloys

Magnesium, aluminum and aluminum alloys

Rubber, plastic electrical insulation

Cork, felt, fabric, leather, fiber

Sound insulation

Thermal insulation

Firebrick and refractory material

Electric windings, electro-magnets, resistance, etc.

Concrete

Brick and stone masonry

Marble, slate, glass porcelain, etc.

Earth

Rock

Sand

Water and other liquids

Across grain — Wood
With grain —

FIG. 17.4 The symbols used for lining parts in section. The cast-iron symbol can be used for any material.

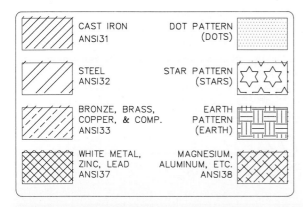

CAST IRON ANSI31	DOT PATTERN (DOTS)	
STEEL ANSI32	STAR PATTERN (STARS)	
BRONZE, BRASS, COPPER, & COMP. ANSI33	EARTH PATTERN (EARTH)	
WHITE METAL, ZINC, LEAD ANSI37	MAGNESIUM, ALUMINUM, ETC. ANSI38	

FIG. 17.5 A few of the sectioning symbols provided by AutoCAD. The pattern scale can be used to vary the spacing of the lines and dashes.

FIG. 17.6 Good section lines are thin and $\frac{1}{16}$″ to $\frac{1}{8}$″ apart. Some typical lining errors are shown.

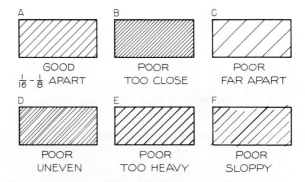

A
GOOD
$\frac{1}{16}$ - $\frac{1}{8}$ APART

B
POOR
TOO CLOSE

C
POOR
FAR APART

D
POOR
UNEVEN

E
POOR
TOO HEAVY

F
POOR
SLOPPY

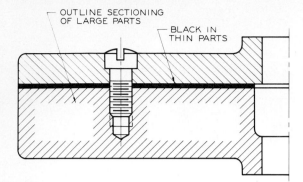

FIG. 17.7 Thin parts are blacked in, and large areas are section-lined around their outlines to save time and effort.

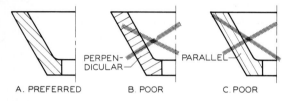

FIG. 17.8 Section lines should be drawn so they are neither parallel nor perpendicular to the outline of a part.

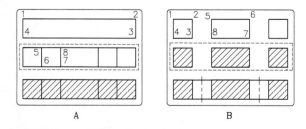

FIG. 17.9 Sections by computer.

A. If lines were drawn from 1 to 2 to 3 to 4 and division lines were drawn from 5 to 6 and from 7 to 8, the resulting section would ignore the intermediate lines (5–6, for example) when the area is windowed. The section lines will extend to the perimeter, 1–2–3–4.

B. If lines are drawn in segments to stop at the corners of the sectioned areas, and then windowed, the section lines will fill the areas as intended. The areas can now be connected and centerlines added to complete the sectional view.

Sectioned areas should be lined with symbols that are neither parallel nor perpendicular to the outlines of the parts lest they be confused for serrations or other machining treatments of the surface. (Fig. 17.8).

COMPUTER METHOD The method of drawing areas to be section-lined using AutoCAD's HATCH command is shown in Fig. 17.9. The areas to be sectioned must be drawn with lines terminating at each corner point, as shown in Fig. 17.9B.

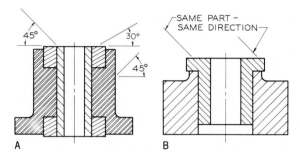

FIG. 17.10 Section lines of the same part should be drawn in the same direction. Section lines of different parts should be drawn at varying angles to distinguish the parts.

17.3
Sectioning assemblies

When an assembly of several parts is sectioned, it is important that the section lines be drawn at varying angles to distinguish the parts (Fig. 17.10). The use of different material symbols in an assembly is also helpful in distinguishing the various materials of the parts. The same part is cross-hatched at the same angle and with the same symbol even though the part may be separated into different areas, (Fig. 17.10B). Section lines are effectively used in Fig. 17.11 to identify the parts of the assembly.

17.4
Full sections

A *full section* is a view formed by passing a cutting plane fully through an object and removing half of it.

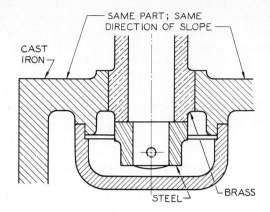

FIG. 17.11 A typical assembly in section with well-defined parts and correctly drawn section lines.

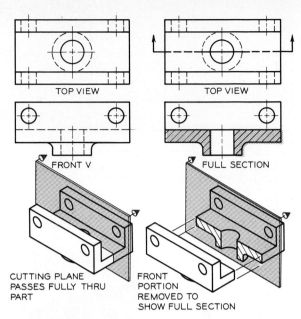

CUTTING PLANE PASSES FULLY THRU PART

FRONT PORTION REMOVED TO SHOW FULL SECTION

FIG. 17.12 A full section is formed by a cutting plane that passes completely through the part. The cutting plane is shown passing through the top view, and the direction of sight is indicated by the arrows at each end. The front view is converted to a sectional view to give a clear understanding of the internal features.

In Fig. 17.12 an object is drawn as two orthographic views in which a number of hidden lines are shown. The front view can be drawn as a full section by passing a cutting plane fully through the top view and removing the front portion. The arrows on the cutting plane indicate the direction of sight, and the front view is then section-lined to give the full section.

A full section through a cylindrical part is shown in Fig. 17.13A, where half the object is removed. A common mistake in constructing sectional views is omitting the visible lines behind the cutting plane, as in Fig. 17.13B. The correctly drawn sectional view is

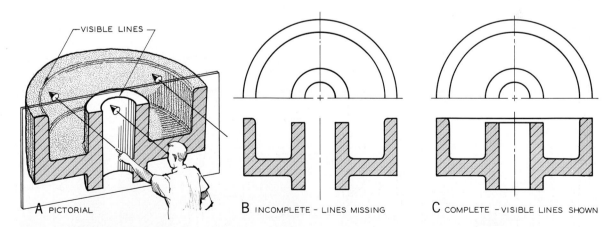

FIG. 17.13 Full section–cylindrical part.

A. When a full section is passed through an object, you will see lines behind the sectioned area.

B. If only the sectioned area were shown, the view would be incomplete.

C. Visible lines behind the sectioned area must be shown also.

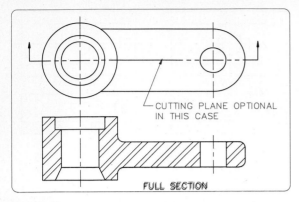

FIG. 17.14 A full section with the cutting plane shown.

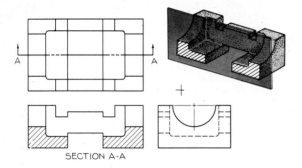

FIG. 17.15 A full section, section A-A, is used to supplement the given views of the object.

shown in Fig. 17.13C. Hidden lines are omitted in all sectional views unless they are considered necessary to provide a clear understanding of the view.

Figure 17.14 is an example of a part whose front view is shown as a full section. Likewise, the part in Fig. 17.15 illustrates a front view that appears as a full section. Lines behind the cutting plane are shown as visible lines.

17.5
Parts not section-lined

Many standard parts like nuts and bolts, rivets, shafts, and set screws, are not section-lined even though the cutting plane passes through them (Fig. 17.16). Since

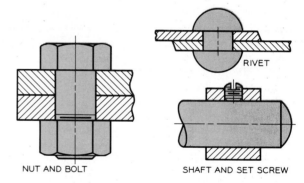

FIG. 17.16 These parts are not section-lined even though the cutting plane passes through them.

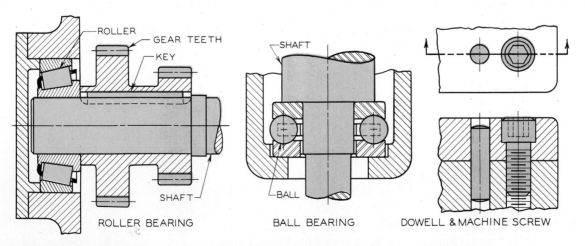

FIG. 17.17 These parts are not section-lined even though cutting planes pass through them.

these parts have no internal features, sections through them would be of no value. Other parts not section-lined are roller bearings, ball bearings, gear teeth, dowels, pins, and washers (Fig. 17.17).

17.6
Ribs in section

Ribs are not section-lined when the cutting plane passes flatwise through them, as in Fig. 17.18A, since this would give a misleading impression of the rib. However, a rib is section-lined when the cutting plane passes through it and shows its true thickness (Fig. 17.18B).

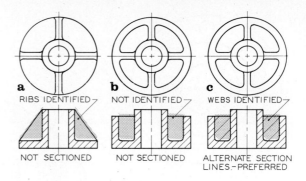

FIG. 17.19 Outside ribs in section are not section-lined. Poorly identified webs, as in part b, should be identified by alternating section lines with cross-hatching, as in part c.

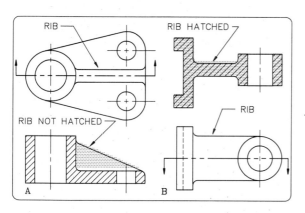

FIG. 17.18 A rib cut in a flatwise direction by a cutting plane is not section-lined. Ribs are section-lined when cutting planes pass perpendicularly through them, as shown in part B.

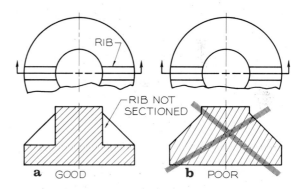

FIG. 17.20 When ribs are not section-lined, the view is more descriptive of the part. Partial views are used in the top views to save space. The front part of the top view is removed when the front view is a section.

An alternative method of section-lining webs and ribs is shown in Fig. 17.19. The ribs are not section-lined since the cutting plane passes through them in a flatwise direction (Fig. 17.19A). The webs are symmetrically spaced about the hub (Fig. 17.19B). As a rule, webs are not cross-hatched, but this would leave them unidentified; therefore it is better to use the *alternate sectioning* technique, which extends every other section line through the webs (Fig. 17.19C).

The ribs in Fig. 17.20A are not section-lined and thus afford a more descriptive view of the part. If the ribs had been section-lined, the section would have given the impression that the part was solid and con-

ical in shape, as shown in Fig. 17.20B. Note that the top views are partial views and that the portion nearest the sectional view has been omitted.

17.7
Half-sections

A *half-section* is a view that results from passing a cutting plane halfway through an object and removing a quarter of it to show external and internal features.

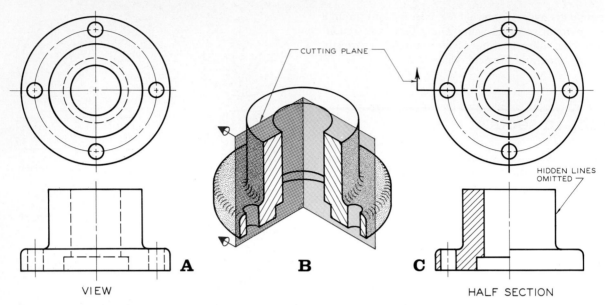

FIG. 17.21 The cutting plane of a half-section passes halfway through the object, which results in a sectional view that shows half the outside and half the inside of the object. Hidden lines are omitted unless they are necessary to clarify the view.

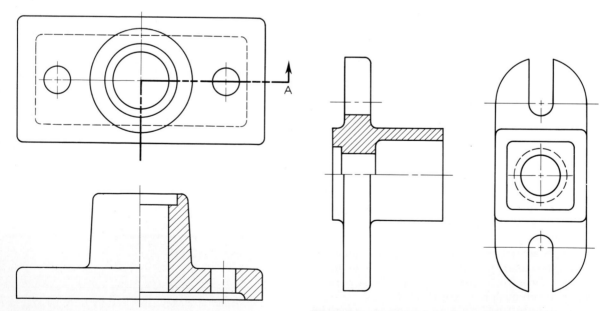

FIG. 17.22 Half-sections. A typical half-section drawing.

When it is obvious where the cutting plane is located, it is unneccessary to show it, as in this example.

A half-section is most often used with symmetrical parts, and cylinders in particular. A cylindrical part in Fig. 17.21A is shown as a pictorial half-section in Fig. 17.21B. The method of drawing the orthographic half-section is shown in Fig. 17.21C, where both the internal and external features can be seen. Hidden lines are omitted in the sectional view.

The half-section in Fig 17.22 has been drawn without showing the cutting plane, which is permissible if it is obvious where the cutting plane was passed through the object. Instead of using an object line, centerlines separate the sectional half from the half that appears as an external view.

17.8
Partial views

A conventional method of representing symmetrical views is shown in Fig. 17.23. A half-view is sufficient when it is drawn adjacent to the sectional view (Fig. 17.23A). In full sections, the removed half is the portion nearest the section (Fig. 17.23B). When drawing half-views associated with views (nonsectional views), the removed half of the partial view is the half away from the adjacent view (Fig. 17.23A).

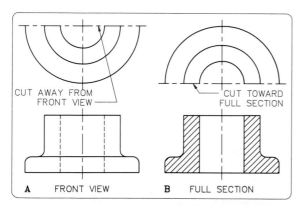

FIG. 17.23 Half-views can be used for symmetrical objects to conserve space and drawing time. In part A, the omitted portion of the view is away from the view. In part B, the omitted portion of the view is toward the full section. The omitted half can be toward or away from the section in the case of a half-section.

When partial views are drawn with half-sections, either the near or the far halves of the partial views can be omitted. Partial views are faster to draw and require less space.

17.9
Offset sections

> An *offset section* is a full section in which the cutting plane is offset to pass through important features.

An offset section is shown pictorially in Fig. 17.24A, where the plane is offset to pass through the large hole and one of the small holes. The cut object is shown in Fig. 17.24B, and the method of drawing an offset section orthographically is given in Fig. 17.24C. The cut formed by the offset is not shown in the section since this is an imaginary cut. The computer-drawn object in Fig. 17.25 also lends itself to representation by an offset section.

17.10
Revolved sections

> A *revolved section* is used to describe a cross section of a part by revolving it about an axis of revolution and placing it on the view on the axis of revolution.

For example, revolved sections are used to indicate cross sections of the parts in Fig. 17.26. One of the revolved sections is positioned within the view, and conventional breaks are drawn on each side of the hexagonal section. The circular cross section is drawn superimposed on the cylindrical portion of the part without using conventional breaks. Either method can be used when drawing revolved sections.

A more advanced type of revolved section is illustrated in Fig. 17.27, where a cutting plane is passed through the object (Step 1). The plane is imagined to be revolved in the top view to give a true-size revolved section in the front view (Step 2). Note that the object lines do not pass through the revolved section in the front view. As in Fig. 17.26, it would have been per-

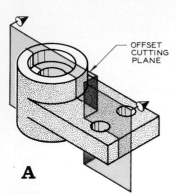

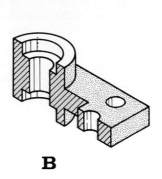

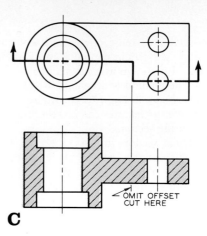

FIG. 17.24 Offset section.

A. An offset cutting plane may be necessary to show all typical features of a sectioned part.

B. When the front portion is removed, the internal features can be seen. Note that the cutting plane has been offset.

C. In an offset section, the offset cut is not shown. The section is shown as if it were a typical full section.

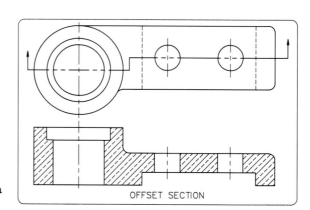

FIG. 17.25 An offset section drawn by a computer.

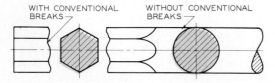

FIG. 17.26 Examples of revolved sections with and without conventional breaks.

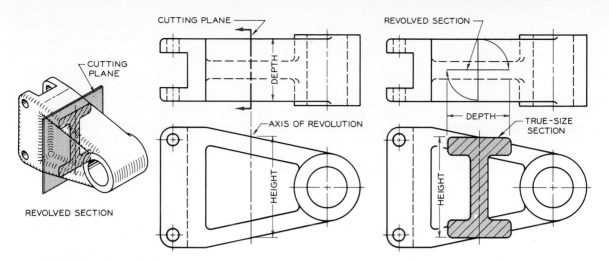

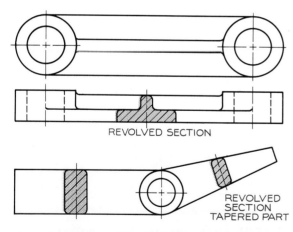

FIG. 17.27 Revolved section.

Step 1 An axis of revolution is shown in the front view. The cutting plane would appear as an edge in the top view if it were shown.

Step 2 The vertical section in the top view is revolved so that the section can be seen true size in the front view. Object lines are not drawn through the revolved section.

FIG. 17.28 The revolved sections given here are helpful in describing the cross-sections of the two parts without using additional orthographic views.

missible to use conventional breaks on each side of the revolved section.

Typical revolved sections are shown in Fig. 17.28. These sections provide a method of giving a part's cross section without relying on another complete orthographic view.

17.11
Removed sections

A *removed section* is a revolved section that has been removed from the view where it was revolved (Fig. 17.29).

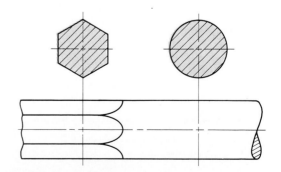

FIG. 17.29 Removed sections are similar to revolved sections, but they have been removed outside the object along an axis of revolution.

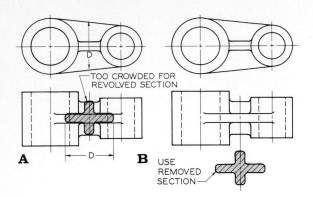

FIG. 17.30 Removed sections can be used where space does not permit the use of revolved sections.

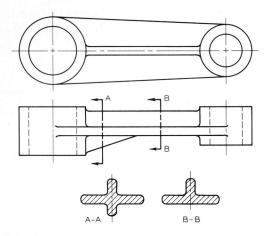

FIG. 17.31 Sections can be lettered at each end of a cutting plane, such as A-A. This removed section can then be shown elsewhere on the drawing and is designated section A-A.

FIG. 17.32 If it is necessary to remove a section to another page in a set of drawings, each end of the cutting plane can be labeled with a letter and a number. The letters refer to section A-A, and the numbers mean this section is located on page 7.

Centerlines are used as axes of rotation to show where the sections were taken from. Removed sections may be necessary where room does not permit revolution on the given view (Fig. 17.30A); instead, the cross section must be removed from the view (Fig. 17.30B).

Removed sections do not have to be positioned directly along an axis of revolution adjacent to the view from where the sections were taken. Instead, cutting planes can be labeled at each end, as shown in Fig. 17.31, to specify the sections. For example, the plane labeled with an A at each end is used to label section A-A; section B-B is similarly found.

When a set of drawings consists of many pages, removed sections may be put on different sheets. In this case, a cutting plane may be labeled as shown in Fig. 17.32: The A at each end indicates section A-A, and the numerals indicate on which page the section is located.

Removed views can also be used to provide inaccessible orthographic views (nonsectional views) as shown in Fig. 17.33.

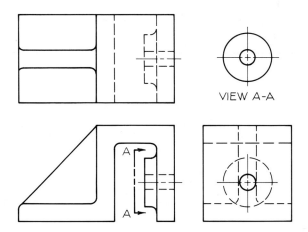

FIG. 17.33 A removed view (not a section) can also be used to view a part from an unconventional direction.

17.12
Broken-out sections

A *broken-out section* is used to show interior features by breaking away a portion of a view.

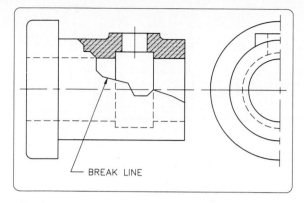

FIG. 17.34 A broken-out section drawn by computer shows internal features by using a conventional break.

A portion of the object in Fig. 17.34 is broken out to reveal details of the wall thickness that better explain the drawing. Using broken-out sections reduces the need for hidden lines; therefore they may be omitted. The irregular lines representing breaks are called *conventional breaks.*

17.13
Phantom (ghost) sections

A *phantom,* or *ghost, section* is used to depict parts as if they were viewed by an x ray.

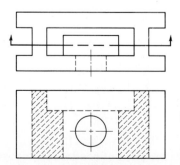

FIG. 17.35 Phantom sections give an "x-ray" view of an object. The section lines are shown as dashed lines, which makes it possible to show the section without removing the hole in the front of the part.

In Fig. 17.35, the cutting plane is drawn in the usual manner, but the section lines are drawn as dashed lines. If the object had been shown as a regular full section, the circular hole through the front surface could not have been shown in the same view.

17.14
Conventional breaks

Some of the previous examples have used *conventional breaks* to indicate the removal of parts of an object. Examples of conventional breaks are illustrated in Fig. 17.36. The "figure 8" breaks, which are used for cylindrical and tubular parts, can be drawn freehand (Fig. 17.37). But when they are drawn to a large scale, they can be drawn with a compass, as shown in Fig. 17.38.

One use of conventional breaks is to shorten a long piece that has a uniform cross section. The long part in Fig. 17.39 has been shortened and drawn at a larger scale for more clarity by using conventional breaks (Fig. 17.39). The dimension specifies the true length of the part, and the breaks indicate a portion of the length has been removed.

17.15
Conventional revolutions

Three conventional sections are shown in Fig. 17.40. The center hole is omitted in Fig. 17.40a, since it does not pass through the center of the circular plate. However, the hole in Fig. 17.40b does pass through the plate's center and is sectioned accordingly. In Fig. 17.40c, although the cutting plane does not pass through one of the symmetrically spaced holes in the top view, the hole is revolved to the cutting plane to show the recommended full section.

When ribs are symmetrically spaced about a hub (Fig. 17.41), it is conventional practice to revolve them to where they will appear true size in either a view or a section. A full section that shows both ribs and holes revolved to their true-size locations can be seen in Fig. 17.42.

The cutting plane can be positioned as shown at Fig. 17.43. Even though the cutting plane does not pass through the ribs and holes in Fig. 17.43a, the sectional view should be drawn as shown in Fig. 17.43b,

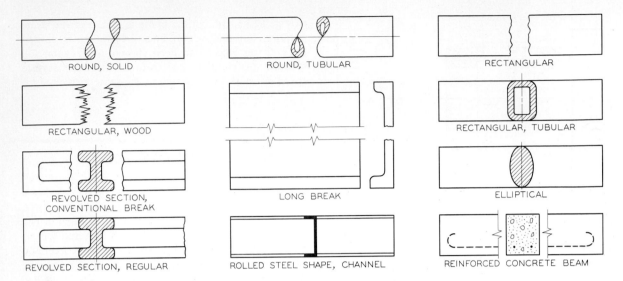

FIG. 17.36 These conventional breaks are used to remove a portion of an object so it can be drawn at a larger scale.

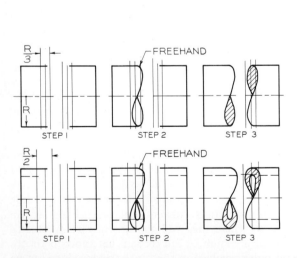

FIG. 17.37 Conventional breaks in cylindrical and tubular sections can be drawn freehand with the aid of the guidelines shown. The radius, *R*, is used to establish the width of both "figure 8s."

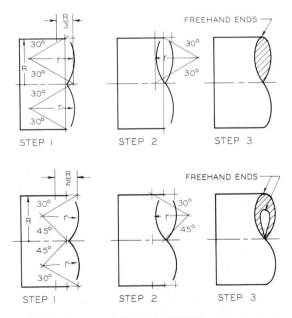

FIG. 17.38 The steps of constructing conventional breaks of solid and tubular shapes with instruments.

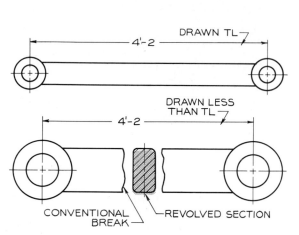

FIG. 17.39 By using conventional breaks and a revolved section, this part can be drawn at a larger scale that is easier to read.

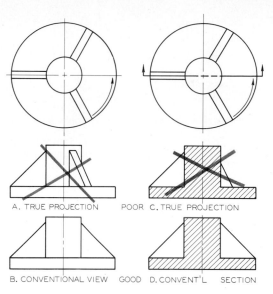

A. TRUE PROJECTION POOR C. TRUE PROJECTION

B. CONVENTIONAL VIEW GOOD D. CONVENT'L SECTION

FIG. 17.41 Symmetrically located ribs are shown revolved in both orthographic and sectional views as a conventional practice.

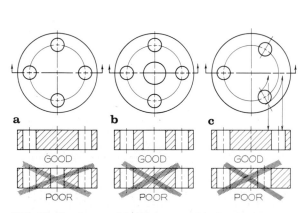

a b c

GOOD GOOD GOOD

POOR POOR POOR

FIG. 17.40 Symmetrically spaced holes in a circular plate should be revolved in their sectional views to show them at their true radial distance from the center. In part a, no hole is shown at the center, but a hole is shown at the center in part b, since the hole is through the center of the plate. In part c, one of the holes is rotated to the cutting plane so that the sectional view will be symmetrical and more descriptive.

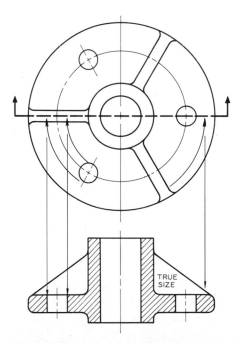

TRUE SIZE

FIG. 17.42 A part with symmetrically located ribs and holes is shown in section with both ribs and holes rotated to the cutting plane.

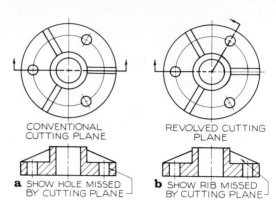

FIG. 17.43 Symmetrically located ribs are shown in section in revolved positions to show the ribs true size. Foreshortened ribs are omitted.

CONVENTIONAL CUTTING PLANE

REVOLVED CUTTING PLANE

a SHOW HOLE MISSED BY CUTTING PLANE

b SHOW RIB MISSED BY CUTTING PLANE

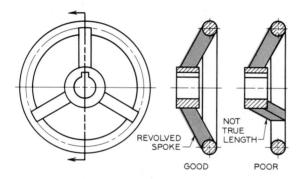

REVOLVED SPOKE

NOT TRUE LENGTH

GOOD   POOR

FIG. 17.44 Symmetrically positioned spokes are revolved to show the spokes true size in section. Spokes are not section-lined in section.

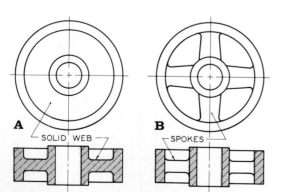

A SOLID WEB

B SPOKES

FIG. 17.45 Solid webs in sections of the type in part A are section-lined. Spokes are not section-lined when the cutting plane passes through them, as in part B.

where the cutting plane is revolved. The cutting plane can be drawn in either position.

In the same manner as ribs in section, symmetrically spaced spokes are rotated and not section-lined. Figure 17.44 illustrates the preferred and the poor methods of representing spokes. Only the revolved, true-size spokes are drawn; the intermediate spokes are omitted.

> If the spokes in Fig. 17.45B had been section-lined, the cross section of the part would be confused with the part in Fig. 17.45A, where there are no spokes but a continuous web.

The lugs symmetrically positoned about the central hub of the object in Fig. 17.46 are revolved to show they are true size in both views and sections. A more complex object involving the same principle of rotation can be seen in Fig. 17.47, where the oblique arm is drawn in the section as if it had been revolved to the centerline in the top view and then projected to the sectional view.

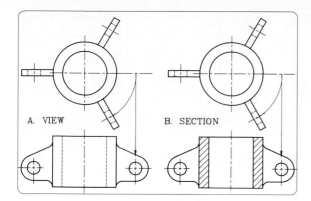

FIG. 17.46 Symmetrically spaced lugs (flanges) are revolved to show the front view and the sectional view as symmetrical.

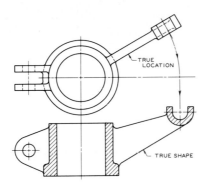

FIG. 17.47 A part with an oblique feature attached to the circular hub is revolved so it will appear true shape in the front view, which is a sectional view.

17.16
Auxiliary sections

Auxiliary sections can be used to supplement the principal views used in orthographic projections, as shown in Fig. 17.48. Auxiliary cutting plane A-A is passed through the front view, and the auxiliary view is projected from the cutting plane as indicated by the sight arrows. Section A-A gives the cross-sectional description of the part.

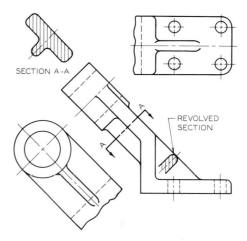

FIG. 17.48 Sectional views can be shown as auxiliary views for added clarity.

Problems

These problems can be solved on Size A or Size B sheets.

1–24. (Fig. 17.49) Full sections: Draw two of these problems per Size A sheet. Each grid is equal to 0.20 inch or 5 mm. Complete the front views as full sections.

25–30. (Figs. 17.50–17.54) Full sections: Complete the drawings as full sections. Draw one problem per Size A sheet. Each grid is equal to 0.20 inch or 5 mm. Show the cutting planes when they are not given.

31–37. (Figs. 17.55–17.56) Half sections: Complete the drawings as half-sections. Draw one problem per Size A sheet. Each grid is equal to 0.20 inch or 5 mm. Show the cutting planes when they are not given.

38–39. (Figs. 17.57–17.58) Offset sections: Complete the drawings as offset sections. Draw one problem per Size A sheet. Each grid is equal to 0.20 inch or 5 mm. Show the cutting planes.

40. (Fig. 17.59) Full section: Complete the partial view as a full section. Draw the views on a Size

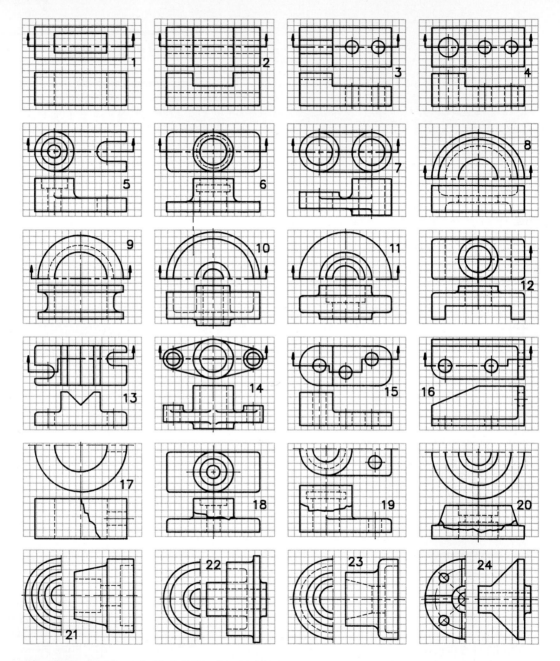

FIG. 17.49 Problems 1–24. Introductory sections.

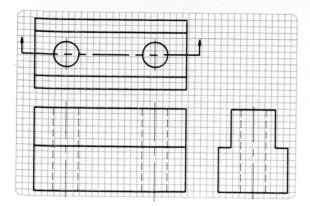

FIG. 17.50   Problem 25. Full section.

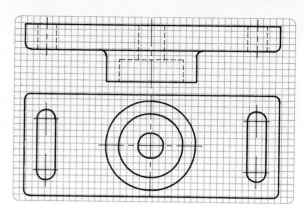

FIG. 17.53   Problem 28. Full section.

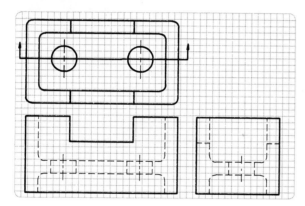

FIG. 17.51   Problem 26. Full section.

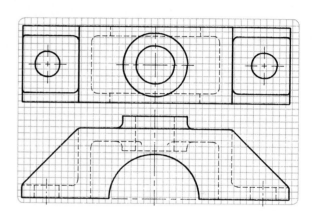

FIG. 17.54   Problem 29. Full section.

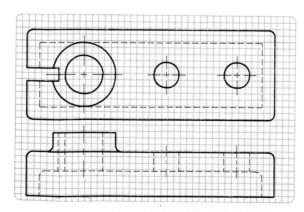

FIG. 17.52   Problem 27. Full section.

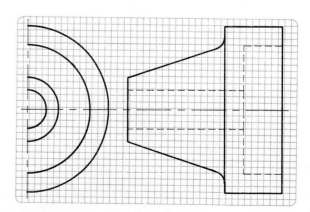

FIG. 17.55   Problem 30. Half-section.

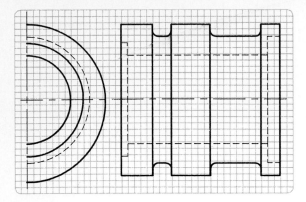

FIG. 17.56 Problem 31. Half-section.

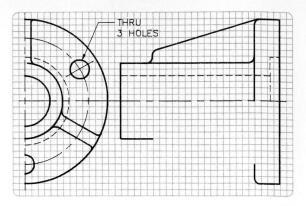

FIG. 17.59 Problem 34. Full section.

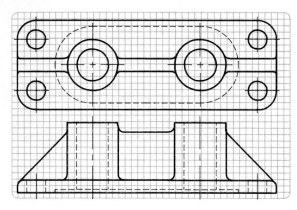

FIG. 17.57 Problem 32. Half-section.

A sheet. Each grid is equal to 0.20 inch or 5 mm. Show the cutting plane.

41. (Fig. 17.60) Assembly: Complete the front view as a full section of the assembled parts. Draw the views on a Size A sheet. Each grid is equal to 0.20 inch or 5 mm. Show the cutting plane.

42–44. (Figs. 17.61–17.63) Sections: Draw the necessary views to describe the parts using sections and conventional practices. Draw one problem per Size B sheet. Omit the dimensions.

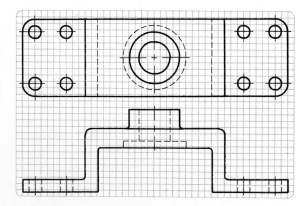

FIG. 17.58 Problem 33. Offset section.

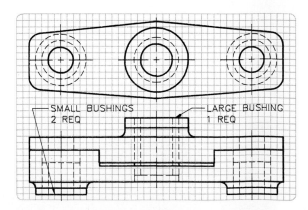

FIG. 17.60 Problem 35. Full section.

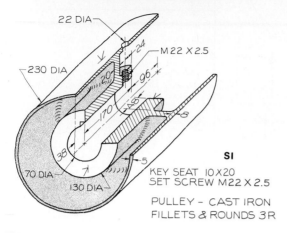

22 DIA

24

M22 X 2.5

230 DIA

20°

96

170

45°

8

38

5

70 DIA

7

130 DIA

SI

KEY SEAT 10 X 20
SET SCREW M22 X 2.5

PULLEY – CAST IRON
FILLETS & ROUNDS 3R

FIG. 17.61 Problem 36. Section.

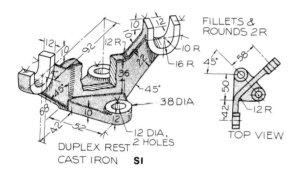

FILLETS &
ROUNDS 2 R

12

10

12

8

12 R

10 R

92

22

16 R

45°

58

36

50

45°

45°

38 DIA

42

68

10

2

12 R

52

12 DIA,
2 HOLES

TOP VIEW

DUPLEX REST
CAST IRON **SI**

FIG. 17.62 Problem 37. Section.

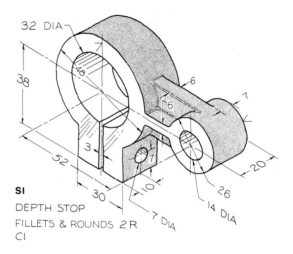

32 DIA

48

38

6

6

7

3

52

7

30

10

26

14 DIA

20

SI

DEPTH STOP

FILLETS & ROUNDS 2 R
CI

7 DIA

FIG. 17.63 Problem 38. Section.

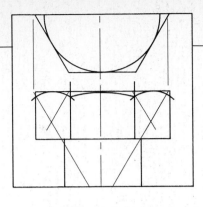

CHAPTER 18

Screws, Fasteners, and Springs

18.1
Threaded fasteners

Screw threads provide a fast and easy method of fastening two parts together and of exerting a force that can be used for adjustment of movable parts. For a screw thread to function, there must be an *internal thread* and an *external thread*. Internal threads may be tapped inside a part such as a motor block or, more commonly, a nut. Whenever possible, the nuts and bolts used in industrial projects should be stock parts that can be obtained from many sources. This reduces manufacturing expenses and improves the interchangeability of parts.

Progress has been made toward establishing standards that will unify threads in this country and abroad by the introduction of metric standards. Other efforts have led to the adoption of the Unified Screw thread by the United States, Britain, and Canada (ABC Standards), which is a modification of both the American Standard thread and the Whitworth thread.

18.2
Definitions of thread terminology

Succeeding sections will discuss the uses and methods of representing screw threads. The terms used and defined below are illustrated in Fig. 18.1.

EXTERNAL THREAD   is a thread on the outside of a cylinder, such as a bolt (Fig. 18.2).

INTERNAL THREAD   is a thread cut on the inside of a part, such as a nut (Fig. 18.2).

MAJOR DIAMETER   is the largest diameter on an internal or external thread.

MINOR DIAMETER   is the smallest diameter that can be measured on a screw thread.

PITCH DIAMETER is the diameter of an imaginary cylinder passing through the threads at the points where the thread width is equal to the space between the threads.

LEAD is the distance a screw will advance when turned 360 degrees.

PITCH is the distance between crests of threads. Pitch is found mathematically by dividing one inch by the number of threads per inch of a particular thread.

CREST is the peak edge of a screw thread.

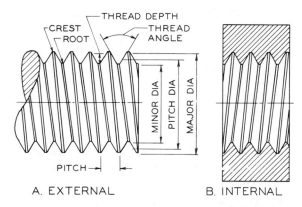

FIG. 18.1 Thread terminology.

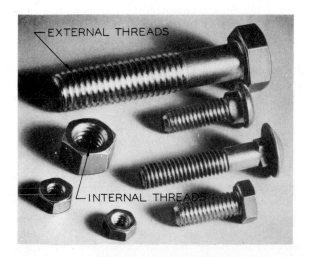

FIG. 18.2 Examples of external threads (bolts) and internal threads (nuts). (Courtesy of Russell, Burdsall & Ward Bolt and Nut Co.)

THREAD ANGLE is the angle between threads cut by the cutting tool.

ROOT is the bottom of the thread cut into a cylinder.

THREAD FORM is the shape of the thread cut into a threaded part.

THREAD SERIES is the number of threads per inch for a particular diameter, grouped into three series: coarse, fine, and extra fine. Coarse series provides rapid assembly, and extra-fine series provides fine adjustment.

THREAD CLASS is a closeness of fit between two mating threaded parts. Class 1 represents a loose fit and Class 3 a tight fit.

RIGHT-HAND THREAD is a thread that will assemble when turned clockwise. A right-hand thread slopes downward to the right on an external thread when the axis is horizontal, and in the opposite direction on an internal thread.

LEFT-HAND THREAD is a thread that will assemble when turned counterclockwise. A left-hand thread slopes downward to the left on an external thread when the axis is horizontal, and in the opposite direction on an internal thread.

18.3
Thread specifications (English system)

Form

Thread form is the shape of the thread cut into a part, as illustrated in Fig. 18.3. The Unified form, a combination of the American National and British Whitworth, is the most widely used, because it is a standard in several countries. The Unified form is signified by UN in abbreviations and thread notes, and the American National form is signified by N.

Another thread form, the Unified National Rolled, abbreviated UNR, was introduced into the 1974 ANSI standards. This designation is specified only for external threads—there is no UNR designation for internal threads. The UN form has a flat root (rounded root is

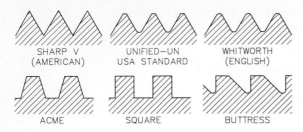

FIG. 18.3 Standard thread forms.

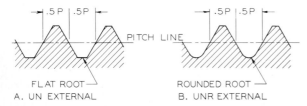

FIG. 18.4 A. The UN external thread has a flat root (rounded root is optional). **B.** The UNR has a rounded root formed by rolling. The UNR form does not apply to internal threads.

optional) in Fig. 18.4A, whereas the UNR thread *must* have a rounded root formed by rolling, as shown in Fig. 18.4B. The rounded root of the UNR thread is designed to reduce the wear of the threading tool and to improve the fatigue strength of the thread.

The transmission of power is achieved by the use of the *Acme, square,* and *buttress* threads that are commonly used in gearing and other pieces of machinery (Fig. 18.4). The *sharp V* thread is used for set screws and in applications where friction in assembly is desired.

Series

Thread series, which is closely related to thread form, designates the type of thread specified for a given application and is abbreviated C, F, and EF.

There are eleven standard series of threads listed under the American National form and the Unified National (UN/UNR) form. There are three series, with abbreviations coarse (C), fine (F), and extra fine (EF), and eight series with constant pitches (4, 6, 8, 12, 16, 20, 28, and 32 threads per inch).

A Unified National form for a coarse-series thread is specified as UNC or UNRC, which is a combination of form and series in a single note. Similarly, an American National form for a coarse thread is written NC. The *coarse-thread* series (UNC/UNRC or NC) is suitable for bolts, screws, nuts, and general use with cast iron, soft metals, or plastics when rapid assembly is desired. The *fine-thread* series (NF or UNF/UNRF) is suitable for bolts, nuts, or screws when a high degree of tightening is required. The *extra-fine*-thread series (UNEF/UNREF or NEF) is suitable for sheet metal, thin nuts, ferrules, or couplings when length of engagement is limited and is used for applications that will have to withstand high stresses.

The 8-thread series (8 UN), 12-thread series, (12 N or 12 UN/UNR), and 16-thread series (16 N or 16 UN/UNR) are threads with a uniform pitch for large diameters. The 8 UN is used as a substitute for the coarse-thread series on diameters larger than 1 inch when a medium-pitch thread is required. The 12 UN is used on diameters larger than $1\frac{1}{2}$ inches, with a thread of a medium-fine pitch as a continuation of the fine-thread series. The 20 UN is used on diameters larger than $1\frac{11}{16}$ inches, with threads of an extra-fine pitch as a continuation of the extra-fine series.

Class of fit

Thread classes are used to indicate the tightness of fit between a nut and a bolt or any two mating threaded parts. This fit is determined by the tolerances and allowances applied to threads. Classes of fit are indicated by the numbers 1, 2, or 3 followed by the letters *A* or *B*. For UN forms, the letter *A* represents an external thread, whereas the letter *B* represents an internal thread. These letters are omitted when the American National form is used.

CLASS 1A AND 1B threads are used on parts that require assembly with a minimum of binding.

CLASS 2A AND 2B threads are general-purpose threads for bolts, nuts, screws, and nominal applications in the mechanical field and are widely used in the mass-production industries.

CLASS 3A AND 3B Threads are used in precision assemblies where a close fit is desired to withstand stresses and vibration.

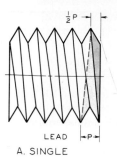
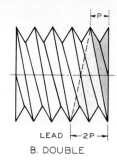
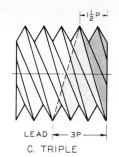

LEAD ⊢P→
A. SINGLE

LEAD ⊢2P→
B. DOUBLE

LEAD ⊢ 3P →
C. TRIPLE

FIG. 18.5 Single and multiple threads.

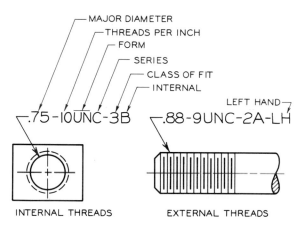

.75 - 10UNC - 3B

- MAJOR DIAMETER
- THREADS PER INCH
- FORM
- SERIES
- CLASS OF FIT
- INTERNAL

LEFT HAND

.88 - 9UNC - 2A - LH

INTERNAL THREADS   EXTERNAL THREADS

FIG. 18.6 Parts of a thread note for an external thread.

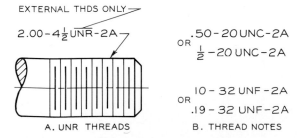

EXTERNAL THDS ONLY

2.00 - 4½ UNR - 2A

A. UNR THREADS

.50 - 20 UNC - 2A
OR
½ - 20 UNC - 2A

10 - 32 UNF - 2A
OR
.19 - 32 UNF - 2A

B. THREAD NOTES

FIG. 18.7 A. The UNR thread notes apply to external threads only. **B.** Notes can be given as decimal fractions or as common fractions.

Single and multiple threads

A *single thread* (Fig. 18.5A) is a thread that will advance the distance of its pitch in one full revolution of 360°; in other words, its pitch is equal to its lead. In the drawing of a single thread, the crest line of the thread will slope $\frac{1}{2}P$, since only 180 degrees of the revolution is visible in a single view. A double thread is composed of two threads, resulting in a lead equal to $2P$, meaning that the threaded part will advance a distance of $2P$ in a single revolution of 360° (Fig. 18.5B). The crest line of a double thread will slope a distance equal to P in the view in which 180 degrees can be seen. Similarly, a triple thread will advance $3P$ in 360° with a crest line slope of $1\frac{1}{2}P$ in the view in which 180 degrees of the cylinder is visible (Fig. 18.5C). The lead of a double thread is $2P$, and the lead of a triple thread is $3P$. Although power on multiple threads is limited, they are used wherever quick assembly is required.

Thread notes

Drawings of threads are only symbolic representations and are inadequate unless accompanying notes give the thread specifications (Fig. 18.6). The major diameter is given first, followed by the number of threads per inch, the form and series, the class of fit, and a letter denoting whether the thread is external or internal. This completes the note for a single right-hand thread. However, for a double or triple thread, the word *DOUBLE* or *TRIPLE* is included in the note; for a left-hand thread, the letters *LH* are included.

The UNR thread note is shown for the external thread in Fig. 18.7A (UNR does not apply to internal threads). When inches are used as the unit of measurement, thread notes can be written as common fractions, but decimal fractions are preferred.

18.4
Using thread tables

The UN/UNR thread table is given in Appendix 15. A portion of this table is shown in Table 18.1. If an external thread (bolt) with a 1.500-inch diameter is to have a "fine" thread, it will have twelve threads per inch. Therefore the thread note can be written

$$1\frac{1}{2}\text{–}12 \text{ UNF–2A} \quad \text{or} \quad 1.500\text{–}12 \text{ UNF–2A}.$$

If the thread had been an internal one (nut), the thread note would have been the same, but the letter B would have been used instead of the letter A.

A constant-pitch thread series can be selected for the larger diameters. The constant-pitch thread notes are written with the abbreviations C, F, and EF omit-

ted. For example, a $1\frac{3}{4}$-inch diameter bolt with a fine thread could be noted in constant-pitch series as

$$1\frac{3}{4}\text{–}12 \text{ UN–2A} \quad \text{or} \quad 1.750\text{–}12 \text{ UN–2A}.$$

This table can also be used for the UNR thread form (for external threads only) by substituting UNR in place of UN; for example, UNEF can be written UNREF.

18.5
Metric thread specifications (ISO)

Metric thread specifications are recommended by the ISO (International Organization for Standardization). Thread specifications can be given with a *basic designation*, which is suitable for general applications, or with the *complete designation*, which is used where detailed specifications are needed.

TABLE 18.1
AMERICAN NATIONAL STANDARD UNIFIED INCH SCREW THREADS (UN AND UNR THREAD FORM)*

						Threads per inch								
			Series with graded pichers			Series with constant pitches								
Sizes		Basic major diameter	Coarser UNC	Fine UNF	Extra fine UNEF	UN	6 UN	8 UN	12 UN	16 UN	20 UN	28 UN	32 UN	Sizes
Primary	Secondary													
1		1,0000	8	12	20	—	—	UNC	UNF	16	UNEF	28	32	1
	$1\frac{1}{16}$	1.0625	—	—	18	—	—	8	12	16	20	28	—	$1\frac{1}{16}$
$1\frac{1}{8}$		1.1250	7	12	18	—	—	8	UNF	16	20	28	—	$1\frac{1}{8}$
	$1\frac{3}{16}$	1.1875	—	—	18	—	—	8	12	16	20	28	—	$1\frac{3}{16}$
$1\frac{1}{4}$		1.2500	7	12	18	—	—	8	UNF	16	20	28	—	$1\frac{1}{4}$
	$1\frac{5}{16}$	1.3125	—	—	18	—	—	8	12	16	20	28	—	$1\frac{5}{16}$
$1\frac{3}{8}$		1.3750	6	12	18	—	UNC	8	UNF	16	20	28	—	$1\frac{3}{8}$
	$1\frac{7}{16}$	1.4375	—	—	18	—	6	8	12	16	20	28	—	$1\frac{7}{16}$
$1\frac{1}{2}$		1.5000	6	12	18	—	UNC	8	UNF	16	20	28	—	$1\frac{1}{2}$
	$1\frac{9}{16}$	1.5625	—	—	18	—	6	8	12	16	20	—	—	$1\frac{9}{16}$
$1\frac{5}{8}$		1.6250	—	—	18	—	6	8	12	16	20	—	—	$1\frac{5}{8}$
	$1\frac{11}{16}$	1.6875	—	—	18	—	6	8	12	16	20	—	—	$1\frac{11}{16}$
$1\frac{3}{4}$		1.7500	5	—	—	—	6	8	12	16	20	—	—	$1\frac{3}{4}$
	$1\frac{13}{16}$	1.8125	—	—	—	—	6	8	12	16	20	—	—	$1\frac{13}{16}$
$1\frac{7}{8}$		1.8750	—	—	—	—	6	8	12	16	20	—	—	$1\frac{7}{8}$
	$1\frac{15}{16}$	1.9375	—	—	—	—	6	8	12	16	20	—	—	$1\frac{15}{16}$

*By using this table, a diameter of $1\frac{1}{2}$ inches that is to be threaded with a fine thread would have the following thread note: $1\frac{1}{2}$-12 UNF-2A. (Courtesy of ANSI; B1.1).

TABLE 18.2
BASIC THREAD DESIGNATIONS FOR COMMERCIAL SERIES OF ISO METRIC THREADS

Nominal size (mm)	Pitch P (mm)	Basic thread designation*	Nominal size (mm)	Pitch P (mm)	Basic thread designation*	Nominal size (mm)	Pitch P (mm)	Basic thread designation*
1.6	0.35	M1.6		1.25	M8		2.5	M22
1.8	0.35	M1.8	8	1	M8 × 1	22	1.5	M22 × 1.5
2	0.4	M2		1.5	M10		3	M24
2.2	0.45	M2.2	10	1.25	M10 × 1.25	24	2	M24 × 2
2.5	0.45	M2.5		1.75	M12		3	M27
3	0.5	M3	12	1.25	M12 × 1.25	27	2	M27 × 2
3.5	0.6	M3.5		2	M14		3.5	M30
4	0.7	M4	14	1.5	M14 × 1.5	30	2	M30 × 2
4.5	0.75	M4.5		2	M16		3.5	M33
5	0.8	M5	16	1.5	M16 × 1.5	33	2	M33 × 2
6	1	M6		2.5	M18		4	M36
7	1	M7	18	1.5	M18 × 1.5	36	3	M36 × 3
				2.5	M20		4	M39
			20	1.5	M20 × 1.5	39	3	M39 × 3

*U.S. practice is to include the pitch symbol even for the coarse pitch series. Basic descriptions shown are as specified in ISO Recommendations.
Source: Courtesy of Greenfield Tap and Die Corp.

Basic designation

Examples of metric screw thread notes are shown in Fig. 18.8. Each note begins with the letter *M*, which designates the note as a metric note, followed by the diameter in millimeters, and the pitch in millimeters separated by ×, the multiplication sign. The pitch can be omitted in notes for coarse threads, but U.S. standards prefer that it be shown. Table 18.2 shows the commercially available ISO threads recommended for general use. Additional ISO specifications are given in Appendix 18.

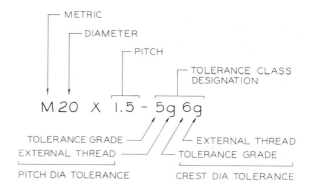

FIG. 18.9 A complete designation note for metric threads.

Complete designation

For some applications it is necessary to show a complete thread designation (Fig. 18.9). The first part of this note is the same as the basic designation; however, the note also has a tolerance class designation separated by a dash. The 5g represents the pitch diameter tolerance, and 6g represents the crest diameter tolerance.

The numbers 5 and 6 are *tolerance grades* (variations from the basic diameter). Grade 6 is commonly used for a medium general-purpose thread that is

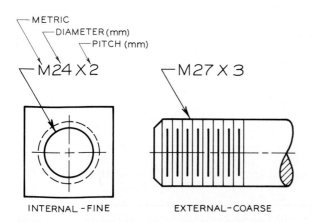

FIG. 18.8 Basic designations for metric threads.

TABLE 18.3
TOLERANCE GRADES, ISO THREADS

External thread		Internal thread	
Major diameter (d_1)	Pitch diameter (d_2)	Minor diameter (D_1)	Pitch diameter (D_2)
—	3	—	—
4	4	4	4
—	5	5	5
6	6	6	6
—	7	7	7
8	8	8	8
—	9	—	—

Grade 6 is medium; smaller numbers are finer and larger numbers are coarser. (Courtesy of ANSI B1).

nearly equal to class 2A and 2B of the Unified system. Grades less than 6 are used for fine-quality fits and short lengths of engagement. Grades greater than 6 are recommended for coarse-quality fits and long lengths of engagement. Table 18.3 gives the tolerance grades for internal and external threads for the pitch diameter and the major and minor diameters.

The letters following the grade numbers designate *tolerance positions*. Lowercase letters represent external threads (bolts), as shown in Fig. 18.10. The lowercase letters *e*, *g*, and *n* represent large allowance, small allowance, and no allowance, respectively. (Allowance is the variation from the basic diameter.) Uppercase letters designate internal threads (nuts); *G* designates small allowance, and *H* designates no allowance. The letters are placed after the tolerance grade number. For example, 5g designates a medium tolerance with small allowance for the pitch diameter of an external thread, and 6H designates a medium tolerance with no allowance for the minor diameter of an internal thread.

Tolerance classes are fine, medium, and coarse, as listed in Table 18.4. These classes of fit are combinations of tolerance grades, tolerance positions, and lengths of engagement—short (S), normal (N), and long (L). The length of engagement can be determined by referring to Appendix 17. Once it has been decided to use a fine, medium, or coarse class of fit for a particular application, the specific designation should be selected first from the classes shown in large print in Table 18.4, second from the classes shown in me-

TOLERANCE POSITIONS

EXTERNAL THREADS (Lower–case Letters)	INTERNAL THREADS (Upper–case Letters)
e = LARGE ALLOWANCE	G = SMALL ALLOWANCE
g = SMALL ALLOWANCE	H = NO ALLOWANCE
n = NO ALLOWANCE	

LENGTH OF ENGAGEMENT

S = SHORT   N = NORMAL   L = LONG

FIG. 18.10 Symbols used to represent tolerance grade, position, and class.

TABLE 18.4
PREFERRED TOLERANCE CLASSES, ISO THREADS*

Quality	External threads (bolts)									Internal threads (nuts)					
	Tolerance position e (large allowance)			Tolerance position g (small allowance)			Tolerance position h (no allowance)			Tolerance position G (small allowance)			Tolerance position H (no allowance)		
	Length of engagement			Length of engagement			Length of engagement			Length of engagement			Length of engagement		
	Group S	Group N	Group L	Group S	Group N	Group L	Group S	Group N	Group L	Group S	Group N	Group L	Group S	Group N	Group L
Fine Medium Coarse		6e 7e6e		5g6g	6g 8g	7g6g 9g8g	3h4h 5h6h	4h 6h	5h4h 7h6h	5G	6G 7G	7G 8G	4H 5H	5H 6H 7H	6H 7H 8H

*In selecting tolerance class, select first from the large bold print, second from the medium-size print, and third from the small-size print. Classes shown in boxes are for commercial threads.

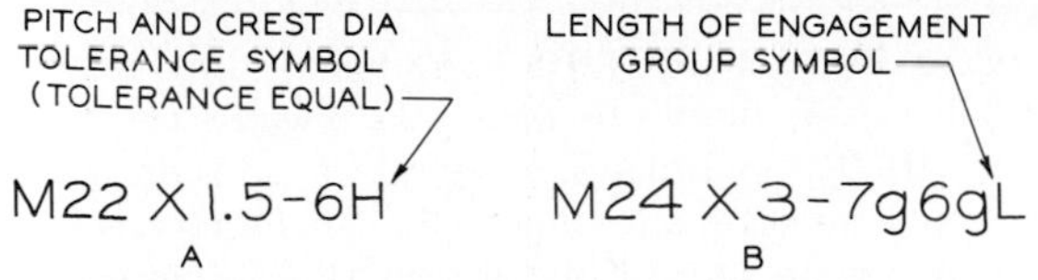

PITCH AND CREST DIA
TOLERANCE SYMBOL
(TOLERANCE EQUAL)

LENGTH OF ENGAGEMENT
GROUP SYMBOL

M22 X 1.5-6H

A

M24 X 3-7g6gL

B

FIG. 18.11 A. When both pitch and crest diameter tolerance grades are the same, the tolerance class symbol is shown only once. **B.** Letters *S*, *N*, and *L* are used to indicate the length of the thread engagement.

dium-size print, and third from the classes shown in small print. Classes shown in boxes are for commercial threads.

Variations in the complete designation thread notes are shown in Fig. 18.11. The tolerance class symbol is written 6H if the crest and pitch diameters have identical grades (Fig. 18.11A). Since an uppercase *H* is used, this is an internal thread. Where considered necessary, the length-of-engagement symbol may be added to the tolerance class designation (Fig. 18.11B).

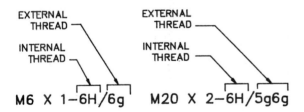

EXTERNAL
THREAD

INTERNAL
THREAD

EXTERNAL
THREAD

INTERNAL
THREAD

M6 X 1-6H/6g   M20 X 2-6H/5g6g

FIG. 18.12 A slash mark is used to separate the tolerance class designations of mating internal and external threads.

Designations for the desired fit between mating threads can be specified as shown in Fig. 18.12. A slash is used to separate the tolerance class designations of the internal and external threads.

Additional information pertaining to ISO threads may be obtained from *ISO Metric Screw Threads*, a booklet of standards published by ANSI. These standards were used as the basis for most of this section.

18.6
Thread representation

Three major types of thread representations are the simplified, schematic, and detailed (Fig. 18.13). The detailed representation is the most realistic approximation of a thread's appearance, and the simplified representation is the least realistic.

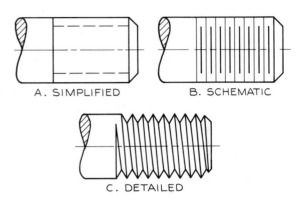

A. SIMPLIFIED   B. SCHEMATIC

C. DETAILED

FIG. 18.13 Three major types of thread representations.

18.7
Detailed UN/UNR threads

Examples of detailed representations of internal and external threads are shown in Figs. 18.14. Instead of helical curves, straight lines are used to indicate crest and root lines.

The construction of a detailed thread representation is shown in Fig. 18.15. The pitch is found by dividing 1 inch by the number of threads per inch. This can be done graphically, as shown in Step 1. However, in most cases this construction is unnecessary, since the pitch can be approximated or laid off with a scale or dividers. Where threads are close, they should be drawn at a larger spacing to be easier to draw. Note in Step 4 that forty-five degree chamfer is used to draw a bevel of the threaded end to improve ease of assembly of the threaded parts.

Metric threads should be drawn in the same manner, using the pitch given in millimeters in the metric thread table, or an expanded pitch, if needed.

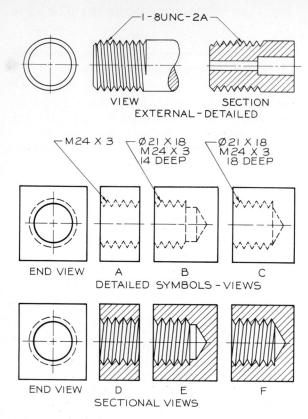

FIG. 18.14 Detailed thread representations of external and internal threads.

18.8
Detailed square threads

Figure 18.17 shows the method of drawing a detailed representation of a square thread to give an approximation of a square thread.

In Step 1, the major diameter is laid off. The number of threads per inch is taken from Appendix 19. The pitch (P) is found by dividing 1 inch by the number of threads per inch, but this pitch can be expanded if needed. Distances of P/2 are marked off with dividers. Steps 2, 3, and 4 are then completed, and a thread note is added.

Square internal theads are drawn in a similar manner (Fig. 18.18). The threads in the section view are drawn in a slightly different way. The thread note

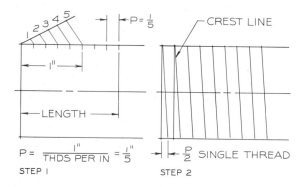
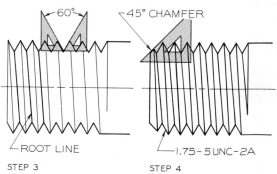

FIG. 18.15 Detailed thread representation.

Step 1   To draw a detailed representation of a 1.75–5 UNC–2A thread, the pitch is determined by dividng 1″ by the number of threads per inch, 5 in this case. The pitch is laid off the length of the thread.

Step 2   Since this is a right-hand thread, the crest lines slope downward to the right equal to ½P. The crest lines will be final lines drawn with an H or F pencil.

Step 3   The root lines are found by constructing 60° vees between the crest lines. The root lines are drawn from the bottom of the vees. Root lines are parallel to each other, but not to crest lines.

Step 4   A 45° chamfer is constructed at the end of the thread from the minor diameter. Strengthen all lines and add a thread note.

310

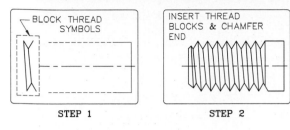

FIG. 18.16 Detailed threads by computer.

Step 1 A typical detailed thread symbol is drawn and BLOCKed by using a window.

Step 2 The BLOCK is repetitively INSERTed along the centerline. A chamfer is drawn at the left end, and the right end is completed.

for an internal thread is placed in the circular view whenever possible, with the leader pointing toward the center. When a square thread is rather long, it need not be drawn continuously but can be represented using the symbol shown in Fig. 18.19.

18.9
Detailed Acme threads

The methods of preparing detailed drawings of Acme threads is shown in Fig. 18.20. The length and the major diameter are laid off with light construction lines. In Step 1, the pitch is found by dividing 1 inch

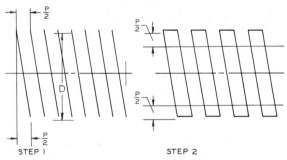

FIG. 18.17 Drawing the square thread.

Step 1 Lay out the major diameter. Space the crest lines $\frac{1}{2}P$ apart. Slope them downward to the right for right-hand threads.

Step 2 Connect every other pair of crest lines. Find the minor diameter by measuring $\frac{1}{2}P$ inward from the major diameter.

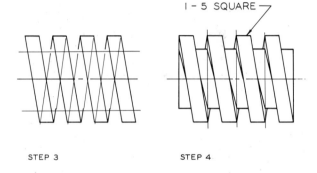

Step 3 Connect the opposite crest lines with light construction lines. This will establish the profile of the thread form.

Step 4 Connect the inside crest lines with light construction lines to locate the points on the minor diameter where the thread wraps around the minor diameter.

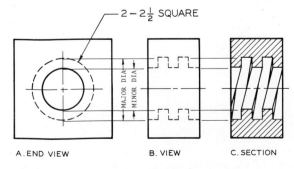

FIG. 18.18 Internal square threads.

by the number of threads per inch. Steps 2, 3, and 4 complete the thread representation, and the thread note is added.

Internal Acme threads are shown in Fig. 18.21. Note that in the section view, left-hand internal threads are sloped so that they look the same as right-hand external threads.

Figure 18.22 shows a shaft that is being threaded with Acme threads on a lathe as the tool travels the length of the shaft.

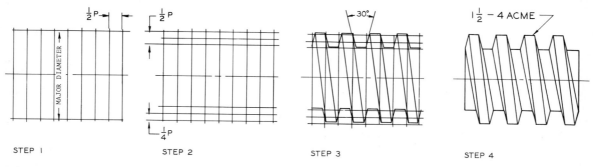

FIG. 18.19 Conventional method of showing square threads without drawing each thread.

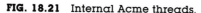

FIG. 18.20 Drawing the Acme thread.

Step 1   Lay out the major diameter and the thread length, and divide the shaft into equal divisions $\frac{1}{2}P$ apart.

Step 2   Locate the minor diameter a distance $\frac{1}{2}P$ inside the major diameter. Locate the pitch diameter between the major and minor diameters.

Step 3   Draw construction lines at 15° with the vertical along the pitch diameter as shown to make a total angle of 30°. Draw the crest lines and the thread profile.

Step 4   Darken the lines, draw the root lines, and add the thread note to complete the drawing.

FIG. 18.21 Internal Acme threads.

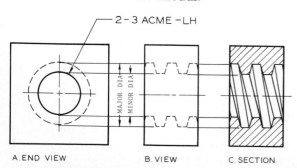

A. END VIEW   B. VIEW   C. SECTION

FIG. 18.22 Cutting an Acme thread on a lathe. (Courtesy of Clausing Corp.)

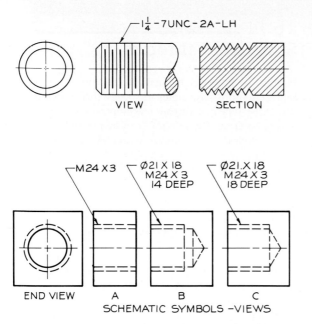

VIEW   SECTION

END VIEW   A   B   C

SCHEMATIC SYMBOLS –VIEWS

END VIEW   D   E   F

SECTIONAL VIEWS

FIG. 18.23 A. Schematic representations of external threads. **B.** Schematic representations of internal threads.

18.10
Schematic threads

Schematic representations of internal and external threads are shown in Figs. 18.23A and B with parallel nonsloping lines. Left-hand threads must be specified in the thread note. Since the schematic representation is easy to draw and gives a good symbolic representation of threads, it is the most often used thread symbol.

The method of constructing schematic threads is illustrated in Fig. 18.24. The method of drawing metric threads using schematic representations is shown in Fig. 18.25. The pitch (in millimeters) given in the metric thread table is used as the approximate distance to separate the crest lines.

> **COMPUTER METHOD** Schematic threads can be drawn by computer as shown in Fig. 18.26. In Step 1, a typical major and minor diameter is drawn and a BLOCK is made of them by a window. In Step 2, the BLOCK of the threads are repetitively inserted to complete the thread symbols.

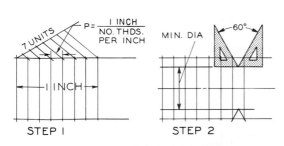

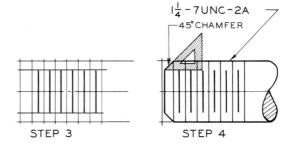

STEP 1   STEP 2   STEP 3   STEP 4

FIG. 18.24 Drawing schematic threads.

Step 1 Lay out the major diameter, and divide the shaft into divisions of a distance P apart. These crest lines should be drawn as thin lines.

Step 2 Find the minor diameter by drawing a 60° angle between two crest lines on each side.

Step 3 Draw heavy root lines between the crest lines.

Step 4 Chamfer the end of the thread from the minor diameter, and give a thread note.

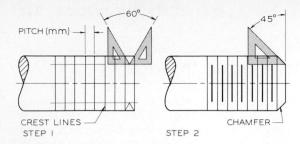

FIG. 18.25 Schematic metric threads.

Step 1 The pitch of metric threads can be taken directly from the metric tables, which can be used to find the minor diameter.

Step 2 The root lines are drawn in heavy between the crest lines. The end of the thread is chamfered.

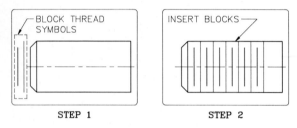

FIG. 18.26 Schematic threads by computer.

Step 1 Draw the outline of the threaded shaft with a chamfered end. Draw typical minor and major diameters, and make them into a BLOCK by using a window.

Step 2 Repetitively INSERT the BLOCKS of the threads along the centerline to complete the thread representation.

18.11
Simplified threads

Figures 18.27A and B illustrate the use of simplified representations with notes to specify thread details. Of the three types of thread representations, this is the easiest to draw. Hidden lines can be positioned by eye to approximate the minor diameter. The steps involved in constructing a simplified thread drawing are shown in Fig. 18.28.

18.12
Drawing small threads

Instead of drawing small threads to exact measurements, minor diameters can be drawn smaller to separate the root and crest lines (Fig. 18.29). This procedure makes the thread easier to draw and read. Since the drawing is only a symbolic representation of a thread, exactness is unnecessary. Schematic threads are drawn with crest lines farther apart to prevent crowding of lines. For both internal and external threads, a thread note completes the symbolic drawing

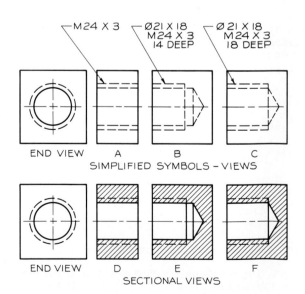

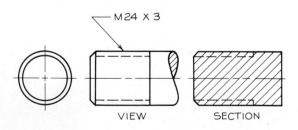

FIG. 18.27 **A.** Simplified thread representations of external threads. **B.** Simplified thread representations of internal threads.

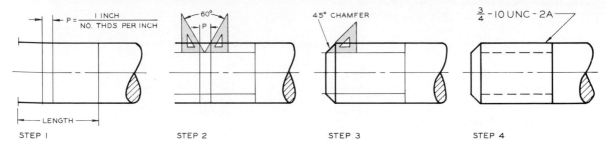

FIG. 18.28 Drawing simplified threads.

Step 1 Lay out the major diameter. Find the pitch (P), and lay out two lines a distance P apart.

Step 2 Find the minor diameter by constructing a 60° angle between the two lines on both sides.

Step 3 Draw a 45° chamfer from the minor diameter to the major diameter.

Step 4 Show the minor diameter as a dashed line. Add a thread note.

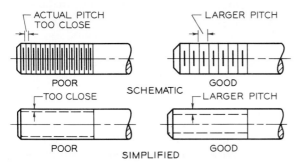

FIG. 18.29 Simplified and schematic threads should be drawn using approximate dimensions if the actual dimensions would result in lines drawn too close together.

FIG. 18.30 Examples of nuts and bolts. (Courtesy of Russell, Burdsall & Ward Bolt and Nut Co.)

by giving necessary specifications and description of the part.

18.13
Nuts and bolts

Nuts and bolts come in many forms and sizes for different applications (Fig. 18.30). The more common types of threaded fasteners are shown in Fig. 18.31. A *bolt* is a threaded cylinder with a head and a *nut* for holding two parts together (Fig. 18.31A). A *stud* does not have a head but is screwed into one part with a nut attached to the other threaded end (Fig. 18.31B). A *cap screw* is similar to a bolt, but it does not have a nut; instead, it is screwed into a member with internal threads (Fig. 18.31C). A *machine screw* is similar to, but smaller than, a cap screw (Fig. 18.31D). A *set screw* is used to adjust one member with respect to another, usually to prevent a rotational movement (Fig. 18.31E).

The types of heads used on standard bolts and nuts are illustrated in Fig. 18.32. These heads are used on both *regular* and *heavy* bolts; the thickness of the head is the primary difference between the two types. Heavy-series bolts have the thicker heads and are used at points where bearing loads are heaviest. Bolts and nuts are either *finished* or *unfinished*. Figure 18.32 shows an unfinished head; that is, none of the surfaces of the head are machined. The finished head has

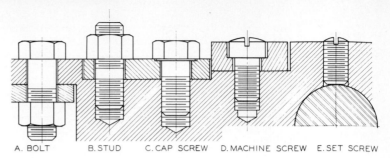

FIG. 18.31 Types of threaded bolts and screws.

A. BOLT   B. STUD   C. CAP SCREW   D. MACHINE SCREW   E. SET SCREW

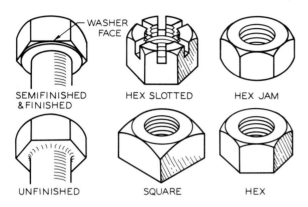

SEMIFINISHED & FINISHED   HEX SLOTTED   HEX JAM

UNFINISHED   SQUARE   HEX

FIG. 18.32 Types of finishes for bolt heads and types of nuts.

a washer face that is $\frac{1}{64}$-inch thick to provide a circular boss on the bearing surface of the bolt head or the nut.

Other standard forms of bolt and screw heads shown in Fig. 18.33 are used primarily on cap screws and machine screws. Standard types of nuts are illustrated in Fig. 18.32. Finished nuts have washer faces

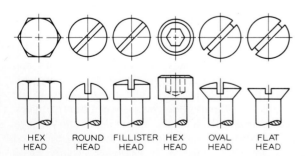

HEX HEAD   ROUND HEAD   FILLISTER HEAD   HEX HEAD   OVAL HEAD   FLAT HEAD

FIG. 18.33 Common types of bolt and screw heads.

for more accurate assembly. A hexagon *jam nut* does not have a washer face, but it is chamfered on both sides.

Although ANSI tables in Appendixes 21–25 indicate the standard bolt lengths and their corresponding thread lengths, the following can be used as a general guide for square- and hexagon-head bolts;

- Hexagon bolt lengths are available in $\frac{1}{4}$-inch increments up to 8-inches long, in $\frac{1}{2}$-inch increments from 8-to-20-inches long, and in 1-inch increments from 20-to-30-inches long.

- Square-head bolt lengths are available in $\frac{1}{8}$-inch increments from $\frac{1}{2}$-to-$\frac{3}{4}$-inch long, in $\frac{1}{4}$-inch increments from $\frac{3}{4}$-to-5-inches long, in $\frac{1}{2}$-inch increments from 5-to-12-inches long, and in 1-inch increments from 12-to-30-inches long.

- The lengths of the threads on both hexagon- and square-head bolts up to 6 inches long can be found by the formula: Thread length = $2D + \frac{1}{4}$ in., where D is the diameter of the bolt. The threaded length for bolts more than 6 inches long can be found by the formula: Thread length = $2D + \frac{1}{2}$ in.

- The threads for bolts can be coarse, fine, or 8-pitch threads. It is understood that the class of fit for bolts and nuts will be 2A and 2B if no class is specified.

- Standard square- and hexagon-head bolts are designated by notes in one of the following forms:

$\frac{3}{8}$–16 × 1$\frac{1}{2}$ SQUARE BOLT—STEEL;
$\frac{1}{2}$–13 × 3 HEX CAP SCREW,
SAE GRADE 8—STEEL;
0.75 × 5.00 HEX LAG SCREW—STEEL.

The numbers represent bolt diameter, threads per inch (omit for lag screws), length, name of screw,

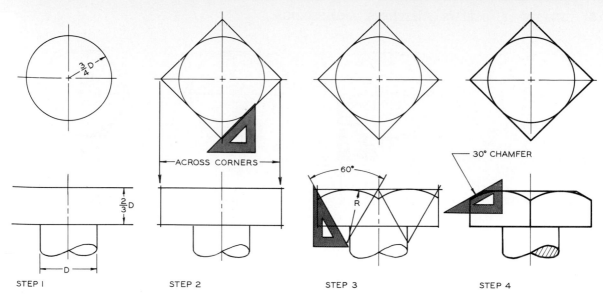

FIG. 18.34 Drawing the square head.

Step 1 Draw the diameter of the bolt. Use the diameter (D) to establish the head diameter and thickness.

Step 2 Draw the top view of the square head with a 45° triangle to give an across-corners view.

Step 3 Show the chamfer in the front view by using a 30°–60° triangle to find the centers for the radii.

Step 4 Show a 30° chamfer tangent to the arcs in the front view. Strengthen the lines.

and material. It is understood that these will have a class 2 fit.

- Nuts are designated by notes in one of the following forms:

  ½–13 SQUARE NUT—STEEL;
  ¾–16 HEAVY HEX NUT, SAE GRADE 5—STEEL;
  1.00–8 HEX THICK SLOTTED NUT—
  CORROSION RESISTANT STEEL

When nuts are not noted as "HEAVY," they are assumed to be REGULAR. The class of fit is assumed to be 2B for nuts when not noted.

18.14
Drawing square bolt heads

Detailed tables are available in Appendix 20 for various types of threaded parts. In most cases it is sufficient to draw nuts and bolts with only general proportions.

The first step in drawing a bolt head or a nut is to determine whether it is to be *across corners* or *across flats*—that is, are the outlines at either side of the view going to represent corners, or edge views of flat sur-

faces of the part? Nuts and bolts should be drawn across corners for the best representation (Fig. 18.34).

18.15
Drawing hexagon bolt heads

An example of constructing the head of a hexagon-head bolt across corners is shown in Fig. 18.35.

Note that the diameter of the bolt is *D*. The thickness of the head is drawn equal to ⅔*D*. The top view of the head is drawn as a circle with a radius of ¾*D*. This proportionality based on *D* is sufficient for drawing bolt heads in most applications.

> **COMPUTER METHOD** A bolt head can be drawn in sequence, as shown in Step 1 of Fig. 18.36, by using a bolt diameter of 1 inch to make a UNIT BLOCK after the drawing has been completed. Step 2 illustrates how the BLOCK can be reduced and rotated to any position. To draw a thread diameter of 0.50 inch, a size factor of 0.50 is assigned when prompted by the BLOCK and INSERT commands.

317

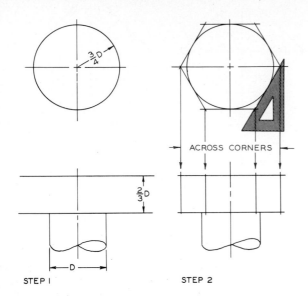
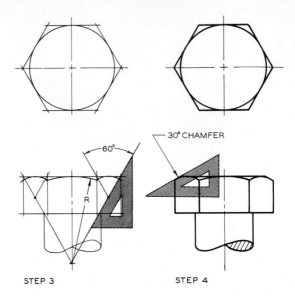

FIG. 18.35 Drawing the hexagon head.

Step 1   Draw the diameter of the bolt and use it to establish the head diameter and thickness.

Step 2   Construct a hexagon with a 30°–60° triangle to give an across-corners view.

Step 3   Find arcs in the front view to show the chamfer of the head.

Step 4   Draw a 30° chamfer tangent to the arcs in the front view. Strengthen the lines.

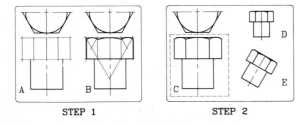

FIG. 18.36 Bolt head by computer.

Step 1   A hexagon head is drawn using the steps of geometric construction covered in the previous figure. The size of the head is based on a bolt diameter of 1 inch to form a UNIT BLOCK.

Step 2   The drawing is made into a BLOCK by using a window. The block can be INSERTed at any size and positioned as shown at D and E.

18.16
Drawing nuts

The drawing of a square and a hexagon nut across corners is the same as the drawing of a bolt head across corners. The only variation is the thickness of the nut. The REGULAR nut thickness is $\frac{7}{8}D$, and for the HEAVY nut, the thickness is $1 \times D$ (D = diameter).

Square and hexagon nuts drawn across corners are shown in Fig. 18.37. Hidden lines are shown in the front view to indicate threads. Since it is understood that nuts are threaded, these hidden lines may be omitted in general applications.

A $\frac{1}{64}$ inch washer face is shown on the hexagon nut. It is usually drawn thicker than $\frac{1}{64}$ inch to make the face more noticeable in the drawing. Thread notes are placed in the circular views where possible. Since the note for square nut is not labeled "HEAVY," it is a REGULAR square nut. The hexagon nut is similar, except that it is a finished hexagon nut. The leader from the note is directed toward the center of the cir-

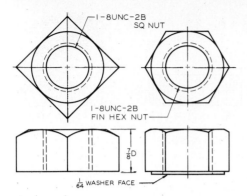

FIG. 18.37 Drawings of hexagon and square nuts are constructed in the same manner as drawings of bolt heads. Standard notes are added to give nut specifications.

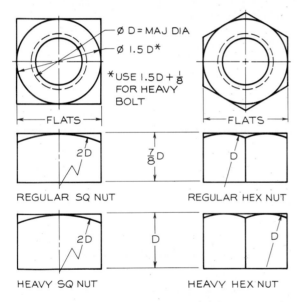

FIG. 18.38 Examples of hexagon and square nuts drawn across flats. Notes are added to give nut specifications.

cular view, but the arrow stops at the first visible circle it makes contact with.

Nuts can be drawn across flats if doing so improves the drawing (Fig. 18.38). For regular nuts, the distance across flats if $1\frac{1}{2} \times D$ (D-major diameter of the thread); for heavy nuts, this distance is $1\frac{5}{8} \times D$. The top views are drawn in the same manner as in across-corners drawings, except that they are positioned to give different front views.

A hexagon nut drawn across corners is a better representation than a nut drawn across flats (Fig. 18.35). To complete the representation of the nuts, notes are added with leaders. A washer face should be added to a nut if it is finished—except for the square nut, which is always unfinished.

18.17
Drawing nuts and bolts in combination

The rules followed when drawing nuts and bolts separately also apply when drawing nuts and bolts in assembly (Fig. 18.39). The diameter, D, of the bolt is used as the basis for other dimensions. The note is added to give the specifications of the nut and bolt. In Fig. 18.39, the bolt heads are drawn across corners, and the nuts across flats. The half-end views show how the front views were found by projection.

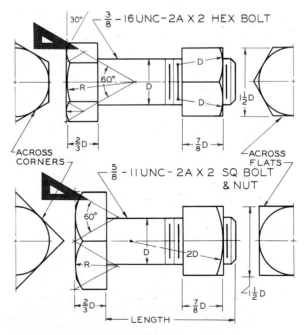

FIG. 18.39 Construction of nuts and bolts in assembly.

18.18

Cap screws

Cap screws are used to hold two parts together without a nut since one of these parts has a threaded cylindrical hole. The other part is drilled with an oversized hole so that the cap screw will pass through it freely. When the cap screw is tightened, the two parts are held securely together.

The standard types of cap screws are illustrated in Fig. 18.40. Appendixes 22 and 23 give the dimensions of several types of cap screws. In Fig. 18.40, the cap screws are drawn on a grid to show their proportions, which can be used for drawing cap screws of all sizes, ranging in diameter from No. 0 (0.060) to No. 12.

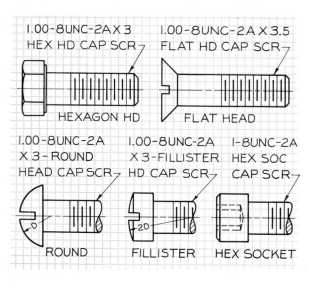

FIG. 18.40 These proportions of the standard types of cap screws can be used for drawing cap screws of all sizes. Notes provide typical specifications.

18.19

Machine screws

Machine screws are smaller than most cap screws, usually less than 1 inch in diameter. The machine screw is used to attach parts together; it is screwed either into another part or into a nut. Machine screws are fully threaded when their length is 2 inches or shorter.

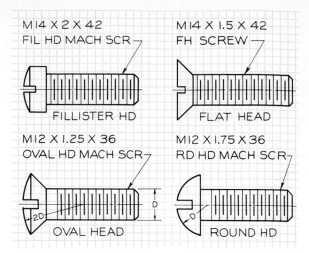

FIG. 18.41 Standard types of machine screws. These proportions can be used for drawing machine screws of all sizes.

Drawings of a few of the common machine screws and their notes are given in Fig. 18.41. The dimensions of round-head machine screws are given in Appendix 24. The four types of machine screws in Fig. 18.41 are drawn on a grid to give proportions of the head in relation to the major diameter of the screw. Machine screws range in diameter from No. 0 (0.060 inch) to $\frac{3}{4}$ inch.

When slotted-head screws are drawn, conventional practice is to show the slots positioned at a forty-five-degree angle in the circular view (Fig. 18.33) and the front view shows the width and depth of the slot.

18.20

Set screws

Set screws or keys are used to affix parts like pulleys and handles on a shaft. Various types of set screws are shown in Fig. 18.42 and Table 18.5 shows their dimensions. Set screws, like the other fasteners, can be drawn as approximations.

Set screws are available in combinations of points and heads. The shaft against which the set screw is tightened may have a flat surface machined to give a good bearing surface for the set screw point; if so, a dog point or flat point would be most effective to press against the flat surface. The cup point gives good friction when applied to a round shaft. Specifications for set screws are given in Appendixes 27–29.

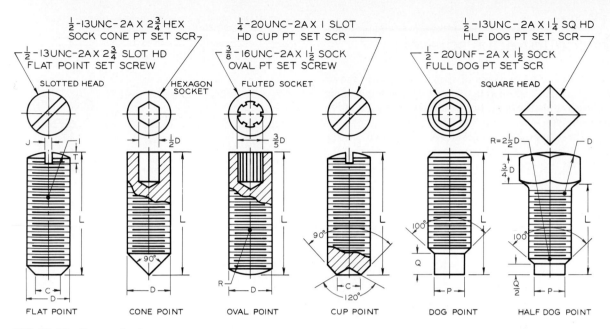

FIG. 18.42 Types of set screws. Set screws are available with various combinations of heads and points. Notes give their specifications. Dimensions are given in Table 18.5.

TABLE 18.5
DIMENSIONS FOR THE SET SCREWS SHOWN IN FIG. 18.42 (ALL DIMENSIONS GIVEN IN INCHES)

D	I	J	T	R	C		P		Q	q
Nominal size	Radius of headless crown	Width of slot	Depth of slot	Oval point radius	Diameter of cup and flat points		Diameter of dog point		Length of dog point	
					Max	Min	Max	Min	Full	Half
5 0.125	0.125	0.023	0.031	0.094	0.067	0.057	0.083	0.078	0.060	0.030
6 0.138	0.138	0.025	0.035	0.109	0.047	0.064	0.092	0.087	0.070	0.035
8 0.164	0.164	0.029	0.041	0.125	0.087	0.076	0.109	0.103	0.080	0.040
10 0.190	0.190	0.032	0.048	0.141	0.102	0.088	0.127	0.120	0.090	0.045
12 0.216	0.216	0.036	0.054	0.156	0.115	0.101	0.144	0.137	0.110	0.055
$\frac{1}{4}$ 0.250	0.250	0.045	0.063	0.188	0.132	0.118	0.156	0.149	0.125	0.063
$\frac{5}{16}$ 0.3125	0.313	0.051	0.076	0.234	0.172	0.156	0.203	0.195	0.156	0.078
$\frac{3}{8}$ 0.375	0.375	0.064	0.094	0.281	0.212	0.194	0.250	0.241	0.188	0.094
$\frac{7}{16}$ 0.4375	0.438	0.072	0.109	0.328	0.252	0.232	0.297	0.287	0.219	0.109
$\frac{1}{2}$ 0.500	0.500	0.081	0.125	0.375	0.291	0.270	0.344	0.344	0.250	0.125
$\frac{9}{16}$ 0.5625	0.563	0.091	0.141	0.422	0.332	0.309	0.391	0.379	0.281	0.140
$\frac{5}{8}$ 0.625	0.625	0.102	0.156	0.469	0.371	0.347	0.469	0.456	0.313	0.156
$\frac{3}{4}$ 0.750	0.750	0.129	0.188	0.563	0.450	0.425	0.563	0.549	0.375	0.188

Source: Courtesy of ANSI; B18.6.2.

18.21

Miscellaneous screws

A few of the many types of specialty bolts are shown in Fig. 18.43, each having its own special application. Figure 18.44 shows further examples of specialty bolts and screws.

Three types of wing screws are shown in Fig. 18.45. They are available in incremental lengths of $\frac{1}{8}$ inch and are used to join parts assembled and disas-

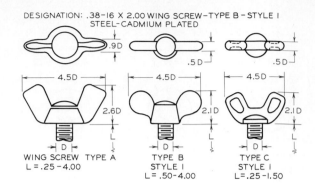

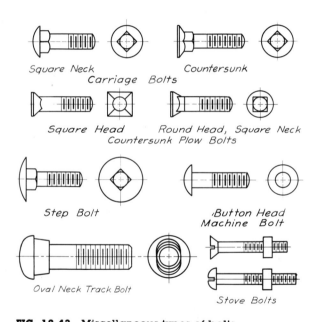

FIG. 18.43 Miscellaneous types of bolts.

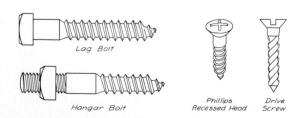

FIG. 18.44 Specialty bolts and screws.

FIG. 18.45 Wing screw proportions for screw diameters of about $\frac{5}{16}''$. These proportions can be used to draw wing screws of any size diameter. Type A is available in screw diameters of 4, 6, 8, 10, 12, 0.25, 0.313, 0.375, 0.438, 0.50, and 0.625; Type B in diameters of 10 to 0.625; and Type C in diameters of 6 to 0.375.

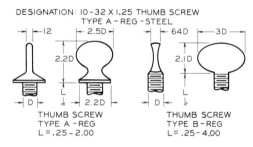

FIG. 18.46 Thumb screw proportions for screw diameters of about $\frac{1}{4}''$. These proportions can be used to draw thumb screws of any screw diameter. Type A is available in diameters of 6, 8, 10, 12, 0.25, 0.313, and 0.375. Type B thumb screws are available in diameters of 6 to 0.50.

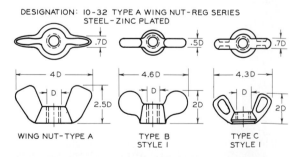

FIG. 18.47 Wing nut proportions for screw diameters of $\frac{3}{8}''$. These proportions can be used to draw thumb screws of any size. Type A wing nuts are available in screw diameters (in inches) of 3, 4, 5, 6, 8, 10, 12, 0.25, 0.313, 0.375, 0.438, 0.50, 0.583, 0.625, and 0.75; Type B nuts are available in sizes from 5 to 0.75; and Type C nuts in sizes from 4 to 0.50.

sembled by hand. Figure 18.46 shows two types of thumb screws, which serve the same purpose as wing screws. Three types of wing nuts are shown in Fig. 18.47 that can be used with various types of screws.

18.22
Wood screws

A wood screw is pointed and has a sharp thread of coarse pitch for insertion into wood. The three most common types of wood screws are shown in Fig. 18.48. These are drawn on a grid to show the proportions of the heads in relation to their major diameters.

Sizes of wood screws are specified by single numbers, such as 0, 6, or 16. From 0 to 10 each digit represents a different size (1, 2, 3, etc). Beginning at 10, only even-numbered sizes are standard; that is, 10, 12, 14, 16, 18, 20, 22, and 24. The following formula relates these numbered sizes to the actual diameter of the screws:

Actual DIA = 0.060 + screw number × 0.013.

For example, the diameter of a No. 5 wood screw is

$$0.060 + 5(0.013) = 0.125.$$

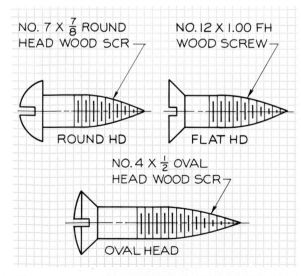

FIG. 18.48 Standard types of wood screws. The proportions shown here can be used for drawing wood screws of all sizes.

18.23
Tapping a hole

A threaded hole is called a *tapped hole,* since the tool used to cut the threads is called a tap. The types of taps available for threading small holes by hand are shown in Fig. 18.49.

The taper, plug, and bottoming hand taps are identical in size, length, and measurement; only the chamfered portion of their ends is different. The taper tap has a long chamfer (8 to 10 threads), the plug tap has a shorter chamfer (3 to 5 threads), and the bottoming tap has the shortest chamfer (1 to 1½ threads).

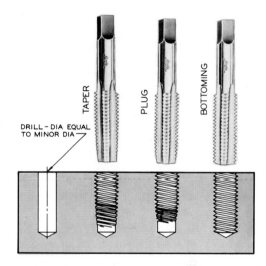

FIG. 18.49 Three types of taps for threading internal holes.

When tapping by hand in open or "through" holes, the taper should be used for coarse threads, since it ensures straighter starting. The taper tap is also recommended for the harder metals. The plug tap can be used in soft metals of for fine-pitch threads. When a hole is tapped to the very bottom, all three taps— taper, plug, and bottoming—should be used in this order.

Notes can be added to specify the depth of the drilled hole and the depth of the threads. For example, a note reading 7/8 DIA–3 DEEP–1–8 UNC–2A–2 DEEP means the hole will be drilled deeper than it is threaded, and the last usable thread will be 2 inches deep in the hole. Note that the drill point has a 120° angle.

18.24
Washers, lock washers, and pins

Washers, called *plain washers,* are used with nuts and bolts to improve the assembly and strength of the fastening. Plain washers are noted on a working drawing in the following manner:

    0.938 × 1.750 × 0.134 TYPE A PLAIN WASHER

These numbers represent the washer's inside diameter, outside diameter, and thickness, in that order (Appendix 36).

    A *lock washer* prevents a nut or cap screw from loosening as a result of vibration or movement. Two of the more common types of lock washers are the *external-tooth lock washer* and the *helical-spring lock washer* (Fig. 18.50). Other types of locking washers and devices are shown in Fig. 18.51.

    Tables for spring lock washers are given in Appendix 37 for regular and extra–heavy-duty helical-spring lock washers and are noted in the following form:

HELICAL-SPRING LOCK WASHER–$\frac{1}{4}$ REGULAR—
        PHOSPHOR BRONZE.

(The $\frac{1}{4}$ is the washer's inside diameter). Tooth lock washers are designated with notes in one of the

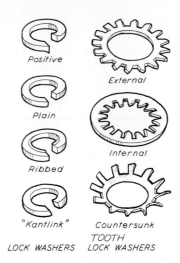

FIG. 18.51 Types of lock washers and locking devices.

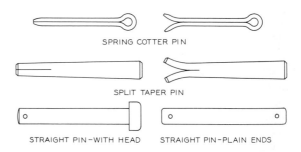

FIG. 18.52 Types of pins used to fix parts together.

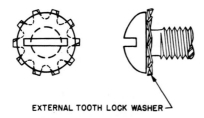
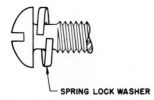

FIG. 18.50 Two types of lock washers for preventing a bolt from unscrewing.

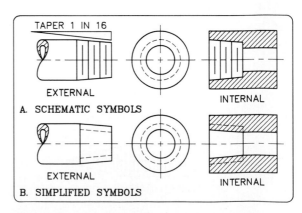

FIG. 18.53 A. Schematic techniques of representing pipe threads. B. Simplified techniques of representing pipe threads.

following forms:

INTERNAL-TOOTH LOCK WASHER–¼–TYPE A–
STEEL,
EXTERNAL-TOOTH LOCK WASHER–562–TYPE B–
STEEL.

Straight pins and taper pins (Fig. 18.52) are used to fix parts together. Dimensions for them are given in Appendix 34. Another locking device is the cotter pin, for which tables of specifications are given in Appendix 38.

18.25
Pipe threads

Pipe threads are used in connecting pipes, tubing, and lubrication fittings. The most commonly used pipe thread is tapered at a ratio of 1 to 16, but straight pipe threads are available. Tapered pipe threads will only engage for an effective length determined by the formula below:

$$L = (0.80D + 6.8)P,$$

where D is the outside diameter of the pipe, and P is the pitch of the thread. Methods of representing tapered threads are shown in Fig. 18.53.

The following abbreviations are associated with pipe threads:

N = National	G = Grease
P = Pipe	I = Internal
T = Taper	M = Mechanical
C = Coupling	L = Locknut
S = Straight	H = Hose coupling
F = Fuel and oil	R = Railing fittings

Combining these abbreviations gives the following ANSI symbols:

NPT	= National pipe taper
NPTF	= National pipe thread (dryseal— for pressure-tight joints)
NPS	= Straight pipe thread
NPSC	= Straight pipe thread in couplings
NPSI	= National pipe straight internal thread
NPSF	= Straight pipe thread (dryseal)
NPSM	= Straight pipe thread for mechanical joints

NPSL	= Straight pipe thread for locknuts and locknut pipe threads
NPSH	= Straight pipe thread for hose couplings and nipples
NPTR	= Taper pipe thread for railing fittings

To specify a pipe thread in note form, the nominal pipe diameter (the internal diameter), the number of threads per inch, and the symbol that denotes the type of thread are given; for example,

$$1\tfrac{1}{4}\text{–}11\tfrac{1}{2}\ \text{NPT} \quad \text{or} \quad 3\text{–}8\ \text{NPTR}$$

These specifications are given in Appendix 10. Examples of external and internal thread notes are shown in Fig. 18.54. Dryseal threads, which may be straight or tapered, are used in applications where a pressure-tight joint is required without the use of a lubricant or sealer. No clearance between the mating parts of the joint is permitted, so the fit is of the highest quality. The tap drill is sometimes given in the internal pipe thread note, but this is optional.

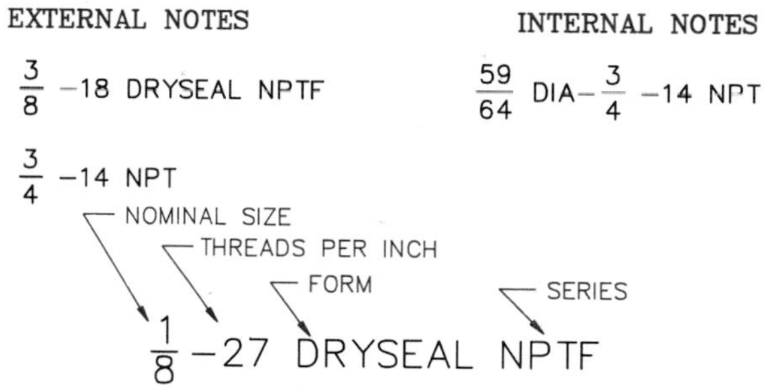

FIG. 18.54 Typical pipe-thread notes.

18.26
Keys

Keys are used to attach parts to shafts in order to transmit power to pulleys, gears, or cranks. The four types of keys shown pictorially and orthographically in Fig. 18.55 are the most commonly used. Notes must be given for the keyway, the key, and the keyseat, as shown in Figs. 18.55A, C, E, and G. These notes are typical of those used to give key specifications. Dimensions for various types of keys are given in Appendixes 31 and 32.

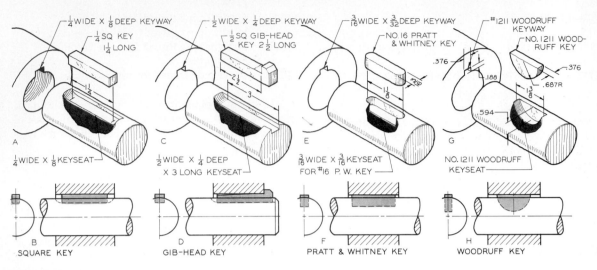

FIG. 18.55 Standard keys used to hold parts on a shaft.

18.27

Rivets

Rivets are fasteners used to join thin overlapping materials in a permanent joint. The rivet is inserted in a hole slightly larger than the diameter of the rivet, and the headless end is formed into the specified shape by applying pressure to the projecting end. This forming operation is done when the rivets are either hot or cold, depending on the application.

Typical shapes and proportions of small rivets are shown in Fig. 18.56. These rivets vary in diameter

from $\frac{1}{16}$ to $1\frac{3}{4}$ inches. Rivets are used extensively in pressure-vessel fabrication, in heavy structures such as bridges and buildings, and in sheet-metal construction.

Three types of lap joints are shown in Fig. 18.57. The joints are held secure by one, two, or three rivets, as shown in the sectional view. Note that the bodies of the rivets are drawn as hidden circles in the top views.

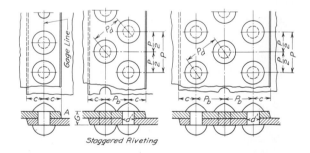

FIG. 18.57 Examples of lap joints using single rivets, double rivets, and triple rivets.

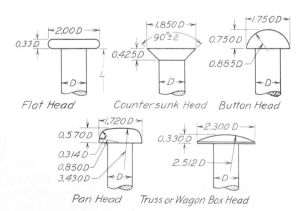

FIG. 18.56 Types and proportions of small rivets. Small rivets have shank diameters up to $\frac{1}{2}''$.

The standard symbols recommended by ANSI for representing rivets are shown in Fig. 18.58. Rivets that are driven in the shop are called *shop rivets*, and those assembled on the job at the site are called *field rivets*.

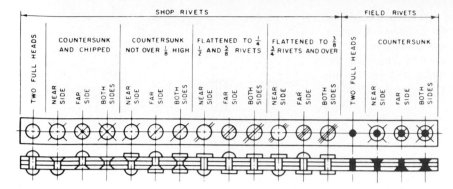

FIG. 18.58 The symbols used to represent rivets in a drawing. (Courtesy of ANSI 14.14.)

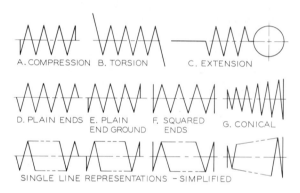

A. COMPRESSION B. TORSION C. EXTENSION

D. PLAIN ENDS E. PLAIN END GROUND F. SQUARED ENDS G. CONICAL

SINGLE LINE REPRESENTATIONS – SIMPLIFIED

FIG. 18.59 Single-line representations of various types of springs.

18.28
Springs

Some of the more commonly used types of springs are *compression, torsion, extension, flat,* and *constant force.* Figure 18.59 shows the single-line conventional representation of the first three types. Also shown are the types of ends that can be used on compression springs and the simplified single-line representation of coil springs.

A typical working drawing of a compression spring is shown in Fig. 18.60. The ends of the spring are drawn by using the double-line representation, and conventional lines are used to indicate the undrawn portion of the spring. The diameter and the free length of the spring are given on the drawing and the remaining specifications are given in tabular form near the drawing.

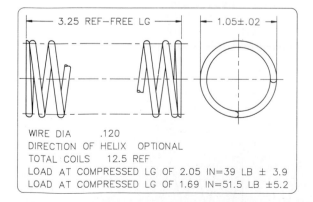

3.25 REF—FREE LG   1.05±.02

WIRE DIA   .120
DIRECTION OF HELIX OPTIONAL
TOTAL COILS   12.5 REF
LOAD AT COMPRESSED LG OF 2.05 IN=39 LB ± 3.9
LOAD AT COMPRESSED LG OF 1.69 IN=51.5 LB ±5.2

FIG. 18.60 A conventional double-line drawing of a compressions spring and its specifications.

A working drawing of an extension spring (Fig. 18.61) is very similar to that of a compression spring. In a drawing of a helical torsion spring (Fig. 18.62), angular dimensions must be shown to specify the initial and final positions of the spring as torsion is applied to it. All springs require a table of specifications to describe their details.

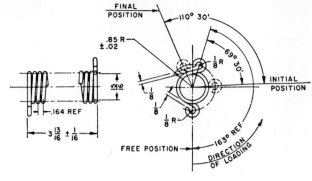

WIRE DIA   .148
DIRECTION OF HELIX   LEFT HAND
TOTAL COILS   20.55 REF
TORQUE   15 LB IN. ± 1.5 LB IN. AT INITIAL POSITION
TORQUE   33 LB IN. ± 3.3 LB IN. AT FINAL POSITION
MAXIMUM DEFLECTION WITHOUT SET BEYOND FINAL POSITION 56°
SPRING RATE   .16 LB IN. / DEG REF

FIG. 18.62   A conventional double-line drawing of a helical torsion spring and its specifications. (Courtesy of the U.S. Dept. of Defense.)

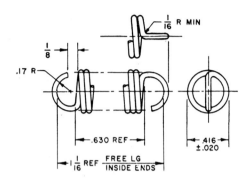

WIRE DIA   .042
DIRECTION OF HELIX   OPTIONAL
TOTAL COILS   14 REF
RELATIVE POSITION OF ENDS 180° ± 20°
EXTENDED LG INSIDE ENDS
WITHOUT PERMANENT SET   2.45 IN. (MAX)
INITIAL TENSION   1.0 LB ± .10 LB
LOAD   4 LB ± .4 LB AT 1.56 IN.
EXTENDED LG INSIDE ENDS
LOAD   6.3 LB ± .63 LB AT 1.95 IN.
EXTENDED LG INSIDE ENDS

FIG. 18.61   A conventional double-line drawing of an extension spring and its specifications. (Courtesy of the U.S. Dept. of Defense.)

18.29
Drawing springs

Springs may be drawn as schematic representations using single lines (Fig. 18.63). Each example is drawn by laying out the diameter of the coils and the lengths of the springs, and then the number of active coils are drawn by using the diagonal-line method. In Fig. 18.63B, the two end coils are "dead" coils, and only four are active. An extension spring is drawn in Fig. 18.63C. Where more realism is desired, a double-line drawing of a thread can be made, as shown in Fig. 18.64.

FIG. 18.63   A. A schematic drawing of a spring with 4 active coils. The diagonal-line method is used to divide it into 4 equally spaced coils. **B.** A spring with 6 coils, but only 4 of them are active. **C.** An extension spring with 5 active coils.

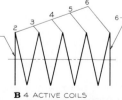

A 4 COILS
COMPRESSION SPRING

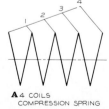

B 4 ACTIVE COILS
COMPRESSION SPRING

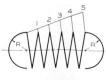

C 5 ACTIVE COILS
EXTENSION SPRING

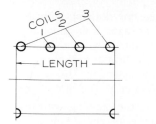
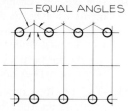

FIG. 18.64 Detailed drawing of a spring.

Step 1 Lay out the diameter and the length of the spring, and locate the coils by the diagonal-line technique.

Step 2 Locate the coils on the lower side along the bisectors of the spaces between the coils on the upper side.

Step 3 Connect the coils on each side. This is a right-hand coil; a left-hand spring would slope in the opposite direction.

Step 4 Construct the back side of the spring and the end coils to complete the detailed drawing of a compression spring.

Problems

These problems are to be completed on Size A sheets. Problems 1–16 are to be laid out one problem per sheet, and Problems 17–30 are to be laid out two problems per sheet. The boxes drawn around the figures representing the problems are approximately 6 by 5 inches, or about half a Size A sheet.

Fasteners

1. The layout in Fig. 18.65 is to be used for constructing a detailed representation of an Acme thread with a major diameter of 3 inches. The thread note specifications are 3–1½ ACME. Show both external and internal thread representations and the thread note. Use inches or millimeters as instructed.

2. Repeat Problem 1, but draw internal and external detailed representations of a square thread that is 3 inches in diameter. The note specifications are 3–½ SQUARE. Apply notes to both parts.

3. Repeat Problem 1, but draw internal and external detailed thread representations of an American National thread form. The major diameter of each part is 3 inches. The note specifications are 3–4 NC–2. Apply notes to both parts.

4. Notes are given in Fig. 18.66 to specify the depth of the holes to be drilled and the threads to be

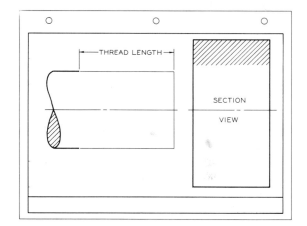

FIG. 18.65 Problems 1–3. Construction of thread symbols.

tapped in the holes. Following these notes, draw detailed representations of the threads as views according to specifications.

5. Repeat Problem 4, but use schematic representations.

6. Repeat Problem 4, but use simplified thread representations.

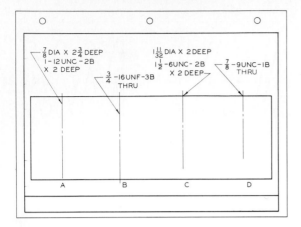

FIG. 18.66 Problems 4–6. Internal threads.

7. Figure 18.67 shows a layout of two external threaded parts and their end views. Also shown is a piece into which the external threads will be screwed. Complete all three views of each of the parts. Use detailed threads, and apply notes to the internal and external threads. Use the table in Appendix 15 for thread specifications. Use UNC threads with a 2A fit.

8. Repeat Problem 7, but use schematic thread representations.

9. Repeat Problem 7, but use simplified thread representations.

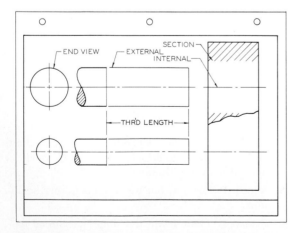

FIG. 18.67 Problems 7–9. Internal and external threads.

10. (Fig. 18.68). Complete the drawing with instruments as a finished hexagon bolt and nut. The bolt head is a heavy nut drawn across corners. Use detailed thread specifications. Show notes to specify the parts of the assembly. Thread specifications are $1\frac{1}{2}$–6 UNC–3 or M36 × 4.

11. (Fig. 18.68). Complete the drawing with instruments as an unfinished squarehead bolt and nut. The bolt head is drawn across corners. The regular nut is to be drawn across corners. Use schematic thread representations. Show notes to specify the parts. Use the table in Appendix 15 or 18 for thread specifications (English or SI).

12. (Fig. 18.68). Complete the drawing with instruments as a finished hexagon nut and bolt. Draw the regular bolt and nut across flats, using simplified thread representations. Show notes to specify the parts. Use the table in Appendix 15 or 18 for specifications.

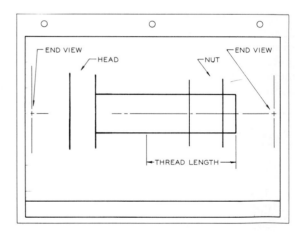

FIG. 18.68 Problems 10–12. Nuts and bolts in assembly.

13. The notes in Fig. 18.69 apply to machine and cap screws that are to be drawn in the section view of the two parts. The holes in which the screws are to be drawn are through holes. Complete the drawings, show the notes and the remaining section lines. Use detailed thread symbols.

14. Repeat Problem 13, but use schematic thread symbols.

15. Repeat Problem 13, but use simplified thread symbols.

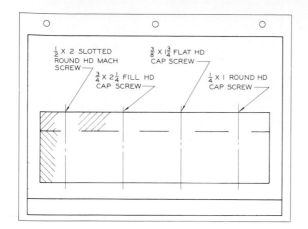

FIG. 18.69   Problems 13–15. Cap screws and machine screws.

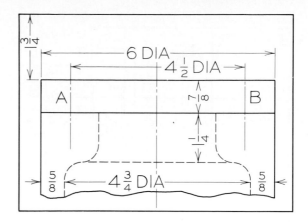

FIG. 18.71   Problem 17.

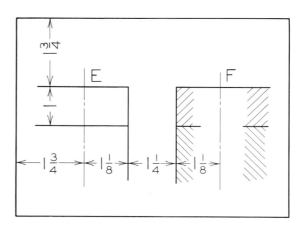

FIG. 18.72   Problem 18.

FIG. 18.70   Problem 16. Design involving threaded parts.

16. The pencil pointer shown in Fig. 18.70 has a $\frac{1}{4}$ in. shaft that fits into a bracket designed to clamp onto a desk top. A set screw holds the shaft in position. Make a drawing of the bracket, estimating its dimensions. Show the details and the method of using the set screw to hold the shaft. Give the specifications for the set screw.

17. (Fig. 18.71). On axes A and B, construct $\frac{5}{8}$ in. (16 mm) hexagonal head bolts across flats with a coarse thread, UNC. Convert the view to a half-section to show how the parts are assembled together.

18. (Fig. 18.72). On axes E and F, draw a stud with a hexagon head nut shown across corners. The stud is to have a diameter of $\frac{7}{8}$ in. (22 mm) and should be a UNC form and series.

19. (Fig. 18.73). Draw a $\frac{7}{8}$ in. (22 mm) DIA special stud with a collar that is $1\frac{1}{4}$ in. DIA. Show a plain washer and a regular nut at end B. On axes CD, draw a machine screw of your selection to hold the two parts together.

20. (Fig. 18.74). Draw a cap screw on axis AB to fasten the parts together. The dimensions are bolt DIA $1\frac{1}{4}$ in. or 32 mm; C, 1.25 in.; E, 1.12 in.; H,

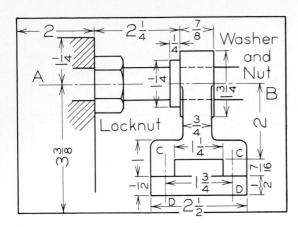

FIG. 18.73 Problem 19.

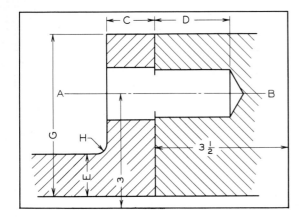

FIG. 18.74 Problem 20.

0.25 in.; hexagon head across corners; *G*, 4.50 in.; and *D*, 2.00 in.

21. Figure 18.75 shows two parts assembled on a cylindrical shaft. These parts are to be held in position by a square key in part A and a gib-head key in part B. Show the necessary notes to specify the key, the keyway, and the keyseat.

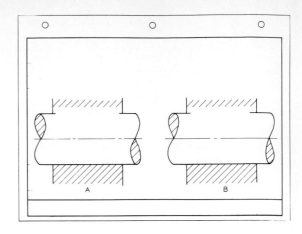

FIG. 18.75 Problems 21–22.

22. Repeat Problem 21, but use a No. 16 Pratt & Whitney key in part A and a No. 1211 Woodruff key in part B. Show the necessary notes to specify the key, the keyway, and the keyseat.

23–26. (Fig. 18.76 and Table 18.6). Using the double-line method, draw helical springs, two per Size A sheet. Use the specifications in the table for each.

27–30. (Fig. 18.76). Same as Problems 23–26, but use single-line representations for the springs.

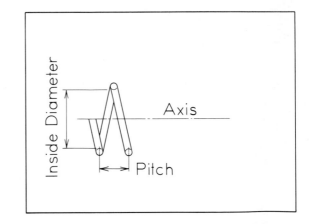

FIG. 18.76 Problems 23–30.

TABLE 18.6
PROBLEMS 23–26

Problem	No. of Turns	Pitch	Size Wire	Inside Dia	Outside Dia	
23	4	1	No. 4 = 0.2253	3		RH
24	5	$\frac{3}{4}$	No. 6 = 0.1920		2	LH
25	6	$\frac{5}{8}$	No. 10 = 0.1350	2		RH
26	7	$\frac{3}{4}$	No. 7 = 0.1770		$1\frac{3}{4}$	LH

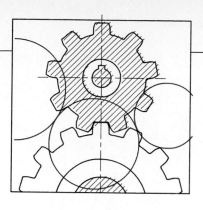

CHAPTER 19

Gears and Cams

19.1
Introduction to gears

Gears are toothed wheels whose circumferences mesh together to transmit force and motion from one gear to the next. The three most common types—spur gears, bevel gears, and worm gears (Fig. 19.1)—are discussed in this chapter.

A   B   C

FIG. 19.1 The three basic types of gears are (A) spur gears, (B) bevel gears, and (C) worm gears. (Courtesy of the Process Gear Co.)

19.2
Spur gear terminology

The spur gear is a circular gear with teeth cut around its circumference. Two mating spur gears can transmit power from a shaft to another, parallel shaft.

> When the two meshing gears are unequal in diameter, the smaller gear is called the *pinion,* and the larger one is called the *gear.*

The following terms are used to describe the parts of a spur gear. Many of these features are illustrated and labeled in Figs. 19.2 and 19.3. The corresponding formulas for each feature are also given.

PITCH CIRCLE (PC) is the imaginary circle of a gear if it were a friction wheel without teeth that contacted the pitch circle of another friction wheel.

PITCH DIAMETER (PD) is the diameter of the pitch circle: $PD = N/DP$, where N is the number of teeth, and DP is the diametrical pitch.

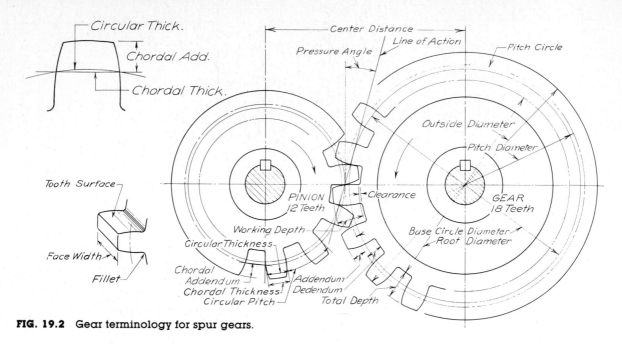

FIG. 19.2 Gear terminology for spur gears.

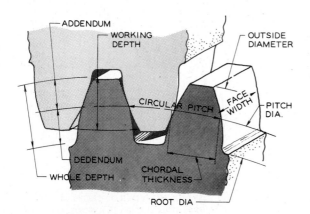

FIG. 19.3 Gear terminology for spur gears.

DIAMETRICAL PITCH (DP) is the ratio between the number of teeth on a gear and its pitch diameter. For example, a gear with twenty teeth and a 4-inch pitch diameter will have a diametrical pitch of 5, which means there are five teeth per inch of a diameter: $DP = N/PD$, where N is the number of teeth.

CIRCULAR PITCH (CP) is the circular measurement from one point on a tooth to the corresponding point on the next tooth measured along the pitch circle: $CP = 3.14/DP$.

CENTER DISTANCE (CD) is the distance from the center of a gear to its mating gear's center: $CD = (N_P + N_S)/(2DP)$, where N_P and N_S are the number of teeth in the pinion and spur, respectively.

ADDENDUM (A) is the height of a gear above its pitch circle: $A = 1/DP$.

DEDENDUM (D) is the depth of a gear below the pitch circle: $D = 1.157/DP$.

WHOLE DEPTH (WD) is the total depth of a gear tooth: $WD = A + D$.

WORKING DEPTH (WKD) is the depth to which a tooth fits into a meshing gear: $WKD = 2/DP$, or $WKD = 2A$.

CIRCULAR THICKNESS (CRT) is the circular distance across a tooth measured along the pitch circle: $CRT = 1.57/DP$.

CHORDAL THICKNESS (CT) is the straight-line distance across a tooth at the pitch circle: $CT = PD (\sin 90°/N)$, where N is the number of teeth.

FACE WIDTH (FW) is the width across a gear tooth parallel to its axis. This is a variable dimension, but it is usually three to four times the circular pitch: $FW = 3$ to $4(CP)$.

OUTSIDE DIAMETER (OD) is the maximum diameter of a gear across its teeth: $OD = PD + 2A$.

ROOT DIAMETER (RD) is the diameter of a gear measured from the bottom of its gear teeth: $RD = PD - (2D)$.

PRESSURE ANGLE (PA) is the angle between the line of action and a line perpendicular to the centerline of two meshing gears. Angles of 14.5 and 20 degrees are standard for involute gears.

BASE CIRCLE (BC) is the circle from which an involute tooth curve is generated or developed: $BC = PD (\cos PA)$.

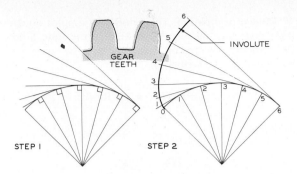

FIG. 19.4 Construction of an involute.

Step 1 The base arc is divided into equal divisions with radial lines from the center. Tangents are drawn perpendicular to the radial lines on the arc.

Step 2 The chordal distance from 1 to 0 is used as the radius and 1 as the center to find point 1 on the involute curve. The distance from 2 to newly found 1 is revolved to the tangent line through 2 to locate a second point. The process is continued.

19.3
Tooth forms

The most common gear tooth is an *involute tooth* with a 14.5° pressure angle. The 14.5° angle is the angle of contact between two gears when the tangents of both gears pass through the point of contact. Gears with pressure angles of 20 and 25° are also used. Gear teeth with larger pressure angles are wider at the base and thus are stronger than the standard 14.5° teeth.

The standard gear face is an involute that keeps the meshing gears in contact as the gear teeth are revolved. The principle of constructing an involute is illustrated in Fig. 19.4.

An involute curve can be thought of as the path of a string that is kept taut as it is unwound from the base arc. It is unnecessary to draw gear teeth as involutes since most detail drawings approximations of teeth, if teeth are shown at all.

19.4
Gear ratios

The diameters of two meshing spur gears establish ratios that are important to the function of the gears. (Fig. 19.5).

If the radius of a gear is twice that of its pinion (the small gear), then the diameter is twice that of the pinion, and the gear has twice as many teeth as the pinion. The pinion must make twice as many turns as the larger gear; in other words, the revolutions per minute (RPM) of the pinion is twice that of the larger gear.

When the diameter of the gear is four times the diameter of the pinion, there must be four times as many teeth on the gear as on the pinion, and the number of revolutions of the pinion will be four times that of the larger gear.

The relationship between two meshing spur gears can be developed in formula form by finding the velocity of a point on the small gear that is equal to πPD

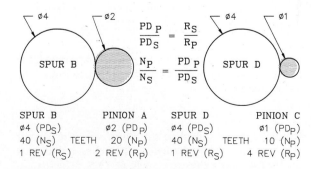

FIG. 19.5 Ratios between meshing spur gears.

× RPM of the pinion. The velocity of a point on the large gear is equal to πPD × RPM of the spur. Since the velocity of points on each gear must be equal, the equation may be written as

$$\pi PD_P(\text{RPM}) = \pi PD_S(\text{RPM});$$

therefore

$$\frac{PD_P}{PD_S} = \frac{\text{RPM}_S}{\text{RPM}_P}.$$

If the radius of the pinion is 1 inch, the radius of the spur 4 inches, and the RPM of the pinion 20, then the RPM of the spur can be found as follows:

$$\frac{2(1)}{2(4)} = \frac{\text{RPM}_S}{20 \text{ RPM}_P}$$

$$\text{RPM}_S = \frac{2(20)}{2(4)} = 5 \text{ RPM}.$$

The RPM of the spur is 5, or one fourth of the RPM of the pinion.

The number of teeth on each gear is proportional to the diameters of a pair of meshing gears. This relationship can be written

$$\frac{N_P}{N_S} = \frac{PD_P}{PD_S},$$

where N_P and N_S are the number of teeth on the pinion and spur, respectively, and PD_P and PD_S are their pitch diameters.

19.5
Spur gear calculations

Before a working drawing of a gear can be started, the drafter must perform calculations to determine the gear's dimensions.

PROBLEM 1   Calculate the dimensions for a spur gear that has a pitch diameter of 5 inches, a diametral pitch of 4, and a pressure angle of 14.5. (The diametral pitch is the same for meshing gears.)

SOLUTION

Number of teeth: $PD \times DP = 5 \times 4 = 20$

Addendum: $\frac{1}{4} = 0.25$

Dedendum: $= 1.157/4 = 0.2893$

Circular thickness: $1.5708/4 = 0.3927$

Outside diameter: $(20 + 2)/4 = 5.50$

Root diameter: $5 - 2(0.2893) = 4.421$

Chordal thickness: $5(\sin 90°/20) = 5(0.079) = 0.392$

Chordal addendum:

$$0.25 + [0.3927^2/(4 \times 5)] = 0.2577$$

Face width: $3.5(0.79) \times 2.75$

Circular pitch: $3.14/4 = 0.785$

Working depth: $0.6366 \times (3.14/4) = 0.4997$

Whole depth: $0.250 + 0.289 = 0.539$

These dimensions can be used to draw the spur gear and to provide specifications necessary for its manufacture.

The method of determining the design information for two meshing gears when their working ratios are known is shown in the following problem.

PROBLEM 2   Find the number of teeth and other specifications for a pair of meshing gears with a driving gear that turns at 100 RPM and a driven gear that turns at 60 RPM. The diametrical pitch for each is 10. The center-to-center distance between the gears is 6 inches.

SOLUTION

STEP 1   Find the sum of the teeth on both gears:

$$\text{Total teeth} = 2 \times (\text{c-to-c distance}) \times DP$$
$$= 2 \times 6 \times 10 = 120 \text{ teeth.}$$

STEP 2   Find the number of teeth for the driving gear:

$$\frac{\text{Driver RPM}}{\text{Driven RPM}} + 1 = \frac{100}{60} + 1 = 2.667;$$

$$\frac{\text{Total teeth}}{\dfrac{100}{60} + 1} = \frac{120}{2.667} = 45 \text{ teeth.}$$

---

*The number of teeth must be a whole number since there cannot be fractional teeth on a gear. It may be necessary to adjust the center distance to yield a whole number of teeth.

STEP 3 Find the number of teeth for the driven gear:

Total teeth − teeth on driver = teeth on driven gear
120 − 45 = 75 teeth.

STEP 4 Other specifications for the gears can be calculated as shown in the first problem by using the formulas in Section 19.2.

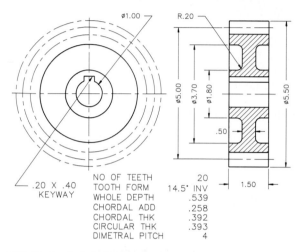

NO OF TEETH	20
TOOTH FORM	14.5° INV
WHOLE DEPTH	.539
CHORDAL ADD	.258
CHORDAL THK	.392
CIRCULAR THK	.393
DIMETRAL PITCH	4

FIG. 19.6 A detail drawing of a spur gear with a table of values to supplement the dimensions shown on the drawing.

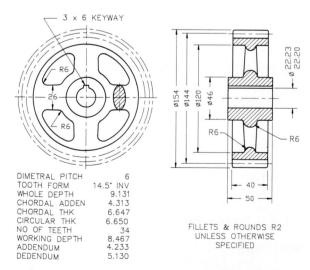

DIMETRAL PITCH	6
TOOTH FORM	14.5° INV
WHOLE DEPTH	9.131
CHORDAL ADDEN	4.313
CHORDAL THK	6.647
CIRCULAR THK	6.650
NO OF TEETH	34
WORKING DEPTH	8.467
ADDENDUM	4.233
DEDENDUM	5.130

FILLETS & ROUNDS R2
UNLESS OTHERWISE
SPECIFIED

FIG. 19.7 A computer-drawn detail drawing of a spur gear.

19.6
Drawing spur gears

A conventional drawing of a spur gear is shown in Fig. 19.6. Since it is so time-consuming, the teeth need not be drawn. It is possible to omit the circular view and to show only a sectional view of the gear with a table of dimensions called *cutting data*. Circular centerlines are drawn to represent the root circle, pitch circle, and outside circle of the gear in the circular view.

A table of dimensions is a necessary part of a gear drawing, as shown in Fig. 19.7. These data can be calculated by formula or taken from tables of standards in gear handbooks, such as *Machinery's Handbook*.

19.7
Bevel gear terminology

Bevel gears are gears whose axes intersect at angles. Although the angle of intersection is usually 90°, other angles are used. The smaller of the two bevel gears is called the *pinion*, as in spur gearing.

The terminology of bevel gearing is illustrated in Fig. 19.8. The corresponding formulas for each feature are given below. Gear handbooks can also be used for finding these dimensions.

PITCH ANGLE OF PINION (SMALL GEAR) (PA$_p$):

$$\tan PA_p = \frac{N_p}{N_g}.$$

Where N_g and N_p are the number of teeth on the gear and pinion, respectively.

PITCH ANGLE OF GEAR (PA$_g$):

$$\tan PA_g = \frac{N_g}{N_p}.$$

PITCH DIAMETER (PD) is the number of teeth (N) divided by the diametrical pitch (DP): $PD = N/P$.

ADDENDUM(A) is measured at the large end of the tooth: $A = 1/DP$.

DEDENDUM (D) is measured at the large end of the tooth: $D = 1.157/DP$.

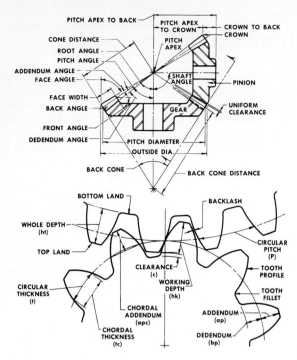

FIG. 19.8 The terminology and definitions of bevel gears. (Courtesy of Philadelphia Gear Corp.)

WHOLE TOOTH DEPTH (WD):

$$WD = 2.157/DP.$$

THICKNESS OF TOOTH (TT) At pitch circle:
$TT = 1.571/DP.$

DIAMETRICAL PITCH: $DP = N/PD$, where N is the number of teeth.

ADDENDUM ANGLE (AA) is the angle formed by the addendum and pitch cone distance:

$$\tan AA = \frac{A}{PCD}.$$

PITCH CONE DISTANCE (PCD):

$$PD/(2 \times \sin PA).$$

DEDENDUM ANGLE (DA) is the angle formed by the dedendum and the pitch cone distance:

$$\tan DA = \frac{D}{PCD}.$$

FACE ANGLE (FA) is the angle between the gear's centerline and the top of its teeth:

$$FA = 90° - (PCD + AA).$$

CUTTING ANGLE (OR ROOT ANGLE) (CA) is the angle between the gear's axis and the roots of the teeth: $CA = PCD - D.$

OUTSIDE DIAMETER (OD) is the greatest diameter of a gear across its teeth: $OD = PD + 2A.$

APEX TO CROWN DISTANCE (AC) is the distance from the crown of the gear to the apex of the cone measured parallel to the axis of the gear: $AC = OD/(2 \tan FA).$

CHORDAL ADDENDUM (CA):

$$A + \frac{TT^2 \cos PA}{4(PD)}.$$

CHORDAL THICKNESS (CT) at the large end of the tooth:

$$CT = PD \times \sin\frac{90°}{N}.$$

FACE WIDTH (FW) can vary, but it is recommended that it be approximately equal to the pitch cone distance divided by 3: $FW = PCD/3.$

19.8
Bevel gear calculations

The following problem demonstrates how the formulas in the previous section are used. Some of the formulas result in the same specifications that apply to both the gear and pinion.

PROBLEM 3 Two bevel gears intersect at right angles. They have a diametrical pitch of 3, 60 teeth on the gear, 45 teeth on the pinion, and a face width of 4 inches. Find the dimensions of the gear.

SOLUTION

Pitch cone angle of gear: $\tan PCA = 60/45 = 1.33$;
$$PCA = 53°7'.$$

Pitch cone angle of pinion: $\tan PCA = 45/60$;
$$PCA = 36°52'.$$

Pitch diameter of gear: $60/3 = 20.00''$.

Pitch diameter of pinion: $45/3 = 15.00''$.

The following formulas are the same for both the gear and the pinion.

Addendum: $\frac{1}{3} = 0.333''$.

Dedendum: $1.157/3 = 0.3857''$.

Whole depth: $2.157/3 = 0.719''$.

Tooth thickness on pitch circle: $1.571/3 = 0.5237$.

Pitch cone distance: $20/(2 \sin 53°7') = 12.5015$.

Addendum angle: $\tan AA = 0.333/12.5015 = 1°32'$.

Dedendum angle:

$$DA = 0.3857/12.5015 = 0.0308 = 1°46'.$$

Face width: $PCD/3 = 4.00''$.

The remainder of the formulas must be applied separately to the gear and pinion.

Chordal addendum of gear:

$$0.333 + \frac{0.5237^2 \times \cos 53°7'}{4 \times 20} = 0.336''.$$

Chordal addendum of pinion:

$$0.333 + \frac{0.5237^2 \times \cos 36°52'}{4 \times 15} = 0.338''.$$

Chordal thickness of gear:

$$\sin \frac{90°}{60} \times 20 = 0.524''.$$

Chordal thickness of pinion:

$$\sin \frac{90°}{45} \times 15 = 0.523''.$$

Face angle of gear:

$$90° - (53°7' + 1°32') = 35°21'.$$

Face angle of pinion:

$$90° - (36°52' + 1°32') = 51°36'.$$

Cutting angle of gear:

$$53°7' - 1°46' = 51°21'.$$

Cutting angle of pinion:

$$36°52' - 1°46' = 35°6'.$$

Angular addendum of gear:

$$0.333 \times \cos 53°7' = 0.1999''.$$

Angular addendum of pinion:

$$0.333 \times \cos 36°52' = 0.2667''.$$

Outside diameter of gear:

$$20 + 2(0.1999) = 20.4000''.$$

Outside diameter of pinion:

$$15 + 2(0.2667) = 15.533''.$$

Apex-to-crown distance of gear:

$$\frac{20.400}{2} \times \tan 35°7' = 7.173''.$$

Apex-to-crown distance of pinion:

$$\frac{15.533}{2} \times \tan 51°36' = 9.800''.$$

19.9
Drawing bevel gears

The dimensions calculated above are used to lay out the bevel gears in a detail drawing. Many of the calculated dimensions would be difficult to measure on a drawing within a high degree of accuracy; therefore it is important to provide a table of cutting data for each gear.

The steps of drawing the bevel gears are shown in Fig. 19.9. The finished drawings are shown with a combination of dimensions and a table of dimensions.

19.10
Worm gears

A worm gear is composed of a threaded shaft called a *worm* and a circular gear called a *spider* (Fig. 19.10). The worm is revolved, which causes the spider to revolve about its axis. The following terminology is illustrated in Figs. 19.10 and 19.11.

Worm specifications and formulas

LINEAR PITCH (P) is the distance from one thread to the next measured parallel to the worm's

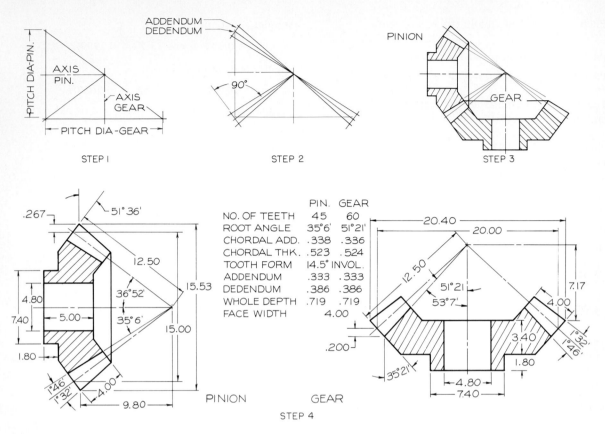

FIG. 19.9 Construction of bevel gears.

Step 1 Lay out the pitch diameters and axes of the two bevel gears.

Step 2 Draw construction lines to establish the limits of the teeth by using the addendum and dedendum dimensions.

Step 3 Draw the pinion and the gear using the specified dimensions or those that were calculated by formula.

Step 4 Complete the detail drawings of both gears and provide a table of cutting data.

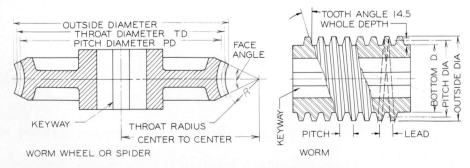

FIG. 19.10 The terminology and definitions of worm gears.

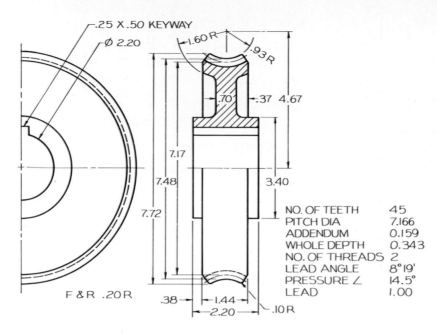

NO. OF TEETH	45
PITCH DIA	7.166
ADDENDUM	0.159
WHOLE DEPTH	0.343
NO. OF THREADS	2
LEAD ANGLE	8°19'
PRESSURE ∠	14.5°
LEAD	1.00

FIG. 19.11 A detail drawing of a worm gear (spider) and the table of cutting data.

axis: $P = L/N$, where N is the number of threads: 1 if a single thread, 2 if a double thread, and so on.

LEAD (L) is the distance a thread advances in a turn of 360°.

ADDENDUM OF TOOTH $AW = 0.3183P$.

PITCH DIAMETER $PDW = OD - 2AW$, where OD is the outside diameter.

WHOLE DEPTH OF TOOTH
$$WDT = 0.6866 \times P.$$

BOTTOM DIAMETER OF WORM
$$BD = OD - 2WDT.$$

WIDTH OF THREAD AT ROOT $WT = 0.31P$.

MINIMUM LENGTH OF WORM $MLW = \sqrt{8PDS \times AW}$, where PDS is the pitch diameter of the spider.

HELIX ANGLE OF WORM

$$\cot \beta = \frac{3.14(PDW)}{L}.$$

OUTSIDE DIAMETER $OD = PD + 2A$.

Spider specifications and formulas

PITCH DIAMETER OF SPIDER

$$PDS = \frac{N(P)}{3.14},$$

where N is the number of teeth on the spider.

THROAT DIAMETER OF SPIDER

$$TD = PDS + 2A.$$

RADIUS OF SPIDER THROAT

$$RST = \frac{OD \text{ of worm}}{2} - 2A.$$

FACE ANGLE (FA) may be selected to be between 60° and 80° for the average application.

CENTER-TO-CENTER DISTANCE (CD) between the worm and spider:

$$CD = \frac{PDW + PDS}{2}.$$

OUTSIDE DIAMETER OF SPIDER

$$ODS = TD + 0.4775P.$$

FACE WIDTH OF GEAR $\quad FW = 2.38(P) + 0.25.$

19.11
Worm gear calculations

The following problem has been solved for a worm gear by using the formulas given above.

PROBLEM 4 Calculate the specifications for a worm and worm gear (spider). The gear has 45 teeth, and the worm has an outside diameter of 2.50 inches. The worm has a double thread and a pitch of 0.5 inch.

SOLUTION

Lead: $L = 0.5'' \times 2 = 1''$.
Worm addendum: $AW = 0.3183P = 0.1592''$.
Pitch diameter of worm:

$$PDW = 2.50'' - 2(0.1592'') = 2.1818''.$$

Pitch diameter of gear:

$$PDS = (45 \times 0.5)/3.14 = 7.166''.$$

Center distance between worm and gear:

$$CD = \frac{(2.182 + 7.166)}{2} = 4.674''.$$

Whole depth of worm tooth:

$$WDT = 0.687 \times 0.5 = 0.3433''.$$

Bottom diameter of worm:

$$BD = 2.50 - 2(0.3433) = 1.813''.$$

Helix angle of worm:

$$\cot \beta = \frac{3.14(2.1816)}{1} = 8°19'.$$

Width of thread at root: $WT = 0.31(1) = 0.155''$.
Minimum length of worm:

$$MLW = \sqrt{8(0.1592)\,(7.1656)} = 3.02''.$$

Throat diameter of gear:

$$TD = 7.1656 + 2(0.1592) = 7.484''.$$

Radius of gear throat:

$$RST = (2.5/2) - (2 \times 0.1592) = 0.9318''.$$

Face Width:

$$FW = 2.38\,(0.5) + 0.25 = 1.44''.$$

Outside diameter of gear:

$$ODS = 7.484 + 0.4775\,(0.5) = 7.723''.$$

19.12
Drawing worm gears

The worm and worm wheel (spider) are drawn and dimensioned as shown in Figs. 19.11 and 19.12. The specifications derived by the formulas in the previous section must be used for scaling, laying out the drawings, and providing cutting data.

19.13
Introduction to cams

Plate cams are irregularly shaped machine elements that produce motion in a single plane, usually up and down (Fig. 19.13). As the cam revolves about its cen-

FIG. 19.12 A detail drawing of a worm using the dimensions calculated.

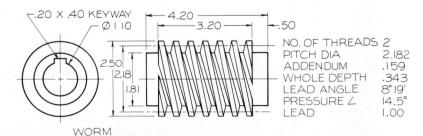

ter, the variation in the cam's shape produces a rise or fall in the follower that is in contact with it. The shape of the cam is determined graphically before the preparation of manufacturing specifications.

Cams utilize the principle of the inclined wedge, with the surface of the cam causing a change in the slope of the plane, thereby producing the desired motion.

FIG. 19.13 Examples of machined cams. (Courtesy of Ferguson Machine Co.)

19.14
Cam motion

Cams are designed primarily to produce (1) uniform or linear motion, (2) harmonic motion, (3) gravity motion, or (4) combinations of these.

Uniform motion

Uniform motion is shown in the displacement diagram in Fig. 19.14A to represent the motion of the cam follower as the cam rotates through 360°. The uniform-motion curve has sharp corners, indicating abrupt changes of velocity, causing the follower to bounce. Therefore uniform motion is usually modified with arcs that smooth this change of velocity. The radius of the modifying arc is varied up to a radius of one half the total displacement, depending on the speed of operation.

Harmonic motion

Harmonic motion, plotted in Fig. 19.14B, is a smooth, continuous motion based on the change of position of points on the circumference of a circle. At moderate speeds, this displacement gives a smooth operation.

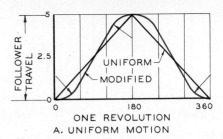

A. UNIFORM MOTION

Uniform-motion diagrams are modified with arcs of one fourth to one third of total displacement to smooth out the velocity of the follower at these points of abrupt change.

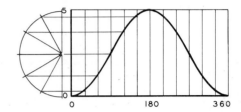

B. HARMONIC MOTION

Harmonic motion is plotted by projecting from a semicircle whose diameter is equal to the rise of the follower. The semicircle must be divided into the same number of sectors as the divisions on the x-axis of the graph to the point of maximum rise.

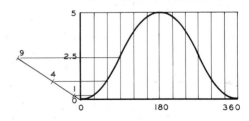

C. GRAVITY MOTION

Gravity-motion diagrams are constructed so that the rise of the follower is relative to the square of the units on the x-axis: 1^2, 2^2, 3^2, etc.

FIG. 19.14 Displacement diagrams.

Gravity motion

Gravity motion (uniform acceleration), (Fig. 19.14C), is used for high-speed operation. The variation of displacement is analogous to the force of gravity, with the difference in displacement being 1, 3, 5, 5, 3, 1, based on the square of the number; for instance, $1^2 = 1$; $2^2 = 4$; $3^2 = 9$. This motion is repeated in reverse

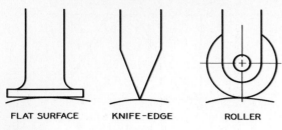

FIG. 19.15 Three basic types of cam followers are the flat surface, the roller, and the knife edge.

order for the remaining half of the motion of the follower. Intermediate points can be found by squaring fractional increments, such as $(2.5)^2$.

Cam followers

Three basic types of cam followers are the flat surface, the roller, and the knife edge, (Fig. 19.15). The flat-surface and knife-edge followers are limited to use with slow-moving cams where minor force will be exerted during rotation. The roller follower is used to withstand higher speeds.

FIG. 19.16 Construction of a plate cam with harmonic motion.

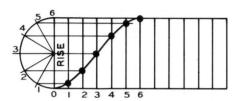

Step 1 Construct a semicircle whose diameter is equal to the rise of the follower. Divide the semicircle into the same number of divisions as there are between 0° and 180° on the horizontal axis of the displacement diagram. Plot half of the displacement curve.

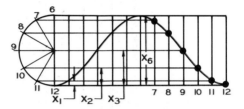

Step 2 Continue the process of plotting points by projecting from the semicircle, starting from the top of the semicircle and proceeding to the bottom. Complete the curve symmetrical half of the curve.

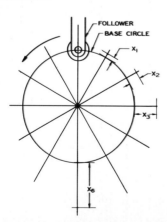

Step 3 Construct the base circle and draw the follower. Divide the circle into the same number of sectors as there are divisions on the displacement diagram. Transfer distances from the displacement diagram to the respective radial lines of the circle, measuring outward from it.

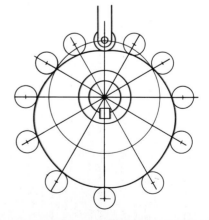

Step 4 Draw circles to represent the positions of the roller as the cam revolves in a counterclockwise direction. Draw the cam profile tangent to all the rollers to complete the drawing.

19.15
Construction of a plate cam

Plate cam—harmonic motion

The steps of constructing a plate cam with harmonic motion are shown in Fig. 19.16. Before designing a cam, the drafter must know the motion of the follower, the rise of the follower, the diameter of the base circle, and the direction of rotation. The specifications for the cam pictured in Fig. 19.16 are given graphically in the displacement diagram.

Plate cam—uniform acceleration

The steps of constructing a cam with uniform acceleration are the same as in the previous example except

for a different displacement diagram and a knife-edge follower. The graphic layout of the problem is shown in Fig. 19.17.

Plate cam—combination

In Fig. 19.18, a knife-edge follower is used with a plate cam to produce a 4-inch rise with harmonic motion from 0° to 180°, a 4-inch fall with a uniform acceleration from 180° to 300°, and dwell (no follower motion) from 300° to 360°. The drafter must draw the cam that will give this motion from the base circle.

The displacement diagram is drawn with a harmonic curve with a full rise of 4 inches. The curve is then drawn with a uniform acceleration drop of 4 inches. The dwell is a horizontal line to complete the diagram of the 360° rotation of the cam.

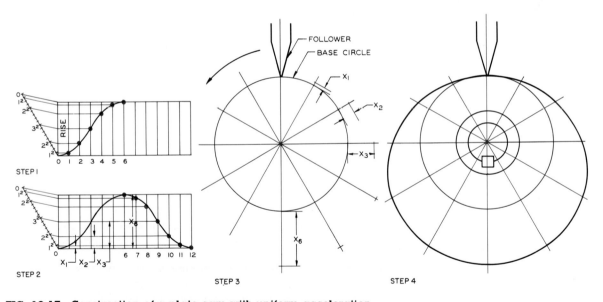

FIG. 19.17 Construction of a plate cam with uniform acceleration.

Step 1 Construct a displacement diagram to represent the rise of the follower. Divide the horizontal axis into angular increments of 30°. Draw a construction line through point 0; locate the 1^2, 2^2, and 3^2 divisions and project them to the vertical axis to represent half of the rise. The other half of the rise is found by laying off distances along the construction line with descending values.

Step 2 Use the same construction to find the right half of the symmetrical curve.

Step 3 Construct the base circle and draw the knife-edge follower. Divide the circle into the same number of sectors as there are divisions in the displacement diagram. Transfer distances from the displacement diagram to the respective radial lines of the base circle, measuring outward from the base circle.

Step 4 Connect the points found in Step 3 with a smooth curve to complete the cam profile. Show also the cam hub and keyway.

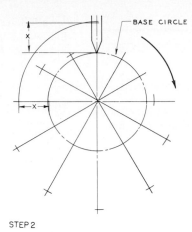

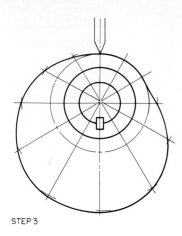

STEP 1   STEP 2   STEP 3

FIG. 19.18 Construction of a plate cam with combination motions.

Step 1   The cam is to rise 4″ in 180° with harmonic motion, fall 4″ in 120° with uniform acceleration, and dwell for 60°. These motions are plotted on the displacement diagram.

Step 2   Construct the base circle and knife-edge follower. Transfer distances from the displacement diagram to the respective radial lines of the base circle, measuring outward from it.

Step 3   Draw a smooth curve through the points found in Step 2 to complete the profile of the cam.

19.16
Construction of a cam with an offset follower

The cam in Fig. 19.19 is required to produce harmonic motion through 360°. This motion is plotted directly from the follower rather than from a displacement diagram, since no combinations of motion are involved.

A semicircle is drawn with its diameter equal to the total motion of the follower. The base circle is drawn to pass through the center or roller of the follower. The centerline of the follower is extended downward, and a circle is drawn tangent to the extension with its center at the center of the base circle. The small circle is divided into 30° intervals to establish points through which construction lines will be drawn tangent to the circle.

The distances from tangent points to the position points along the path of the follower are laid out along the tangent lines drawn at 30° intervals. These points can be located by measuring from the base circle, as shown in the figure, where point 3 was located distance X from the base circle. The circular roller is drawn in all views, and the profile of the cam is constructed tangent to the rollers at all positions.

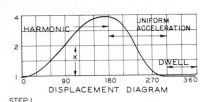

FIG. 19.19 Construction of a plate cam with an offset roller follower.

Problems

Gears

Use Size A sheets (8½ by 11 inches) for the following gear problems. Select the most appropriate scale so that the drawings will use the available space.

1–5. Calculate the dimensions for the following spur gears, and make a detail drawing of each. Give the dimensions and cutting data for each gear. Provide any other dimensions that are needed.

Problem	Gear Teeth	Diametrical Pitch	14.5° Involute
1	20	5	"
2	30	3	"
3	40	4	"
4	60	6	"
5	80	4	"

6–10. Calculate the gear sizes and number of teeth using the ratios and data below.

Problem	RPM Pinion	RPM Gear	Center to Center	Diametrical Pitch
6	100 (driver)	60	6.0″	10
7	100 (driver)	50	8.0″	9
8	100 (driver)	40	10.0″	8
9	100 (driver)	35	12.0″	7
10	100 (driver)	25	14.0″	6

11–20. Make detail drawings of each of the gears for which calculations were made in Problems 6–10. Provide a table of cutting data and other dimensions that are needed to complete the specifications.

21–25. Calculate the specifications for the bevel gears that intersect at 90°, and make detail drawings of each with the necessary dimensions and cutting data.

Problem	Diametrical Pitch	No. of Teeth on Pinion	No. of Teeth on Gear
21	3	60	15
22	4	100	40
23	5	100	60
24	6	100	50
25	7	100	30

26–30. Calculate the specifications for worm gears and make detail drawings of each, providing the necessary dimensions and cutting data.

Problem	No. of Teeth in Spider Gear	Outside DIA of Worm	Pitch of Worm	Thread of Worm
26	45	2.50	0.50	double
27	30	2.00	0.80	single
28	60	3.00	0.80	double
29	30	2.00	0.25	double
30	80	4.00	1.00	single

Cams—problems 31–41

Draw the following on Size B sheets (11 by 17 inches) with the following standard dimensions: base circle, 3.50 in.; roller follower, 0.60-in. diameter; shaft, 0.75-in. diameter; hub, 1.25-in. diameter; direction of rotation, clockwise. The follower is positioned vertically over the center of the base circle except in Problems 40 and 41. Lay out the problems and displacement diagrams as shown in Fig. 19.20.

31. Draw a plate cam with a knife-edge follower for uniform motion and a rise of 1.00 in.

32. Draw a displacement diagram and a cam that will give a modified uniform motion to a knife-edge follower with a rise of 1.7 in. Modify the uniform motion with an arc of one quarter of the rise in the displacement diagram.

33. Draw a displacement diagram and a cam that will give a harmonic motion to a roller follower with a rise of 1.60 in.

34. Draw a displacement diagram and a cam that will give a harmonic motion to a knife-edge follower with a rise of 1.00 in.

35. Draw a displacement diagram and a cam that will give uniform acceleration to a knife-edge follower with a rise of 1.70 in.

36. Draw a displacement diagram and a cam that will give a uniform acceleration to a roller follower with a rise of 1.40 in.

37. Draw a displacement diagram and a cam that will give the following motion to a knife-edge follower: rise 1.25 in. with harmonic motion in 120°; and fall 1.25 in. with uniform acceleration.

38. Draw a displacement diagram and a cam that will give the following motion to a knife-edge follower: dwell for 70°; rise 1 in. with a modified uniform motion in 100°; fall 1 in. with a harmonic motion in 100°; and dwell for 90°.

39. Draw a displacement diagram and a cam that will give the following motion to a roller follower: rise 1.25 in. with a harmonic motion in 120°; dwell for 120°; and fall 1.25 in. with a uniform acceleration in 120°.

40. Repeat Problem 32, but offset the follower 0.60 in. to the right of the vertical centerline.

41. Repeat Problem 33, but offset the follower 0.60 in. to the left of the vertical line.

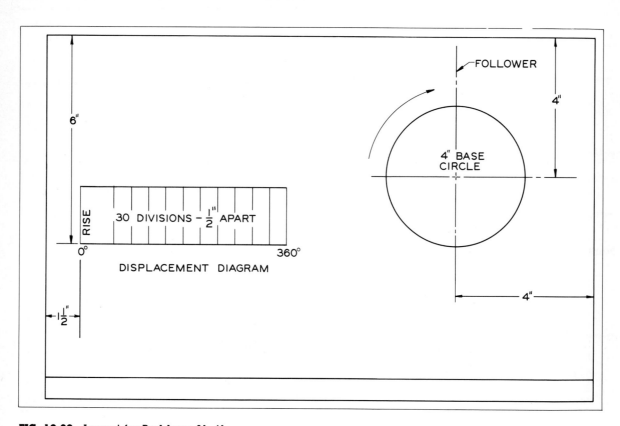

FIG. 19.20 Layout for Problems 31–41.

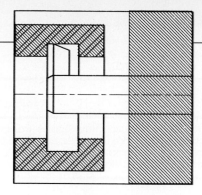

CHAPTER 20

Materials and Processes

20.1
Introduction

Metallurgy, the study of metals, is a complex area that is constantly changing as new processes and alloys are developed (Fig. 20.1) The guidelines for designating various types of metals have been standardized by three associations: the American Iron and Steel Institute (AISI), the Society of Automotive Engineers (SAE), and the American Society for Testing Materials (ASTM).

20.2
Iron*

Metals that contain iron, even in small quantities, are called *ferrous metals*. The three types of iron are *gray iron*, *white iron*, and *ductile iron*. *Cast iron*, iron that is

*This section on iron was developed by Dr. Tom Pollock, a metallurgist at Texas A&M University.

FIG. 20.1 This furnace operator is pouring an aluminum alloy of manganese into ingots (shown at the right) that will be remelted and cast. (Courtesy of the Aluminum Co. of America.)

TABLE 20.1

NUMBERING AND APPLICATIONS OF GRAY IRON

ATSM Grade (1000 psi)	SAE Grade	Typical Uses
ASTM 25 CI	G 2500 CI	Small engine blocks, pump bodies, transmission cases, clutch plates
ASTM 30 CI	G 3000 CI	Auto engine blocks, flywheels, heavy casting
ASTM 35 CI	G 3500 CI	Diesel engine blocks, tractor transmission cases, heavy and high-strength parts
ASTM 40 CI	G 4000 CI	Diesel cylinders, pistons, camshafts

melted and poured into a mold to form it, is used in the production of machine parts. Though cheaper and easier to machine than steel, iron does not have its ability to withstand shock and force.

GRAY IRON contains flakes of graphite, which results in low strength and ductility but makes the material easy to machine. Gray iron resists vibrations better than other types of iron. Types of gray iron with two designations and their typical applications are given in Table 20.1.

WHITE IRON contains carbide particles that are very hard and brittle, which enables it to withstand wear and abrasion. There are no designated grades of white iron, but there are differences in composition from one supplier to another. White iron is used for parts on grinding and crushing machines, digging teeth on earthmovers and mining equipment, and wear plates on reciprocating machinery used in textile mills.

DUCTILE IRON (also called *nodular* or *spheroidized* iron) contains tiny spheres of graphite, making it stronger and tougher than most types of gray iron also but this makes it more expensive to produce. The numbering system for ductile iron is given by three sets of numbers, as shown below:

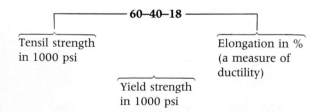

Commonly used alloys of ductile iron and their applications are shown in Table 20.2.

TABLE 20.2

NUMBERING AND APPLICATIONS OF DUCTILE IRON

Grade	Typical Uses
60–40–18 CI	Valves, steam fittings, chemical plant equipment, pump bodies
65–45–12 CI	Machine components that are shock loaded, disc brake calipers
80–55–6– CI	Auto crankshafts, gears, rollers
100–70–3 CI	High-strength gears and machine parts
120–90–2 CI	Very high-strength gears, rollers, and slides

MALLEABLE IRON is made from white iron by a heat-treatment process that converts carbides into carbon nodules (similar to ductile iron). The numbering system for designating the grades of malleable iron is shown below:

A. Tensile strength in MPa $\left.\right\}$ **325** — **10** $\left\{\right.$ Elongation in %

B. Stands for "martensite," which is not essential $\left.\right\}$ **M** **32** **10** $\left\{\right.$ Elongation in %; Tensil strength in MPa × 0.1

Some of the commonly used grades of malleable iron and their applications are given in Table 20.3.

TABLE 20.3
NUMBERING AND APPLICATIONS OF
MALLEABLE IRON

ASTM Grade	Typical Uses
35018 CI	Marine and railroad valves and fittings, "black-iron" pipe fittings (similar to 60–40–18 ductile CI)
45006 CI	Machine parts (similar to 80–55–6 ductile CI)
M3210 CI	Low-stress components, brackets
M4504 CI	Crankshafts, hubs
M7002 CI	High-strength parts, connecting rods, universal joints
M8501 CI	Wear-resistant gears and sliding parts

20.3
Steel

Steel is an alloy of iron and carbon and often contains other constituents such as manganese, chromium, or nickel. Carbon (usually between 0.20 and 1.50 percent) is the ingredient that has the greatest effect on the grade of the steel. Broadly, the three types of steel are *plain carbon steels, free-cutting carbon steels,* and *alloy steels.* The types of steels and their designations by four-digit numbers are shown in Table 20.4. The first digit indicates the type of steel: 1 is carbon steel, 2 is nickel steel, and so on. The second digit gives the percentage content of the material represented by the first digit. The last two digits give the percentage of carbon in the alloy, where 100 is equal to 1 percent, and 50 is equal to 0.50 percent.

Some frequently used SAE (Society of Automotive Engineers) steels are 1010, 1015, 1020, 1030, 1040, 1070, 1080, 1111, 1118, 1145, 1320, 2330, 2345, 2515, 3135, 3130, 3240, 3310, 4023, 4042, 4063, 4140, and 4320.

20.4
Copper

Copper, one of the first metals discovered, can be easily formed and bent without breakage. Because it is highly resistant to corrosion and highly conductive, it is used in the manufacture of pipes, tubing, and electrical wiring. It is also an excellent roofing and screening material since it withstands the weather well.

Copper has a number of alloys, including brasses, tin bronzes, nickel silvers, and copper nickels. Brass is an alloy of copper and zinc, and bronze is an alloy of copper and tin. Copper and copper alloys can be easily finished by buffing or plating. These alloys can be joined by soldering, brazing, or welding and can be easily machined and used for casting.

Wrought copper has properties that permit it to be formed by hammering. A few of the numbered designations of wrought copper are C11000, C11100, C11300, C11400, C11500, C11600, C10200, C12000, and C12200.

TABLE 20.4
NUMBERING AND APPLICATIONS OF STEEL

Type of Steel	Number	Application
Carbon steels		
Plain carbon	10XX	Tubing, wire, nails
Resulphurized	11XX	Nuts, bolts, screws
Manganese steel	13XX	Gears, shafts
Nickel steel	23XX	Keys, levers, bolts
	25XX	Carburized parts
Nickel-chromium	31XX	Axles, gears, pins
	32XX	Forgings
	33XX	Axles, gears
Molybdenum steel	40XX	Gears, springs
Chromium-molybdenum	41XX	Shafts, tubing
Nickel-chromium	43XX	Gears, pinions
Nickel-molybdenum	46XX	Cams, shafts
	48XX	Roller bearings, pins
Chromium steel	51XX	Springs, gears
	52XX	Ball bearings
Chromium vanadium	61XX	Springs, forgings
Silicon manganese	92XX	Leaf springs

Source: Courtesy of the Society of Automotive Engineers.

20.5
Aluminum

Aluminum is a corrosion-resistant, lightweight metal that has applications for many industrial products. Most materials called aluminum are actually aluminum alloys, which are stronger than pure aluminum.

The types of wrought aluminum alloys are designated by four digits, as shown in Table 20.5. The first digit, from 2 through 9, indicates the alloying element

TABLE 20.5

NUMBERING DESIGNATIONS FOR WROUGHT
ALUMINUM AND ALUMINUM ALLOYS

Composition	Alloy Number	Applications
Aluminum (99% pure)	1XXX	Tubing, tank cars
Aluminum alloys		
Copper	2XXX	Aircraft parts, screws, rivets
Manganese	3XXX	Tanks, siding, gutters
Silicon	4XXX	Forging, wire
Magnesium	5XXX	Tubes, welded vessels
Magnesium and silicon	6XXX	Auto body, pipe
Zinc	7XXX	Aircraft structures
Other elements	8XXX	

that is combined with aluminum. The second digit indicates modifications of the original alloy or impurity limits. The last two digits identify the other alloying materials or indicate the aluminum purity.

A four-digit numbering system, as shown in Table 20.6, is used to designate types of cast aluminum and aluminum alloys. The first digit indicates the alloy group. The next two digits identify the aluminum alloy or the aluminum purity. The numeral 1 to the

TABLE 20.6

ALUMINUM CASTING AND INGOT
DESIGNATIONS

Composition	Alloy Number
Aluminum (99 % pure)	1XX.X
Aluminum alloys	
Copper	2XX.X
Silicon with copper and/or magnesium	3XX.X
Silicon	4XX.X
Magnesium	5XX.X
Magnesium and silicon	6XX.X
Zinc	7XX.X
Tin	8XX.X
Other elements	9XX.X

Source: Society of Automotive Engineers.

right of the decimal point represents the aluminum form: XX.0 indicates castings, XX.1 indicates ingots with a specified chemical composition, and XX.2 indicates ingots with a specified chemical composition other than the XX.1 ingot whereas 0 represents aluminum for casting. *Ingots* are blocks of cast metal that are to be remelted, and *billets* are castings of aluminum that are to be formed by forging.

20.6
Magnesium

Magnesium is a light metal available in an inexhaustible supply since it is extracted from seawater and natural brines. Approximately half the weight of aluminum, magnesium is an excellent material for aircraft parts, clutch housing, crankcases for air-cooled engines, and applications where lightness is desirable.

Magnesium is used for die and sand castings, extruded tubing, sheet metal, and forging. Magnesium and its alloys can be joined by bolting, riveting, or welding. Some numbered designations of magnesium alloys are M10100, M11630, M11810, M11910, M11912, M12390, M13320, M16410, and M16620.

20.7
Properties of materials

All materials have properties that designers must use to their best advantage. The following terms are used to describe these properties.

DUCTILITY is a softness present in some materials, such as copper and aluminum, that permits them to be formed by stretching (drawing) or hammering without breaking. Wire is made of ductile materials that can be drawn through a die.

BRITTLENESS is a characteristic of metals that will not stretch without breaking, such as cast irons and hardened steels.

MALLEABILITY is the ability of a metal to be rolled or hammered without breaking.

HARDNESS is a metal's ability to resist being dented when it receives a blow.

TOUGHNESS is the property of being resistant to cracking and breaking while remaining malleable.

ELASTICITY   is the ability of a metal to return to its original shape after being bent or stretched.

20.8
Heat treatment of metals

The properties of metals can be changed by various forms of heat treating. Steels are affected to a greater extent by heat treating than are other materials.

HARDENING   of steel is performed by heating the material to a prescribed temperature and then quenching it in oil or water.

QUENCHING   is the process of rapidly cooling heated metal by immersing it in liquids, gases, or solids (such as sand, limestone, or asbestos).

TEMPERING   is the process of reheating previously hardened steel and then cooling it, usually by air. This increases the steel's toughness.

ANNEALING   is the process of heating and cooling metals to soften them, release their internal stresses, and make them easier to machine.

NORMALIZING   is achieved by heating metals and letting them cool in air to relieve their internal stresses.

CASE HARDENING   is the process of hardening a thin outside layer of a metal. The outer layer is placed in contact with carbon or nitrogen compounds that are absorbed by the metal as it is heated; afterward, the metal is quenched.

FLAME HARDENING   is the method of hardening by heating a metal to within a prescribed temperature range with a flame and then quenching the metal.

20.9
Castings

Two major methods of forming shapes are *casting* and *pressure forming*. Casting involves the preparation of a mold into which is poured molten metal that cools and forms the part. The types of casting, which differ

FIG. 20.2   A large casting of an aircraft's landing-gear mechanism is being removed from its mold. (Courtesy of Cameron Iron Works.)

in the way the molds are made, are *sand casting, permanent-mold casting, die casting,* and *investment casting.*

Sand casting

In the first step of sand casting, a wood or metal form or pattern is made that is representative of the final part to be cast. The pattern is placed in a metal box called a *flask,* and molding sand is packed around the pattern. When the pattern is withdrawn from the sand, it leaves a void forming the mold. Molten metal is poured into the mold through sprues, or gates. After cooling, the casting is removed and cleaned (Fig. 20.2).

Cores, parts formed in sand, are placed within a mold to leave holes or hollow portions within the finished casting (Fig. 20.3). Once the casting has been formed, the cores can be broken apart and removed, leaving behind the desired void within the casting. Cores add to the cost of a casting and should not be used unless their expense is offset by savings in materials.

Since the patterns are placed in sand and then withdrawn before the metal is poured, the sides of the patterns must be tapered for ease of withdrawal from the sand (Fig. 20.4). This taper is called *draft.* The amount of draft depends on the depth of the pattern in the sand; for most applications, it varies from 2 to

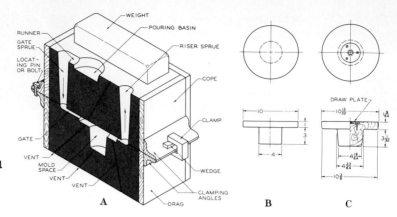

FIG. 20.3 A typical sand mold. The opening, which will be filled with molten metal, is formed by a wood or metal pattern that is pressed into the sand.

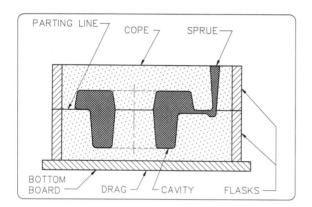

FIG. 20.4 A two-section sand mold.

8°. Also, to compensate for shrinkage that will occur when the metal cools, patterns are made oversize.

Because the sand casting has a rough surface that ought not be a contact surfaces with other moving parts, it is common practice to machine portions of a casting by drilling, grinding, or shaping (Fig. 20.5). The pattern should be made larger in these areas to compensate for removal of metal.

Fillets and rounds are used at intersections to increase the strength of a casting and because it is difficult to form square corners by the sand-casting process.

Permanent-mold casting

Permanent molds are made for the mass production of parts. They are generally made of cast iron and are

FIG. 20.5 This casting of the outer cylinder of an aircraft's landing gear is being bored on a horizontal boring mill. (Courtesy of Cameron Iron Works.)

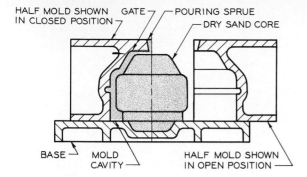

FIG. 20.6 Permanent molds are made of metal for repetitive usage. Here, a sand core, made from another mold, is placed in the permanent mold to give a void within the casting.

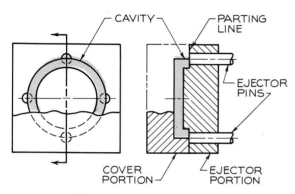

FIG. 20.7 A die for casting a simple part. Unlike the sand casting, the metal is forced into the die to form the casting.

FIG. 20.8 Three stages of manufacturing a turbine fan: **A.** The blank is formed by forging. **B.** It is machined. **C.** The fan blades are attached to their machined slots. (Courtesy of Avco Lycoming.)

coated to prevent fusing with the molten metal that is poured into them (Fig. 20.6).

Die casting

Die castings are used for the mass production of parts made of aluminum, magnesium, zinc alloys, copper, or other materials. Made by forcing molten metal into dies (or molds) under pressure, die castings can be produced at a low cost, at close tolerances, and with good surface qualities. The same general principles recommended for sand castings—using fillets and rounds, allowing for shrinkage, and specifying draft angles—apply to die castings also (Fig. 20.7).

Investment casting

Investment casting is a process used to produce complicated parts that would be difficult to form by any other method. This technique is used to form the intricate shapes of artistic sculptures.

Since a new pattern must be used for each investment casting, a mold or die is made to cast a wax master pattern. The wax pattern, which will be identical to the finished casting, is placed inside a container, and plaster or sand is poured (or invested) around the pattern. Once the investment has cured, the wax pattern is melted, leaving a hollow cavity that will serve as the mold for the molten metal. When filled and set, the plaster or sand is broken away from the finished investment casting.

20.10
Forgings

Forging is the process of shaping or forming heated metal by hammering or squeezing it into a die. Drop forges and press forges are used to hammer the metal (called billets) into the forging dies by multiple blows. The resulting forging possesses high strength and a resistance to loads and impacts.

When preparing forging drawings, the following must be considered: (1) draft angles and parting lines, (2) fillets and rounds, (3) forging tolerances, (4) allowance of extra material for machining, and (5) heat treatment of the finished forging (Fig. 20.8). Some of the standard steels used for forging are designated by the SAE numbers 1015, 1020, 1025, 1045, 1137, 1151, 1335, 1340, 4620, 5120, and 5140. Iron, copper, and aluminum can also be forged.

Examples of dies are shown in Fig. 20.9. A single-impression die gives an impression on one side of the parting line between the mating dies; a double-impression die gives an impression on both sides of the parting line; and the interlocking dies gives an impression that may cross the parting line on either side. An object that is forged with auxiliary rams to hollow the forging is shown in Fig. 20.10. A drawing of a forged part is illustrated in Fig. 20.11.

Rolling

Rolling is a type of forging in which the stock is rolled between two rollers to give it a desired shape. Rolling can be done at right angles to the axis of the part or

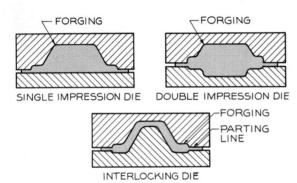

FIG. 20.9 Three types of forging dies: **A.** A single-impression die. **B.** A double-impression die. **C.** An interlocking die.

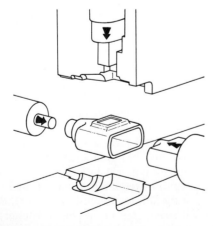

FIG. 20.10 Auxiliary rams can be used to form internal features on a part. (Courtesy of General Motors Corp.)

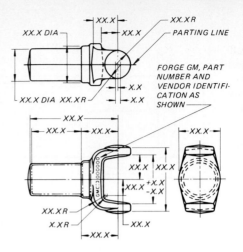

UNLESS OTHERWISE SPECIFIED:
DRAFT ANGLES X°.
ALL FILLETS X.XR, CORNERS X.XR.
+X.X– X.X TOLERANCES ON
FORGING DIM.

SNAG AND REMOVE SCALE.

SAMPLE FORGINGS ARE TO BE
APPROVED BY METALLURGICAL
AND ENGRG DEPTS FOR GRAIN
FLOW STRUCTURE.

FORGING DRAWING

FIG. 20.11 A drawing of a forging. The blank is forged oversize to allow for machining operations that will remove metal from it. (Courtesy of General Motors Corp.)

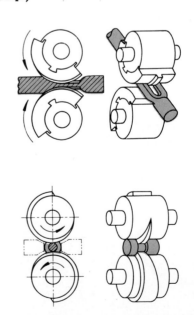

FIG. 20.12 Features on parts may be formed by rolling. In these examples, parts are being rolled parallel and perpendicular to their axes. (Courtesy of General Motors Corp.)

parallel to its axis (Fig. 20.12). If a high degree of shaping is required, the stock is usually heated before rolling. If the forming requires only a slight change in configuration, the rolling can be performed when the metal is cold; this is called *cold rolling*.

20.11
Stamping

Stamping is a method of forming flat metal stock into three-dimensional shapes. The first step of stamping is to cut out the shapes, called *blanks*, that are to be bent. Blanks are formed into shape by bending and pressing them against forms. Examples of box-shaped parts are shown in Fig. 20.13 and a flange stamping is illustrated in Fig. 20.14. Holes in stampings are made by punching, extruding, or piercing (Fig. 20.15).

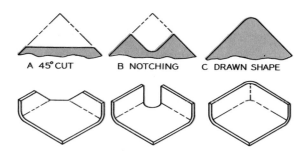

FIG. 20.13 Box-shaped parts formed by stamping: **A.** A corner cut of 45° permits folding flanges and may require no further trim. **B.** Notching has the same effect as the 45° cut, but it is often more attractive. **C.** A continuous corner flange requires that the blank be developed so that it can be drawn into shape.

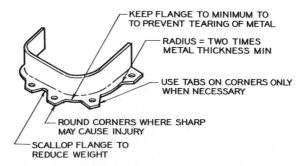

FIG. 20.14 A sheet metal flange design, with notes calling attention to design details.

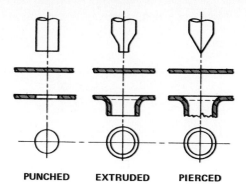

FIG. 20.15 Three methods of forming holes in sheet metal. (Courtesy of General Motors Corp.)

20.12
Machining operations

After the metal has been formed, machining operations have to be performed to complete the part, which involve the following machines: *lathe, drill press, milling machine, shaper,* and *planer*.

The lathe

The lathe shapes cylindrical parts while rotating the work piece between the centers of the lathe (Fig. 20.16). The more fundamental operations performed

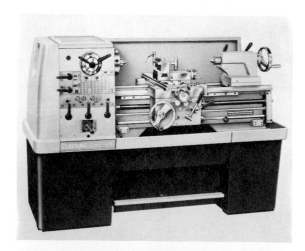

FIG. 20.16 A typical metal lathe that holds and rotates the work piece between the centers of the lathe. (Courtesy of the Clausing Corp.)

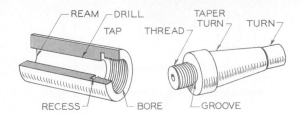

FIG. 20.17 The fundamental operations that are performed on a lathe are illustrated on the two parts above.

on the lathe are *turning, facing, drilling, boring, reaming, threading,* and *undercutting* (Fig. 20.17).

TURNING forms a cylinder by a tool that advances against and moves parallel to the cylinder being turned (Fig. 20.18).

FACING forms flat surfaces perpendicular to the axis of rotation of the part being rotated by the lathe.

DRILLING is performed by mounting a drill in the tail stock of the lathe and rotating the work while the bit is advanced into the part (Fig. 20.19).

BORING makes large holes that are too big to be drilled. Large holes are bored by enlarging smaller drilled holes (Fig. 20.20).

REAMING removes only a few thousandths of an inch of material inside a drilled hole to bring it to its required level of tolerance. Conical and cylindrical reaming can be performed on the lathe (Fig. 20.21).

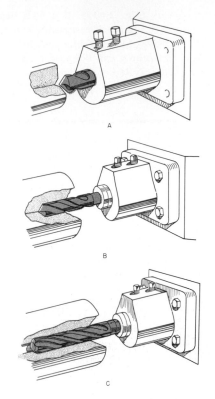

FIG. 20.19 Three steps of drilling a hole in the end of a cylinder: **A.** start drilling, **B.** twist drilling, and **C.** core drilling, which enlarges the previously drilled hole to the required size.

FIG. 20.18 Turning is the most basic of all operations performed on the lathe. A continuous chip is removed by a cutting tool as the part is rotated.

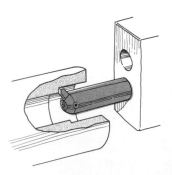

FIG. 20.20 Boring is the method of enlarging holes that are larger than available drill bits. The cutting tool is attached to the boring bar on the lathe.

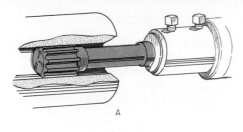

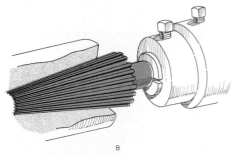

FIG. 20.21 Fluted reamers can be used to finish inside cylindrical and conical holes within a few thousandths of an inch.

THREADING of external and internal holes can be done on the lathe. The die used for cutting internal holes is called a *tap* (Fig. 20.22).

UNDERCUTTING cuts a recess inside a cylindrical hole with a tool mounted on a boring bar. The groove is cut as the tool advances from the center of the axis of revolution into the part (Fig. 20.23).

The turret lathe is a programmable lathe that can perform sequential operations, such as drilling a series

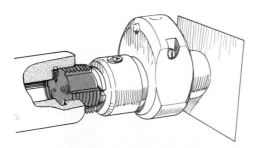

FIG. 20.22 External and internal threads (shown here) can be cut on a lathe. The die used to cut the threads is called a tap. Note that a recess has been formed at the end of the threaded hole prior to threading.

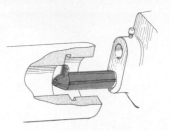

FIG. 20.23 A recess (undercut) can be formed by using the boring bar with the cutting tools attached as shown. As the boring bar is moved off center of the axis, the tool will form the recess.

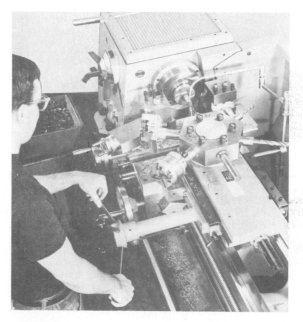

FIG. 20.24 A turret lathe that performs a sequence of operations.

of holes, boring them, and then reaming them in succession. The turret is mounted to rotate each tool into position for its particular operation (Fig. 20.24).

The drill press

The drill press is used to drill small- and medium-sized holes into stock that is held on the bed of the press by a fixture or clamp (Fig. 20.25). The drill press can be

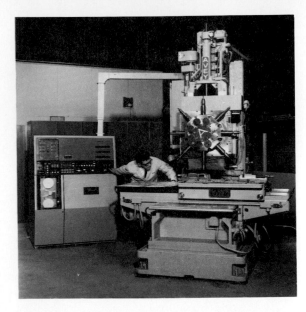

FIG. 20.25 A multiple-head drill press that can be programmed to perform a series of drill press operations in a desired sequence.

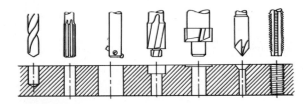

FIG. 20.26 The basic operations that can be performed on the drill press are (left to right): drilling, reaming, boring, counterboring, spotfacing, countersinking, and tapping (threading).

used for *counterdrilling, countersinking, counterboring, spotfacing,* and *threading* (Fig. 20.26).

BROACHING Cylindrical holes can be converted into square holes or hexagonal holes by using a tool called a *broach*. The broach has a series of teeth along its axis, beginning with teeth that are nearly the size of the hole to be broached and tapering to the size of the finished hole that is to be broached. The broach is forced through the hole, with each tooth cutting more from the hole as it passes through.

The milling machine

The millng machine uses a variety of cutting tools, rotated about a shaft (Fig. 20.27), to form different grooved slots, threads, and gear teeth. The milling machine can cut irregular grooves in cams and finish surfaces on a part within a high degree of tolerance.

FIG. 20.27 This small milling machine is being used to cut a slot in the work piece. (Courtesy of the Clausing Corp.)

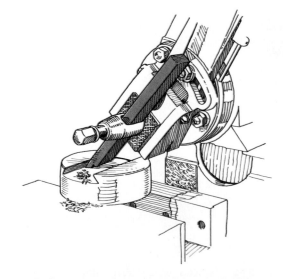

FIG. 20.28 The shaper moves back and forth across the part, removing metal as it advances. It can be used to finish surfaces, cut slots, and for many other operations.

The shaper

The shaper is a machine that holds the work piece stationary while the cutter passes back and forth across the work to finish the surface or to cut a groove, one stroke at a time (Fig. 20.28). With each stroke of the cutting tool, the material is shifted slightly so as to align the part for the next overlapping stroke.

The planer

Unlike the shaper, which holds the work piece stationary, the planer passes the work under the cutters (Fig. 20.29). Like the shaper, the planer can cut grooves or slots and finish surfaces that must meet tolerance specifications.

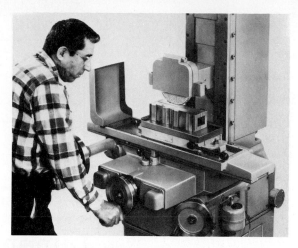

FIG. 20.30 The upper surface of this part is being ground to a smooth finish by a grinding wheel. (Courtesy of the Clausing Corp.)

FIG. 20.29 The planer has stationary cutters, and the work is fed past them to finish larger surfaces. This planer has a 30-foot bed. (Courtesy of Simmons Machine Tool Corp.)

20.13
Surface finishing

Surface finishing is the process of finishing a surface to the desired uniformity. It may be accomplished by several methods, including *grinding, polishing, lapping, buffing,* and *honing*.

GRINDING finishes of a flat surface by holding it against a rotating abrasive wheel (Fig. 20.30). Grinding is used to smooth surfaces and to sharpen edges used for cutting, such as drill bits.

POLISHING is performed in the same manner as grinding, except the polishing wheel is flexible since it is made of felt, leather, canvas, or fabric.

LAPPING produces very smooth surfaces. The surface to be finished is held against large, flat surface, called a *lap*, which has been coated with a fine abrasive powder. As the lap rotates, the surface is finished. Lapping is done only after the surface has been previously finished by a less accurate technique, such as grinding or polishing. Cylindrical parts can be lapped by using a lathe with the lap.

BUFFING removes scratches from a surface with a rotating buffer wheel made of wool, cotton, or other fabric. Sometimes the buffer is a cloth or felt belt that is applied to the surface being buffed. To enhance the buffing, an abrasive mixture is applied to the buffed surface from time to time.

HONING finishes the outside or the inside of holes within a high degree of tolerance. As it is passed through the holes, the honing tool is rotated to produce the sort of finishes found in gun barrels, engine cylinders, and products where a high degree of smoothness is required.

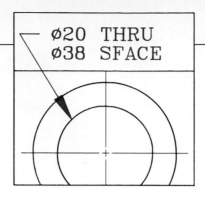

CHAPTER 21

Dimensioning

21.1
Introduction

Working drawings are dimensioned drawings used to describe the details of a part or a project so that construction can be performed in accordance with specifications. When properly applied, dimensions and notes will supplement the drawings so they can be used as legal contracts for construction.

The techniques of dimensioning presented in this chapter are based primarily on the standards of the American National Standards Institute (ANSI), especially Y14.5M, *Dimensioning and Tolerancing for Engineering Drawings*. Various industrial standards from corporations, such as the General Motors Corperation, have also been used.

21.2
Dimensioning terminology

The guide slide in Fig. 21.1 is used as an example to identify some of the terms of dimensioning.

DIMENSION LINES are thin lines (2H–4H pencil) with arrows at each end. Numbers placed near their midpoints specify a part's size.

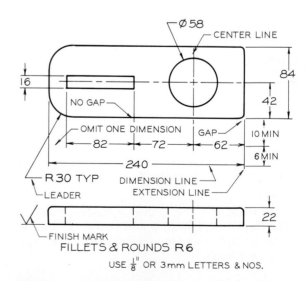

FIG. 21.1 This typical working drawing is dimensioned in millimeters.

EXTENSION LINES are thin lines (2H–4H pencil) that extend from a view of an object for dimensioning the part. The arrowheads of dimension lines end at these lines.

CENTERLINES are thin lines (2H–4H pencil) used to locate the centers of cylindrical parts, such as cylindrical holes.

LEADERS are thin lines (2H–4H pencil) drawn from a note to a feature to which the note applies.

ARROWHEADS are placed at the ends of dimension lines and leaders to indicate their endpoints. Arrowheads are drawn the same length as the height of the letters or numerals, one-eighth inch in most cases. The form of the arrowhead is shown in Fig. 21.2.

DIMENSION NUMBERS are placed near the middle of the dimension line and are usually ⅛ inch high; units of measurement (″, IN, or mm) are omitted.

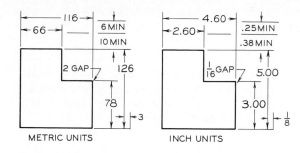

FIG. 21.3 Dimension lines should be placed at least ⅜″ (10 mm) from an object. Other rows of dimensions should be located at least ¼″ (6 mm) apart.

form. A comparison of dimensions in millimeters with those in inches is shown in Fig. 21.3.

Examples are shown in Fig. 21.4, where the units are given in millimeters, decimal inches, and fractional inches.

Dimensions in millimeters are usually rounded off to whole numbers without decimal fractions.

When a metric dimension is less than a millimeter, a zero precedes the decimal point. When using decimal inches, show all dimensions with two-place decimal fractions even if the last numbers are zeros. For di-

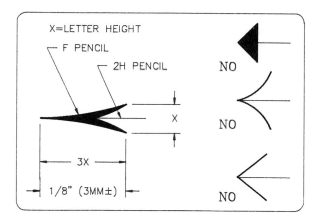

FIG. 21.2 Arrowheads are drawn as long as the height of the letters used on a drawing. They are one third as wide as they are long.

21.3
Units of measurement

The two most commonly used units of measurement are the decimal inch, in the English (imperial) system, and the millimeter, in the metric (SI) system.

The inch in its common fraction form can be used, but it is preferable to give fractions in decimal form. Common fractions make arithmetic hard to per-

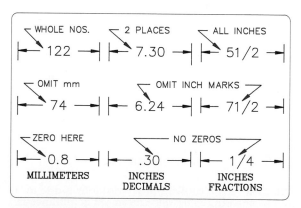

FIG. 21.4 When using SI units, dimensions are usually given to the nearest whole millimeter. When decimal inches are used, the fractions are carried to two decimal places. If common fractions are used, they should be twice as tall as whole numbers.

mensions of less than an inch, no zero ˙s the decimal.

> Units of measurement are omitted from the dimension numbers, since they are normally understood to be in millimeters or inches. For example, 112, not 112 mm; and 67, not 67″ or 5′–7″.

Architects do use a combination of feet and inches, but the inch units are omitted (e.g., 7′–2). Engineers use feet and decimal fractions of feet to dimension large-scale projects such as road designs (e.g., 252.7′ where feet units *are* shown).

21.4
English/metric conversions

> Dimensions in inches can be converted to millimeters by multiplying by 25.4. Similarly, dimensions in millimeters can be converted to inches by dividing by 25.4.

For most applications, the millimeter does not need more than a one-place decimal when it is found by conversion from inches:

- The last digit retained in a conversion of either mm or inches is unchanged if it is followed by a number less than 5; for example, 34.43 is rounded off to 34.4.
- The last digit to be retained is increased by one if it is followed by a number greater than 5; 34.46 is rounded off to 34.5.
- The last digit to be retained is unchanged if it is even and is followed by exactly 5; 34.45 is rounded off to 34.4.
- The last digit to be retained is increased by one if it is odd and is followed by exactly 5; 34.75 is rounded off to 34.8.

21.5
Dual dimensioning

Some drawings require that both metric and English units be given on each dimension. This is called *dual dimensioning* and can be shown by placing the inch equivalent of millimeters either under or over the other units, (Fig. 21.5). If the drawing was originally dimensioned in inches, then the inch dimensions are placed on top, and the equivalent millimeters are given underneath. If the drawing was originally dimensioned in millimeters and then converted to inches, the millimeters would be placed over the equivalent in inches.

A second method of dual dimensioning uses brackets placed on either side of the converted dimensions (Fig. 21.5). Do not mix these two methods on the same drawing.

21.6
Metric designation

The metric system is the Système Internationale d'Unités and is denoted by the abbreviation SI (Fig. 21.6). This system uses the first-angle of projection, which locates the front view over the top view and the right-side view to the left of the front view.

When drawings are made for international circulation, it is customary to use one of the symbols shown in Figs. 21.6C and D to designate the angle of projection used. Either the letters *SI* or the word *MET-*

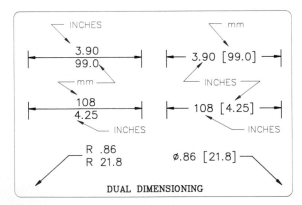

FIG. 21.5 If the drawing was originally made in inches, the equivalent measurement in millimeters is placed under the inches or to the right in brackets in dual-dimensioning. If the drawing was originally made in millimeters, the inch equivalents are placed under or to the right. When inches are converted to millimeters, the millimeters may need to be written as decimal fractions.

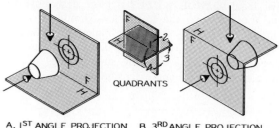

FIG. 21.6 **A.** The European system of orthographic projection places the top view under the front view. **B.** The American system places the front view under the top view. **C.** These symbols indicate first-angle projection. **D.** These symbols indicate third-angle projection. SI indicates that metric units are used.

RIC, written prominently on the drawing or in the title block, is used to indicate the measurements are metric.

21.7
Aligned and unidirectional numbers

The two methods of positioning dimension numbers on a dimension line are the *aligned* and *unidirectional* methods. The unidirectional system is more widely ac-

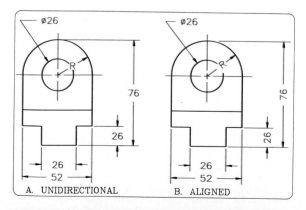

FIG. 21.7 **A.** When dimensions are positioned so all of them read from the bottom of the page, they are unidirectional. **B.** Aligned dimensions are positioned to read from the bottom and right side of the sheet.

cepted since it is easier to apply numerals positioned horizontally (Fig. 21.7A). The aligned system aligns the numerals with the dimension lines (Fig. 21.7B). The numbers must be readable from the bottom or the right side of the page.

Examples of aligned dimensions on angular dimension lines are shown in Fig. 21.8. Avoid placing aligned dimensions in the "trouble zone," since these numerals would read from the left instead of the right side of the sheet.

COMPUTER METHOD A mode of DIMTIH (dimensioning text inside dimension lines is horizontal) under DIM Vars of the Dim: command of AutoCAD must be set to OFF for aligned dimensions and to ON for unidirectional dimensions (Fig. 21.7). The DIMTOH mode controls the positioning of text that lies outside the dimension line in cases where the numerals do not fit within a short dimension line. When DIMTOH is ON, the dimensions will be horizontal; when OFF, the dimensions will be aligned with the direction of the dimension line.

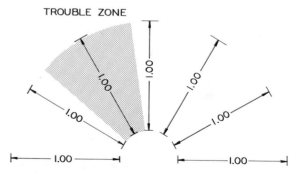

FIG. 21.8 Numbers on angular dimension lines should not be placed in the trouble zone. To do so would cause the numbers to be read from the left side of the sheet rather than from the right side and bottom.

21.8
Placement of dimensions

It is good practice to dimension the most descriptive views.

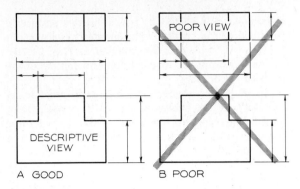

FIG. 21.9 Place the dimensions on the most descriptive views where the true contour of the object can be seen.

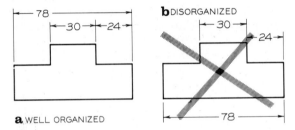

FIG. 21.10 Dimensions should be placed on the views in a well-organized manner to make them as readable as possible.

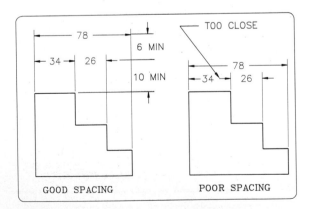

FIG. 21.11 The first row of dimensions should be placed at least 0.40 in. (10 mm) from the view, and successive rows should be at least 0.25 in. (6 mm) from the first row. If greater spaces are used, these proportions should still be maintained.

The front view of Fig. 21.9 is more descriptive than the top view, so the front view should be dimensioned. Dimensions should be applied to views in an organized manner (Fig. 21.10). Locate the dimension lines by beginning with the smaller ones to avoid crossing dimension and extension lines.

> Always leave at least 0.40 in. (10 mm) between the object and the first row of dimensions (Fig. 21.11). Successive rows of dimensions should be at least 0.25 in. (0.6 mm) apart.

If greater spaces are used, apply these same general proportions. The Braddock-Rowe lettering guide triangle can be used to space the dimension lines. (Fig. 21.12).

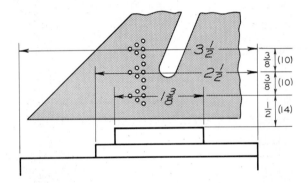

FIG. 21.12 When common fractions are used, the center holes of the triangular arrangements on the Braddock-Rowe triangle are aligned with the dimension lines to automatically space the lines.

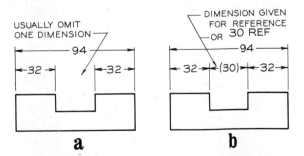

FIG. 21.13 **A.** One intermediate dimension is customarily omitted since the overall dimension provides this measurement. **B.** If all the intermediate dimensions are given, one should be placed in parentheses to indicate that it has been given as a reference dimension.

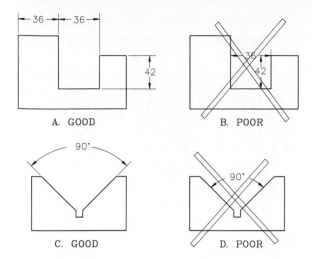

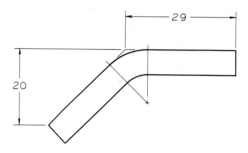

FIG. 21.14 Dimensioning rules.

A. Placing the dimensions outside a part is the preferred practice.

B. Dimension lines should not be used as extension lines.

C. Angles are dimensioned with arcs and extension lines.

D. The angle should not be placed inside the angular cut.

When a row of dimensions is placed on a drawing, one of the dimensions is omitted, since the overall dimension supplements the omitted dimension (Fig. 21.13A). A dimension that needs to be given as a reference dimension is either placed in parentheses or is followed by the abbreviation REF to indicate it is a reference dimension.

The recommended techniques of dimensioning features are shown in Fig. 21.14. Examples of the placement of extension lines are shown in Fig. 21.15. Extension lines may cross other extension lines or object lines; they are also used to locate theoretical points outside curved surfaces (Fig. 21.16).

FIG. 21.16 A curved surface is dimensioned by locating the theoretical point of intersection with extension lines.

21.9
Dimensioning in limited spaces

Several examples of dimensioning in limited spaces are shown in Fig. 21.17. Regardless of space limitations, the numerals should not be drawn smaller than they appear elsewhere on the drawing. Where dimen-

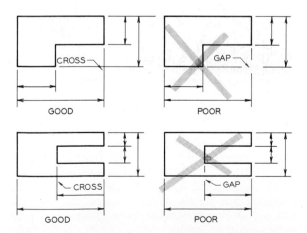

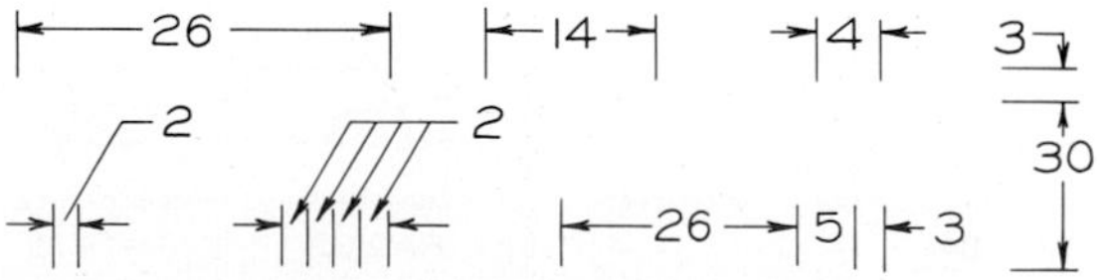

FIG. 21.15 Extension lines extend from the edge of an object, leaving a small gap. They do not have gaps where they cross object lines or other extension lines.

FIG. 21.17 Where room permits, numerals and arrows should be placed inside the extension lines. Other placements are shown as the spacing becomes smaller.

sion lines are closely grouped, the numerals should be staggered to make them more readable (Fig. 21.18).

21.10
Dimensioning symbology

A number of symbols used in dimensioning are shown in Fig. 21.19. The sizes of the symbols are based on the height of the lettering used in dimensioning an object, usually one-eighth inch. Symbols are used to reduce time in preparing notes while adding to the clarity of a dimension.

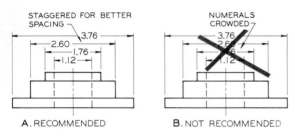

FIG. 21.18 Dimensioning numerals should be staggered when close spacing tends to crowd them.

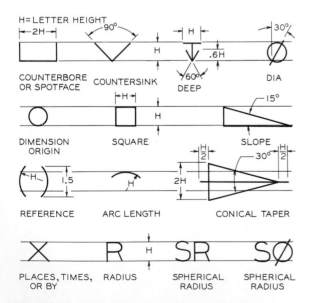

FIG. 21.19 These symbols can be used to dimension parts. The proportions of the symbols are based on the letter height, *H,* which is usually $\frac{1}{8}$ inch.

21.11
Computer dimensioning

The rules of dimensioning must be followed whether you are using computer graphics or manual techniques. The AutoCAD program offers a number of features to use in applying the rules. For example, several combinations of DIM VARS are shown in Fig. 21.20, which are modes of the Dim: command.

Text can be placed inside the dimension line (DIMTAD: OFF) or above the dimension line (DIMTAD: ON). Arrowheads can be placed at the ends of dimension lines DIMASZ>0, or tick marks (slashes) can be used when DIMTSZ is set to a value greater than 0, usually about half the letter height of text. UNITS can be selected to be architectural (feet and inches), metric (no decimal fractions), decimal inches (two or more decimal fractions), or engineering units (feet and decimal inches).

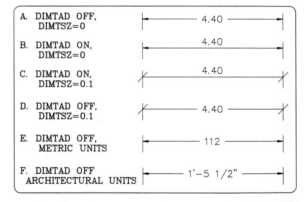

FIG. 21.20 The dimension lines illustrate the effects of using the different DIM VARS (dimensioning variables) in AutoCAD.

An important DIM VARS in dimensioning is DIMSCALE, which can be used to change all variables of dimensioning to allow for changes in scales (Fig. 21.21). DIMSCALE is set at a default of 1.00, which contains the lengths of arrows, text size, and extension line offsets. When invoked, DIMSCALE will not change previously drawn dimensions, only those applied afterwards.

The text STYLE that was last used will be used as the dimensioning text. If you had set the text to a specified height, then this height would be used in di-

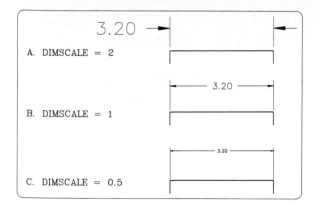

FIG. 21.21 DIMSCALE, α subcommand under Dim:, can be used to change dimensioning variables by inputting a single factor. The text sizes, arrows, offsets, and extensions are all changed at one time.

mensioning whether DIMSCALE was used or not. For DIMSCALE to change text height, you should set the text height for the STYLE being used at 0 (zero). You will be prompted for the letter height each time, which allows you either to accept the default last used or to assign another height. In this case, DIMSCALE will enlarge or reduce text height along with the other variables.

You must apply DIMSCALE to a dimensioned drawing when it will be plotted at a different scale (Fig. 21.22). For example, if a drawing is to be plotted

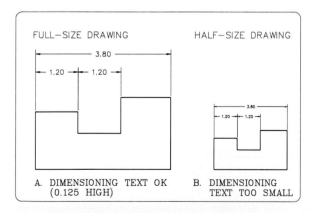

FIG. 21.22 When dimensioning a drawing, you should be aware of its final plotted size in order for the dimensioning variables to be sized to compensate for reduction or enlargement. DIMSCALE is the most efficient command for assigning a scale to dimensioning variables.

half size, you would want to use a DIMSCALE of 2.00 so the variables would appear full size when reduced.

AutoCAD performs many dimensioning operations automatically, but occasionally you will want to modify the placement of arrows and text. An example of an edited dimension is shown in Fig. 21.23, where MOVE and ERASE commands are used.

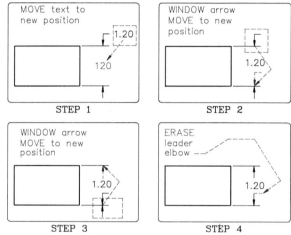

FIG. 21.23 Editing dimensions.

Step 1 You may wish to MOVE α dimension text within the extension lines.

Step 2 Arrows can be MOVED inside extension lines also.

Step 3 A second arrow is windowed and MOVED inside the extension lines.

Step 4 The unneeded leader extension is ERASED to complete the modified dimension.

21.12
Dimensioning prisms

The following rules for dimensioning prisms are illustrated in Fig. 21.24.

1. Dimensions should extend from the most descriptive views (Fig. 21.24A).

2. Dimensions that apply to two views should be placed between them (Fig. 21.24A).

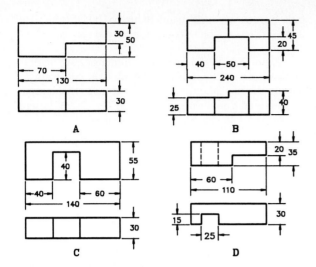

FIG. 21.24 Dimensioning prisms (metric).

A. Dimensions should extend from the most descriptive veiw and be placed between the views to which they apply.

B. One intermediate dimension is not given. Extension lines may cross object lines.

C. It is permissible to dimension a notch inside the object if this improves clarity.

D. Whenever possible, dimensions should be placed on visible lines, not hidden lines.

3. The first row of dimension lines should be placed a minimum of 0.40 in. (10 mm) from the view. Successive rows are placed at least 0.25 in. (6 mm) apart.

4. Extension lines may cross each other and other lines, but dimension lines should not cross unless absolutely necessary.

5. To dimension a part in its most descriptive view, dimensions may have to be placed in more than one view (Fig. 21.24B).

6. Dimension lines may be placed inside a notch to improve clarity.

7. Whenever possible, dimensions should be applied to visible lines rather than to hidden lines (Fig. 21.24D).

8. Dimensions should not be repeated, nor should unnecessary information be given.

21.13
Dimensioning angles

Angles can be dimensioned either by using coordinates to locate the ends of angular lines or planes or by using angular measurements in degrees (Fig. 21.25). The two methods should not be mixed when dimensioning the same angle, since they may not agree.

Units for angular measurements are degrees, minutes, and seconds, as shown in Fig. 21.25C. There are 60 minutes in a degree and 60 seconds in a minute. Seldom will angular measurements need to be measured to the nearest second.

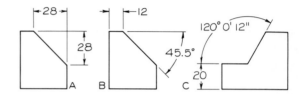

FIG. 21.25 **A.** Angular planes can be dimensioned by using coordinates. **B.** Angular planes can also be dimensioned by using an angle measured in degrees from the located vertex. Angles can be measured in decimal fractions. **C.** Angles can also be measured in degrees, minutes, and seconds.

21.14
Dimensioning cylinders

Cylinders that are dimensioned may be either solid cylinders or cylindrical holes.

> Solid cylinders are dimensioned in their rectangular views by using diameters, not radii.

All diametral dimensions should be preceded by the symbol $\varnothing$ (Fig. 21.26), which indicates the dimension is a diameter. The English system often uses the abbreviation DIA following the diametral dimension.

Parts having several cylinders, which are concentric unless otherwise noted, are dimensioned with di-

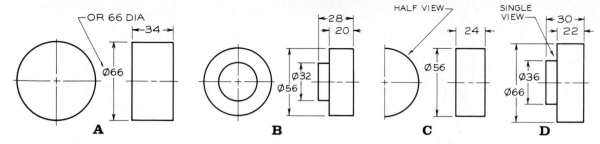

FIG. 21.26 **A.** It is preferred that cylinders be dimensioned in their rectangular views using a diameter rather than a radius. **B.** Dimensions should be placed between the views when possible. **C.** The circular view can be omitted if DIA or ∅ is used with the diametral dimensons.

ameters, beginning with the smallest cylinder (Fig. 21.26B). A cylindrical part may be sufficiently dimensioned with only one view if ∅ or DIA is used with the diametral dimension (Fig. 21.26C).

FIG. 21.27 An internal micrometer caliper for measuring internal cylindrical diameters.

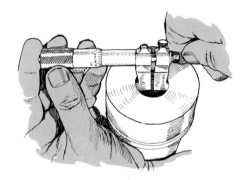

21.15
Measuring cylindrical parts

Cylindrical parts are dimensioned with diameters rather than radii because diameters are easier to measure. An internal cylindrical hole is measured with an internal micrometer caliper (Fig. 21.27). Likewise, an external micrometer caliper can be used for measuring the outside diameters of a part (Fig. 21.28). Measuring a diameter rather than a radius makes it possible to measure during machining when the part is held between centers on a lathe.

21.16
Cylindrical holes

Cylindrical holes may be dimensioned by one of the methods shown in Fig. 21.29. The preferred method of dimensioning cylindrical holes is to draw a leader from the circular view, and then add the dimension, preceded by ∅ (Fig. 21.30) or followed by DIA. Sometimes the note DRILL or BORE is added to specify the shop operation, but current standards prefer the use of DIA instead.

To illustrate various methods of dimensioning, a part containing cylindrical features is dimensioned in

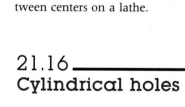

FIG. 21.28 An external micrometer caliper for measuring the diameter of a cylinder.

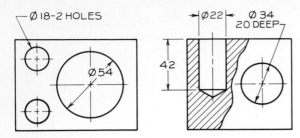

FIG. 21.29 Several acceptable methods of dimensioning cylindrical holes and shapes. The symbol ⌀ is always placed in front of the diametral dimension.

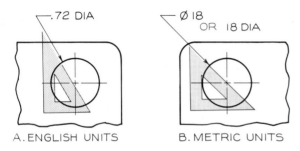

A. ENGLISH UNITS   B. METRIC UNITS

FIG. 21.30 The preferable method of dimensioning cylindrical holes is with a leader, a dimension, and the symbol ⌀ to indicate that the dimension is a diameter. Previous standards recommended the abbreviation DIA after the dimension.

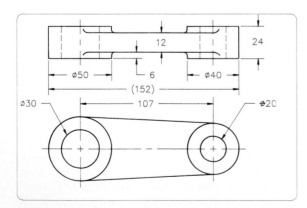

FIG. 21.31 This part, composed of cylindrical features, has been dimensioned using several approved methods. (F&R R6 means that fillets and rounds have a 6.6 radius.)

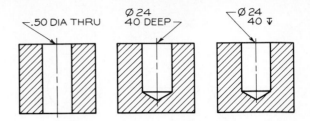

FIG. 21.32 Three methods of dimensioning holes with combinations of notes and symbols.

Fig. 21.31 in both the circular and the rectangular views. Three methods of dimensioning holes are shown in Fig. 21.32. The depth of a hole is its *usable* depth, not the depth to the point left by the drill bit.

21.17
Pyramids, cones, and spheres

Three methods of dimensioning pyramids are shown in Figs. 21.33A, B, and C. The pyramids in Figs. 21.33B and C are truncated (that is, the apex has been cut off and replaced by a plane). Note that in all three examples, the apex of the pyramid is located in the rectangular view.

Two acceptable methods of dimensioning cones are shown in Figs. 21.33D and E.

A sphere, if it is complete, is dimensioned by using its diameter (Fig. 21.33F); a radius is used if it is not a complete sphere (Fig. 21.33G). Only one view is necessary to describe a sphere.

21.18
Leaders

Leaders are used to apply notes and dimensions to a feature they describe. As illustrated in Fig. 21.30, leaders are drawn at a standard angle of a triangle. Examples of notes using leaders are shown in Fig. 21.34. The leader should be drawn from either the first word of the note or the last word of the note and begin with a short horizontal line from the note.

Applications of leaders on a part are shown in Fig. 21.35. A dot is used instead of an arrowhead

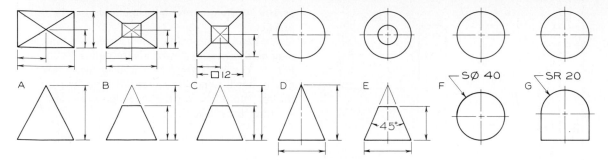

FIG. 21.33 Methods of dimensioning pyramids, cones, and spheres.

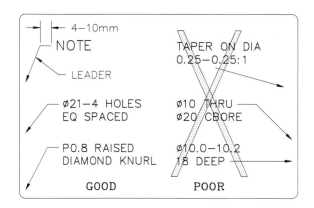

FIG. 21.34 Leaders from notes should begin with a horizontal bar from the first or last word of the note, not from the middle of the note.

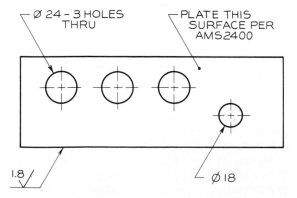

FIG. 21.35 Examples of notes with leaders applied to a part.

when the note applies to a surface that does not appear as an edge.

21.19
Dimensioning arcs

Cylindrical parts less than a full circle are dimensioned with radii, as shown in Fig. 21.36. Current standards recommend that radii be dimensioned with an *R* preceding the dimension, such as R10. Previous standards recommended that the *R* follow the dimension, such

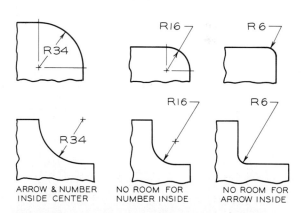

FIG. 21.36 When space permits, the dimension and arrow should be placed between the center and the arc. If room is not available for the number, the arrow is placed between the center and the arc with the number on the outside. If there is no room for the arrow, then both the dimension and the arrow are placed outside the arc with a leader.

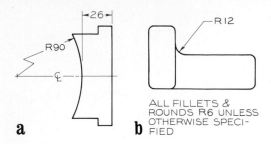

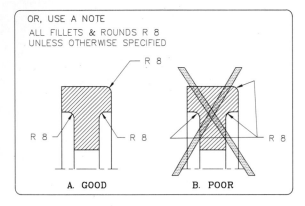

FIG. 21.37 a. When a radius is very long, it may be shown with a false radius and a "zigzag" to indicate that it is not true length. It should end on the centerline of the true center. **b.** Fillets and rounds may be noted to reduce repetitive dimensions of small arcs.

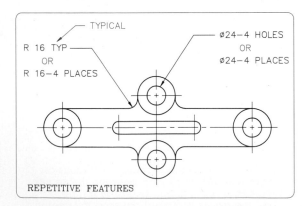

FIG. 21.38 When serveral arcs are dimensioned, it is preferable that separate leaders be used rather than extending the leaders.

FIG. 21.39 Notes can be used to indicate that similar features and dimensions are repeated on drawings without having to dimension them individually.

as 10R. Consequently, both methods will be seen in practice.

When the arc being dimensioned is a long arc, it may be dimensioned with a false radius, as shown in Fig. 21.37a. A "zigzag" is placed in the radius to indicate that it is not a true radius. The false center of the arc should lie on the extended centerline where the true center lies.

21.20
Fillets and rounds, and TYP

Fillets and *rounds* are rounded corners conventionally used on castings. A fillet is an internal rounding, and a round is an external rounding.

A note may be placed on the drawing to eliminate the need for repetitive dimensioning of fillets and rounds. The note may read, ALL FILLETS AND ROUNDS R6. If most, but not all, of the fillets and rounds have equal radii, the note may read ALL FILLETS AND ROUNDS R6 UNLESS OTHERWISE SPECIFIED. In this case, only the fillets and rounds of different radii are dimensioned (Fig. 21.37b). The notes may be abbreviated, such as F&R10.

Fillets and rounds should be dimensioned with short, simple leaders (Fig. 21.38A), rather than with long, confusing leaders (Fig. 21.38B).

Repetitive features on a drawing may be noted as shown in Fig. 21.39. The note TYPICAL or TYP means that although only one of these features is dimensioned, the dimensions are typical of those that are undimensioned. The note PLACES is sometimes used to specify the number of places that a similar feature appears. The number of holes sharing the same dimension may be similarly indicated.

21.21
Curved surfaces

An irregular shape composed of a number of tangent arcs of varying sizes (Fig. 21.40) can be dimensioned by using a series of radii.

When the curve is irregular rather than composed of arcs (Fig. 21.41), the coordinate method can be used to locate a series of points along the curve from two datum lines. The drafter must use judgment to determine the proper spacing for the points. Extension

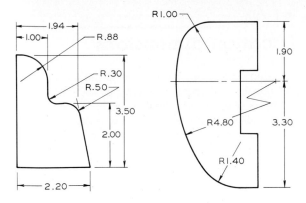

FIG. 21.40 Examples of dimensioned parts composed of a series of tangent parts.

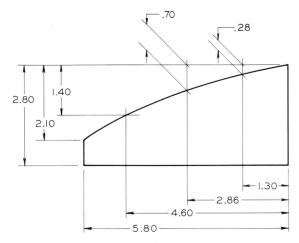

FIG. 21.41 This object with an irregular curve is dimensioned by using coordinates.

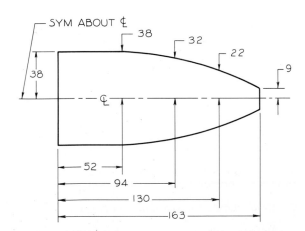

FIG. 21.42 This symmetrical part is dimensioned about its centerline.

lines may be placed at an angle to provide additional space for showing dimensions.

An irregular curve that is symmetrical about an axis is shown dimensioned in Fig. 21.42. Dimension lines are used as extension lines, which is an acceptable violation of rules.

21.22
Symmetrical objects

Symmetrical objects may be dimensioned as shown in Figure. 21.43A, where it is assumed that the dimensions are each centered about the centerline, abbreviated CL. The better method is shown in Fig. 21.43B, where the assumption is eliminated. All dimensions are located with respect to each other.

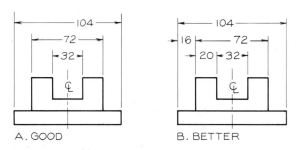

FIG. 21.43 **A.** Symmetrical parts may be dimensioned about their centerlines as shown here. **B.** But the better way of dimensioning symmetrical parts is shown here.

21.23
Finished surfaces

Many parts are formed as castings in a mold that gives the parts' exterior surfaces a rough finish. If the part is designed to come in contact with another surface, the rough finish must be machined by grinding, shaping, lapping, or similar process.

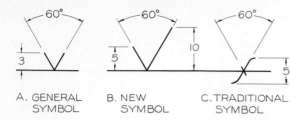

A. GENERAL SYMBOL   B. NEW SYMBOL   C. TRADITIONAL SYMBOL

FIG. 21.44 Finish marks indicate that a surface has been machined to a smooth surface. **A.** The traditional V can be used for general applications. **B.** The new finish mark is related to surface texture. **C.** The f is also used to indicate finished surfaces.

To indicate that a surface is to be finished, finish marks are drawn on the surface where it appears as an edge (Fig. 21.44).

> Finish marks should be repeated in every view where the finished surface appears as an edge, even if it is a hidden line.

Three methods of drawing finish marks are shown in Fig. 21.44. The simple V mark is preferred in general cases. The uneven V (Fig. 21.44B), a newly recommended symbol, is related to surface texture and will be discussed in the next chapter. The steps of constructing the traditional F finish mark are shown in Fig. 21.45. When an object is finished on all surfaces, the note FINISHED ALL OVER (abbreviated FAO) is placed on the drawing.

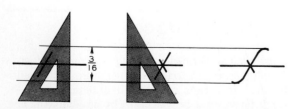

FIG. 21.45 The steps of drawing the f-finish mark.

21.24
Location dimensions

Location dimensions are used to locate the positions, not the sizes, of geometric elements, such as cylindrical holes (Fig. 21.46). The sizes of the holes are omitted for clarity. The centers of the holes are located with coordinates in the circular view when possible.

> Holes should be located from finished surfaces since holes can be located more accurately from a smooth, machined surface than from a rough, unfinished one.

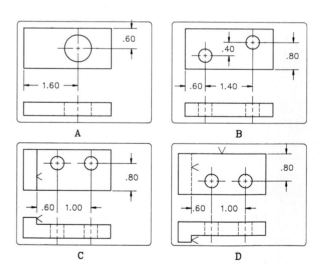

FIG. 21.46 Location dimensions.

A. Cylindrical holes should be located in their circular views from two surfaces of the object.

B. When more than one hole is to be located, they should be located from center to center.

C. Holes should be located from finished surfaces.

D. Holes should be located in the circular view and from finished surfaces even if the finished surfaces are hidden.

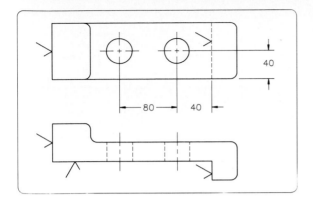

FIG. 21.47 Cylindrical holes are located from center to center in their circular view and from a finished surface if one is available.

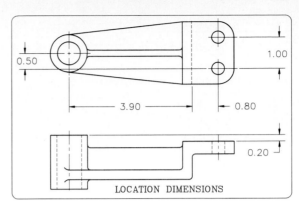

FIG. 21.49 Location dimensions applied to a part to locate its geometric features.

This rule is followed even if the finished surface is a hidden line, as in Fig. 21.47.

Location dimensions should be placed on views where both dimensions can be shown (Fig. 21.48). Cylinders are located in their circular views (Figs. 21.48B and 21.49).

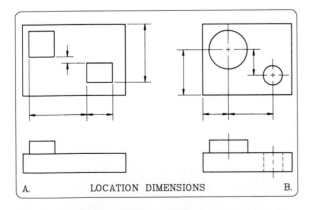

FIG. 21.48 Prisms and cylinders are located with coordinates in the view where both coordinates can be seen.

21.25
Location of holes

When holes must be located accurately, the dimensions should originate from a common datum plane on the part to reduce errors in measurement (Figs. 21.50A and B). When several holes in a series are to be equally spaced, as in Fig. 21.50C, a note specifying as much can be used to locate the holes. The first and last holes of the series are determined by the usual location dimensions.

Holes through circular plates may be located by coordinates or by a note (Fig. 21.51). When a note is used, the diameter of the circle passing through the centers of the holes must be given. This circle is referred to as the *bolt circle* or *circle of centers*.

A similar method of locating holes is the polar system illustrated in Fig. 21.52. The radial distances from the point of concurrency and their angular measurements (in degrees) between the holes are used to locate the centers.

21.26
Objects with rounded ends

Objects with rounded ends should be dimensioned from end to end (Fig. 21.53A). The radius, shown as R without a dimension, specifies the end is formed by

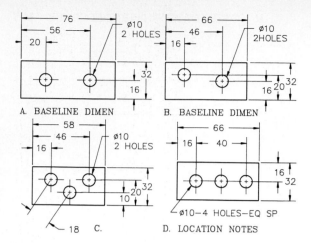

A. BASELINE DIMEN   B. BASELINE DIMEN

C.   D. LOCATION NOTES

FIG. 21.50 Location of holes.

A and B. Holes can be more accurately located if measured from common datum.

C. A diagonal dimension can be used to locate a hole of this type from another hole's center.

D. A note can be used to specify the spacing between the centers of equally spaced holes.

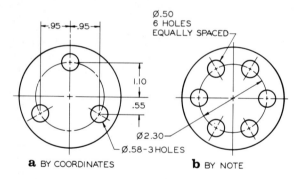

a BY COORDINATES   **b** BY NOTE

FIG. 21.51 Holes may be located in circular plates by coordinates or by notes.

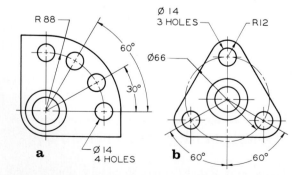

a   **b**

FIG. 21.52 Methods of locating cylindrical holes on concentric arcs.

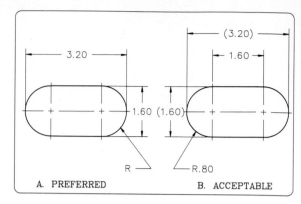

A. PREFERRED   B. ACCEPTABLE

FIG. 21.53 A. The preferred method of dimensioning objects with rounded ends. **B.** The less desirable method.

an arc. Since the height is given, the radius is understood to be half the height.

If the object is dimensioned from center to center (Fig. 21.53B), the overall dimension should be given as a reference dimension (3.40) to eliminate calculating the overall dimension. In this case, the radius must be given.

A part with partially rounded ends is dimensioned in Fig. 21.54A. The overall dimension and the radii are given so that their centers may be located. When an object has a rounded end that is less than a

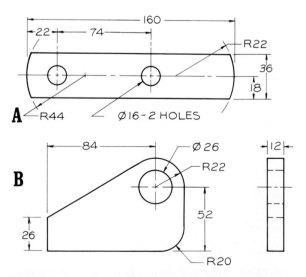

FIG. 21.54 Examples of dimensioned parts with rounded ends and cylindrical features.

semicircle (Fig. 21.54B), location dimensions must be used to locate the center of the arc.

Slots with rounded ends are dimensioned in Fig. 21.55. Only one slot is dimensioned in Fig. 21.55A, with a note indicating that there are two slots. The slot in Fig. 21.55B is dimensioned giving the overall dimension and the two arcs, which are understood to apply to both ends. The distance between the centers is given as a reference dimension.

The dimensioned views of the tool holder table in Fig. 21.56 show examples of arcs and slots. To prevent dimension lines from crossing, it is often necessary to place dimensions in a view less descriptive than might be desired.

21.27
Machined holes

Machined holes are made or refined by a machine operation, such as drilling or boring (Fig. 21.57). It is preferable to give the diameter of the hole with the symbol $\varnothing$ in front of the dimension ($\varnothing$ 32); however, the note 32 DRILL may be used in some cases. You will also see diameters dimensioned as XX DIA, since this was the standard previously recommended.

DRILLING is the most common method of machining holes. The depth of a drilled hole can be specified

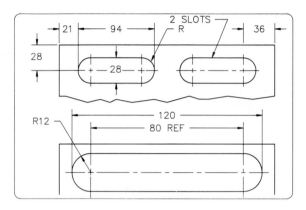

FIG. 21.55 Methods of dimensioning slots.

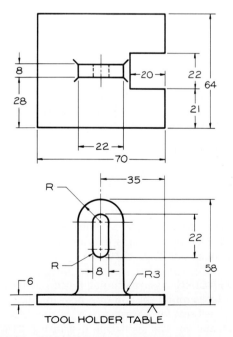

TOOL HOLDER TABLE

FIG. 21.56 Examples of arcs and slots.

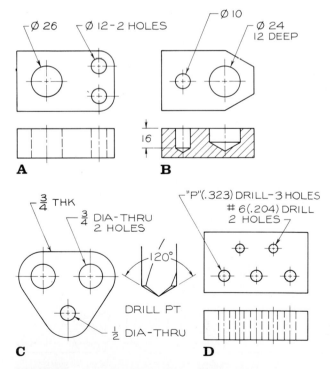

FIG. 21.57 **A.** and **B.** Cylindrical holes may be dimensioned by either of these methods. **C.** and **D.** When only one view is given, it is necessary to note the "THRU" holes.

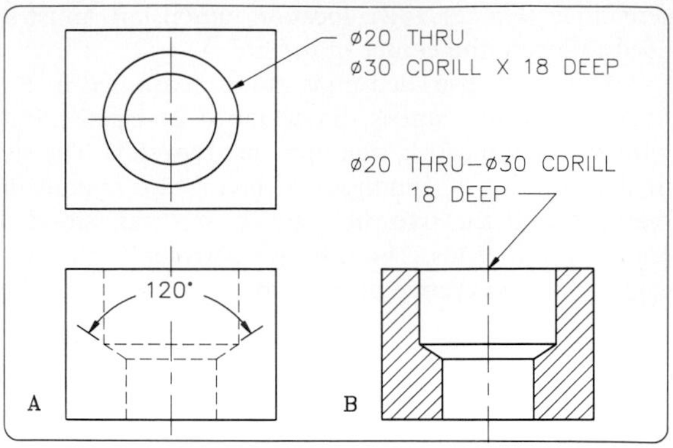

FIG. 21.58 Counterdrilling notes give the specifications for a larger hole drilled inside a smaller hole. The 120° angle is not required as a dimension since this is the standard angle of a drill point.

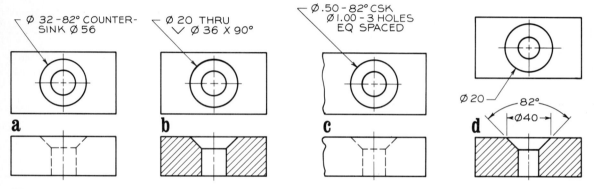

FIG. 21.59 Examples of notes and symbols specifying countersunk holes.

in the note or it may be dimensioned in the rectangular view. The depth of a drilled hole is dimensioned as the usable part of the hole; the conical point is disregarded (Fig. 21.57B).

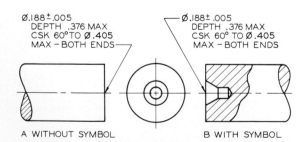

FIG. 21.60 Methods of specifying countersinks in the ends of cylinders for mounting them on a lathe between centers.

COUNTERDRILLING is drilling a large hole inside a smaller drilled hole to enlarge it (Fig. 21.58). The 120-degree dimension indicates the angle of the drill point; it need not be dimensioned on the drawing.

COUNTERSINKING is the process of forming a conical hole often used with flat head screws (Fig. 21.59). The diameter of the countersunk hole (the maximum diameter on the surface) and the angle of the countersink are given in the notes. Countersinking is also used to provide center holes in shafts, spindles, and other cylindrical parts that are held between the centers of a lathe (Fig. 21.60).

SPOTFACING is a machining process used to finish the surface around the top of a hole in order to provide a level seat for a washer or a fastener head (Figs. 21.61A and B). The spotfacing tool in Fig. 21.62 has spotfaced a boss (a raised cylindrical element).

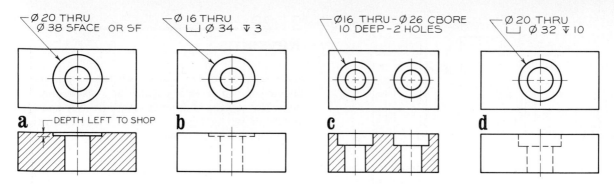

FIG. 21.61. **a.** and **b.** Spotfaces can be specified as shown here. The depth of the spotface can be specified if needed. **c.** and **d.** Counterbores are dimensioned by giving the diameters of both holes and the depth of the larger hole.

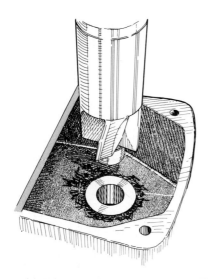

FIG. 21.62 This tool is spotfacing the cylindrical boss to provide a smooth seat for a bolt head.

FIG. 21.63 Boring a large hole on a lathe with a boring bar. (Courtesy of Clausing Corp.)

BORING is a machine operation for making large holes. It is usually performed on a lathe with a boring bar (Fig. 21.63).

COUNTERBORING is the proces of enlarging the diameter of a drilled hole (Fig. 21.64). The bottoms of counterbored holes are flat with no taper as in counterdrilled and countersunk holes.

REAMING is the operation of finishing or slightly enlarging holes that have been drilled or bored. This operation uses a ream similar to a drill bit.

21.28
Chamfers

Chamfers are beveled edges that are made on cylindrical parts, such as shafts and threaded fasteners. They eliminate sharp edges and facilitate the assembly of parts.

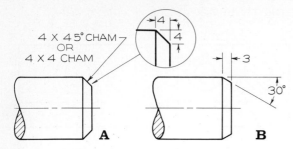

FIG. 21.64 **A.** Chamfers can be dimensioned by using either type of note when the angle is 45°. **B.** If the angle is other than 45°, it should be dimensioned as shown here.

When a chamfer angle is 45°, a note can be used in either of the forms shown in Fig. 21.64A. When the chamfer angle is other than forty-five degrees, the angle and the length are given, as shown in Fig. 21.64B. Chamfers can also be specified at the openings of holes (Fig. 21.65).

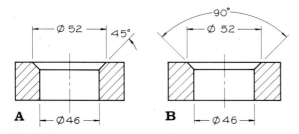

FIG. 21.65 Insider chamfers are dimensioned by using one of these methods.

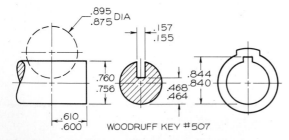

FIG. 21.66 Methods of dimensioning slots in a shaft and a slot for a Woodruff key that will hold the part on the shaft. The dimensions for these features are given in Appendix 33.

21.29
Keyseats

A *keyseat* is a slot cut into a shaft for the purpose of aligning the shaft with a pulley or a collar mounted on it. The method of dimensioning a keyway and keyseat is shown in Fig. 21.66. The double dimensions are tolerances, which will be discussed in the next chapter.

21.30
Knurling

Knurling is the operation of cutting diamond-shaped or parallel patterns on cylindrical surfaces for gripping, decoration, or a press fit between two parts that will be permanently assembled.

A *diamond knurl* and a *straight knurl* are drawn and dimensioned in Fig. 21.67. Knurls should be dimensioned with specifications that give type, pitch, and diameter.

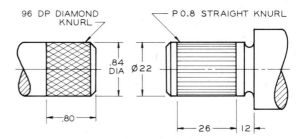

FIG. 21.67 **A.** A diamond knurl with a diametral pitch of 96. **B.** A straight knurl where the linear pitch *(P)* is 0.8 mm. Pitch is the distance between the grooves on the circumference.

The abbreviation DP in Fig. 21.67 means diametral pitch, the ratio of the number of grooves on the circumference *(N)* to the diameter *(D)*, which is found by the equation $DP = N/D$. The preferred diametral pitches for knurling are 64 DP, 96 DP, 128 DP, and 160 DP. For diameters of one inch, knurling of 64 DP, 96 DP, 128 DP, and 160 DP will have 64, 96, 128, and 160 teeth, respectively on the circumference. The

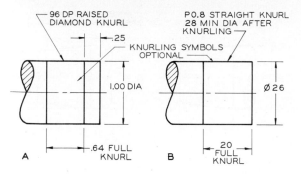

FIG. 21.68 Knurls need not be drawn if they are dimensioned as shown here.

note P 0.8 means that the knurling grooves are 0.8 mm apart. Calculations for knurling must be made using inches with conversion to millimeters made afterward.

Knurls for press fits are specified with diameters before knurling and with the minimum diameter after knurling. A simplified method of representing knurls is shown in Fig. 21.68, where notes are used and the knurls are not drawn.

21.31
Necks and undercuts

A *neck* is a recess cut into a cylindrical part. Where cylinders of different diameters join (Fig. 21.69), a neck ensures that the part assembled on the smaller shaft will fit flush against the shoulder of the larger cylinder.

Undercuts, which are similar to necks (Fig. 21.70A), ensure that a part fitting in the corner of the part will fit flush against both surfaces; they also permit space for trash to drop out of the way when entrapped in the corner. An undercut could also be a recessed neck on the inside of a cylindrical hole. A thread relief (Fig. 21.70B) is used to square the threads where they intersect a larger cylinder.

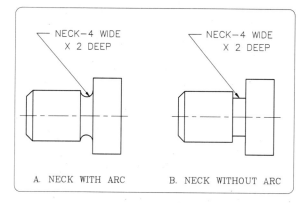

FIG. 21.69 Necks are recesses in cylinders that are used where cylinders of different sizes join together.

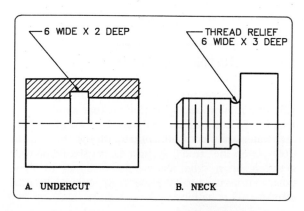

FIG. 21.70 **A.** An undercut can be dimensioned as shown here. **B.** A thread relief, which is a type of neck, is dimensioned here.

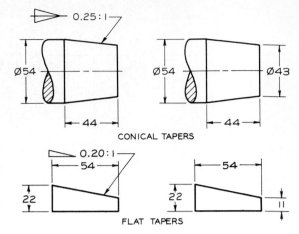

FIG. 21.71 Examples of flat and conical tapers. Taper is the ratio of the diameters (or heights) at each end of a sloping surface to the length of the taper.

21.32
Tapers

Tapers can be either conical surfaces or flat planes (Fig. 21.71). A taper can be dimensioned by (1) the diameter or width at each end of the taper, (2) the length of the tapered feature, or (3) the rate of taper.

Taper is the ratio of the difference in the diameters of a cone to the distance between two diameters (Fig. 21.71). Taper can be expressed as inches per inch (.25 per inch), inches per foot (3.00 per foot), or millimeters per millimeter (0.25:1).

Flat taper is the ratio of the difference in the heights at each end of a feature to the distance between the heights. Flat taper can be expressed as inches per inch (.20 per inch), inches per foot (2.40 per foot), or millimeters per millimeter (0.20:1), as shown in Fig. 21.71.

21.33
Dimensioning sections

Sections are dimensioned in the same manner as regular views (Fig. 21.72). Most of the principles of dimensioning covered in this chapter have been applied to the part shown in this figure.

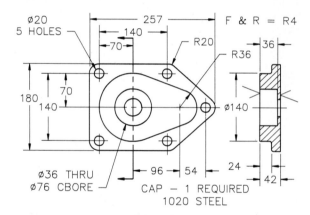

FIG. 21.72 An example of a computer-drawn part dimensioned with a variety of dimensioning principles.

21.34
Miscellaneous notes

A variety of notes are used on detail drawings to provide information and specifications that would otherwise be difficult to represent (Figs. 21.73 and 21.74).

Notes are placed horizontally on the sheet; if the notes lie on the same line short dashes should be used between them. The abbreviations shown in these notes can be used to save space. Appendix 1 lists the standard abbreviations.

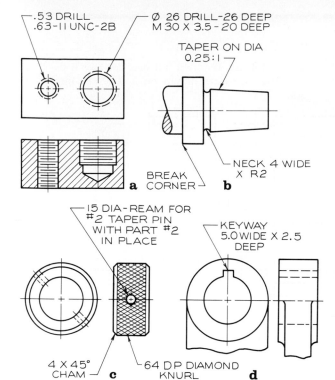

.53 DRILL
.63-11 UNC-2B

Ø 26 DRILL-26 DEEP
M 30 X 3.5 - 20 DEEP

TAPER ON DIA
0.25:1

NECK 4 WIDE
X R2

BREAK
CORNER

a

b

15 DIA-REAM FOR
#2 TAPER PIN
WITH PART #2
IN PLACE

KEYWAY
5.0 WIDE X 2.5
DEEP

4 X 45°
CHAM

c

64 DP DIAMOND
KNURL

d

FIG. 21.73 **a.** and **b.** Threaded holes are sometimes dimensioned by giving the tap drill size in addition to the thread specifications. The tap drill size is not required, but it is permissible. **b.** This part is dimensioned to indicate a neck, a taper, and a break corner, which is a slight round to remove the sharpness from a corner. **c.** This collar has a knurl note, a chamfer note, and a note indicating the insertion of a #2 taper pin. **d.** This part has a note for dimensioning a keyway.

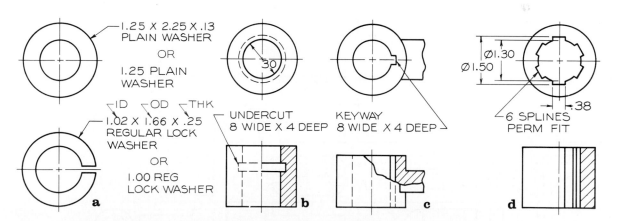

1.25 X 2.25 X .13
PLAIN WASHER

OR

1.25 PLAIN
WASHER

ID   OD   THK
1.02 X 1.66 X .25
REGULAR LOCK
WASHER

OR

1.00 REG
LOCK WASHER

a

UNDERCUT
8 WIDE X 4 DEEP

b

KEYWAY
8 WIDE X 4 DEEP

c

Ø1.30
Ø1.50

.38

6 SPLINES
PERM FIT

d

FIG. 21.74 **a.** and **b.** Methods of dimensioning washers and lock washers. These dimensions can be found in Appendixes 36 and 37. **b.** An undercut is dimensioned. **c.** A keyway is dimensioned **d.** A spline inside a hole is dimensioned.

Problems

1–24 (a) and (b) (Figs. 21.75–21.76) These problems are to be solved on Size A paper if they are drawn full size. If drawn full size, use Size B paper. The views are drawn on a 0.20″ (5 mm) grid.

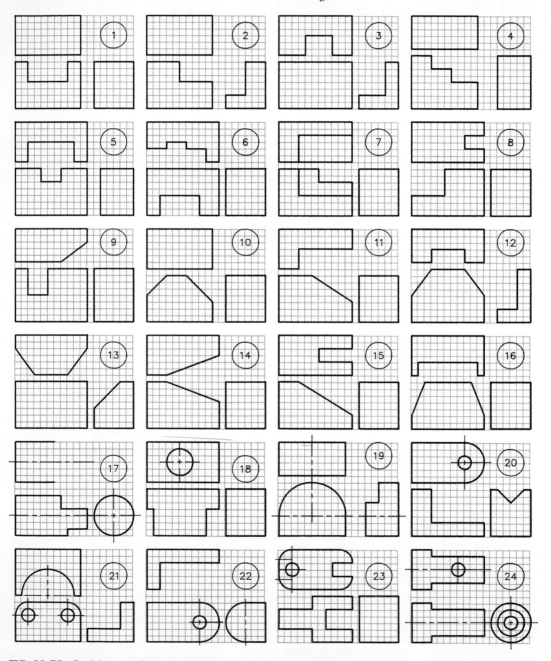

FIG. 21.75 Problems 1–24(a). Lay out the views and dimension them.

You will need to vary the spacing between the views to provide adequate room for the dimensions. It would be a good idea to sketch the views and the dimensions to determine the required spacing before laying out the problems with instruments.

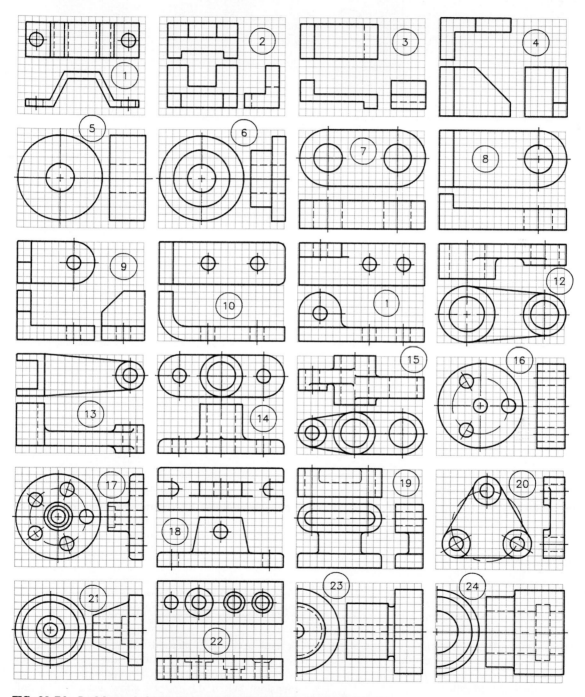

FIG. 21.76 Problems 1–24(b). Lay out the views and dimension them.

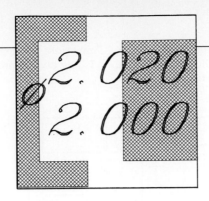

CHAPTER 22

Tolerances

22.1
Introduction

Today's technologies require increasingly exact dimensions. What is more, many of today's parts are made by different companies in different locations; therefore these parts must be specified so they will be interchangeable.

The techniques of dimensioning parts to ensure interchangeability is called *tolerancing*. Each dimension is allowed a certain degree of variation within a specified zone, called a *tolerance*. For example, a part's dimension might be expressed as 100 ± 0.50, which yields a tolerance of 1.00 mm.

Dimensions should be given as large a tolerance as possible without interfering with the function of the part to reduce production costs. Manufacturing to close tolerances is expensive.

22.2
Tolerance dimensions

Several acceptable methods of specifying tolerances are shown in Fig. 22.1. When "plus-and-minus" tolerancing is used, tolerances are applied to a *basic* dimension. When dimensions allow variation in only one direction, the tolerancing is *unilateral.* Tolerancing that permits variation in either direction from the basic dimension is *bilateral.*

Tolerances may also be given in the form of *limits;* that is, two dimensions are given that represent the largest and smallest sizes permitted for a feature of the part.

The customary methods of indicating toleranced dimensions are shown in Fig. 22.2. The positioning

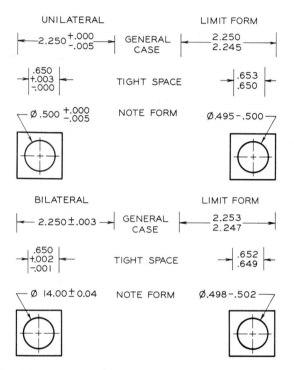

FIG. 22.1 Methods of positioning and indicating tolerances in unilateral, bilateral, and limit forms.

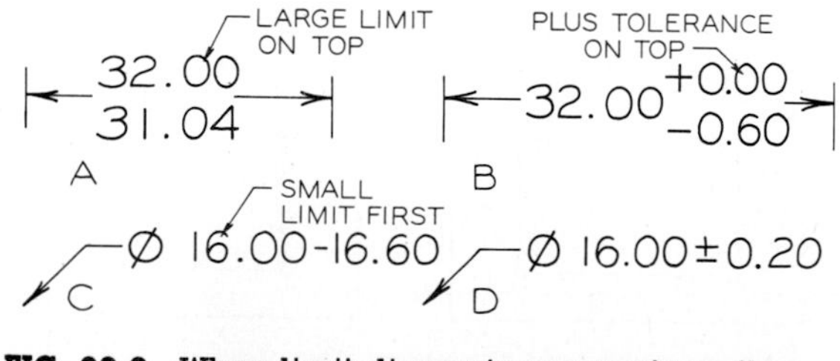

LARGE LIMIT ON TOP

32.00
31.04

A

PLUS TOLERANCE ON TOP

32.00 +0.00 -0.60

B

SMALL LIMIT FIRST

∅ 16.00-16.60

C

∅ 16.00±0.20

D

FIG. 22.2 When limit dimensions are given, the large limits are placed either above or to the right of the small limits. In plus-and-minus tolerancing, the plus limits are placed above the minus limits.

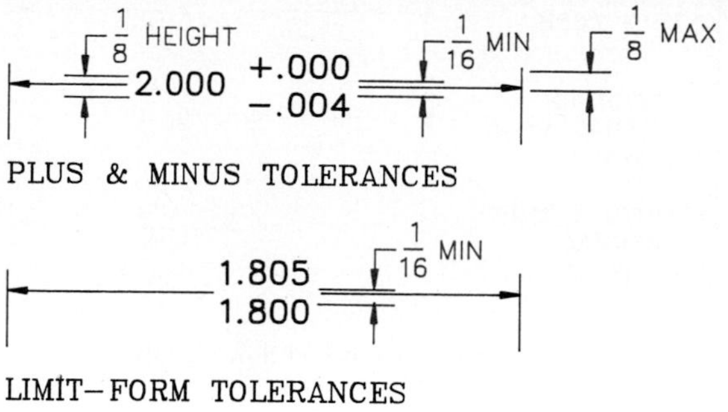

1/8 HEIGHT 2.000 +.000 -.004 1/16 MIN 1/8 MAX

PLUS & MINUS TOLERANCES

1.805
1.800 1/16 MIN

LIMIT–FORM TOLERANCES

FIG. 22.3 Positioning and spacing of numerals used to specify tolerances.

and spacing of numerals of toleranced dimensions are shown in Fig. 22.3.

TOLERANCES BY COMPUTER Using AutoCAD, toleranced dimensions can be shown in limit form or plus-and-minus form (Fig. 22.4). Both DIMLIM and DIMTOL must be turned ON to obtain dimensions in limit form. The tolerances are assigned to DIMTM and DIMTP modes under the DIM: command.

In addition to linear dimensions, diametral and radial dimensions are automatically given with either a circle symbol or an R preceding the dimensions as shown in Fig. 22.5 and 22.6.

You will want to edit a toleranced dimension when you change the limits given by the automatic process. By a series of erasures and moves (Fig. 22.7), you will be able to make the desired changes.

22.3
Mating parts

Mating parts are parts that fit together within a prescribed degree of accuracy (Fig. 22.8). The upper piece is dimensioned with two measurements that indicate the upper and lower limits of the size. The notch is slightly larger, allowing the parts to be assembled with a clearance fit.

An example of mating cylindrical parts is shown in Fig. 22.9A; Fig. 22.9B illustrates the meaning of the tolerance dimensions. The size of the shaft can vary in

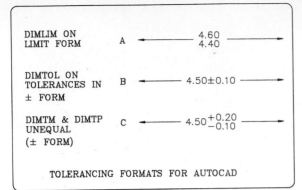

DIMLIM ON LIMIT FORM	A	4.60 4.40
DIMTOL ON TOLERANCES IN ± FORM	B	4.50±0.10
DIMTM & DIMTP UNEQUAL (± FORM)	C	4.50 +0.20 −0.10

TOLERANCING FORMATS FOR AUTOCAD

FIG. 22.4 AutoCAD will give toleranced dimensions in the forms shown here. In all cases, the DIMTOL mode must be ON. If DIMLIM is ON, tolerances will be applied in limit form; when OFF, tolerances will be in plus-and-minus form.

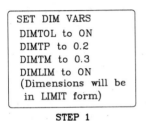
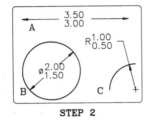

FIG. 22.5 Limit tolerances by computer.

Step 1 Set DIM: variables (DIM VARS) as shown for tolerances to be given in LIMIT form.

Step 2 Linear dimensions will be given in limit form, diametrical dimensions will be given as limits preceded by ∅, and radial dimensions will be preceded by R.

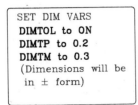
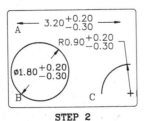

FIG. 22.6 Plus-and-minus tolerances by computer.

Step 1 Set DIM: variables (DIM VARS) as shown with DIMLIM set to OFF.

Step 2 Linear dimensions will be given as a basic diameter followed by plus-and-minus tolerances, diametral dimensions will be given as linear dimensions preceded by ∅, and radial dimensions will be preceded by R.

diameter from 1.500 in. (maximum size) to 1.498 in. (minimum size). The difference between these limits on a single part is a tolerance of 0.002 in. The dimensions of the hole in Fig. 22.9A are given with limits of 1.503 and 1.505, for a tolerance of 0.002 (the difference between the limits as illustrated in Fig. 22.9B).

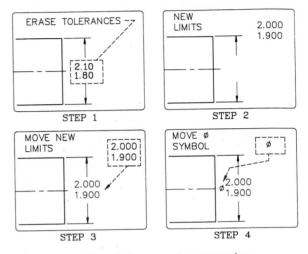

FIG. 22.7 Editing tolerances by computer.

Step 1 When you wish to change a toleranced dimension, begin by erasing the given tolerances.

Step 2 Using TEXT, draw the new limits of tolerance in a convenient location.

Step 3 Using a window, MOVE the new limits to the dimension line.

Step 4 Locate a diameter symbol, ∅, by typing %%C, and move it in front of the two new limits.

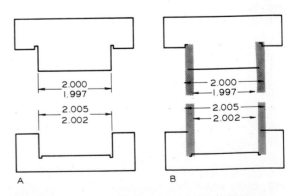

FIG. 22.8 Each of these mating parts has a tolerance of 0.003″ (variation in size). The allowance between the assembled parts (tightest fit) is 0.002″.

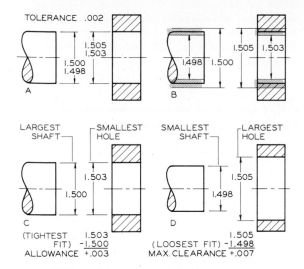

FIG. 22.9 The allowance (tightest fit) between these assembled parts is +0.003″. The maximum clearance is 0.007″.

22.4
Terminology of tolerancing

The meaning of most of the terms used in tolerancing can be seen by referring to Fig. 22.9.

TOLERANCE is the difference between the limits prescribed for a single part. The tolerance of the shaft in Fig. 22.9 is 0.002 in.

LIMITS OF TOLERANCE are the extreme measurements permitted by the maximum and minimum sizes of a part. The limits of tolerance of the shaft in Fig. 22.9 are 1.500 and 1.498.

ALLOWANCE is the tightest fit between two mating parts. The allowance between the largest shaft and the smallest hole in Fig. 22.9 is 0.003 (negative for an interference fit).

NOMINAL SIZE is an approximate size that is usually expressed with common fractions. The nominal sizes of the shaft and hole in Fig. 22.9 are 1.50 in. or $1\frac{1}{2}$ in.

BASIC SIZE is the exact theoretical size from which limits are derived by the application of plus-and-minus tolerances. There is no basic diameter if this is expressed with limits.

ACTUAL SIZE is the measured size of the finished part.

FIT signifies the type of fit between two mating parts when assembled. There are four types of fit: clearance, interference, transition, and line.

CLEARANCE FIT is a fit that gives a clearance between two assembled mating parts. The fit between the shaft and the hole in Fig. 22.9 is a clearance fit that permits a minimum clearance of 0.003 in. and a maximum clearance of 0.007 in.

INTERFERENCE FIT is a fit that results in an interference between the two assembled parts. The shaft in Fig. 22.10A is larger than the hole, so it requires a force or press fit, which has an effect similar to welding the two parts.

TRANSITION FIT can result in either an interference or a clearance. The shaft in Fig. 22.10B can be either smaller or larger than the hole and still be within the prescribed tolerances.

LINE FIT can result in a contact of surfaces or a clearance between them. The shaft in Fig. 22.10C can have contact or clearance when the limits are approached.

SELECTIVE ASSEMBLY is a method of selecting and assembling parts by trial and error. Using this method, parts can be assembled that have greater tolerances and can be produced at a reduced cost. This hand-assembly process is a compromise between a

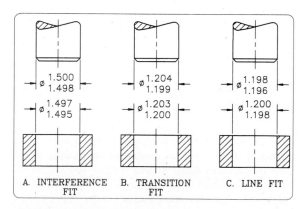

FIG. 22.10 Types of fits between mating parts. The clearance fit is not shown.

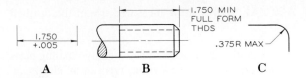

FIG. 22.11 Single tolerances can be given in some applications in MAX or MIN form.

high degree of manufacturing accuracy and an ease of assembly of interchangeable parts.

SINGLE LIMITS are dimensions designated by either MIN (minimum) or MAX (maximum), not by both (Fig. 22.11). Depths of holes, lengths, threads, corner radii, chamfers, and so on are sometimes dimensioned in this manner. Caution should be taken to prevent substantial deviations from the single limit.

22.5
Basic hole system

Widely used by industry, the basic hole system of dimensioning holes and shafts gives the required allowance between two assembled parts. The smallest hole is taken as the basic diameter from which the limits of tolerance and allowance are applied. It is advantageous to use the hole diameter as the basic dimension because many of the standard drills, reamers, and machine tools are designed to give standard hole sizes.

If the smallest diameter of a hole is 1.500 in., the allowance (0.003 in this example) can be subtracted from this diameter to find the diameter of the largest shaft (1.497 in.). The smallest limit for the shaft can then be found by subtracting the tolerance from 1.497 in.

22.6
Basic shaft system

Some industries use the basic shaft system of applying tolerances to dimensions of shafts, since many shafts come in standard sizes. In this system, the largest diameter of the shaft is used as the basic diameter from which the tolerances and allowances are applied.

If the largest permissible shaft is 1.500 in., the allowance can be added to this dimension to yield the

smallest possible diameter of the hole into which the shaft must fit. Therefore if the parts are to have an allowance of 0.004 in., the smallest hole would have a diameter of 1.504 in.

22.7
Metric limits and fits

This section will cover the metric system as recommended by the International Standards Organization (ISO), which has been presented in ANSI B4.2. These fits usually apply to cylinders—holes and shafts. However, these tables can also be used to determine the fits between any parallel surfaces, such as a key in a slot.

Metric definitions of limits and fits

Some of the definitions given below are illustrated in Fig. 22.12.

BASIC SIZE is the size from which the limits or deviations are assigned. Basic sizes, usually diameters, should be selected from Table 22.1 under the column heading ''First Choice.''

DEVIATION is the difference between the hole or shaft size and the basic size.

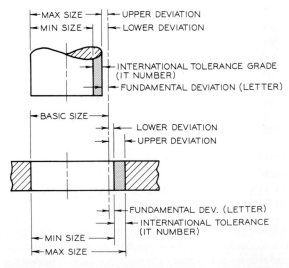

FIG. 22.12 Terms related to metric fits and limits.

TABLE 22.1
PREFERRED SIZES

Basic Size, (mm)		Basic Size, (mm)		Basic Size, (mm)	
First Choice	Second Choice	First Choice	Second Choice	First Choice	Second Choice
1		10		100	
	1.1		11		110
1.2		12		120	
	1.4		14		140
1.6		16		160	
	1.8		18		180
2		20		200	
	2.2		22		220
2.5		25		250	
	2.8		28		280
3		30		300	
	3.5		35		350
4		40		400	
	4.5		45		450
5		50		500	
	5.5		55		550
6		60		600	
	7		70		700
8		80		800	
	9		90		900
				1000	

UPPER DEVIATION is the difference between the maximum permissible size of a part and its basic size.

LOWER DEVIATION is the difference between the minimum permissible size of a part and its basic size.

FUNDAMENTAL DEVIATION is the deviation closest to the basic size. In the note 40H7, the *H* (an uppercase letter) represents the fundamental deviation for a hole. In the note 40g6, the *g* (a lowercase letter) represents the fundamental deviation for a shaft.

TOLERANCE is the difference between the maximum and minimum allowable sizes of a single part.

INTERNATIONAL TOLERANCE (IT) GRADE is a group of tolerances that vary in accordance with the basic size and that provide a uniform level of accuracy within a given grade. In the note 40H7, the 7 represents the IT grade. There are eighteen IT grades: IT01, IT0, IT1, . . . , IT16.

TOLERANCE ZONE is the zone that represents the tolerance grade and its position in relation to the basic size. This is a combination of the fundamental deviation (represented by a letter) and the International Tolerance grade (IT number). In note 40H8, the H8 indicates the tolerance zone.

HOLE BASIS is a system of fits based on the minimum hole size as the basic diameter. The fundamental deviation for a hole basis system is an uppercase letter, *H*, for example (Fig. 22.13).

SHAFT BASIS is a system of fits based on the maximum shaft size as the basic diameter. The fundamental deviation for a shaft basis system is a lowercase letter, *f*, for example (Fig. 22.13).

CLEARANCE FIT is a fit that results in a clearance between two assembled parts under all tolerance conditions.

INTERFERENCE FIT is a fit between two parts that requires they be forced together when assembled.

TRANSITION FIT is a fit that results in either a clearance or an interference fit between two assembled parts.

TOLERANCE SYMBOLS are notes used to communicate the specifications of tolerance and fit (Fig. 22.13). The basic size is the primary dimension from

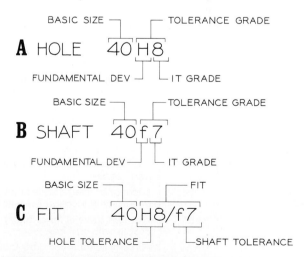

FIG. 22.13 Symbols and their definitions as applied to holes and shafts.

40H8   40H8$\left(\dfrac{40.039}{40.000}\right)$ $\dfrac{40.039}{40.000}$ (40H8)

A   **B**   **C**

FIG. 22.14 Three methods of giving tolerance symbols. The numbers in parentheses are for reference.

which the tolerances are determined; therefore it is the first part of the symbol. It is followed by the fundamental deviation letter and the IT number to give the tolerance zone. Uppercase letters are used to indicate the fundamental deviation for holes, and lowercase letters are used for shafts.

Three methods of specifying tolerance information are shown in Fig. 22.14. Parenthetical information is for reference only. The upper and lower limits are found in Appendix 44–48.

22.8
Preferred sizes and fits

The preferred basic sizes for computing tolerances are shown in Table 22.1. Under the "First Choice" heading, each number increases by about twenty-five percent of the preceding number. Each number in the "Second Choice" column increases by about twelve percent. To reduce expenses, you should, where possible, select basic diameters from the first column since these correspond to standard stock sizes for round, square, and hexagonal metal products.

Preferred fits for clearance, transition, and interference fits are shown in Table 22.2 for hole basis and shaft basis fits. The tables in Appendixes 45–48 correspond to these fits.

PREFERRED FITS—HOLE BASIS SYSTEM Figure 22.15 illustrates the symbols used to show the

TABLE 22.2
DESCRIPTION OF PREFERRED FITS

ISO Symbol		Description
Hole Basis	Shaft Basis	
H11/c11	C11/h11	*Loose running* fit for wide commercial tolerances or allowances on external members.
H9/d9	D9/h9	*Free running* fit—not for use where accuracy is essential, but good for large temperature variations, high running speeds, or heavy journal pressures.
H8/f7	F8/h7	*Close running* fit for running on accurate machines and for accurate location at moderate speeds and journal pressures.
H7/g6	G7/h6	*Sliding* fit—not intended to run freely, but to move and turn freely and locate accurately.
H7/h6	H7/h6	*Locational clearance* fit provides snug fit for locating stationary parts but can be freely assembled and disassembled.
H7/k6	K7/h6	*Locational transisition* fit for accurate location, a compromise between clearance and interference.
H7/n6	N7/h6	*Locational transition* fit for more accurate location where greater interference is permissible.
H7/p6[1]	P7/h6	*Locational interference* fit for parts requiring rigidity and alignment with prime accuracy of location but without special bore pressure requirements.
H7/s6	S7/h6	*Medium drive* fit for ordinary steel parts or shrink fits on light sections, the tightest fit usable with cast iron.
H7/u6	U7/h6	*Force* fit suitable for parts that can be highly stressed or for shrink fits where the heavy pressing forces required are impractical.

Clearance fits → (more clearance)
Transition fits
Interference fits ↑ (more interference)

[1]Transition fit for basic sizes in range from 0 through 3 mm.

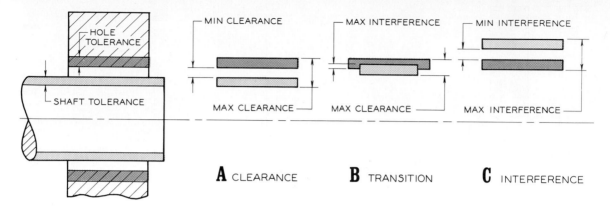

FIG. 22.15 Types of fits.

A. A clearance fit. **B.** A transition fit where there can be an interference or a clearance.
C. An interference fit where the parts must be forced together.

possible combinations of fits when using the hole basis system. There is a clearance fit between the two parts, a transition fit, and an interference fit. This technique of representing fits is used in Fig. 22.16 to show a series of fits for a hole basis system. Note that the lower deviation of the hole is zero; in other words, the smallest size of the hole is the basic size. The different sizes of the shafts give a variety of fits from c11 to u6, where there is a maximum of interference. These fits correspond to those given in Table 22.2.

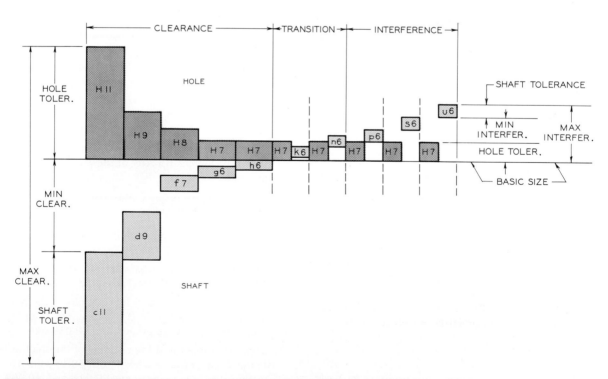

FIG. 22.16 The preferred fits for a hole basis system. These fits correspond to those given in Table 22.2

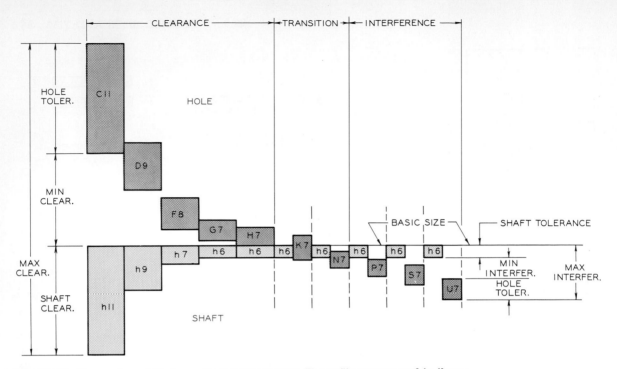

FIG. 22.17 The preferred fits for a shaft basis system. These fits correspond to those given in Table 22.2.

PREFFERED FITS—SHAFT BASIS SYSTEM Figure 22.17 illustrates the preferred fits based on the shaft basis system, where the largest shaft size is the basic diameter. The variation in the fit between the parts is caused by varying the size of the holes, which results in a range from a clearance fit of C11/h11 to an interference fit of U7/h6.

22.9
Example problems—
metric system

The following problems are given and solved as examples of determining the sizes and limits and the applications of the proper symbols to mating parts. The solution of these problems requires the use of the tables in Appendix 45, the table of preferred sizes (Table 22.1), and the table of preferred fits (Table 22.2).

EXAMPLE 1 (Fig. 22.18)

Given: Hole basis system, close running fit, basic diameter = 39 mm.

Solution: Use a basic diameter of 40 mm (Table 22.1) and fit of H8/f7 (Table 22.2).

Hole: Find the upper and lower limits of the hole in Appendix 45 under H8 and across from 40 mm. These limits are 40.000 and 40.039 mm.

Shaft: The upper and lower limits of the shaft are found under f7 and across from 40 mm in Appendix 45. These limits are 39.950 and 39.975 mm.

Symbols: The methods of noting the drawings are shown in Fig. 22.18. Any of these methods is appropriate.

EXAMPLE 2 (Fig. 22.19)

Given: Hole basis system, locational transition fit, basic diameter = 57 mm.

Solution: Use a basic diameter of 60 mm (Table 22.1) and a fit of H7/k6 (Table 22.2).

Hole: Find the upper and lower limits of the hole in Appendix 46 under H7 and across from 60 mm. These limits are 60.000 and 60.030 mm.

Shaft: The upper and lower limits of the shaft are found under k6 and across from 60 mm in Appendix 46. These limits are 60.021 and 60.002 mm.

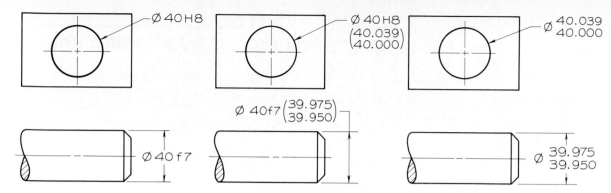

FIG. 22.18 Any of these three methods can be used to apply symbols to a detail drawing of two mating parts.

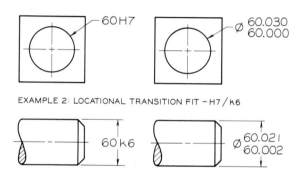

EXAMPLE 2: LOCATIONAL TRANSITION FIT – H7 / K6

FIG. 22.19 Two methods of applying tolerance symbols to a transition fit.

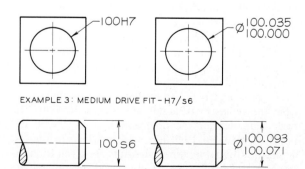

EXAMPLE 3: MEDIUM DRIVE FIT – H7/s6

FIG. 22.20 Two methods of applying tolerance symbols to an interference fit.

Symbols: Two methods of applying the tolerance symbols to a drawing are shown in Fig. 22.19.

EXAMPLE 3 (Fig. 22.20)

Given: Hole basis system, medium drive fit, basic diameter = 96 mm.

Solution: Use a basic diameter of 100 mm (Table 22.1) and a fit of H7/s6 (Table 22.2).

Hole: Find the upper and lower limits of the hole in Appendix 46 under H7 and across from 100 mm. These limits are 100.035 and 100.000 mm.

Shaft: The upper and lower limits of the shaft are found under s6 and across from 100 mm in Appendix 46. These limits are found to be 100.093 and 100.071 mm. From the appendix, the tightest fit is an interference of 0.093 mm, and the loosest fit is an interference of 0.036 mm. An interference is indicated by a minus sign in front of the numbers.

EXAMPLE 4 (Fig. 22.21)

Given: Shaft basis system, loose running fit, basic diameter = 116 mm.

Solution: Use a basic diameter 120 mm (Table 22.1) and a fit of C11/h11 (Table 22.2).

Hole: Find the upper and lower limits of the hole in Appendix 47 under C11 and across from 120 mm. These limits are 120.400 and 120.180 mm.

Shaft: The upper and lower limits of the shaft are found under h11 and across from 120 mm in Appendix 47. These limits are 119.780 and 120.000 mm.

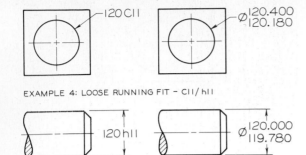

EXAMPLE 4: LOOSE RUNNING FIT – C11/ h11

FIG. 22.21 Two methods of applying tolerance symbols to a clearance fit.

22.10
Preferred metric fits— nonpreferred sizes

Limits of tolerances for preferred fits, shown in Table 22.2, can be calculated for nonstandard sizes. Limits of tolerances appear in Appendix 49 for nonstandard hole sizes and in Appendix 50 for nonstandard shaft sizes.

The hole and shaft limits for an H8/f7 fit and a 45-mm DIA are calculated in Fig. 22.22. The tolerance limits of 0.000 and 0.039 mm for an H8 hole are taken from Appendix 49 across from the size range of 40–50 mm. The tolerance limits of 0.000 and 0.050 mm are taken from Appendix 50. The limits of sizes for the hole and shaft are calculated by applying these limits of tolerance to the 45-mm basic diameter.

FIT: H8/f7   Ø 45 BASIC

FROM APPENDIX			
HOLE H8	SHAFT f7	HOLE LIMITS	45.039 45.000
0.039 0.000	−0.025 −0.050	SHAFT LIMITS	44.975 44.950

FIG. 22.22 The limits of a nonstandard diameter, 45 mm, and an H8/f7 fit are calculated by using values from Appendixes 49 and 50.

22.11
Standard fits—English units

The ANSI B4.1 standard specifies a series of fits between cylindrical parts that are based on the basic hole system in inches. The types of fit covered in this standard are

RC—Running and sliding fits

LC—Clearance locational fits

LT—Transition locational fits

LN—Interference locational fits

FN—Force and shrink fits

These five types of fit, for which each has several classes, are listed in Appendixes 39–43.

RUNNING AND SLIDING FITS (RC) are fits for which limits of clearance to provide a similar running performance, with suitable lubrication allowance throughout the range of sizes. The clearance for the first two classes (RC 1 and RC 2), used chiefly as slide fits, increases more slowly with diameter size than that of other classes, so that accurate location is maintained even at the expense of free relative motion.

LOCATIONAL FITS (LC, LT, LN) are intended to determine only the location of the mating parts; they may provide rigid or accurate location (interference fits) or some freedom of location (clearance fits). Locational fits are divided into three groups: clearance fits (LC), transition fits (LT), and interference fits (LN).

FORCE FITS (FN) are special types of interference fits, typically characterized by maintenance of constant bore pressures throughout the range of sizes. The interference therefore varies almost directly with diameter, and the difference between its minimum and maximum values is small to maintain the resulting pressures within reasonable limits.

Figure 22.23 illustrates how to use the values from the tables in Appendix 39 for an RC 9 fit. The basic diameter for the hole and shaft is 2.5000 in., which is between 1.97 and 3.15 in. given in the last column of the table. Since all limits are given in thousandths, the values can be converted by moving the decimal point three places to the left; for example, $+0.7$ is $+0.0007$ in.

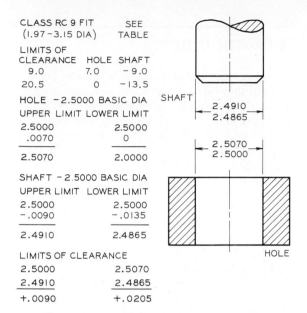

CLASS RC 9 FIT (1.97-3.15 DIA) SEE TABLE

LIMITS OF

CLEARANCE	HOLE	SHAFT
9.0	7.0	- 9.0
20.5	0	-13.5

HOLE - 2.5000 BASIC DIA

UPPER LIMIT	LOWER LIMIT
2.5000	2.5000
.0070	0
2.5070	2.0000

SHAFT - 2.5000 BASIC DIA

UPPER LIMIT	LOWER LIMIT
2.5000	2.5000
-.0090	-.0135
2.4910	2.4865

LIMITS OF CLEARANCE

2.5000	2.5070
2.4910	2.4865
+.0090	+.0205

SHAFT 2.4910 2.4865

2.5070 2.5000

HOLE

FIG. 22.23 The method of calculating limits and allowances for an RC 9 fit between a shaft and a hole. The basic diameter is 2.5000 inches.

The upper and lower limits of the shaft (2.4910 and 2.4865 in.) are found by subtracting the two limits (−0.0090 and −0.0135 in.) from the basic diameter. The upper and lower limits of the hole (2.5007 and 2.5000 in.) are found by adding the two limits (+0.007 and 0.000 in.) to the basic diameter.

When the two parts are assembled, the tightest fit (+0.0205 in.) and the loosest fit (+0.0090 in.) are found by subtracting the maximum and minimum sizes of the holes and shafts. These values are provided in the second column of the table as a check on the limits.

The same method (but different tables) is used for calculating the limits for all types of fit. Plus values indicate clearance, and minus values indicate interference between the assembled parts.

When using the metric system, convert these values to millimeters either by multiplying inches by 25.4 or by using metric tables.

22.12

Chain dimensions

When parts are dimensioned to locate surfaces or geometric features by a chain of dimensions (Fig.

22.24A), variations may occur that exceed the tolerances specified. As successive measurements are made, with each based on the preceding one, the tolerances may accumulate, as shown in Fig. 22.24A. For example, the tolerance between surfaces *A* and *B* is 0.002; between *A* and *C*, 0.004; and between *A* and *D*, 0.006.

This accumulation of tolerances can be eliminated by measuring from a single plane called a *datum plane*. A datum plane is usually on the object, but it could be on the machine used to make the part. Since each of the planes in Fig. 22.24B was located with respect to a single datum the tolerances between the intermediate planes are a uniform 0.002, which represents the maximum tolerance.

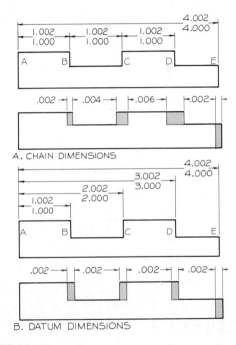

FIG. 22.24 When dimensions are given as chain dimensions, the tolerances can accumulate to give a variation of 0.006″ at *D* instead of 0.002″. When dimensioned from a single datum, the variations of *X* and *Y* cannot deviate more than the specified 0.002″ from the datum.

22.13

Origin selection

Sometimes there is a need to specify a surface as the origin for locating another surface. An example is il-

FIG. 22.25 This method is used to indicate the origin surface for locating one feature of a part with respect to another.

lustrated in Fig. 22.25, where the origin surface is the shorter one at the base of an object. The result is that the angular variation permitted is less for the longer surface than it would be if the origin plane had been the longer surface.

22.14
Conical tapers

Taper is a ratio of the difference in the diameters of two circular sections of a cone to the distance between the sections. A method of specifying a conical taper by giving a basic diameter and a basic taper is shown in Fig. 22.26. The basic diameter of 20 mm is located midway in the length of the cone with a toleranced dimension. Figure 22.26 also shows how to find the radial tolerance zone.

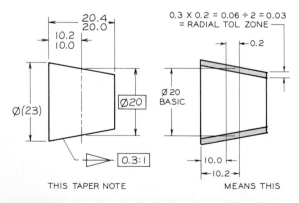

FIG. 22.26 Taper is indicated with a combination of tolerances and taper symbols. The variation in diameter at any point is 0.06 mm, or 0.03 mm in radius.

22.15
Tolerance notes

All dimensions on a drawing are toleranced either by the rules previously discussed or by a note placed in or near the title block. For example, the note TOLERANCE $\pm \frac{1}{64}$ (or its decimal equivalent, 0.40 mm) might be given on a drawing for less critical dimensions.

Some industries may give dimensions in inches where decimals are carried out to two, three, and four places. A note for dimensions with two and three decimal places might be given on the drawing as TOLERANCES XX.XX ± 0.10; XX.XXX ± 0.005. Tolerances of four places would be given directly on the dimension lines.

The most common method of noting tolerances is to give as large a tolerance as feasible in a note, such as TOLERANCES ± 0.05 (± 1 mm when using metrics), and to give the tolerances for the mating dimensions that require smaller tolerances on the dimension lines.

Angular tolerances should be given in a general note in or near the title block, such as ANGULAR TOLERANCES ±0.5° or ±30'. Angular tolerances less than this should be given on the drawing where these angles are dimensioned. Techniques of tolerancing angles are shown in Fig. 22.27.

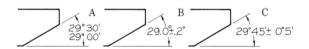

FIG. 22.27 Angles can be toleranced by any of these techniques using limits or the plus-and-minus method.

22.16
General tolerances—metric

All dimensions on a drawing are understood to have tolerances, and the amount of tolerance must be noted. This section, which is based on the metric system as outlined in ANSI B4.3, uses the millimeter as the unit of measurement.

Tolerances may be specified by (1) applying them directly to the dimensions, (2) giving them in specifi-

FIG. 22.28 International Tolerance (IT) grades and their applications.

cation documents, or (3) giving them in a general note on the drawing.

LINEAR DIMENSIONS may be toleranced by indicating ± one half of an International Tolerance (IT) grade as given in Appendix 44. The appropriate IT grade can be selected from the graph in Fig. 22.28.

IT grades for mass produced items range from IT12 through IT16. When the machining process is known, the IT grades can be selected from Table 22.3.

General tolerances using IT grades may be expressed in a note as follows:

UNLESS OTHERWISE SPECIFIED ALL

UNTOLERANCED DIMENSIONS ARE $\pm\dfrac{IT14}{2}$.

This means that a tolerance of ±0.700 mm is allowed for a dimension between 315 and 400 mm. The value of the tolerance is listed in Appendix 44 as 1.400.

Table 22.4 shows recommended tolerances for fine, medium, and coarse series for dimensions of graduated sizes. A medium tolerance, for example, can

TABLE 22.3
IT GRADES AND THEIR RELATIONSHIP TO MACHINING PROCESSES

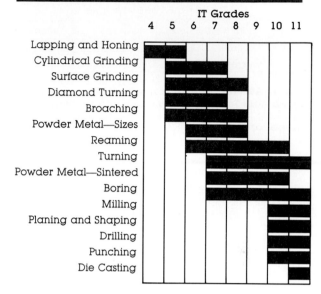

be specified by the following note:

GENERAL TOLERANCES SPECIFIED IN
ANSI B4.3 MEDIUM SERIES APPLY.

This same information can be given on the drawing in table form by selecting the grade—medium in this example—from Table 22.4 and giving it as shown in Fig. 22.29.

TABLE 22.4
FINE, MEDIUM, AND COARSE SERIES: GENERAL TOLERANCE—LINEAR DIMENSIONS

		Variations in mm						
Basic Dimensions (mm)		0.5 to 3	Over 3 to 6	Over 6 to 30	Over 30 to 120	Over 120 to 315	Over 315 to 1000	Over 1000 to 2000
Permissible Variations	Fine series	±0.05	±0.05	±0.1	±0.15	±0.2	±0.3	±0.5
	Medium Series	±0.1	±0.1	±0.2	±0.3	±0.5	±0.8	±1.2
	Coarse Series		±0.2	±0.5	±0.8	±1.2	±2	±3

DIMENSIONS IN mm							
GENERAL TOLERANCE UNLESS OTHERWISE SPECIFIED THE FOLLOWING TOLERANCES ARE APPLICABLE							
LINEAR	OVER TO	0.5 6	6 30	30 120	120 315	315 1000	1000 2000
TOL	±	0.1	0.2	0.3	0.5	0.8	1.2

FIG. 22.29 This table is a *medium* series of values taken from Table 22.4. It is placed on the working drawing to provide the tolerances for various ranges of sizes.

DIMENSIONS IN mm						
GENERAL TOLERANCE UNLESS OTHERWISE SPECIFIED THE FOLLOWING TOLERANCES ARE APPLICABLE						
LINEAR		OVER TO	− 120	120 315	315 1000	1000 −
TOL	ONE DECIMAL ±		0.3	0.5	0.8	1.2
	NO DECIMALS ±		0.8	1.2	2	3

FIG. 22.30 This table of tolerances can be placed on a drawing to indicate the tolerances for dimensions with one or no decimal places.

General tolerances can be expressed in a table that gives the tolerances for dimensions with one or no decimal places (Fig. 22.30).

Another method of giving general tolerances is a note in this form:

UNLESS OTHERWISE SPECIFIED ALL
UNTOLERANCED
DIMENSIONS ARE ±0.8 mm.

This method should be used only where the dimensions on a drawing have slight differences in size.

ANGULAR TOLERANCES are expressed (1) as an angle in decimal degrees or in degrees and minutes, (2) as a taper expressed in percentage (number of millimeters per 100 mm), or (3) as milliradians. A milliradian is found by multiplying the degrees of an angle by 17.45. The suggested tolerances for each of these units are shown in Table 22.5 and are based on the length of the shorter leg of the angle.

General angular tolerances may be given on the drawing with a note in the following form:

UNLESS OTHERWISE SPECIFIED THE GENERAL
TOLERANCES IN ANSI B4.3 APPLY.

A second method shows a portion of Table 22.5 on the drawing using the units desired (Fig. 22.31). A third method is a note with a single tolerance such as:

UNLESS OTHERWISE SPECIFIED THE GENERAL
ANGULAR TOLERANCES ARE ±0°30′ (or ±0.5°).

22.17
Geometric tolerances

Geometric tolerancing is a term used to describe tolerances that specify and control form, profile, orientation, location, and runout on a dimensioned part. The basic principles of this area of tolerancing are standardized by the ANSI Y14.5M–1982 Standards and the Military Standards (Mil-Std) of the U.S. Department of Defense.

TABLE 22.5
GENERAL TOLERANCE—ANGLES AND TAPERS

	Length of the Shorter Leg (mm)	Up to 10	Over 10 to 50	Over 50 to 120	Over 120 to 400
Permissible Variations	In Degrees and Minutes	±1°	±0°30′	±0°20′	±0°10′
	In Millimeters per 100 mm	±1.8	±0.9	±0.6	±0.3
	In Milliradians	±18	±9	±6	±3

ANGULAR TOLERANCE				
LENGTH OF SHORTER LEG - mm	UP TO 10	OVER 10 TO 50	OVER 50 TO 120	OVER 120 TO 400
TOL	±1°	±0° 30'	±0° 20'	±0° 10'

FIG. 22.31 This table can be placed on a drawing to indicate the general tolerances for angles that were extracted from Table 22.5.

	TOLERANCE	CHARACTERISTIC	SYMBOL
INDIVIDUAL FEATURES	FORM	STRAIGHTNESS	—
		FLATNESS	▱
		CIRCULARITY	○
		CYLINDRICITY	⌭
INDIVIDUAL OR RELATED FEATURES	PROFILE	PROFILE OF A LINE	⌒
		PROFILE OF A SURFACE	⌓
RELATED FEATURES	ORIENTATION	ANGULARITY	∠
		PERPENDICULARITY	⊥
		PARALLELISM	//
	LOCATION	POSITION	⌖
		CONCENTRICITY	◎
	RUNOUT	CIRCULAR RUNOUT	↗
		TOTAL RUNOUT	⌰

FIG. 22.32 These symbols are used to specify the geometric characteristics of a dimensioned part.

These standards are based on the metric system with the millimeter as the unit of measurement. However, inch units with decimal fractions can be used instead of millimeters if needed.

22.18
Symbology of geometric tolerances

The various symbols used to specify geometric characteristics of dimensioned drawings are shown in Fig. 22.32. Additional features and their proportions are shown in Fig. 22.33, where the letter height *(H)* is used as the basis of the proportions. On most draw-

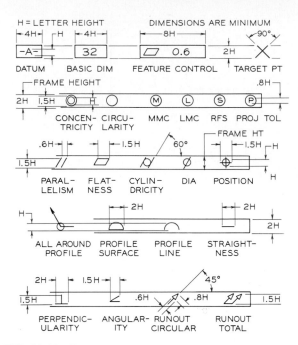

FIG. 22.33 The general proportions of notes and symbols used in feature control symbols.

ings, a $\frac{1}{8}$-in. or 3-mm letter height is recommended. Other examples of feature control symbols are given in Fig. 22.34.

22.19
Limits of size

Three terms used to specify the limits of size of a part when applying geometric tolerances are *maximum ma-*

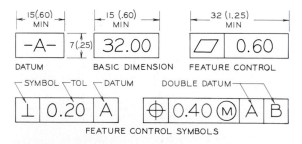

FIG. 22.34 Examples of symbols used to indicate datum planes, basic dimensions, and feature control symbols.

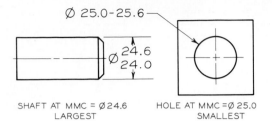

Ø 25.0-25.6

SHAFT AT MMC = Ø 24.6
LARGEST

HOLE AT MMC =Ø 25.0
SMALLEST

FIG. 22.35 A shaft is at MMC when it is at the largest size permitted by its tolerance. A hole is at MMC when it is at the smallest size.

terial condition (MMC), least material condition (LMC), and regardless of feature size (RFS).

MMC indicates when a part is made with the maximum amount of material. For example, the shaft in Fig. 22.35 is at MMC when it has the largest permitted diameter of 24.6 mm. The hole is at MMC when it has the most material or has the smallest diameter of 25.0 mm.

LMC indicates that a part has the least amount of material. The shaft in Fig. 22.35 is at LMC when it has the smallest diameter of 24.0 mm. The hole is at LMC when it has the largest diameter of 25.6 mm.

RFS indicates that tolerances apply to a geometric feature regardless of the size it may be, from MMC to LMC.

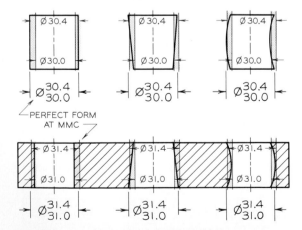

FIG. 22.36 When only a tolerance of size is specified on a part, the limits prescribe the form of the part, as shown in these shafts and holes with the same limits of tolerance.

22.20
Three rules of tolerances

There are three general rules of tolerancing geometric features that should be followed in this type of dimensioning.

RULE 1 (INDIVIDUAL FEATURE SIZE) When only a tolerance of size is specified on a part, the limits of size prescribe the amount of variation permitted in its geometric form. In Fig. 22.36, the forms of the shaft and hole are permitted to vary within the tolerance of size indicated by the dimensions.

RULE 2 (TOLERANCES OF POSITION) When a tolerance of position is specified on a drawing, RFS, MMC, or LMC must be specified with respect to the tolerance, the datum, or both. The specification of symmetry of the part in Fig. 22.37 is based on a tolerance at RFS from a datum at RFS.

RULE 3 (ALL OTHER GEOMETRIC TOLERANCES) RFS applies for all other geometric tolerances for individual tolerances and datum references if no modifying symbol is given in the feature control symbol. If a feature is to be at MMC, it must be specified.

22.21
Three-datum plane concept

A datum plane is used as the origin of a part's dimensioned features that have been toleranced. Datum planes are usually associated with manufacturing equipment, such as machine tables, or with locating pins.

Three mutually perpendicular datum planes are used to dimension a part accurately. For example, the part in Fig. 22.38 is placed in contact with the primary datum plane at its base where three points must make contact with the datum. The part is further related to the secondary plane with two contacting points. The third (tertiary) datum is contacted by a single point on the object.

The priority of these datum planes is noted on the drawing of the part by feature control symbols as shown in Fig. 22.39. The primary datum is surface *P*, the secondary is surface *S*, and the tertiary is surface *T*. Examples of feature control symbols are given in

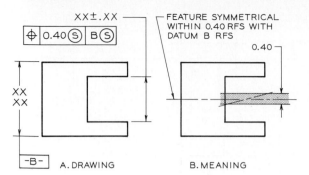

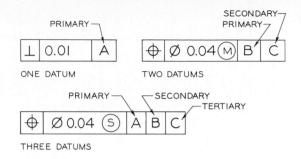

FIG. 22.37 Tolerances of position should include a note of M, S, or L to indicate maximum material condition, regardless of feature size, or least material condition.

FIG. 22.40 Feature control symbols may indicate from one to three datum planes listed in order of priority.

Fig. 22.40, where the primary, secondary, and tertiary datum planes are listed in order of priority.

22.22
Cylindrical datum features

Figure 22.41 illustrates a part with a cylindrical datum feature that is the axis of a true cylinder. Primary datum *K* establishes the first datum. Datum *m* is associated with two theoretical planes—the second and third in a three-plane relationship.

The two theoretical planes are represented on the drawing by perpendicular centerlines. The intersection of the centerlines coincides with the datum axis. All dimensions originate from the datum axis, which is perpendicular to datum *K*; the two intersecting datum

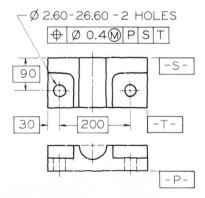

FIG. 22.38 When an object is referenced to a primary datum plane, it contacts the plane with its three highest points. The vertical surface contacts the secondary vertical datum plane with two points. The third datum plane is contacted by one point on the object. The datum planes are listed in this order in the feature control symbol.

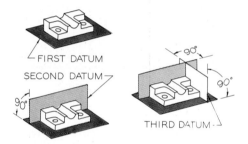

FIG. 22.39 The sequence of the three-plane reference system (shown in Fig. 22.38) is labeled where the planes appear as edges. Note that the primary datum plane, *P*, is listed first in the feature control symbol; the secondary plane, *S*, is next; and the tertiary plane, *T*, is last.

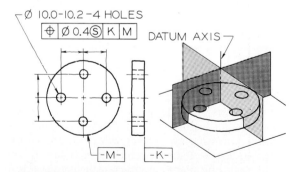

FIG. 22.41 These true-position holes are located with respect to primary datum *K* and datum *M*. Since datum *M* is a circle, this implies that the holes are located about two intersecting datum planes formed by the crossing centerlines in the circular view satisfying the three-plane concept.

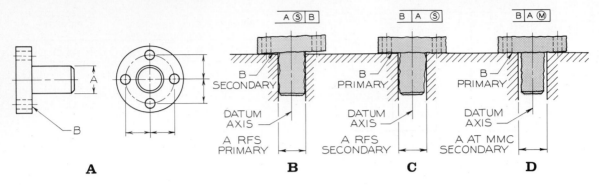

A   **B**   **C**   **D**

FIG. 22.42 Three examples that illustrate the effects of selection of the datum planes in order of priority and the effect of RFS and MMC.

planes indicate the direction of measurements in an *x* and *y* direction.

The sequence of the datum reference in the feature control symbol is significant to the manufacturing and inspection processes. The part in Fig. 22.42 is dimensioned with an incomplete feature control symbol; it does not specify the primary and secondary datum planes. The schematic drawing in Fig. 22.42B illustrates the effect of specifying the diameter *A* as the primary datum plane and surface *B* as the secondary datum plane. This means that the part is centered about cylinder *A* by mounting the part in a chuck, mandrel, or centering device on the processing equipment, which centers the part at RFS. Surface *B* is assembled to contact at least one point of the third datum plane.

If surface *B* were specified as the primary datum feature, it would be assembled to contact datum plane *B* in at least three points. The axis of datum feature *A* will be gauged by the smallest true cylinder that is perpendicular to the first datum that will contact surface *A* at RFS.

In Fig. 22.42D, plane *B* is specified as the primary datum feature, and cylinder *A* is specified as the secondary datum feature at MMC. The part is mounted on the processing equipment where at least three points on feature *B* are in contact with datum *B*. The second and third planes intersect at the datum axis to complete the three-plane relationship. Using the modifier to specify MMC gives a more liberal tolerance zone than otherwise would be acceptable when RFS was specified.

22.23
Datum features at RFS

When dimensions of size are applied to a part at RFS, the datum is established by contacting surfaces on the processing equipment with surfaces of the part. Variable machine elements, such as chucks or center devices, are adjusted to fit the external or internal features of a part and thereby establish datums.

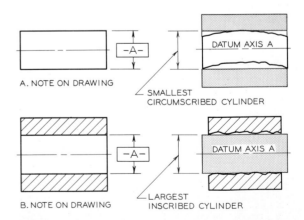

FIG. 22.43 The datum axis of a shaft is the smallest circumscribed cylinder that contacts the shaft. The datum axis of a hole is the centerline of the largest inscribed cylinder that contacts the hole.

PRIMARY DIAMETER DATUMS For an external cylinder (shaft), the datum axis is the axis of the smallest circumscribed cylinder that contacts the cylindrical feature of the part. In other words, the largest diameter of the part will make contact with the smallest contacting cylinder of the machine element that holds the part (Fig. 22.43).

For an internal cylinder (hole), the datum axis is the axis of the largest inscribed cylinder that contacts the inside of the hole. In other words, the smallest diameter of the hole will make contact with the largest cylinder of the machine element inserted in the hole (Fig. 22.43).

PRIMARY EXTERNAL PARALLEL DATUMS The datum for external features is the center plane between two parallel planes, at their minimum separation, that contact the planes of the object (Fig. 22.44A). These are planes of a viselike device that holds the part; therefore the planes of the part are at maximum separation, whereas the planes of the device are at minimum separation.

PRIMARY INTERNAL PARALLEL DATUMS The datum for internal features is the center plane between two parallel planes, at their maximum separa-

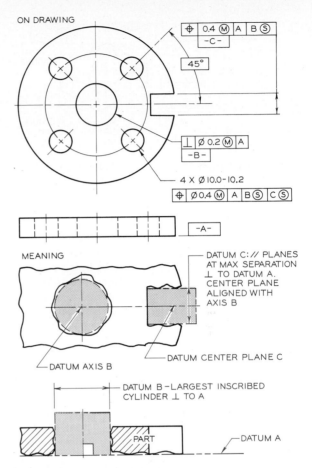

FIG. 22.45 A part located with respect to primary, secondary, and tertiary datum planes.

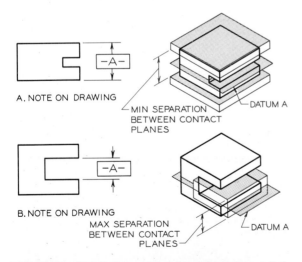

FIG. 22.44 The datum plane for external parallel surfaces is the center plane between two contacting parallel planes at their minimum separation. The datum plane for internal parallel surfaces is the center plane between two contacting parallel surfaces at their maximum separation.

tion, that contact the planes of the object (Fig. 22.44B). This is the condition in which the slot is at its smallest opening size.

SECONDARY DATUMS The secondary datum (axis or center plane) for both external and internal diameters or distances between parallel planes is found as covered in the previous two paragraphs but with an additional requirement: The contacting cylinder of the contacting parallel planes must be oriented perpendicular to the primary datum. Fig. 22.45 illustrates how datum *B* is the axis of a cylinder. This principle also can be applied to parallel planes.

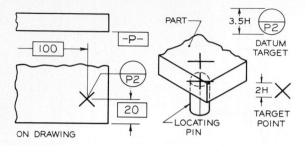

FIG. 22.46 Target points from which a datum point is established are located with an X and a target symbol.

TERTIARY DATUMS The third datum (axis or center plane) for both external and internal features is found in the same manner as covered in the previous three paragraphs but with an additional requirement: The contacting cylinder or parallel planes must be angularly oriented with respect to the secondary datum. Datum *C* in Fig. 22.45 is the tertiary datum plane.

22.24
Datum targets

Instead of using a plane surface as a datum, specified datum targets are indicated on the surface of a part where the part is supported by spherical or pointed locating pins. The symbol X indicates target points that are supported by locating pins at specified points (Fig. 22.46). Datum target symbols are placed outside the outline of the part with a leader directed toward the target. When the target is on the near (visible) surface, the leaders are solid lines; when the target is on the

far (invisible) surface (Fig. 22.48) the leaders are hidden.

Three target points are required to establish the primary datum plane, two for the secondary, and one for the tertiary. The target symbol in Fig. 22.46 is labeled *P2* to match the designation of the primary datum, *P*. Were the other two points shown, they would be labeled *P1* and *P2* to establish the primary datum.

A datum target line is specified in Fig. 22.47 for a part supported on a datum line instead of on a datum point. An X and a phantom line are used to locate the line of support.

Target areas are specified for cases where spherical or pointed locating pins are inadequate to support a part. The diameters of the targets are specified with crosshatched circles surrounded by phantom lines as shown in Fig. 22.48. The target symbols give both the diameter of the targets and their number designations. The X symbol could also be used to indicate target areas.

The part is located on its datum plane by placing it on the three locating pins with 30-mm diameters, as shown in Fig. 22.48. Leaders from the target areas to the target symbols are hidden, indicating that the targets are on the hidden side of the object.

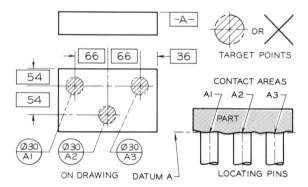

FIG. 22.48 Target points with areas are located with basic dimensions and target symbols that give the diameters of the targets. Hidden leaders indicate that the targets are on the hidden side of the plane.

22.25
Tolerances of location

Tolerances of location deal with *position, concentricity,* and *symmetry.*

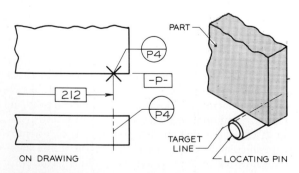

FIG. 22.47 An X and a phantom line are used to locate target lines on a drawing.

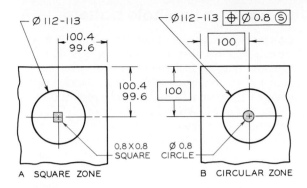

FIG. 22.49 **A.** These dimensions give a square tolerance zone for the axis of the hole. **B.** Basic dimensions locate the true center of the circle about which a circular tolerance zone of 0.8 mm is specified.

True-position tolerancing

Whereas toleranced location dimensions give a square tolerance zone for locating the center of a hole, *true-position dimensions* locate the exact position of a hole's center, about which a circular tolerance zone is given (Fig. 22.49). *Basic dimensions* are exact untoleranced dimensions used to locate true positions indicated by boxes drawn around them (Fig. 22.49B).

In both methods, the diameters of the holes are toleranced by notes. The true-position method (Fig. 22.49B) uses a feature control symbol to specify the diameter of the circular tolerance zone inside which the center of the hole must lie. More than a square A circular zone gives a more uniform tolerance of the hole's true position than a square.

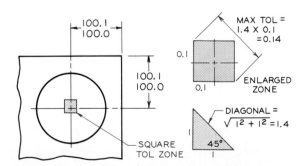

FIG. 22.50 The toleranced-coordinate method of dimensioning gives a square tolerance zone. The diagonal of the square exceeds the specified tolerance by a factor of 1.4.

In Fig. 22.50, you can see an enlargement of the square tolerance zone that results from using toleranced coordinates to locate a hole's center. The diagonal across the square zone is greater than the specified tolerance by a factor of 1.4. The true-position method shown in Fig. 22.51 can have a larger circular tolerance zone by a factor of 1.4 and still have the same degree of accuracy of position as the specified 0.1 square zone.

If the toleranced coordinate method could accept a variation of 0.014 across the diagonal of the square tolerance zone, then the true-position tolerance should be acceptable with a circular zone of 0.014, which is a greater tolerance than the square zone permitted (Fig. 22.50). True-position tolerances can be applied by symbol, as in Fig. 22.51.

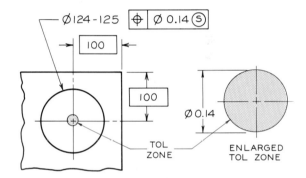

FIG. 22.51 The true-position method of tolerancing gives a circular tolerance zone with equal variations in all directions from the true axis of the hole.

The circular tolerance zone specified in the circular view of a hole is assumed to extend the full depth of the hole. Therefore the tolerance zone for the centerline of the hole is a cylindrical zone inside which the axis must lie. Since the size of the hole and its position are both toleranced, these two tolerances are used to establish the diameter of a gauge cylinder used to check for the conformance of the holes to the specifications (Fig. 22.52).

By subtracting the true-position tolerance from the hole at MMC (the smallest permissible hole), the circle is found. This zone represents the least favorable condition when the part is gauged or assembled with a mating part. When the hole is not at MMC, it is larger and permits greater tolerance and easier assembly.

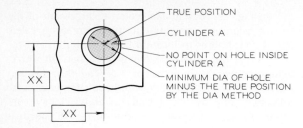

FIG. 22.52 When a hole is located at true position at MMC, no element of the hole will be inside the imaginary cylinder *A*.

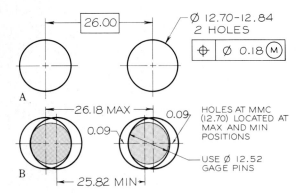

FIG. 22.53 When two holes are located at true position at MMC, they may be gauged with pins 12.52 mm in diameter that are located 26.00 mm apart.

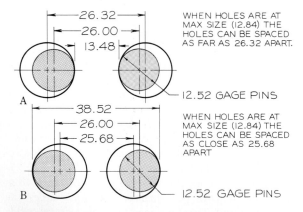

FIG. 22.54 **A.** When two holes are at their maximum size, the centers of the holes can be spaced as far as 26.32 mm apart and still be acceptable. **B.** The holes can be placed as close as 25.68 mm apart when they are at maximum size.

Gauging a two-hole pattern

Gauging is a technique of checking dimensions to determine whether they have met the specifications of tolerance (Fig. 22.53). The two holes, which are positioned 26.00 mm apart with a basic dimension, have limits of 12.70 and 12.84 for a tolerance of 0.14, and are located at true position within a diameter of 0.18. The gauge pin diameter is calculated to be 12.52 mm (the smallest hole's size minus the true-position tolerance), as illustrated in Fig. 22.53B. This means that two pins with diameters of 12.52 mm spaced exactly 26.00 mm apart could be used to check the diameters and positions of the holes at MMC, the most critical size. If the pins can be inserted in the holes, the holes are properly sized and located.

When the holes are not at MMC—that is, when they are larger than their minimum size—these gauge pins will permit a greater range of variation (Fig. 22.54). When the holes are at their maximum size of 12.84 mm, they can be located as close as 25.68 mm from center to center or as far apart as 26.32 mm from center to center. When not specified, true-position tolerances are assumed to apply at MMC.

Concentricity

Concentricity is a feature of location because it specifies the relationship of one cylinder with another since both share the same axis. In Fig. 22.55, the large cylinder is "flagged" as datum *A*, which means the large diameter is used as the datum for measuring the variation of the smaller cylinder's axis.

Feature control symbols will be used to specify concentricity and other geometric characteristics throughout the remainder of this chapter (Fig. 22.56).

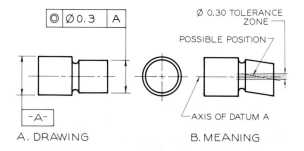

FIG. 22.55 Concentricity is a tolerance of location. The feature control symbol specifies that the smaller cylinder should be concentric to cylinder *A*, within 0.3 mm about the axis of *A*.

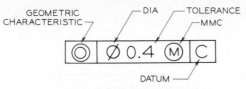

FIG. 22.56 A typical feature control symbol. This one indicates that a surface is concentric to datum C within a diameter of 0.4 mm at MMC.

Symmetry

Symmetry is also a feature of location. A part or a feature is symmetrical when it has the same contour and size on opposite sides of a central plane. A symmetry tolerance locates features with respect to a datum plane (Fig. 22.57). The feature control symbol notes that the notch is symmetrical about datum B within a zone of 0.6 mm.

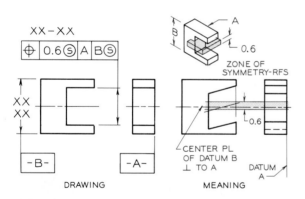

FIG. 22.57 Symmetry is a tolerance of position that specifies a part's feature be symmetrical about the center plane between parallel surfaces of the part.

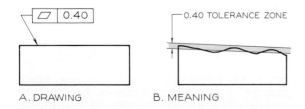

FIG. 22.58 Flatness is a tolerance of form that specifies two parallel planes inside of which the object's surface must lie.

22.26
Tolerances of form

Flatness

A surface is flat when all of its elements are in one plane. A feature control symbol is used to specify flatness within a 0.4 mm zone RFS in Fig. 22.58. No point on the surface may vary more than 0.40 from the highest to the lowest point on the surface.

Straightness

A surface is straight if all of its elements are straight lines. A feature control symbol is used to specify straightness of a cylinder in Fig. 22.59. A total of 0.12 mm RFS is permitted as the elements are gauged in a vertical plane parallel to the axis of the cylinder.

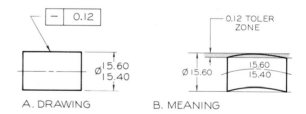

FIG. 22.59 Straightness is a tolerance of form that indicates the elements of a surface are straight lines. The symbol must be applied where the elements appear as straight lines.

Roundness

A surface of revolution (a cylinder, cone, or sphere) is round when all points on the surface intersected by a plane are equidistant from the axis. A feature control symbol is used to specify roundness of a cone and cylinder in Fig. 22.60. This symbol permits a tolerance of 0.34 mm RFS on the radius of each part. The roundness of a sphere is specified in Fig. 22.61.

Cylindricity

A surface of revolution is cylindrical when all of its elements form a cylinder. A cylindricity tolerance zone is specified in Fig. 22.62, where a tolerance of 0.54 mm RFS is permitted on the radius of the cylinder.

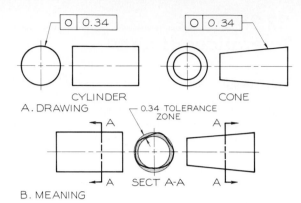

FIG. 22.60 Roundness is a tolerance of form that indicates a cross section through a surface of revolution is round and lies within two concentric circles.

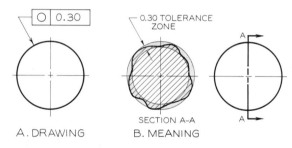

FIG 22.61 Roundness of a sphere is indicated in this manner, which means that any cross section through it is round within the specified tolerance in the symbol.

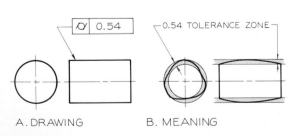

FIG. 22.62 Cylindricity is a tolerance of form that indicates the surface of a cylinder lies within an envelope formed by two concentric cylinders.

Cylindricity is a combination of tolerances of roundness and straightness.

22.27
Tolerances of profile

Profile tolerancing is used to specify tolerances about a contoured shape formed by arcs or irregular curves. Profile can apply to a surface or to a single line.

The surface in Fig. 22.63 is given a unilateral profile tolerance because it can only be smaller than the points located. Examples of specifying bilateral

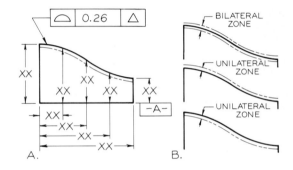

FIG. 22.63 Profile is a tolerance of form that is used to tolerance irregular curves of planes. **A.** The curving plane is located by coordinates. **B.** The tolerance is located by any of these methods.

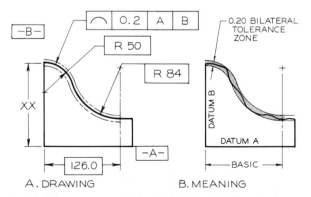

FIG. 22.64 Profile of a line is a tolerance of form that specifies the variation allowed from the path of a line. Here, the line is formed by tangent arcs. The tolerance zone may be either bilateral or unilateral, as shown in Fig. 22.63B.

and unilateral tolerance zones are shown in Fig. 22.63B.

A profile tolerance for a single line can be specified as shown in Fig. 22.64. In this example, the curve is formed by tangent arcs whose radii are given as basic dimensions. The radii are permitted to vary by plus or minus 0.10 mm about the basic radii.

22.28
Tolerances of orientation

Tolerances of orientation include *parallelism, perpendicularity,* and *angularity.*

Parallelism

A surface or a line is parallel when all of its points are equidistant from a datum plane or axis. Two types of parallelism are:

1. A tolerance zone between planes parallel to a datum plane within which the axis or surface of the feature must lie (Fig. 22.65). This tolerance also controls flatness.

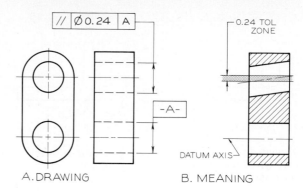

A. DRAWING   B. MEANING

FIG. 22.66 Parallelism of one centerline to another can be specified by using the diameter of one of the holes as the datum.

ply RFS when not specified. Specifying parallelism at MMC means that the axis of the cylindrical hole must vary no more than 0.20 mm when the holes are the smallest permissible size.

As the hole approaches its upper limit of 30.30, the tolerance zone increases until it reaches 0.50 DIA. Therefore a greater variation is given at MMC than at RFS.

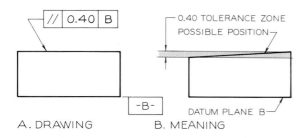

A. DRAWING   B. MEANING

FIG. 22.65 Parallelism is a tolerance of form that specifies a plane is parallel to another within specified limits. Plane *B* is the datum plane in this figure.

2. A cylindrical tolerance zone parallel to a datum feature within which the axis of a feature must lie (Fig. 22.66).

The effect of specifying parallelism at MMC can be seen in Fig. 22.67, where the modifier *M* is given in the feature control symbol. Tolerances of form ap-

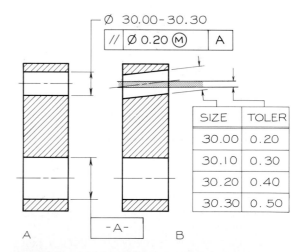

SIZE	TOLER
30.00	0.20
30.10	0.30
30.20	0.40
30.30	0.50

FIG. 22.67 The most critical tolerance will exist when features are at MMC. Here, the upper hole must be parallel to the hole used as datum *A* within 0.20 DIA. As the hole approaches its maximum size of 30.30 mm, the tolerance zone approaches 0.50 mm.

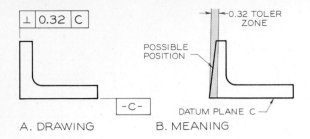

A. DRAWING   B. MEANING

FIG. 22.68 Perpendicularity is a tolerance form that gives a tolerance zone for a plane perpendicular to a specified datum plane.

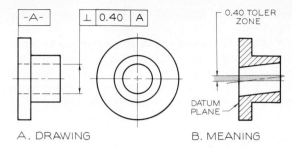

A. DRAWING   B. MEANING

FIG. 22.69 Perpendicularity can apply to the axis of a feature, such as the centerline of a cylinder.

Perpendicularity

The perpendicularity of two planes is specified in Fig. 22.68. Note that datum plane *C* is "flagged," and the feature control symbol is applied to the perpendicular surface. A hole is specified as perpendicular to a surface in Fig. 22.69, where surface *A* is indicated as the datum plane.

Angularity

A surface or line is angular when it is at a specified angle (other than 90°) from a datum or an axis. The angularity of a surface is specified in Fig. 22.70, where the angle is given a basic dimension of thirty degrees. The angle is permitted to vary within a tolerance zone of 0.25 mm about the angle.

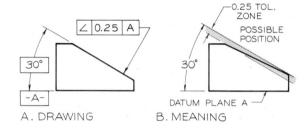

A. DRAWING   B. MEANING

FIG. 22.70 Angularity is a tolerance of form that specifies the tolerance for an angular surface with respect to a datum plane. The 30° angle is a true angle, a basic angle. The tolerance of 0.25 mm is applied to this basic angle.

22.29
Tolerances of runout

Runout tolerance is a means of controlling the functional relationship between one or more parts to a common datum axis. The features controlled by runout are surfaces of revolution about an axis and surfaces perpendicular to the axis.

The datum axis is established by using a functional cylindrical feature that rotates about the axis, such as diameter *B* in Fig. 22.71. When the part is rotated about this axis, the features of rotation must fall within the prescribed tolerance at *full indicator movement (FIM)*.

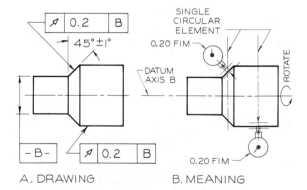

A. DRAWING   B. MEANING

FIG. 22.71 Runout tolerance of a surface, a composite of several tolerance-of-form characteristics, is used to specify concentric cylindrical parts. The part is mounted on the datum axis and is gauged as it is rotated about it.

The two types of runout are circular runout and total runout. One arrow in the feature control symbol indicates circular runout, and two arrows indicate total runout.

CIRCULAR RUNOUT (one arrow) is measured by rotating an object about its axis for 360° to determine whether a circular cross section at any point exceeds the permissible runout tolerance. This same technique is used to measure the amount of wobble existing in surfaces perpendicular to the axis of rotation.

TOTAL RUNOUT (two arrows) is used to specify cumulative variations of circularity, straightness, coaxiality, angularity, taper, and profile of a surface (Fig. 22.72). Total runout is applied to all circular and profile positions as the part is rotated 360°. When applied to surfaces perpendicular to the axis, total runout controls variations in perpendicularity and flatness.

The part shown in Fig. 22.73 is dimensioned by using a composite of several techniques of geometric tolerancing.

FIG. 22.72 The runout tolerance in this example is measured by mounting the object on the primary datum plane, surface C, and the secondary datum plane, cylinder D. The cylinder and conical surface is gauged to determine if it conforms to a tolerance zone of 0.03 mm. The end of the cone could have been noted to specify its runout (perpendicularity to the axis).

22.30
Surface texture

Because the surface texture of a part will affect its function, it must be precisely specified. The finish mark V does not elaborate on the finish desired. Most of the terms of surface texture defined below are illustrated in Fig. 22.74.

SURFACE TEXTURE is the variation in the surface, including roughness, waviness, lay, and flaws.

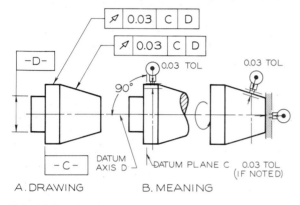

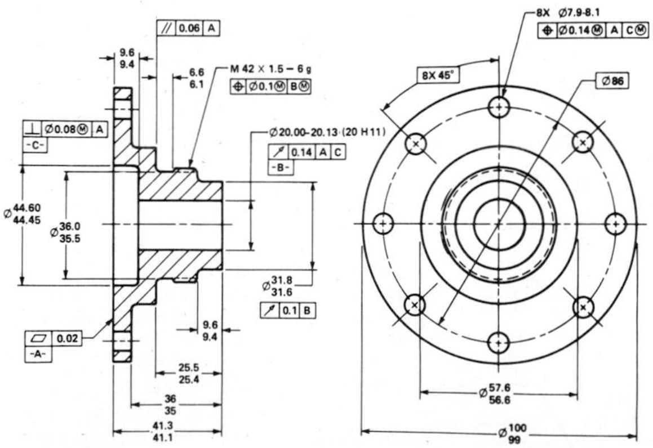

FIG. 22.73 A part dimensioned with a combination of notes and symbols to describe its geometric features. (Courtesy of ANSI Y14.5M–1982.)

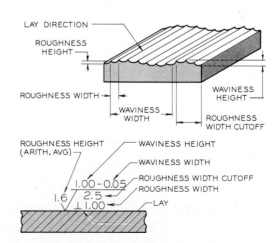

FIG. 22.74 Characteristics of surface texture.

A. BASIC SURFACE TEXTURE SYMBOL: Surface may be produced by any method except when the bar or circle (B or D) is specified.

B. MATERIAL REMOVAL BY MACHINING REQUIRED: The horizontal bar indicates that material removal by machining is required to produce the surface and that material must be provided for that purpose.

C. MATERIAL REMOVAL ALLOWANCE: The number indicates the amount of stock to be removed by machining in millimeters (or inches). Tolerances may be added to the basic value shown in a general note.

3.5

D. MATERIAL REMOVAL PROHIBITED: The circle in the vee indicates that the surface must be produced by processes such as casting, forging, hot finishing, cold finishing, die casting, powder metallurgy, or injection molding without subsequent removal of material.

E. SURFACE TEXTURE SYMBOL: To be used when any surface characteristics are specified above the horizontal line or to the right of the symbol. Surface may be produced by any method except when the bar or circle (B or D) is specified.

X = LETTER HEIGHT

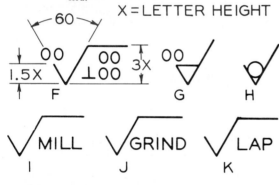

FIG. 22.75 Surface control symbols for specifying surface finish. General notes (I, J, and K) can be used to specify the recommended machining operation for finishing a surface.

ROUGHNESS describes the finest of the irregularities in the surface. These are usually caused by the manufacturing process used to smooth the surface.

ROUGHNESS HEIGHT is the average deviation from the mean plane of the surface. It is measured in microinches (μin) or micrometers (μm), which are millionths of an inch and of a meter, respectively.

ROUGHNESS WIDTH is the width between successive peaks and valleys that forms the roughness measured in microinches or micrometers.

ROUGHNESS WIDTH CUTOFF is the largest spacing of repetitive irregularities that includes average roughness height (measured in inches or millimeters). When not specified, a value of 0.8 mm (0.030 in.) is assumed.

WAVINESS is a widely spaced variation that exceeds the roughness width cutoff. Roughness may be regarded as superimposed on a wavy surface. Waviness is measured in inches or millimeters.

 1.6

Roughness average rating is placed at the left of the long leg. One rating indicates the maximum value in micrometers or microinches.

 1.6 0.8

The maximum and minimum roughness average values are specified in this manner. (Microinches or micrometers.)

0.005-5

Maximum waviness height rating is the first rating above the horizontal extension in millimeters or inches.

Maximum waviness spacing is the second rating above the horizontal extension and to the right of the waviness height rating. This is specified in millimeters or inches.

 1.6 3.5

The amount of stock provided for material removal is specified at the left of the short leg of the symbol in millimeters or inches.

 1.6

Removal of material is prohibited

 0.8

Lay is designated by the symbol at the right of the long leg. This means that the lay is perpendicular to this edge of the surface.

 0.8 2.5

Roughness sampling length or cutoff rating is placed below the horizontal extension. When no value is shown, 0.80 mm (0.030 inches) applies. Specify in mm or inches.

 0.8

Maximum roughness spacing shall be placed at the right of the lay symbol. Specify in mm or inches.

FIG. 22.76 Values can be added to surface control symbols for more precise specifications. These may be in combinations other than those shown here.

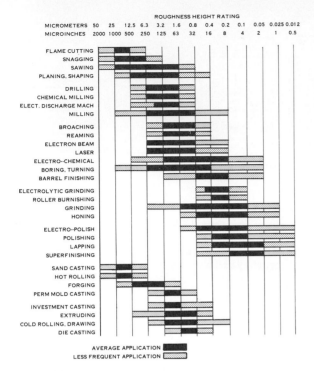

ROUGHNESS HEIGHT RATING

| MICROMETERS | 50 | 25 | 12.5 | 6.3 | 3.2 | 1.6 | 0.8 | 0.4 | 0.2 | 0.1 | 0.05 | 0.025 | 0.012 |
| MICROINCHES | 2000 | 1000 | 500 | 250 | 125 | 63 | 32 | 16 | 8 | 4 | 2 | 1 | 0.5 |

AVERAGE APPLICATION
LESS FREQUENT APPLICATION

FIG. 22.77 The surface roughness heights produced by various types of production methods are shown here in micrometers (microinches). (Courtesy of the General Motors Corp.)

The symbols used to specify surface texture are given in Fig. 22.75. The point of the V must touch the edge view of the surface, an extension line from the surface, or a leader pointing to the surface being specified.

In Fig. 22.76, values of surface texture that can be applied to surface texture symbols, individually or in combination, are given. The roughness height values are related to manufacturing processes used to finish the surface (Fig. 22.77).

Lay symbols that indicate the direction of texture of a surface (Fig. 22.78) can be incorporated into sur-

WAVINESS HEIGHT is the peak-to-valley distance between waves. It is measured in inches or millimeters.

WAVINESS WIDTH is the spacing between peaks or wave valleys measured in inches or millimeters.

LAY is the direction of the surface pattern and is determined by the production method used.

FLAWS are irregularities or defects that occur infrequently or at widely varying intervals on a surface. These include cracks, blow holes, checks, ridges, scratches, and the like. Unless otherwise specified, the effect of flaws is not included in roughness height measurements.

CONTACT AREA is the surface that will make contact with its mating surface.

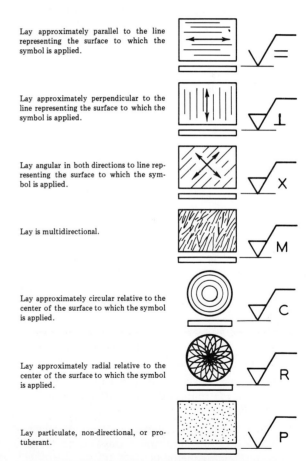

Lay approximately parallel to the line representing the surface to which the symbol is applied.

Lay approximately perpendicular to the line representing the surface to which the symbol is applied.

Lay angular in both directions to line representing the surface to which the symbol is applied.

Lay is multidirectional.

Lay approximately circular relative to the center of the surface to which the symbol is applied.

Lay approximately radial relative to the center of the surface to which the symbol is applied.

Lay particulate, non-directional, or protuberant.

FIG. 22.78 These symbols are used to indicate the direction of lay with respect to the surface where the control symbol is placed. (Courtesy of ANSI B46.1.)

face texture symbols as shown in Fig. 22.79. A part with a variety of surface texture symbols is shown in Fig. 22.80.

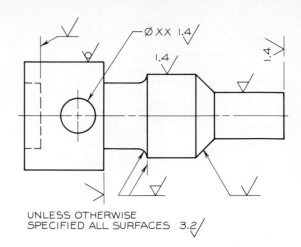

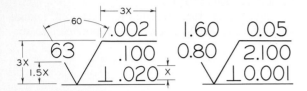

FIG. 22.79 Examples of fully specified surface control symbols.

FIG. 22.80 This drawing illustrates the techniques of applying surface texture symbols to a part.

Problems

These problems can be solved on Size A sheets. The problems are laid out on a grid of 0.20 inches (5 mm).

Cylindrical fits

1. (Fig. 22.81) Construct the drawing of a shaft and hole as shown (it need not be drawn to scale), give the limits for each diameter, and complete the table of values. Use a basic diameter of 1.00 in. (25 mm) and a class RC 1 fit, or a corresponding metric fit.

2. Same as Problem 1, but use a basic diameter of 1.75 in. (45 mm) and a class RC 9 fit, or a corresponding metric fit.

3. Same as Problem 1, but use a basic diameter of 2.00 in. (51 mm) and a class RC 5 fit, or a corresponding metric fit.

4. Same as Problem 1, but use a basic diameter of 12.00 in. (305 mm) and a class LC 11 fit, or a corresponding metric fit.

5. Same as Problem 1, but use a basic diameter of 3.00 in. (76 mm) and a class LC 1 fit, or a corresponding metric fit.

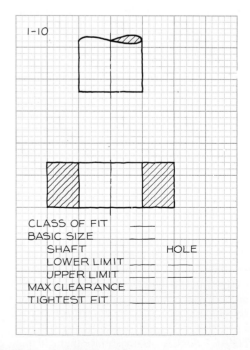

FIG. 22.81 Problems 1–10.

6. Same as Problem 1, but use a basic diameter of 8.00 in. (203 mm) and a class LC 1 fit, or a corresponding metric fit.

7. Same as Problem 1, but use a basic diameter of 102 in. (2591 mm) and a class LN 3 fit, or a corresponding metric fit.

8. Same as Problem 1, but use a basic diameter of 11.00 in. (279 mm) and a class LN 2 fit, or a corresponding metric fit.

9. Same as Problem 1, but use a basic diameter of 6.00 in. (152 mm) and a class FN 5 fit, or a corresponding metric fit.

10. Same as Problem 1, but use a basic diameter of 2.60 in. (66 mm) and a class FN 1 fit, or a corresponding metric fit.

Tolerances of position

11. (Fig. 22.82) On Size A paper, make an instrument drawing of the part shown. Locate the two holes

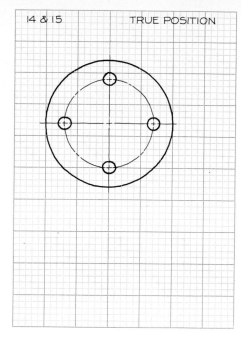

FIG. 22.83 Problems 14 and 15.

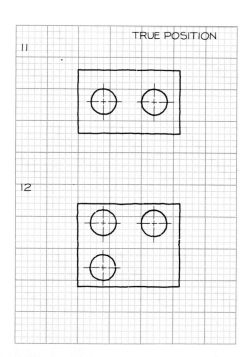

FIG. 22.82 Problems 11–13.

with a size tolerance of 1.00 mm and a true-position tolerance of 0.50 DIA. Show the proper symbols and dimensions for this arrangement.

12. Same as Problem 11, except locate three holes using the same tolerances for size and position.

13. Give the specifications for a two-pin gauge that can be used to gauge the correctness of the two holes specified in Problem 11. Make a sketch of the gauge and show the proper dimensions on it.

14. (Fig. 22.83) Using true positioning, locate the holes and properly note them to provide a size tolerance of 1.50 mm and a locational tolerance of 0.60 DIA.

15. Same as Problem 14, except locate six equally spaced, equally sized holes using the same tolerances of position.

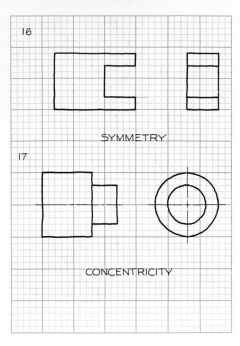

FIG. 22.84 Problems 16 and 17.

16. (Fig. 22.84) Using a feature control symbol and the necessary dimensions, indicate that the notch is symmetrical to the left-hand end of the part within 0.60 mm.

17. (Fig. 22.84) Using a feature control symbol and the necessary dimensions, indicate that the small cylinder is concentric with the large one (the datum cylinder) within a tolerance of 0.80.

18. (Fig. 22.85) Using a feature control symbol and the necessary dimensions, indicate that the elements of the cylinder are straight within a tolerance of 0.20 mm.

19. (Fig. 22.85) Using a feature control symbol and the necessary dimensions, indicate the upper surface of the object is flat within a tolerance of 0.08 mm.

20–22. (Fig. 22.86) Using feature control symbols and the necessary dimensions, indicate that the cross sections of the cylinder, cone, and sphere are round within a tolerance of 0.40 mm.

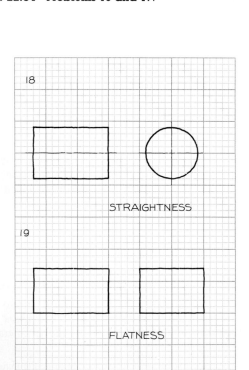

FIG. 22.85 Problems 18 and 19.

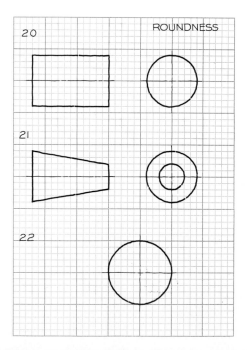

FIG. 22.86 Problems 20–22.

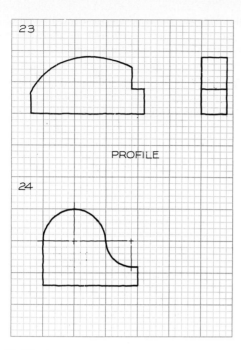

FIG. 22.87 Problems 23 and 24.

23. (Fig. 22.87) Using a feature control symbol and the necessary dimensions, indicate that the profile of the irregular surface of the object lies within a bilateral or unilateral tolerance zone of 0.40 mm.

24. (Fig. 22.87) Using a feature control symbol and the necessary dimensions, indicate that the profile of the line formed by tangent arcs lies within a bilateral or unilateral tolerance zone of 0.40 mm.

25. (Fig. 22.88) Using a feature control symbol and the necessary dimensions, indicate that the cylindricity of the cylinder is 0.90 mm.

26. (Fig. 22.88) Using a feature control symbol and the necessary dimensions, indicate that the angularity tolerance of the inclined plane is 0.7 mm from the bottom of the object, the datum plane.

27. (Fig. 22.89) Using a feature control symbol and the necessary dimensions, indicate that the upper surface of the object is parallel to the lower surface, the datum, within 0.30 mm.

28. (Fig. 22.89) Using a feature control symbol and the necessary dimensions, indicate that the small hole is parallel to the large hole, the datum, within a tolerance of 0.80 mm.

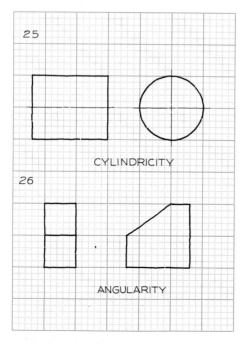

FIG. 22.88 Problems 25 and 26.

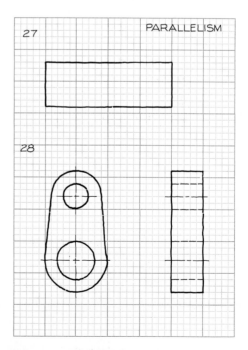

FIG. 22.89 Problems 27 and 28.

29. (Fig. 22.90) Using a feature control symbol and the necessary dimensions, indicate that the vertical surface is perpendicular to the bottom of the object, the datum, within a tolerance of 0.20 mm.

30. (Fig. 22.90) Using a feature control symbol and the necessary dimensions, indicate that the hole is perpendicular to datum *A* within a tolerance of 0.08 mm.

31. (Fig. 22.91) Using a feature control symbol and cylinder *A* as the datum, indicate that the conical feature has a runout of 0.80 mm.

32. (Fig. 22.91) Using a feature control symbol with cylinder *B* as the primary datum and surface *C* as the secondary datum, indicate that surfaces *D, E,* and *F* have a runout of 0.60 mm.

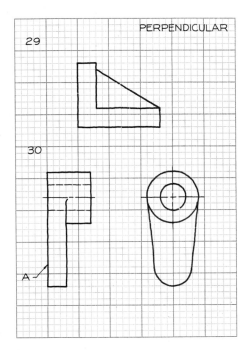

FIG. 22.90 Problems 29 and 30.

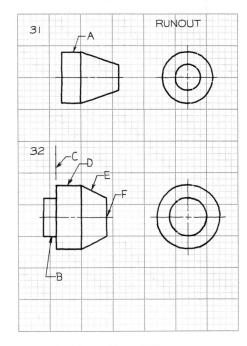

FIG. 22.91 Problems 31 and 32.

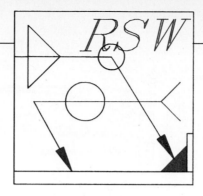

CHAPTER 23

Welding

23.1

Introduction

Welding is the process of permanently joining metal by heating a joint to a suitable temperature with or without the application of pressure and with or without the use of filler material.

The welding practices presented in this chapter are in compliance with the standards developed by the American Welding Society and the American National Standards Institute (ANSI). Reference is also made to the drafting standards used by General Motors Corporation.

Advantages of welding over other methods of fastening include (1) simplified fabrication, (2) economy, (3) increased strength and rigidity, (4) ease of repair, (5) creation of gas- and liquid-tight joints, and (6) reduction in weight and size.

Various types of welding processes are given in Fig. 23.1 but the four main types are *gas welding, arc welding, flash welding,* and *resistance welding.*

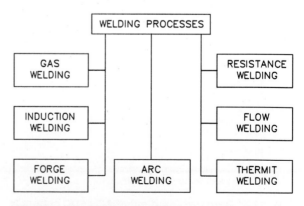

FIG. 23.1 The four major types of welding processes are gas welding, arc welding, flash welding, and resistance welding.

423

GAS WELDING PRINCIPLE

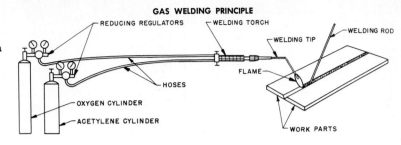

FIG. 23.2 The gas welding process burns gases like oxygen and acetylene in a torch to apply heat to a joint. The welding rod supplies the filler material. (Courtesy of General Motors Corp.)

GAS WELDING is a process in which gas flames are used to melt and fuse metal joints. Gases like acetylene or hydrogen are mixed in a welding torch and burned with air or oxygen (Fig. 23.2). The oxyacetylene method is widely used for repair work and field construction.

Most oxyacetylene welding is done manually with a minimum of equipment. Filler material in the form of welding rods is used to deposit metal at the joint as it is heated. Most metals, except for low- and medium-carbon steels, require fluxes to aid the process of melting and fusing the metals.

ARC WELDING is a process that uses an electric arc to heat and fuse the joints (Fig. 23.3). Pressure is sometimes required in addition to heat. The filler material is supplied by a consumable or a nonconsumable electrode through which the electric arc is transmitted. Metals well-suited to arc welding are wrought iron, low- and medium-carbon steels, stainless steel, copper, brass, bronze, aluminum, and some nickel alloys.

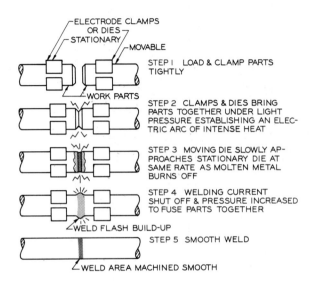

STEP 1 LOAD & CLAMP PARTS TIGHTLY

STEP 2 CLAMPS & DIES BRING PARTS TOGETHER UNDER LIGHT PRESSURE ESTABLISHING AN ELECTRIC ARC OF INTENSE HEAT

STEP 3 MOVING DIE SLOWLY APPROACHES STATIONARY DIE AT SAME RATE AS MOLTEN METAL BURNS OFF

STEP 4 WELDING CURRENT SHUT OFF & PRESSURE INCREASED TO FUSE PARTS TOGETHER

STEP 5 SMOOTH WELD

FIG. 23.4 Flash welding, a type of arc welding, uses a combination of electric current and pressure to fuse two parts.

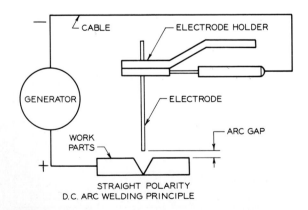

STRAIGHT POLARITY
D.C. ARC WELDING PRINCIPLE

FIG. 23.3 In arc welding either AC or DC current is passed through an electrode to heat the joint.

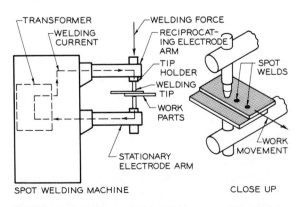

SPOT WELDING MACHINE   CLOSE UP

FIG. 23.5 Resistance spot welding can be used to join lap and butt joints.

FLASH WELDING is a form of arc welding, but it is similar to resistance welding in that both pressure and an electric current are used to join two pieces (Fig. 23.4). The two pieces are brought together, and an electric current causes heat to build up between them. As the metal burns, the current is turned off and the pressure between the pieces is increased to fuse them together.

RESISTANCE WELDING is a group of processes where metals are fused together both by the heat produced from the resistance of the parts to the flow of an electric current and by the pressure applied. Fluxes and filler materials are normally not used. All resistance welds are either lap- or butt-type welds.

Fig. 23.5 illustrates how resistance spot welding is performed on a lap joint. The two parts are lapped and pressed together, and an electric current fuses the parts where they join. A series of spots spaced at intervals, called *spot welds,* are used to secure the parts. Table 23.1 suggests the welding processes that can be used for different materials.

23.2
Weld joints

Five standard weld joints are illustrated in Fig. 23.6. The *butt joint* can be joined with the following types of welds: square groove, V-groove, bevel groove, U-groove, and J-groove. The *corner joint* can be joined with these welds and with the fillet weld. The *lap joint* can be joined with the bevel groove, J-groove, fillet, slot, plug, spot, projection, and seam welds. The *edge joint* uses the same welds as the lap joint along with

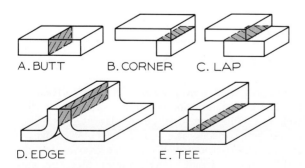

FIG. 23.6 The standard weld joints.

TABLE 23.1
RECOMMENDED RESISTANCE WELDING PROCESSES

Material	Spot Welding	Flash Welding
Low-carbon mild steel		
SAE 1010	R	R
SAE 1020	R	R
Medium-carbon steel		
SAE 1030	R	R
SAE 1050	R	R
Wrought alloy steel		
SAE 4130	R	R
SAE 4340	R	R
High-alloy austenitic stainless steel		
SAE 30301–30302	R	R
SAE 30309–30316	R	R
Ferritic and martensitic stainless steel		
SAE 51410–51430	S	S
Wrought heat-resisting alloys*		
19–9–DL	S	S
16–25–6	S	S
Cast iron	NA	NR
Gray iron	NA	NR
Aluminum and aluminum alloys	R	S
Nickel and nickel alloys	R	S

S—Satisfactory   NA—Not applicable
R—Recommended   NR—Not recommended
*For composition see *American Society of Metals Handbook.*
Source: Courtesy of General Motors Corp.

the square groove, V-groove, U-groove, and seam welds. The *tee joint* can be joined by the bevel groove, J-groove, and fillet welds.

23.3
Welding symbols

The specification of welds on a working drawing is done by symbols. If a drawing has a general note such as ALL JOINTS WELDED or WELDED THROUGHOUT, the designer has transferred the design responsibility to the welder. Welding is too important to be left to chance; it must be thoroughly specified.

A welding symbol is used to provide specifications on a drawing (Fig. 23.7). This example gives the sym-

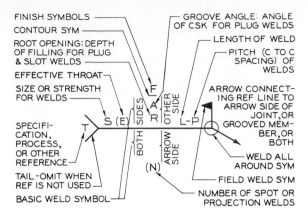

FIG. 23.7 The welding symbol. Usually it is modified to a simpler form.

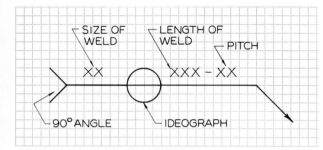

FIG. 23.8 When the grid is drawn full size ($\frac{1}{8}$" or 3 mm), the size of the welding symbol can be determined.

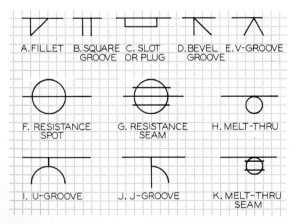

FIG. 23.9 The sizes of the ideographs are shown on the $\frac{1}{8}$" (3 mm) grid. These sizes are proportionately equal to the size of the welding symbol shown in Fig. 23.8.

bol in its complete form, which is seldom needed. The symbol is usually modified to a simpler form when not all the specifications are necessary.

The scale of the welding symbol is shown in Fig. 23.8, where it is drawn on a $\frac{1}{8}$ in. (3 mm) grid. Its size can be scaled using these same proportions. The lettering used is the standard height of $\frac{1}{8}$ in. or 3 mm.

The *ideograph* is the symbol that denotes the type of weld desired. Generally, the ideograph depicts the cross section of the weld.

Figure 23.9 illustrates the most often used ideographs. They are drawn to scale on the $\frac{1}{8}$ in. (3 mm) grid to represent their full size when added to the welding symbol.

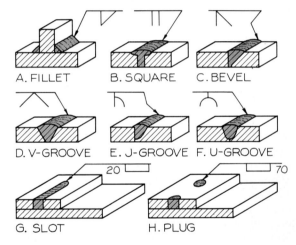

FIG. 23.10 The standard welds and their corresponding ideographs.

23.4
Types of welds

Commonly used welds, along with their corresponding ideographs, are shown in Fig. 23.10. The fillet weld is a built-up weld at the angular intersection between two surfaces. The square, bevel, V-groove, J-groove, and U-groove welds all have grooves, and the weld is placed inside these grooves. Slot and plug welds have intermittent holes or openings where the

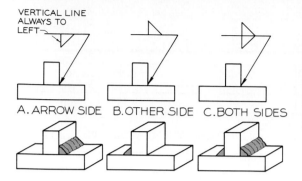

VERTICAL LINE ALWAYS TO LEFT

A. ARROW SIDE   B. OTHER SIDE   C. BOTH SIDES

FIG. 23.11 Fillet welds are indicated by abbreviated symbols. When the ideograph is below the horizontal line, it refers to the arrow side; when it is above the line, it refers to the other side.

parts are welded. Holes are unnecessary when resistance welding is used.

23.5
Application of symbols

In Fig. 23.11A, the fillet ideograph is placed below the horizontal line of the symbol, indicating that the weld is to be at the joint on the *arrow side,* the side of the arrow.

> The vertical leg of the ideograph is *always* on the left side.

Placing the ideograph above the horizontal line indicates that the weld is to be on the *other side*—that is, the joint on the other side of the part away from the arrow. When the part is to be welded on both sides, the ideograph shown in Fig. 23.11C is used. It is permissible to omit the tail and other specifications from the symbol when detailed specifications are given elsewhere.

A single arrow is often used to specify a weld that is to be all around two joining parts (Fig. 23.12); a circle, 6 mm in diameter, placed at the bend in the leader of the symbol denotes this. If the welding is to be done in the field (on the site rather than in the shop), a black circle, 3 mm in diameter, can be used to denote this joint.

A fillet weld that is to be the full length of the two parts may be specified as shown in Fig. 23.13.

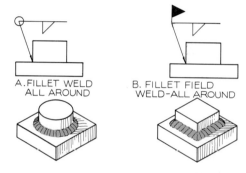

A. FILLET WELD ALL AROUND   B. FILLET FIELD WELD-ALL AROUND

FIG. 23.12 Symbols for indicating fillet welds all around two types of parts.

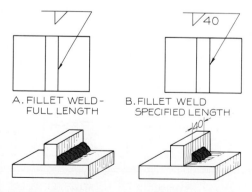

A. FILLET WELD- FULL LENGTH   B. FILLET WELD SPECIFIED LENGTH

FIG. 23.13 **A.** Symbol for indicating full-length fillet welds. **B.** Symbol for indicating fillet welds less than full length.

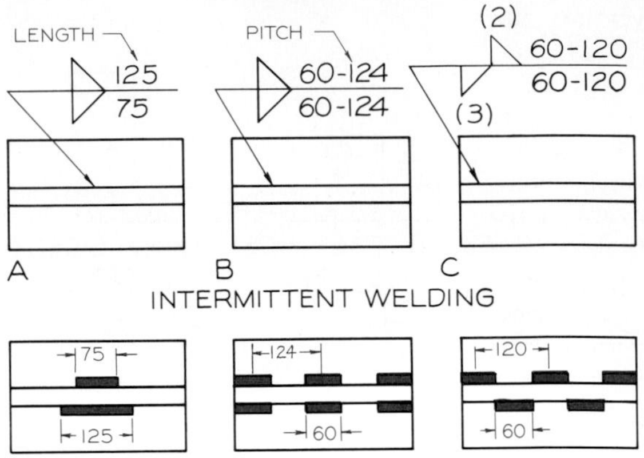

FIG. 23.14 Symbols for specifying varying and intermittent welds.

INTERMITTENT WELDING

Since the ideograph is on the lower side of the horizontal line, the weld will be on the arrow side. A fillet weld that is to be less than full length may be specified as shown in Fig. 23.13B, where 40 represents the weld's length in millimeters.

Fillet welds of different lengths and positioned on both sides may be specified as shown in Fig. 23.14A. The dimension on the lower side of the horizontal gives the length of the weld on the arrow side, and the dimension on the upper side gives the length on the other side.

Intermittent welds are welds of a given length that are spaced uniformly apart from center to center by a distance called the *pitch*. In Fig. 23.14B, the welds

are on both sides, are 60 mm long, and have pitches of 124 mm; this can be indicated by the symbol shown. Intermittent welds that are to be staggered to alternate positions on both sides can be specified by the symbol shown in Fig. 23.14C.

23.6
Groove welds

Figure 23.15 illustrates the more standard groove welds. When the depth of the grooves, the angle of the chamfer, and the root openings are not given on a

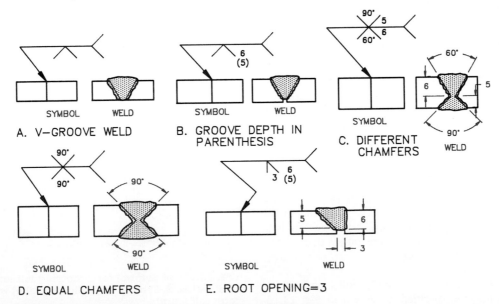

A. V-GROOVE WELD   B. GROOVE DEPTH IN PARENTHESIS   C. DIFFERENT CHAMFERS

D. EQUAL CHAMFERS   E. ROOT OPENING=3

FIG. 23.15 Types of groove welds and their general specifications.

symbol, they must be specified elsewhere on the drawings or in supporting documents. In Figs. 23.15B and E, the depth of the chamfer of the prepared joint is given in parentheses under the size dimension of the weld, which takes into account the penetration of the weld beyond this chamfer. If the size of the joint is equal to the depth of the prepared joint, only one number is given.

When the chamfer is different on each side of the joint, it can be noted with the symbol shown in Fig. 23.15C. If the spacing between the two parts, the *root opening,* is to be specified, this is done by placing its dimensions between the groove angle number and the weld ideograph as shown in Fig. 23.15E.

A bevel weld is a groove beveled from one of the parts being joined; therefore the symbol must indicate which part is to be beveled (Fig. 23.15E). To call attention to this operation, the leader from the symbol is bent and aimed toward the part to be beveled. This practice also applies to J-welds where one side is grooved and the other is not (Fig. 23.16).

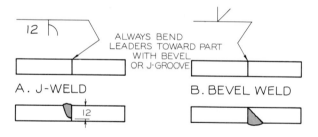

FIG. 23.16 J-welds and bevel welds are specified by bent arrows pointing to the side of the joint to be grooved.

TYPE	SYMBOL	EXAMPLE
FLUSH		
CONCAVE		
CONVEX		

23.7
Surface contoured welds

Contour symbols are used to indicate which of the three types of contours is desired on the surface of the weld: *flush, concave,* or *convex.* Flush contours are smooth with the surface or flat across the hypotenuse of a fillet weld. Concave contours bulge inward with a curve, and convex contours bulge outward with a curve (Fig. 23.17).

It is often necessary to finish the weld by a supplementary process to bring it to the desired contour. These processes, which may be indicated by their abbreviations, are chipping (C), grinding (G), hammering (H), machining (M), rolling (R), and peening (P). Examples of these are shown in Fig. 23.18.

23.8
Seam welds

A seam weld joints two lapping parts with either a continuous weld or a series of closely spaced spot welds. The process used for seam welds must be given by abbreviations placed in the tail of the weld symbol. The ideograph for a resistance weld is about 12 mm in diameter, and it is placed with the horizontal line of the symbol through its center. The weld's width, length, and pitch are indicated (Fig. 23.19).

When the seam weld is made by arc welding, the diameter of the ideograph is about 6 mm, and it is placed on the upper or lower side of the symbol's horizontal line to indicate whether the seam will be ap-

FIG. 23.17 The contour symbols used to specify the surface finish of a weld.

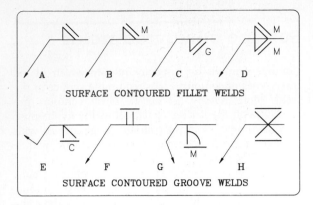

SURFACE CONTOURED FILLET WELDS

SURFACE CONTOURED GROOVE WELDS

FIG. 23.18 Examples of contoured symbols and letters of finishing applied to them.

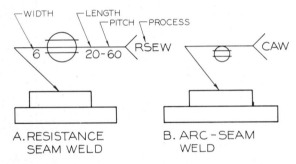

A. RESISTANCE SEAM WELD

B. ARC–SEAM WELD

FIG. 23.19 The process used for resistance seam welds and arc seam welds is indicated in the tail of the symbol. The arc weld must specify arrow side or other side in the symbol.

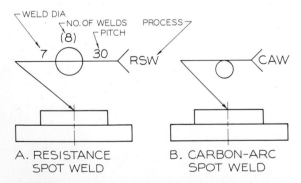

A. RESISTANCE SPOT WELD

B. CARBON-ARC SPOT WELD

FIG. 23.20 The process used for resistance spot welds and arc spot welds is indicated in the tail of the symbol. The arc weld must specify arrow side or other side in the symbol.

plied to the arrow side or the other side (Fig. 23.19B). When the length of the weld is omitted from the symbol, it is understood that the seam weld extends between abrupt changes in the seam or as it is dimensioned.

Spot welds are similarly specified with ideographs, and specifications are given by diameter, number of welds, and pitch between the welds. The process, resistance spot welding (RSW), is noted in the tail of the symbol (Fig. 23.20A). For arc welding, the arrow side or other side must be indicated by a symbol (Fig. 23.20B). Also note the abbreviation of the welding process. (See Table 23.2 for the abbreviations of various welding processes.)

23.9
Built-up welds

When the surface of a part is to be enlarged, or *built-up*, by welding, this can be indicated by a symbol (Fig. 23.21). The width of the built-up weld is dimensionsed in the view, and the height of the weld above the surface is specified in the symbol to the left of the ideograph. The radius of the circular segment is 6 mm.

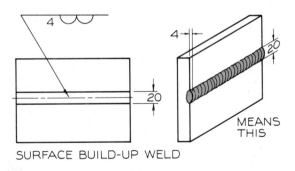

SURFACE BUILD-UP WELD

MEANS THIS

FIG. 23.21 The method of applying a symbol to a built-up weld on a surface.

23.10
Welding standards

Figure 23.22 (pages 432 and 433) gives an overview of the welding symbols and specifications that have been discussed in the previous paragraphs. The chart can be used as a reference for most general types of welding and their symbols.

TABLE 23.2
WELDING PROCESS SYMBOLS

CAW	Carbon arc welding	FRW	Friction welding	PGW	Pressure gas welding
CW	Cold welding	FW	Flash welding	RB	Resistance brazing
DB	Dip brazing	GMAW	Gas metal arc welding	RPW	Projection welding
DFW	Diffusion welding	GTAW	Gas tungsten welding	RSEW	Resistance seam welding
EBW	Electron beam welding	IB	Induction brazing	RSW	Resistance spot welding
ESW	Electroslag welding	IRB	Infrared brazing	RW	Resistance welding
EXW	Explosion welding	OAW	Oxyacetylene welding	TB	Torch brazing
FB	Furnace brazing	OHW	Oxyhydrogen welding	UW	Upset welding
FOW	Forge welding				

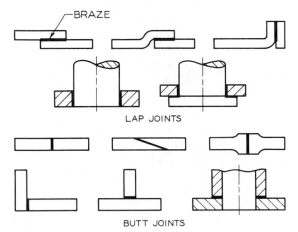

FIG. 23.23 Two basic types of brazing joints: lap joints and butt joints.

23.11
Brazing

Like welding, brazing is a method of joining pieces of metal. The process entails heating the joints above 800°F, and distributing by capillary action a nonferrous filler material, with a melting point below the base materials, between the closely fit parts.

Before brazing, the parts must be cleansed and the joints fluxed. The brazing filler is added before or just as the joints are heated beyond the filler's melting point. After the filler material has melted, it is allowed to flow between the parts to form the joint. As shown in Fig. 23.23, there are two basic brazing joints: lap joints and butt joints.

Brazing is used to hold parts together, to provide gas- and liquid-tight joints, to ensure electrical conductivity, and to aid in repair and salvage. Brazed joints will withstand more stress, higher temperature, and more vibration than will soft-soldered joints.

23.12
Soft soldering

Soldering is the process of joining two metal parts with a third metal that melts below the temperature of the metals being joined. Solders are alloys of nonferrous metals that melt below 800°F. Widely used in the automotive and electrical industries, soldering is one of the basic techniques of welding and is often done by hand with a soldering iron like the one shown in Fig. 23.24. The iron is placed on the joint to heat it and to melt the solder. The method of indicating a soldered joint is shown in Fig. 23.24.

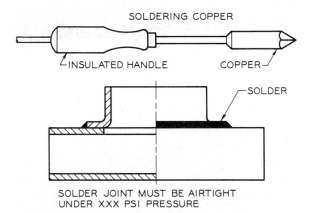

FIG. 23.24 A typical hand-held soldering iron used to soft-solder two parts together, and the method of indicating a soldered joint on a drawing.

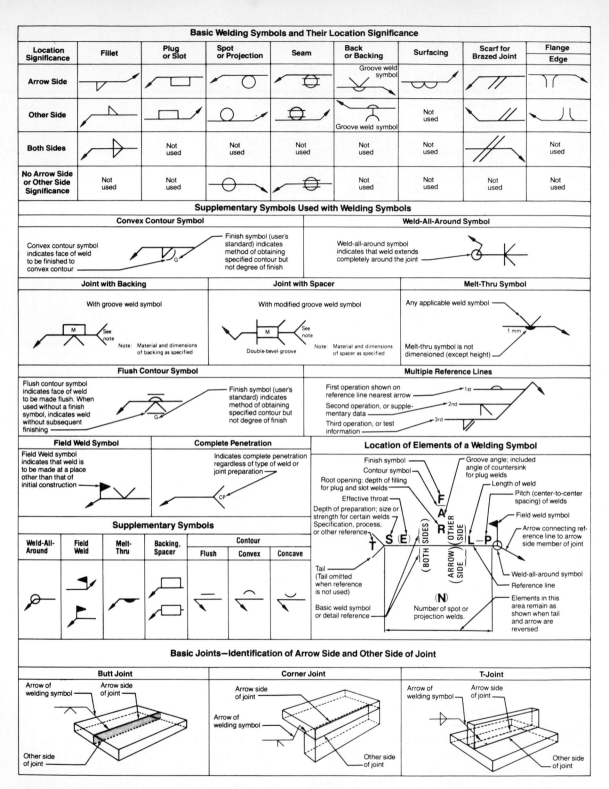

FIG. 23.22 The American Welding Society Standard Welding Symbols. (Courtesy of the American Welding Society of Miami, FL.)

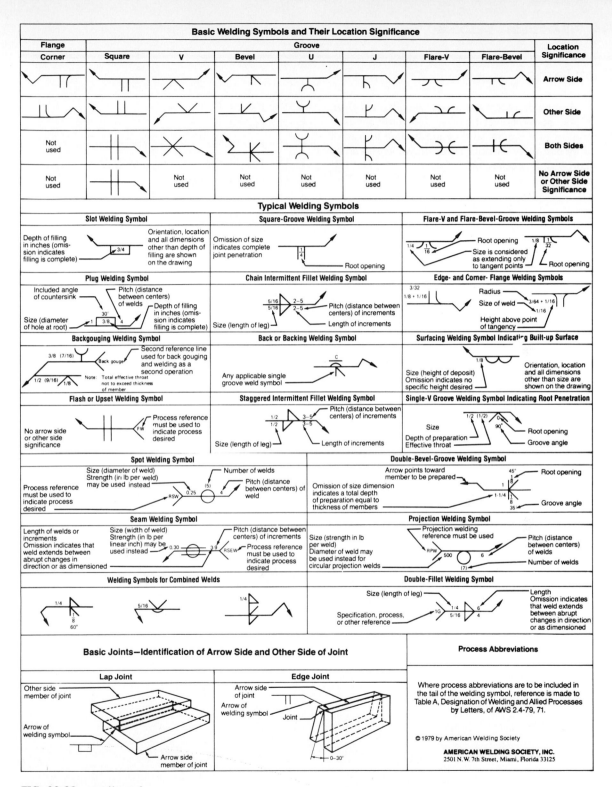

FIG. 23.22 continued

Problems

Make working drawings of the following figures given in Chapter 24. Wherever feasible, change the joints of the features of the parts so they are welded instead of joined by one-piece casting. These drawings should be made on Size B sheets.

1. Use Fig. 24.36.

2. Use Fig. 24.40.

3. Use Fig. 24.45.

4. Use Fig. 24.47.

5. Use Fig. 24.51.

6. Use Fig. 24.53.

7. Use Fig. 24.67.

8. Use Fig. 24.78, Part 1.

9. Use Fig. 24.84, Part 2.

10. Use Fig. 24.84, Part 1.

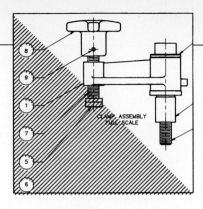

CHAPTER 24

Working Drawings

24.1

Introduction

Working drawings are drawings from which a design is constructed. Depending on the complexity of the project, a set of working drawings may contain any number of sheets, from only one to more than one hundred. It is important to give the number of sheets in the set on each sheet; for example, sheet 2 of 6, sheet 3 of 6, and so on.

The written instructions that accompany working drawings are called *specifications*. When the project can be represented on several sheets, the specifications are often written on the drawings to consolidate the information into a single format. Much of the work in preparing working drawings is done by the drafter, but the designer, who is most often an engineer, is responsible for their correctness.

A working drawing is often called a *detail drawing* because it describes and gives the dimensions of the details of the parts being presented.

All the principles of orthographic projection and all the techniques of graphical presentation are used to communicate the details in a working drawing.

24.2

Working drawings— inch system

The inch is the basic unit of the English system, which is slowly being superseded by the metric system. An example of a computer-drawn working drawing is

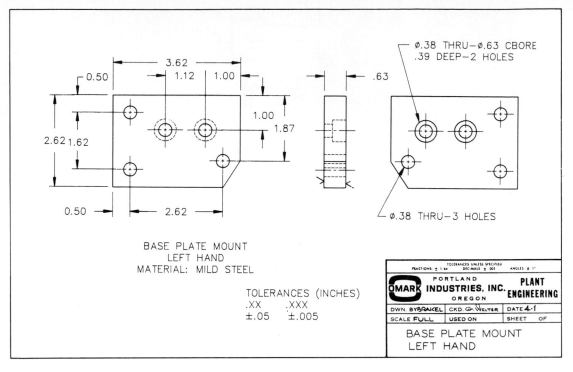

FIG. 24.1 A computer-drawn working drawing of a single part dimensioned in inches.

FIG. 24.2 A revolving clamp assembly manufactured to hold parts while they are being machined.

shown in Fig. 24.1, where the base-plate mount is detailed in three orthographic views. Dimensions and notes give the information necessary to construct the piece.

This particular drawing is dimensioned with decimal inches, which are preferable to common fractions, although both systems are still in widespread use. Using decimal inches makes it possible to handle arithmetic with greater ease than is possible with fractions. Inch marks are omitted from dimensions on a working drawing since it is understood that the units are in inches.

An example of working drawings dimensioned with decimal fractions using the inch as the unit of measurement is shown in Fig. 24.2. The computer-drawn detail drawings of the parts of the assembly are shown in Figs. 24.3–24.5. Several dimensioned parts are shown on each sheet as orthographic views. The arrangement of these parts on the sheet has no relationship to how they fit together; they are simply positioned to take advantage of the available space. Each part is given a number and a name for identification. The material that each part is made of is indicated

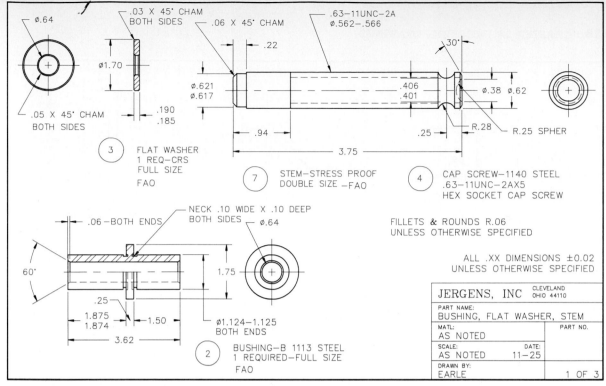

FIG. 24.3 A set of three computer-drawn detail drawings showing parts of the clamp assembly. (Courtesy of Jergens, Inc.)

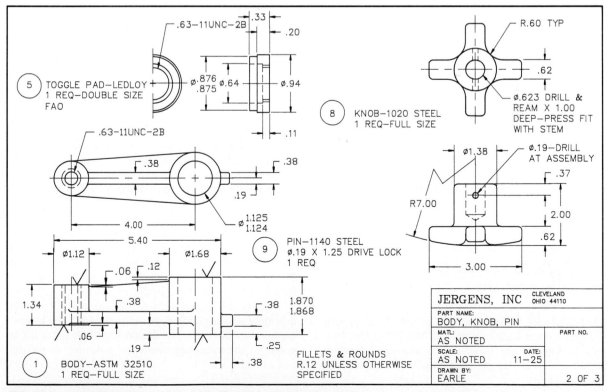

FIG. 24.4 A continuation of Fig. 24.3. (Courtesy of Jergens, Inc.)

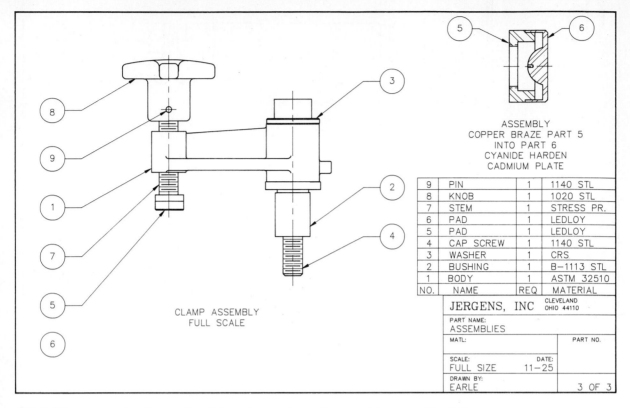

FIG. 24.5 A computer-drawn detail drawing and an orthographic assembly of the clamp assembly.

along with other notes to explain any necessary manufacturing procedures.

The orthographic assembly given on sheet 3 (Fig. 24.5) explains how the parts fit together. The parts are numbered to correspond to the part numbers in the parts list, which serves as a bill of materials.

24.3
Working drawings—metric system

The millimeter is the basic unit of the metric system. A single part, a back tool post, is detailed in Fig. 24.6. All dimensions are measured to the nearest whole millimeter, with no decimal fractions except where

tolerances are shown. Metric abbreviations are omitted from dimensions on a working drawing since it is understood from the title block that the units are in millimeters. Remember that the fingernail of your index finger is about 10 millimeters wide. A single part is dimensioned in Fig. 24.7 by using the metric system.

The lifting device shown in Fig. 24.8 is detailed in working drawings shown in Figs. 24.9 and 24.10. All dimensions are given in millimeters, indicated by the SI symbol near the title block. Also, note the symbol for the third angle of projection.

In Fig. 24.10, the parts are shown in a full-section assembly that explains how the parts are to be assembled. Assemblies are not dimensioned. A parts list is given on the same page with the assembly above the title block.

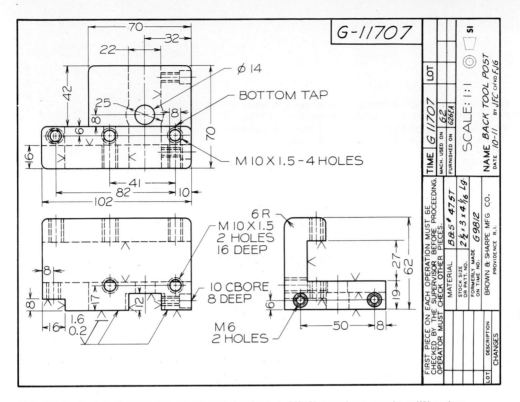

FIG. 24.6 A detail drawing of a back tool post. All dimensions are in millimeters.

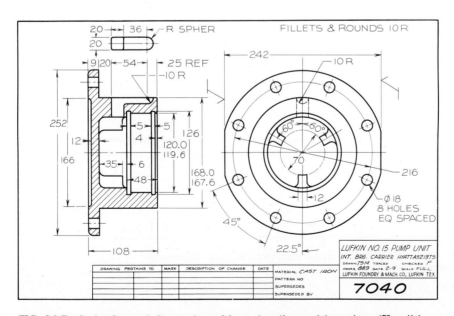

FIG. 24.7 A single part dimensioned by using the metric system (SI units).

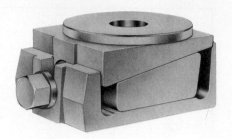

FIG. 24.8 A Lev-L-ine lifting device used to level heavy machinery. This product is the basis of the working drawings in Figs. 24.9 and 24.10. (Courtesy of Unisorb Machinery Installation Systems.)

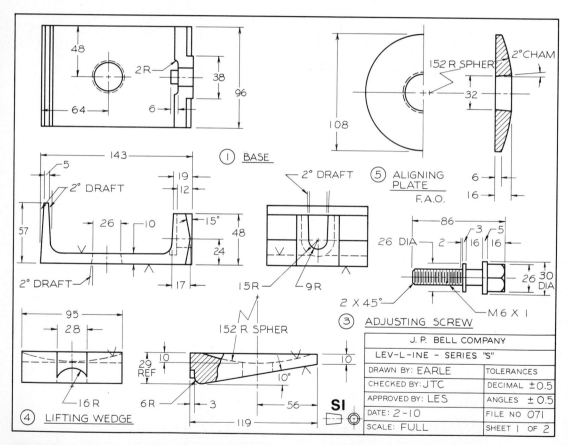

FIG. 24.9 A working drawing of the lifting device dimensioned in SI units. (Courtesy of Unisorb Machinery Installation Systems.)

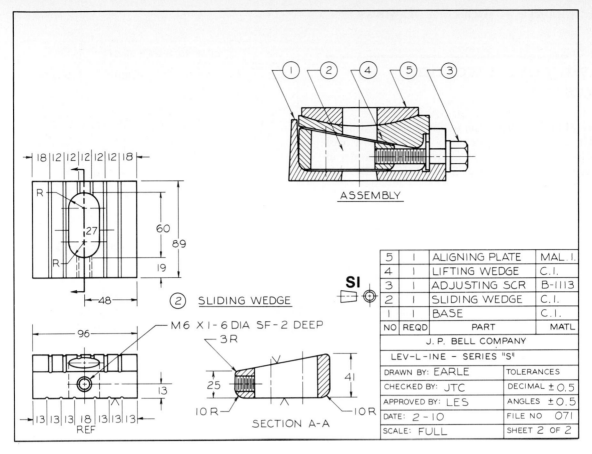

FIG. 24.10 A working drawing and assembly of the lifting device dimensioned in SI units. (Courtesy of Unisorb Machinery Installation Systems.)

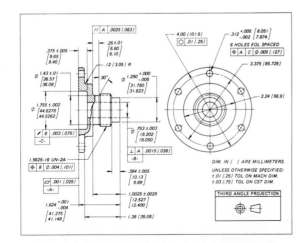

FIG. 24.11 A dual-dimensioned drawing. The dimensions are given in millimeters, with their equivalents in inches given in parentheses. (Courtesy of General Motors Corp.)

24.4
Working drawings— dual dimensions

Some working drawings are dimensioned in both inches and millimeters. An example is shown in Fig. 24.11, where the dimensions in parentheses are millimeters. The units may also be given in inches first and then converted to millimeters. Converting from one unit to the other will result in fractional units that must be rounded off. An explanation of the system used should be noted in the title block.

24.5
Laying out a working drawing

The working drawing is laid out by beginning with the border, if a printed border is not provided. At least 0.25 in. (7 mm) should be allowed for the border at the edge of the sheet (Fig. 24.12, Step 1). The title

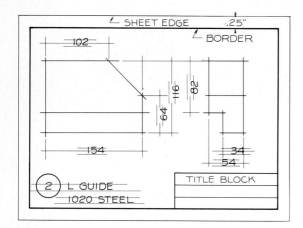

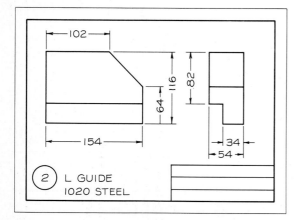

FIG. 24.12 Laying out a detail drawing.

Step 1 The border and title block are drawn on a preliminary sheet that will be traced. The views are positioned to allow adequate room for their dimensions. Guidelines are constructed for all dimensions and notes.

Step 2 The layout is overlaid with vellum or film. Then the lines are drawn to their proper weights, the dimensions and notes are lettered, and the title block is completed.

LETTER SIZE	DRAWING SHEET SIZES		
	X WIDTH	Y HEIGHT	Z MARGIN
A (HORIZ)	8.5	11	.38
A (VERT)	11	8.5	.38
B	11	17	.50
C	17	22	.50
D	22	34	.50
E	34	44	.50
F	28	40	.50

FIG. 24.13 The standard sheet sizes for working drawings dimensioned in inches.

block is drawn to size at the lower right corner of the sheet, and views and dimensions are drawn lightly to take advantage of the available space. Properly positioning the views to provide adequate space is one of the major concerns of the drafter in laying out a drawing.

When using tracing paper or film, it is more efficient to lay out the views and dimensions on a different sheet of paper and then overlay this drawing with vellum or film for tracing the final working drawing. Guidelines for lettering are drawn for each dimension line. The lines are darkened to their proper weight, and the drawing is completed as shown in Step 2.

The standard sheet sizes of working drawings are shown in Fig. 24.13. Papers, films, cloths, and reproduction materials are available in these modular sizes.

	5" APPROXIMATELY			
2	SHAFT	2	1020 STL
1	BASE	1	CI
NO	PART NAME	REQ	MATERIAL

.125" LETTERS

TITLE	
BY:	SECT:
DATE:	SHEET:
SCALE:	OF SHEETS

FIG. 24.14 A typical parts list and title block suitable for most student assignments.

24.6
Title blocks and parts lists

A parts list and title block suitable for student assignments are shown in Fig. 24.14. Title blocks are usually located in the lower right corner of the drawing sheet against the borders. Title blocks usually contain the title or part name, drafter, date, scale, company, and sheet number. Other information such as tolerances, checkers, and materials may be shown. The parts list should be placed directly over and in contact with the title block.

A title block used by General Motors is shown in Fig. 24.15. A note to the left of the title block lists John F. Brown as the inventor. Two associates were

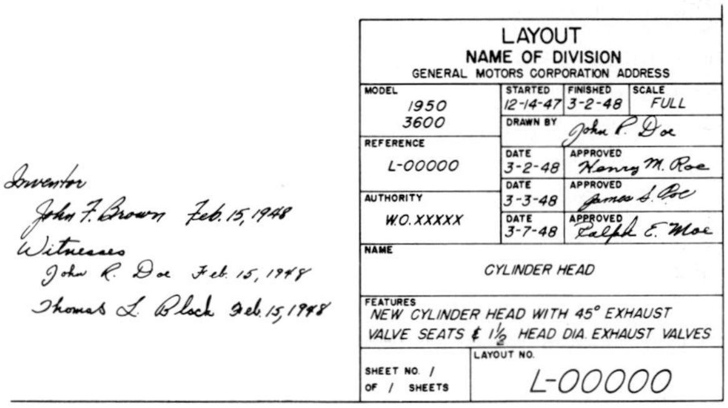

FIG. 24.15 A title block typical of those used in industry. (Courtesy of General Motors Corp.)

asked to date and sign the drawings as witnesses of his work. This procedure establishes the ownership of the ideas and dates of their development in case this becomes an issue in obtaining a patent.

Another example of a title block, shown in Fig. 24.16, is typical of those used by various industries. Revision blocks list any modifications that will improve the design.

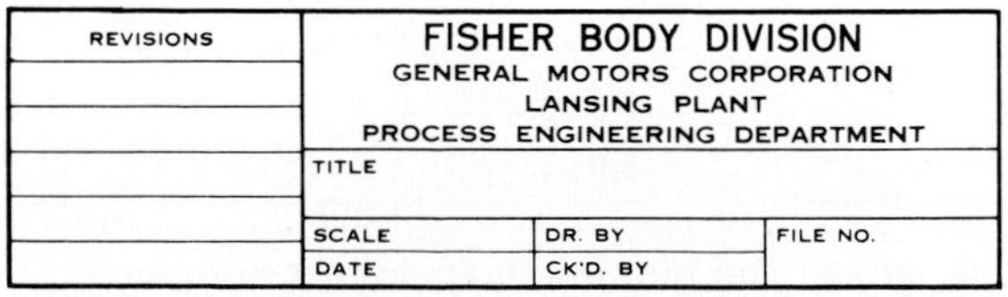

FIG. 24.16 An example of a title block with a revision block. (Courtesy of General Motors Corp.)

24.7
Scale specification

If all drawings are the same scale, the scale of a working drawing should be indicated in or near the title block. If several drawings are made at different scales, the scales should be indicated on the drawings.

Several methods of indicating scales are shown in Fig. 24.17. When the colon is used (e.g., 1:2) the metric system is implied; whereas when the equal sign is used (e.g., 1 = 2), the English system is implied. The SI symbol or the METRIC designation on a drawing specifies that the units of measurement are millimeters.

In some cases, a graphical scale with calibrations is given that permits the interpretation of linear units by transferring dimensions, with your dividers from the drawing to the scale.

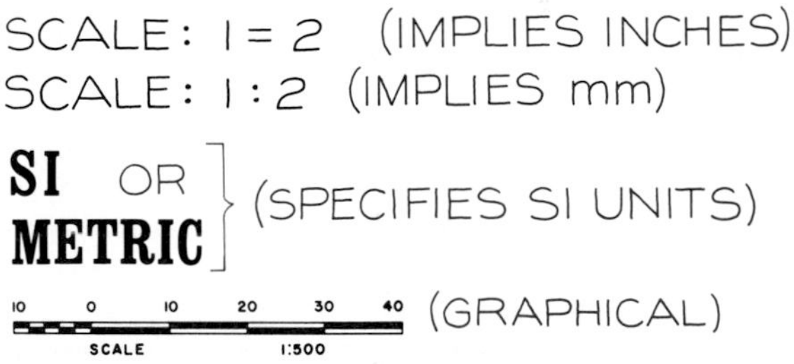

FIG. 24.17 Methods of specifying scales and metric units on working drawings.

24.8
Tolerances

General notes can be given on working drawings to specify the tolerances of dimensions. A table of values is shown in Fig. 24.18, where a check can be made to indicate whether the units will be in inches or in millimeters. As the figure shows, plus-or-minus tolerances are given in the blanks, under the number of digits, and under each decimal fraction. For example, this table specifies that each dimension with two-place decimals will have a tolerance of ± 0.01 in.

Angular tolerances can be given in general notes also. Refer to Chapter 22 for more detailed examples of using tolerance notes.

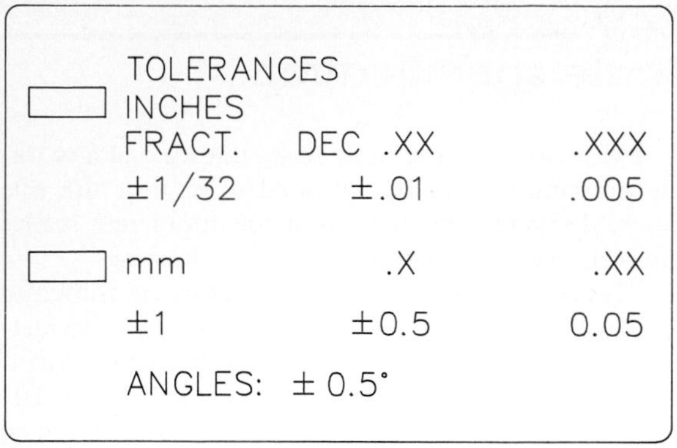

TOLERANCES
INCHES
FRACT.   DEC .XX   .XXX
±1/32   ±.01   .005

mm   .X   .XX
±1   ±0.5   0.05

ANGLES: ± 0.5°

FIG. 24.18 General tolerance notes given on working drawings to specify the tolerances permitted on dimensions.

NAME OF PART
PLACE NEAR PART
BALLOON 4 X LETTER HEIGHT

FIG. 24.19 Each part of a working drawing should be named and numbered for listing in the parts list.

24.9
Part names and numbers

Each part should be given a name and a number (Fig. 24.19). The letters and numbers should be ⅛ in. (3 mm) high. The part numbers are placed inside circles, called *balloons,* which are drawn approximately four times the height of the numbers.

On the working drawings, the part numbers should be placed near the parts to clarify which part they are associated with. On assembly drawings, balloons are especially important since the numbers of the parts refer to the same numbers in the parts list.

24.10
Checking a drawing

All drawings must be checked before they are released for production, since a mistake could prove expensive. The people who check drawings have special qualifi-

cations enabling them to suggest revisions and modifications that will result in a better product at less cost. The checker may be a chief drafter experienced in this type of work or the engineer or designer who originated the project. In larger companies, the drawings are reviewed by the various shops involved to determine whether the most efficient production methods are specified for each part.

Checkers never check the original drawing; instead, they note corrections with a colored pencil on a *diazo print* (a blue-line print). The print is returned to the drafter for revision of the original drawing, and another print is made for approval.

In Fig. 24.20, the various modifications made by checkers are labeled with letters that are circled and placed near the revisions. Changes are listed and dated in the revision record by the drafter.

Checkers check for the soundness of the design and its functional characteristics. They are also responsible for the drawing's completeness and quality, its readability and clarity, and its lettering and drafting techniques. A poorly drawn view must be redrawn so that it will reproduce well and be understood by those using it. Quality of lettering is especially important since the shop person must rely on lettered notes and dimensions.

The best method for students to check their drawings is to make a scale drawing of the part from the working drawings. It is easier to find another's mistakes than one's own. A grading scale for checking

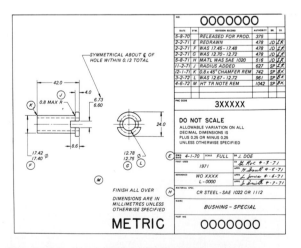

FIG. 24.20 The modifications to this working drawing are noted near the details to be revised. The letters in balloons are cross-referenced in the revision table. (Courtesy of General Motors Corp.)

	Max Value	Points Earned
TITLE BLOCK		
Student's name	1	_____
Checker	1	_____
Date	1	_____
Scale	1	_____
Sheet number	1	_____
REPRESENTATION OF DETAILS		
Selection of views	5	_____
Assembly drawings	10	_____
Positioning of views	5	_____
DRAFTING PRACTICES		
Line quality	8	_____
Lettering	8	_____
Proper dimensioning	10	_____
Proper use of sections	5	_____
Proper use of auxiliary views	5	_____
DESIGN INFORMATION		
Indication of tolerances	5	_____
General tolerance notes	5	_____
Fillets and rounds notes	5	_____
Finish marks	5	_____
Parts list	8	_____
Thread notes and symbols	5	_____
PRESENTATION		
Properly trimmed	2	_____
Properly folded	2	_____
Properly stapled	2	_____
	100	

grade

FIG. 24.21 A checklist for evaluating a working-drawing assignment.

FIG. 24.22 An assembly drawing is used to explain how the parts of a product such as this Ford tractor are assembled. (Courtesy of Ford Motor Co.)

working drawings prepared by students is given in Fig. 24.21. This list can be used as an outline for reviewing working drawings to ensure that the major requirements have been met.

24.11
Drafter's log

Drafters should keep a record called a *log* to show all changes made during the project. As the project progresses, changes, dates, and the people involved should be recorded for reference and for later review of the project. Calculations are often made during a drawing's preparation. If they are lost or poorly done, it may be necessary to do them again; therefore they should be made a permanent part of the log.

24.12
Assembly drawings

After the parts have been made according to the specifications of the working drawings, they will be assembled (Fig. 24.22). This requires an *assembly drawing*.

Two general types of assembly drawings are prepared by using either pictorial techniques or orthographic projection.

A pictorial assembly drawing is shown in Fig. 24.23. The parts are numbered and cross-referenced to the

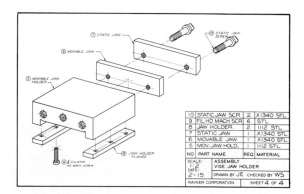

FIG. 24.23 An exploded pictorial assembly. Each part is listed by number in the parts list.

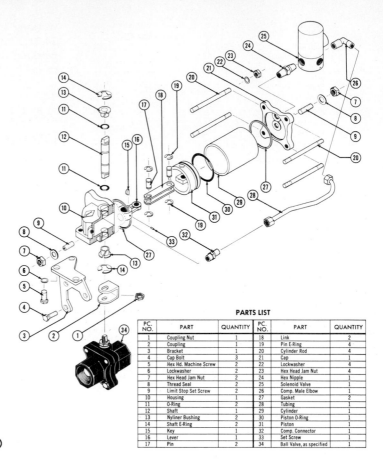

PARTS LIST

PC. NO.	PART	QUANTITY	PC. NO.	PART	QUANTITY
1	Coupling Nut	1	18	Link	2
2	Coupling	1	19	Pin E-Ring	4
3	Bracket	1	20	Cylinder Rod	4
4	Cap Bolt	3	21	Cap	1
5	Hex Hd. Machine Screw	2	22	Lockwasher	4
6	Lockwasher	2	23	Hex Head Jam Nut	4
7	Hex Head Jam Nut	2	24	Hex Nipple	1
8	Thread Seal	2	25	Solenoid Valve	1
9	Limit Stop Set Screw	2	26	Comp. Male Elbow	1
10	Housing	1	27	Gasket	2
11	O-Ring	2	28	Tubing	1
12	Shaft	1	29	Cylinder	1
13	Nyliner Bushing	2	30	Piston O-Ring	1
14	Shaft E-Ring	2	31	Piston	1
15	Key	1	32	Comp. Connector	1
16	Lever	1	33	Set Screw	1
17	Pin	2	34	Ball Valve, as specified	1

FIG. 24.24 An exploded pictorial assembly drawing of the parts of a solenoid control valve. (Courtesy of Jenkins Bros.)

parts list, where more information about each part is given. This assembly shows the parts separated, or exploded, along their centerlines of assembly. The exploded pictorial assembly offers the most understandable view of the relationship of the parts. This is especially important when the assemblies are complex and difficult to visualize in orthographic views.

An exploded assembly drawing of a solenoid control valve is shown in Fig. 24.24. Drawings of this type are used in catalogs and maintenance manuals.

It is usually unnecessary to dimension assemblies since the parts have been dimensioned individually in the working drawings. An example of a dimensioned assembly is shown in Fig. 24.25, where a helicopter frame is drawn as a pictorial. More clarity is provided by this assembly than would be possible in an orthographic assembly.

An orthographic assembly of a special puller is shown in Fig. 24.26. These parts are completely assembled with two end views and a half-section to explain their relationship. A similar partially exploded orthographic assembly is the gear-cutting fixture in Fig. 24.27.

The outline assembly drawing illustrated in Fig. 24.28 is used to show how various components are connected. Each part is composed of subassemblies that are not shown in detail. A computer-drawn assembly is shown as a half-section in Fig. 24.29.

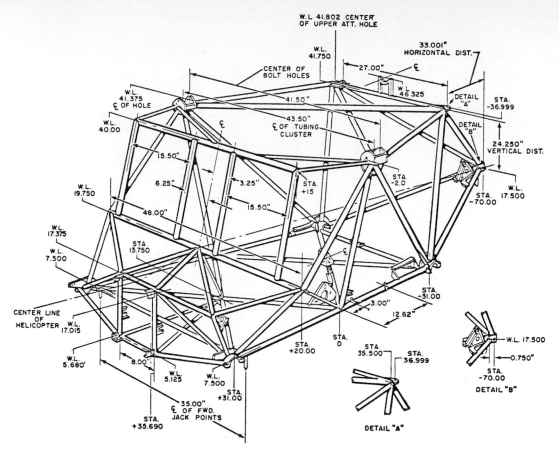

FIG. 24.25 A dimensioned pictorial assembly of a helicopter frame. (Courtesy of Bell Helicopter Co.)

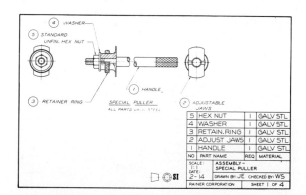

Fig. 24.26 An assembled orthographic assembly of a special puller. Note that sections can be used to clarify the assembly.

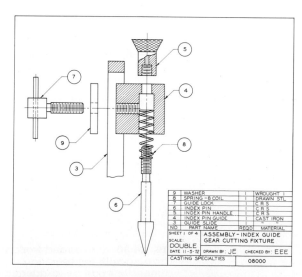

FIG. 24.27 A partially exploded orthographic assembly drawing of an index guide of a cutting fixture.

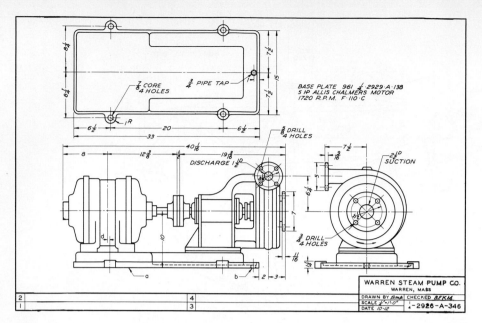

FIG. 24.28 An outline assembly showing the general relationship of the parts of the assembly and their overall dimensions.

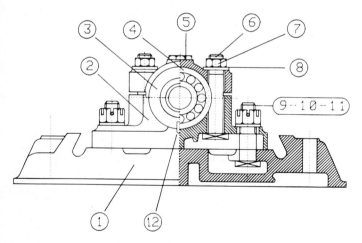

FIG. 24.29 A computer-drawn half-section of an assembly.

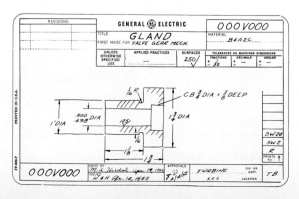

FIG. 24.30 A freehand working drawing with the essential dimensions can be as adequate as a instrument-drawn detail drawing. (Courtesy of General Electric Co.)

24.13
Freehand working drawings

A freehand sketch can serve the same purpose as an instrument drawing, provided the part is simple and the essential dimensions are given (Fig. 24.30). The same principles of working-drawing construction should be followed as when instruments are used.

24.14
Castings and forged parts

The two parts shown in Fig. 24.31 illustrate the difference between a forged part and a machined part. A *forging* is a rough, oversize form made by hammering the metal into shape or by pressing it between two forms (called *dies*). The forging is then machined to its finished dimensions and tolerances.

FIG. 24.31 The top part is a "blank" that has been forged. When the forging has been machined, it will look like the bottom part.

A *casting*, like a forging, is a general shape that must be machined so it will fit with other parts in an assembly. The casting is formed by pouring molten metal into a mold formed by a pattern made slightly larger than the finished part (Fig. 24.32). For the pattern to be removable from the sand that forms the mold, its sides must be tapered. This taper of from five to ten degrees is called the *draft* (Fig. 24.32).

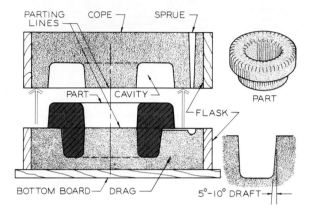

FIG. 24.32 A two-part sand mold is used to produce a casting. A draft of from 5° to 10° is necessary to permit withdrawal of the pattern from the sand. The casting must be machined to size it within specified tolerances.

In some industries, casting and forging drawings are separately made from machine drawings. These are more often combined into one drawing, with the understanding that the forgings and castings must be made with additional material to allow for the removal of excess by machining to meet the final design specifications. The body-wheel cylinder drawing in Fig. 24.33 dimensions both the casting and the machining operations in one drawing.

24.15
Sheet metal drawings

Parts made of sheet metal are formed by bending or stamping. The flat metal patterns for these parts must be developed graphically. Figure 24.34 shows an example of a sheet metal part, where the angles and the radii of the bends are given. The note B.D. means to bend downward, and B.U. means to bend upward.

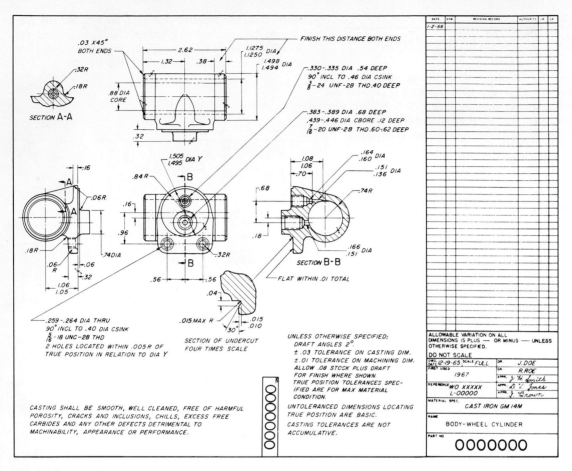

FIG. 24.33 A detail drawing of a casting that shows both the machining operations and the specifications. (Courtesy of General Motors Corp.)

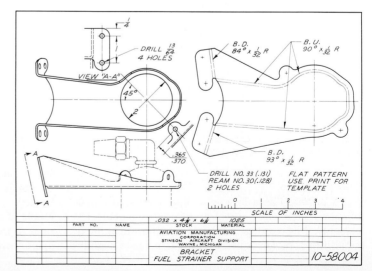

FIG. 24.34 A detail drawing of a sheet metal part shown as a flat pattern and as orthographic views when bent into shape.

Problems

The following problems (Figs. 24.35–24.85) are to be drawn on the sheet sizes assigned or those suggested. Each problem should be drawn with the appropriate dimensions and notes to fully describe the parts and assemblies being drawn.

Working drawings may be made on film or tracing vellum in ink or pencil. Select a suitable title block, and complete it using good lettering practices. Some problems will require more than one sheet to show all the parts properly.

Assemblies should be prepared with a parts list where there are several parts. These may be ortho-

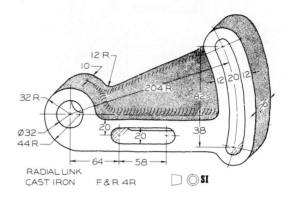

FIG. 24.36 Make a detail drawing on a Size B sheet.

graphic or pictorial assemblies, either exploded, partially exploded, or assembled.

The dimensions given in the problems do not always represent good dimensioning practices because of space limitations; however, the dimensions given are usually adequate for you to complete the detail drawings. In some cases, there may be omitted dimensions that you must approximate using your own judgment. When making the detail drawings, strive to provide all the necessary information, notes, and dimensions to describe the views completely. Use any of the previously covered principles, conventions, and techniques to present the views with the maximum clarity and simplicity.

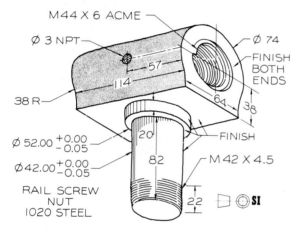

FIG. 24.35 Make a detail drawing on a Size B sheet.

FIG. 24.37 Make a detail drawing on a Size C sheet.

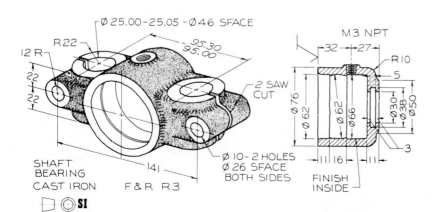

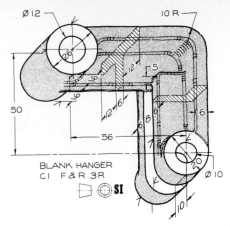

FIG. 24.38 Make a detail drawing on a Size B sheet.

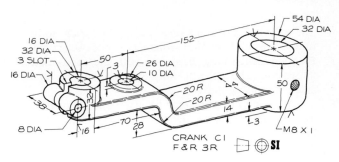

FIG. 24.41 Make a detail drawing on a Size B sheet.

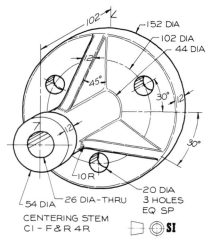

FIG. 24.39 Make a detail drawing on a Size B sheet.

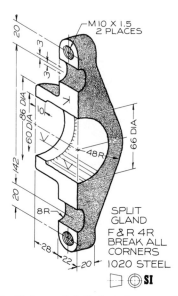

FIG. 24.42 Make a detail drawing on a Size B sheet.

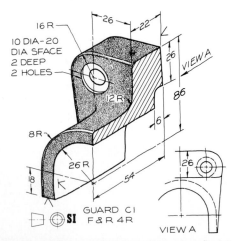

FIG. 24.40 Make a detail drawing on a Size B sheet.

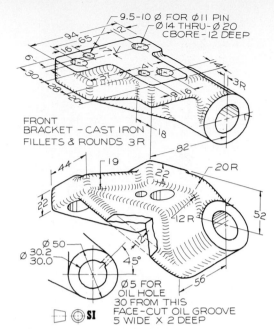

9.5-10 Ø FOR Ø11 PIN
Ø 14 THRU - Ø 20
CBORE - 12 DEEP

94
65
12

6
30
28
20

FRONT
BRACKET - CAST IRON
FILLETS & ROUNDS 3 R

18

3R

82

44
19
22
20R

22

22

12 R

52

Ø 50
Ø 30.2
30.0

45°

56

Ø 5 FOR
OIL HOLE
30 FROM THIS
FACE-CUT OIL GROOVE
5 WIDE X 2 DEEP

SI

FIG. 24.43 Make a detail drawing on a Size B sheet.

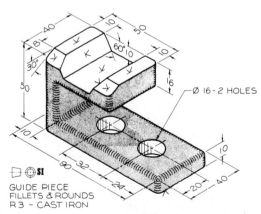

40
10
50
8
60°
30°
10
10

50
16

10
Ø 16 - 2 HOLES

10
90
32
24
20
40
10

SI

GUIDE PIECE
FILLETS & ROUNDS
R 3 - CAST IRON

FIG. 24.44 Make a detail drawing on a Size B sheet.

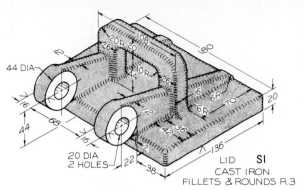

44 DIA

12
20 R 20
108
180
36
10 R
44
26
6
6 R 70
6
20

16
68
16
12
36
42
20

20 DIA
2 HOLES

22
38
136

LID   SI
CAST IRON
FILLETS & ROUNDS R 3

FIG. 24.45 Make a detail drawing on a Size B sheet.

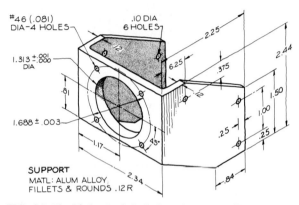

#46 (.081)
DIA-4 HOLES

.10 DIA
6 HOLES

2.25

.12

2.44

1.313 ±.001
DIA

6.25
.375

1.688 ±.003

.81

2

1.50

1.00

.25
.25

1.17
45°

.84

SUPPORT
MATL: ALUM ALLOY
FILLETS & ROUNDS .12 R

2.34

FIG. 24.46 Make a detail drawing on a Size B sheet using (A) decimal inches or (B) inches converted to millimeters.

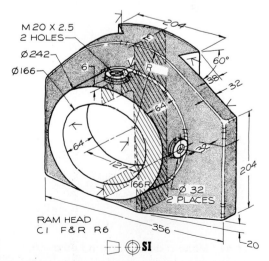

M 20 X 2.5
2 HOLES

Ø242
Ø166

6
R
60°
38
32

64
64
32

166 R
12

Ø 32
2 PLACES

204

204

356
20

RAM HEAD
C I   F & R   R 6

SI

FIG. 24.47 Make a detail drawing on a Size B sheet.

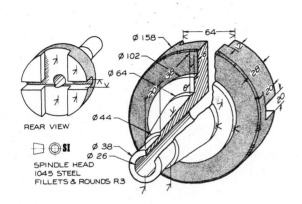

Ø 158
Ø 102
Ø 64
Ø 44
64
REAR VIEW
Ø 38
Ø 26
SPINDLE HEAD
1045 STEEL
FILLETS & ROUNDS R3

FIG. 24.48 Make a detail drawing on a Size B sheet.

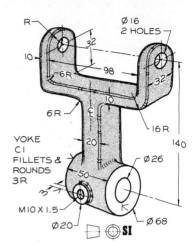

R
Ø 16
2 HOLES
10
32
98
6R
32
10
6R
16R
140
YOKE
CI
FILLETS &
ROUNDS
3R
20
50
Ø 26
Ø 68
M10 X 1.5
Ø 20

FIG. 24.50 Make a detail drawing on a Size B sheet.

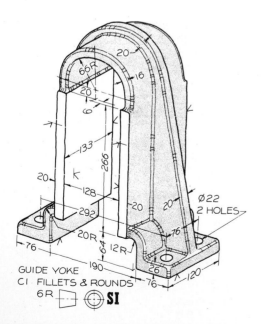

20
66R
16
20
6
133
266
20
128
292
20
20
Ø 22
2 HOLES
76
20R
64
12R
76
190
26
120
GUIDE YOKE
CI FILLETS & ROUNDS
6R SI

FIG. 24.49 Make a detail drawing on a Size B sheet.

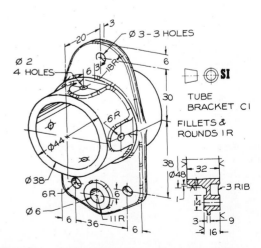

3
20
Ø 3 – 3 HOLES
6
Ø 2
4 HOLES
6
18R
6
30
TUBE
BRACKET CI
FILLETS &
ROUNDS 1R
6R
Ø44
38
Ø48
32
Ø 38
3 RIB
6R
6
Ø 6
6
36
11R
6
4
3
9
16

FIG. 24.51 Make a detail drawing on a Size B sheet.

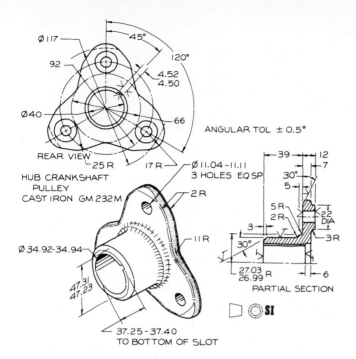

Ø117
45°
92
120°
4.52
4.50
Ø40
66
ANGULAR TOL ± 0.5°
REAR VIEW
25 R
17 R
Ø11.04 - 11.11
3 HOLES EQ SP
HUB CRANKSHAFT
PULLEY
CAST IRON GM 232M
2 R
11 R
Ø34.92 - 34.94
47.31
47.23
37.25 - 37.40
TO BOTTOM OF SLOT

39   12
7
30°
5
5 R   22
2 R   DIA
3   3 R
30°
27.03   6
26.99 R
PARTIAL SECTION

FIG. 24.52 Make a detail drawing on a Size B sheet.

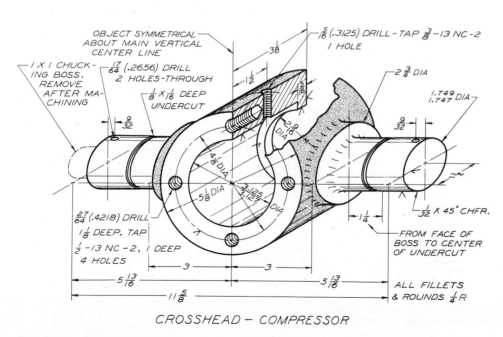

OBJECT SYMMETRICAL
ABOUT MAIN VERTICAL
CENTER LINE
$\frac{5}{16}$ (.3125) DRILL - TAP $\frac{3}{8}$ -13 NC -2
1 HOLE
$3\frac{1}{8}$
$2\frac{3}{4}$ DIA
1 X 1 CHUCK-
ING BOSS.
REMOVE
AFTER MA-
CHINING
$\frac{17}{64}$ (.2656) DRILL
2 HOLES-THROUGH
$1\frac{1}{2}$
$\frac{1}{8}$ X $\frac{1}{16}$ DEEP
UNDERCUT
1.749
1.747 DIA
$\frac{9}{32}$
$\frac{9}{32}$
$\frac{1}{32}$ X 45° CHFR.
$1\frac{1}{4}$
$\frac{27}{64}$ (.4218) DRILL
$1\frac{1}{8}$ DEEP. TAP
$\frac{1}{2}$ -13 NC -2, 1 DEEP
4 HOLES
$5\frac{1}{8}$ DIA
FROM FACE OF
BOSS TO CENTER
OF UNDERCUT
3   3
$5\frac{13}{16}$
$5\frac{13}{16}$
ALL FILLETS
& ROUNDS $\frac{1}{4}$ R
$11\frac{5}{8}$

CROSSHEAD — COMPRESSOR

FIG. 24.53 Make a detail drawing on a Size B sheet using (A) decimal inches or (B) inches converted to millimeters.

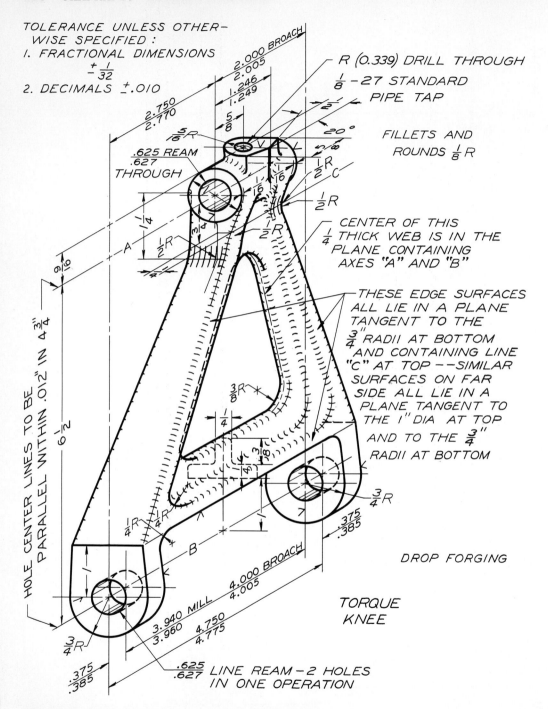

TOLERANCE UNLESS OTHER-
 WISE SPECIFIED :
1. FRACTIONAL DIMENSIONS
   $\pm\frac{1}{32}$
2. DECIMALS $\pm .010$

2.000 BROACH
2.005

1.246
1.249

2.750
2.770

5/8

20°

$\frac{5}{16}R$

.625 REAM
.627
THROUGH

R (0.339) DRILL THROUGH
$\frac{1}{8}$ -27 STANDARD
PIPE TAP

FILLETS AND
ROUNDS $\frac{1}{8}R$

$\frac{1}{2}R$ C

$\frac{1}{2}R$

$\frac{1}{16}$

$\frac{1}{2}R$

A

$\frac{1}{2}R$

$\frac{9}{16}$

HOLE CENTER LINES TO BE
PARALLEL WITHIN .012" IN 4 $\frac{3}{4}$"

$6\frac{1}{2}$

CENTER OF THIS
$\frac{1}{4}$ THICK WEB IS IN THE
PLANE CONTAINING
AXES "A" AND "B"

THESE EDGE SURFACES
ALL LIE IN A PLANE
TANGENT TO THE
$\frac{3}{4}$"RADII AT BOTTOM
AND CONTAINING LINE
"C" AT TOP --SIMILAR
SURFACES ON FAR
SIDE ALL LIE IN A
PLANE TANGENT TO
THE 1" DIA AT TOP
AND TO THE $\frac{3}{4}$"
RADII AT BOTTOM

$\frac{3}{8}R$

$\frac{1}{4}$

$\frac{3}{4}R$

.375
.385

DROP FORGING

$\frac{1}{4}R$ $\frac{1}{4}R$

B

4.000 BROACH
4.005

3.940 MILL
3.960

4.750
4.775

TORQUE
KNEE

$\frac{3}{4}R$

.375
.385

.625
.627   LINE REAM - 2 HOLES
IN ONE OPERATION

FIG. 24.54 Make a detail drawing on a Size B sheet
using (A) decimal inches or (B) inches converted to millimeters.

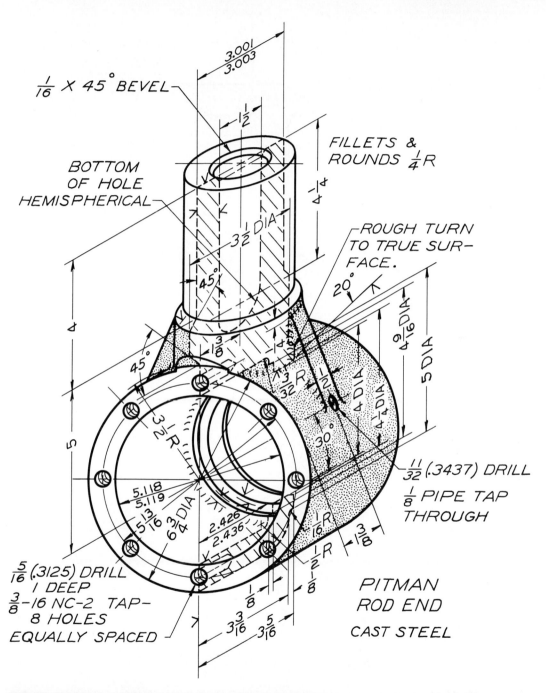

$\frac{1}{16}$ X 45° BEVEL

BOTTOM
OF HOLE
HEMISPHERICAL

$\frac{3.001}{3.003}$

$1\frac{1}{2}$

FILLETS &
ROUNDS $\frac{1}{4}$R

$1\frac{1}{4}$

$3\frac{1}{2}$ DIA

45°

ROUGH TURN
TO TRUE SUR-
FACE.

20°

$\frac{3}{8}$

$4\frac{9}{16}$ DIA

5 DIA

45°

$\frac{3}{32}$ R

$\frac{1}{2}$

$\frac{3}{4}$ R

$4\frac{1}{4}$ DIA

$4\frac{1}{4}$ DIA

30°

$3\frac{1}{2}$ R

$\frac{11}{32}$ (.3437) DRILL
$\frac{1}{8}$ PIPE TAP
THROUGH

5.118
5.119
$5\frac{13}{16}$

$6\frac{3}{4}$ DIA

2.426
2.436

$\frac{1}{16}$ R

$\frac{1}{2}$ R

$1\frac{3}{8}$

$\frac{5}{16}$ (.3125) DRILL
1 DEEP
$\frac{3}{8}$-16 NC-2 TAP-
8 HOLES
EQUALLY SPACED

$\frac{1}{8}$

$\frac{1}{8}$

$3\frac{3}{16}$

$3\frac{5}{16}$

PITMAN
ROD END

CAST STEEL

FIG. 24.55 Make a detail drawing on a Size B sheet using
(A) decimal inches or (B) inches converted to millimeters.

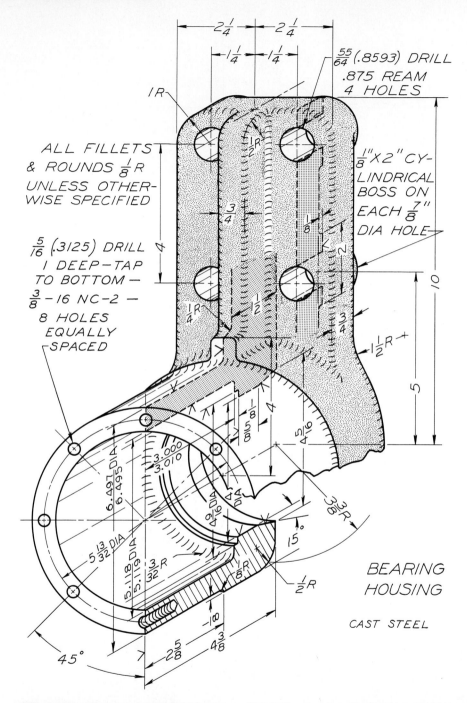

FIG. 24.56 Make a detail drawing on a Size B sheet using (A) decimal inches or (B) inches converted to millimeters.

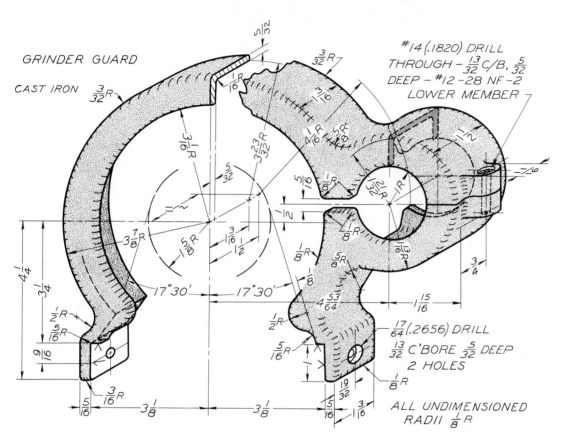

FIG. 24.57 Make a detail drawing on a Size B sheet using (A) decimal inches or (B) inches converted to millimeters.

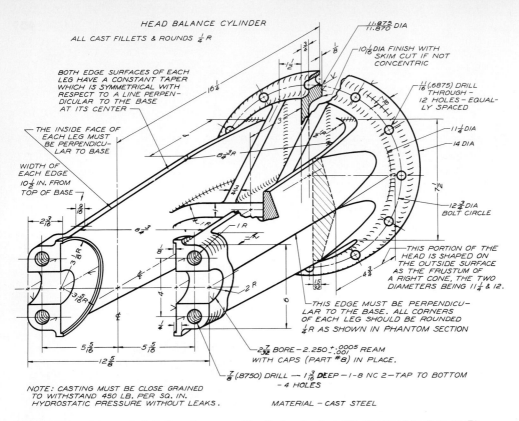

HEAD BALANCE CYLINDER

ALL CAST FILLETS & ROUNDS ¼ R

BOTH EDGE SURFACES OF EACH LEG HAVE A CONSTANT TAPER WHICH IS SYMMETRICAL WITH RESPECT TO A LINE PERPENDICULAR TO THE BASE AT ITS CENTER

THE INSIDE FACE OF EACH LEG MUST BE PERPENDICULAR TO BASE

WIDTH OF EACH EDGE 10½ IN. FROM TOP OF BASE

11.875 / 11.870 DIA

10½ DIA FINISH WITH SKIM CUT IF NOT CONCENTRIC

1 1/16 (.6875) DRILL THROUGH – 12 HOLES – EQUALLY SPACED

11¼ DIA

14 DIA

12¾ DIA BOLT CIRCLE

THIS PORTION OF THE HEAD IS SHAPED ON THE OUTSIDE SURFACE AS THE FRUSTUM OF A RIGHT CONE, THE TWO DIAMETERS BEING 11¼ & 12.

THIS EDGE MUST BE PERPENDICULAR TO THE BASE. ALL CORNERS OF EACH LEG SHOULD BE ROUNDED ¼ R AS SHOWN IN PHANTOM SECTION

2 7/32 BORE – 2.250 +.0005/–.001 REAM WITH CAPS (PART #8) IN PLACE.

7/8 (.8750) DRILL – 1 3/16 DEEP – 1–8 NC 2–TAP TO BOTTOM – 4 HOLES

NOTE: CASTING MUST BE CLOSE GRAINED TO WITHSTAND 450 LB. PER SQ. IN. HYDROSTATIC PRESSURE WITHOUT LEAKS.

MATERIAL – CAST STEEL

FIG. 24.58 Make a detail drawing on a Size B sheet using (A) decimal inches or (B) inches converted to millimeters.

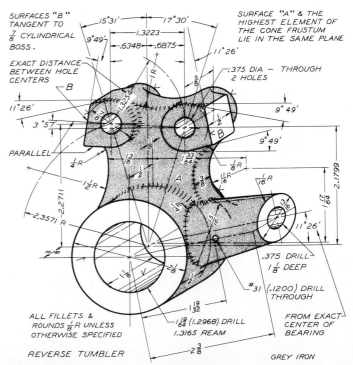

SURFACES "B" TANGENT TO 7/8 CYLINDRICAL BOSS.

EXACT DISTANCE BETWEEN HOLE CENTERS

SURFACE "A" & THE HIGHEST ELEMENT OF THE CONE FRUSTUM LIE IN THE SAME PLANE

15°31'   17°30'

9°49'

1.3223

.6348   .6875

11°26'

.375 DIA – THROUGH 2 HOLES

11°26'

3°57'

9°49'

9°49'

PARALLEL

2.3571 R   2.2711 R

2.1799

1 17/64

11°26'

.375 DRILL 1⅛ DEEP

#31 (.1200) DRILL THROUGH

FROM EXACT CENTER OF BEARING

ALL FILLETS & ROUNDS ⅛ R UNLESS OTHERWISE SPECIFIED

1 19/64 (1.2968) DRILL 1.3165 REAM

1 19/32

2⅜

REVERSE TUMBLER

GREY IRON

FIG. 24.59 Make a detail drawing on a Size C sheet using (A) decimal inches or (B) inches converted to millimeters.

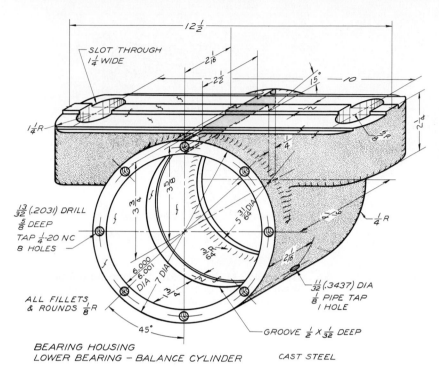

SLOT THROUGH 1¼ WIDE

$12\frac{1}{2}$

$2\frac{1}{16}$

$2\frac{1}{2}$

15°

10

$2\frac{1}{4}$

1¼R

$5\frac{3}{8}R$

$\frac{1}{4}$

$\frac{13}{32}$ (.203I) DRILL
$\frac{5}{8}$ DEEP
TAP ¼-20 NC
8 HOLES

$3\frac{3}{8}$

$3\frac{3}{4}$

$5\frac{31}{64}$ DIA

$4\frac{1}{16}$

$\frac{1}{4}$ R

$\frac{3}{64}$R

$\frac{2}{16}$

ALL FILLETS
& ROUNDS ⅛R

6.000
6.001
DIA 7 DIA

$3\frac{3}{4}$

$\frac{11}{32}$ (.3437) DIA
⅛ PIPE TAP
I HOLE

45°

GROOVE ½ X $\frac{1}{32}$ DEEP

BEARING HOUSING
LOWER BEARING – BALANCE CYLINDER   CAST STEEL

FIG. 24.60 Make a detail drawing on a Size C sheet using (A) decimal inches or (B) inches converted to millimeters.

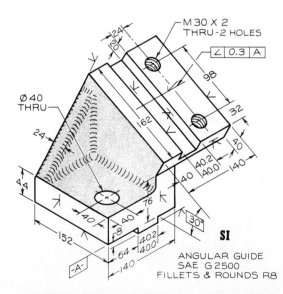

M30 X 2
THRU - 2 HOLES

∠ 0.3 A

24

20

Ø40
THRU

98

32

24

162

40

40.2
40.0

140

40

152

40

8

76

40.2
40.0

64

140

30°

SI

ANGULAR GUIDE
SAE G 2500
FILLETS & ROUNDS R8

-A-

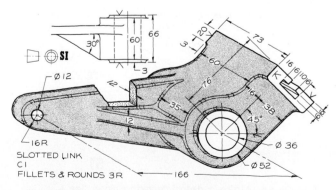

30°

60

66

3

SI

Ø 12

12

16R

35

12

SLOTTED LINK
CI
FILLETS & ROUNDS 3R

20

73

60

76

16 16 10 6

16

38

45°

Ø 36

Ø 52

166

FIG. 24.61 Make a detail drawing on a Size B sheet using (A) decimal inches or (B) millimeters.

FIG. 24.62 Make a detail drawing on a Size B sheet using (A) decimal inches or (B) millimeters.

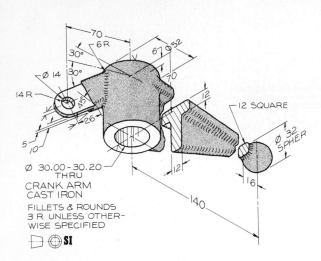

Ø 30.00 - 30.20
THRU
CRANK ARM
CAST IRON

FILLETS & ROUNDS
3 R UNLESS OTHER-
WISE SPECIFIED

FIG. 24.63 Make a detail drawing on a Size B sheet.

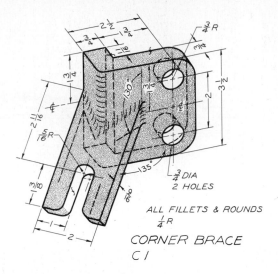

ALL FILLETS & ROUNDS
¼ R

CORNER BRACE
C 1

FIG. 24.65 Make a detail drawing on a Size B sheet. Convert the fractional inches to (A) decimal inches or (B) millimeters.

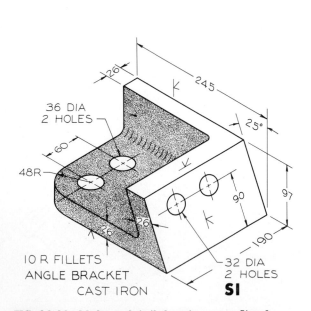

36 DIA
2 HOLES

48R

10 R FILLETS
ANGLE BRACKET
CAST IRON
SI

32 DIA
2 HOLES

FIG. 24.64 Make a detail drawing on a Size A sheet.

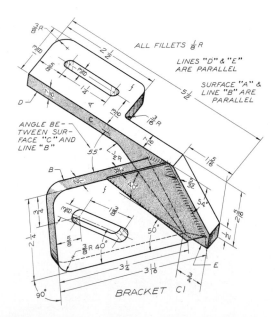

ALL FILLETS ⅛ R

LINES "D" & "E"
ARE PARALLEL

SURFACE "A" &
LINE "B" ARE
PARALLEL

ANGLE BE-
TWEEN SUR-
FACE "C" AND
LINE "B"

BRACKET C1

FIG. 24.66 Make a detail drawing on a Size C sheet. Convert the fractional inches to (A) decimal inches or (B) millimeters.

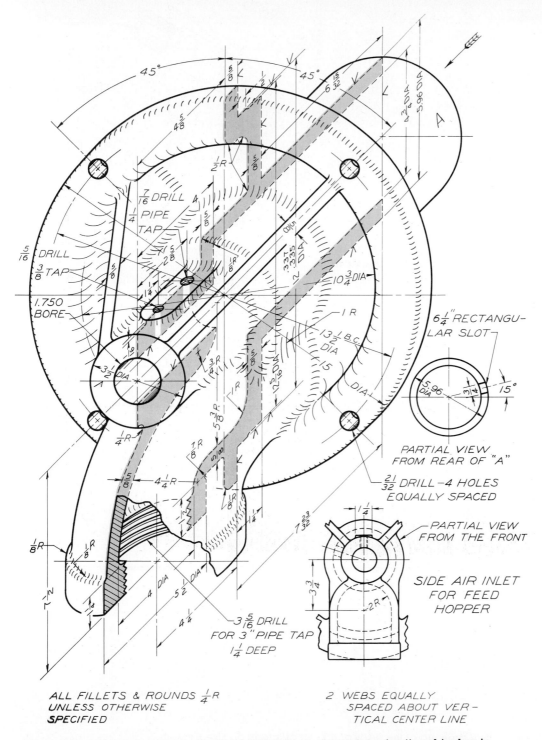

PARTIAL VIEW
FROM REAR OF "A"

PARTIAL VIEW
FROM THE FRONT

SIDE AIR INLET
FOR FEED
HOPPER

ALL FILLETS & ROUNDS $\frac{1}{4}$ R
UNLESS OTHERWISE
SPECIFIED

2 WEBS EQUALLY
SPACED ABOUT VER-
TICAL CENTER LINE

FIG. 24.67 Make a detail drawing on a Size C sheet. Convert the fractional inches to
(A) decimal inches or (B) millimeters.

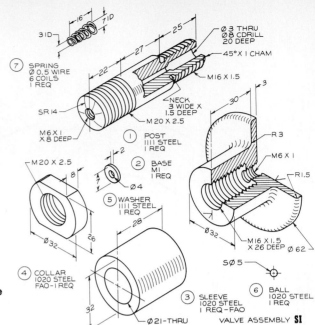

FIG. 24.68 Make a detail drawing of the parts of the valve assembly on Size B sheets. Draw an assembly and provide a parts lists.

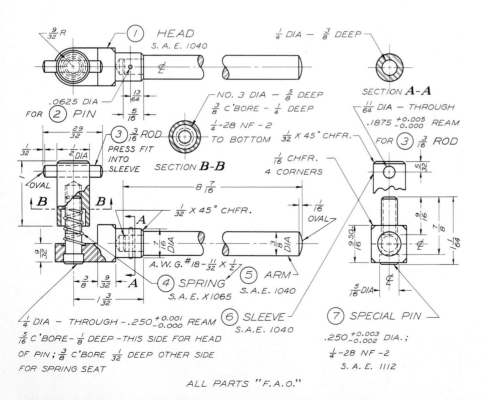

FIG. 24.69 Make a detail drawing of the parts of the cut-off crank on Size B sheets. Convert the fractional inches to (A) decimal inches or (B) millimeters. Draw an assembly and provide a parts list.

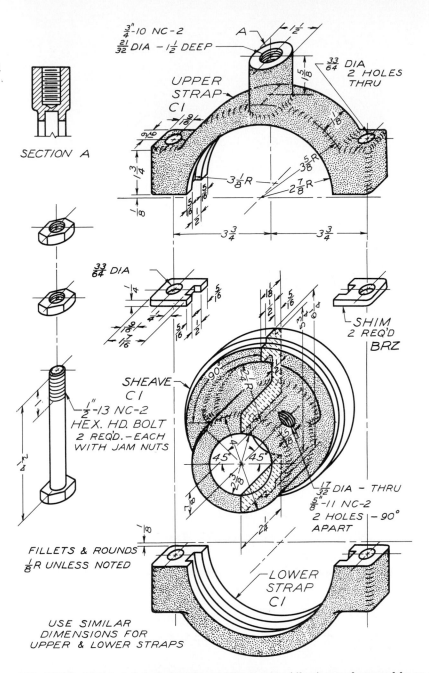

$\frac{3}{4}"$-10 NC-2
$\frac{21}{32}$ DIA - $1\frac{1}{2}$ DEEP

A

$1\frac{1}{2}$

$\frac{33}{64}$ DIA
2 HOLES
THRU

UPPER
STRAP
CI

$\frac{5}{8}$

$\frac{9}{16}$

$\frac{9}{16}$

SECTION A

$\frac{3}{4}$

$\frac{1}{8}$

$\frac{5}{16}$ $\frac{1}{2}$ $\frac{5}{16}$

$3\frac{1}{8}R$

$3\frac{5}{8}R$

$2\frac{7}{8}R$

$3\frac{3}{4}$ $3\frac{3}{4}$

$\frac{33}{64}$ DIA

$\frac{1}{4}$

$\frac{5}{16}$

$\frac{9}{16}$

$\frac{7}{16}$

$\frac{5}{16}$ $\frac{1}{2}$

SHIM
2 REQ'D
BRZ

SHEAVE
CI

$\frac{1}{2}"$-13 NC-2
HEX. HD. BOLT
2 REQ'D.-EACH
WITH JAM NUTS

90°

$2\frac{3}{4}R$

45 45

$\frac{17}{32}"$ DIA - THRU
$\frac{5}{8}"$-11 NC-2
2 HOLES - 90°
APART

$\frac{1}{8}$

FILLETS & ROUNDS
$\frac{1}{8}R$ UNLESS NOTED

LOWER
STRAP
CI

USE SIMILAR
DIMENSIONS FOR
UPPER & LOWER STRAPS

FIG. 24.70 Make a detail drawing of the parts of the journal assembly on Size B sheets. Convert the fractional inches to (A) decimal inches or (B) millimeters. Draw an assembly and provide a parts lists.

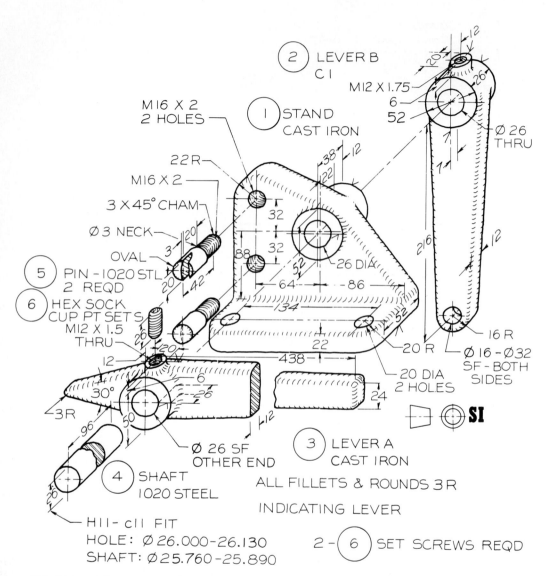

FIG. 24.71 Make a detail drawing of the parts of the indicating lever on Size B sheets. Draw an assembly and provide a parts list.

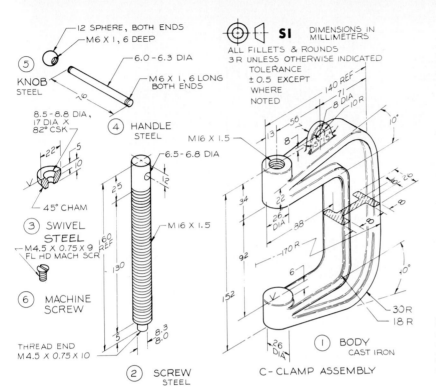

12 SPHERE, BOTH ENDS
M6 X 1, 6 DEEP

⑤ KNOB STEEL

6.0-6.3 DIA

④ HANDLE STEEL

M6 X 1, 6 LONG BOTH ENDS

SI DIMENSIONS IN MILLIMETERS

ALL FILLETS & ROUNDS
3 R UNLESS OTHERWISE INDICATED
TOLERANCE
± 0.5 EXCEPT WHERE NOTED

8.5-8.8 DIA,
17 DIA X
82° CSK

22
5
10

45° CHAM

③ SWIVEL STEEL

M4.5 X 0.75 X 9
FL HD MACH SCR

⑥ MACHINE SCREW

THREAD END
M4.5 X 0.75 X 10

6.5-6.8 DIA

25

12

M16 X 1.5

60 REF

130

5

8.3
8.0

② SCREW STEEL

140 REF
71
8 DIA
10 R
M16 X 1.5

13
56
8
22
26 DIA
88
34
170 R
92
6
152

30 R
18 R

① BODY CAST IRON

26 DIA

C- CLAMP ASSEMBLY

FIG. 24.72 Make a detail drawing of the parts of the C-clamp assembly on Size B sheets. Draw an assembly and provide a parts lists.

FIG. 24.73 Make a detail drawing of the parts of the step bearing on Size B sheets. Draw an assembly and provide a parts lists.

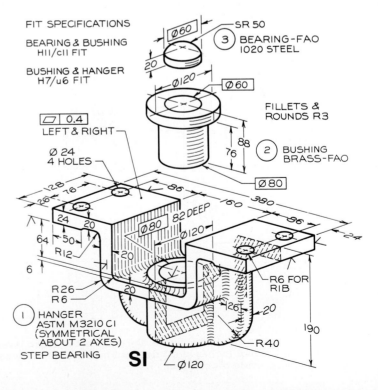

FIT SPECIFICATIONS

BEARING & BUSHING
H11/c11 FIT

BUSHING & HANGER
H7/u6 FIT

SR 50

Ø60

③ BEARING-FAO
1020 STEEL

20

Ø120
Ø60

FILLETS & ROUNDS R3

88
76

② BUSHING BRASS-FAO

0.4
LEFT & RIGHT

Ø80

Ø 24
4 HOLES

128
76
26
24
50
R12

86
160
380
96
24

82 DEEP
Ø80
Ø120

64
6
20

R26
R6

20

① HANGER
ASTM M3210 C1
(SYMMETRICAL ABOUT 2 AXES)

STEP BEARING

126
20
R6 FOR RIB

190

R40

SI
Ø120

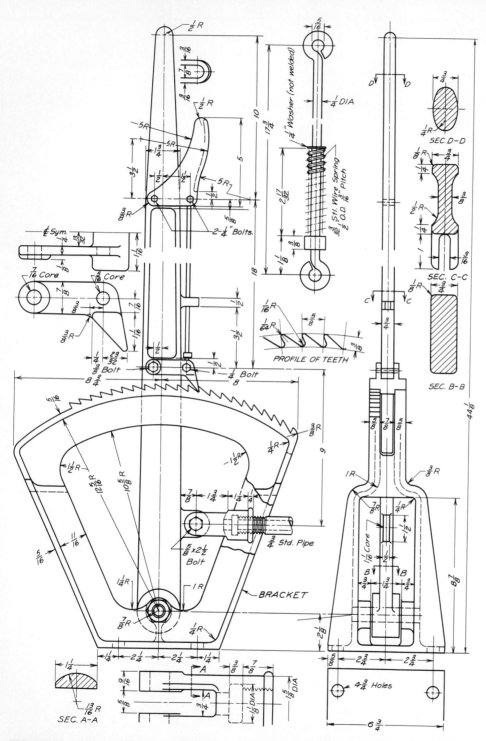

FIG. 24.74 Make a detail drawing of the parts of the brake lever on Size B sheets. Convert the fractional inches to (A) decimal inches or (B) millimeters. Draw an assembly and provide a parts lists.

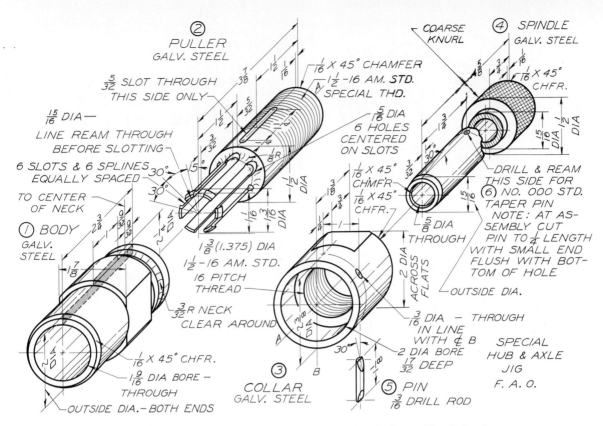

FIG. 24.75 Make detail drawings of the parts of the hub and axle jig on Size B sheets. Convert the fractional inches to (A) decimal inches or (B) millimeters. Draw an assembly of the parts and provide a parts lists.

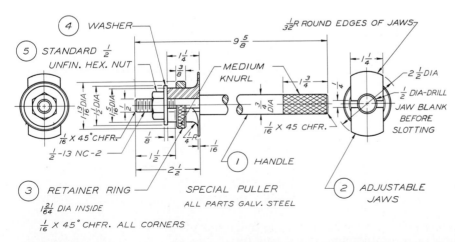

FIG. 24.76 Make a detail drawing of the parts of the special puller on Size B sheets. Convert the fractional inches to (A) decimal inches or (B) millimeters. Draw a pictorial assembly of the parts and provide a parts lists.

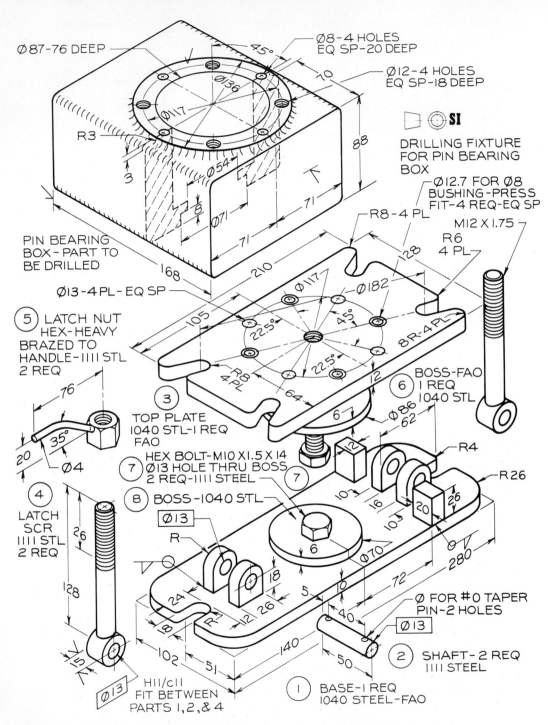

FIG. 24.77 Make detail drawings of the parts of the drilling jig and crank pin bearing box. Draw an assembly of the parts and provide a parts lists.

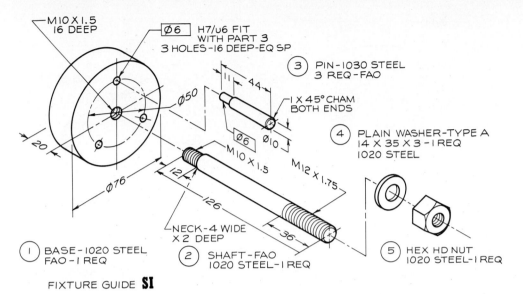

FIXTURE GUIDE **SI**

FIG. 24.78 Make detail drawings of the parts of the fixture guide. Draw an assembly and provide a parts lists.

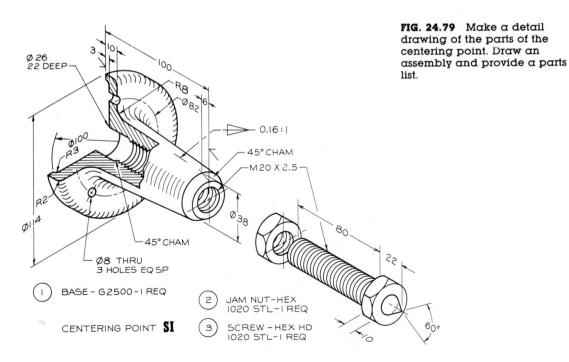

FIG. 24.79 Make a detail drawing of the parts of the centering point. Draw an assembly and provide a parts list.

CENTERING POINT **SI**

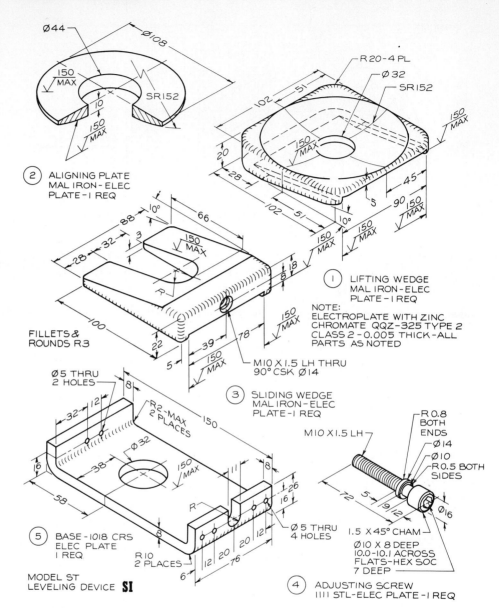

Ø44

Ø108

150/MAX

10

150/MAX

SR152

150/MAX

② ALIGNING PLATE
MAL IRON – ELEC
PLATE – 1 REQ

R20-4 PL

Ø32

SR152

51

102

150/MAX

150/MAX

20

28

45

5

90 150/MAX

102

51

10°

150/MAX

150/MAX

10°

10°

88

66

3

150/MAX

28

32

18

8

100

R

22

5

39

78

150/MAX

150/MAX

FILLETS &
ROUNDS R3

① LIFTING WEDGE
MAL IRON – ELEC
PLATE – 1 REQ

NOTE:
ELECTROPLATE WITH ZINC
CHROMATE QQZ-325 TYPE 2
CLASS 2 – 0.005 THICK – ALL
PARTS AS NOTED

M10 X 1.5 LH THRU
90° CSK Ø14

③ SLIDING WEDGE
MAL IRON – ELEC
PLATE – 1 REQ

Ø5 THRU
2 HOLES

8

32   12

R2-MAX
2 PLACES

150

Ø32

150/MAX

11

8

16

38

26

58

R

16

M10 X 1.5 LH

R0.8
BOTH
ENDS

Ø14

Ø10

R0.5 BOTH
SIDES

72

5

9

2

Ø16

1.5 X 45° CHAM

Ø10 X 8 DEEP
10.0-10.1 ACROSS
FLATS-HEX SOC
7 DEEP

⑤ BASE – 1018 CRS
ELEC PLATE
1 REQ

R10
2 PLACES

8

12   20   20   12

6

76

Ø5 THRU
4 HOLES

MODEL ST
LEVELING DEVICE **SI**

④ ADJUSTING SCREW
1111 STL-ELEC PLATE – 1 REQ

FIG. 24.80 Make a detail drawing of the parts of the leveling device. Draw an assembly and provide a parts list.

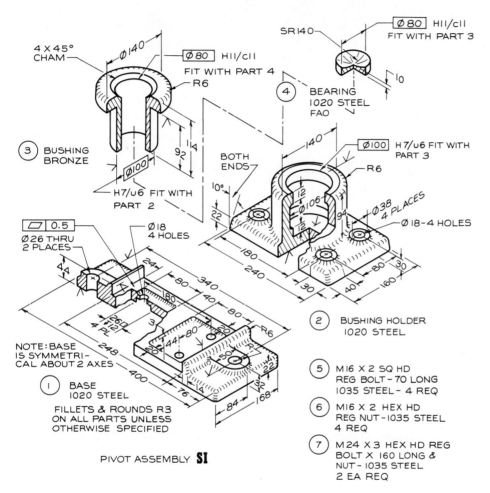

4 X 45°
CHAM

Ø140

Ø 80 H11/c11
FIT WITH PART 4

R6

3 BUSHING
BRONZE

Ø100

92

14

H7/υ6 FIT WITH
PART 2

BOTH
ENDS

10°

SR140

Ø 80 H11/c11
FIT WITH PART 3

10

4 BEARING
1020 STEEL
FAO

140

Ø100 H7/υ6 FIT WITH
PART 3

R6

12
Ø106
12

94

Ø38
4 PLACES

Ø18-4 HOLES

22

180

240

30

80 30

40 160

2 BUSHING HOLDER
1020 STEEL

0.5

Ø 26 THRU
2 PLACES

Ø18
4 HOLES

24

80

340

40

80

260
12

3

180

80

44

50

R6

R

NOTE: BASE
IS SYMMETRI-
CAL ABOUT 2 AXES

248

400

14

76

84 168

22

35

1 BASE
1020 STEEL

FILLETS & ROUNDS R3
ON ALL PARTS UNLESS
OTHERWISE SPECIFIED

PIVOT ASSEMBLY **SI**

5 M16 X 2 SQ HD
REG BOLT - 70 LONG
1035 STEEL - 4 REQ

6 M16 X 2 HEX HD
REG NUT - 1035 STEEL
4 REQ

7 M24 X 3 HEX HD REG
BOLT X 160 LONG &
NUT - 1035 STEEL
2 EA REQ

FIG. 24.81 Make detail drawings of the parts of the pivot assembly. Draw an
assembly and provide a parts list.

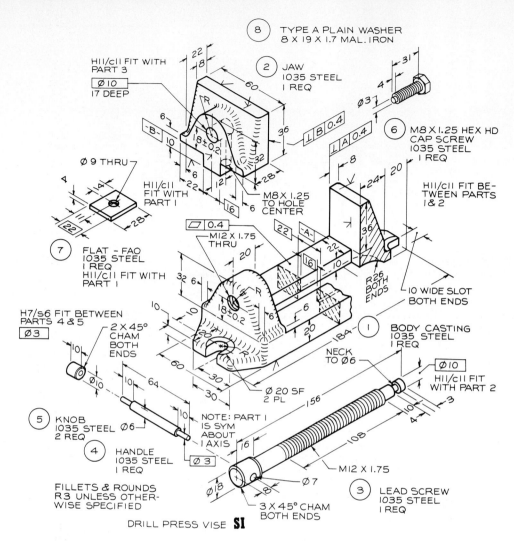

FIG. 24.82 Make detail drawings of the parts of the drill press vise. Draw an assembly and provide a parts list.

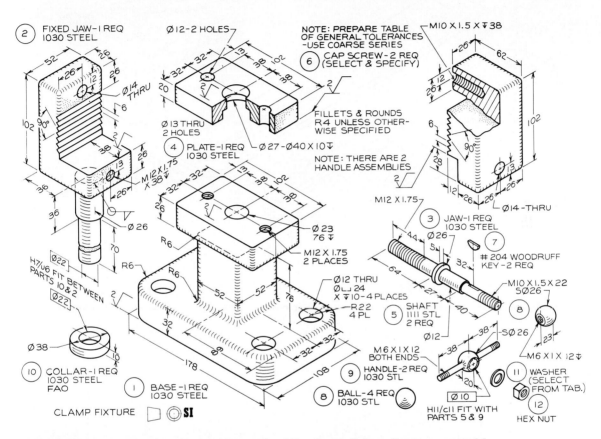

FIG. 24.83 Make detail drawings of the parts of the clamp fixture. Draw an assembly and give a parts list.

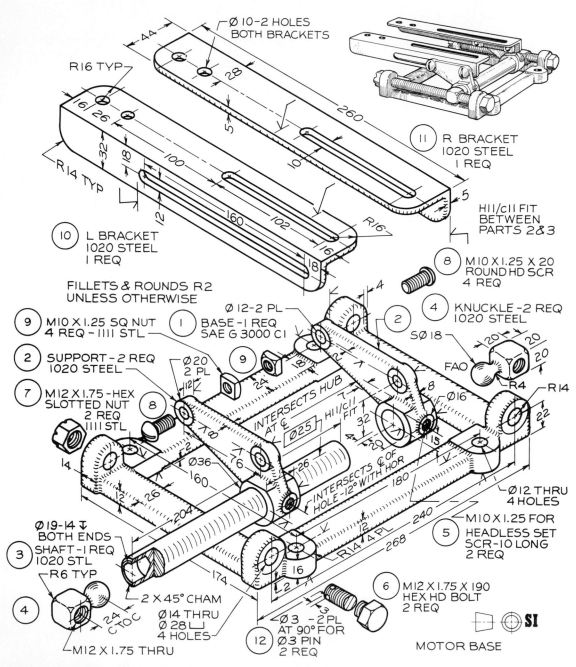

FIG. 24.84 Make a detail drawing of the parts of the motor base. Draw an assembly and provide a parts lists.

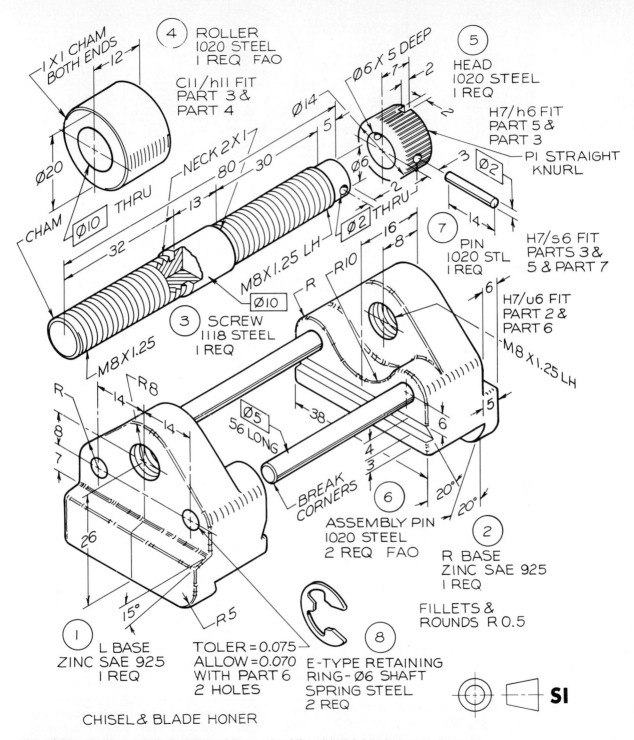

I X I CHAM
BOTH ENDS

12

4 ROLLER
1020 STEEL
I REQ FAO

CII/hII FIT
PART 3 &
PART 4

Ø6 X 5 DEEP

7

2

5 HEAD
1020 STEEL
I REQ

H7/h6 FIT
PART 5 &
PART 3

PI STRAIGHT
KNURL

Ø20

Ø14

5

2

Ø6

3

Ø2

CHAM

Ø10 THRU

NECK 2 X I

80

30

14

Ø2 THRU

2

H7/s6 FIT
PARTS 3 &
5 & PART 7

13

32

M8 X 1.25 LH

Ø6

16

7 PIN
1020 STL
I REQ

6

H7/u6 FIT
PART 2 &
PART 6

Ø10

8

R

R10

6

M8 X 1.25 LH

3 SCREW
1118 STEEL
I REQ

M8 X 1.25

R

R8

14

Ø5
56 LONG

14

38

5

8

7

6

Ø5
56 LONG

26

15°

R5

BREAK
CORNERS

4

3

6

ASSEMBLY PIN
1020 STEEL
2 REQ FAO

6

20°

20°

2 R BASE
ZINC SAE 925
I REQ

FILLETS &
ROUNDS R 0.5

I L BASE
ZINC SAE 925
I REQ

TOLER = 0.075
ALLOW = 0.070
WITH PART 6
2 HOLES

8 E-TYPE RETAINING
RING–Ø6 SHAFT
SPRING STEEL
2 REQ

SI

CHISEL & BLADE HONER

FIG. 24.85 Make a detail drawing of the parts of the chisel & blade honer. Draw an
assembly and give a parts lists.

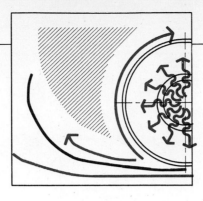

CHAPTER 25

Reproduction Methods and Drawing Shortcuts

25.1
Introduction

So far this text has discussed the processes of preparing drawings and specifications through the working-drawing stage where a detailed drawing is completed on tracing film or paper. Now the drawing must be reproduced, folded, and prepared for filing or transmittal to the drawing's users.

25.2
Reproduction of working drawings

A drawing made by a drafter is of little use in its original form. It would be impractical for the original to be handled by checkers and, even more so, by workers in the field or shop. The drawing would quickly be damaged or soiled, and no copy would be available as a permanent record of the job. Therefore reproduction of drawings is necessary so that copies can be available for use by the people concerned.

The most often used processes of reproducing engineering drawings are (1) *diazo printing*, (2) *blueprinting*, (3) *microfilming*, (4) *xerography*, and (5) *photostatting*.

Diazo printing

The diazo print is more correctly called a "whiteprint" or a "blue-line print" than a blueprint, since it has a white background and blue lines. Other colors of lines are available depending on the type of paper used. The white background makes notes and corrections drawn on the drawing more clearly visible than does the blue background of the blueprint.

> Both blueprinting and diazo printing require that the original drawing be made on semitransparent tracing paper, cloth, or film that will allow light to pass through the drawing.

The paper on which the copy is made, the diazo paper, is chemically treated so that it has a yellow tint on one side. To prevent spoilage, this paper must be stored away from heat and light.

The tracing paper or film drawing is placed face up on the yellow side of the diazo paper and is run through the diazo-process machine, which exposes the drawing to a built-in light. The light passes through the tracing paper and burns out the yellow chemical on the diazo paper except where the drawing lines have shielded the paper from the light. After exposure, the diazo paper is a duplicate of the original drawing except that the lines are light yellow and are not permanent. The diazo paper is then passed through the developing unit of the diazo machine where the yellow lines are developed into permanent blue lines by exposure to ammonia fumes. A typical diazo printer-developer, sometimes called a whiteprinter, is shown in Fig. 25.1.

FIG. 25.2 The Micro-Master 35-mm camera and copy table are used for microfilming engineering drawings. (Courtesy of Keuffel & Esser Co., Morristown, N.J.)

FIG. 25.1 A typical wheelprinter that operates on the diazo process. (Courtesy of Blu-Ray, Inc., Essex, Conn.)

The speed at which the drawing passes under the light determines the darkness of the copy. A slow speed burns out more of the yellow and produces a clear white background; however, some of the lighter lines of the drawing may be lost. Most diazo copies are made at a somewhat faster speed to give a light tint of blue in the background and stronger lines in the copy. Ink drawings give the best reproductions.

Blueprinting

Blueprints are made with paper that is chemically treated on one side. As in the diazo process, the tracing-paper drawing is placed in contact with the chemically treated side of the paper and exposed to light. The exposed blueprint paper is washed in clear water and coated with a solution of potassium dichromate. Then the print is washed again and dried. The wet sheets can be hung on a line to dry, or they can be dried by equipment made for this purpose.

Microfilming

Microfilming is a photographic process that converts large drawings into film copies—either aperture cards or roll film. Drawings must be photographed on either 16-mm or 35-mm film. A camera and copy table are shown in Fig. 25.2.

The roll film or aperture cards can be placed in a microfilm enlarger-printer (Fig. 25.3), where the individual drawings can be viewed on a built-in screen. The selected drawings can then be printed from the film to give standard-size drawings. Microfilm copies are usually smaller than the original drawings to save paper and make the drawings easier to use.

FIG. 25.3 The Bruning 1200 microfilm enlarger-printer makes drawings up to 18″ × 24″ from aperture cards and roll film. (Courtesy of Bruning Co.)

Microfilming makes it possible to eliminate large, bulky files of drawings, since hundreds of drawings can be stored in miniature size on a small amount of film. The aperture cards shown in Fig. 25.3 are data processing cards that can be cataloged and recalled by a computer to make them accessible with a minimum of effort.

FIG. 25.4 This Xerox 7080 Engineering Print System accepts original drawings sized A to E; makes prints sized A to C as fast as 58 per minute; and stamps, folds, and sorts prints automatically.

Xerography

Xerography is an electrostatic process of duplicating drawings on ordinary, unsensitized paper. This process was originally developed for business and clerical uses, but it has more recently been used for the reproduction of engineering drawings.

One advantage of the xerographic process is its ability to make reduced copies of drawings (Fig. 25.4). The Xerox 2080 can reduce a 24-by-36-inch drawing to 8 by 10 inches.

Photostatting

Photostatting is a method of enlarging or reducing drawings using a camera. The combination camera and processor shown in Fig. 25.5 is used for photographing drawings and producing high-contrast copies.

FIG. 25.5 A camera-processor for enlarging and reducing drawings to be reproduced as photostats. (Courtesy of the Duostat Corp.)

The drawing or artwork is placed under the glass of the exposure table (Fig. 25.6), which is lit by built-in lamps. The image can be seen on the glass inside the darkroom where it is exposed on photographically sensitive paper. The negative paper that has been exposed to the image is placed in contact with receiver paper, and the two are fed through the developing solution to obtain a photostatic copy.

These high-contrast reproductions are often used to prepare artwork that is to be printed by offset printing presses. Photostatting can also be used to make reproductions on transparent films and for reproducing halftones (photographs with tones of gray).

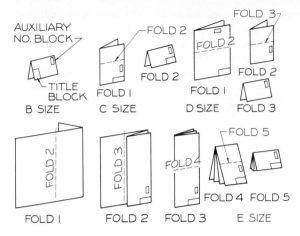

FIG. 25.6 Standard folds for engineering drawing sheets. The final size in each case is $8\frac{1}{2}'' \times 11''$.

25.3
Folding the drawing

Once the prints have been finished, the original drawings should be stored in a flat file for future use and updating. The original drawings should not be folded, and their handling should be kept to a minimum.

The printed drawings, on the other hand, are usually folded for transmittal from office to office. The methods of folding Sizes B, C, D, and E sheets are shown in Fig. 25.6; in each case, the final size after folding is $8\frac{1}{2}$ by 11 inches (or 9 by 12 inches). Note that the drawings are folded so that the title blocks are positioned at the top and the lower right of the drawing to allow them to be easily retrieved from a file.

25.4
Overlay drafting techniques

Valuable drafting time can be saved by taking advantage of current processes and materials that use a series of overlays to separate parts of a single drawing. Engineers and architects often work from a single site plan or floor plan. For example, the floor plan of a building will be used for the electrical plan, furniture arrangement plan, air-conditioning plan, floor materials plan, and so on. It would be expensive to retrace the plan for each application.

A series of overlays can be used in a system referred to as *pin drafting,* where accurately spaced holes are punched in the polyester drafting film at the top edge of the sheets. These holes are aligned on pins attached to a metal strip that match the holes punched in the film (Fig. 25.7). The pins ensure accurate alignment of registration of a series of sheets, and the polyester ensures stability of the material since it does not stretch or sag with changes in humidity.

FIG. 25.7 In the pin system, separate overlays are aligned by seven pins mounted on metal strips. (Courtesy of Keuffel & Esser Co., Morristown, N.J.)

The set of overlays could be attached by the alignment pins or taped together and run through a diazo machine for full-size prints. Another reproduction operation is the use of a flat-bed process camera (Fig. 25.8), which photographs and reduces the drawings to $8\frac{1}{2}$ by 11 inches.

When a large number of prints or multicolor prints are needed, offset lithography is the reproduction process frequently used.

25.5
Paste-on photos

When a number of repetitive drawings are necessary, it is often more economical to use photographic reproductions of the drawings on transparent film. These features can be pasted into position on the master drawing. (Although the term "pasted" is commonly

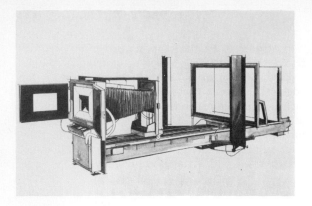

FIG. 25.8 The process camera used to reduce and enlarge engineering drawings is the heart of the pin system. (Courtesy of Keuffel & Esser Co., Morristown, N.J.)

FIG. 25.9 When drawings are to be used repetitively, it may be more economical to reproduce them than to redraw them. (Courtesy of Eastman Kodak Co.)

FIG. 25.10 Photographically reproduced drawings are taped in position to complete the overall drawing. (Courtesy of Eastman Kodak Co.)

used to describe this attachment, tape is actually used when the drawings are reproduced on transparent film. If the reproductions are made as opaque photostats, then rubber cement may be used.)

The office arrangements shown in Fig. 25.9 are transparencies that have been photographically duplicated from a single drawing. The architect shown in Fig. 25.10 is composing an entire drawing sheet with paste-on images of repetitive features that were previously drawn.

25.6
Photo revisions

When a previously made drawing is in need of revision, drawing time can be saved by photographically modifying the drawing. If you want to change the original drawing in Fig. 25.11 to look like the revision shown, you would make a clear film reproduction of the parts, cut them out, and tape them into position on a new form. The new drawing is then photographed onto a new film on which additional notes and lines can be provided to complete the drawing. This new drawing can be used as a master for making diazo prints.

25.7
Stick-on materials

A number of companies market stick-on symbols, screens, and lettering that can be applied to drawings to save time and improve the appearance of drawings. The three standard types of materials are stick-ons, burnish-ons, and tape-ons.

Stick-on symbols or letters are printed on thin plastic sheets. The symbols are cut out with a razor-sharp blade and affixed to the drawing. (Fig. 25.12). This material is available in glossy and matte finishes.

Burnish-on symbols are applied by placing the entire sheet over the drawing and burnishing the desired symbol into place with a rounded-end object such as the end of a pen cap.

Tape-ons are colored adhesive tapes in varying widths that are used in the preparation of graphs and charts. Tapes are also used to represent wide lines on large drawings.

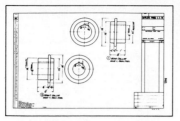

Say you have an existing drawing:

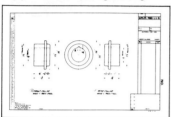

and you want to revise it like this.

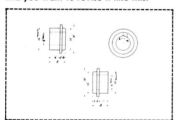

Step 1: First you make a *clear film* reproduction of the original and cut out the elements.

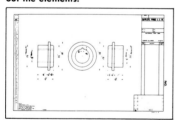

Step 2: Then tape the elements in their new positions on a new form.

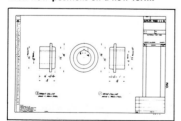

Step 3: Photograph it on film with a matte finish (the tapes and film edges will disappear). Draw in whatever extra detail you want—and you have a new original drawing.

FIG. 25.11 The steps in photographically revising an engineering drawing. (Courtesy of Eastman Kodak Co.)

FIG. 25.12 Stick-on lettering can be applied to a drawing by cutting the letters from the plastic sheet, applying them to the drawing in alignment with a guideline, and burnishing them to the drawing. (Courtesy of Graphic Products Corp.)

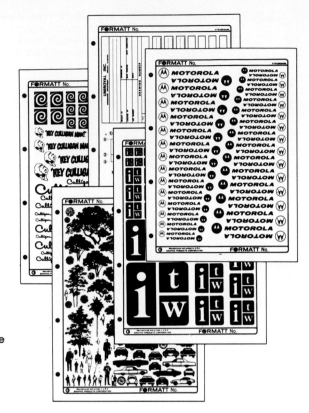

FIG. 25.13 Stick-on symbols are available in a wide range of styles, and they can be custom-designed to suit the client's needs. (Courtesy of Graphic Products Corp.)

Sheets of symbols can be custom-printed for users who have repetitive needs for trademarks and other often-used symbols (Fig. 25.13). Title blocks are sometimes printed in this manner to reduce drawing time. A number of symbols that are used on architectural plans are shown in (Fig. 25.14).

Also available is a matte finished sheet with an adhesive back that can be typed on with a standard typewriter and then transferred to the drawing.

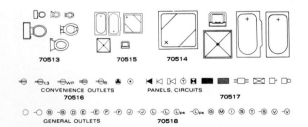

FIG. 25.14 Examples of stick-on architectural symbols. (Courtesy of Zip-a-Tone Inc.)

25.8
Photo drafting

To clarify assembly details, it is sometimes worthwhile to build a model for photographing. This is especially true in the piping industry, where complex refineries are built as models, then photographed, noted, and reproduced as photodrawings.

The steps of preparing a photodrawing are shown in Fig. 25.15, where it is desired to specify certain parts of a sprocket-and-chain assembly. In Step 1, a halftone print of the photograph is made. (This is the process of screening the photograph, or representing it by a series of dots that give varying tones of gray.) In Step 2, the halftone is taped to a white drawing sheet and then photographed to give a negative. The negative is used to produce a positive on polyester drafting film. The notes can be lettered on this drawing film to complete the master drawing (Step 4). The master drawing can then be used to make diazo prints, or it can be microfilmed or reproduced photographically to the desired size.

Step 1: Make a halftone print of the photograph.

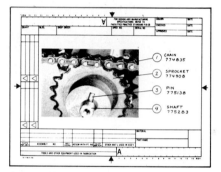

Step 3: Make a positive reproduction on matte film.

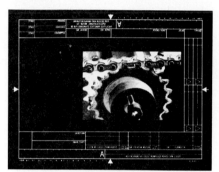

Step 2: Tape the halftone print to a drawing form, and photograph it to produce a negative.

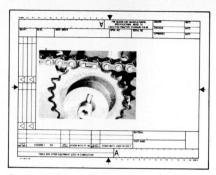

Step 4: Now draw in your callouts—and the job is done.

FIG. 25.15 The steps in making a photodrawing. (Courtesy of Eastman Kodak Co.)

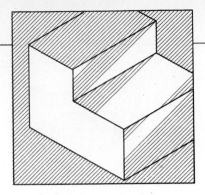

CHAPTER 26

Pictorials

26.1
Introduction

A *pictorial* is an effective means of communicating an idea. Pictorials are especially helpful if a design is unique or if the person to whom it is being explained has difficulty interpreting multiview drawings.

Pictorials, sometimes called *technical illustrations*, are widely used to describe various products in catalogs, parts manuals, and maintenance publications.

26.2
Types of pictorials

The four commonly used types of pictorials are (1) obliques, (2) isometrics, (3) axonometrics, and (4) perspectives (Fig. 26.1).

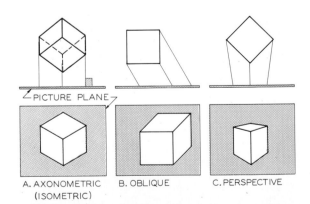

FIG. 26.1 Types of projection systems for pictorials. **A.** Axonometric pictorials are formed by parallel projectors perpendicular to the picture plane. **B.** Obliques are formed by parallel projectors oblique to the picture plane. **C.** Perspectives are formed by converging projectors that make varying angles with the picture plane.

OBLIQUE PICTORIALS are three-dimensional pictorials made on a plane of paper by projecting from the object with parallel projectors that are oblique to the picture plane (Fig. 26.1B).

ISOMETRIC AND AXONOMETRIC PICTORIALS are three-dimensional pictorials on a plane of paper drawn by projecting from the object to the picture plane (Fig. 26.1A). The parallel projectors are perpendicular to the picture plane.

PERSPECTIVE PICTORIALS are drawn with projectors that converge at the viewer's eye and make varying angles with the picture plane (Fig. 26.1C).

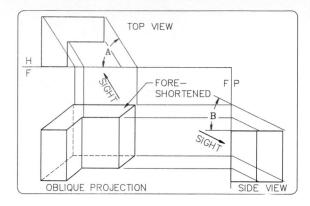

FIG. 26.3 An oblique projection can be drawn at any angle of sight to obtain an oblique pictorial. However, the line of sight should not make an angle less than 45° with the picture plane. This would result in a receding axis longer than true length, thereby distorting the pictorial.

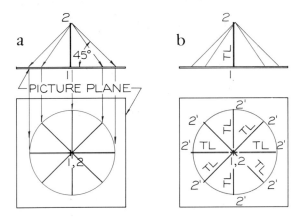

FIG. 26.2 The underlying principle of the cavalier oblique can be seen here, where a series of projectors form a cone. Each element makes a 45° angle with the picture plane. Thus, the projected lengths of 1–2′ are equal in length to line 1–2, which is perpendicular to the picture plane.

26.3
Oblique pictorials

Although seldom used, *oblique pictorials* are the basis of oblique drawings; In Fig. 26.2a, a number of lines of sight are drawn through point 2 of line 1–2. Each line of sight makes a 45° angle with the picture plane, which creates a cone with its apex at 2, and each element on the cone makes a 45° angle with the plane. A variety of projections of line 1–2′ on the picture plane can be seen in the front view (Fig. 26.2b). Each

of these projections of 1–2′ is equal in length to the true length of 1–2. This is a *cavalier oblique projection* because the projectors make a 45° angle with the picture plane, and measurements along the receding axis can be made true length and in any direction.

The top and side views are given as orthographic views, and the front view is drawn as an oblique projection by using the two views of a selected line of sight (Fig. 26.3). Projectors are drawn from the object parallel to the lines of sight to locate their respective points in the oblique view.

26.4
Oblique drawings

Oblique projections are seldom used in the manner just illustrated.

Instead, based on these principles are three basic types of *oblique drawings*: (1) cavalier, (2) cabinet, and (3) *general* (Fig. 26.4).

In each case, the angle of the receding axis can be at any angle between 0° and 90° (Fig. 26.4). Measurements along the receding axes of the cavalier oblique

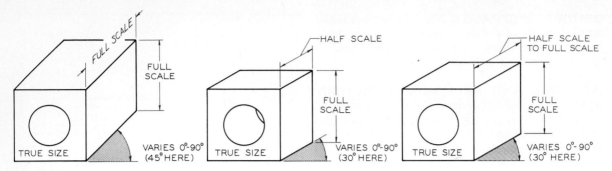

FIG. 26.4 Types of obliques.

A. The *cavalier oblique* can be drawn with a receding axis at any angle, but the measurements along this axis are true length.

B. The *cabinet oblique* can be drawn with a receding axis at any angle, but the measurements along this axis are half-size.

C. The *general oblique* can be drawn with a receding axis at any angle, but the measurements along this axis can vary from half-to full size.

are true length (full scale). The cabinet oblique has measurements along the receding axes reduced to half-length. The general oblique has measurements along the receding axes reduced to between half- and full length.

Three examples of cavalier obliques of a cube are shown in Fig. 26.5. Each has a different angle for the receding axes, but the measurements along the receding axes are true length. A comparison of cavalier and cabinet obliques is given in Fig. 26.6.

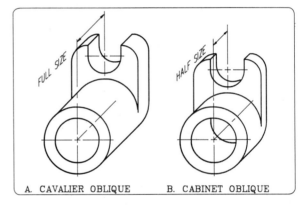

A. CAVALIER OBLIQUE   B. CABINET OBLIQUE

FIG. 26.6 Measurements along the receding axis of a cavlier oblique are full size; those in a cabinet oblique are half-size.

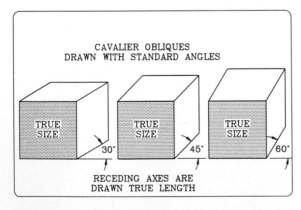

CAVALIER OBLIQUES
DRAWN WITH STANDARD ANGLES

RECEDING AXES ARE
DRAWN TRUE LENGTH

FIG. 26.5 A cavalier oblique is usually drawn with the receding axis at the standard angles of the drafting triangles. Each gives a different view of a cube.

26.5
Constructing obliques

An oblique should be drawn by constructing a box using the overall dimensions of height, width, and depth with light construction lines. In Fig. 26.7, the front view is drawn true size in Step 1. This will be a cavalier oblique. True measurements can be made parallel to the three axes. To complete the oblique, the notches are removed from the blocked-in construction box These measurements are transferred from the given orthographic views with your dividers.

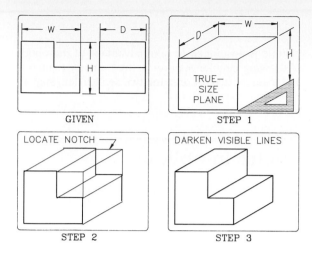

GIVEN

STEP 1

LOCATE NOTCH

STEP 2

DARKEN VISIBLE LINES

STEP 3

FIG. 26.7 Oblique drawing construction.

Step 1 The front surface of the oblique is drawn as a true-size plane. The receding axis is drawn at a convenient angle, and depth is found by using the true distance of *D* taken from the given views.

Step 2 The notch in the front plane is drawn and projected to the rear plane.

Step 3 The lines are darkened to complete the cavalier oblique.

26.6
Angles in oblique

Angular measurements can be made on the true-size plane of an oblique. However, angular measurements will not be true size on the other two planes of the oblique.

To construct an angle in an oblique, coordinates must be used (Fig. 26.8). The sloping surface of 30° must be found by locating the vertex of the angle *H* distance from the bottom. The inclination is found by measuring the distance of *D* along the receding axis to establish the slope. This angle is not equal to the 30° angle that was given in the orthographic view.

You can see in Fig. 26.9 that a true angle can be measured on a true-size surface. In Fig. 26.9B, angles along the receding planes are either smaller or larger than their true angles.

It requires less effort and gives a better appearance when obliques are drawn where angles will appear true size, as shown in Fig. 26.9A, rather than in Fig. 26.9B.

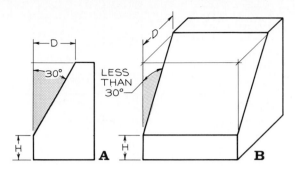

FIG. 26.8 Angles in oblique must be located by using coordinates; they cannot be measured true size except on a true-size plane.

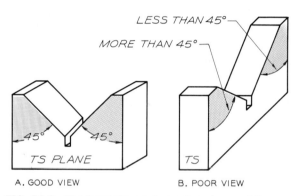

A. GOOD VIEW

B. POOR VIEW

FIG. 26.9 The best view takes advantage of the ease of construction offered by oblique drawings. The view in part B is less descriptive and more difficult to construct than the view in part A.

26.7
Cylinders in oblique

The major advantage of obliques is that circular features can be drawn as true circles when parallel to the picture plane (Fig. 26.10).

The centerlines of the circular end at *A* are drawn and the receding axis is drawn at any desired angle. The end at *B* is located by measuring along the axis. Circles drawn at each end using centers *A* and *B* are connected with tangent lines parallel to the axis.

These same principles are used to construct an object with semicircular features (Fig. 26.11). The

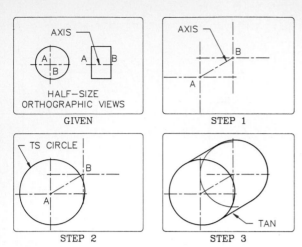

FIG. 26.10 An oblique drawing of a cylinder.

Step 1 Draw axis *AB*, and locate the centers of the circular ends of the cylinder at *A* and *B*. Since the axis is drawn true length, this will be a cavalier oblique.

Step 2 Draw a true-size circle with its center at *A* using a compass or computer-graphics techniques.

Step 3 Draw the other circular end with its center at *B*, and connect the circles with tangent lines parallel to the axis, *AB*.

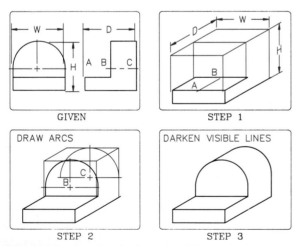

FIG. 26.11 Circular features in oblique.

Step 1 Block in the overall dimensions of the cavalier oblique with light construction lines, and ignore the semicircular features.

Step 2 Locate centers *B* and *C*, and draw arcs with a compass or computer tangent to the sides of the construction boxes.

Step 3 Connect the arcs with lines tangent to each arc and parallel to axis *BC*; then darken the lines.

oblique is positioned to take advantage of the option of drawing circular features as true circles. Centers *B* and *C* are located for drawing two semicircles (Step 3).

26.8
Circles in oblique

Although circular features will be true size on a true-size plane of an oblique pictorial, circular features on the other two planes will appear as ellipses (Fig. 26.12).

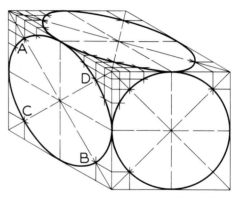

FIG. 26.12 Circular features on the faces of a cavalier oblique of a cube appear as two ellipses and one true circle.

A more frequently used technique of drawing elliptical views in oblique is the four-center ellipse technique shown in Fig. 26.13. A rhombus is drawn that would be tangent to the circle at four points. Perpendicular construction lines are drawn from where the centerlines cross the sides of the rhombus. As shown in Steps 2 and 3, this construction is made to locate four centers that are used to draw four arcs that form the ellipse in oblique.

The four-center ellipse method will not work for the cabinet or the general oblique. In these cases, coordinates must be used to locate a series of points on the curve. Figure 26.14 shows coordinates in orthographic view and in cabinet oblique. The coordinates parallel to the receding axis are reduced to half-size; horizontal coordinates are drawn full size. The ellipse can be drawn with an irregular curve or with an ellipse template that approximates the plotted points.

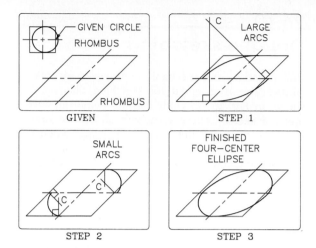

FIG. 26.13 Four-center ellipse in oblique.

Step 1 The circle that is to be drawn in oblique is blocked in with a square that is tangent to the circle at four points. This square will appear as a rhombus on the oblique plane.

Step 2 Construction lines are drawn perpendicularly from the points of tangency to locate the centers for drawing two segments of the ellipse.

Step 3 The centers for the two remaining arcs are located with perpendiculars drawn from adjacent tangent points.

Step 4 When the four arcs have been drawn, the final result is an approximate ellipse.

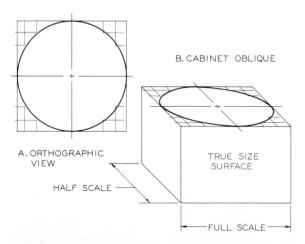

FIG. 26.14 The four-center ellipse technique cannot be used to locate circular shapes on the foreshortened surface of a cabinet oblique. These ellipses must be plotted with coordinates.

Whenever possible, oblique drawings of objects with circular features should be positioned with circles on true-size planes so they can be drawn as true circles. The view in Fig. 26.15A is better than the one in Fig. 26.15B because it gives a more descriptive view of the part and was easier to draw.

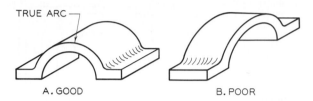

FIG. 26.15 An oblique should be positioned to enable circular features to be drawn most easily.

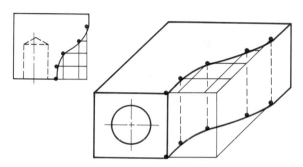

FIG. 26.16 Coordinates are used to establish irregular curves in oblique. The lower curve is found by projecting the points downward a distance equal to the height of the oblique.

26.9
Curves in oblique

Irregular curves in oblique must be plotted point by point using coordinates (Fig. 26.16). The coordinates are transferred from the orthographic view to the oblique view, and the curve is drawn through these points with an irregular curve.

If the object has a uniform thickness, the lower curve can be found by projecting vertically downward from the upper points a distance equal to the height of the object.

In Fig. 26.17, the elliptical feature on the inclined surface was found by using a series of coordinates to locate the points along its curve. These points are then connected by using an irregular curve or an ellipse template of approximately the same size.

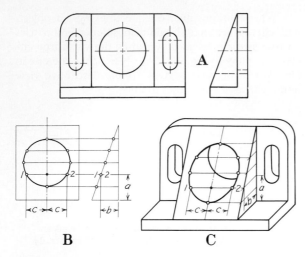

FIG. 26.17   The construction of a circular feature on an inclined surface must be found by plotting points using three coordinates, *a*, *b*, and *c*, to locate points 1 and 2 in this example. The plotted points are connected to complete the elliptical feature.

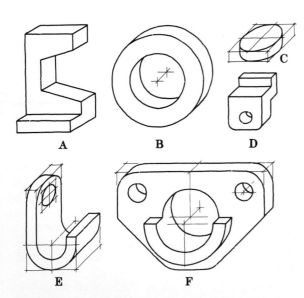

FIG. 26.18   Obliques may be drawn as freehand sketches by using the same principles used for instrument pictorials. Use light construction lines to locate the more complex features.

26.10
Oblique sketching

Understanding the mechanical principles of oblique construction is essential for sketching obliques. As shown in Fig. 26.18, guidelines are helpful in developing a sketch. These guidelines should be drawn lightly so they will not need to be erased when the finished lines of the sketch are darkened. When sketching on tracing vellum, it is helpful if a printed grid is placed under the vellum to provide guidelines.

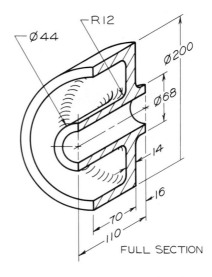

FIG. 26.19   Oblique pictorials can be drawn as sections and dimensioned to serve as working drawings.

26.11
Dimensioned obliques

A full-section dimensioned oblique is given in Fig. 26.19, where the interior features and dimensions are shown.

In oblique pictorials, numerals and lettering should be applied using either the aligned method, in which the numerals are aligned with the dimensioned lines, or the unidirectional method, in which the numerals are all positioned in a single direction regardless of the direction of the dimension lines (Fig. 26.20). Notes connected with leaders are positioned horizontally in both methods.

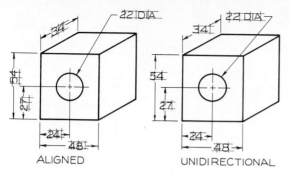

FIG. 26.20 Oblique pictorials can be dimensioned by using either of these methods of applying numerals to the dimension lines.

ALIGNED   UNIDIRECTIONAL

26.12
Isometric pictorials

An *isometric pictorial* is a type of axonometric projection in which parallel projectors are perpendicular to the picture plane, and the diagonal of a cube is seen as a point (Fig. 26.21). The three axes are spaced 120° apart, and the sides are foreshortened to 82% of their true length.

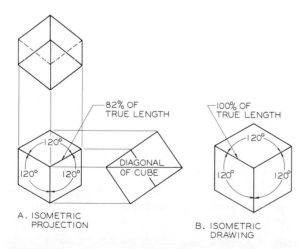

A. ISOMETRIC PROJECTION   B. ISOMETRIC DRAWING

FIG. 26.21 A true isometric projection is found by constructing a view that shows the diagonal of a cube as a point. An isometric drawing is not a true projection since the dimensions are drawn true size rather than reduced as in a projection.

Isometric, which means "equal measurement," is used to describe this type of pictorial since the planes are equally foreshortened.

An *isometric drawing* is similar to an *isometric projection* except that it is not a true axonometric projection but an approximate method of drawing a pictorial. Instead of reducing the measurements along the axes 82%, they are drawn true length (Fig. 26.21B).

A comparison between an isometric projection and an isometric drawing is shown in Fig. 26.22. By using the isometric drawing instead of the isometric projection, pictorials can be measured using standard scales, the only difference being the 18% increase in size. Isometric drawings are used more often than isometric projections.

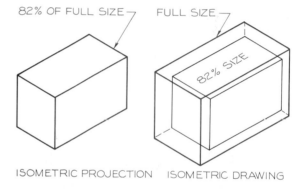

ISOMETRIC PROJECTION   ISOMETRIC DRAWING

FIG. 26.22 The isometric projection is foreshortened to 82% of full size. The isometric drawing is drawn full size for convenience.

The axes of isometric drawings are separated by 120° (Fig. 26.23). Although one of the axes is usually drawn vertically, this is not necessary.

Constructing isometric drawings

An isometric drawing is begun by drawing three axes 120° apart. Lines parallel to these axes are called *isometric lines* (Fig. 26.24A). True measurements can be made along isometric lines but not along nonisometric lines (Fig. 26.24B).

The three surfaces of a cube in isometric are called *isometric planes* (Fig. 26.24C). Planes parallel to these planes are isometric planes, and planes not parallel to them are called nonisometric planes.

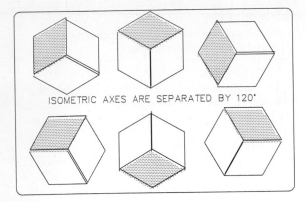

FIG. 26.23 Isometric axes are spaced 120° apart, but they can be revolved into any position.

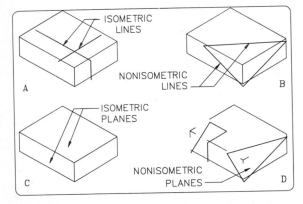

FIG. 26.24 **A.** True measurements can be made along isometric lines (parallel to the three axes). **B.** Nonisometric lines cannot be measured true length. **C.** Three isometric planes are indicated; there are no nonisometric planes in this drawing. **D.** Nonisometric planes are planes inclined to any of the three isometric planes of a cube.

To draw an isometric, you will need a scale and a 30°–60° triangle (Fig. 26.25). Begin by constructing a plane of the isometric using the dimensions of height (*H*) and depth (*D*). In Step 3, the third dimension, width (*W*), is used to complete the isometric drawing.

All isometric drawings should be blocked in using light guidelines (Fig. 26.26) and overall dimensions of *W*, *D*, and *H*. Other dimensions can be taken from the given views and measured along the isometric axes to locate notches and portions removed from the blocked-in drawing.

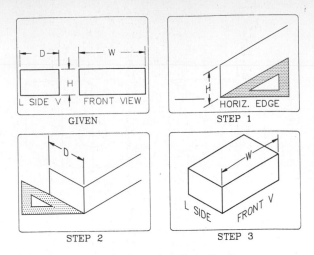

FIG. 26.25 An isometric drawing of a box.

Step 1   Use a 30°–60° triangle and a horizontal straight edge to construct a vertical line equal to the height (*H*), and draw two isometric lines through each end.

Step 2   Draw two 30° lines, and locate the depth (*D*) by transferring this dimension from the given views.

Step 3   Locate the width (*W*) of the object, and complete the surfaces of the isometric box.

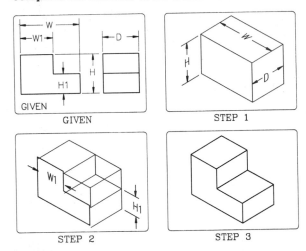

FIG. 26.26 Layout of a simple isometric drawing.

Step 1   Construct an isometric drawing of a box by using the overall dimensions *W*, *D*, and *H* taken from the given view.

Step 2   Locate the notch in the box by using dimensions *W*1 and *H*1 taken from the given orthographic views.

Step 3   Darken the lines to complete the isometric drawing.

A more complex isometric is shown in Fig. 26.27. Again, the object is blocked in using *H, W,* and *D,* and portions of the block are removed to complete the isometric drawing.

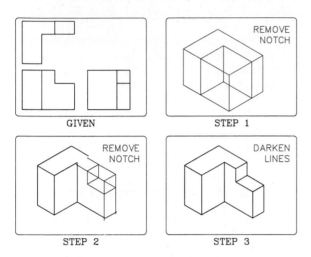

GIVEN

STEP 1 — REMOVE NOTCH

STEP 2 — REMOVE NOTCH

STEP 3 — DARKEN LINES

FIG. 26.27 Layout of an isometric drawing.

Step 1 The overall dimensions of the object are used to block in the object with light lines. The notch is removed.

Step 2 The second notch is removed.

Step 3 The lines are darkened to complete the isometric drawing.

26.13
Angles in isometric

Angles cannot be measured true size in an isometric drawing since the surfaces of an isometric are not true size. Angles must be located by using coordinates measured along isometric lines (Fig. 26.28). Lines *AD* and *BC* are equal in length in the orthographic view, but they are shorter and longer than true length in the isometric drawing.

A similar example can be seen in Fig. 26.29, where two angles are drawn in isometric. The equal-size angles in orthographic are less than and greater than true size in the isometric drawing.

An isometric drawing of an object with an inclined surface is drawn in three steps in Fig. 26.30. The object is blocked in pictorially, and portions are removed.

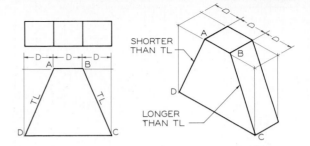

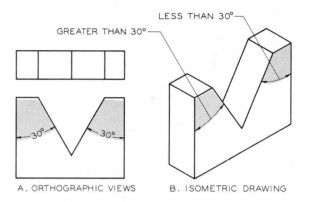

FIG. 26.28 Inclined surfaces must be found by using coordinates measured along the isometric axes. The lengths of angular lines will not be true length in isometric.

FIG. 26.29 Angles in isometric must be found by using coordinates. Angles will not appear true size in isometric.

A. ORTHOGRAPHIC VIEWS   B. ISOMETRIC DRAWING

26.14
Circles in isometric

Three methods of constructing circles in isometric drawings are (1) point plotting, (2) four-center ellipse construction, and (3) ellipse templates

Point plotting

The point-plotting technique is a method where a series of points on a circle are located by coordinates in the X- and Y-directions. The coordinates are transferred to the isometric drawing to locate the points on the ellipse one at a time. A series of points located on a circle can be located in an isometric drawing by us-

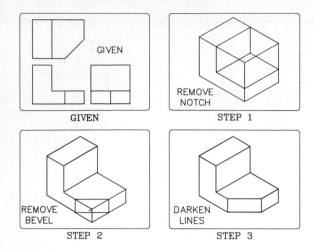

FIG. 26.30 Inclined planes in isometric.

Step 1 The object is lightly blocked in using the overall dimensions, and the notch is removed.

Step 2 The ends of the inclined plane are located using measurements parallel to the isometric axes.

Step 3 The lines are darkened to complete the isometric drawing.

ing *coordinates* parallel to the isometric axes (Fig. 26.31).

The cylinder is blocked in and drawn pictorially, with the centerlines added in Step 1. Coordinates *A, B, C,* and *D* are used in Step 2 to locate points on the ellipse and then are connected using an irregular curve.

The lower ellipse located on the bottom plane of the cylinder can be found by using a second set of coordinates. The most efficient method is by measuring the distance, *E,* vertically beneath each point that was located on the upper ellipse (Step 3). A plotted ellipse is a true ellipse; it is equivalent to a 35° ellipse on an isometric plane. An example of a design composed of circular features drawn in isometric is the handwheel shown in Fig. 26.32.

Four-center ellipse construction

The *four-center ellipse* method can be used to construct an approximate ellipse in isometric by using four arcs drawn with a compass (Fig. 26.33). The four-center ellipse is drawn by blocking in the orthographic view of the circle with a square tangent to the circle at four points. The four centers are found by constructing perpendiculars to the sides of the rhombus at the midpoints of the sides (Step 2). The four arcs are drawn to give the completed four-center ellipse (Step 3). This method can be used to draw ellipses on any of the

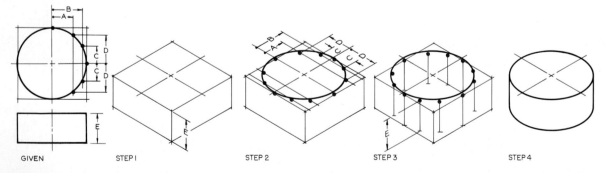

FIG. 26.31 Plotting circles.

Step 1 The cylinder is blocked in using the overall dimensions. The centerlines locate the points of tangency of the ellipse.

Step 2 Coordinates are used to locate points on the circumference of the circle.

Step 3 The lower ellipse is found by dropping each point a distance equal to the height of the cylinder, *E.*

Step 4 The two ellipses can be drawn with an irregular curve and connected with tangent lines to complete the cylinder.

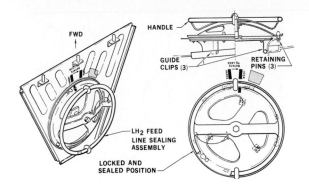

FIG. 26.32 An example of parts drawn by using ellipses in isometric to represent circles. This is a handwheel proposed for use in an orbital workshop to be launched into space.

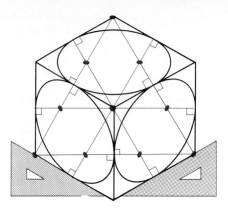

FIG. 26.34 Four-center ellipses can be drawn on all three surfaces of an isometric drawing.

three isometric planes, since each is equally foreshortened (Fig. 26.34).

You can see in Fig. 26.35 that the four-center ellipse is only an approximate ellipse when compared with a true ellipse.

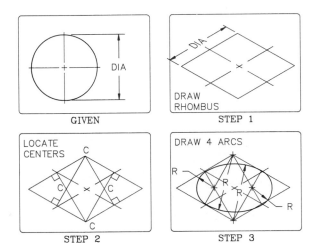

FIG. 26.33 The four-center ellipse.

Step 1 The diameter of the given circle is used to draw a rhombus and centerlines.

Step 2 Light construction lines are drawn perpendicularly from the midpoints of each side to locate four centers.

Step 3 Four arcs are drawn from each center to represent an ellipse tangent to the rhombus.

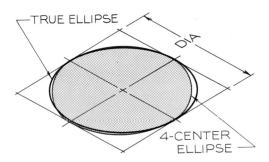

FIG. 26.35 The four-center ellipse is not a true ellipse but an approximate ellipse.

Ellipse templates

A specially designed *ellipse template* can be purchased for drawing ellipses in isometric (Fig. 26.36).

The diameters of the ellipses on the template are measured along the direction of the isometric lines, since this is how diameters are measured in an isometric drawing (Fig. 26.37). The maximum diameter across the ellipse is the *major diameter*, which is a true diameter. Thus, the size of the diameter marked on the template is less than the ellipses' major diameter, the true diameter.

The isometric ellipse template can be used to draw an ellipse by constructing the centerlines of the ellipse in isometric and aligning the ellipse template with these isometric lines (Fig. 26.37A).

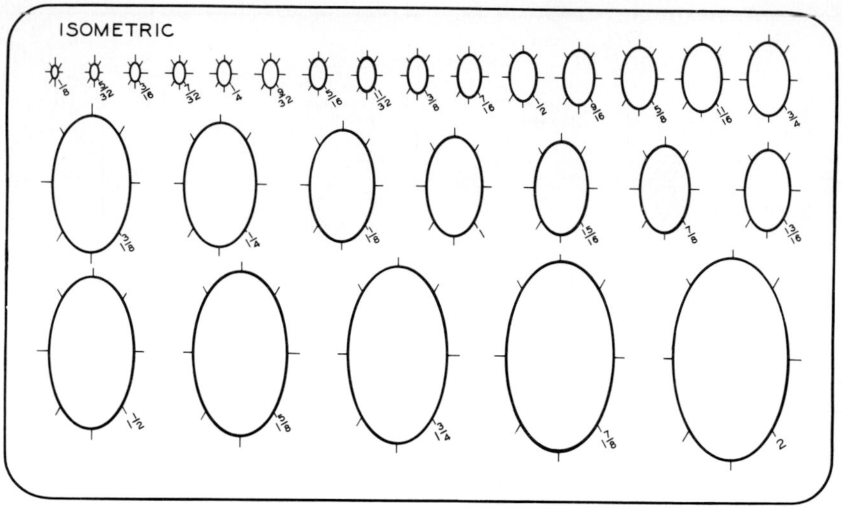

ISOMETRIC

FIG. 26.36 The isometric template is designed to reduce drafting time. The isometric diameters of the ellipses are not the major diameters of the ellipses but are axes parallel to the isometric axes.

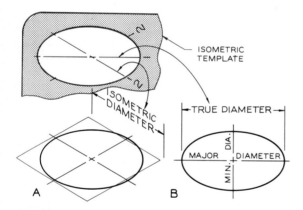

FIG. 26.37 The diameter of a circle in isometric is measured along the direction of the isometric axes. Therefore the major diameter of an isometric ellipse is greater than the measured diameter. The minor diameter is perpendicular to the major diameter.

Ellipses by computer

AutoCAD can draw ellipses by BLOCKing a circle and INSERTing it and by varying the x-scale factor or the y-scale factor for vertical and horizontal ellipses, respectively. The line of sight makes an angle of 35° with a circle on an isometric surface (Fig. 26.38). The ratio of the minor diameter to the major diameter is

the sine of 35°, or 0.57. When the circle BLOCK is inserted, respond as follows to the INSERT prompts:

```
x scale factor <1>/Corner?XYZ: 1
y scale factor <default=x>: 0.57
Rotation angle <0>:0
```

The resulting ellipse will be an isometric ellipse (Fig. 26.38). Examples of various ellipses found by varying

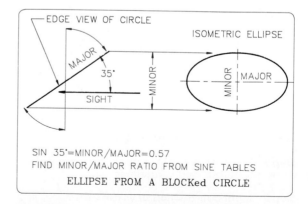

SIN 35°=MINOR/MAJOR=0.57
FIND MINOR/MAJOR RATIO FROM SINE TABLES
ELLIPSE FROM A BLOCKed CIRCLE

FIG. 26.38 A circle on an isometric plane appears as an ellipse. The ratio between the minor and major diameters is equal to the sine of the angle that the line of sight makes with the plane of the circle, 35°.

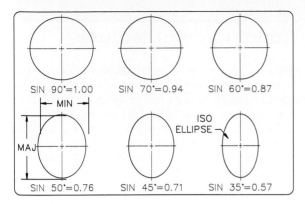

FIG. 26.39 Various ellipses may be drawn by inserting circles as BLOCKS and varying the ratios of the minor diameters to the major diameters based on the sine of the angle of the line of sight.

26.15
Cylinders in isometric

A cylinder can be drawn in isometric by using the four-center ellipse method (Fig. 26.41). A rhombus is drawn at each end of the cylinder's axis with the centerlines drawn as isometric lines (Step 1). The ellipses are drawn using the four-center ellipse method at each end (Step 2). The ellipses are then connected with tangent lines and the lines are darkened (Step 3).

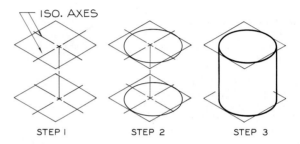

FIG. 26.41 Cylinder: four-center method.

Step 1 A rhombus is drawn in isometric at each end of the cylinder's axis.

Step 2 A four-center ellipse is drawn within each rhombus.

Step 3 Lines are drawn tangent to each rhombus to complete the isometric drawing.

the ratio of the minor diameter to the major diameters can be seen in Fig. 26.39.

The circle BLOCK can be INSERTed and rotated 120° and −120° and the ratio of the minor diameter to the major diameter can be set to 0.57 to represent the circle in isometric on each of the isometric planes (Fig. 26.40). Another method of constructing an isometric ellipse is the four-center ellipse method. A four-center BLOCK can be inserted, scaled, and rotated as needed.

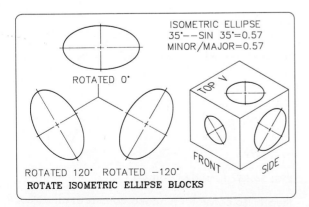

FIG. 26.40 Isometric ellipses with minor/major diameter ratios of 0.57 can be inserted at rotation angles of 120° and −120° to orient ellipses on each of the three isometric planes.

A cylinder can also be drawn using the ellipse template (Fig. 26.42). The axis of the cylinder is drawn and perpendiculars are constructed at each end (Step 1). Since the axis of a right cylinder is perpendicular to the major diameter of its elliptical end, the ellipse template is positioned as shown (Step 2). The ellipses are drawn at each end and are connected with tangent lines (Step 3).

To construct a cylindrical hole in the block (Fig. 26.43), begin by locating the center of the hole on the isometric plane. The axis of the cylinder is drawn parallel to the isometric axis that is perpendicular to this plane through its center (Step 2). The ellipse template is aligned with the major and minor diameters to complete the elliptical view of the cylindrical hole (Step 3).

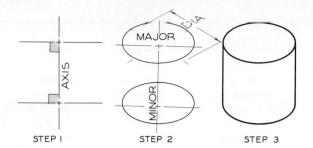

FIG. 26.42 Cylinder: ellipse template.

Step 1 The axis of the cylinder is drawn to its proper length, and perpendiculars are drawn at each end.

Step 2 The elliptical ends are drawn by aligning the major diameter with the perpendiculars at the ends of the axis. The isometric diameter of the isometric ellipse template is given along the isometric axis.

Step 3 The ellipses are connected with tangent lines to complete the isometric drawing. Hidden lines are omitted.

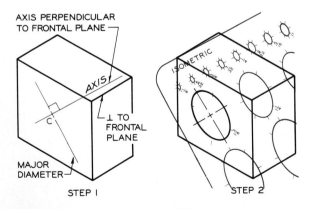

FIG. 26.43 Cylinders in isometric.

Step 1 The center of the hole with a given diameter is located on a face of the isometric drawing. The axis of the cylinder is drawn from the center parallel to the isometric axis perpendicular to the plane of the circle. The major diameter is drawn perpendicular to the axis.

Step 2 The 1⅝″ ellipse template is used to draw the ellipse by aligning the major and minor diameters with the guidelines on the template.

26.16
Partial circular features

When an object has a semicircular end, as in Fig. 26.44, the four-center ellipse method can be used with only two centers to draw half of the circle (Step 2). To draw the lower ellipse at the bottom of the object, the centers are projected downward a distance of *H*, the height of the object. These centers are used with the same radii that were used on the upper surface to draw the arcs.

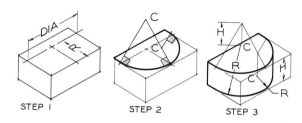

FIG. 26.44 Semicircular features.

Step 1 Objects with semicircular features can be drawn by blocking in the objects as if they had square ends. The centerlines are drawn to locate the centers and tangent points.

Step 2 Perpendiculars are drawn from each point of tangency to locate two centers. These are used to draw half of a four-center ellipse.

Step 3 The lower surface can be drawn by lowering the centers by the distance of *H*, the thickness of the part. The same radii are used with these new centers to complete the isometric.

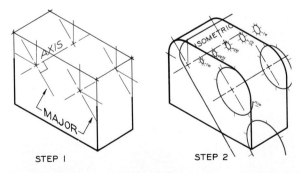

FIG. 26.45 Rounded corners.

Step 1 To construct rounded corners of an object, the centerlines of the ellipse are drawn.

Step 2 The elliptical corners are drawn with an ellipse template.

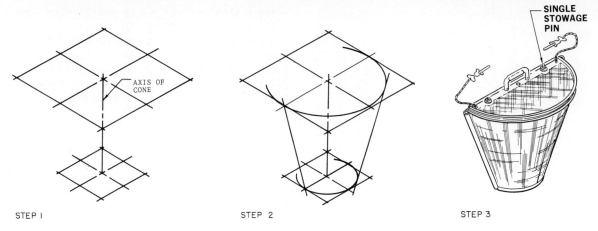

STEP I   STEP 2   STEP 3

FIG. 26.46 Construction of a cone in isometric.

Step 1 The axis of the cone is constructed. Each circular end of the cone is blocked in.

Step 2 An ellipse guide is used for constructing the circles in isometric at each end. These ends are connected to give the outline of the object.

Step 3 The remaining details of the screen storage provisions are added to complete the isometric. (Courtesy of the National Aeronautics and Space Administration.)

In Fig. 26.45, an object with rounded corners is blocked in, and the centerlines are located at each rounded corner (Step 1). An ellipse template is used to construct the rounded corners (Step 2). The rounded corners could have been constructed by the four-center ellipse method or by plotting points on the arcs.

A similar drawing involving the construction of ellipses is the conical shape in Fig. 26.46. The ellipses are blocked in at the top and bottom surfaces (Step 1), and by using a template or the four-center method, the half ellipses are drawn (Step 2).

26.17
Measuring angles

Angles in isometric may be located by coordinates (Fig. 26.47), since angles will not appear true size in an isometric.

A second method of measuring and locating angles is the ellipse template method shown in Fig. 26.48. Since the hinge line is perpendicular to the path of revolution, an ellipse is drawn in Step 1 with the major diameter perpendicular to the hinge line. A

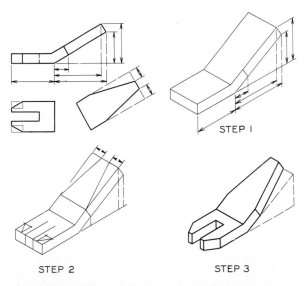

STEP I

STEP 2   STEP 3

FIG. 26.47 Inclined surfaces in isometric must be located by using coordinates laid off parallel to the isometric axes. True angles cannot be measured in isometric drawings.

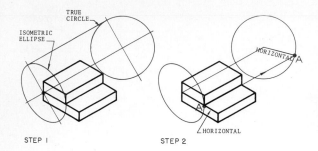

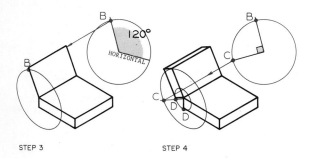

FIG. 26.48 Measuring angles with an ellipse template.

Step 1 An ellipse is drawn with the major diameter perpendicular to the hinge line of the two parts. Any size of ellipse could be used. A true circle is drawn with its center on the projection of the hinge line and with a diameter equal to the major diameter of the ellipse.

Step 2 Point A is projected to the circle to locate the direction of the horizontal in the circular view.

Step 3 The position of rotation is measured 120° from the horizontal to locate point B, which is then projected to the ellipse to locate the position of the revolved surface.

Step 4 To locate the perpendicular to the surface, a 90° angle is drawn in the circular view, and point C is projected to the ellipse; a line is drawn from point C on the ellipse to the center of the ellipse. A smaller ellipse is drawn to pass through point D on the lower part of the object. The point where this ellipse intersects the line from C to the center of the ellipse establishes the thickness of the revolved part.

true circle is drawn with a diameter that is equal to the major diameter of the ellipse.

In Step 2, point A is located on the ellipse and is projected to the circle to locate the direction of a horizontal line. From this line, the angle of revolution of the hinged part can be measured true size, 120° in this

example (Step 3). Point B is projected to the ellipse to locate a line at 120° in isometric.

The thickness of the revolved part is found by drawing line C perpendicular to line A and projecting back to the ellipse. A smaller ellipse through point D is drawn to locate the thickness of the revolved part in the isometric. The remaining lines are drawn parallel to these key lines.

26.18
Curves in isometric

Irregular curves must be plotted point by point, using coordinates to locate each point. Points A through F are located in the orthographic view with coordinates of width and depth (Fig. 26.49). These coordinates are transferred to the isometric view of the blocked-in part (Step 1) and are connected with an irregular curve.

Each point on the upper curve is projected downward for a distance of H, the height of the part, to locate points on the lower curve. The points are connected with an irregular curve to complete the isometric.

26.19
Ellipses on nonisometric planes

When an ellipse lies on a nonisometric plane, such as the one shown in Fig. 26.50, points on the ellipse can be plotted to locate it. Coordinates are located in the orthographic views and then transferred to the isometric as shown in Steps 1 and 2. The plotted points can be connected with an irregular curve, or an ellipse template can be selected that will approximate the plotted points.

26.20
Surfaces of revolution

A surface of revolution is a solid form made by revolving a plane about an axis of revolution (Fig. 26.51). To draw these in isometric, the axes were drawn through point 0 to point 4. Circular cross sections were located along this axis using the correct elliptical diameters taken from the orthographic views for each.

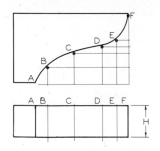
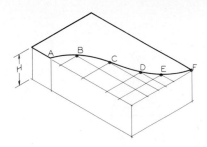
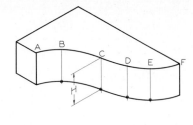

FIG. 26.49 Plotting irregular curves.

Step 1 Coordinates are established in the orthographic views.

Step 2 One set of coordinates is transferred to the isometric view.

Step 3 The second set of coordinates is transferred to the isometric to establish points on the ellipse.

Step 4 The plotted points are connected with an elliptical curve. An ellipse template can usually serve as a guide for connecting the points.

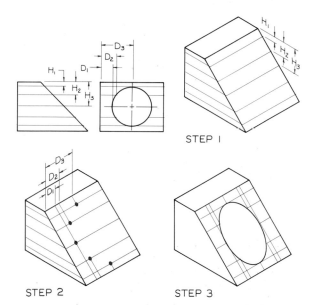

STEP 1

STEP 2 STEP 3

FIG. 26.50 Construction of ellipses on an inclined plane.

Step 1 Draw two coordinates to locate a series of points on the irregular curve. These coordinates must be parallel to the standard W, D, and H dimensions.

Step 2 Block in the shape using overall dimensions. Locate points on the irregular curve using the coordinates from the orthographic views.

Step 3 Since the object has a uniform thickness, the lower curve can be found by projecting downward the distance H from the upper points.

The elliptical cross sections are connected with tangent lines to find the outlines of the objects in isometrics. The more cross sections used, the more accurate the location of the tangent lines.

26.21
Machine parts in isometric

Orthographic and isometric views of a spotface, countersink, and boss are shown in Fig. 26.52. The isometric drawings of these features can be drawn by point-plotting the circular features, by using the four-center method, or by using an ellipse template, which is by far the easiest method.

A threaded shaft can be drawn in isometric, as shown in Fig. 26.53, by first drawing the cylinder in isometric (Step 1). Next, the major diameters of the crest lines of the thread are drawn separated at a distance of P, the pitch of the thread (Step 2). Ellipses are then drawn by aligning the major diameter of the ellipse template with the perpendiculars to the cylinder's axis (Step 3). Note that the 45° chamfered end is drawn using a smaller ellipse at the end.

A hexagon head nut (Fig. 26.54) is drawn in three steps using an ellipse template. The nut is blocked in, and an ellipse drawn tangent to the rhombus. The hexagon is constructed by locating the distance across a flat, W, parallel to the isometric axes. The other sides of the hexagon are found by drawing lines tangent to the ellipse (Step 2). The distance H is

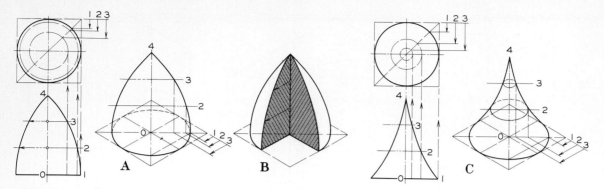

FIG. 26.51 Surfaces of revolution must be constructed by drawing a series of cross sections that are connected to give their overall shapes. The cross sections in these examples are circles drawn as ellipses in isometric.

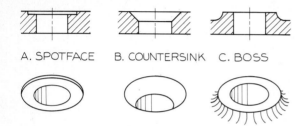

FIG. 26.52 Examples of circular features drawn in isometric. These can be drawn by using ellipse templates.

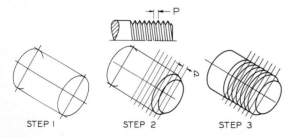

FIG. 26.53 Threads in isometric.

Step 1 Draw the cylinder to be threaded by using an ellipse template.

Step 2 Lay off perpendiculars that are spaced by a distance equal to the pitch of the thread, *P*.

Step 3 Draw a series of ellipses to represent the threads. The chamfered end is drawn by using an ellipse whose major diameter is equal to the root diameter of the threads.

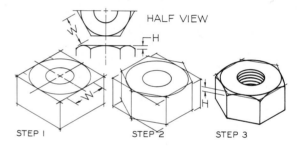

FIG. 26.54 Construction of a nut.

Step 1 The overall dimensions of the nut are used to block in the nut.

Step 2 The hexagonal sides are constructed at the top and bottom.

Step 3 The chamfer is drawn with an irregular curve. Threads are drawn to complete the isometric.

laid off at each corner to establish the amount of chamfer at each corner (Step 3).

A hexagon-head bolt is drawn in two positions in Fig. 26.55. The washer face can be seen on the lower side of the head, and the chamfer on the upper side of the bolt head.

To draw a sphere in isometric three ellipses are drawn as isometric planes with a common center (Fig. 26.56). The center is used to construct a circle that will be tangent to each ellipse, as shown in Step 3. The sphere is larger in isometric than in a true axonometric projection.

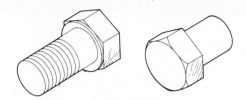

FIG. 26.55 Isometric drawings of the upper and lower sides of a hexagon-head bolt.

A portion of a sphere is used to draw a round-head screw in Fig. 26.57. A hemisphere is constructed in Step 1. The centerline of the slot is located along one of the isometric planes. The thickness of the head is measured as distance of E from the highest point on the sphere.

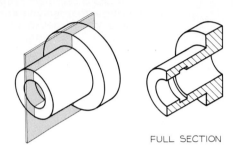

FULL SECTION

FIG. 26.58 Parts can be shown in isometric sections to clarify internal features.

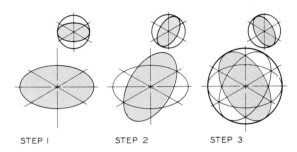

STEP I   STEP 2   STEP 3

FIG. 26.56 An isometric sphere.

Step 1 The three intersecting isometric axes are drawn. The ellipse template is used to draw the horizontal elliptical section.

Step 2 The isometric ellipse template is used to draw one of the vertical elliptical sections.

Step 3 The third vertical elliptical section is drawn, and the center is used to draw a circle tangent to the three ellipses.

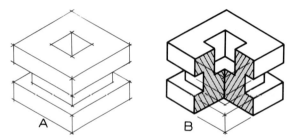

A   B

FIG. 26.59 An isometric drawing of a half-section can be constructed.

26.22
Isometric sections

A full section can be drawn in isometric to clarify internal details that might otherwise be overlooked (Fig. 26.58). Half-sections can also be used, as illustrated in Fig. 26.59. The same part is shown as a full section and a half-section in Fig. 26.60.

26.23
Dimensioned isometrics

When it is advantageous to dimension and note a part shown in isometric, either the aligned or the unidirectional method can be used to apply the notes (Fig.

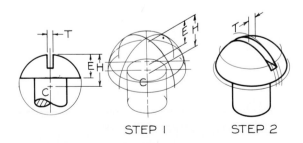

STEP I   STEP 2

FIG. 26.57 Spherical features.

Step 1 An isometric ellipse template is used to draw the elliptical features of a round-head screw.

Step 2 The slot in the head is drawn, and the lines are darkened to complete the isometric of the head.

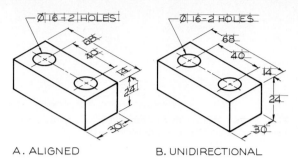

FIG. 26.60 Dimensions can be placed on isometric drawings by using either technique shown here. Guidelines should always be used for lettering.

A. ALIGNED

B. UNIDIRECTIONAL

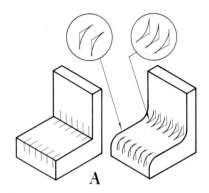 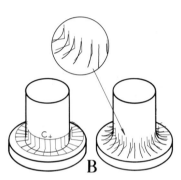 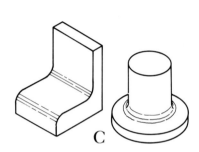

FIG. 26.61 Representation of fillets and rounds.

A. Fillets and rounds can be represented by segments of an isometric ellipse if guidelines are constructed at intervals.

B. Fillets and rounds can be represented by elliptical arcs by constructing radial guidelines.

C. Fillets and rounds can be represented by lines that run parallel to the fillets and rounds.

26.61). In both cases, notes connected with leaders are positioned horizontally. Always use guidelines for your lettering and numerals.

26.24
Fillets and rounds

Fillets and rounds in isometric can be represented by either of the methods shown in Fig. 26.61 to give added realism to a pictorial drawing. Figures 26.61A and B show how intersecting guidelines are drawn equal in length to the radii of the fillets and rounds, and how arcs are drawn tangent to these lines. These arcs can be drawn freehand or with an ellipse template. The method in Fig. 26.61C uses freehand lines drawn parallel to or concentric with the directions of the fillets and rounds. An example of these two meth-

ods of showing fillets and rounds is shown in Fig. 26.62. The stipple shading was applied by using an adhesive overlay film (see Section 26.34).

When fillets and rounds are illustrated as shown in Fig. 26.63 and dimensions are applied, it is much easier to understand the features of the part than when it is represented by orthographic views.

26.25
Isometric assemblies

Assemblies are used to explain how parts are assembled. Common mistakes in applying leaders to an assembly are shown in Fig. 26.64A; whereas, the more acceptable techniques are shown in Fig. 26.64B. The numbers in the circles ("balloons") refer to the number given to each part listed in the parts list.

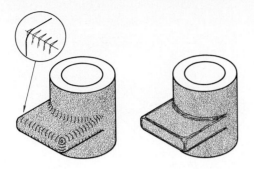

FIG. 26.62 Two methods of representing fillets and rounds on a part.

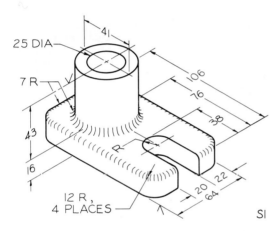

FIG. 26.63 An isometric drawing with complete dimensions and fillets and rounds represented.

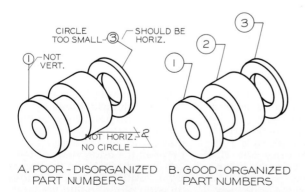

A. POOR - DISORGANIZED PART NUMBERS

B. GOOD-ORGANIZED PART NUMBERS

FIG. 26.64 **A.** Common mistakes in applying part numbers to an assembly. **B.** Acceptable techniques of applying part numbers.

An assembly from a parts manual, along with its parts list, is shown in Fig. 26.65. This assembly is "exploded" so it is more clear how the parts would be assembled.

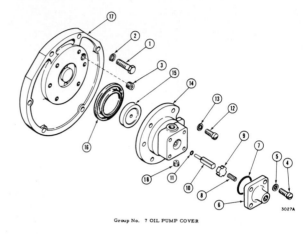

Group No. 7 OIL PUMP COVER

FIG. 26.65 An exploded isometric assembly.

26.26
Axonometric pictorials

An *axonometric pictorial* is a form of orthographic projection in which the pictorial view is projected perpendicularly onto the picture plane with parallel projectors. The object is positioned in an angular position with the picture plane so that its pictorial projection will be a three-dimensional view. Three types of axonometric pictorials are possible: (1) isometric, (2) dimetric, or (3) trimetric.

The *isometric projection* is the view where the diagonal of a cube is viewed as a point. The planes will be equally foreshortened and the axes equally spaced 120° apart (Fig. 26.66A). The measurements along the

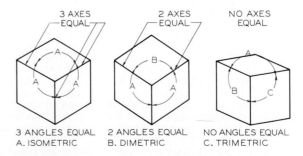

FIG. 26.66 Three types of axonometric projection.

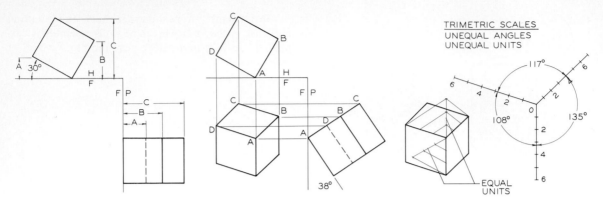

FIG. 26.67 Trimetric-scale construction.

Step 1 Revolve the top view 30° clockwise. Find the side view by transferring dimensions *A*, *B*, and *C* from the top view. If the revolution had been 45° in the top view, the resulting projection would be either a dimetric or an isometric.

Step 2 Tilt the side view 34°. This will change the projection of the top view but not its width dimension; Thus, it is unnecessary to change the top view. Find the trimetric projection of the cube by projecting from the top and side views.

Step 3 The sides of a cube are equal; therefore divide each axis into an equal number of units by proportional division even though the three axes in a trimetric have different lengths. The axes can be extended and scaled into as many units as desired.

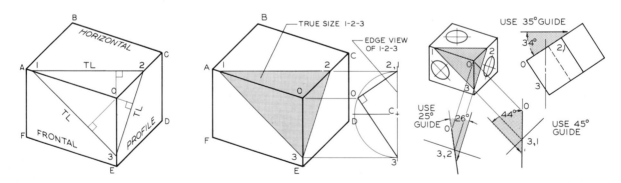

FIG. 26.68 Ellipse template angles for a trimetric scale.

Step 1 The planes of a cube are mutually perpendicular; therefore line 0*A* is perpendicular to plane 0*CDE*, and line 0*E* is perpendicular to plane *ABC*0. On each plane, construct true-length lines, which are located perpendicular to the axis lines *A*0, *C*0, and *E*0. The true-length lines 1–2, 2–3, and 3–1 will intersect at points on the axes forming triangle 1–2–3.

Step 2 Since plane 1–2–3 is composed of true-length lines, it is true size in the trimetric projection. Determine the side view of the plane, which is a frontal plane, by projection. Find the 90° angle of the cube in the side view by constructing a semicircle, using the edge view of plane 1–2–3 as the diameter. Project point 0 to the semicircle where the 90° angle is inscribed.

Step 3 Determine the edge view of each principal plane by locating the point views of lines 1–2, 2–3, and 3–1. The ellipse guide angle is the angle between the edge views of 1–0–2, 2–0–3, and 1–0–3 and the line of sight. Position the ellipse guides on each plane so that the major diameter is parallel to the true-length lines on that plane.

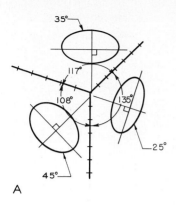

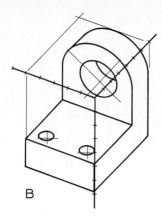

A   B

FIG. 26.69 The complete trimetric scales showing the units along the axes and the ellipse template angles. The object in part B was drawn using the axonometric scales.

three axes will be equal but less than true length, since this is true projection.

A *dimetric projection* is the view where two planes are equally foreshortened, and two of the axes are separated by equal angles (Fig. 26.66B). The measurements along two of the axes are equal.

A *trimetric projection* is the view where all three planes are unequally foreshortened, and the angles between the three axes are different (Fig. 26.66C).

26.27
Axonometric construction

All axonometric constructions can be made in the same manner as the trimetric constructed in Fig. 26.67. A cube should always be used instead of the object to be drawn. By using a cube, you can construct axonometric scales that can be used to draw trimetrics of many objects, not just a single part. The trimetric scales found in Step 3 can be lengthened to accommodate any size part.

Trimetric scales are not complete unless the ellipse angles are found for each plane so that ellipse templates can be used. In a trimetric, a different ellipse angle will be used for each plane since each plane is foreshortened differently.

When a trimetric is given (Fig. 26.68), a true-size plane can be found by drawing lines 1–2, 1–3, and 3–4 perpendicular to the three axes (Step 1). Next, the side view of this true-size plane is found as a vertical edge (Step 2). Using these two views, the angles between the lines of sight and each plane can be found (Step 3). These angles are the ellipse angles for each surface.

The ellipses are positioned on each plane perpendicular to the axes intersecting each. This gives you a set of trimetric scales with calibrations and the ellipse angles for each plane (Fig. 26.69A). These scales can be used to construct a trimetric by overlaying the scales with tracing vellum (Fig. 26.69B).

A second technique of constructing an axonometric is shown in Fig. 26.70. The three axes are located in a convenient position, and plane 1–2–3 is constructed with three true-length lines (Step 1). Semicircles are drawn with diameters equal to each of these true-length lines. Angles are inscribed inside each semicircle to give a 90° angle at point 0 (Step 2). The top, front, and side views of the object are located at point 0 for each view, and each view is projected back where the three projectors converge at common points.

26.28
Perspective pictorials

A *perspective pictorial* is a view normally seen by the eye or a camera; it is the most realistic form of pictorial. In a perspective, all parallel lines converge at infinite vanishing points as they recede from the observer.

The three basic types of perspectives are (1) one-point, (2) two-point, and (3) three-point, depending on the number of vanishing points used in their construction (Fig. 26.71).

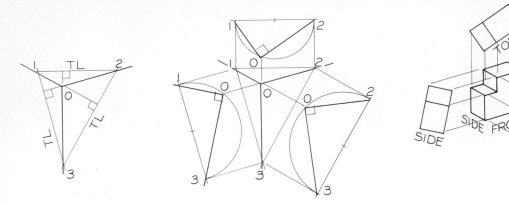

FIG. 26.70 Axonometric projection.

Step 1 The three axes of an axonometric projection can be selected and drawn at the angle of your choice. True-length perpendiculars are drawn so they intersect at 1, 2, and 3.

Step 2 Semicircles are projected from each true-length line. Right angles are inscribed in the semicircle, 2–0–3 for example.

Step 3 The three views of the object are located with the same corner placed at the right angles found in Step 2. These views are projected back to intersect and form a trimetric.

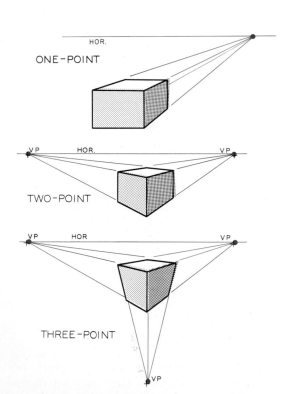

FIG. 26.71 A comparison of one-point, two-point, and three-point perspectives.

The *one-point perspective* has one surface of the objective parallel to the picture plane; therefore it is true shape. The other sides vanish to a single point on the horizon, called a *vanishing point* (Fig. 26.71A).

A *two-point perspective* is a pictorial positioned with two sides at an angle to the picture plane; this requires two vanishing points (Fig. 26.71B). All horizontal lines converge at the vanishing points, but vertical lines remain vertical and have no vanishing point.

The *three-point perspective* has three vanishing points since the object is positioned so that all sides of it are at an angle with the picture plane (Fig. 26.71C). The three-point perspective is used in drawing larger objects such as buildings.

26.29
One-point perspectives

The steps of drawing a one-point perspective are shown in Fig. 26.72, which shows the top and side views of the object, the picture plane, the station point, the horizon, and the ground line.

PICTURE PLANE is the plane on which the perspective is projected. It appears as an edge in the top view.

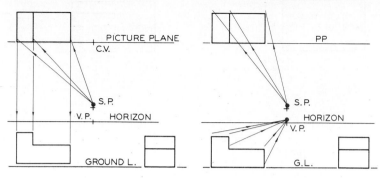
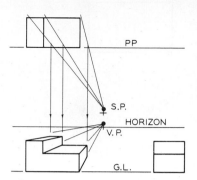

FIG. 26.72 Construction of a one-point perspective.

Step 1 Since the object is parallel to the picture plane, there will be only one vanishing point, located on the horizon below the station point. Projections from the top and side views establish the front plane. This surface is true size, since it lies in the picture plane.

Step 2 Draw projectors from the station point to the rear points of the object in the top view and from the front view to the vanishing point on the horizon. In a one-point perspective, the vanishing point is the front view of the station point.

Step 3 Construct vertical projectors from the top view to the front view from the points where the projectors cross the picture plane. These projectors intersect the lines leading to the vanishing point. This is a one-point perspective, since the lines converge at a single VP.

STATION POINT (SP) is the location of the observer's eye in the plan view. The front view of the station point will always lie on the horizon.

HORIZON is a horizontal line in the front view that represents an infinite horizontal, such as the surface of the ocean.

GROUND LINE is an infinite horizontal line in the front view that passes through the base of the object being drawn.

CENTER OF VISION (CV) is a point that lies on the picture plane in the top view and on the horizon in the front view. In both cases, it is on the line from the station point that is perpendicular to the picture plane.

When drawing any perspective, the station point should be located far enough away from the object so that the perspective can be contained in a cone of vision not more than 30° (Fig. 26.73). If a larger cone of vision is required, the perspective will be distorted.

Measuring points

An additional vanishing point used to locate measurements along the receding lines that vanish to the horizon is the *measuring point*. In Fig. 26.74, the measuring point of a one-point perspective is found by revolving line 0–2 into the picture plane to 0–2' and drawing a construction line from the station point to the picture plane parallel to 2–2'. The measuring point is located on the horizon by projection from the picture plane (Step 1).

Since the distance 0–2 is equal to 0–2', depth dimensions can be laid off along the ground line and

FIG. 26.73 The station point (SP) should be placed far enough away from the object to permit the cone of vision to be less than 30° to reduce distortion.

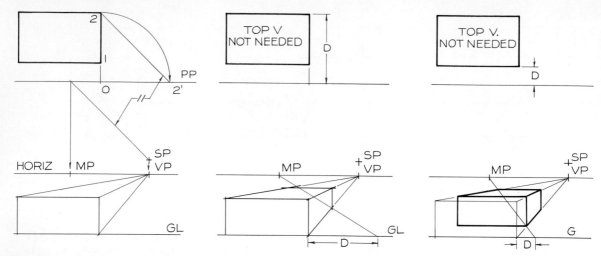

FIG. 26.74 One-point perspective—measuring points.

Step 1 Line 0–2 is revolved into the picture plane to locate point 2′. A line is drawn parallel to 2′–2 through SP to the PP. The measuring point is located on the horizon.

Step 2 Distance D is laid off along the ground line from the front corner of the perspective. This distance is projected to the measuring point to locate the rear corner of the perspective.

Step 3 The front of the object is located by laying off distance D from the corner of the perspective and projecting to the measuring point. This locates the front surface of the object.

then projected to the measuring point. This locates the real corner of the one-point perspective (Step 2). The depth from the picture plane to the front of the object is found in the same manner.

The use of the measuring-point method eliminates the need for placing the top view in the customary top-view position; instead, the dimensions can be more conveniently transferred to the ground line.

26.30
Two-point perspectives

If two surfaces of an object are positioned at an angle to the picture plane, two vanishing points will be required to draw it as a perspective. Different views can be obtained by changing the relationship between the horizontal and the ground line (Fig. 26.75).

An *aerial view* is obtained when the horizon is placed above the ground line and the top of the object in the front view. When the ground line and the horizon coincide in the front view, a *ground-level view* is obtained. A *general view* is obtained when the horizon

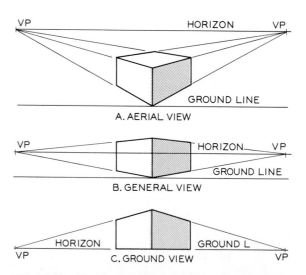

FIG. 26.75 Different perspectives can be obtained by locating the horizontal over, under, and through the object.

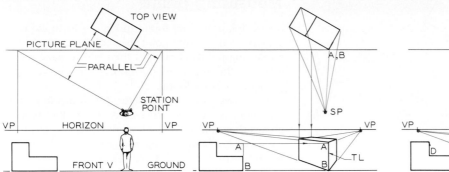

FIG. 26.76 Construction of a two-point perspective.

Step 1 Construct projectors that extend from the top view of the station point to the picture plane parallel to the forward edges of the object. Project these points vertically to the horizon in the front view to locate vanishing points. Draw the ground line below the horizon, and construct the side view on the ground line.

Step 2 Since all lines in the picture plane are true length, line *AB* is true length. Thus, line *AB* is projected from the side view to determine its height. Then project each end of *AB* to the vanishing points. Draw projectors from the SP to the exterior edges of the top view. Project the intersections of these projectors with the picture plane to the front view.

Step 3 Determine point *C* in the front view by projecting from the side view to line *AB*. Draw a projector from point *C* to the left vanishing point. Point *D* will lie on this projector beneath the point where a projector from the station point to the top view of point *D* crosses the picture plane. Complete the notch by projecting to the respective vanishing points.

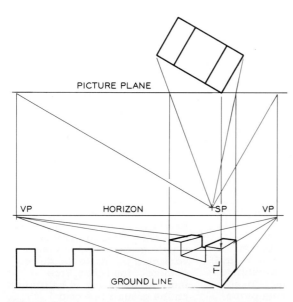

FIG. 26.77 A two-point perspective of an object.

is placed above the ground line and through the object, usually equal to the height of a person.

The steps of constructing a two-point perspective are shown in Fig. 26.76.

Since line *AB* lies in the picture plane, it will be true length in the perspective. All height dimensions must originate at this vertical line because it is the only true-length line in the perspective. Points *C* and *D* are found by projecting to *AB*, and then projecting toward the vanishing points.

A typical two-point perspective is shown in Fig. 26.77. By referring to Fig. 26.76, you will be able to understand the development of the construction used.

The object in Fig. 26.78 does not contact the picture plane in the top view as in the previous examples. To draw a perspective of this object, the lines of the object must be measured where the extended plane intersects the picture plane. The height is measured, and the infinite plane is drawn to the vanishing point. The corner of the object can be located on this infinite plane by projecting the corner to the picture plane in the top view with a projector from the station point.

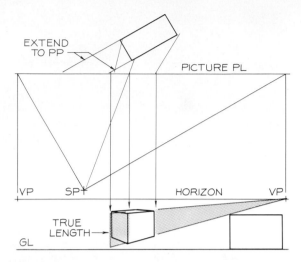

FIG. 26.78 A two-point perspective of an object that is not in contact with the picture plane.

Measuring points

Measuring points are found in Fig. 26.79 to aid in the construction of two-point perspectives. The use of measuring points eliminates the need to have the top view in the top-view position, after the vanishing points have been found.

For two-point perspectives, two vanishing points are used to locate dimensions along the receding planes that vanish to the horizon (Step 2). Depth dimensions are laid off along the ground line from point *B*, and secondary planes are passed from these points to their measuring points.

Measuring points are used in Fig. 26.80 to construct a two-point perspective. No top view is needed since this system is used to locate measurements along the receding lines.

Arcs in perspective

Arcs in two-point perspectives must be found by using coordinates to locate points along the curves in perspective (Fig. 26.81). Points 1 through 7 are located along the semicircular arc in the orthographic view.

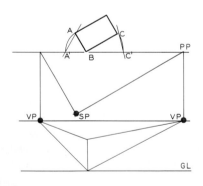 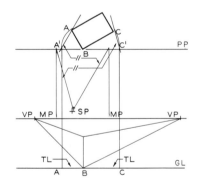 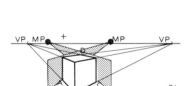

FIG. 26.79 Construction of measuring points.

Step 1 In the top view, revolve lines *AB* and *BC* into the picture plane using point *B* as the center of revolution. Draw construction lines through points *A–A'* and points *C–C'*. These lines represent edge views of vertical planes passing through the corners *A* and *C*.

Step 2 Draw lines from the station point parallel to lines *A–A'* and *C–C'* to the picture plane. Project these points of intersection to the horizon to locate two measuring points. Distances *AB* and *BC* can be laid off true length on the ground line (GL), since they have been revolved into the picture plane in the top view.

Step 3 Extend planes from points *A* and *C* on the ground line to their respective measuring points. These planes intersect the infinite planes, which are extended to their vanishing points, to locate two corners of the block. True measurements can to be laid off on the ground line and projected to measuring points to find corner points.

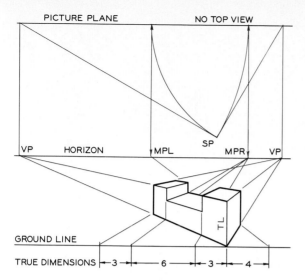

FIG. 26.80 A two-point perspective drawn with the use of measuring points.

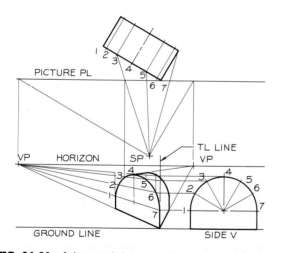

FIG. 26.81 A two-point perspective of an object with semicircular features.

These same points are found in the perspective by projecting coordinates from the top and side views. The points do not form a true ellipse, but an egg-shaped oval. An irregular curve is used to connect the points.

Sloping planes

A nonhorizontal sloping plane will have its vanishing point below or above the horizon, not on the horizon.

In Fig. 26.82, a roof slopes 25° with the horizontal in two directions. The vanishing points of these two planes are found in Step 2 by revolving *AB* into the picture plane to *AB′*, and *B′* is projected to the horizon. From this point, two lines are drawn at 25° with the horizon to the vertical projector through the vanishing point found in Step 1. This locates the vanishing points of the two planes of the roof. The planes of the roof are then located in the perspective and projected to their respective vanishing points.

26.31
Three-point perspectives

A three-point perspective can be constructed by positioning the object so that all its sides make angles with the picture plane. In Fig. 26.83, the top and side views are orthographic views that have been tilted with respect to the picture plane. The station point is the same distance from the picture plane in the side view.

Three vanishing points are located by constructing lines from the two views of the station point that are parallel to the sides of the object. The corner points in perspective are found by projecting to the picture plane and then to the perspective view, where the lincs vanish to their respective vanishing points.

In laying out a three-point perspective, you must begin by positioning the top and side views. The vanishing point for vertical lines will lie along a vertical construction line through the station point (SP).

A short-cut method of drawing a three-point perspective is shown in Fig. 26.84. An equilateral triangle with 60° angles is drawn, and a convenient point 0 is located within the triangle. The vertexes of the triangle are used as the three vanishing points.

Width and depth are laid off on either side of point 0. Height is laid off along a construction line through 0 that is parallel to one on the other sides of the triangle. Height is projected back to the vertical line from 0 to the lower vanishing point.

26.32
Axonometric pictorials
by computer

As previously covered in Sections 10.35 and 10.36, an ISOMETRIC style grid and snap can be used as an aid in making isometric drawings. The STYLE sub-

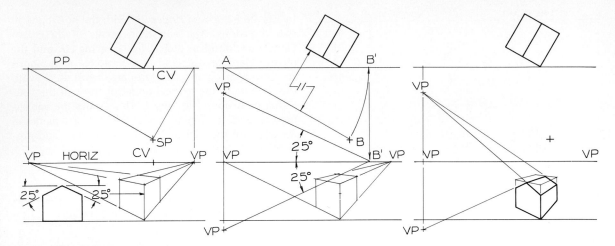

FIG. 26.82 Perspectives of sloping surfaces.

Step 1 The vanishing points are located and the perspective of the object is drawn as if it had no sloping planes. The point where the sloping plane begins is found on the true-length vertical line by projecting from the side view.

Step 2 Line *AB* is rotated in the top view to locate *B'* on PP which is projected to the horizon. Since both planes slope 25°, their vanishing points are drawn at 25° with the horizon from *B'* to a vertical line that passes through the left VP.

Step 3 Each sloping plane is found by using the two vanishing points found in Step 2 and projecting from points found on the vertical lines where the slopes begin.

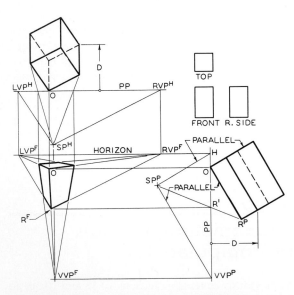

FIG. 26.83 The construction of a three-point perspective.

command of the SNAP command can be used to change the rectangular GRID (called STANDARD-(S) to ISOMETRIC (I) where the dots are plotted vertically and at 30° to the horizontal (Fig. 26.85). While in this mode, the cursor's cross hairs, which can be made to snap to the grid points, will align with two axes of the isometric.

Although an isometric drawing can be made by using the grid and snap features, this method is not truly a three-dimensional system; instead, points must be located manually, and elliptical features must be inserted individually, as shown in Fig. 26.86.

AutoCAD 3-D level 1

AutoCAD contains features within its ADE-3 package that provide limited three-dimensional capabilities for drawing axonometric pictorials with three commands: ELEV, VPOINT, and HIDE. These commands enable you to draw the plan view of an object, specify its height, select a desired viewpoint, obtain its axonometric projection, and remove hidden lines from it. Each point of the object is rotated as a three-dimensional point in space when each viewpoint is selected, and circular features are converted to ellipses.

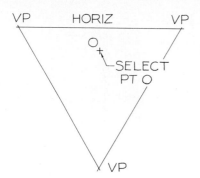
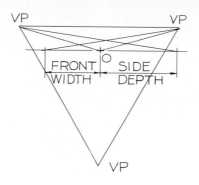
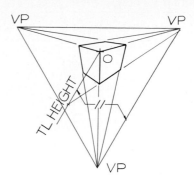

FIG. 26.84 Three-point perspective.

Step 1 Draw an equilateral triangle. The vanishing points lie at each vertex. Locate point *0* anywhere within this triangle, and it will be the beginning point of the perspective.

Step 2 Draw a horizontal through point *0*. The width and depth can be laid off from point *0*. Projectors are drawn to the two upper vanishing points to locate the top surface.

Step 3 The height is laid off along a line parallel to one of the triangle's sides. The height is projected back to where it intersects a line from point *0* to the lower VP.

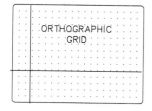
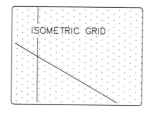

A. ORTHOGRAPHIC GRID   B. ISOMETRIC GRID

FIG. 26.85 The SNAP command permits you to rotate the orthographic grid (Standard) or to select the isometric grid option that can be used as an aid in drawing isometric pictorials.

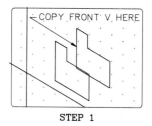
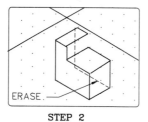

STEP 1   STEP 2

FIG. 26.86 Isometric drawings-computer.

Step 1 When the isometric grid is on the screen, the cursor can be made to SNAP to it. The front view is drawn as an isometric plane and is duplicated at the back of the object by using the COPY command.

Step 2 The hidden lines are ERASEd to complete the isometric drawing.

A box with a cylindrical hole through it is drawn as follows:

```
Command: ELEV
New current elevation <.20
current elevation>: 0
New current thickness <0.000>: 4
```

Thickness is the distance the object extends above its current elevation. A minus thickness results in downward extrusion of the plane of the object drawn on elevation 0. We can now draw the box:

```
Command: LINE
From point: (Draw plan view of box)
```

Since the elevation and thickness of the cylindrical hole will be the same as that of the box, these values need not be changed, and the cylinder is located in the plan view as a circle:

```
Command: CIRCLE
3P/2P/<Center point>: (Specify center)
Diameter/<Radius>: 1.0
```

VPOINT command can now be used to select a viewpoint of the object that you want to see in three dimensions. Respond by entering the desired location of your viewpoint, such as 1, -1, 1, which locates the point of vision at $X=1$, $Y=-1$, and $Z=1$ from the

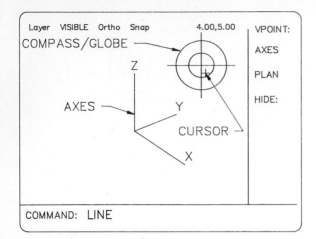

FIG. 26.87 When using the AXES option under the VPOINT command, the screen will show the three axes and compass/globe that can be used to select a viewpoint for an axonometric projection of the object being drawn. As the cursor is moved about the globe, the axes will change on the screen to show the current relationship of the axes.

origin of 0, 0, 0. By not giving coordinates but selecting AXES from the Root Menu, the X-, Y-, and Z-axes will be shown on the screen (Fig. 26.87). The circles represent the compass on a globe. The center point is the north pole (0, 0, 1), the inner circle is the equator (n, n, 1), and the outer circle is the south pole (0, 0, −1). When the cursor is moved about the compass, the tripod axes will change accordingly to show the axes of the axonometric that will be drawn.

The viewpoint can be selected in the following manner:

```
Command: VPOINT
Enter view point <current X, Y, Z
view points>: 1, −1, 1
```

Press CR and a three-dimensional wire diagram will appear on the screen (Fig. 26.88).

The Root Menu will show subcommands PLAN, AXES, and HIDE under the VPOINT command. By selecting the PLAN command, the drawing will return to its plan view for modification if needed.

The HIDE command is illustrated in Fig. 26.89, where two boxes with cylindrical holes through them have been drawn. The one at the left was drawn using the DRAW and CIRCLE commands, and the one at the right was drawn the same way but the SOLID

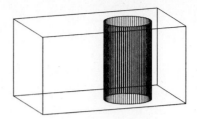

FIG. 26.88 This axonometric drawing is a wire diagram of a cylindrical hole in a box.

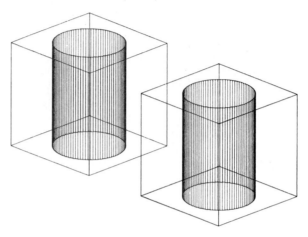

FIG. 26.89 Two cubes with cylindrical holes through them.

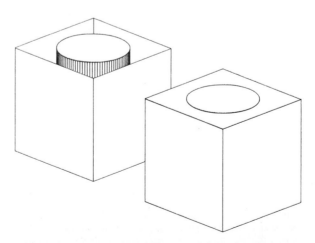

FIG. 26.90 When the HIDE command is applied to the cubes drawn in Fig. 26.89, visibility for each is shown differently. The cube at the right was filled in with the SOLID command when it was drawn in the PLAN mode.

command was used to fill in the plan view (over the circle). The viewpoint was selected, CR pressed, and a wire diagram of each object was drawn on the screen. The HIDE command is selected from the Root Menu, and AutoCAD calculates the visibility for each line and redraws the objects on the screen (Fig. 26.90). The object at the left shows the box as an empty box, whereas the one at the right shows the box as a solid.

To plot or print the axonometric projections with the hidden lines removed, you must respond during plotter configuration when prompted as follows:

Remove hidden lines? <N>: Y

This response will remove the hidden lines at the time of plotting.

Irregular shapes can also be drawn as axonometrics by using the PLINE command (Fig. 26.91). The Root Menu will show subcommands PLAN, AXES, and HIDE under the VPOINT command.

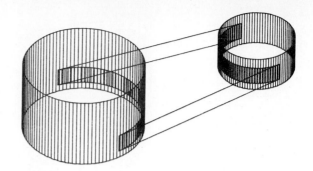

FIG. 26.92 A three-dimensional drawing with varying elevations can be constructed with AutoCAD's 3-D option.

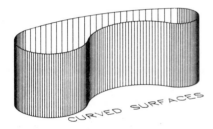

FIG. 26.91 This object with curved surfaces was drawn using a combination of ARC and PLINE commands.

Objects drawn with TRACE lines, wide POLY-LINES, CIRCLES, or SOLIDS will have solid tops and bottoms when they are extruded. Therefore when the HIDE command is used, the objects will appear as solids rather than as empty perimeters. A "wire-diagram drawing" is often better than one where visibility has been shown, since all features can be seen better.

Surfaces can be drawn on multiple elevations and thicknesses to represent a part (Fig. 26.92). The cylinders have the same base elevation, 0, and thicknesses of 2 and 1, respectively. The rib has an elevation of 0.2 and a thickness of 0.5.

All pictorials using this option of AutoCAD will be axonometrics with projection lines parallel to the picture plane. Perspective drawings cannot be drawn with this system. However, you may obtain an infinite number of views of each part, and the view that you wish to have, by selecting VPOINTS.

TEXT and BLOCKS can be inserted in three-dimensional drawings when the drawing is in its PLAN mode. When converted to axonometric, the text will lie on the ELEVation at which it was inserted. BLOCKS will be inserted on the current layer and will be extruded to the thickness of the part.

26.33
Perspectives by computer

Design Board Professional® a software package produced by MegaCADD, Inc., is covered in this section to illustrate how perspectives can be constructed in three-dimensional space. The MegaCADD program offers three basic modes for developing perspective drawings of objects: CREATE, VIEW, and MODIFY. The CREATE mode (Fig. 26.93) gives a screen on the monitor on which a perspective can be drawn by establishing three-dimensional shapes in their plan views. The X- and Y-axes are given at the perimeters of the plan view with a grid for locating measurements.

The VIEW mode gives a screen on which you may select a viewpoint from which to look at the orthographic views (Fig. 26.94). The orthographic views of the drawing will appear at the right of the screen, and the corresponding perspective view will be shown in the large area at the left. By selecting CREATE, the screen will return to the mode shown in Fig. 26.93, where the perspective axes are shown.

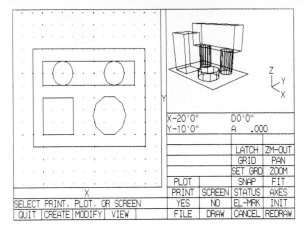

FIG. 26.93 The CREATE screen of MegaCADD's Design Board Professional® is used to create the orthographic views of the objects that will be converted into perspective views.

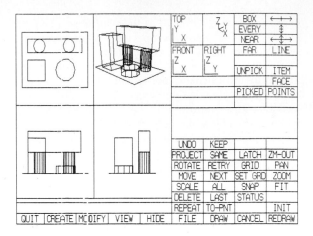

FIG. 26.95 The MODIFY screen of DBPro®. shows three orthographic views and the perspective of the model created. Modifications of the model can be made in this mode.

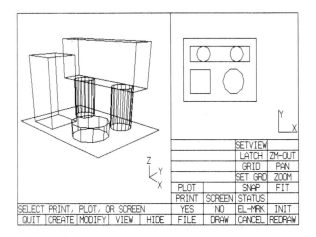

FIG. 26.94 The VIEW screen of DBPro®. shows the plan view of the model and its corresponding perspective view. The viewpoint can be changed by selecting the observer's position and the target position in the plan view.

The MODIFY mode (Fig. 26.95) gives three orthographic views (top, front, and right-side views) of the model as well as its perspective. Parts of the drawing can be deleted, changed, moved, enlarged, and edited in a number of ways. Once the model is completed and the viewpoint selected, the DRAW command can be used to obtain a drawing on the

screen or a hard copy by printer or plotter (Fig. 26.96). The plotted drawing in Fig. 26.96 is a wire diagram. To show visibility, hidden lines can be omitted before plotting (Fig. 26.97). Another wire-diagram view of the model from a different position is shown in Fig. 26.98. Views can be obtained from any position, from any distance around the model, and from any height.

MegaCADD's Prohouse is viewed from different locations in Fig. 26.99, where the hidden lines have been removed. A series of views can be generated to

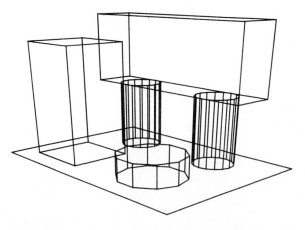

FIG. 26.96 A wire diagram of several three-dimensional solids drawn by DBPro®.

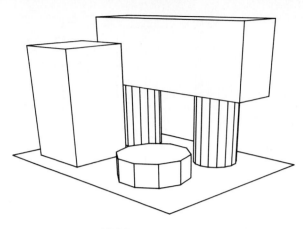

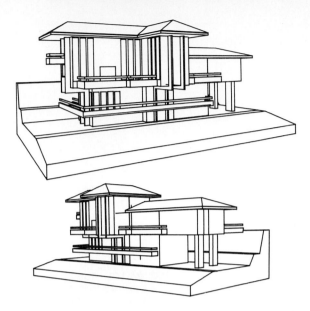

FIG. 26.97 The HIDE command is used to remove hidden lines from the wire diagram plotted in the previous figure.

FIG. 26.99 An example of different views obtained from the same house developed by DBPro.

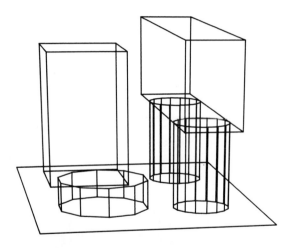

FIG. 26.98 An infinite number of views may be obtained from each model.

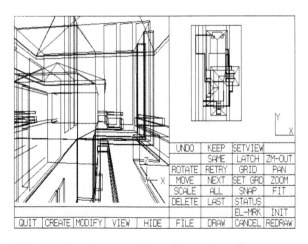

FIG. 26.100 A viewpoint and target position can be located in the plan view at the right to obtain the perspective shown at the left when the screen is in the VIEW mode.

represent "walk-through" views of the house by locating viewpoints and target points within the house (Fig. 26.100). The hidden lines can be removed and the view can be enlarged to give a perspective from this vantage point (Fig. 26.101). Since the house was drawn in six hours, some details were omitted; however, the true value of the views lies in their giving a quick means of analyzing the major shapes within the structure.

A different application of this powerful software program is the representation of an entire downtown area for the comparison and study of a group of build-

ings (Fig. 26.102). Like the house, viewpoints can be selected at sky level or street level to simulate the view from any position. Views also can be generated that represent a walk through the town.

MegaCADD offers methods of drawing curved surfaces, domes, inclined surfaces, free forms, ortho-

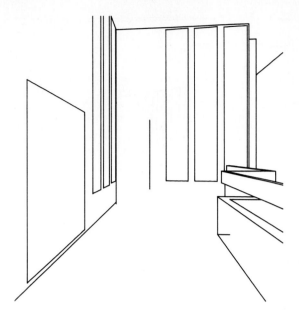

FIG. 26.101 The hidden lines are omitted to give a "walk-through" view of the house.

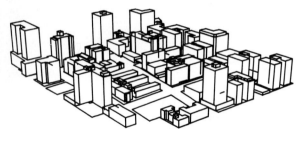

FIG. 26.102 A perspective view of MegaCADD's Urban City.

graphic views, and stereographic views that can be used for representing perspectives for viewing models in three dimensions. This software package provides an excellent means of drawing three-dimensional pictorials.

26.34
Overlay film

Overlay film is an acetate film on which is printed a pattern that can be applied to a drawing to shade an illustration. These films have adhesive backings and can be burnished permanently to the surface of a drawing. Patterns, symbols, letters, numbers, arrow-

heads, and other designs are available on overlay film. Arrowheads are applied to an assembly in Fig. 26.103.

Overlay films are available in dull and glossy surfaces as well as in percentage screens labeled in percent or lines per inch (Fig. 26.104). The percentage indicates the percentage of solid black; the lines per inch represent the number of rows of dots per inch. A 27.5-line screen is an open screen, whereas a 55-line screen is composed of smaller dots twice as close together.

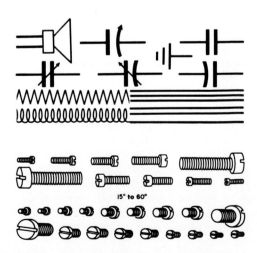

FIG. 26.103 Examples of a few of the symbols available in 9″ × 12″ sheets of overlay film. (Courtesy of Artype, Inc.)

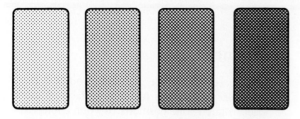

FIG. 26.104 Overlay film is available in screens at 10% increments. A 10% screen is 10% solid black, and a 70% screen is 70% solid black.

26.35
Photographic illustrations

Technical illustrations can be made from photographs as shown in Fig. 26.105. Drawings can emphasize certain aspects of the photograph that might not be otherwise clear. The photograph can be converted to a drawing by projecting it into the drawing surface by an opaque projector, where it can be outlined and then rendered.

Large-size photographs can be overlaid with tracing vellum or polyester film and the drawing made by tracing from the photograph. A bold illustration is required if it is to be reduced to a much smaller size for publication.

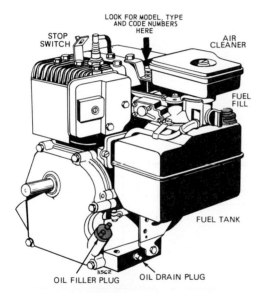

LOOK FOR MODEL, TYPE AND CODE NUMBERS HERE

STOP SWITCH

AIR CLEANER

FUEL FILL

FUEL TANK

OIL FILLER PLUG   OIL DRAIN PLUG

FIG. 26.105 An example of a line drawing made from a photograph to improve its reproduction when reduced to a small size. (Courtesy of Briggs & Stratton Corp.)

Problems

The following problems (Fig. 26.106) are to be shown on Size A or B sheets as assigned. Select the appropriate scale that will take advantage of the space available on each sheet. By setting each square to 0.20 inch (about 5 mm) two problems can be drawn on each Size A sheet. When each square is set to 0.40 inch (about 10 mm), one problem can be drawn on each Size B sheet.

Obliques

1–24. Construct either cavalier, cabinet, or general obliques of the parts assigned in Fig. 26.106.

Isometrics

1–24. Construct isometric drawings of the objects assigned in Fig. 26.106.

Axonometrics

1–24. Construct axonometric scales, and find the ellipse template angles for each surface by using a cube rotated into positions of your choice. Calibrate the scales to represent ½-in. intervals. Overlay the scales, and draw axonometrics of the objects assigned from Fig. 26.106.

Perspectives

1–24. Lay out perspective views of the assigned parts in Fig. 26.106 on Size B sheets.

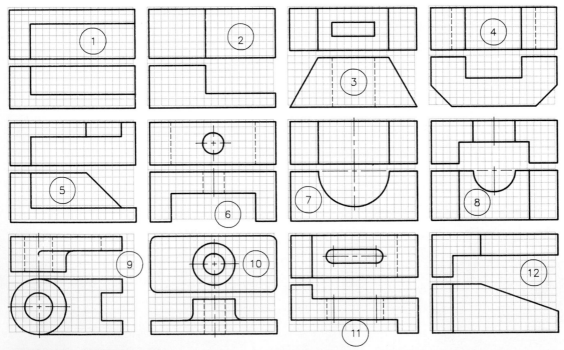

FIG. 26.106 Problems 1–24.

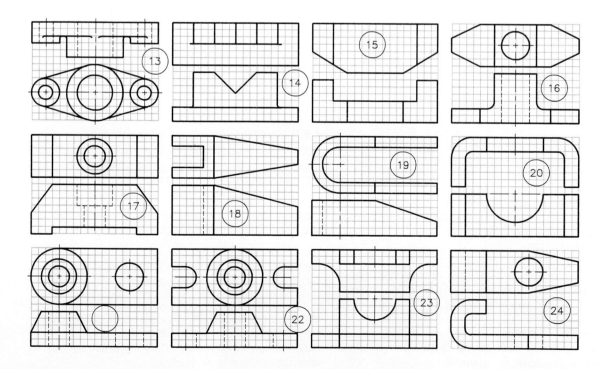

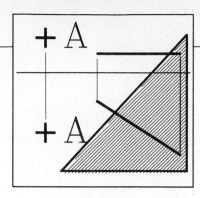

CHAPTER 27

Points, Lines, and Planes

27.1

Introduction

Points, lines, and planes are the basic geometric elements used in descriptive geometry. The following rules of solving and labeling descriptive problems are illustrated in Fig. 27.1.

LETTERING should be labeled using $\frac{1}{8}$-in. letters with guidelines. Lines should be labeled at each end and planes at each corner. Either letters or numbers can be used.

POINTS should be indicated by two short perpendicular dashes that form a cross, *not* a dot. Each dash should be approximately $\frac{1}{8}$-in. long.

POINTS ON A LINE should be indicated with a short perpendicular dash on the line, *not* a dot. Label the point with a letter or numeral.

REFERENCE LINES are thin black lines that should be labeled in accordance with the instructions given in Section 15.3.

OBJECT LINES are used to represent points, lines, and planes. They should be drawn heavier than reference lines, with an H or F pencil. Hidden lines are drawn thinner than visible lines.

TRUE-LENGTH LINES should be labeled by the full note, TRUE LENGTH, or by the abbreviation TL.

TRUE-SIZE PLANES should be labeled by a note, TRUE SIZE, or by the abbreviation TS.

PROJECTION LINES should be precisely drawn with a 4H pencil. These should be thin, gray lines, just dark enough to be visible. They need not be erased after the problem is completed.

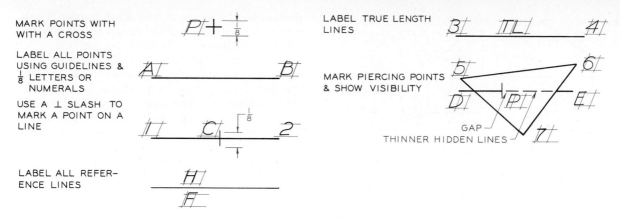

MARK POINTS WITH
WITH A CROSS

LABEL ALL POINTS
USING GUIDELINES &
$\frac{1}{8}$ LETTERS OR
NUMERALS

USE A ⊥ SLASH TO
MARK A POINT ON A
LINE

LABEL ALL REFER-
ENCE LINES

LABEL TRUE LENGTH
LINES

MARK PIERCING POINTS
& SHOW VISIBILITY

GAP
THINNER HIDDEN LINES

FIG. 27.1 Standard practices for labeling points, lines, and planes.

27.2
Orthographic projection of a point

A point is a theoretical location in space; it has no dimension. However, a series of points can establish areas, volumes, and lengths.

A point must be projected perpendicularly onto at least two principal planes to establish its position (Fig. 27.2). Note that when the planes of the projection box (Fig. 27.2A) are opened into the plane of the drawing surface (Fig. 27.2B), the projectors from each view of point 2 are perpendicular to the reference lines be-

tween the planes. The letters *H, F,* and *P* are used to represent the horizontal, frontal, and profile planes, the three principal projection planes.

A point can be located from verbal descriptions with respect to the principal planes. For example, point 2 in Fig. 27.2 can be described as being (1) 4 units left of the profile plane, (2) 3 units below the horizontal plane, and (3) 2 units behind the frontal plane.

When looking at the front view, the horizontal and profile planes appear as edges. The frontal and profile planes appear as edges in the top view, and the frontal and horizontal planes appear as edges in the side view.

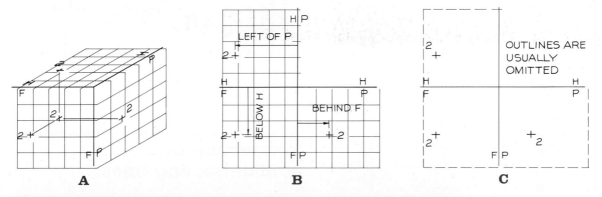

FIG. 27.2 A. The three projections of point 2 are shown pictorially. **B.** The three projections are shown orthographically. The projection planes are opened into the plane of the drawing paper. Point 2 is 4 units to the left of the profile, 3 below the horizontal, and 2 behind the frontal. **C.** The outlines of the projection are usually omitted in orthographic projection.

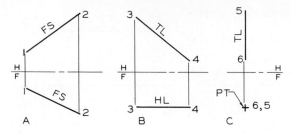

FIG. 27.3 A line in orthographic projection can appear as a point (PT), foreshortened (FS), or true length (TL).

27.3
Lines

A line is a straight path between two points in space. A line can appear as (1) a foreshortened line, (2) a true-length line, or (3) a point (Fig. 27.3).

OBLIQUE LINES are neither parallel nor perpendicular to a principal projection plane (Fig. 27.4). When line 1–2 is projected onto the horizontal, frontal, and profile planes, it appears foreshortened in each view.

PRINCIPAL LINES are parallel to at least one of the principal projection planes. A principal line is true length in the view where the principal plane to which it is parallel appears true size. The three types of principal lines are horizontal, frontal, and profile lines.

A *horizontal line* is shown in Fig. 27.5A, where it appears true length in the horizontal view, the top

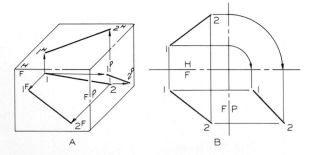

FIG. 27.4 A. A pictorial of the orthographic projection of a line. **B.** A standard orthographic projection of a line.

view. It may be shown in an infinite number of positions in the top view and still appear true length provided it is parallel to the horizontal plane.

An observer cannot tell whether the line is horizontal when looking at the top view. This must be determined from looking at the front or side views where a horizontal line will be parallel to the edge view of the horizontal, which is the *H-F* fold line. A line that projects as a point in the front view is a combination horizontal and profile line.

A *frontal line* is parallel to the frontal projection plane, and it appears true length in the front view since the observer's line of sight is perpendicular to it in this view. Line 3–4 in Fig. 27.5B is determined to be a frontal line by observing its top and side views where the line is parallel to the edge view of the frontal plane.

A *profile line* is parallel to the profile projection planes and it appears true length in the side views, the profile views. To tell whether or not a line is a profile line, it is necessary to look at a view adjacent to the profile view. In Fig. 27.5C, line 5–6 is parallel to the edge view of the profile plane in the top and side views.

27.4
Location of a point on a line

The top and front views of line 1–2 are shown in Fig. 27.6. Point 0 is located on the line in the top view, and it is required that the front view of the point be found.

Since the projectors between the views are perpendicular to the *H-F* fold line in orthographic projection, point 0 is found by projecting in this same direction from the top view to the front view of the line.

If a point is to be located at the midpoint of a line, it will be at the line's midpoint whether the line appears true length or foreshortened.

27.5
Intersecting and nonintersecting lines

Lines that intersect have a point of intersection that lies on both lines and is common to both. Point 0 in Fig. 27.7a is a point of intersection since it projects to a common crossing point in the three views.

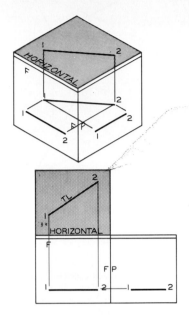

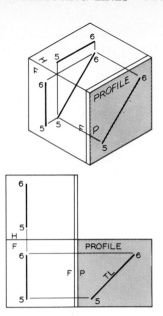

FIG. 27.5 Principal lines.

A. The horizontal line is true length in the horizontal view (the top view). It will appear parallel to the edge view of the horizontal plane in the front and side views.

B. The frontal line is true length in the front view. It will appear parallel to the edge view of the frontal plane in the top and side views.

C. The profile line is true length in the profile view (the side view). It will appear parallel to the edge view of the profile plane in the top and front views.

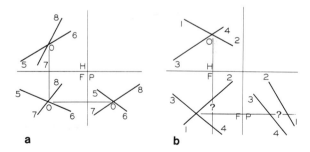

FIG. 27.7 **a.** The lines intersect because the point of intersection, 0, projects as a point of intersection in all views. **b.** The lines cross in the top and front views, but they do not intersect. A common point of intersection does not project from view to view.

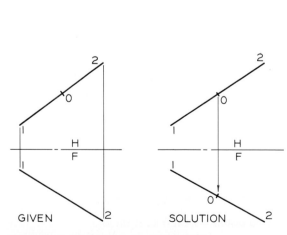

FIG. 27.6 A point on a line shown orthographically can be found on the front view by projection. The direction of the projection is perpendicular to the reference line between the two views.

On the other hand, the crossing point of the lines in Fig. 27.7b in the front view is not a point of intersection. Point 0 does not project to a common crossing point in the top view. Therefore the lines do not intersect although they do cross.

27.6
Visibility of crossing lines

Lines *AB* and *CD* in Fig. 27.8 do not intersect; however, it is necessary to determine the visibility of the lines by analysis. Select a crossing point in one of the views, the front view in Step 1, and project it to the top view to determine which line is in front of the other. This process of determining visibility is done by analysis rather than by visualization. If only one view were available, it would be impossible to determine visibility.

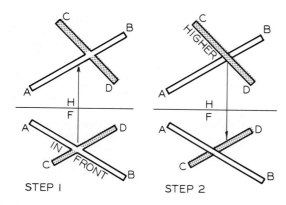

FIG. 27.8 Visibility of lines.

Required   Find the visibility of the lines in both views.

Step 1   Project the point of crossing from the front to the top view. This projector strikes *AB* before *CD*; therefore, line *AB* is in front and is visible in the front view.

Step 2   Project the point of crossing from the top view to the front view. This projector strikes *CD* before *AB*; therefore, *CD* is above *AB* and is visible in the top view.

27.7
Visibility of a line
and a plane

The principle of visibility of intersecting lines is used in determining the visibility for a line and a plane. In Step 1 of Fig. 27.9, the intersections of *AB* and lines 1–3 and 2–3 are projected to the top view to deter-

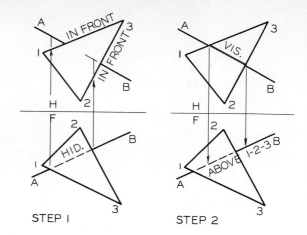

FIG. 27.9 Visibility of line and a plane.

Required   Find the visibility of the plane and the line in both views.

Step 1   Project the points where *AB* crosses the plane in the front view to the top view. These projectors encounter lines 1–3 and 2–3 of the plane first; therefore, the plane is in front of the line, making the line invisible in the front view.

Step 2   Project the points where *AB* crosses the plane in the top view to the front view. These projectors encounter *AB* first; therefore, the line is higher than the plane, and the line is visible in the top view.

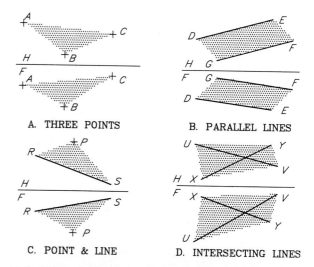

FIG. 27.10 Representations of a plane.

A. Three points not on a straight line.

B. Two parallel lines.

C. A line and a point not on the line or its extension.

D. Two intersecting lines.

mine that the lines of the plane are in front of *AB* in the front view. Therefore the line is shown as a dashed line in the front.

Similarly, the two intersections on *AB* in the top view are projected to the front view, where line *AB* is found to be above the two lines of the plane, 2–3 and 1–3. Since it is above the plane, *AB* is drawn as visible in the top view.

27.8
Planes

A plane can be represented by any of the four methods shown in Fig. 27.10. Planes in orthographic pro-

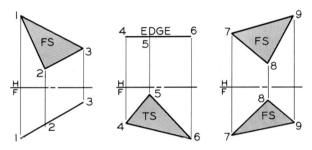

FIG. 27.11 A plane in orthographic projection can appear as an edge, true size (TS), or foreshortened (FS). If a plane is foreshortened in all principal views, it is an oblique plane.

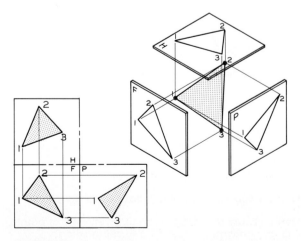

FIG. 27.12 An oblique plane is one that is neither parallel nor perpendicular to a projection plane; it can be called a general-case plane.

jection can appear as (1) an edge, (2) true-size plane, (3) a foreshortened plane (Fig. 27.11).

OBLIQUE PLANES are not parallel to principal projection planes in any view (Fig. 27.12).

PRINCIPAL PLANES are parallel to projection planes (Fig. 27.13). The three types of principal planes are frontal, horizontal, and profile planes.

A *frontal plane* is parallel to the frontal projection plane, and it appears true size in the front view. To tell that the plane is frontal, you must observe the top or side views where its parallelism to the edge view of the frontal plane can be seen.

A *horizontal plane* is parallel to the horizontal projection plane, and it is true size in the top view. To tell that the plane is horizontal, you must observe the front or side views where its parallelism to the edge view of the horizontal plane can be seen.

A *profile plane* is parallel to the profile projection plane, and it is true size in the side view. To tell that it is a profile plane, you must observe the top or front views where its parallelism to the edge view of the profile plane can be seen.

27.9
A line on a plane

Line *AB* is given on the front view of the plane in Fig. 27.14. It is required to find the top view of the line. Points *A* and *B* that lie on lines 1–4 and 2–3 of the plane can be projected to the top view to the same lines of the plane. Points *A* and *B* are found in the top view and are connected to complete the top view of line *AB*. This is an extension of the principle covered in Section 27.4.

27.10
A point on a plane

Point 0 is given on the front view of plane 4–5–6 in Fig. 27.15. It is required to locate the point on the plane in the top view. In Step 1, a line in any direction but vertical is drawn through the point to establish a line on the plane. The line is projected to the top view in Step 2, and the point is projected from the front view to the top view of the line.

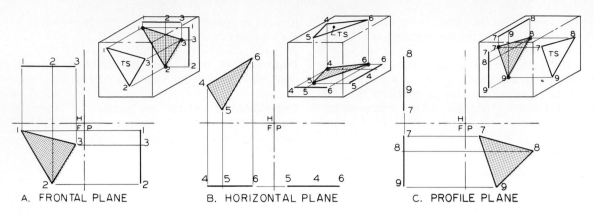

FIG. 27.13 Principal planes.

A. The horizontal line is true length in the horizontal view (the top view). It will appear parallel to the edge view of the horizontal plane in the front and side views.

B. The frontal line is true length in the front view. It will appear parallel to the edge view of the frontal plane in the top and side views.

C. The profile line is true length in the profile view (the side view). It will appear parallel to the edge view of the profile plane in the top and front views.

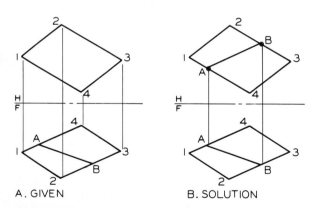

FIG. 27.14 If line *AB* lying on the plane is given, the top view of the line can be found. Points *A* and *B* are projected to lines 1–4 and 2–3, respectively, and are connected to form line *AB*.

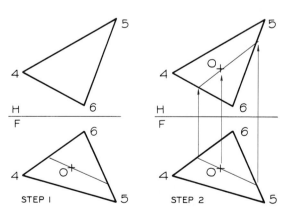

FIG. 27.15 Location of a point on a plane.

Find the top view of point 0 that lies on the plane.

Step 1 Draw a line through the given view of point 0 in any convenient direction except vertical.

Step 2 Project the ends of the line to the top view, and draw the line. Point 0 is projected to the line.

27.11
Principal lines on a plane

Principal lines may be found in any view of a plane when at least two views of the plane are given. An infinite number of principal lines can be drawn on a single plane.

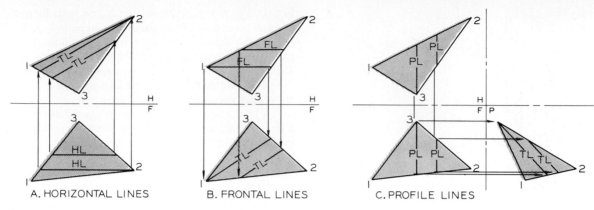

FIG. 27.16 Principal lines on a plane.

A. Horizontal lines are drawn first in the front view parallel to the edge view of the horizontal plane. These lines are found true length when they are projected to the top view.

B. Frontal lines are drawn first in the top view parallel to the edge view of the frontal plane. These lines are found true length when they are projected to the front view.

C. Profile lines are drawn first in the front view parallel to the edge view of the profile plane. These lines are found true length when they are projected to the profile view (side view).

Horizontal lines parallel to the edge view of the horizontal projection in Fig. 27.16A are drawn in the front view of the plane. These horizontal lines are projected to the top view of the plane, where they are horizontal and true length.

Frontal lines are drawn parallel to the frontal projection plane in the top view in Fig. 27.16B. When projected to the front view, the lines are true length.

Profile lines are drawn parallel to the profile projection plane in the top and front views in Fig. 27.16C. When projected to the side view, the lines will appear true length.

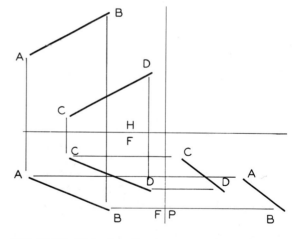

FIG. 27.17 When two lines are parallel, they will project as parallel in all orthographic views.

27.12
Parallelism of lines

If two lines are parallel, they will appear parallel in all views in which they are seen, except where both appear as points.

Lines *AB* and *CD* appear parallel in three views in Fig. 27.17. Parallelism of lines in space cannot be determined if only one view is given; two or more views are required. Using this principle, a line can be drawn parallel to a given line through a specified point (Fig. 27.18).

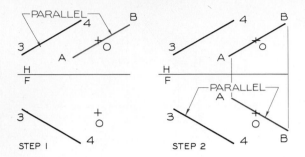

FIG. 27.18 A line parallel to a line.

Draw a line through 0 parallel to the given line.

Step 1 Draw line *AB* parallel to the top view of line 3–4 with its midpoint at 0.

Step 2 Draw the front view of line *AB* parallel to the front view of 3–4 through point 0.

27.13

Parallelism of a line and a plane

A line is parallel to a plane if it is parallel to any line in the plane.

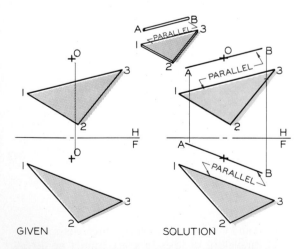

FIG. 27.19 A line can be drawn through point 0 that is parallel to the given plane if the line is parallel to any line in the plane. Line *AB* is drawn parallel to line 1–3 in the front and top views, making it parallel to the plane.

In Fig. 27.19, it is required that a line with its midpoint at point 0 be drawn parallel to plane 1–2–3. This is done by drawing line *AB* parallel to a line in the plane, line 1–3 in this case, in the top and front views. The line could have been drawn parallel to *any* line in the plane; therefore for lines parallel to a given plane, an infinite number of positions exist.

A similar example is shown in Fig. 27.20, where a line with its midpoint at 0 and parallel to the plane formed by two intersecting lines is required.

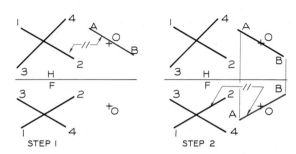

FIG. 27.20 A line parallel to plane.

Draw a line parallel to the plane represented by two intersecting lines.

Step 1 Line *AB* is drawn parallel to line 1–2 through point 0.

Step 2 Line *AB* is drawn parallel to the same line, line 1–2, in the front view, which makes *AB* parallel to the plane.

27.14

Parallelism of planes

Two planes are parallel when intersecting lines in one plane are parallel to intersecting lines in the other (Fig. 27.21).

It is easy to determine that planes are parallel when both appear as parallel edges in a view. In Fig. 27.22, it is required that a line be drawn through point 0 and parallel to plane 1–2–3. In Step 1, *EF* is drawn through point 0 and parallel to line 1–2 in both the top and front views. In Step 2, a second line is drawn through point 0 parallel to line 2–3 of the plane in the front and top views. These two intersecting lines form a plane parallel to plane 1–2–3.

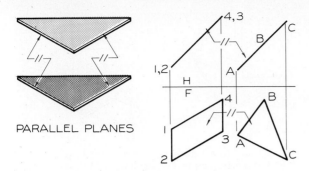

FIG. 27.21 Two planes are parallel when intersecting lines in one are parallel to intersecting lines in the other. When parallel planes appear as edges, the edges will be parallel.

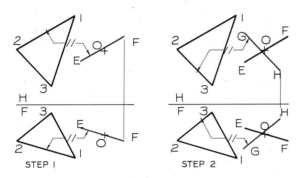

FIG. 27.22 A plane through a point parallel to a plane.

Draw a plane through point 0 parallel to the given plane.

Step 1   Draw line *EF* parallel to any line in the plane, line 1–2 in this case. Show the line in both views.

Step 2   Draw a second line parallel to line 2–3 in the top and front views. These two intersecting lines represent a plane parallel to 1–2–3.

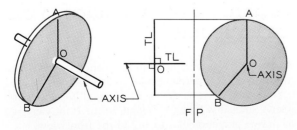

FIG. 27.23 Perpendicular lines will intersect at 90° angles in a view where one or both of the lines appear true length.

27.15
Perpendicularity of lines

> When two lines are perpendicular, they will project at true 90° angles if one or both are true length (Fig. 27.23). When two lines are perpendicular but neither is true length, they will not project with a true 90° angle.

In Fig. 27.23, it can be seen that the axis is true length in the front view; therefore any spoke will be shown perpendicular to the axis in the front view. Spokes 0*A* and 0*B* are examples where one is true length and the other foreshortened.

27.16
A line perpendicular to a principal line

In Fig. 27.24, it is required that a line be constructed through point 0 perpendicular to frontal line 5–6, which is true length in the front view. In Step 1, 0*P* is drawn perpendicular to 5–6 since it is true length. In

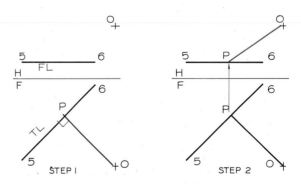

FIG. 27.24 A line perpendicular to a frontal line.

Draw a line from point 0 perpendicular to line 5–6.

Step 1   Since line 5–6 is a frontal line and is true length in the front view, a perpendicular from point 0 will make a true 90° angle with it.

Step 2   Project point *P* to the top view and connect it to point 0. Since neither line is true length, they will not intersect at 90° in the top view.

Step 2, point *P* is projected to the top view of 5–6. Line 0*P* in the top view cannot be drawn as a true 90° angle since neither of the lines is true length in this view.

27.17

A line perpendicular to an oblique line

In Fig. 27.25, it is required that a line be constructed from point 0 perpendicular to oblique line 1–2. A line could have been found perpendicular to 1–2 in the front view by drawing a frontal line in the top view.

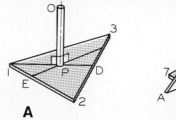
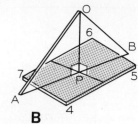

FIG. 27.26   **A.** A line is perpendicular to a plane if it is perpendicular to two intersecting lines on the plane. **B.** A plane is perpendicular to another plane if the plane contains a line perpendicular to the other plane.

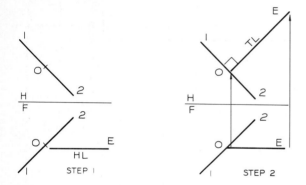

FIG. 27.25   A line perpendicular to an oblique line.

Draw a line from point 0 perpendicular to line, 1–2.

Step 1   Draw a horizontal line from point 0 in the front view.

Step 2   Horizontal line 0*E* will be true length in the top view; therefore, it can be drawn perpendicular to line 1–2 in this view.

27.18

Perpendicularity involving planes

A line can be drawn perpendicular to a plane if it is drawn perpendicular to any two intersecting lines in the plane (Fig. 27.26A). A plane is perpendicular to another plane if a line in one is perpendicular to the other (Fig. 27.26B).

27.19

A line perpendicular to a plane

In Fig. 27.27, it is required that a line be drawn from point 0 on the plane perpendicular to the plane.

In Step 1, a frontal line is drawn on the plane in the top view and is projected to the front view, where the line is true length. Line 0*P* is drawn perpendicular to the true-length line.

In Step 2, a horizontal line is drawn in the front view and then in the top view of the plane through point 0 perpendicular to the true-length line. This results in a line perpendicular to the plane since the line is perpendicular to two intersecting lines in the plane, a horizontal and a frontal line.

27.20

A plane perpendicular to an oblique line

In Fig. 27.28, it is required that a plane be constructed through point 0 perpendicular to line 1–2.

In Step 1, *AB* is drawn as a frontal line in the top view. Since it will be true length in the front view, *AB* is drawn perpendicular to 1–2 in the front view.

In Step 2, *CD* is drawn as a horizontal line in the front view. Line *CD* is drawn perpendicular to 1–2 in the top view since *CD* is true length in this view. These

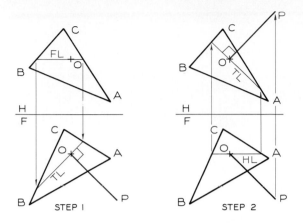

FIG. 27.27 A line perpendicular to a plane.

Draw a line from point 0 perpendicular to the plane.

Step 1 Construct a frontal line on the plane through 0 in the top view. This line is true length in the front view; therefore, line 0P can be drawn perpendicular to the true-length line.

Step 2 Construct a horizontal line through point 0 in the front view. This line is true length in the top view; therefore, line 0P can be drawn perpendicular to it.

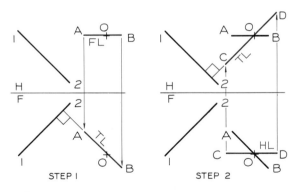

FIG. 27.28 A plane through a point perpendicular to a line.

Draw a plane through 0 perpendicular to line 1–2.

Step 1 Draw line AB as a frontal line in the top view. Since it will be true length in the front view, it can be drawn perpendicular to line 1–2.

Step 2 Draw line CD as a horizontal line in the front view. Since it will be true length in the top view, it can be drawn perpendicular to line 1–2. The intersecting lines form a plane perpendicular to the line.

two intersecting lines, AB and CD, form a plane perpendicular to line 1–2.

27.21
Perpendicularity of planes

In Fig. 27.29, it is required that a plane be constructed through line AB perpendicular to plane 1–2–3.

In Step 1, a true-length frontal line is found on the plane. Line CD is drawn through AB and perpendicular to the frontal line in the plane.

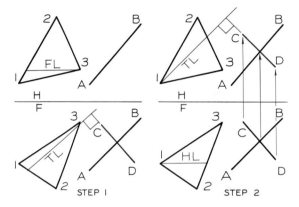

FIG. 27.29 A plane through a line perpendicular to a plane.

Draw a plane through the line perpendicular to the plane.

Step 1 Draw a frontal line on the plane in the top view, and find its true-length view in the front view. Line CD is drawn through line AB and perpendicular to the true-length line.

Step 2 Draw a horizontal line on the plane in the front view, and find its true-length view in the top view. Draw line CD perpendicular to the true-length line and through line AB at the point of intersection projected from the front view.

In Step 2, the intersection point between AB and CD is projected to the top view. A true-length point between AB and CD is projected to the top view. A true-length horizontal line is found on the plane. The top view of CD is drawn through the point of intersection and perpendicular to the true-length line in the plane. These two intersecting lines, AB and CD, form a plane perpendicular to the plane since a single line in the plane, CD, is perpendicular to the plane.

Problems

Use Size A sheets for the following problems, and lay out the problems using instruments. Each square on the grid is equal to 0.20 inch (about 5 mm). The problems can be laid out on grid paper or plain paper. Label all reference planes and points in each problem with $\frac{1}{8}$-inch letters or numbers, using guidelines

1. (Fig. 27.30) **A.–D.** Find the missing third views of the given points. **E.** and **F.** Find the missing third views of the lines between the given points.

2. (Fig. 27.31) **A.** Find the front and top views of the profile line. **B.** Find the front and side views of the horizontal line. **C.** Find the side view of 5–6. **D.** Find the front and side views of 7–8. **E.** and **F.** Find the top and side views of the given lines.

3. (Fig. 27.32) **A.** Find the side view of the line. **B.** Find the front and side views of the horizontal line. **C.** Draw three views of a line from point 2 to point 0 on the line. **D.** Find the side view of the line and locate the midpoint of the line in each view. **E.** and **F.** Find the piercing points between the lines and planes, show visibility, and complete the missing third view.

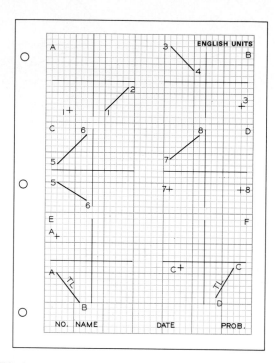

FIG. 27.31 Problems 2A–2F.

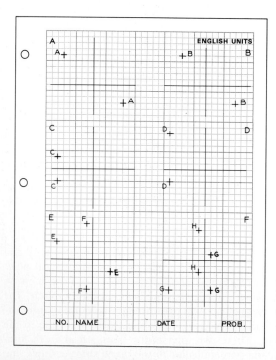

FIG. 27.30 Problems 1A–1F.

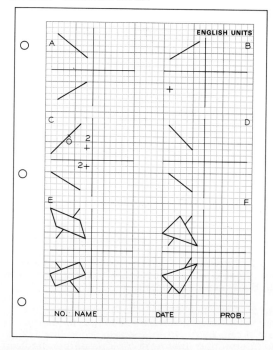

FIG. 27.32 Problems 3A–3F.

4. (Fig. 27.33) **A.** Find the front and side views of the horizontal plane. **B.** Draw the top and side views of the frontal plane. **C.** Draw the front and top views of the profile plane. **D.** Draw the top view of the plane. **E.** Find the side view of the plane, and locate the line on it in each view. **F.** Find the missing view of the plane, and locate point 0 on each view of the plane.

5. (Fig. 27.34) **A.** Draw a line with its midpoint at 0 parallel to the line. **B.** Construct the front view of a second plane parallel to the given plane. **C.** and **D.** In each problem, draw lines through point 0 parallel to the given planes.

6. (Fig. 27.35) In each problem, draw lines through point 0 perpendicular to the given lines.

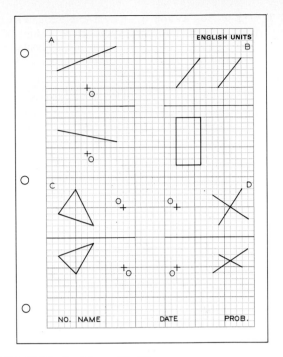

FIG. 27.34 Problems 5A–5D.

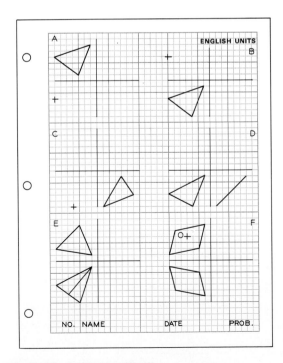

FIG. 27.33 Problems 4A–4F.

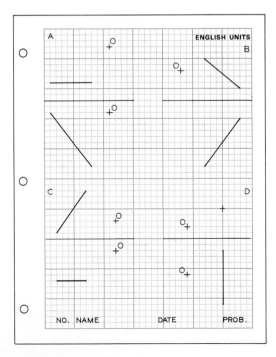

FIG. 27.35 Problems 6A–6D.

7. (Fig. 27.36) **A.** and **B.** Draw a line through each point 0 perpendicular to the given planes. **C.** Draw a plane through point 0 perpendicular to the given line. **D.** Draw a plane through the given line perpendicular to the plane.

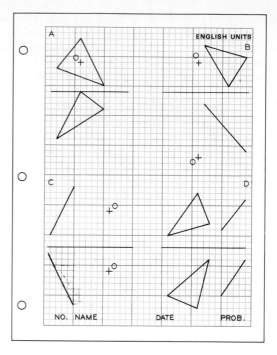

FIG. 27.36 Problems 7A–7D.

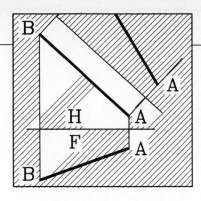

CHAPTER 28

Primary Auxiliary Views in Descriptive Geometry

28.1
Introduction

Descriptive geometry is the projection of three-dimensional figures onto a two-dimensional plane of paper in such manner as to allow geometric manipulations to determine lengths, angles, shapes, and other geometric information by means of graphics. Orthographic projection is the system used for laying out descriptive geometry problems.

A powerful tool of descriptive geometry is the *primary auxiliary view*, which permits the analysis of three-dimensional geometry.

28.2
Primary auxiliary view of a line

The top and front views of line 1–2 are shown pictorially and orthographically in Fig. 28.1. Since the line is not a principal line, it is not true length in the prin-

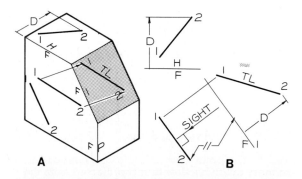

FIG. 28.1 **A.** A pictorial of line 1–2 is shown inside a projection box where a primary auxiliary plane is established perpendicular to the frontal plane and parallel to the line. **B.** The orthographic arrangement shows the auxiliary view projected from the front view to find 1–2 true length.

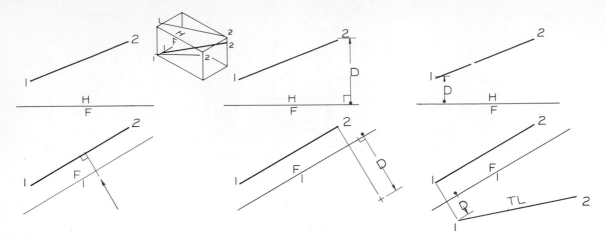

FIG. 28.2 True length of a line by a primary auxiliary view.

Step 1 To find 1–2 true length, construct the line of sight perpendicular to the line. Draw reference line F-1 parallel to the front view of the line.

Step 2 Project point 2 perpendicularly from the front view. Dimension D from the top view locates point 2.

Step 3 Point 1 is located by transferring dimension D from the top view. Line 1–2 is true length in the auxiliary view.

cipal views. A primary auxiliary view must be used to find its true-length view.

In Fig. 28.1B, the line of sight is drawn perpendicular to the front view of the line, and reference line F-1 is drawn parallel to its frontal view. You can see in the pictorial that the auxiliary plane is parallel to the line and perpendicular to the frontal plane, which accounts for its being labeled as F-1.

The auxiliary view is found by projecting parallel to the line of sight and perpendicular to the F-1 reference line. Point 2 is found by transferring distance D with your dividers to the auxiliary view, since the frontal plane appears as an edge in both the top and auxiliary views. Point 1 is located in the same manner, and the points are connected to find the true-length view of the line.

Figure 28.2 separates the sequential steps required to find the true length of an oblique line. It is best to letter all reference planes using the notation suggested in the various steps with the exception of dimensions such as D, which are transferred from one view to another with your dividers.

A primary auxiliary view can result in a point view of the line if projected from a true-length view of the line in a principal view (Fig. 28.3). The auxil-

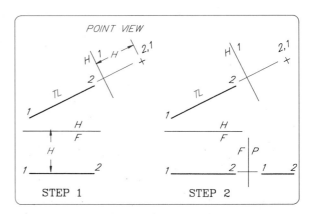

FIG. 28.3 Point view of a line.

Step 1 The point view of a line can be found in a primary auxiliary view that is projected from the true-length view of the line.

Step 2 An auxiliary view projected from a foreshortened view of a line will result in a foreshortened view of the line, not a point view.

iary view projected from the front view of the line does not give a point view since the line is foreshortened in the front view.

28.3
True length by analytical geometry

You can see in Fig. 28.4 that the length of a frontal line can be found in the front view by the application of analytical geometry (mathematics) and the Pythagorean theorem, which states that the hypotenuse of a right triangle is equal to the square root of the sum of the squares of the other two sides. The length of the line in Fig. 28.4 can be measured in the front view since it is true length in this view.

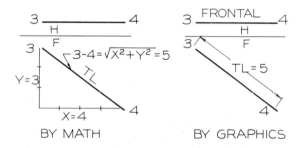

FIG. 28.4 A line that appears true length in a view (the front view in this case) can have its length calculated by application of the Pythagorean theorem. Since the line is a frontal line and is true length in the front view, its length can be measured graphically.

The line shown pictorially in Fig. 28.5A can be found true length by analytical geometry by determining the length of the front view where the X- and Y-coordinates form a right triangle. In Fig. 28.5B, a second right triangle, 1–0–2, is solved to find its hypotenuse, which is the true length of 1–2. The true length of an oblique line is the square root of the sum of the squares of the X-, Y-, and Z-coordinates that correspond to width, height, and depth. The steps in determining the true length of line 1–2 by analytical geometry are shown in Fig. 28.6.

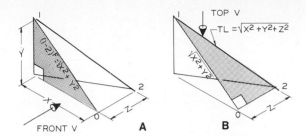

FIG. 28.5 A three-dimensional line that is foreshortened in a principal view can be found true length by the Pythagorean theorem in two steps. **A.** The frontal projection, 1–0, is found using the X- and Y-coordinates. **B.** The hypotenuse of the right triangle 1–0–2 is found using X-, Y-, and Z-coordinates.

28.4
The true-length diagram

A true-length diagram is constructed with two perpendicular lines to find a line of true length (Fig. 28.7). This method does not give a direction for the line but merely its true length.

The two measurements laid out on the true-length diagram can be transferred from any two adjacent orthographic views. One measurement is the distance between the endpoints in one of the views. The other measurement, taken from the adjacent view, is measured between the endpoints in a direction perpendicular to the reference line between the two views.

28.5
Angles between lines and principal planes

> To measure the angle between a line and a plane, the line must appear true length in the view where the plane appears as an edge.

Since a principal plane will appear as an edge in a primary auxiliary view, the angle a line makes with this principal plane can be measured if the line is found true length in this view (Fig. 28.8).

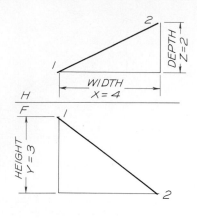

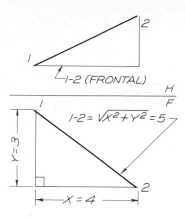

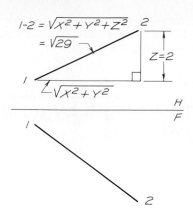

FIG. 28.6 True length of a line—analytical method.

Step 1 Right triangles are drawn with line 1–2 as the hypotenuse in the top and front views. The coordinates, or legs, of the right triangles are drawn parallel and perpendicular to the H-F line.

Step 2 The true length of the front view of 1–2 is found by the Pythagorean theorem. The resulting length is found to be 5 units.

Step 3 The true length of the line is found by combining the length of the front projection with the true length of the Z-coordinate in the top view. The length is $\sqrt{29}$.

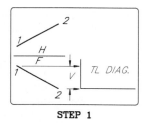

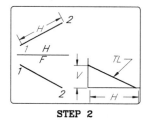

FIG. 28.7 True-length diagram.

Step 1 Transfer the vertical distance between the ends of 1–2 to the vertical leg of the TL diagram.

Step 2 Transfer the horizontal length of the line in the top view to the horizontal leg of the TL diagram.

28.6
Slope of a line

Slope is the angle a line makes with the horizontal plane.

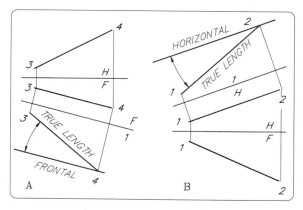

FIG. 28.8 Angles between lines and principal planes.

A. When an auxiliary view is projected from the front view, the frontal plane appears as an edge and the line is true length. The angle with the frontal plane can be measured in this view.

B. An auxiliary view projected from the top view shows the horizontal plane as an edge and the line true length. The angle with the horizontal can be measured here.

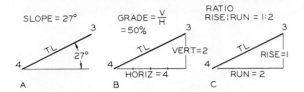

FIG. 28.9 The inclination of a line with the horizontal can be measured and expressed by any of three methods. **A.** Slope angle. **B.** Percent grade. **C.** Slope ratio.

Slope may be specified by any of the three methods shown in Fig. 28.9: *slope angle, percent grade,* or *slope ratio.*

Slope angle

The slope of a line can be measured in a view where the line is true length and the horizontal plane appears as an edge. Thus, the slope of *AB* in Fig. 28.10 can be measured in the front view, where Θ is found to be 31°. Using the trigonometric tables, this angle can also be found by converting its tangent of 0.60 to 31°.

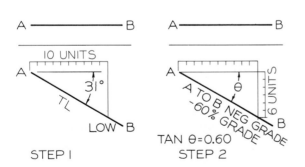

FIG. 28.10 Percent grade of a line.

Step 1 The percent grade of a line can be measured in the view where the horizontal appears as an edge and the line is true length (the front view here). Ten units are laid off parallel to the horizontal from the end of the line.

Step 2 A vertical distance from *A* to the line is measured to be 6 units. The percent grade is 6 divided by 10 or 60%. This is negative when the direction is from *A* to *B*. The tangent of this slope angle is $\frac{6}{10}$ or 0.60.

Percent grade

The percent grade of a line is found in the view where the line is true length and the horizontal plane appears as an edge.

> *Grade* is the ratio of the vertical *(rise)* divided by the horizontal *(run)* between the ends of a line expressed as a percentage.

The percent grade of *AB* is found in Fig. 28.10 by using a combination of mathematics and graphics. Ten units are laid off parallel to the horizontal from *A* using a convenient scale with decimal units (Step 1). The vertical drop of the line after 10 units along the horizontal is measured as 6 units (Step 2). Since 10 units were used along the horizontal, the tangent of the angle is 0.60, which is easily converted into a −60% grade. The grade is negative from *A* to *B* since this is downhill; it would be positive from *B* to *A* if it were uphill. Line *CD* has a +50% grade from *C* to *D* in Fig. 28.11A.

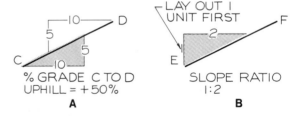

FIG. 28.11 A. The percent grade of a line is positive if uphill, and negative if downhill. **B.** The slope ratio is expressed as 1:*XX*, where 1 is the rise and *XX* is the horizontal distance. The 1 unit must be drawn before the horizontal distance.

Slope ratio

The first number of the slope ratio is the rise, and the second number is the run (see Fig. 28.9). In the slope ratio the rise is always 1 (for example, 1:10, 1:200).

The graphical method of finding the slope ratio is shown in Fig. 28.11B, where the rise of 1 is laid off on the true-length view of *EF*. The corresponding horizontal is found to be 2, which results in a slope ratio of 1:2.

OBLIQUE LINES When a line is oblique and does not appear true length in the front view, it must be found true length by an auxiliary view projected from the top view. This auxiliary view shows the horizontal as an edge and the line true length, making it possible to measure the slope angle (Fig. 28.12A).

Similarly, an auxiliary view projected from the top view must be used to find the percent grade of an oblique line (Fig. 28.12B). Ten units are laid off horizontally, parallel to the H-1 reference line, and the vertical distance is found to be 6, or a −60% grade downhill from 3 to 4.

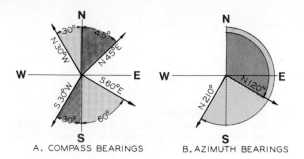

FIG. 28.13 A. Compass bearings are measured with respect to north and south directions on the compass. **B.** Azimuth bearings are measured with respect to north in a clockwise direction up to 360°.

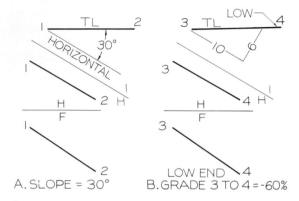

FIG. 28.12 Slope of an oblique line.

A. The slope angle can be measured in a view where the horizontal appears as an edge and the line is true length. The slope of 30° is found in an auxiliary view projected from the top view.

B. The percent grade can be measured in a true-length view of the line projected from the top view. Line 3–4 has a −60% grade from 3 to the low end at 4.

The line in Fig. 28.13A that makes 30° with north has a bearing of N 30° W. A line making 60° with south toward the east has a compass bearing of south 60° east, or S 60° E. Since a compass can be read only when held horizontally, the compass bearings of a line can be determined only in the top, or the horizontal view.

> An azimuth bearing is measured from north in clockwise direction to 360° (Fig. 28.13B).

Azimuth bearings of a line are written N 120°, N 210°, and so forth, with this notation indicating that the measurements are made from the north.

The compass bearing (direction) of a line is assumed to be toward the low end of the line unless

28.7
Compass bearing of a line

Two types of bearings of a line's direction are *compass bearings* and *azimuth bearings* (Fig. 28.13).

> Compass bearings always begin with the north or south directions, and the angles with north and south are measured toward east or west.

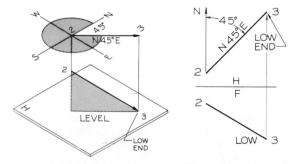

FIG. 28.14 The compass bearing of a line is measured in the top view toward its low end (unless specified toward the high end). Line 2–3 has a bearing of N 45° E toward the low end at 3.

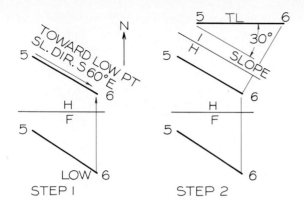

FIG. 28.15 Slope and bearing of a line.

Step 1 Slope bearing can be found in the top view toward its low end. Direction of slope is S 60° E.

Step 2 The slope angle of 30° is found in an auxiliary view projected from the top view where the line is found true length.

otherwise specified. For example, line 2–3 in Fig. 28.14 has a bearing of N 45° E since the line's low end is point 3. It can be seen in the front view that point 3 is the lower end.

The compass bearing and slope of a line are found in Fig. 28.15. This information can be used to verbally describe the line as having a compass bearing of S 60° E and a slope of 30° from 5 to 6. When given verbal information and one point in the top and front views, a line can be drawn as shown in Fig. 28.16.

28.8
Contour maps and profiles

The *contour map* uses contour lines to show variations in elevation on irregular surfaces of the earth. A pictorial view of a sectional plane and a portion of the earth are shown in Fig. 28.17, along with the conventional orthographic views of the contour map and profiles.

CONTOUR LINES are horizontal lines that represent constant elevations from a horizontal datum such as sea level. Contour lines can be thought of as the intersections of horizontal planes with the surface of

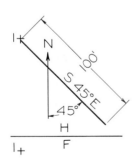

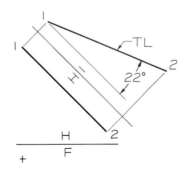

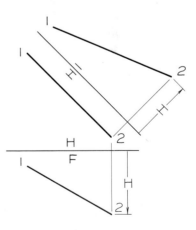

FIG. 28.16 A line from slope specifications.

Step 1 It is required to draw a line through point 1 that bears S 45° E for 100′ horizontally and slopes 22°. The bearing and horizontal are drawn in the top view.

Step 2 An auxiliary view is projected from the top view where the 22° slope angle is measured.

Step 3 The front view of 1–2 is found by locating point 2 in the front view.

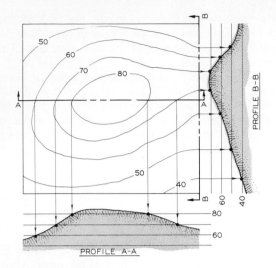

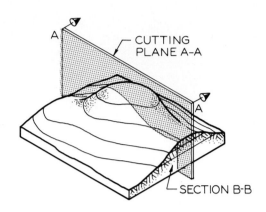

FIG. 28.17 A contour map uses contour lines to show variations in elevation on an irregular surface. Vertical sections taken through a contour map are called profiles.

the earth. The vertical interval of spacing between the contours in Fig. 28.17 is 10 feet.

CONTOUR MAPS contain contour lines that are drawn to represent irregularities of the surface (Fig. 28.17). The closer the contour lines are to each other, the steeper the terrain.

PROFILES are vertical sections through a contour map that are used to show the earth's surface at any desired location (Fig. 28.17). Contour lines represent edge views of equally spaced horizontal planes in profiles. The true representation of a profile is drawn with the vertical scale equal to the scale of the contour map; however, this vertical scale is often increased to emphasize changes in elevation that would otherwise not be apparent.

CONTOURED SURFACES are also depicted on the drawing board by using contour lines. When applied to objects other than the earth's surface—such as airfoils, automobile bodies, ship hulls, and household appliances—this technique of representing contours is called *lofting*.

STATION NUMBERS are used to locate distances on a contour map. Since the civil engineer uses a chain (metal tape) 100 feet long, stations are located 100 feet apart (Fig. 28.18). Station 7 is 700 feet from

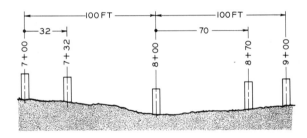

FIG. 28.18 Station points are located 100 ft apart. For example, station 7 is 700 ft from the beginning point. A point 32 ft beyond station 7 is labeled station 7 + 32. A point 870 ft from the origin is labeled station 8 + 70.

the beginning point, station 0. A point 32 feet from station 7 toward station 8 is labeled station 7 + 32.

28.9
Vertical sections

In Fig. 28.19, a vertical section (a *profile*) is passed through the top view of an underground pipe that is known to have an elevation of 90 feet at point 1 and 60 feet at point 2. An auxiliary view is projected perpendicularly from the top view, contour lines are lo-

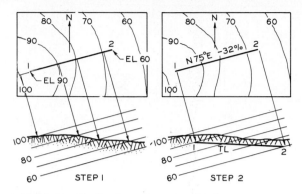

FIG. 28.19 Vertical sections (profiles).

Step 1 An underground pipe is known to have elevations of 90 ft and 60 ft at each end. An auxiliary view is projected perpendicularly from the top view, and contours are drawn at 10-ft intervals to correspond to the plan view. The top of the ground is found by projecting from the contour lines in the plan view.

Step 2 Points 1 and 2 are located at elevations of 90 ft and 60 ft in the vertical section (profile). Since 1–2 is TL in the section, its slope or percent grade can be measured. The compass direction and percent grade are labeled in the top view of the line.

cated, and the top of the earth over the pipe is found in a profile.

To measure the true lengths and angles of slope in the profile, the same scale used to draw the contour map was used to draw the profile. The percent grade and compass bearing of the line are labeled on the contour map.

28.10
Plan-profiles

A *plan-profile* is a drawing that includes a plan with contours and a vertical section called a profile. A plan-profile is used to show an underground drainage system from manhole 1 to manhole 3 in Figs. 28.20 and 28.21.

The profile is drawn with an exaggerated vertical scale to emphasize the variations in the earth's surface and the grade of the pipe. Although the vertical scale is usually increased, it can be drawn at the same scale used in the plan if desired.

Manhole 1 is projected to the profile using orthographic projection, but the other points are not orthographic projections (Fig. 28.20). The points where the

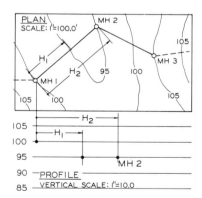
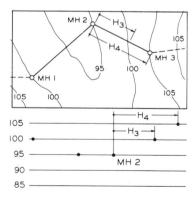
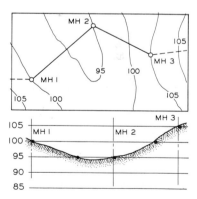

FIG. 28.20 Plan-profile.

Required Find the profile of the earth over the underground drainage system.

Step 1 Distances H_1 and H_2 from manhole 1 are transferred to their respective elevations in the profile. This is not an orthographic projection.

Step 2 Distances H_3 and H_4 are measured from manhole 2 in the plan and are transferred to their respective elevations in the profile. These points represent elevations of points on the earth above the pipe.

Step 3 The five points are connected with a freehand line and the drawing is crosshatched to represent the earth's surface. Centerlines are drawn to show the locations of the three manholes.

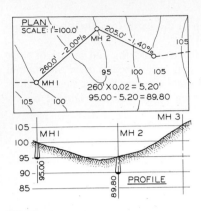
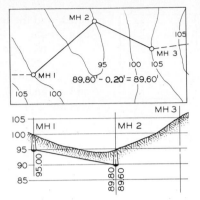
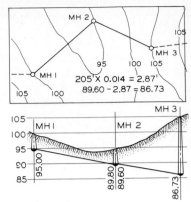

FIG. 28.21 Plan-profile—manhole location.

Step 1 The horizontal distance from MH1 to MH2 is multiplied by the percent grade. The elevation of the bottom of manhole 2 is calculated by subtracting from the elevation of manhole 1.

Step 2 The lower side of manhole 2 is 0.20' lower than the inlet side to compensate for loss of head (pressure) due to the turn in the pipeline. The lower side is found to be 89.60' and is labeled.

Step 3 The elevation of manhole 3 is calculated to be 86.73' since the grade is 1.40% from manhole 2 to manhole 3. The flow line of the pipeline is drawn from manhole to manhole, and the elevations are labeled.

contour lines cross the top view of the pipe are transferred to their respective elevations in the profile with your dividers. These points are connected to show the surface of the earth over the pipe and the location of manhole centerlines (Fig. 28.21). The drop from manhole 1 to manhole 2 is found to be 5.20 feet by multiplying the horizontal distance of 260.00 feet by a −2.00% grade (Fig. 28.21).

Since the pipes intersect at manhole 2 at an angle, the flow of the drainage is disrupted at the turn; thus, a drop of 0.20 foot is given from the inlet across the floor of the manhole to compensate for the loss of pressure (head) through the manhole.

The true lengths of the pipes cannot be accurately measured in the profile when the vertical scale is different from the horizontal scale. For this computation, trigonometry must be used.

28.11

Edge view of a plane

The edge view of a plane can be found in a view where a line on the plane appears as a point.

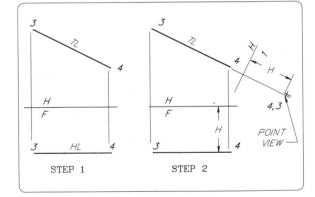

FIG. 28.22 Point view of a line.

Step 1 Line 3–4 is horizontal in the front view and is therefore true length in the top view.

Step 2 The point view of 3–4 is found by projecting parallel lines from the top view to the auxiliary view.

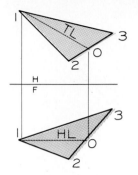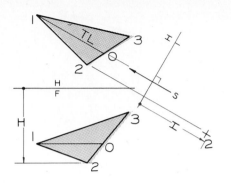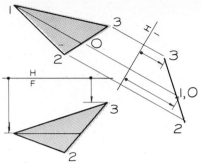

FIG. 28.23 Edge view of a plane.

Step 1 To find plane 1–2–3 as an edge, horizontal line 1–10 is drawn on the plane in the front view. Line 1–0 is projected to the top view, where it is true length.

Step 2 A line of sight is constructed parallel to the true-length line 1–0. H-1 is drawn perpendicular to the line of sight. Point 2 is found in the auxiliary view by transferring dimension *H* from the front view.

Step 3 Points 1 and 3 are found by transferring their height dimensions from the front view. These points will lie in a straight line, which is the edge of the plane. Line 1–10 will appear as a point in this view.

A line can be found as a point by projecting from a true-length view of the line. (Fig. 28.22).

A true-length line can be found on any plane by drawing the line parallel to one of the principal planes and projecting it to the adjacent view, as shown in Fig. 28.23 (Step 1). Since line 3–4 is true length in the top view, its point view may be found, and the plane will appear as an edge in this auxiliary view.

28.12
Dihedral angles

The angle between two planes, called a *dihedral angle,* can be found in the view where the line of intersection between two planes appears as a point.

The line of intersection, 1–2, between the two planes in Fig. 28.24 is true length in the top view. This makes it possible to find the point view of line 1–2 and the edge view of both planes in a primary auxiliary view.

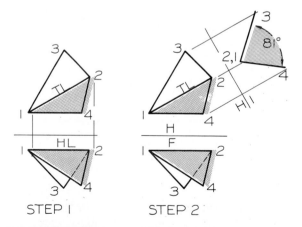

FIG. 28.24 Dihedral angle.

Step 1 The line of intersection between the planes, 1–2, is true length in the top view.

Step 2 The angle between the planes (the dihedral angle) can be found in the auxiliary view where the line of intersection appears as a point.

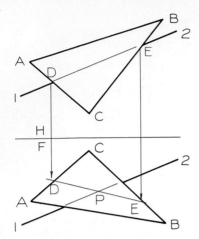

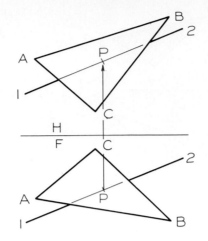

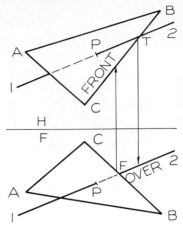

FIG. 28.25 Piercing points by projection.

Step 1 Assuming a vertical cutting plane is passed through the top view of 1–2, the plane intersects *AC* and *BC* at *D* and *E*. The intersection of *DE* with 1–2 in the front view locates the piercing point *P*.

Step 2 Point *P* is projected to the top view of line 1–2 from the front view, where it was first located.

Step 3 Lines *CB* and *P2* cross at point *T* in the top view. By projecting downward from *T*, *P2* is found to be over *CB*; therefore, *PT* is visible in the top view. Intersection *F* can be used to find that line *CB* is in front of *P2*; therefore, *PF* is hidden in the front view.

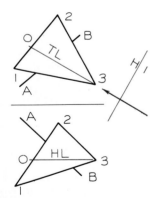

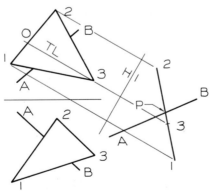

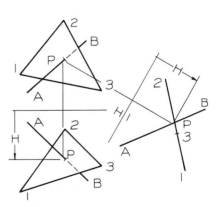

FIG. 28.26 Piercing points by auxiliary view.

Step 1 Draw a horizontal line on the plane in the front view, project it to the top view to find TL line 0–3 on the plane. The line of sight is drawn parallel to the TL line.

Step 2 Find the edge view of the plane and project line *AB* to this view. Point *P* is the piercing point in the auxiliary view.

Step 3 Point *P* is projected to the top and front views. Point *A* of the auxiliary view is the highest point and *AP* will be visible in the top view. Point *B* in the top view is the farthest back and is therefore hidden in the front view.

28.13
Piercing points by projection

Figure 28.25 shows the steps necessary to find the piercing point of line 1–2 that passes through the plane. Cutting planes are passed through the line and plane in the top view. The trace of this cutting plane, line, *DE*, is then projected to the front view, where piercing point *P* is found (Step 1). The top view of *P* is located (Step 2), and the visibility of the line is found (Step 3).

28.14
Piercing points by auxiliary views

The piercing point of a line and a plane can be found by an auxiliary view (Fig. 28.26). Piercing point *P* can be seen in Step 2, where the plane is found as an edge. Point *P* is projected back to the line in the top and front views from the auxiliary view in Step 3. The location of *P* in the front view is checked by transferring di-

mension *H* from the auxiliary view with your dividers.

Visibility is easily determined for the top view since it can be seen that *AP* is higher than the plane in the auxiliary view and is therefore visible in the top view. Analysis of the top view shows the endpoint *A* is the most forward point; therefore *AP* is visible in the front view.

28.15
Perpendicular to a plane

In Fig. 28.27, it is required that a line be drawn from point 0 perpendicular to the plane.

> A perpendicular line will appear true length and perpendicular to a plane where the plane appears as an edge.

The true-length perpendicular is drawn in Step 2 to locate piercing point *P*. Point *P* is found in the top

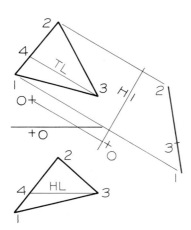
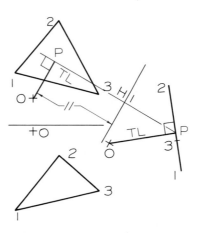
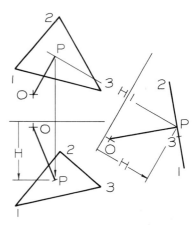

FIG. 28.27 Line perpendicular to a plane.

Step 1 Find the edge view of the plane by finding the point view of a line on it. Project from either view. Project point 0.

Step 2 Line 0*P* is drawn perpendicular to the edge view of the plane. Since 0*P* is TL in the auxiliary view, it will be parallel to the H-1 line in the top view and perpendicular to a TL line on the plane.

Step 3 Piercing point *P* is found in the front view by projecting from the top view. Point *P* is accurately located by transferring dimension *H* from the auxiliary view to the front view.

view by drawing line 0P parallel to the H-1 reference line. It must lie in this direction since 0P is true length in the auxiliary view. Line 0P will also be perpendicular to a true-length line in the top view of the plane. The front view of point P, along with its visibility, is found by projection in Step 3.

28.16
Intersections by projection

The line of intersection between two planes can be found by locating the piercing points of two lines on one plane with the other. In Fig. 28.28, a cutting plane is used in Step 1 to find piercing point P by projection. In Step 2, piercing point T is found by the same method. Line PT is the line of intersection.

The two lines selected to be analyzed for their piercing points should be lines of a plane that cross the other plane. Lines AB and 1–2 would be poor selections since they lie outside the other plane. Each selected line is then analyzed to find its piercing point and visibility as if it were a single line.

28.17
Intersections by auxiliary view

The intersection between planes can be found by finding the edge view of one of the planes, as shown in Fig. 28.29 (Step 1). Piercing points L and M are projected from the auxiliary view to their respective lines, 5–6 and 4–6, in the top view (Step 2).

The visibility of plane 4–5–6 in the top view is apparent by inspection of the auxiliary view, where sight line S_1 has an unobstructed view of the 4–L–M portion of the plane. Plane 4–5–L–M is visible in the front view, since sight line S_2 has an unobstructed view of the top view of this portion of the plane.

28.18
Slope of a plane

Planes can be established by using verbal specifications of slope and direction of slope of a plane.

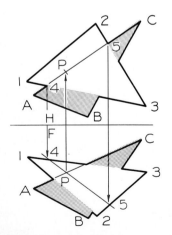 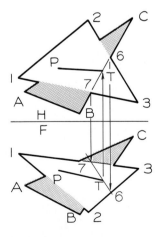 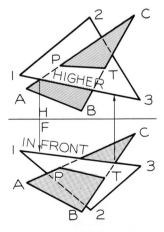

FIG. 28.28   Intersection by projection.

Step 1 Pass a cutting plane through AC in the top view to locate points 4 and 5 that are projected to the front view. Point P is found on 4–5 in the front view, and point P is projected to the top view.

Step 2 Pass a cutting plane through the top view of BC to locate points 6 and 7 that are projected to the front view. Piercing point T is located on 6–7 in the front view, and point T is projected to the top view.

Step 3 The intersection between AC and 1–3 is projected to the front view to determine that 1–3 is higher in the top view. Line 1–3 is found to be in front of BC in the front view to establish the visibility of the plane.

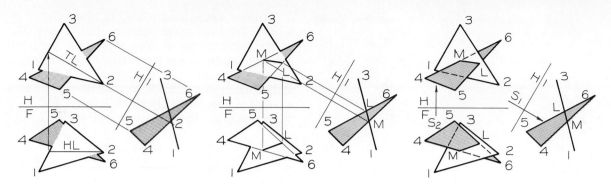

FIG. 28.29 Intersection by auxiliary view.

Step 1 Find the edge view of one of the planes, and project the other plane to this view also.

Step 2 Piercing points *L* and *M* can be seen on the edge view of the plane. *LM* is projected to the top and front views.

Step 3 The line of sight from the top view strikes 1–5 first in the auxiliary view, which makes 1–5 visible in the top view. Line 4–5 is farthest forward in the top view and is visible in the front view.

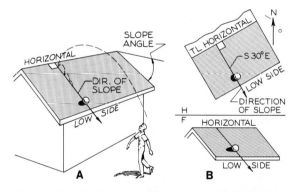

FIG. 28.30 The direction of slope of a plane is the compass bearing of a line on the plane. This is measured in the top view toward the low side of the plane.

SLOPE is the angle the plane's edge view makes with the edge of the horizontal plane.

DIRECTION OF SLOPE is the compass bearing of a line perpendicular to a true-length line in the top view of a plane toward its low side. This is the direction in which a ball would roll on the plane. It can be seen in Fig. 28.30A that a ball would roll perpendicular to all horizontal lines of the roof toward the low side. The slope is seen when the roof and the horizontal are edges in a single view.

Figure 28.31 gives the steps of finding the direction of slope and the slope angle of a plane. An understanding of these terms enables you to verbally describe a sloping plane.

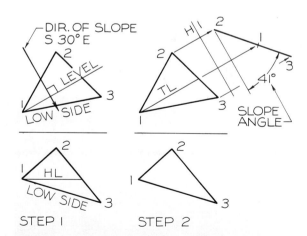

FIG. 28.31 Slope and direction of slope of a plane.

Step 1 Slope direction can be found as perpendicular to a true-length level line in the top view toward the low side of the plane, S 30° E in this case.

Step 2 Slope is measured in an auxiliary view where the horizontal is an edge and the plane is an edge, 41° in this case.

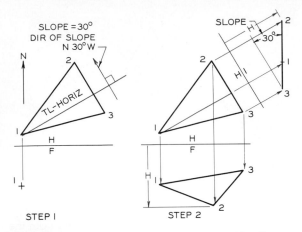

FIG. 28.32 Construction from slope specifications.

Step 1 If the top view, front view of 1, and slope specifications are given, the front view can be completed. Draw the direction of slope in the top view. A true-length horizontal line on the plane is drawn perpendicular to the slope direction.

Step 2 A point view of the TL line is found in the auxiliary view to locate point 1. The edge view of the plane is drawn through 1 at a slope of 30°, according to the specifications. The front view is completed by transferring height dimensions from the auxiliary view to the front view.

FIG. 28.33 The road across the top of this dam was built by applying the principles of cut and fill. (Courtesy of the Bureau of Reclamation, U.S. Department of the Interior.)

A three-dimensional plane can be established by working from slope and direction specifications (Fig. 28.32). The direction of slope is drawn in the given top view, which locates the direction of a true-length horizontal line on the plane (Step 1). The edge view of the plane is then found by locating point 1 and constructing a slope of 30° (Step 2). Points 3 and 2 are transferred to the front view to complete that view.

28.19
Cut and fill

A level roadway routed through irregular terrain or the embankment of a fill used to build a dam involves the principles of cut and fill (Fig. 28.33). *Cut and fill* is the process of cutting away equal amounts of the high ground to fill the lower areas.

In Fig. 28.34, it is required that a level roadway of an elevation of 60 feet be constructed about the given centerline in the contour map using the specified angles of cut and fill.

In Step 1, the roadway is drawn in the top view, and the contour lines in the profile view are drawn 10 feet apart, since the contours in the top view are this far apart. In Step 2, the cut angles are measured and drawn on both sides of the roadway on the upper side. The points on the elevation lines crossed by the cut embankments are projected to the top view to find the limits of cut in this view.

In Step 3, the fill angles are laid off in the profile from given specifications. The crossing points on the profile view of the elevation lines are projected to the top view to find the limits of fill. New contour lines are drawn inside the areas of cut and fill to indicate they have been changed.

28.20
Design of a dam

Some of the terms used in the drawing of a dam are (1) *crest,* the top of the dam; (2) *water level;* and (3) *freeboard,* the height of the crest above the water level (Fig. 28.35).

An earthen dam is located on the contour map in Fig. 28.36. It makes an arc with its center at point *C.* The top of the dam is to be level to provide a roadway. The method of drawing the top view of the dam and

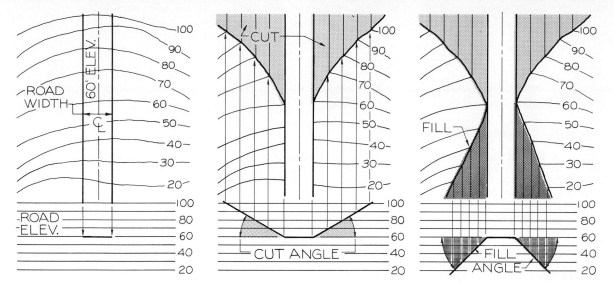

FIG. 28.34 Cut and fill of a level roadway.

Step 1 Draw a series of elevation planes in the front view at the same scale as the map, and label them to correspond to the contours on the map. Draw the width of the roadway in the top view and in the front view at the given elevation, 60′ in this case.

Step 2 Draw the cut angles on the upper sides of the road in the front view according to the given specifications. The points of intersection between the cut angles and the contour planes in the front view are projected to their respective contour lines in the top view to determine the limits of cut.

Step 3 Draw the fill angles on the lower sides of the road in the front view. The points in the front view where the fill angles cross the contour lines are projected to their respective contour lines in the top view to give the limits of the fill. Contour lines are changed in the cut-and-fill areas to indicate the new contours parallel to the roadway.

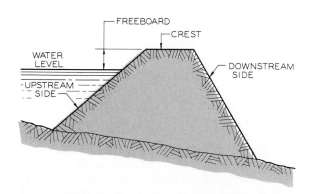

FIG. 28.35 Terms and symbols used in the construction of a dam.

indicating the level of the water held by the dam are shown in Fig. 28.36.

These same principles were used in the design and construction of the 726-foot Hoover Dam, built in the 1930s. Since this dam was made of concrete instead of earth, the dam was built in the shape of an arch bowed toward the water to take advantage of the compressive strength of concrete (Fig. 28.37).

28.21
Strike and dip

Strike and dip are terms used in geological and mining engineering to describe strata of ore under the surface of the earth.

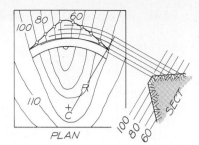 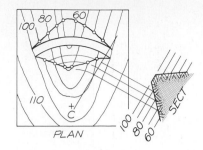 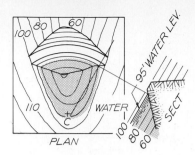

FIG. 28.36 Graphical design of a dam.

Step 1 A dam in the shape of an arc with its center at *C* has an elevation of 100'. Draw radius *R* from *C* and project perpendicularly from this line, and draw a section through the dam from specifications. The downstream side of the dam is projected to radial line *R*. Using the radii from *C*, locate points on their respective contour lines.

Step 2 The elevations of the dam on the upstream side of the section are projected to the radial line, *R*. Using center *C* and your compass, locate points on their respective contour lines in the plan view as they are projected from the section.

Step 3 The elevation of the water level is 95' and is drawn in the section. The point where the water intersects the dam is projected to the radial line in the plan view and is drawn as an arc using center *C*. The limit of the water is drawn between the 90' and 100' contour lines in the top view.

FIG. 28.37 The Hoover Dam and Lake Mead, which were built from 1931 to 1935. (Courtesy of the Bureau of Reclamation, U.S. Department of the Interior.)

STRIKE is the compass bearing of a level line in the top view of a plane. It has two possible compass bearings since it is the direction of a level line.

DIP is the angle the edge view of a plane makes with the horizontal plus its general compass direction, such as NW or SW. The dip angle is found in the primary auxiliary view projected from the top view, and its dip direction is measured perpendicular to a level line in the top view toward the low side.

The steps of finding the strike and dip of a plane are given in Fig. 28.38. Strike can be measured in the top view by finding a true-length line on the plane in this view. The dip angle is found in an auxiliary view projected from the top view that shows the horizontal and the plane as edges.

A plane can be constructed from strike and dip specifications (Fig. 28.39). The strike is drawn to represent a true-length horizontal line on the plane. Dip direction is perpendicular to the strike (Step 1). The edge view of the plane is then drawn through point 1 at a dip of 300° (Step 2). Points 2 and 3 are located in the front view.

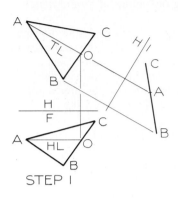

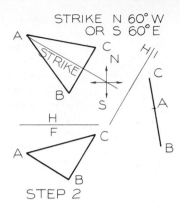

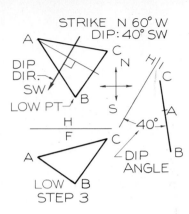

FIG. 28.38 Strike and dip of a plane.

Step 1 Find the edge view of the plane by projecting from the top view.

Step 2 Strike is the compass direction of a level line on the plane and is measured in the top view. The strike of the plane is either N 60° W or S 60° E.

Step 3 Dip is the angle a plane makes with the horizontal (40° in the auxiliary view) plus the general compass direction toward the low side in the top view (SW). Dip direction is perpendicular to a TL line in the top view. Dip is written 40° SW.

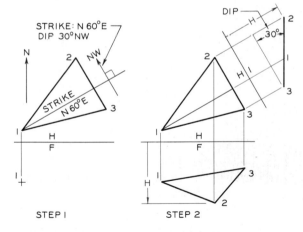

FIG. 28.39 Working from strike and dip specifications.

Step 1 Strike is drawn on the top view of the plane, which is a TL horizontal line. The direction of dip is drawn perpendicular to the strike toward the NW, as specified.

Step 2 A point view of the strike line is found in the auxiliary view to locate point 1. The edge view of the plane is drawn through 1 at a dip of 30°, according to the specifications. The front view is completed by transferring height dimensions from the auxiliary to the front view.

28.22
Distances from a point to a plane

Descriptive geometry principles can be used to find various distances from a point to a plane. An example is shown in Fig. 28.40, where the distance from point 0 on the ground to an underground ore vein is found.

Three points are located on the top plane of an ore vein. Point 0 is the point on the earth from which the tunnels are to be drilled to the vein. Point 4 is a point on the lower plane of the vein.

The edge view of plane 1–2–3 is found by projecting from the top view. The lower plane is drawn parallel to the upper plane through point 4. The horizontal distance from point 0 to the plane is drawn parallel to the H-1 reference line. The vertical distance is perpendicular to the H-1 line. The shortest distance to the plane is perpendicular to the plane. Each of these lines is true length in this view where the ore vein appears as an edge.

The process of finding the distance from a point to a plane or a vein is a technique often used in solving mining and geological problems; for example, test

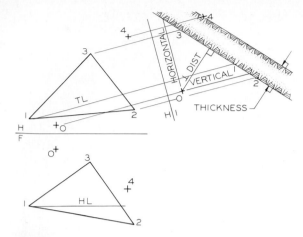

FIG. 28.40 The vertical, horizontal, and perpendicular distances from a point to an ore vein can be found in an auxiliary view projected from the top view. The thickness of an ore vein is perpendicular to the upper and lower planes of the vein.

wells are drilled into coal zones to learn more about them (Fig. 28.41).

28.23
Outcrop

Strata of ore formations usually approximate planes of a uniform thickness. This assumption is employed in analyzing the orientation of underground ore veins. A vein of ore may be inclined to the surface of the earth and may outcrop on its surface. Outcrops permit

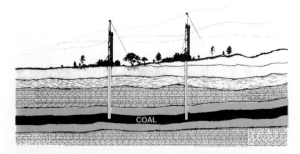

FIG. 28.41 Test wells are drilled into coal zones to determine which coal seams will contribute significantly to the production of gas. (Courtesy of Texas Eastern News.)

open-surface mining operations at a minimum of expense.

The steps of finding the outcrop of an ore vein are given in Fig. 28.42. The locations of sample drillings, *A, B,* and *C,* are shown on the contour map, and their elevations are located on the contours of the profile. These points are known to lie on the upper surface of the vein; point *D* is known to lie on the lower plane of the vein.

The edge view of the ore vein can be found in an auxiliary view projected from the top view (Step 1). The points on the upper surface are projected back to their respective contour lines in the top view (Step 2). The points on the lower surface of the vein are then projected to the top view (Step 3). If the ore vein continues uniformly at its angle of inclination to the surface, the space between these two lines will be the outcrop of the vein on the surface of the earth.

28.24
Intersection between planes—cutting plane method

The top and front views of two planes are given in Fig. 28.43, where it is required to find the line of intersection between them if the planes are infinite in size. Cutting planes are passed through either view at any angle and projected to the adjacent view. The two points, *L* and *M,* found in the top view form the line of intersection. The compass direction of the line describes its direction of slope toward its low end.

The front view of the line of intersection is found by projecting its end points from the top view to their respective planes in the front view.

28.25
Intersection between planes—auxiliary method

In Fig. 28.44, two planes have been located and specified using strike and dip. Since the given strike lines are true-length level lines in the top view, the edge view of the planes can be found in the view where the strike appears as a point (Step 1). The edge views are drawn using the given dip angles and directions.

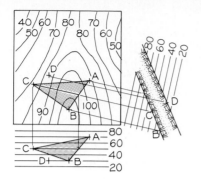

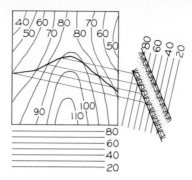

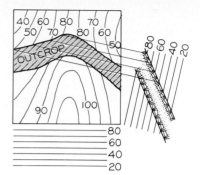

FIG. 28.42 Ore vein outcrop.

Step 1 Using points *A*, *B*, and *C* on the upper surface of the plane, find its edge view by projecting an auxiliary off the top view. The lower surface of the plane is drawn parallel to the upper surface through point *D*, a point on the lower surface.

Step 2 Points of intersection between the upper surface and the contour lines in the auxiliary view are projected to their respective contour lines in the top view to find one line of the outcrop.

Step 3 Points from the lower surface in the auxiliary view are projected to their respective contour lines in the top view to find the second line of outcrop. Crosshatch this area to indicate the outcrop of the vein.

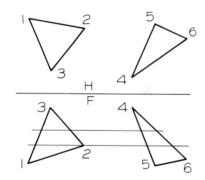

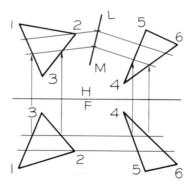

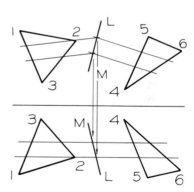

FIG. 28.43 Intersection of planes by cutting-plane method.

Step 1 Pass cutting planes through the front view of the planes. These planes can be drawn in any direction.

Step 2 Project the intersections of the cutting planes in the front view to the top view of the planes. where line of intersection *LM* is found.

Step 3 Points *L* and *M* are projected to their respective cutting planes in the front view to complete the solution.

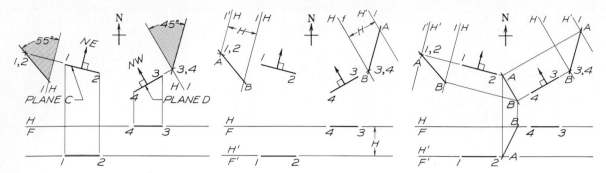

FIG. 28.44 Intersection between ore veins by auxiliary view.

Step 1 Lines 1–2 and 3–4 are strike lines and are true length in the horizontal view. The point view of each strike line is found by auxiliary views, using a common reference plane. The edge views of the ore veins can be found by constructing the dip angles with the H-1 line through the point views. The low side is the side of the dip arrow.

Step 2 A supplementary horizontal plane, H′-F′, is constructed at a convenient location in the front view. This plane is shown in both auxiliary views located *H* distance from the H-1 reference line. The H′-1′ plane cuts through each ore vein edge in the auxiliary views to locate points *A* and *B* on each plane.

Step 3 Points *A*, on each auxiliary view of the H′-1′ plane, are projected to the top view where they intersect at *A*. Points *B* on the H-1 plane are projected to their intersection in the top view *B*. Points *A* and *B* are projected to their respective planes in the front view. Line *AB* is the line of intersection between the two planes.

Horizontal datum planes H-F and H′-F′ are used to find lines on each plane that will intersect when projected from the auxiliary views to the top view. Points *A* and *B* are connected to determine the line of intersection between the two planes in the top view. These points are projected to the front view to find line *AB*.

28.26
Solution of descriptive geometry problems

Figure 28.45 illustrates the techniques of labeling and solving a descriptive geometry problem. Some of the lettering and numbering is aligned with inclined lines and reference lines to which the labeling applies, whereas other lettering is parallel to the edge of the paper. You may use either technique or a combination of the two. Always use guidelines and $\frac{1}{8}$ inch lettering is best for most drawings. Observe the difference in the line qualities used in the problem solution.

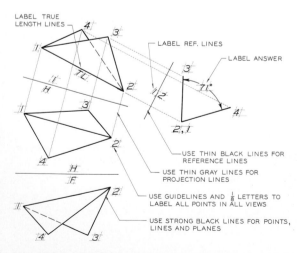

FIG. 28.45 Rules that should be followed in solving descriptive geometry problems.

Problems

Use Size A sheets for the following problems, and lay out the problems using instruments. Each square on the grid is equal to 0.20 inch (about 5 mm). The problems can be laid out on grid or plain paper. Label all reference planes and points in each problem with ⅛-inch (3 mm) letters or numbers, using guidelines.

1. (Fig. 28.46) **A.–D.** Find the true-length views of the lines as indicated by the given lines of sight by an auxiliary view.

2. (Fig. 28.47) **A.–D.** Find the angles that these lines make with the respective principal planes indicated by the given auxiliary reference lines.

3. (Fig. 28.48) **A.** and **B.** Find the lines' true length by the true-length diagram method, using the same diagram for both lines. **C.** and **D.** Find the point views of the lines.

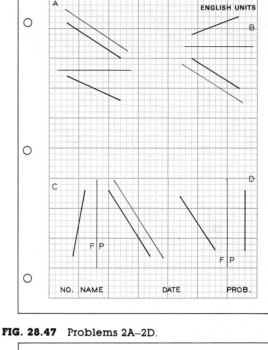

FIG. 28.47 Problems 2A–2D.

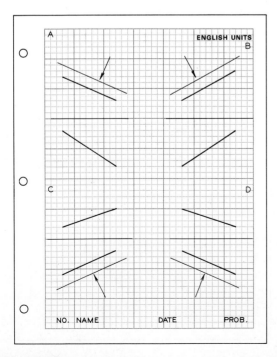

FIG. 28.46 Problems 1A–1D.

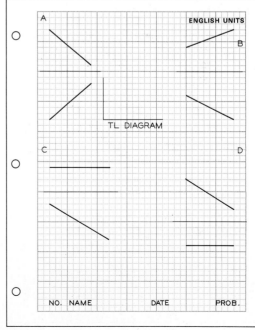

FIG. 28.48 Problems 3A–3D.

4. (Fig. 28.49) **A.** and **B.** Find the slope angle, tangent of the slope angle, and the percent grade of the four lines.

5. (Fig. 28.50) **A.** and **B.** Find the edge views of the two planes.

6. (Fig. 28.51) **A.** Find the angle between the planes. **B.** Find the piercing point by projection. **C.** Find the piercing point by an auxiliary view.

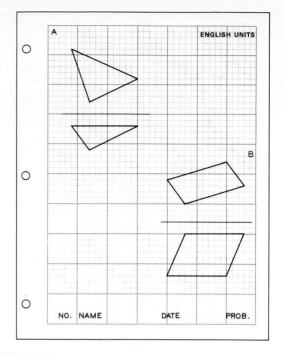

FIG. 28.50 Problems 5A and 5B.

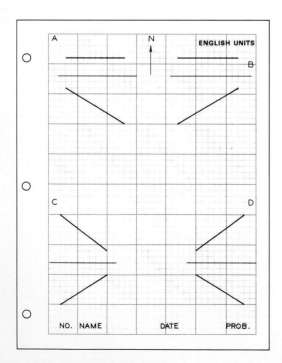

FIG. 28.49 Prolems 4A and 4B.

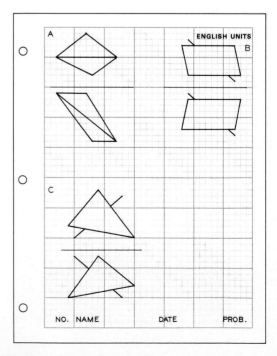

FIG. 28.51 Problems 6A–6C.

7. (Fig. 28.52) **A.** Construct a line perpendicular to the plane and through point 0 on the plane by an auxiliary view. **B.** Construct a line perpendicular to the plane from point 0 by an auxiliary view.

8. (Fig. 28.53) **A.** Find the line of intersection between the planes by projection. **B.** Find the line of intersection by an auxiliary view. Show visibility.

9. (Fig. 28.54) **A.** and **B.** Find the direction of slope and slope of angle of the planes.

10. (Fig. 28.54) **A.** and **B.** Find the strike and dip of the planes.

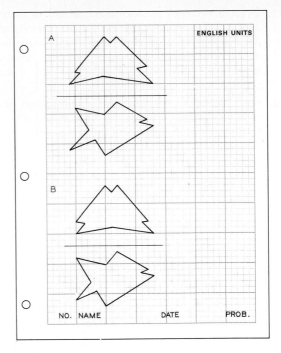

FIG. 28.53 Problems 8A and 8B.

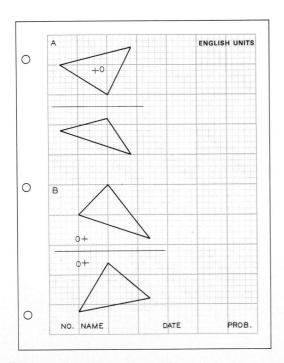

FIG. 28.52 Problems 7A and 7B.

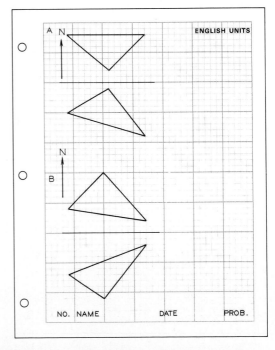

FIG. 28.54 Problems 9A and 9B; 10A and 10B.

11. (Fig. 28.55) Find the shortest distance, the horizontal distance, and the vertical distance from point 0 on the ground to the underground ore vein represented by the triangle. Point *B* is on the lower plane of the vein. Find the thickness of the vein.

12. (Fig. 28.56) **A.** Find the line of intersection between the two planes by the cutting-plane method. **B.** Find the line of intersection between the two planes indicated by strike lines 1–2 and 3–4. The plane with strike 1–2 has a dip of 30°, and the one with strike 3–4 has a dip of 60°.

13. (Fig. 28.57) Find the limits of cut and fill in the plan view of the roadway. The roadway has a cut angle of 35° and a fill angle of 40°.

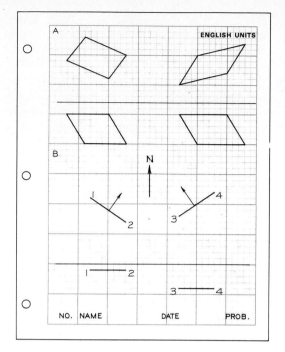

FIG. 28.56 Problems 12A and 12B.

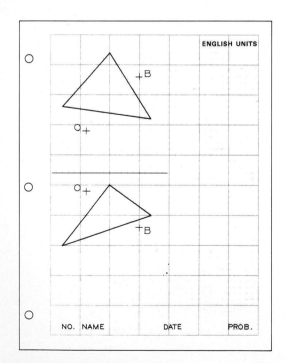

FIG. 28.55 Problem 11.

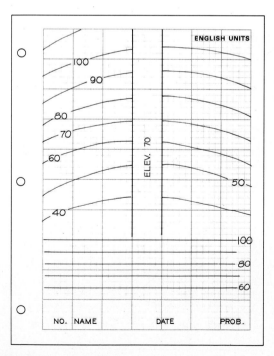

FIG. 28.57 Problem 13.

14. (Fig. 28.58) Find the outcrop of the ore vein represented by the triangle (upper surface). Point *B* is on the lower surface.

15. (Fig. 28.59) Complete the plan-profile drawing of the drainage system from manhole 1, through manhole 2, to manhole 3, using the grades indicated. Allow a drop of 0.20 foot across each manhole to compensate for loss of pressure.

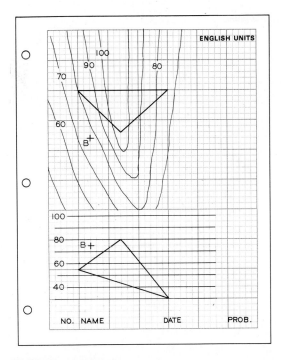

FIG. 28.58 Problem 14.

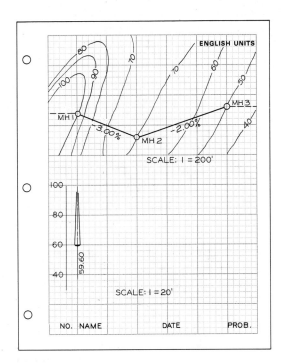

FIG. 28.59 Problem 15.

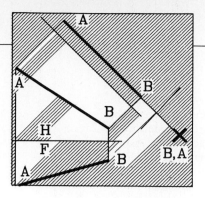

CHAPTER 29

Successive Auxiliary Views

29.1
Introduction

A design cannot be detailed with complete specifications unless its complete geometry has been determined, which usually requires the application of descriptive geometry. The Comsat satellite (Fig. 29.1) is an example of a design where various problems of geometry were solved by the use of successive auxiliary views.

A *secondary auxiliary view* is an auxiliary view projected from a primary auxiliary view. A *successive auxiliary view* is an auxiliary view of a secondary auxiliary view.

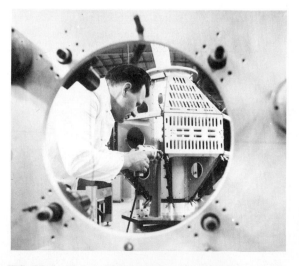

FIG. 29.1 The angles between the planes and the true-size views of the surfaces of this Comsat satellite were determined by applying the principles of descriptive geometry. (Courtesy of TRW Systems.)

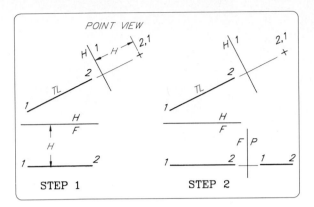

FIG. 29.2 The point of view of a line can be found by projecting an auxiliary view from the true-length view of the line.

29.2
Point view of a line

When a line appears true length, its point view can be found by projecting an auxiliary view from it. In Fig. 29.2, line 1–2 is true length in the top view since it is horizontal in the front view. Its point view is found in

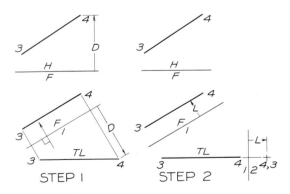

FIG. 29.3 Point view of an oblique line.

Step 1 A line of sight is drawn perpendicular to one of the views, the front view in this example. Line 3–4 is found true length by projecting perpendicularly from the front view.

Step 2 A secondary reference line, 1–2, is drawn perpendicular to the true-length view of 3–4. The point view is found by transferring dimension L from the front view to the secondary auxiliary view.

the primary auxiliary view by constructing reference line H-1 perpendicular to the true-length line. The height dimension, H, is transferred to the auxiliary view to locate the point view of 1–2.

Since the line in Fig. 29.3 is not true length in either view, the line must be found true length by a primary auxiliary view. By projecting from the front view, the line is found true length. This view could have been projected from the top as well. The point view of the line is found by projecting from the true-length line to a secondary auxiliary view and is labeled 2,1 since point 2 is seen first.

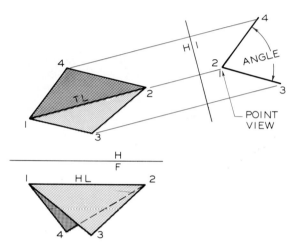

FIG. 29.4 The angle between two planes can be found in the view where the line of intersection between them projects as a point. Since the line of intersection, 1–2, is true length in the top view, it can be found as a point in a view that is projected from the top view.

29.3
Angle between planes

> The angle between two planes is called a *dihedral angle*.

The dihedral angle can be found in the view where the line of intersection appears as a point. Since this view results in the point view of a line that lies on both planes, both will appear as edges.

The two planes in Fig. 29.4 represent a special case since the line of intersection, 1–2, is true length

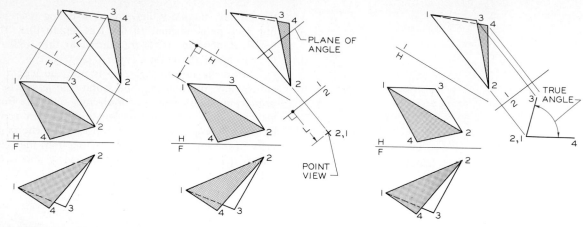

FIG. 29.5 Angle between two planes.

Step 1 The angle between two planes can be measured in a view where the line of intersection appears as a point. The line of intersection is first found true length by projecting a primary auxiliary view perpendicularly from the top view, in this case.

Step 2 The point view of the line of intersection is found in the secondary auxiliary view by projecting parallel to the true length view of 1–2. The plane of the dihedral angle is an edge and is perpendicular to the true-length line of intersection.

Step 3 The edge views of the planes are completed in the secondary auxiliary view by locating points 3 and 4. The angle between the planes, the dihedral angle, can be measured in this view.

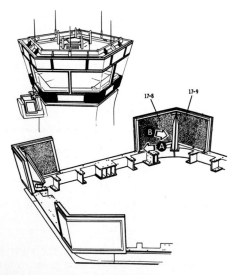

FIG. 29.6. The determination of the angle between the planes of the corner panels of the control tower used principles of descriptive geometry. (Courtesy of the Federal Aviation Agency.)

in the top view. This permits you to find its point view in a primary auxiliary view where the true angle can be measured.

A more typical case is given in Fig. 29.5, where the line of intersection between the two planes is not true length in either view. The line of intersection, 1–2, is found true length in a primary auxiliary view, and the point view of the line is then found in the secondary auxiliary view, where the dihedral angle is measured.

This principle must be used to determine the angles between side panels of a control tower (Fig. 29.6) so that corner braces can be designed that will hold the structure together.

29.4 _____
True size of a plane

A plane can be found true size in a view projected perpendicularly from an edge view of a plane.

The front view of plane 1–2–3 appears as an edge in the front view (Fig. 29.7). It can be found true size in a primary auxiliary view projected perpendicularly from the edge view.

In Fig. 29.8, the true size of plane 1–2–3 is found by first finding the edge view of the plane (Step 1). The secondary auxiliary view is then projected per-

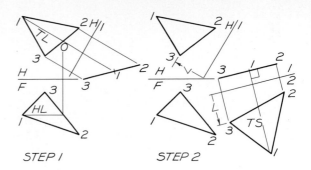

FIG. 29.8 True size of a plane.

Step 1 The edge view of plane 1–2–3 is found by finding the point view of TL line, 1–0, in a primary auxiliary view.

Step 2 A true-size view is found by projecting a secondary auxiliary view perpendicularly from the edge view of the plane. Dimension L is shown as a typical measurement used to complete the TS view.

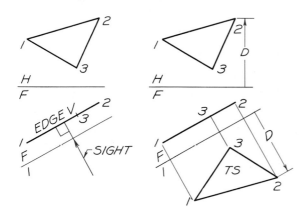

FIG. 29.7 True size of a plane.

Step 1 Since plane 1–2–3 appears as an edge in the front view, the line of sight is drawn perpendicular to the edge. The F-1 reference line is drawn parallel to the edge.

Step 2 The plane will appear true size in the primary auxiliary by locating the vertex points with depth (D) dimensions.

pendicularly from the edge view (Step 2) to find a true-size view of the plane where each angle is true size.

This principle can be used to find the angle between lines, such as bends in a fuel line of an aircraft engine (Fig. 29.9). This type of problem is shown in Fig. 29.10, where the top and front views of intersecting lines are given. It is required that the angles of bend be determined and a radius of curvature be shown.

FIG. 29.9 The angles of bend in the fuel line were found by applying the principle of finding the angle between two lines. (Courtesy of Avco Lycoming.)

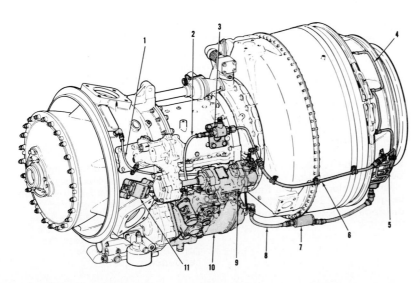

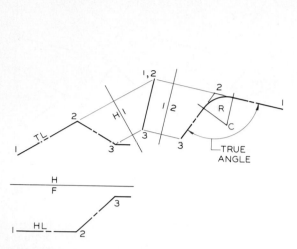

FIG. 29.10. The angle between two lines can be found by finding the plane of the lines true size.

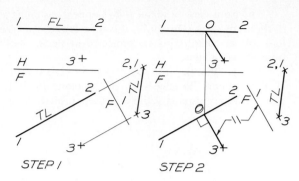

FIG. 29.11 Shortest distance from a point to a line.

Step 1   The shortest distance from a point to a line will appear TL where the line appears as a point. The TL line is found in the primary auxiliary view.

Step 2   For the connecting line to be TL in the auxiliary view, it must be parallel to the F-1 reference line in the preceding view, the front view. Line 3–0 is found and projected to the top view.

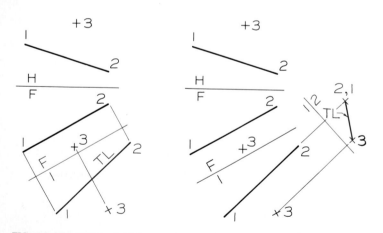

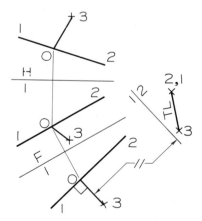

FIG. 29.12 Shortest distance from a point to a line.

Step 1   The shortest distance from a point to a line can be found in the view where the line appears as a point. Line 1–2 is found true length by projecting from the front view.

Step 2   Line 1–2 is found as a point in a secondary auxiliary view projected from the true-length view of 1–2. The shortest distance appears true length in this view.

Step 3   Since 3–0 is true length in the secondary auxiliary view, it is parallel to the 1–2 line in the primary auxiliary view and perpendicular to the line. The front and top views of 3–0 are found by projecting from the primary auxiliary view in sequence.

Angle 1–2–3 is found as an edge in the primary auxiliary view and true-size in the secondary view, where the angle can be measured and the radius of curvature drawn.

29.5
Shortest distance from a point to a line

> The shortest distance from a point to a line can be measured in the view where the line appears as a point.

The shortest distance from point 3 to line 1–2 is found in a primary auxiliary view in Fig. 29.11 (Step 1). Since the distance from 3 to the line is true length where the line is a point, it must be parallel to reference line F-1 in the front view.

This type of problem is solved in Fig. 29.12 by finding the line 1–2 true length in a primary auxiliary. The point view of the line is found in the secondary auxiliary view, where the distance from point 3 is true length. Since line 0–3 is true length in this view, it must be parallel to the 1–2 reference line in the preceding view, the primary auxiliary view. It is also perpendicular to the true-length view of line 1–2 in the primary auxiliary view.

29.6
Shortest distance between skewed lines—line method

> Randomly positioned (non-parallel) lines are called *skewed lines*. The shortest distance between two skewed lines can be measured in the view where one of the lines appears as a point.

The shortest distance between two lines is perpendicular to both lines. The location of the shortest distance is both functional and economical, as demonstrated by the connector between two pipes in Fig. 29.13, since a standard connector is a 90° tee.

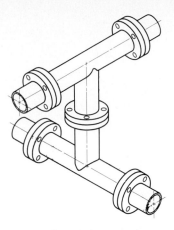

FIG. 29.13 The shortest distance between two lines, or planes, is a line perpendicular to both. This is the most economical and functional connection since perpendicular connectors are standard.

A problem of this type is solved by the *line method*. In Fig. 29.14, line 3–4 is found as a point in the secondary auxiliary view, where the shortest distance is drawn perpendicular to line 1–2. Since the distance is true length in the secondary auxiliary view, it must be parallel to the 1–2 reference line in the primary auxiliary view. Point 0 is found by projection, and 0P is drawn perpendicular to line 3–4. The line is projected back to the given principal views.

29.7
Shortest distance between skewed lines—plane method

The distance between skewed lines can be solved using the alternative *plane method* which involves the construction of a plane through one of the lines parallel to the other (Fig. 29.15). The top and front views of 0–2 are drawn parallel to their respective views of line 3–4. Plane 1–2–0 is parallel to line 3–4. Both lines appear parallel in a view where 1–2–0 is an edge.

In Fig. 29.16, plane 3–4–0 is constructed, its edge view is found, and both lines appear parallel (Step 1). A secondary auxiliary view is projected perpendicularly from these parallel lines to find the view where the lines cross (Step 2). This crossing point is the point

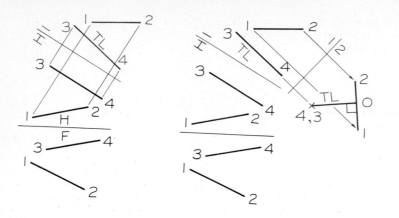

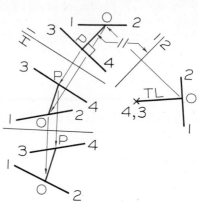

FIG. 29.14 Shortest distance between skewed lines—line method.

Step 1 The shortest distance between two skewed lines can be found in the view where one of the lines appears as a point. Line 3–4 is found true length by projecting from the top view along with line 1–2.

Step 2 The point view of line 3–4 is found in a secondary auxiliary view projected from the true-length view of 3–4. The shortest distance between the lines is drawn perpendicular to line 1–2.

Step 3 Since the shortest distance is TL in the secondary auxiliary view, it must be parallel to the reference line in the preceding view. Points 0 and P are projected to the given view.

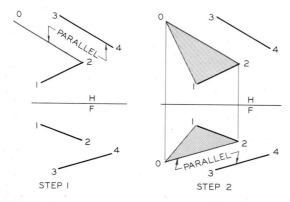

FIG. 29.15 A plane can be constructed through a line and parallel to another line.

Step 1 Line 0–2 is drawn parallel to line 3–4 to a convenient length.

Step 2 The front view of line 0–2 is drawn parallel to the front view of line 3–4. The length of the front view of 0–2 is found by projecting from the top view of 0. Plane 1–2–0 is parallel to line 3–4.

view of the shortest distance between the lines. It is true length and perpendicular to both lines when projected to the primary auxiliary view, where it is labeled line *LM*. It is projected back to the given views to complete the solution.

This principle of the shortest distance between two skewed lines was applied to the design of the separation of power lines, where clearance is critical (Fig. 29.17).

29.8
Shortest level distance between skewed lines

To find the shortest level (horizontal) distance between two skewed lines, the plane method must be used rather than the line method. Also the plane method is used to find a view where the horizontal plane appears as an edge.

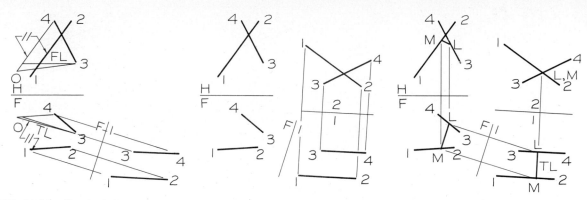

FIG. 29.16 Shortest distance between skewed lines—plane method.

Step 1 Construct a plane through line 3–4 that is parallel to line 1–2. When this plane is found as an edge by projecting from the front view, the two lines will appear parallel.

Step 2 The shortest distance will appear TL in the primary auxiliary view, where it will be perpendicular to both lines. The secondary auxiliary view is projected perpendicularly from the lines in the primary auxiliary.

Step 3 The crossing point of the two lines is the point view of perpendicular distance *LM* between them. This distance is projected to the primary auxiliary view, where it is TL, and back to the given views.

FIG. 29.17 The determination of clearance between power lines is an electrical engineering problem that is an application of skewed-line principles. (Courtesy of the Tennessee Valley Authority.)

In Fig. 29.18, plane 3–4–0 is drawn parallel to line 1–2, and an edge view of the plane is found in the primary auxiliary view. The lines appear parallel in this view, and the horizontal (H-1) appears as an edge. A line of sight is drawn parallel to H-1, and the secondary reference line, 1–2, is drawn perpendicular to H-1. The crossing point of the lines in the secondary auxiliary view locates the point view of the shortest horizontal distance between the lines. This line, *LM*, is true length in the primary auxiliary view and parallel to the H-1 plane. Line *LM* is projected back to the given views. As a check, this line must be parallel to the H–F line in the front view since it is a level or horizontal line.

29.9

Shortest grade distance between skewed lines

Many lines representing highways, power lines, or conveyors are connected by lines at a specified grade other than horizontal or perpendicular. Conveyors,

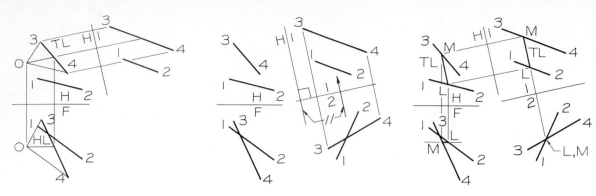

FIG. 29.18 Shortest level distance between skewed lines—plane method.

Step 1 Construct plane 0–3–4 parallel to line 1–2 by drawing line 0–4 parallel to 1–2. Find the edge view of 0–3–4 by projecting off the top view. The lines will appear parallel in this view. The auxiliary view must be projected from the top view to find the horizontal plane as an edge.

Step 2 An infinite number of horizontal (level) lines can be drawn parallel to H-1 between the lines in the auxiliary view, but only the shortest level line will appear true length. Construct the secondary auxiliary view by projecting parallel to the horizontal (H-1) to find the point view of the shortest level line.

Step 3 The crossing point of the two lines in the secondary auxiliary view establishes the point view of the level connector, *LM*. Project *LM* back to the given views. *LM* is parallel to the H-plane in the front view, verifying that it is a level line.

FIG. 29.19 These conveyors represent the application of skewed-line problems.

such as the one shown in Fig. 29.19, are used to transport aggregates or grain for mixing.

If a 50% grade connector between two lines must be found (Fig. 29.20), the plane method is used. A view where the lines appear parallel is constructed, and a 50% grade line is drawn from the edge view of the horizontal (H-1). To have an edge view of the horizontal from which the 50% grade is constructed the auxiliary views must be projected from the top view.

The grade line can be constructed in two directions from the H-1 line, but the shortest distance will be the one most nearly perpendicular to both lines (Step 2). The secondary auxiliary view is projected parallel to this 50% grade line to find the crossing point of the lines to locate the shortest connector, *LM*, at a 50% grade. Line *LM* is projected back to all views. *LM* is true length in the primary auxiliary view, where the lines appear parallel.

The shortest connector between skewed lines will appear true length in the view where the lines appear parallel. Perpendicular, horizontal, and grade lines are true length in the this view.

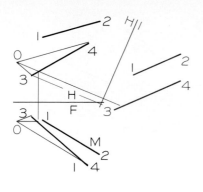 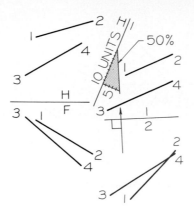 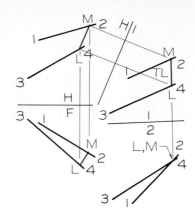

FIG. 29.20 Grade distance between skewed lines.

Step 1 To find a level line or a line on a grade between two skewed lines, the primary auxiliary must be projected from the top view. Plane 3–4–0 is constructed parallel to line 1–2. The edge view of the plane is found and the lines appear parallel.

Step 2 Construct a 50% grade line from the edge view of the H-1 line in the primary auxiliary view that is most nearly perpendicular to the lines. Project the secondary auxiliary view parallel to the grade line. The shortest grade distance will appear TL in the primary auxiliary.

Step 3 The point of crossing of the two lines in the secondary auxiliary view establishes the point view of the 50% grade line, LM. This line is projected back to the previous views in sequence.

29.10
Angular distance to a line

Standard connectors used to connect pipes and structural members are available in two standard angles—90° and 45°. It is far more economical to incorporate these angles into a design rather than to design specially made connectors.

In Fig. 29.21, it is required to locate the point of intersection on line 1–2 of a line drawn from point 0 at a 45° angle to the line. The plane of the line and point, 1–2–0, is found as an edge in the primary auxiliary view and as a true-size plane in the secondary auxiliary view. The angle can be measured in this view where the plane of the line and point is true size.

The 45° connector is drawn from 0 to the line toward point 2 if it slopes downhill, or toward point 1 if it slopes uphill. Slope can be determined by referring to the front view where height can be easily seen. Line 0P is projected back to the given views.

29.11
Angle between a line and a plane—plane method

> The angle between a line and a plane can be measured in the view where the plane appears as an edge and the line appears true length.

In Fig. 29.22, the edge view of the plane is found in a primary auxiliary view projected from any primary view. The plane is then found true size in Step 2, where the line is foreshortened. Line *AB* can be found true length in a third auxiliary view projected perpendicularly from the secondary auxiliary view of *AB*. The line appears true length, and the plane appears as an edge in the third successive auxiliary view.

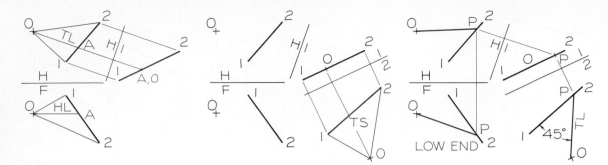

FIG. 29.21 Line through a point with a given angle to a line.

Step 1 Connect 0 to each end of the line to form a plane 1–2–0 in both views. Draw a horizontal line in the front view of the plane and project it to the top view, where it is TL. Find the edge view of the plane by obtaining the point view of A0.

Step 2 Find the true size of plane 1–2–0 projecting perpendicularly from the edge view of the plane in the primary auxiliary view. The plane can be omitted in this view and only line 1–2 and point 0 are shown.

Step 3 Line *OP* is constructed at the specific angle with line 1–2, 45°. If the angle is toward point 2, the line slopes downhill; if toward point 1, it slopes uphill. Point *P* is projected back to the other views in sequence.

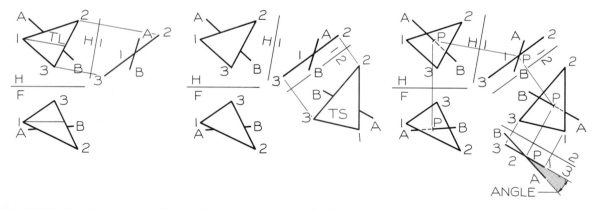

FIG. 29.22 Angle between a line and plane—plane method.

Step 1 The angle between a line and a plane can be measured in the view where the plane is an edge and the line is TL. The plane is found as an edge by projecting off the top view. The line is not true length in this view.

Step 2 The plane is found true size by projecting perpendicularly from the edge view of the plane. A view projected in any direction from a TS plane will show the plane as an edge.

Step 3 A third successive auxiliary view is projected perpendicularly from line *AB*. The line appears TL and the plane appears as an edge in this view where the angle is measured. The piercing points and visibility are shown by projecting back in sequence to all views.

The piercing point is projected back to the views in sequence, and the visibility is determined for each view.

29.12
Angle between a line and a plane—line method

To find the angle between a line and a plane by the line method, the line is first found as a point, and then true length is found (Fig. 29.23). The plane is foreshortened in Step 2 where the line *AB* appears as a point.

The plane is found as an edge in a third auxiliary view by finding the point view of a line on the plane. Since the view is projected from the point view of line *AB*, the line will appear true length. This view satisfies the condition that the line be true length and the plane be an edge. The angle is measured in the third view. The piercing point is projected back to previous views, and the visibility is determined to complete the solution.

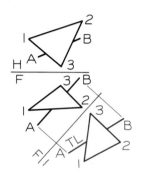 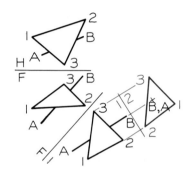 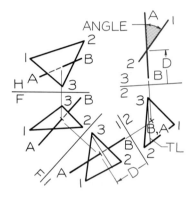

FIG. 29.23 Angle between a line and a plane—line method.

Step 1 The angle between a line and a plane can be measured in the view where the plane appears as an edge and the line is true length. Find a true length view of line *AB* and project the plane to this view also.

Step 2 Find a point view of line *AB* in the secondary auxiliary view. Plane 1–2–3 does not appear true size in this view unless the line is perpendicular to the plane. The point view of the line in this view is the piercing point on the plane.

Step 3 An edge view of the plane is found by finding the point view of a line on the plane. Line *AB* will appear TL since it was a point in the secondary auxiliary view. Measure the angle, project back to every view to locate the piercing point, and determine visibility.

Problems

Use Size A sheets for the following problems, and lay out the problems using instruments. Each square on the grid is equal to 0.20 in. (about 5 mm). The problems can be laid out on grid or plain paper. Label all reference planes and points in each problem with $\frac{1}{8}$-inch letters or numbers, using guidelines.

The crosses marked "1" and "2" are to be used for placing the primary and secondary reference lines. The primary reference line should pass through "1" and the secondary through "2".

1–2. (Fig. 29.24) Find the point views of the line.

3–4. (Fig. 29.24) Find the angles between the planes.

5–6. (Fig. 29.25) Find the true-size views of the planes. Project from the front view in Problem 5 and from the top view in Problem 6.

7–8. (Fig. 29.26) Find the angles between the lines. Project from the top views of both problems.

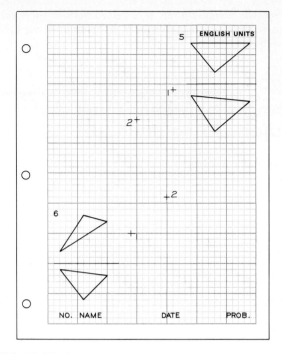

FIG. 29.25   Problems 5 and 6.

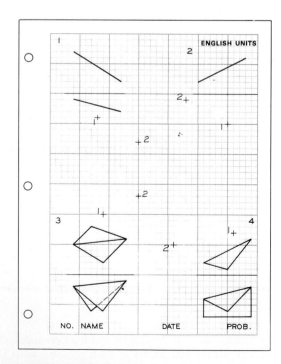

FIG. 29.24   Problems 1–4.

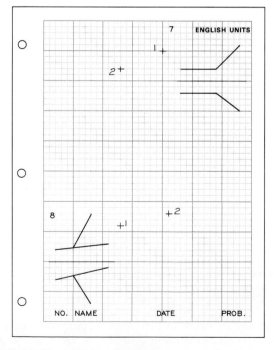

FIG. 29.26   Problems 7 and 8.

9. (Fig. 29.27) Find the shortest distance from the point to the line, and show the distance in all views. Use the plane method, and project from the left-side view. Scale: full size.

10. (Fig. 29.27) Find the shortest distance from point 0 to the line, and show the distance in all views. Use the line method, and project from the top view. Scale: full size.

11–12. (Fig. 29.28) Find the shortest distances between the two skewed lines using the line method, and show the distances in all views. Begin problem twelve by finding line 3–4 true length, using the cross marks given. Scale: full size.

13. (Fig. 29.29) Find the shortest horizontal distance between the two lines by the plane method. Scale: full size.

14. On a separate sheet of paper, redraw Problem 11, and find the shortest 20% grade between the two lines. Scale: full size.

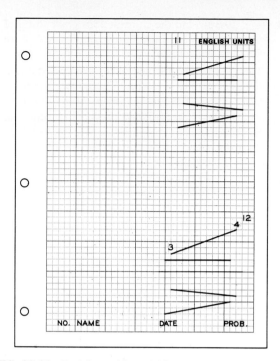

FIG. 29.28 Problems 11 and 12.

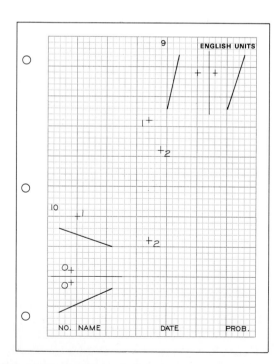

FIG. 29.27 Problems 9 and 10.

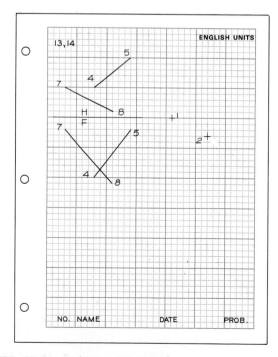

FIG. 29.29 Problems 13 and 14.

15. (Fig. 29.30) Find the shortest 25% grade distance between the two lines. Show the distance in all views. Scale: full size.

16. (Fig. 29.31) Find the connector from point 0 that will intersect line 1–2 at 60°. Show this line in all views. Project from the top view. Scale: full size.

17. (Fig. 29.32) Find the angle between the line and the plane by using the plane method. Project from the front view, and show the visibility in all views. Scale: full size.

18. Same as Problem 17, except use the line method.

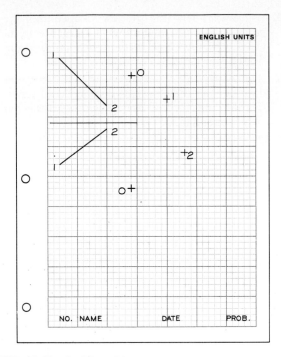

FIG. 29.31 Problem 16.

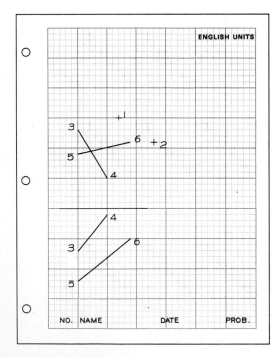

FIG. 29.30 Problem 15.

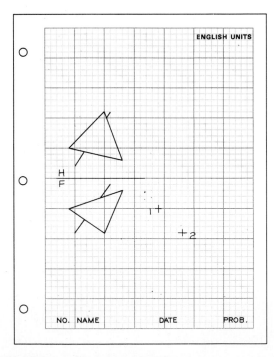

FIG. 29.32 Problems 17 and 18.

19. (Fig. 29.33) Construct a circle that will pass through each vertex of the triangle. Project from the top view, and show the elliptical views of the circle in all views.

20. (Fig. 29.33) Find the front view of the elliptical path of a circular section through the sphere. The edge view of the section is shown in the top view.

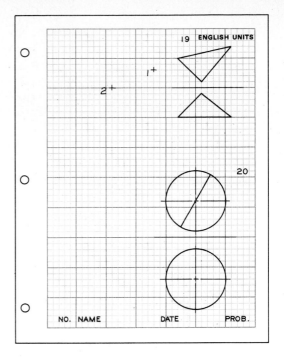

FIG. 29.33. Problems 19 and 20

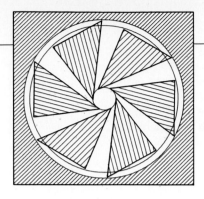

CHAPTER 30

Revolution

30.1
Introduction

Figure 30.1 shows an orthicon camera. It was designed to revolve about three axes; thus, it is possible to aim the camera in any direction to track space vehicles. This is just one of many designs based on the principles of revolution.

Revolution is a technique of revolving an orthographic view into a new position to yield a true-size view of a surface or a line. For example, revolution techniques were used in early descriptive geometry solutions prior to the alternate method of auxiliary views.

30.2
True length of a line
in the front view

The object in Fig. 30.2 demonstrates how an inclined surface can be found true size by auxiliary view and revolution. When the auxiliary-view method is used,

FIG. 30.1 This orthicon camera is an example of a design that uses principles of revolution. Its cradle was designed to permit the camera to be revolved to any position. (Courtesy of ITT Industrial Laboratories.)

584

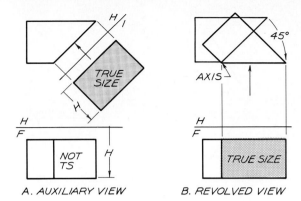

A. AUXILIARY VIEW   B. REVOLVED VIEW

FIG. 30.2   A. The surface is found true size by an auxiliary view. **B.** The surface is found true size by revolving the top view.

the observer changes position to an auxiliary vantage point and looks perpendicularly at the inclined surface.

When the revolution method is used, the top view of the object is revolved about the axis until the edge view of the inclined plane is perpendicular to the standard line of sight from the front view. In other

words, the observer's line of sight does not change, but the conventional lines of sight between adjacent orthographic views are used.

A single line can be found true length in the front view by revolution, as shown in Fig. 30.3. By establishing the point view of an axis in the top view, Line *AB* is revolved into a position parallel to the frontal plane. The top view represents the circular base of a right cone, and the front view is the triangular view of a cone. Line *AB'* is the outside element of the cone and is true length.

Figure 30.4 illustrates the technique of finding line 1–2 true length in the front view. When in its first position, the observer's line of sight is perpendicular neither to the triangular plane nor to line 1–2. But when it is revolved to be perpendicular to the line of sight, the triangle appears true size, and line 1–2 is true length.

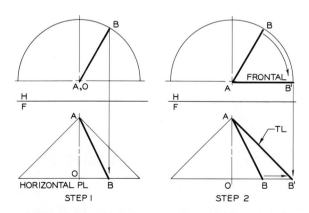

STEP 1   STEP 2

FIG. 30.3   True length in the front view.

Step 1   The top view of line *AB* is used as a radius to draw the base of a cone with point *A* as the apex. The front view of the cone is drawn with a horizontal base through point *B*. Line *A0* is the axis of the cone.

Step 2   The top view of line *AB* is revolved to be parallel to the frontal plane. When projected to the front view, frontal line *AB'* is the outside element of the cone and is true length.

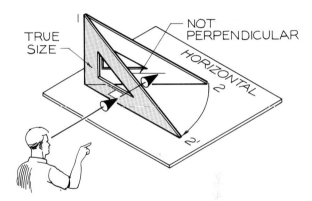

FIG. 30.4   Line 1–2 of the triangle does not appear true length in the front view because your line of sight is not perpendicular to it. When the triangle is revolved to a position where your line of sight is perpendicular to it, line 1–2' can be seen true length.

30.3
True length of a line in the top view

A surface that appears as an edge in the front view can be found true size in the top view by a primary auxiliary view or by a single revolution (Fig. 30.5).

The axis of revolution is located as a point in the front view and is true length in the top view. The edge

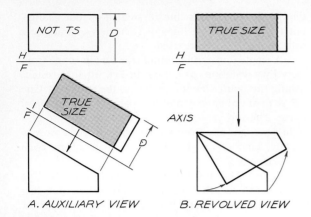

A. AUXILIARY VIEW   B. REVOLVED VIEW

FIG. 30.5 **A.** The inclined plane is found true size by an auxiliary view with a line of sight perpendicular to the surface. **B.** The surface is found true size by revolving the front view until it is perpendicular to the line of sight from the top view.

view of the plane is revolved until it is a horizontal edge in the front view (Fig. 30.5A). It is projected to the top view to find the surface true size. As in the auxiliary-view method, the depth dimension *(D)* does not change.

Line *CD* is found true length in the top view by revolving the line into a horizontal position in Fig. 30.6 (Step 2). The arc of revolution in the front view represents the base of a cone of revolution. Line *CD'* is true length in the top view since it is an outside element of the cone. Note that the depth dimension in the top view does not change.

30.4

True length of a line in the profile view

Line *EF* in Fig. 30.7 is found true length by revolving it in the front view until it is parallel to the edge view of the profile plane (Step 1). The circular view of the cone is projected to the side view, where the triangular shape of the cone is seen. Since *EF'* is a profile line in (Step 2), it is true length in the side view, where it is the outside element of the cone.

In the previous examples, each line has been revolved about one of its ends. However, a line can be revolved about any point on its length. Line 5–6 in

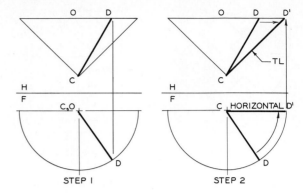

FIG. 30.6 True length of a line in the top view.

Step 1 The front view of line *CD* is used as a radius to draw the base of a cone with point *C* as the apex. The top view of the cone is drawn with the base shown as a frontal plane.

Step 2 The front view of line *CD* is revolved into position *CD'* where it is horizontal. When projected to the top view, *CD'* is the outside element of the cone and is true length.

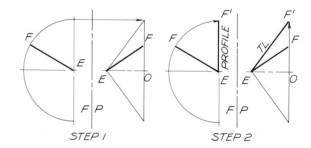

FIG. 30.7 True length of a line in the side view.

Given The front and side views of line *EF*.

Required Find the true-length view of line *EF* in the profile review by revolution.

Step 1 The front view of line *EF* is used as a radius to draw the circular view of the base of a cone. The side view of the cone is drawn with a base through point *F* that is a frontal edge.

Step 2 Line *EF* in the frontal view is revolved to position *EF'* where it is a profile line. Line *EF'* in the profile view is true length, since it is a profile line and the outside element of the cone.

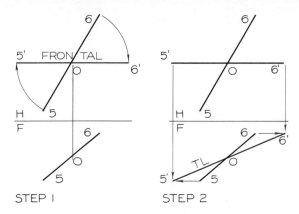

FIG. 30.8 In the preceding figures, the lines have been revolved about their ends, but they can be found true length by revolving them about any point on them. Line 5–6 is revolved into a frontal position in the top view and is found true length in the frontal view.

Fig. 30.8 is an example of a line that is found true length by revolving it about point 0.

30.5
Angles with a line and principal planes

The angle between a line and plane will appear true size in the view where the plane is an edge and the line is true length. Two principal planes appear as edges in all principal views. Therefore when a line appears true length in a principal view, the angle between the line and the two principal planes can be measured.

The angle between the horizontal and the profile planes can be measured in Fig. 30.9A in the front view. The angle with horizontal and profile planes can be measured in the top view in Fig. 30.9B.

30.6
True size of a plane

When a plane appears as an edge in a principal view (Fig. 30.10), it can be revolved to be parallel to the reference line (Step 1). The new front view is true size.

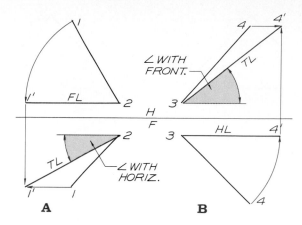

FIG. 30.9 Angles with principal planes.

A. The angle with the horizontal plane can be measured in the front view if the line appears true length.

B. The angle with the frontal plane can be measured in the top view if the line appears true length.

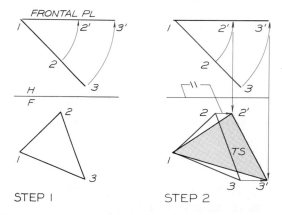

FIG. 30.10 True size of a plane.

Step 1 The edge view of a plane is revolved to be parallel to the frontal plane.

Step 2 Points 2' and 3' are projected to the horizontal projectors from points 2 and 3 in the front view.

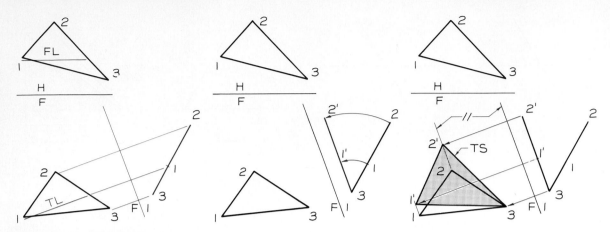

FIG. 30.11   True size of a plane by revolution.

Step 1   Find the edge view of the plane by finding the point view of a true-length line on it in the front view.

Step 2   Revolve the edge view of the plane about one of its points (point 3 in this example) until it is parallel to the F-1 line.

Step 3   Project points 1' and 2' to the front view to the projectors from points 1 and 2. These projectors must be parallel to the F-1 line.

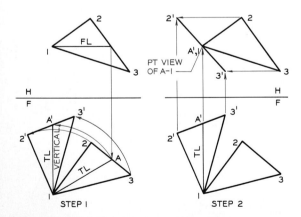

FIG. 30.12   Edge view of a plane.

Step 1   A frontal line is found true length on the front view of the plane. The front view is revolved until the true-length line is vertical.

Step 2   Since the TL line (1–A') is vertical, it will appear as a point in the top view, and the plane will appear as an edge, 1–2'–3'.

The plane in Fig. 30.11 is found true size by the combination of an auxiliary view and a single revolution. The plane is found as an edge projected from a true-length line in the plane. The edge view is revolved to be parallel to the F-1 reference line (Step 2). The true size of the plane is found by projecting the original points, 1, 2, and 3, in the front view parallel to the F-1 line to intersect the projectors from 1' and 2'. The true size of the plane could have been found by projecting from the top view to find the edge view as well.

30.7
True size of a plane by double revolution

The edge view of a plane can be found by revolution without the use of auxiliary views (Fig. 30.12). A frontal line is drawn on plane 1–2–3, and the line appears true length in the front view. The plane is revolved until the true-length line is vertical in the front view (Step 1). The true-length line will project as a point in the top view; therefore the plane will appear as an edge in this view (Step 2). Projectors from the

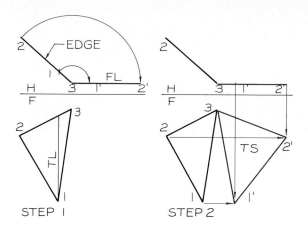

FIG. 30.13 True size of a plane.

Step 1 When a plane appears as an edge in the principal view, it can be revolved to a position parallel to a reference line, the frontal line in this case.

Step 2 Points 1' and 2' are projected to the front view to intersect with the horizontal projectors from the original points 1 and 2. The plane is true size in this view.

top view of points 2 and 3 are parallel to the H-F reference line.

A second revolution, called a *double revolution,* can be made to revolve this edge view of the plane until it is parallel to the frontal plane, as shown in Fig. 30.13. The top views of points 1' and 2' are projected to the front view where plane 1–2–3 is true size.

This second revolution could have been performed in Fig. 30.12, but this would have resulted in an overlapping of views, making it difficult to observe the separate steps.

Double revolution is used in Fig. 30.14 to find the oblique plane of the object, plane 1–2–3, true size. In Step 1, the true-length line 1–2 on the plane is revolved in the top view until it is perpendicular to the frontal plane. Thus, line 1–2 appears as a point in the front view, and the plane appears as an edge. This changes the width and depth dimensions, but the height dimension does not change.

In Step 2, the edge view of the plane is revolved into a vertical position parallel to the profile plane. The plane is found true size by projecting to the profile view, where the depth dimension is unchanged and the height dimension has been increased.

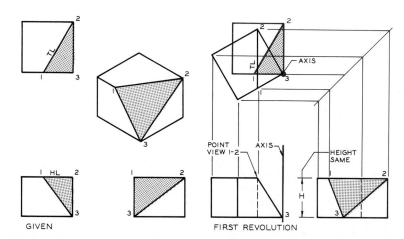
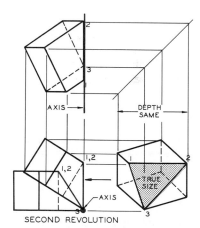

FIG. 30.14 True size by double revolution.

Given Three views of a block with an oblique plane across one corner.

Required Find the plane true size by revolution.

Step 1 Since line 1–2 is horizontal in the frontal view, it is true length in the top view. The top view is revolved into a position where line 1–2 can be seen as a point in the front view.

Step 2 Since plane 1–2–3 was found as an edge in Step 1, this plane can be revolved into a vertical position in the front view, to appear true size in the side view. The depth dimension does not change.

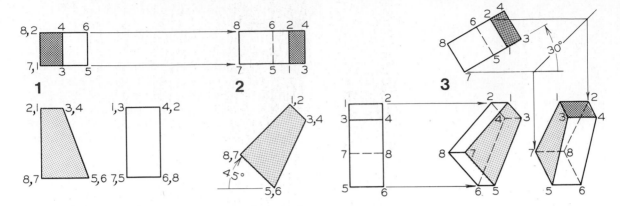

FIG. 30.15 Double revolution.

Given Three views of a surface with an inclined plane.

Step 1 The front view is revolved 45°, and the new width is projected to the top view where the depth is unchanged.

Step 2 The top view is revolved 30° to change the width and depth, but the height remains the same.

A second example of double revolution of a solid is shown in Fig. 30.15 with dimensions transferred from view to view. The front view is revolved clockwise 45°, and new top and side views are drawn with the depth dimension remaining constant. The top view is revolved 30° counterclockwise in Step 2. A new front view and side view are constructed by using projectors from the side view found in Step 1 and the top view found in Step 2.

30.8
Angle between planes

The engine mount frame of a helicopter is an application where the angle between two intersecting planes must be found to provide its design specifications (Fig. 30.16).

In Fig. 30.17, the angle between two planes is found by drawing the edge view of the dihedral angle (the angle between the planes) perpendicular to the line of intersection, and the plane of the angle is projected to the front view (Step 1). The edge view of the angle is revolved until it is a frontal plane in the top view; then, it is projected to the front view where its true-size view is found (Step 2).

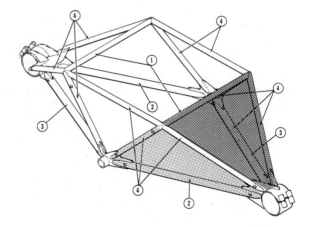

FIG. 30.16 By using revolution principles, the angle between any two planes of the helicopter engine mount can be found. (Courtesy of Bell Helicopter Corp.)

A similar problem is solved in Fig. 30.18. In this example, the line of intersection does not appear true length in the given views; therefore an auxiliary view is used to find its true length (Step 1). The plane of the angle between the planes can be drawn as an edge perpendicular to the true-length line of intersection.

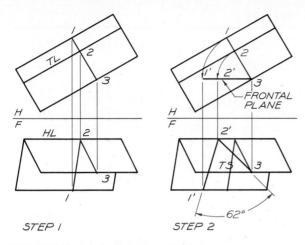

STEP 1

STEP 2

FIG. 30.17 Angle between planes.

Step 1   A right section is drawn perpendicular to the TL line of intersection between the planes in the top view and is projected to the front view. The section is not true size in the front view.

Step 2   The edge view of the right section is revolved to position 1'–2'–3 in the top view to be parallel to the frontal plane. This section is projected to the front view, where it is true size.

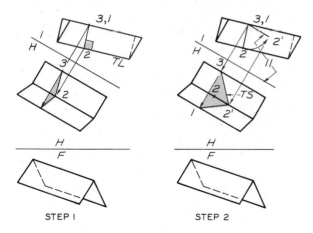

STEP 1

STEP 2

FIG. 30.18 Angle between oblique planes.

Step 1   A true-length view of the line of intersection is found in an auxiliary view projected from the top view. The right section is constructed perpendicular to the true length of the line of intersection and is projected to the top view.

Step 2   The edge view of the right section is revolved to be parallel to the H-1 reference line so the plane will appear true size in the top view after being revolved. The angle between the planes can be found by measuring angle 1–2'–3.

The foreshortened view of plane 1–2–3 is projected to the top view (Step 1). The edge view of plane 1–2–3 is then revolved in the primary auxiliary view until it is parallel to the H-1 line (Step 2).

30.9
Location of directions

To solve more advanced problems of revolution, you must be able to locate the basic directions of up, down, forward, and backward in any given view. In Fig. 30.19A, the directions of backward and up are located by first drawing directional arrows in the given top and front views.

Line 4–5 is drawn pointing backward in the top view, and its front view appears as a point. Arrow 4–5 is projected to the auxiliary view as any other line to locate the direction of backward. By drawing the arrow on the other end of the line, you would find the direction of forward.

The direction of up is located in Fig. 30.19B by drawing line 4–6 in the direction of up in the front view and as a point in the top view. The arrow is found in the primary auxiliary by the usual projection method. The direction of down would be in the opposite direction.

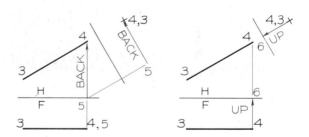

FIG. 30.19 To find the direction of back, forward, up, and down in an auxiliary view, construct an arrow pointing in the desired direction in the given principal views, and project this arrow to the auxiliary view. The directions of back and up are shown here.

The location of directions in secondary auxiliary views are found in the same manner. The direction of up is found in Fig. 30.20 by beginning with an arrow pointing upward in the front view and appearing as a point in the top view (Step 1). The arrow, *AB,* is pro-

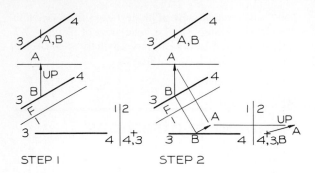

FIG. 30.20 Direction in a secondary auxiliary view.

Step 1 To find the direction of up in the secondary auxiliary view, arrow *AB* is drawn pointing upward in the front view. It appears as a point in the top view.

Step 2 Arrow *AB* is projected to the primary and secondary auxiliary views like any other line. The direction of up is located in the secondary auxiliary view.

jected from the front view to the primary, and then to a secondary auxiliary view to give the direction of up in all views. The other directions can be found in the same manner by beginning with the two given principal views of a known directional arrow.

30.10
Revolution of a point about an axis

In Fig. 30.21, it is required that point 0 be revolved about axis 3–4 to its most forward position. The circular path of revolution is drawn in the primary auxiliary view where the axis is a point (Step 1). The direction of forward is drawn (Step 2), and the new location of point 0 is found at 0'. By projecting back through the successive views, point 0' is found in each view. Note that 0' lies on the line in the front view, verifying that 0' is in its most forward position.

The problem in Fig. 30.22 requires an additional auxiliary view since axis 3–4 is not true length in the given views. Therefore the line must be found true length before it can be found as a point where the path of revolution can be drawn as a circle. Point 0 is revolved into its highest position, 0', where the "up" arrow, 3–5, is found in the secondary auxiliary view.

By projecting back to the given views, 0 is located in each view. Its position in the top view is over the axis, which verifies that the point is located at its highest position.

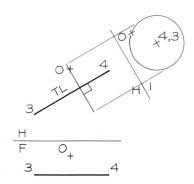 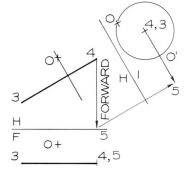 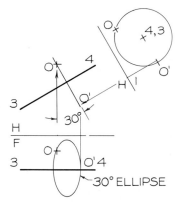

FIG. 30.21 Revolution about an axis.

Step 1 To rotate point 0 about axis 3–4, it is necessary to find the point view of the axis in a primary auxiliary view. The circular path is drawn and the path of revolution is shown in the top view as an edge perpendicular to the axis.

Step 2 If it is required to rotate point 0 to its most forward position, draw an arrow pointing forward in the top view. It will appear as a point in the front view. The arrow, 4–5, is found in the auxiliary view to locate point 0'.

Step 3 Point 0' is projected back to the given views. The path of revolution appears as an ellipse in the front view since the axis is not true length in this view. A 30° ellipse is drawn since this is the angle your line of sight makes with the circular path in the front view.

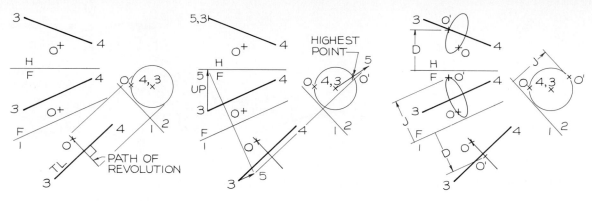

FIG. 30.22 Revolution of a point about an axis.

Step 1 To rotate 0 about axis 3–4, the axis is found as a point in a secondary auxiliary view where the circular path is drawn. The path appears as an edge in the primary auxiliary view where the axis is true length.

Step 2 If it is required to rotate 0 to its highest position, construct an arrow 3–5 in the front and top views that points upward. The direction of 3.5 in the secondary auxiliary view locates the highest position, 0'.

Step 3 Point 0' is projected back to the given views by transferring the dimensions J and D using your dividers. The highest point lies over the line in the top view to verify its position. The path of revolution is elliptical wherever the axis is not true length.

The paths of revolution will appear as edges when their axes are true length, and as ellipses when their axes are not true length. The angle of the ellipse template for drawing the ellipse in the front view is the angle the projectors from the front view make with the edge view of the revolution in the primary auxiliary view. To find the ellipse in the top view, an auxiliary view must be used to find the path of revolution as an edge perpendicular to the true-length axis projected from the top view.

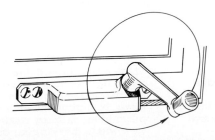

FIG. 30.23 The handcrank on a casement window is an example of a problem solved by using revolution principles. The handle must be properly positioned so not to interfere with the windowsill or wall.

The handcrank of a casement window (Fig. 30.23) is an example of a problem solved using revolution to determine the clearances between the sill and the window frame.

30.11
Revolution of a right prism about its axis

A coal chute between two buildings (Fig. 30.24) is used to convey coal at a continuous rate. For the coal to be transported efficiently, the sides of the enclosed chute must be vertical and the bottom of the chute's right section must be horizontal.

In Fig. 30.25, it is required that the right section be positioned about centerline *AB* so that two of its sides will be vertical. This is done by finding the point view of the axis (Step 1), and the direction of up is projected to this view. The right section is drawn about the axis (Step 2) so that two of its sides are parallel to the upward arrow. The right section is found in the other views. The sides of the chute are then constructed parallel to the axis, (Step 3). The bottom of the chute's right section will be horizontal and properly positioned for conveying coal.

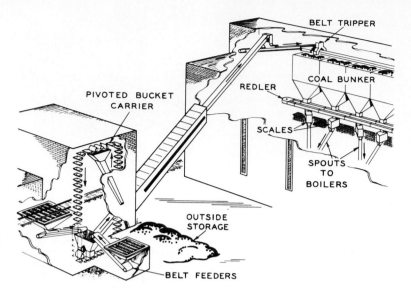

FIG. 30.24 A conveyor chute must be installed so that two edges of its right section are vertical for the conveyors to function properly. This requires the application of the revolution of a prism about its axis. (Courtesy of Stephens-Adamson Manufacturing Co.)

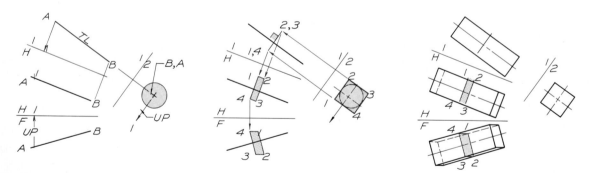

FIG. 30.25 Revolution of a prism about its axis.

Step 1 Locate the point view of centerline *AB* in the secondary auxiliary view by drawing a circle about the axis with a diameter equal to one side of the square right section. Draw a vertical arrow in the front and top views, and project it to the secondary auxiliary view to indicate the direction of vertical.

Step 2 Draw the right section, 1–2–3–4, in the secondary auxiliary view with two sides parallel to the vertical directional arrow. Project this section back to the successive views by transferring measurements with dividers. The edge view of the section could have been located in any where along centerline *AB* in the primary auxiliary view.

Step 3 Draw the lateral edges of the prism through the corners of the right section so they are parallel to the centerline in all views. Terminate the ends of the prism in the primary auxiliary view where they appear as edges perpendicular to the centerline. Project the corner points of the ends to the top and front views to establish the ends in these views.

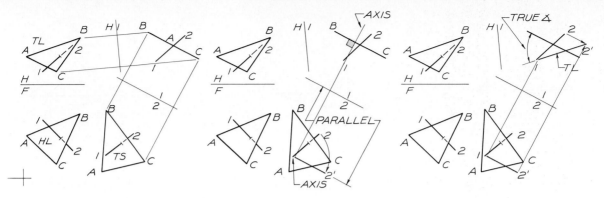

FIG. 30.26 Angle between a line and a plane.

Step 1 Construct plane *ABC* as an edge in a primary auxiliary view, which can be projected from either view. Determine the true size of the plane in the secondary auxiliary view, and project line 1–2 to each view.

Step 2 Revolve the secondary auxiliary view of the line until it is parallel to the 1–2 reference line. The axis of revolution appears as a point through point 1 in the secondary auxiliary view. The axis appears true length and is perpendicular to the 1–2 line and plane *ABC* in the primary auxiliary view.

Step 3 Point 2' is projected to the primary auxiliary view where the true length of line 1–2' is found by projecting the primary auxiliary view of point 2 parallel to the 1–2 line as shown. Since the plane appears as an edge and the line appears true length in this view, the true angle between them found.

30.12
Angle between a line and a plane

The angle between a line and a plane is found by a combination of auxiliary views and revolution in Fig. 30.26. The plane is found true size in a secondary auxiliary view (Step 1).

The line is revolved until it is parallel to the 1–2 reference line (Step 2). The line can then be found true length in the primary auxiliary view (Step 3). Since the line appears true length and the plane appears as an edge in this view, the true angle can be measured here.

30.13
A line at a specified angle with two principal planes

In Fig. 30.27, it is required that a line be drawn through point 0 that will make angles of 35° with the frontal plane and 44° with the horizontal plane, and that will slope forward and downward.

The cone containing elements making 35° with the frontal plane is drawn (Step 1). The cone with elements making 44° with the horizontal plane is drawn (Step 2). The length of the elements of both cones must be equal, so they will intersect with equal elements. Lines 0–1 and 0–2, which are elements that lie on each cone and make the specified angles with the principal planes, are then found (Step 3).

30.14
Revolution of parts on detail drawings

It is standard practice to revolve parts, such as the one shown in Fig. 30.28, to make the views more descriptive. The front view of the part is true size because the top view has been revolved. This technique, called *conventional practice*, should be used to conserve effort while gaining additional clarity in the preparation of working drawings.

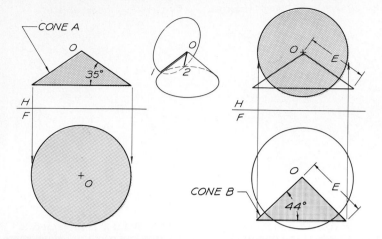

FIG. 30.27 A line at specified angles.

Step 1 Draw a triangular view of a cone in the top view such that the extreme elements make an angle of 35° with the frontal plane. Construct the circular view of the cone in the front view, using point 0 as the apex. All elements of this cone make an angle of 35° with the frontal plane.

Step 2 Draw a triangular view of a cone in the front view such that the elements make an angle of 44° with the horizontal plane. Draw the elements of this cone equal in length to element E of cone A. All elements of cone B make an angle of 44° with the horizontal plane.

Step 3 Since the elements of cones A and B are equal in length, there will be two common elements that lie on the surface of each cone, elements 0–1 and 0–2. Locate points 1 and 2 at the point where the bases of the cone intersect in both views. Either of these lines will satisfy the problem requirements.

FIG. 30.28 Revolving features of parts to show the features true size in the adjacent orthographic views is conventional practice. When this is done, it is not necessary to show the arrows of rotation.

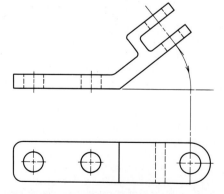

Problems

Use Size A sheets for the following problems, and lay out the problems using instruments. Each square on the grid is equal to 0.20 in. (about 5mm). The problems can be laid out on grid or plain paper. Label all reference planes and points in each problem with $\frac{1}{8}$ inch letters or numbers, using guidelines.

The crosses marked "1" and "2" are to be used for placing primary and secondary reference lines. The primary reference line should pass through "1" and the secondary through "2".

1–4. (Fig. 30.29) Find the true-length views of the lines by revolution.

5. (Fig. 30.30) Find the true-size view of the plane by an auxiliary view and a single revolution.

6. (Fig. 30.30) Find the true-size view of the plane by revolution.

7–8. (Fig. 30.31) Find the true-size views of the planes by double revolution.

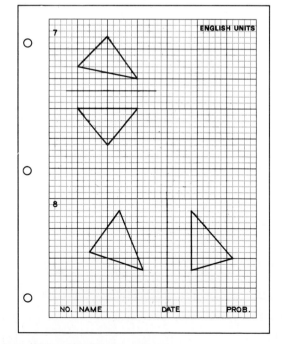

FIG. 30.30 Problems 5 and 6.

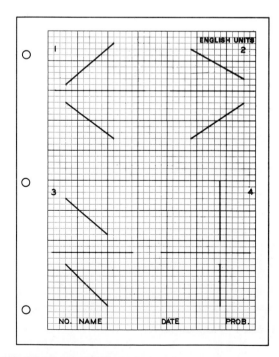

FIG. 30.29 Problems 1–4.

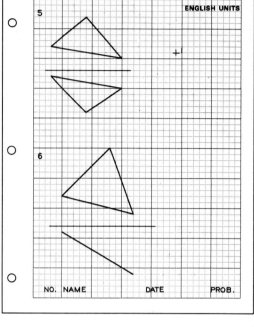

FIG. 30.31 Problems 7 and 8.

9–10. (Fig. 30.32) Find the dihedral angles between the planes.

11–12. (Fig. 30.33) Revolve the points about the given axes, and show the points in all views. In Problem 11, revolve the point into its most forward position, and in Problem 12 into its highest position.

13. The centerline of a conveyor chute is given in the top and front views of Fig. 30.34. The chute has a 10-foot-square cross section. Construct the necessary views to revolve the 10-foot square into a position where two sides of the right section will be vertical planes. Show the chute in all views. Scale: $1'' = 10'$.

14–19. Lay out these problems on a Size B sheet using the horizontal format. Position the crossing division lines at the center of the sheet. The grids are spaced 0.25 in. (approximately 6mm) apart. Problem 14 requires that you lay out the prism in area 1 (Fig. 30.35) and rotate it as specified about its corner point 0 in the remaining areas of the sheet. Problems 15–19 require that you replace

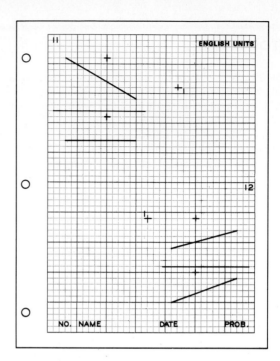

FIG. 30.33 Problems 11 and 12.

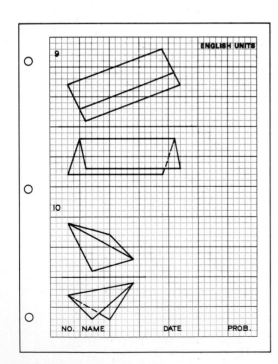

FIG. 30.32 Problems 9 and 10.

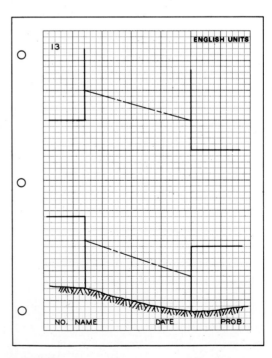

FIG. 30.34 Problem 13.

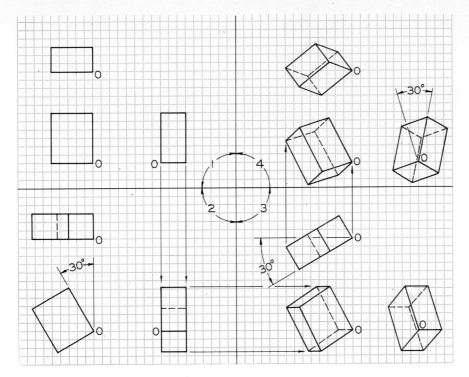

FIG. 30.35 Problem 14.

the object in Problem 14 with one of those given in Fig. 30.36, and rotate these objects through the angles specified in each step.

20. Draw the object in Fig. 30.37, but complete the top view by showing the inclined surface revolved into a true-size position. This will eliminate the need for an auxiliary view as presently shown.

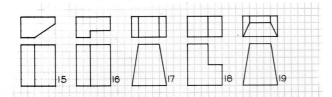

FIG. 30.36 Problems 15–19.

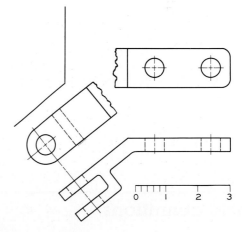

FIG. 30.37 Problem 20.

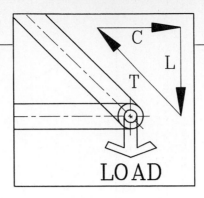

LOAD

CHAPTER 31

Vector Graphics

31.1
Introduction

A system cannot be analyzed for strength without considering the forces of tension and compression within the system. These forces are represented by *vectors*. Other quantities, such as distance, velocity, and electrical properties, may also be represented by vectors.

Graphical methods are useful in the solution of vector problems as alternative methods to conventional trigonometric and algebraic methods.

31.2
Basic definitions

To understand the techniques of problem solving with vectors, it is necessary to know the terminology of graphical vectors.

FORCE is a push or pull that tends to produce motion. All forces have (1) magnitude, (2) direction, and (3) a point of application. A force is represented by the rope being pulled in Fig. 31.1A.

VECTOR is a graphical representation of a quantity of force; it is drawn to scale to indicate magnitude, direction, and point of application. The vector shown in Fig. 31.1B represents the force of the rope pulling the weight, *W*.

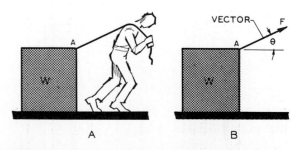

FIG. 31.1 Representation of a force by a vector.

MAGNITUDE is the amount of push or pull; it is represented by the length of the vector line. Magnitude is usually measured in pounds or kilograms of force.

DIRECTION is the inclination of a force (with respect to a reference coordinate system); it is indicated by a line with an arrow at one end.

POINT OF APPLICATION is the point through which the force is applied on the object or member (point A in Fig. 31.1A).

COMPRESSION is the state created in a member by subjecting it to opposite pushing forces. A member tends to be shortened by the compression (Fig. 31.2A). Compression is represented by a plus sign (+) or the letter C.

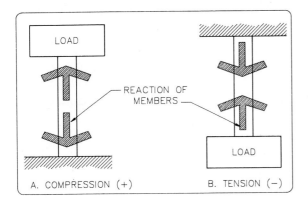

FIG. 31.2 Comparison of tension and compression in a member.

TENSION is the state created in a member by subjecting it to opposite pulling forces. A member tends to be stretched by tension, as shown in Fig. 31.2B. Tension is represented by a minus sign (−) or the letter T.

FORCE SYSTEM is the combination of all forces acting on a given object. Figure 31.3 shows a force system.

RESULTANT is a single force that can replace all the forces of a force system and have the same effect as the combined forces.

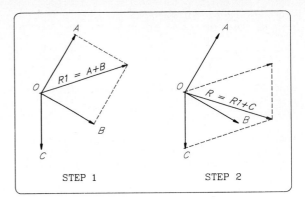

FIG. 31.3 Resultant by the parallelogram method.

Step 1 Draw a parallelogram with its sides parallel to vectors A and B. The diagonal R_1, drawn from point P to point 0, is the resultant of forces A and B.

Step 2 Draw a parallelogram using vectors R_1 and C to find diagonal R from P to Q. This is the resultant that can replace forces A, B, and C.

EQUILIBRANT is the opposite of a resultant; it is the single force that can be used to counterbalance all forces of a force system.

COMPONENTS are any individual forces that, if combined, would result in a given single force. For example, forces A and B are components of resultant R_1 in Step 1 of Fig. 31.3.

SPACE DIAGRAM is a diagram depicting the physical relationship between structural members. The force system in Fig. 31.3 is given as a space diagram.

VECTOR DIAGRAM is a diagram composed of vectors scaled to their appropriate lengths to represent the forces within a given system. The vector diagram is used to solve for unknowns.

STATICS is the study of forces and force systems in equilibrium.

METRIC UNITS are standard units for indicating weights and measures. The kilogram (kg) is the standard unit for indicating mass (loads). A comparison of kilograms with pounds is shown in Fig. 31.4. The metric ton is 1000 kilograms. One kilogram = 2.2 pounds.

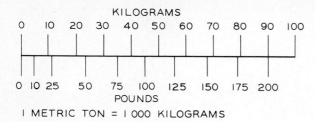

FIG. 31.4 The kilogram is the standard metric unit for measuring forces that are represented by pounds in the English system: 1 kilogram = 2.2 pounds.

31.3
Coplanar, concurrent force systems

When several forces, represented by vectors, act through a common point of application, the system is said to be *concurrent*. Vectors *A*, *B*, and *C* act through a single point in Fig. 31.3; therefore this is a concurrent system. When only one view is necessary to show the true length of all vectors, as in Fig. 31.3, the system is *coplanar*.

The *resultant* represents the composite effect of all forces on the point of application. The resultant is found graphically by (1) the parallelogram method and (2) the polygon method.

31.4
Resultant of a coplanar, concurrent system— parallelogram method

In Fig. 31.3, all vectors lie in the same plane and act through a common point. The vectors are scaled to a known magnitude.

To apply the parallelogram method to determine the resultant, the vectors for a force system must be known and drawn to scale. Two vectors are used to find a parallelogram; the diagonal of the parallelogram is the resultant of these two vectors and has its point of origin at point *P* (Fig. 31.3). Resultant R_1 can be called the *vector sum* of vectors *A* and *B*.

Since vectors *A* and *B* have been replaced by R_1, they can be disregarded in the next step of the solu-

tion. Again, resultant R_2 and vector *C* are resolved by completing a parallelogram (i.e., by drawing a line parallel to each vector). The diagonal of this parallelogram is the resultant of the entire system and is the vector sum of R_1 and *C*. Resultant *R* can be analyzed as though it were the only force acting on the point, thereby simplifying the process.

31.5
Resultant of a coplanar, concurrent system— polygon method

The system of forces shown in Fig 31.3 is shown again in Fig. 31.5, but in this case the resultant is found by the polygon method. The forces are drawn to scale and in their true directions, with each force being drawn head-to-tail to form the polygon. The vectors are drawn in a clockwise sequence, beginning with vector *A*. The polygon does not close, which means the system is not in *equilibrium*; in other words, it would tend to be in motion, since the forces are not balanced. The resultant *R* is drawn from the tail of vector *A* to the head of vector *C* to close the polygon.

An *equilibrant* has the same magnitude, orientation, and point of application as the resultant in a system of forces.

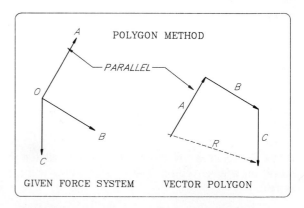

FIG. 31.5 Resultant of a coplanar, concurrent system as determined by the polygon method, in which the vectors are drawn head-to-tail.

The difference between them is their direction. The resultant of the system of forces shown in Fig. 31.6 is found by the parallelogram method. The equilibrant can be applied at point 0 to balance the forces A and B and thereby cause the system to be in equilibrium.

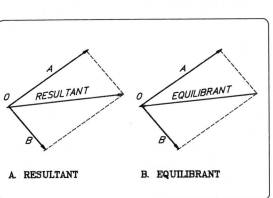

A. RESULTANT   B. EQUILIBRANT

FIG. 31.6 The resultant and equilibrant are equal in all respects except in sense (position of arrowhead).

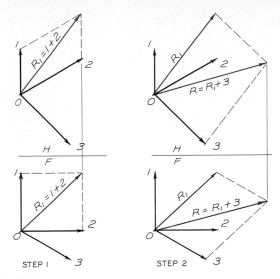

FIG 31.7 Resultant by the parallelogram method.

Step 1   Vectors 1 and 2 are used to construct a parallelogram in the top and front views. The diagonal, R_1, is the resultant of these two vectors.

Step 2   Vectors 3 and R_1 are used to construct a second parallelogram to find the views of the overall resultant, R_2.

31.6
Resultant of noncoplanar, concurrent forces— parallelogram method

When vectors lie in more than one plane of projection, they are said to be *noncoplanar*; therefore more than one view is necessary to analyze their spatial relationships. The resultant of a system of noncoplanar forces can be found by the parallelogram method if their true projections are given in two adjacent orthographic views.

Vectors 1 and 2 in Fig.31.7 are used to construct the top and front views of a parallelogram. The diagonal of the parallelogram, R_1, is found in both views. As a check, the front view of R_1 must be an orthographic projection of its top view.

In Step 2, resultant R_1 and vector 3 are resolved to form resultant R_2 in both views. The top and front views of R_2 must project orthographically. Resultant R_2 can be used to replace vectors 1, 2, and 3. Since R is an oblique line, its true length must be found by auxiliary view, as shown in Fig. 31.8, or by revolution.

31.7
Resultant of noncoplanar, concurrent forces— polygon method

The same system of forces given in Fig. 31.7 is solved in Fig. 31.8 for the resultant of the system by the polygon method.

Each vector is laid head-to-tail in a clockwise direction, beginning with vector 1 (Step 1). In each view, the vectors are drawn to be orthographic projections (Step 2). Since the vector polygon did not close, the system is not in equilibrium. The resultant, R, is constructed from the tail of vector 1 to the head of vector 3 in both views. Resultant R is an oblique line and requires an auxiliary view to find its true length.

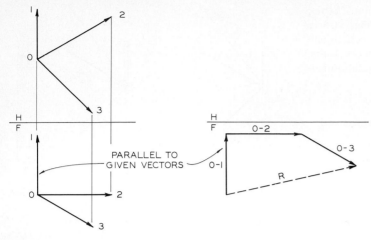

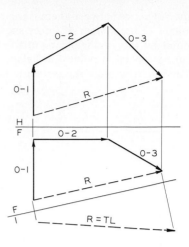

FIG. 31.8 Resultant by the polygon method.

Required Find the resultant of this system of concurrent, noncoplanar forces by the polygon method.

Step 1 Each vector is laid off head-to-tail in the front view. The front view of the resultant is the vector found.

Step 2 The same vectors are drawn head-to-tail in the top view to complete the 3-D polygon. The resultant is found true length by an auxiliary view.

31.8
Forces in equilibrium

An example of a coplanar, concurrent structure in equilibrium can be seen in the loading cranes in Fig. 31.9.

FIG. 31.9 The cargo cranes on this ship are examples of coplanar, concurrent force systems designed to remain in equilibrium. (Courtesy of Exxon Corp.)

The coplanar, concurrent structure given in Fig. 31.10 is designed to support a load of $W = 1000$ kg. The maximum loading in each structural member determines the type and size of the members used in the design.

In Step 1, the only known force, $W = 1000$ kg, is laid off parallel to the given direction. Unknown forces A and B are drawn head-to-tail as vectors to close the force polygon.

In Step 2, vectors A and B are analyzed to determine whether they are in tension or compression. Vector B points upward to the left, which is toward point O when transferred to the structural diagram. A vector that acts toward a point of application is in *compression*. Vector A points away from point A when transferred to the structural diagram and is therefore in *tension*.

In Fig. 31.11, a similar example of a force system involving a pulley is solved to determine the loads in the structural members caused by the weight of 100 lb. The only difference between this solution and the previous one is the construction of two equal vectors to represent the loads in the cable on both sides of the pulley.

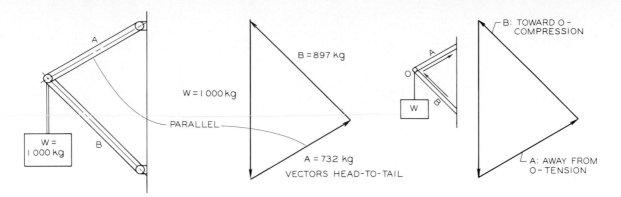

FIG. 31.10 Coplanar forces in equilibrium.

Required Find the forces in the two structural members caused by the load of 1000 kg.

Step 1 Draw the known load of 1000 kg as a vector. Draw the vectors A and B parallel to their directions. Arrowheads are drawn head-to-tail.

Step 2 Vector A points away from the point 0 when transferred to the structural diagram and is in tension. Vector B points toward point 0 and is in compression.

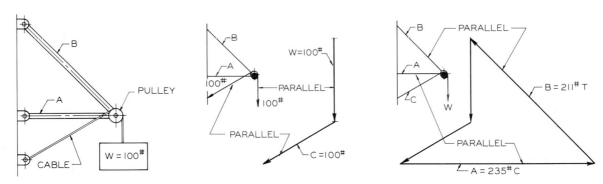

FIG. 31.11 Determination of forces in equilibrium.

Required Find the forces in the members caused by the load of 100 lb (denoted by #) supported by the pulley.

Step 1 The force in the cable is equal to 100 lb on both sides of the pulley. These two forces are drawn as vectors head-to-tail parallel to their directions in the space diagram.

Step 2 A and B are drawn to close the polygon, and arrowheads are drawn head-to-tail. The direction of A is toward the point of application and is in compression; B is away from the point and is in tension.

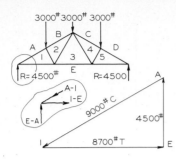
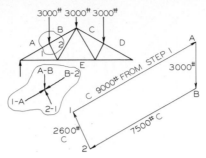
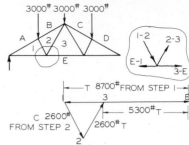

FIG. 31.12 Joint analysis of a truss.

Step 1 The truss is labeled using Bow's notation, with letters between the exterior loads and numbers between interior members. The lower left joint can be analyzed since it has only two unknowns, A–1 and 1–E. These vectors are found by drawing them parallel to their directions from both ends of the 4500 lb reaction in a head-to-tail order.

Step 2 Using vector 1–A found in step 1 and load AB, the two unknowns B–2 and 2–1 can be found. The known vectors are laid out beginning with vector 1–A and moving clockwise about the joint. Vectors B–2 and 2–1 close the polygon. If a vector points toward the point of application, it is in compression; if away from the point, it is in tension.

Step 3 The third joint can be analyzed by laying out the vectors E–1 and 1–2 from the previous steps. Vectors 2–3 and 3–E close the polygon and are parallel to their directions in the space diagram. The directions of 2–3 and 3–E are away from the point of application; these vectors are in tension.

31.9
Truss analysis

Vector polygons can be used to analyze structural trusses to determine the loads in each member by two graphical methods: (1) joint-by-joint analysis and (2) Maxwell diagrams.

Joint-by-joint analysis

The Fink truss in Fig. 31.12 is loaded with 3000 lb. forces concentrated at joints of the structural members. A method of designating forces, called *Bow's notation*, is used. The exterior forces applied to the truss are labeled with letters placed between the forces, and numerals are placed between the interior members.

Each vector is referred to by the number on each of its sides by reading in a clockwise direction. For example, the first vertical load at the left is called *AB*, with *A* at the tail and *B* at the head of the vector.

We first analyze the joint at the left where the reaction of 4500 lb is known. This force, reading in a clockwise direction about the joint, is called *EA* with an upward direction. The tail is labeled *E* and the head *A*. Continuing in a clockwise direction, the next force is *A*–1, and the next 1–*E*, which closes the polygon and ends with the beginning letter, *E*. The arrows are placed, beginning with the known vector *EA*, in a head-to-tail arrangement.

Tension and compression can be determined by relating the direction of each vector to the original joint. For example, *A*–1 points toward the joint and is in compression, whereas 1–*E* points away and is in tension.

Since the truss is symmetrical and equally loaded, the loads in the members on the right will be equal to those on the left.

The other joints are analyzed in the same manner in Steps 2 and 3. The direction of the vectors is opposite at each end. Vector *A*–1 is toward the left in Step 1, and toward the right in Step 2.

Maxwell diagrams

The Maxwell diagram is identical to the joint-by-joint analysis except that the polygons overlap, with some vectors common to more than one polygon. Again, Bow's notation is used to good advantage.

The first step (Fig. 31.13) is to lay out the exterior loads beginning clockwise about the truss—*AB, BC, CD, DE,* and *EA*—head-to-tail. A letter is placed at each end of the vectors. Since they are parallel, this polygon will be a straight line.

The structural analysis begins at the joint through which reaction *EA* acts. For easier analysis, a free-body diagram is drawn to isolate this joint. The two unknowns, members *A*–1 and 1–*E*, are drawn parallel to their direction in the truss in Step 1, with *A*–1 beginning at point *A* and 1–*E* beginning at point *E*. To locate point 1, these directions are extended. Because resultant *EA* points upward, vector *A*–1 must have its tail at *A*, giving it a direction toward point 1. By referring to the free-body diagram, we can see that the direction is toward the point of application, which means *A*–1 is in compression. Vector 1–*E* points away from the joint, which means it is in tension. The vectors are coplanar and can be scaled to determine their loads.

In Step 2, vectors 1–*A* and *AB* are known, whereas vectors *B*–2 and 2–1 are unknown, making it possible to solve for them. Vector *B*–2 is drawn parallel to the structural member through point *B* in the Maxwell diagram, and the line of vector 2–1 is extended from point 1 until it intersects with *B*–2 at point 2. The arrows of each vector are found by laying off each vector head-to-tail. Both vectors *B*–2 and 2–1 point toward the joint in the free-body diagram; therefore they are in compression.

In Step 3, the next joint is analyzed in sequence to find the stresses in 2–3 and 3–*E*. The truss will have equal forces on each side, since it is symmetrical and is loaded symmetrically. The total Maxwell diagram is drawn in Step 3.

If all the polygons in the series do not close at every point with perfect symmetry, there is an error in construction. A slight error of closure can be disregarded, since safety factors are generally applied in derivation of working stresses of structural systems to assure safe construction. Arrowheads are omitted on

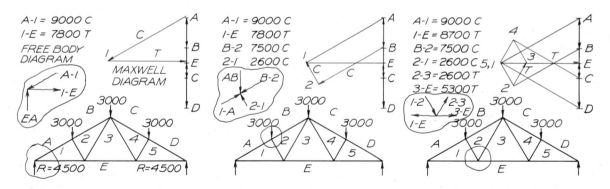

FIG. 31.13   Truss analysis.

Step 1   Label the spaces between the outer forces of the truss with letters and the internal spaces with numbers, using Bow's notation. Add the given load vectors in a Maxwell diagram, and sketch a free-body diagram of the first joint. Using vectors *EA, A*–1, and 1–*E* drawn head-to-tail, draw a vector diagram to find their magnitudes. Vector *A*–1 is in compression (+) because it points toward the joint, and 1–*E* is in tension (−) because it pionts away from the joint.

Step 2   Draw a sketch of the next joint to be analyzed. Since *AB* and *A*–1 are known, we have to determine only two unknowns, 2–1 and *B*–2. Draw these parallel to their direction, head-to-tail, in the Maxwell diagram using the existing vectors found in Step 1. Vectors *B*–2 and 2–1 are in compression since each points toward the joint. Note that vector *A*–1 becomes 1–*A* when read in a clockwise direction.

Step 3   Sketch a free-body diagram of the next joint to be analyzed. The unknowns in this case are 2–3 and 3–*E*. Determine the true length of these members in the Maxwell diagram by drawing vectors parallel to given members to find point 3. Vectors 2–3 and 3–*E* are in tension because they act away from the joint. This process is repeated to find the loads of the members on the opposite side.

Maxwell diagrams, since each vector will have opposite directions when applied to different joints.

31.10
Noncoplanar structural analysis—special case

Three-dimensional structure systems require the use of descriptive geometry, since it is necessary to analyze the system in more than one plane. The manned flying system (MFS) in Fig. 31.14 can be analyzed to determine the forces in the support members (Fig. 31.15). Weight on the moon can be found by multiplying earth weight by a factor of 0.165. A tripod that must support 182 lb on earth has to support only 30 lb on the moon.

This example in Fig. 31.15 is a special case, since members *B* and *C* lie in the same plane that appears as an edge in the front view. A vector polygon is constructed in the front view in Step 1 by drawing force *F* as a vector and using the other vectors as the other

FIG. 31.14 The structural members of this tripod support for a moon vehicle can be analyzed graphically to determine design loads. (Courtesy of NASA.)

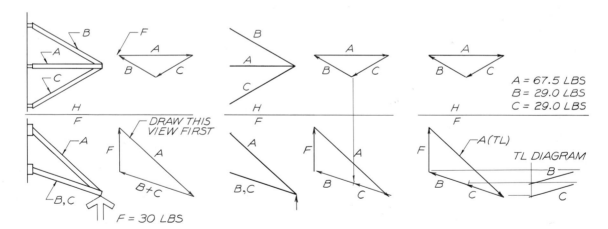

FIG. 31.15 Noncoplanar structural analysis—special case.

Step 1 Forces, *B* and *C*, coincide in the front view, resulting in only two unknowns in this view. Vector *F* (30 lb) and the two unknowns are drawn parallel to their front view to complete the front view of the vector polygon. The top view of *A* can be found by projection, from which vectors *B* and *C* can be found.

Step 2 The point of intersection of vectors *B* and *C* in the top view is projected to the front view to separate these vectors. All vectors are drawn head-to-tail. Vectors *B* and *C* are in tension because they act away from the point in the space diagram, whereas *A* is in compression.

Step 3 The completed top and front views found in Step 2 do not give the true lengths of vectors *B* and *C* since they are oblique. The true lengths of these lines are determined by a true-length diagram where they are scaled to find the forces in each member.

sides of the polygon. One of these vectors is actually a summation of vectors B and C. The top view is drawn using vectors B and C to close the polygon from each end of vector A. In Step 2, the front view of vectors B and C is found.

The true lengths of the vectors are found in a true-length diagram in Step 3. The vectors are measured to determine their loads. Vector A is found to be in compression because it points toward the point of concurrency. Vectors B and C are in tension since they point away from the point of concurrency.

31.11
Noncoplanar structural analysis—general case

The structural frame shown in Fig. 31.16 is attached to a vertical wall to support a load of W = 600 lb. Since there are three unknowns in each of the views, an auxiliary view must be drawn that will give the edge view of a plane containing two of the vectors, thereby reducing the number of unknowns to two. We no longer need to refer to the front view.

A vector polygon is drawn by constructing vectors parallel to the members in the auxiliary view (Step 1). An adjacent orthographic view of the vector polygon is also drawn by constructing its vectors parallel to the members in the top view (Step 2). A true-length diagram is used to find the true length of the vectors to determine their magnitudes (Step 3).

A three-dimensional vector system is the side-boom tractors used for lowering pipe into a ditch during pipeline construction (Fig. 31.17).

31.12
Nonconcurrent, coplanar vectors

Forces *may* be applied in such a manner that they are *nonconcurrent*, as illustrated in Fig. 31.18. Bow's notation can be used to locate the resultant of this type of nonconcurrent system.

In Step 1, the vectors are laid off to form a vector polygon in which the closing vector is the resultant, R = 68 lb. Each vector is resolved into two components by randomly locating point 0 on the interior or exterior of the polygon and connecting point 0 with

the end of each vector. The components, or strings, from point 0 are equal and opposite components of adjacent vectors. For example, component 0b is common to vectors AB and BC. Since the strings from point 0 are equal and opposite, the system has not changed statically.

In Step 2, each string is transferred to the space diagram of the vectors, where it is drawn between the respective vectors to which it applies. (The figure thus produced is called a *funicular diagram*.) For instance, string 0b is drawn in the area between the vectors AB and BC, and string 0c is drawn in the C area to connect at the intersection of 0b and vector BC. The point of intersection of the last two strings, 0a and 0d, locates a point through which the resultant R will pass.

31.13
Nonconcurrent systems resulting in couples

A *couple* is the name given to two parallel, equal, and opposite forces separated by a distance and applied to a member in a way to cause the member to rotate.

An important quantity associated with a couple is its *moment*. The moment of any force is a measure of its rotational effect. An example is shown in Fig. 31.19, in which two equal and opposite forces are applied to a wheel. The forces are separated by the distance D. The moment of the couple is found by multiplying one of the forces by the perpendicular distance between it and the other: $F \times D$. If the force is 20 lb and the distance is 3 ft, the moment of the couple would be given as 60 ft-lb.

A series of parallel forces is applied to a beam in Fig. 31.20. The spaces between the vectors are labeled with letters that follow Bow's notation. We are required to determine the resultant.

After constructing a vector diagram, we have a straight line that is parallel to the direction of forces and that closes at point A. We then locate pole point 0 and draw the strings of a funicular diagram.

The strings are transferred to the space diagram and are drawn in their respective spaces. For example, 0c is drawn in the C-space between vectors BC and CD. The last two strings, 0d and 0a, do not close at a common point but are found to be parallel; the result is therefore a couple. Using the scale of the space diagram, the distance between the forces of the couple is the perpendicular distance, E, between strings 0a and

FIG. 31.16 Noncoplanar structural analysis—general case.

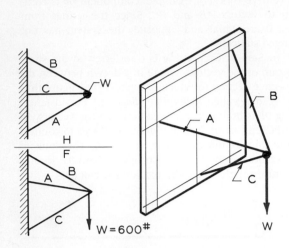

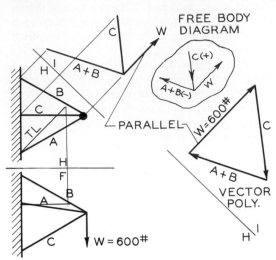

Given The top and front views of a three-member frame which is attached to a vertical wall and supports a weight of 600 lb.

Required Find the loads in the structural members.

Step 1 To limit the unknowns to two, construct an auxiliary view to find two vectors lying in the edge of a plane. Use the auxiliary view and top view in the remainder of the problem. Draw a vector polygon parallel to the members in the auxiliary view in which $W = 600$ lb is the only known vector. Sketch a free-body diagram for preliminary analysis.

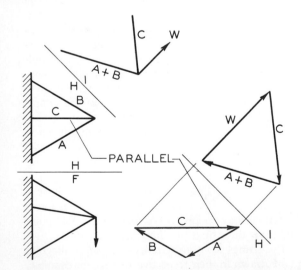

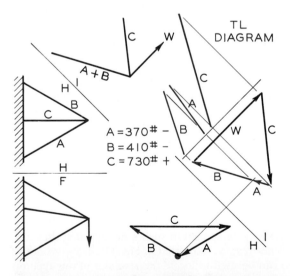

Step 2 Construct an orthographic projection of the view of the vector polygon found in Step 1 so that its vectors are parallel to the members in the top view. The reference plane between the two views is parallel to the H–1 plane. This portion of the problem is closely related to the problem in Fig. 31.15.

Step 3 Project the intersection of vectors A and B in the horizontal view of the vector polygon to the auxiliary view polygon to establish the lengths of vectors A and B. Determine the true lengths of all vectors in a true-length diagram to determine their magnitudes. Analyze for tension or compression.

Fig. 31.17 Tractor sidebooms represent noncoplanar, concurrent systems of forces that can be solved graphically. (Courtesy of Trunkline Gas Co.)

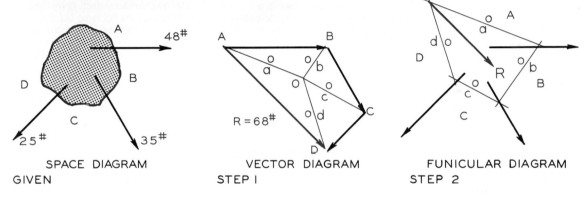

SPACE DIAGRAM	VECTOR DIAGRAM	FUNICULAR DIAGRAM
GIVEN	STEP 1	STEP 2

Fig. 31.18 Resultant of nonconcurrent forces.

Required Find the resultant of the known forces applied to the above object. The forces are nonconcurrent.

Step 1 The vectors are drawn head-to-tail to find resultant *R*. Point 0 is conveniently located for the construction of strings to the ends of each vector.

Step 2 Each string is drawn between the two vectors to which it applies. *Example:* 0c between *BC* and *CD*. These strings are connected in sequence until the strings 0*a* and 0*d* establish the position of *R*.

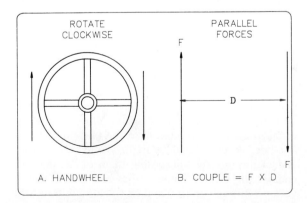

A. HANDWHEEL   B. COUPLE = F X D

Fig. 31.19 Representation of a couple or moment.

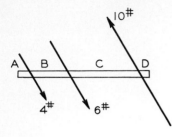

SPACE DIAGRAM

VECTOR DIAGRAM

COUPLE = 7.5$^{\#}$ (E)

FUNICULAR DIAGRAM

Fig. 31.20 Couple resultants.

Required Find the resultant of these nonconcurrent forces applied to this beam.

Step 1 The spaces between each force are labeled in Bow's notation.

Step 2 The vectors are laid out head-to-tail; they will lie in a straight line since they are parallel. Pole point 0 is located in a convenient location and the ends of each vector are connected with point 0.

Step 3 Strings 0a, 0c, and 0d are successively drawn between the vectors to which they apply. Since strings 0a and 0d are parallel, the resultant will be a couple equal to 7.5 lb × E, where E is the distance between 0a and 0d.

0d in the space diagram. Using the scale of the vector diagram, the magnitude of the force is the scaled distance from point 0 to points A and D in the vector diagram. The moment of the couple is equal to 7.5 lb × E in a counterclockwise direction.

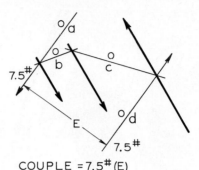

Fig. 31.21 The boom of this crane can be analyzed for its resultant as a parallel, nonconcurrent system of forces when the cables have been disregarded.

31.14
Resultant of parallel, nonconcurrent forces

Forces applied to beams, such as those shown in Fig. 31.21, are parallel and nonconcurrent in many instances, and they may have the effect of a couple, tending to cause a rotational motion.

The beam in Fig. 31.22 is on a rotational crane used to move building materials. The magnitude of the weight W is unknown, but the counterbalance weight is 2000 lb; column R supports the beam as shown. Assuming the support cables have been omitted, we desire to find the weight W that would balance the beam.

The graphical solution (Fig. 31.22B) is found by constructing a line to represent the total distance between the forces F and W. Point 0, the point of balance where the summation of the moments will be equal to zero, is projected from the space diagram to this line. Vectors F and W are drawn to scale at each end of the line by transposing them to the opposite ends of the beam. A line is drawn from the end of vector F through point 0 and extended to intersect the

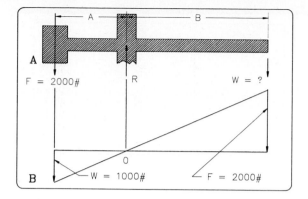

Fig. 31.22 Determining the resultant of parallel, nonconcurrent forces.

direction of vector W. This point represents the end of vector W, which can be scaled to have a magnitude of 1000 lb.

31.15
Resultant of parallel, noncurrent forces on a beam

The beam given in Fig. 31.23 is supported at each end and must carry three given loads. We are required to determine the magnitude of each support, R_1 and R_2, the resultant of the loads, and its location. In Step 1, the spaces between all vectors are labeled in a clockwise direction with Bow's notation, and a force diagram is drawn.

In Step 2, the lines of force in the space diagram are extended, and the strings from the vector diagram are drawn in their respective spaces, parallel to their original direction. For example, string $0a$ is drawn parallel to string $0A$ in space A between forces EA and AB, and string $0b$ is drawn in space B beginning at the intersection of $0a$ with vector AB. The last string, $0e$, is drawn to close the funicular diagram. The direction of string $0e$ is transferred to the force diagram, where it is laid off through point 0 to intersect the load line at point E. Vector DE represents support R_2 (refer to Bow's notation as it was applied in Step 1), and vector EA represents support R_1.

In Step 3, the magnitude of the resultant of the loads is the summation of the vertical downward forces, or the distance from A to D. The location of the resultant is found by extending the extreme outside strings in the funicular diagram, $0a$ and $0d$, to their point of intersection. The resultant is found to have a magnitude of 500 lb, a vertical downward direction, and a point of application established by $\overline{X}$.

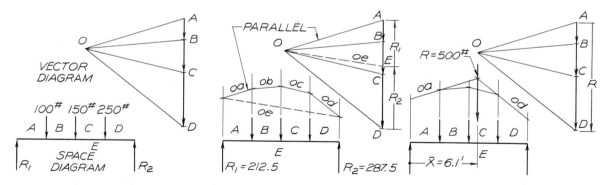

Fig. 31.23 Beam analysis with parallel loads.

Step 1 Letter the spaces between the loads with Bow's notation. Find the graphical summation of the vectors by drawing them head-to-tail in a vector diagram at a convenient scale. Locate pole point 0 at a convenient location and draw strings from point 0 to each end of the vectors.

Step 2 Extend the lines of force in the space diagram, and draw a funicular diagram with string $0a$ in the A-space, $0b$ in the B-space, $0c$ in the C-space, etc. The last string, which is drawn to close the diagram, is $0e$. Transfer $0e$ to the vector polygon to locate point E, thus establishing the lengths of R_1 and R_2, which are EA and DE, respectively.

Step 3 The resultant of the three downward forces will be equal to their graphical summation, line AD. Locate the resultant by extending strings $0a$ and $0d$ in the funicular diagram to a point of intersection. The resultant $R = 500$ lb will act through this point in a downward direction. $\overline{X}$ is a locating dimension.

Problems

Problems should be presented in instrument drawings on Size A paper, grid or plain. Each grid square represents 0.20 in. All notes, sketches, drawings, and graphical work should be neatly prepared in keeping with good design practices. Written matter should be legibly lettered using $\frac{1}{8}$-inch guidelines.

1. (Fig. 31.24) **A**. Determine the resultant of the force system by the parallelogram method at the left of the sheet. Solve the same system using the vector-polygon method at the right of the sheet. Scale $1'' = 100$ lb. **B**. Determine the resultant of the concurrent, coplanar force system by the parallelogram method. Solve the same system using the polygon method at the right of the sheet. Scale $1'' = 100$ lb.

2. (Fig. 31.25) **A**. and **B**. Solve for the resultant of each of the concurrent, noncoplanar force systems by the parallelogram method at the left of the sheet. Solve for the resultant of the same systems by the vector-polygon method at the right of the sheet. Find the true length of the resultant in both problems. Letter all construction. Scale $1'' = 600$ lb.

3. (Fig. 31.26) **A**. and **B**. The concurrent, coplanar force systems are in equilibrium. Find the loads in each structural member. Use a scale of $1'' = 300$ lb in part A and a scale of $1'' = 200$ lb in part B. Show and label all construction.

4. (Fig. 31.27) Solve for the loads in the structural members of the truss. Vector polygon scale $1'' = 2000$ lb. Label all construction.

5. (Fig.31.28) Solve for the loads in the structural members of the concurrent, noncoplanar force system. Find the true length of all vectors. Scale $1'' = 300$ lb.

6. (Fig. 31.29) Solve for the loads in the structural members of the concurrent, noncoplanar force system. Find the true length of all vectors. Scale $1'' = 400$ lb.

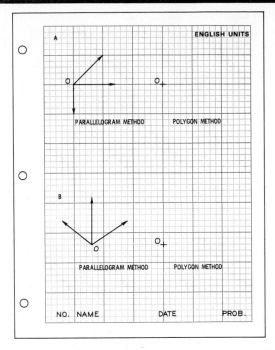

Fig. 31.24 Problems 1A and 1B. Resultant of concurrent, coplanar vectors.

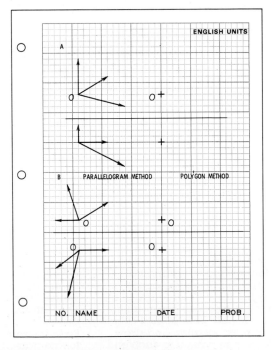

Fig. 31.25 Problems 2A and 2B. Resultant of concurrent, noncoplanar vectors.

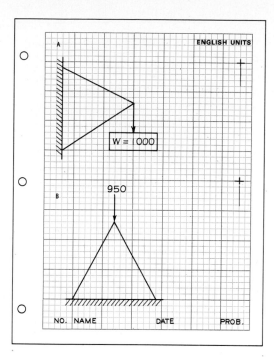

Fig. 31.26 Problems 3A and 3B. Coplanar, concurrent forces in equilibrium.

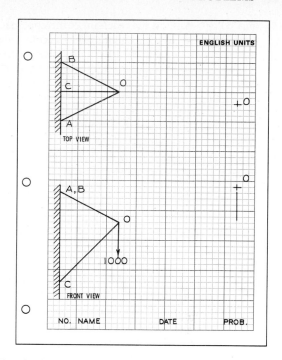

Fig. 31.28 Problem 5. Noncoplanar, concurrent forces in equilibrium.

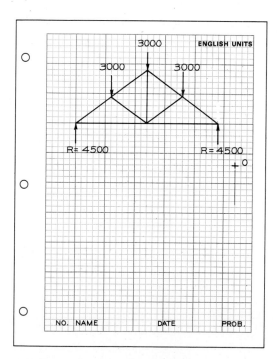

Fig. 31.27 Problem 4. Truss analysis.

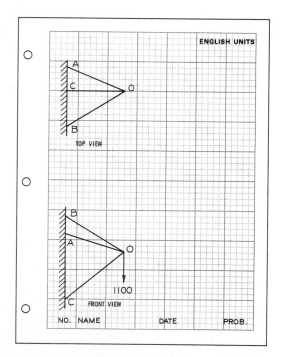

Fig.31.29 Problem 6. Noncoplanar, concurrent forces in equilibrium.

7. (Fig. 31.30) **A.** Find the resultant of the coplanar, nonconcurrent force system. The vectors are drawn to a scale of 1″ = 100 lb. **B.** Solve for the resultant of the coplanar, nonconcurrent force system. The vectors are given in their true positions and at the true distances from each other. The space diagram is drawn to a scale of 1″ = 1.0′. Draw the vectors to a scale of 1″ = 30 lb. Show all construction.

8. (Fig. 31.31) **A.** Determine the force that must be applied at *A* to balance the horizontal member supported at *B*. Scale 1″ = 100 lb. **B.** Find the resultants at each end of the horizontal beam. Find the resultant of the downward loads, and determine where it would be positioned. Scale 1″ = 600 lb.

9. (Fig. 31.32) Determine the forces in the three members of the tripod, which supports a load of *W* = 250 lb. Find the true lengths of all vectors.

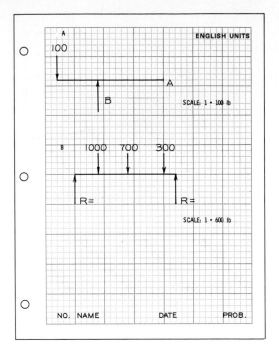

Fig. 31.31 Problems 8A and 8B. Beam analysis.

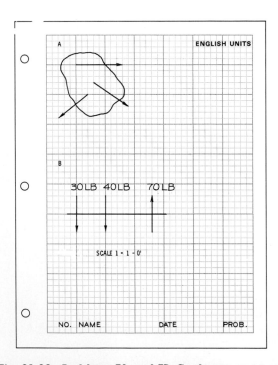

Fig. 31.30 Problems 7A and 7B. Coplanar, nonconcurrent forces.

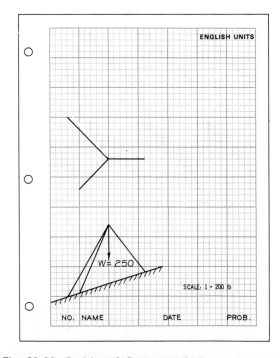

Fig. 31.32 Problem 9. Beam analysis.

10. (Fig. 31.33) The vectors each make an angle of 60° with the structural member on which they are applied. Find the resultant of this force system.

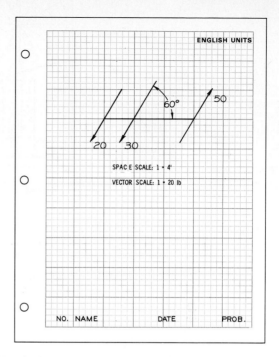

Fig. 31.33 Problem 10. Noncoplanar, concurrent forces in equilibrium.

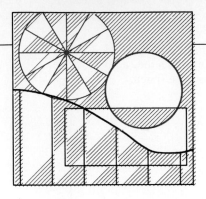

CHAPTER 32

Intersections and Developments

32.1
Introduction

This chapter deals with the methods of finding lines of *intersections* between parts that join. Usually these parts are made of sheet metal or plywood, if used to form concrete.

Once the intersections have been determined, *developments* can be found. These flat patterns can be laid out on the sheet metal and cut to conform to the desired shape. You can see many examples of intersections and developments in Fig. 32.1, where a refinery is under construction.

32.2
Intersections of lines and planes

The steps of finding the intersection between a line and a plane are illustrated in Fig. 32.2. This is a special case where the point of intersection can be easily seen

since the plane appears as an edge (Step 1). It is projected to the front view, and the visibility of the line is found (Step 2).

This principle is used in Fig. 32.3 to find the line of intersection between two planes. Since *EFGH* appears as an edge in the side view, points of intersection 1 and 2 can be found, projected to the front view, and the visibility determined (Step 2). The intersection was found by locating the piercing points of lines *AB* and *DC* and connecting these points.

The intersection of a plane at a corner of a prism results in a line of intersection that bends around the corner (Fig. 32.4). Piercing points 2' and 1' and found in Step1.

Corner point 3 is seen in the side view of Step 2, where the vertical corner pierces the plane. Point 3 is projected to the corner in the front view. Point 2' is hidden in the front view since it is on the back side.

The intersection of a plane and a prism is found in Fig. 32.5, where the plane appears as an edge. The points of intersection are found for each corner line and are connected; visibility is shown to complete the line of intersection.

An intersection between a plane and a prism is shown in Fig. 32.6, where the vertical corners of the

FIG. 32.1 This refinery installation illustrates many applications of the principles of intersections and developments.

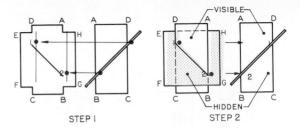

FIG. 32.3 Intersection between planes.

Step 1 The points where lines *AB* and *DC* intersect plane *EFGH* are found where the plane appears as an edge. These points are projected to the front view.

Step 2 Line 1–2 is the line of intersection. Visibility is determined by looking from the front view to the right-side view.

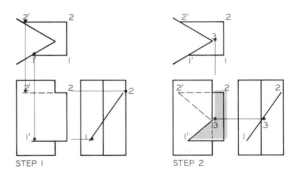

FIG. 32.4 Intersection of a plane at a corner.

Step 1 The intersecting plane appears as an edge in the side view. Intersection points 1' and 2' are projected from the top and side views to the front view.

Step 2 The line of intersection from 1' to 2' must bend around the vertical corner at 3' in the top and side views. Point 3' is projected to the front view to locate line 1'–2'–3'.

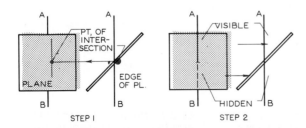

FIG. 32.2 Intersection of a line and a plane.

Step 1 The point of intersection can be found in the view where the plane appears as an edge, the side view in this example.

Step 2 Visibility in the front view is determined by looking from the front view to the right-side view.

prism are true length in the front view and the plane appears foreshortened in both views. Vertical cutting planes are passed through the planes of the prism in the top view to find the piercing points of the corners in the front view. The points are connected and the visibility is determined to complete the solution.

The intersection between a foreshortened plane and an oblique prism is found in Fig. 32.7. The plane

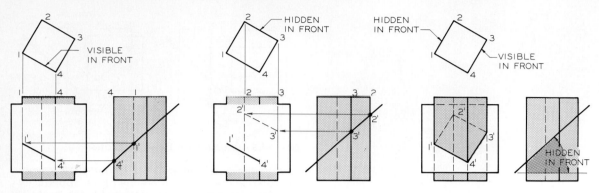

FIG. 32.5 Intersection of a plane and a prism.

Step 1 Vertical corners 1 and 4 intersect the edge view of the plane in the side view at 1' and 4'. These points are projected to the front view and are connected with a visible line.

Step 2 Vertical corners 2 and 3 intersect the edge of the inclined plane at 2' and 3' in the side view. These points are connected in the front view with a hidden line.

Step 3 Lines 1'–2' and 3'–4' are drawn as hidden and visible lines, respectively. Visibility is determined by inspection of the top and side views.

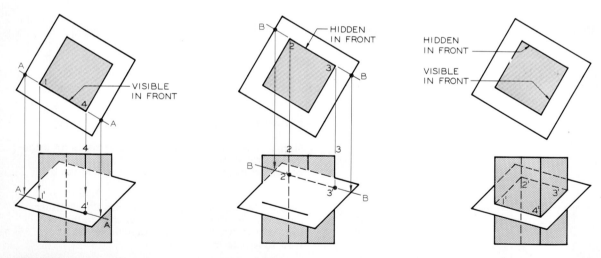

FIG. 32.6 Intersection of an oblique plane and a prism.

Step 1 Vertical cutting plane A–A is passed through the vertical plane, 1–4, in the top view and is projected to the front view. Piercing points 1' and 4' are found in this view.

Step 2 Vertical plane B–B is passed through the top view of plane 2–3 and is projected to the front view where piercing points 2' and 3' are found. Line 2'–3' is a hidden line.

Step 3 The line of intersection is completed by connecting the four points in the front view. Visibility in the front view is found by inspection of the top view.

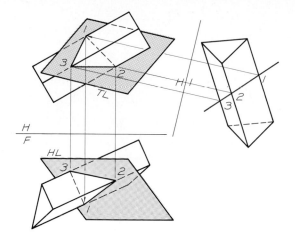

FIG. 32.7 The intersection between a plane and a prism can be found by constructing a view in which the plane appears as an edge.

is found as an edge in a primary auxiliary view. The piercing points of the corners of the prism are located in the auxiliary view and are projected back to the given views.

Points 1, 2, and 3 are projected from the auxiliary view to the given views as examples. Visibility is determined by analysis of crossing lines.

32.3
Intersections between prisms

The same principles used to find the intersection between a plane and a line are used to find the intersection between two prisms in Fig. 32.8. Piercing points 1, 2, and 3 are found in the front view by projecting from the side and top views. Point X is located in the side view where line of intersection 1–2 bends around the vertical corner of the other prism. Points 1, X, and 2 are connected in sequence, and the visibility is determined.

In Fig. 32.9, an inclined prism intersects a vertical prism. The end view of the inclined prism is found by an auxiliary view (Step 1). In the auxiliary view, you can see where plane 2–3 bends around corner AB at point X (Step 2). Points of intersection 1' and 2' are projected from the top to the front view. The line of intersection 2'–X–3' can be drawn to complete this portion of the line of intersection. The remaining lines, 1'–3' and 2'–3', are connected to complete the solution (Step 3).

An alternative method of solving a problem of this type is shown in Fig. 32.10. Piercing points 1' and 2' are found in the front view by projecting from the top view (Step 1). Point 5, the point where line 1'–5–2' bends around vertical corner AB, is then found (Step 2). A cutting plane is passed through corner AB in the top view. The front view of point 5 is found, and the lines of intersection are completed.

The conduit connector in Fig. 32.11 is an example of intersecting planes and prisms.

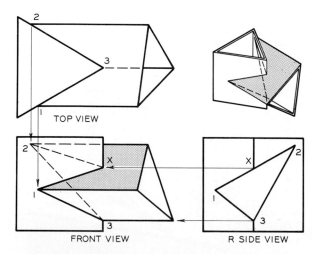

TOP VIEW

FRONT VIEW   R SIDE VIEW

FIG. 32.8 Three views of intersecting prisms. The points of intersection can be seen where intersecting planes appear as edges.

32.4
Intersection of a plane and cylinder

The intersections of the components of the gas transmission system shown in Fig. 32.12 offer numerous applications of the principles of intersections.

The intersection between a plane and a cylinder is found in Fig. 32.13, where the plane appears as an edge in one of the views. Cutting planes are passed vertically through the top view of the cylinder to establish elements on the cylinder and their piercing points. The piercing points are projected to each view to find the line of intersection, which is an ellipse.

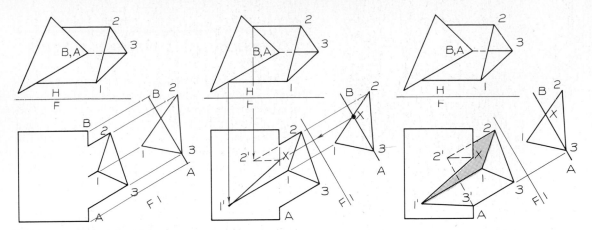

FIG. 32.9 Intersection between two prisms.

Step 1   Construct the end view of the inclined prism by projecting an auxiliary view from the front view. Show only line *AB* of the vertical prism in the auxiliary view.

Step 2   Locate piercing points 1' and 2' in the top and front views. Intersection line 1'–2' will bend around corner *AB* at point *X*, which is projected from the auxiliary view.

Step 3   Intersection lines from 2' and 1' to 3' do not bend around the corner. Therefore these are drawn as straight lines. Line 1'–3' is visible, and line 2'–3' is invisible.

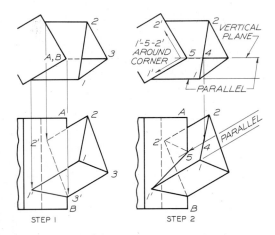

FIG. 32.10   Intersection between prisms.

Step 1   The piercing points of lines 1, 2, and 3 are found in the top view and are projected to the front view where piercing points 1', 2', and 3' are found.

Step 2   A cutting plane is passed through corner *AB* in the top view to locate point 5, where line of intersection 1'–2' bends around the vertical prism. Point 5 is found in the front view, and 1'–5–2' is drawn.

FIG. 32.11   This conduit connector was designed using the principles of the intersection of a plane and prism. (Courtesy of the Federal Aviation Administration.)

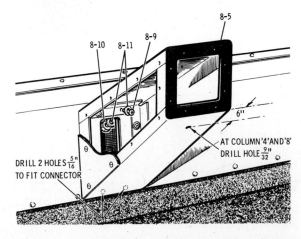

FIG. 32.12 This complex of pipes and vessels contains many applications of intersections. (Courtesy of Trunkline Gas Co.)

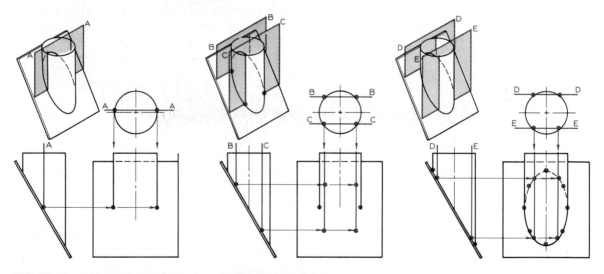

FIG. 32.13 Intersection between a cylinder and a plane.

Step 1 A vertical cutting plane, *A–A* is passed through the cylinder parallel to its axis to find two points of intersection.

Step 2 Cutting planes, *B–B* and *C–C*, are used to find four additional points in the top and left side views; these points are projected to the front view.

Step 3 Additional cutting planes are used to find more points. These points are connected to give an elliptical line of intersection.

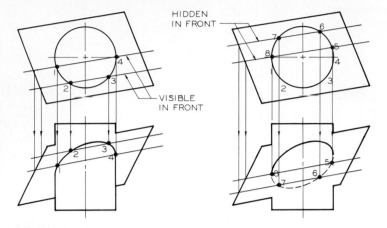

FIG. 32.14 Intersection of a cylinder and an oblique plane.

Step 1 Vertical cutting planes are passed through the cylinder in the top view to establish elements on its surface and lines on the oblique plane. Piercing points 1, 2, 3, and 4 are projected to the front view of their respective lines and are connected with a visible line.

Step 2 Additional cutting planes are used to find other piercing points—5, 6, 7, and 8— which are projected to the front view of their respective lines on the oblique plane. These are connected with a hidden line.

Step 3 Visibility of the plane and cylinder is completed in the front view. Line *AB* is found to be visible by inspection of the top view, and line *CD* is found to be hidden.

A more general problem is solved in Fig. 32.14, where the cylinder is vertical but the plane is oblique. Vertical cutting planes are passed through the cylinder and the plane in the top view to find piercing points of the cylinder's elements on the plane. These points are projected to the front view to complete the line of intersection, an ellipse. The more cutting planes used, the more accurate the line of intersection will be.

The general case of the intersection between a plane and cylinder is solved in Fig. 32.15, where both are oblique in the given views. The edge view of the plane is found in an auxiliary view. Cutting planes are passed through the cylinder parallel to the cylinder's axis in the auxiliary view to find the piercing points. The piercing points of the elements are connected to give elliptical lines of intersection in the given views. Visibility is determined by analysis.

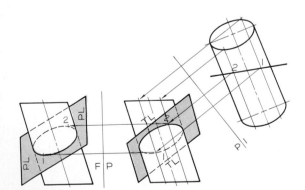

FIG. 32.15 The intersection between an oblique cylinder and an oblique plane can be found by constructing a view that shows the plane as an edge.

32.5
Intersections between cylinders and prisms

A series of vertical cutting planes is used in Fig. 32.16 to establish lines that lie on the surfaces of the cylinder and prism. A primary auxiliary view is drawn to show the end view of the inclined prism. Also shown in this

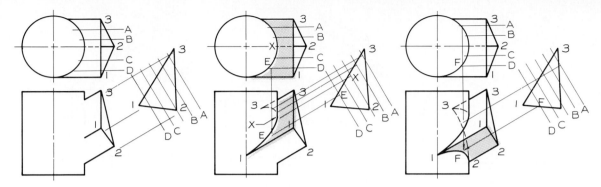

FIG. 32.16 Intersection between a cylinder and a prism.

Step 1 Project an auxiliary view of the triangular prism from the front view to show three of its surfaces as edges. Pass frontal cutting planes through the top view of the cylinder and project them to the auxiliary view. The spacing between the planes is equal in both views.

Step 2 Locate points along the line of intersection 1–3 in the top view and project them to the front view. *Example:* Point *E* on cutting plane *D* is found in the top and auxiliary views and projected to the front view where the projectors intersect. Visibility changes in the front view at point *X*.

Step 3 Determine the remaining points of intersection by using the same cutting planes. Point *F* is shown in the top and auxiliary views and is projected to the front view of 1–2. Connect the points and determine visibility. Space the cutting planes so that they will produce the most accurate line of intersection.

view are the vertical cutting planes, which are spaced the same distance apart as in the top view (Step 1).

The line of intersection from 1 to 3 is projected from the auxiliary view to the front view (Step 2), where the intersection is an elliptical curve. The change of visibility of this line is found at point *X* in the top and auxiliary views, and it is projected to the front view. The process is continued to find the lines of intersection of the other two planes of the prism. (Step 3).

32.6

Intersections between two cylinders

The line of intersection between two perpendicular cylinders can be found by passing cutting planes through the cylinders parallel to the centerlines of each (Fig. 32.17). Each cutting plane locates the piercing points of two elements of one cylinder on an element of the other cylinder. The points are connected and visibility is determined to complete the solution.

The intersection between nonperpendicular cylinders is found in Fig. 32.18 by a series of vertical cutting planes. Each cutting plane is passed through the cylinders parallel to the centerline of each. As examples of points on the line of intersection, points 1 and 2 are labeled on cutting plane *D*. Other points are found in the same manner. Although the auxiliary view is not required for the solution, it assists you in visualizing the problem. Points 1 and 2 are shown on cutting plane *D* in the auxiliary view, where they can be projected to the front view as a check on the solution found when projecting from the top view.

32.7

Intersections between planes and cones

To find points of intersection on a cone, cutting planes can be used that are (1) perpendicular to the cone's axis or (2) parallel to the cone's axis. Horizontal cutting planes are shown in Fig. 32.19A, where they are

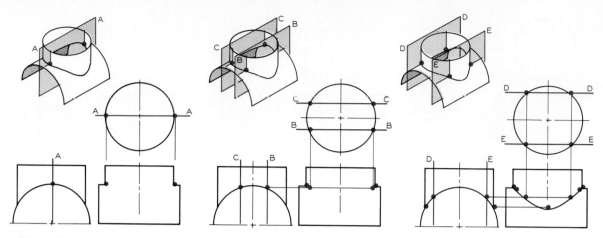

FIG. 32.17 Intersection between two cylinders.

Step 1 Cutting plane *A–A* is passed through the cylinders parallel to the axes of both. Two points of intersection are found.

Step 2 Cutting planes *C–C* and *B–B* are used to find four additional points of intersection.

Step 3 Cutting planes *D–D* and *E–E* locate four more points. Points found in this manner give the line of intersection.

FIG. 32.18 The intersection between these cylinders is found by locating the end view of the inclined cylinder in an auxiliary view. Vertical cutting planes are used to find the piercing points of the elements of the cylinder and the line of intersection.

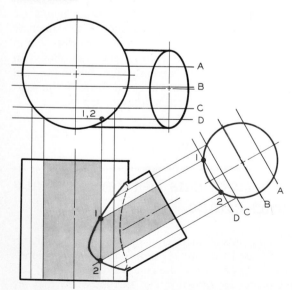

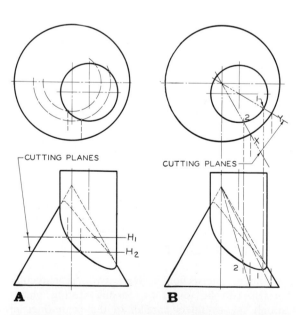

FIG. 32.19 A. Intersections on conical surfaces can be found by cutting planes that pass through the cone parallel to its base. **B.** A second method shows radial cutting planes that pass through the cone's centerline and perpendicular to its base.

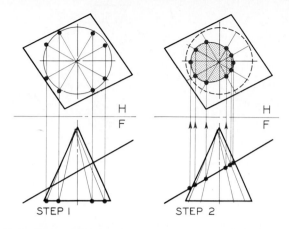

FIG. 32.20 Intersection of a plane and a cone.

Step 1 Divide the base into even divisions in the top view and connect these points with the apex to establish elements on the cone. Project these elements to the front view.

Step 2 The piercing point of each element on the edge view is projected to the top view to the same elements, where they are connected to form the line of intersection.

labeled H_1 and H_2. The horizontal planes cut circular sections that appear true size in the top view.

The cutting planes in Fig. 32.19B are passed radially through the top view to establish elements on the surface of the cone that are projected to the front view. Points 1 and 2 are found on these elements in both views by projection.

A series of radial cutting planes is used to find elements on the cone in Fig. 32.20. These elements cross the edge view of the plane in the front view to locate piercing points that are projected to the top view of the same elements to form the line of intersection.

A cone and an oblique plane intersect in Fig. 32.21, and the line of intersection is found by using a series of horizontal cutting planes. The sections cut by these imaginary planes will be circles in the top view. In addition, the cutting planes locate lines on the oblique plane that intersect the same circular sections cut by each respective cutting plane. The points of intersection are found in the top view and are projected to the front view.

The horizontal cutting-plane method also could have been used to solve the example in Fig. 32.20.

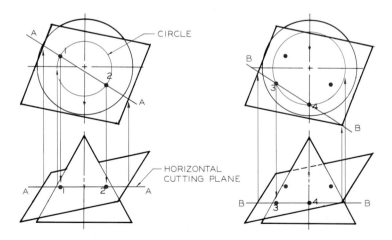

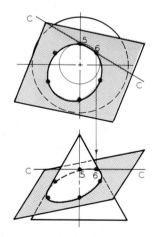

FIG. 32.21 Intersection of an oblique plane and a cone.

Step 1 A horizontal cutting plane is passed through the front view to establish a circular section on the cone and a line on the plane in the top view. The piercing point of this line lies on the circular section. Piercing points 1 and 2 are projected to the front view.

Step 2 Horizontal cutting plane B–B is passed through the front view in the same manner to locate piercing points 3 and 4 in the top view. These points are projected to the horizontal plane in the front view from the top view.

Step 3 Additional horizontal planes are used to find sufficient points to complete the line of intersection. Determination of visibility completes the solution.

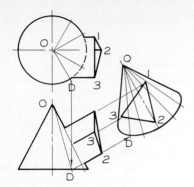
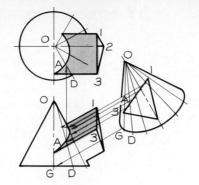
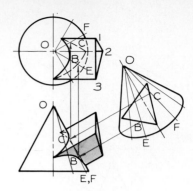

FIG. 32.22 Intersection between a cone and a prism.

Step 1   Construct an auxiliary view to obtain the edge views of the lateral surfaces of the prism. In the auxiliary view, pass cutting planes through the cone that radiates from the apex to establish elements on the cone. Project the elements to the front and auxiliary views.

Step 2   Locate the piercing points of the cone's elements with the edge view of plane 1–3 in the primary view and project them to the front and top views. *Example:* Point A lies on element 0D in the primary auxiliary view, so it is projected to the front and top views of elements 0D.

Step 3   Locate the piercing points where the conical elements intersect the edge views of the planes of the prism in the auxiliary view. *Example:* Point B is found on 0E in the primary auxiliary view and is projected to the front and top views of 0E.

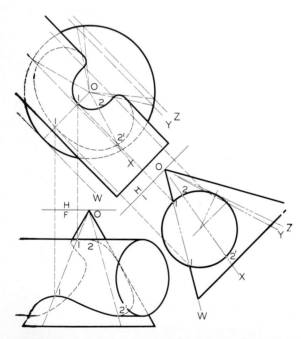

FIG. 32.23 The intersection between the cone and the cylinder is found by projecting an auxiliary view from the top of the cone to find the circular view of the cylinder. Radial cutting planes are passed through the cone and the cylinder in the auxiliary view to locate piercing points of the cylinder's elements.

FIG. 32.24 Horizontal cutting planes are used to find the intersection between the cone and the cylinder. The cutting planes form circles in the top view.

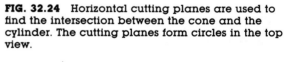
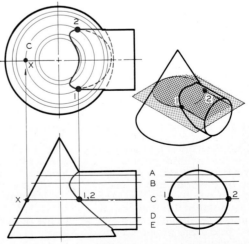

32.8
Intersections between cones and prisms

A primary auxiliary view is used to find the end view of the inclined prism that intersects the cone in Fig. 32.22 (Step 1). Cutting planes that radiate from the apex of the cone in the top view are drawn in the auxiliary view to locate elements on the cone's surface that intersect the prism. These elements are drawn in the front view by projection.

Wherever the edge view of plane 1–3 intersects an element in the auxiliary view, the piercing points are projected to the same element in the front and top views (Step 2). An extra cutting plane is passed through point 3 in the auxiliary view to locate an element that is projected to the front and top views. Piercing point 3 is projected to this element in sequence from the auxiliary view to the top view.

This same procedure is used to find the piercing points of the other two planes of the prism (Step 3). All projectons of points of intersection originate in the auxiliary view, where the planes of the prism appear as edges.

A similar problem is solved in Fig. 32.23, where a cylinder intersects a cone. The circular view of the cylinder is found in a primary auxiliary view. Points 2 and 2' are found in the auxiliary view on element 0X, where a radial cutting plane is passed through apex 0. Element 0X is found in the top and front views by projecting from the auxiliary view to locate points 2 and 2'.

In Fig. 32.24, horizontal cutting planes are passed through the cone and the intersecting perpendicular cylinder to locate the line of intersection. A series of circular sections are found in the top view. Points 1 and 2 are found on cutting plane C in the top view as examples and are projected to the front view. Other points are found in this same manner.

This method is feasible only when the centerline of the cylinder is perpendicular to the axis of the cone, so that circular sections can be found in the top view, rather than elliptical sections that would be difficult to draw.

The distributor housing in Fig. 32.25 is an example of an intersection between cylinders and a cone.

32.9
Intersections between pyramids and prisms

The intersection between an inclined prism and a pyramid is solved in Fig. 32.26. The end view of the inclined prism is found in a primary auxiliary view; the pyramid is shown in this view also (Step 1). Radial lines 0B and 0A are passed through corners 1 and 3 in the auxiliary view (Step 2). The radial lines are projected from the auxiliary view to the front and top views. Intersecting points 1 and 3 are located on 0B and 0A in each of these views by projection. Point 2 is the point where line 1–3 bends around corner 0C. Lines of intersection 1–4 and 4–3 are then found (Step 3). The visibility is determined and the solution is completed.

A prism parallel to the base of a pyramid is shown in Fig. 32.27. Its lines of intersection are found by using a series of horizontal cutting planes that pass through the pyramid parallel to its base to form triangular sections in the top view.

The same cutting planes are passed through the corner lines of the prism in the front and auxiliary views. Each corner edge is extended in the top view to intersect the triangular section formed by the cutting plane passing through it. Point 1 is given as an example.

FIG. 32.25 This electrically operated distributor is an application of intersections between a cone and a series of cylinders. (Courtesy of GATX).

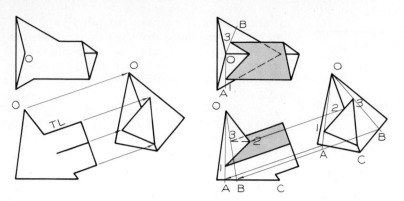
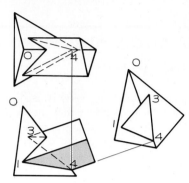

FIG. 32.26 Intersection between a prism and a pyramid.

Step 1 Find the edge view of the surfaces of the prism by projecting an auxiliary view from the front view. Project the pyramid into this view also. Only the visible surfaces need be shown in this view.

Step 2 Pass planes A and B through apex 0 and points 1 and 3 in the auxiliary view. Project lines 0A and 0B to the front and top views. Project points 1 and 3 to 0A and 0B in the principal views. Point 2 lies on line 0C. Connect points 1, 2, and 3 to give the intersection of the upper plane.

Step 3 Point 4 lies on line 0C in the auxiliary view. Project this point to the principal views. Connect point 4 to points 3 and 1 to complete the intersections. Visibility is indicated. These geometric shapes are assumed to be hollow, as though constructed of sheet metal.

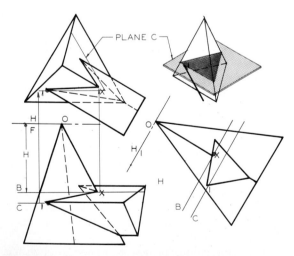
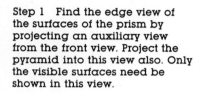

FIG. 32.27 The intersection between this pyramid and prism is found by locating the end view of the prism in an auxiliary view. Horizontal cutting planes are passed through the fold lines of the prism to find the piercing points and the line of intersection.

Corner point *X* is found by passing cutting plane *B* through it in the auxiliary view where it crosses the corner line. This is where the line of intersection of this plane bends around the corner.

32.10

Intersections between spheres and planes

The line of intersection between a sphere and a plane is found in Fig. 32.28, where the plane appears as an edge in the front view. Horizontal cutting planes are passed through the front view to form circular sections in the top view. Two points are located on each cutting plane in the top view by projecting from the front view, where the cutting plane crosses the edge view of the plane. The points are connected and visibility is shown in the top view.

The elliptical intersection could have been drawn with an ellipse template selected by measuring the angle between the edge view of the plane in the front view and the projectors from the top view. Since the plane passes through the center of the sphere, the ma-

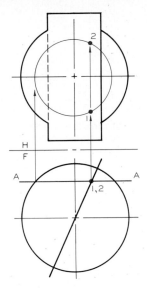 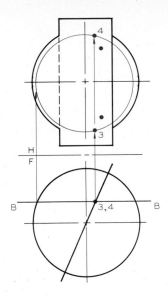 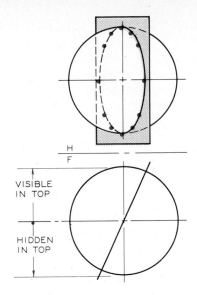

FIG. 32.28 Intersection of a sphere and a plane.

Step 1 Horizontal cutting plane A–A is passed through the front view of the sphere to establish a circular section in the top view. Piercing points 1 and 2 are projected from the front view to the top view.

Step 2 Horizontal cutting plane B–B is used to locate piercing points 3 and 4 in the top view by projecting to the circular section cut by the plane in the top view. Additional horizontal planes are used to find sufficient points.

Step 3 Visibility of the top view is found by inspection of the front view. The upper portion of the sphere will be visible in the top view and the lower portion will be hidden.

jor diameter of the ellipse would be equal to the true diameter of the sphere.

A general case of the intersection between a plane and a sphere is given in Fig. 32.29, where three points, 1, 2, and 3, are located on the sphere's surface. A circle that lies on the surface of the sphere is to be drawn through these three points.

The edge view of the plane is found by projecting from the top view to the primary auxiliary view. The circle on the sphere passing through 1, 2, and 3 will have an elliptical line of intersection in the top and front views. The ellipse template angle for the top view is the angle between the edge of the plane and the projector from the top view, 25°. The major diameter is drawn parallel to the true-length lines on plane 1–2–3 in the top view.

The ellipse for the front view of the intersection is found in the same manner by finding the edge view of the plane in an auxiliary view projected from the front view. The ellipse template angle is 50°.

32.11
Intersections between spheres and prisms

The intersection between a sphere and a prism is found in Fig. 32.30 by drawing a series of vertical cutting planes in the top and side views. The planes form circular sections in the front view.

The intersections of the edges with the cutting planes in the side view are projected to their respective circles in the front view. Points 1 and 2 are located on cutting plane A in the side view and on the circular path of A in the front view. Point X in the side view locates the point where the visibility changes in the front view. Point Y in the side view is the point where the visibility of the intersection changes in the top view. Both points lie on the centerlines of the sphere on the side view.

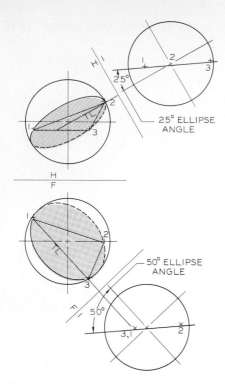

25° ELLIPSE ANGLE

50° ELLIPSE ANGLE

FIG. 32.29 Three points are given on the surface of the sphere through which a circle passes. This plane is found as an edge by projecting from the top and front views. Ellipse angles of 25° and 50° are found for drawing the top and front views of the elliptical intersections.

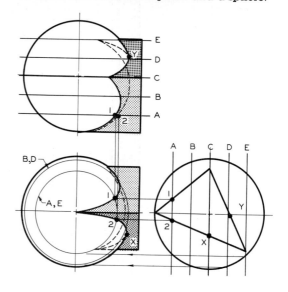

FIG. 32.30 Vertical cutting planes are used to find the intersection between a prism and a sphere.

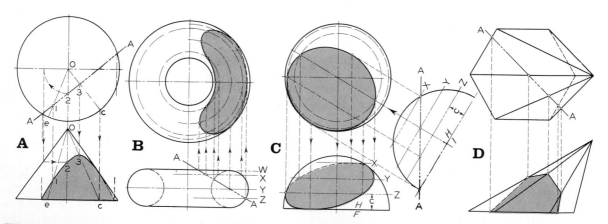

FIG. 32.31 Examples of cutting-plane intersections through four types of figures.

A. Radial lines are used to find a hyperbolic section through a cone.

B. An inclined plane is shown intersecting a torus.

C. The section cut by a plane through a hemisphere.

D. A section cut by a vertical plane through a pyramid.

32.12
Miscellaneous intersections

A series of intersections of cutting planes passed through different figures is shown in Fig. 32.31. Radial lines are used to find a hyperbolic section on a cone in Fig. 32.31A. Horizontal cutting planes are used to locate a section in the top view of a *torus* (a donut-shaped object) in Fig. 32.31B. Horizontal cutting planes are drawn to locate an elliptical section through a hemisphere in Fig. 32.31C with a supplementary auxiliary view. The section of a cutting plane through an oblique pyramid is shown in Fig. 32.31D.

The construction of runouts formed by fillets that intersect cylinders is shown in Fig. 32.32. Horizontal cutting planes are passed through the fillets in the front view and are projected to the top view to locate arcs formed by the cutting planes. Points along the runout in the front view are found by projectng from the top view. Points 1 and 2 are shown as examples.

FIG. 32.33 Almost all the surfaces shown in this refinery were made from flat stock fabricated to form these irregular shapes. These flat patterns are called developments.

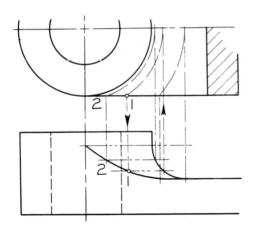

FIG. 32.32 Horizontal cutting planes are used to find the runout (intersection) where surfaces intersect the cylindrical ends of these points.

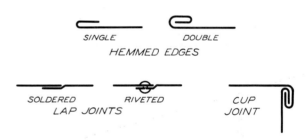

FIG. 32.34 Examples of the types of seams used to join developments.

32.13
Principles of developments

The processing plant shown in Fig. 32.33 illustrates examples of sheet metal shapes designed using the principles of developments. In other words, their pat-terns were laid out on a flat stock and then formed to the proper shape by bending and seaming the joints.

Examples of standard hemmed edges and joints are shown in Fig. 32.34. The application of the sheet metal design will determine the best method of connecting the seams.

The development of the surfaces of three typical shapes into a flat pattern is shown in Fig. 32.35. The sides of a box are imagined to be unfolded into a common plane. The cylinder is rolled out for a distance equal to its circumference. The pattern of a right cone is developed using the length of an element as a radius.

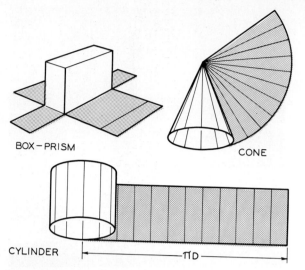

FIG. 32.35 Three standard types of developments: The box, cylinder, and cone.

An inside pattern (development) is more desirable than an outside pattern because most bending machines are designed to fold metal so that markings are folded inward and because markings and scribings are better hidden when the pattern is assembled.

The method of denoting a pattern is labeled by a series of lettered or numbered points about its layout. All lines on a development must be true length.

Seam lines (lines where the pattern is joined) should be the shortest lines so that the expense of riveting or welding the pattern is the least possible.

Patterns of shapes with parallel elements, such as the prisms and cylinders shown in Figs. 32.36a and b, are begun by constructing stretch-out lines parallel to the edge view of the right section of the parts. The distance around the right section is laid off along the stretch-out line. The prism and cylinder in Figs. 32.36 c and d are inclined; thus, the right sections must be drawn perpendicular to their sides, not parallel to their bases.

32.14
Development of prisms

A flat pattern for a prism is developed in Fig. 32.37. Since the edges of the prism are vertical in the front view, its right section is perpendicular to these sides. The top view shows the right section true size. The stretch-out line is drawn parallel to the edge view of the right section, beginning with point 1.

If an inside pattern is drawn and it is to be laid out to the right, point 2 will be to the right of point 1. This is determined by looking from the inside of the top view, where 2 is seen to the right of 1.

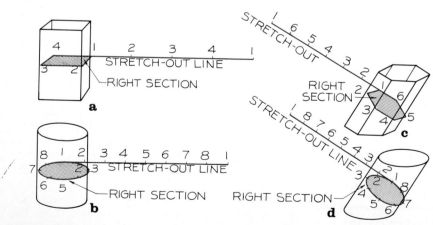

FIG. 32.36 **A.** and **B.** The developments of right prisms and cylinders are found by rolling out the right sections along a stretch-out line. **C.** and **D.** When these figures are oblique, the right sections are found to be perpendicular to the sides of the prism and cylinder. The development is laid out along the stretch-out line parallel to the edge view of the right section.

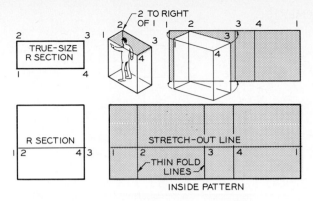

TRUE-SIZE R SECTION

R SECTION

STRETCH-OUT LINE

THIN FOLD LINES

INSIDE PATTERN

FIG. 32.37 The development of a rectangular prism to give an inside pattern. The stretch-out line is parallel to the edge view of the right section.

To locate the fold lines on the pattern, lines 2–3, 3–4, and 4–1 are transferred with your dividers from the right section in the top view to the stretch-out line. The length of each fold line is found by projecting its true length from the front view. The ends of the fold lines are connected to form the limits of the developed surface. Fold lines are drawn as thin lines, and the outside lines are drawn as visible object lines.

The complex installation in Fig. 32.38 is composed of many developments ranging from simple prisms to complicated shapes.

FIG. 32.38 Numerous examples of developments fabricated from sheet metal are shown in this plant. (Courtesy of *Heat Engineering.*)

The development of the prism in Fig. 32.39 is similar to the example in Fig. 32.37 except that one end is beveled rather than square. The stretch-out line is drawn parallel to the edge view of the right section in the front view. The true-length distances around the right section are laid off along the stretch-out line, and the fold lines are located. The lengths of the fold lines are found by projecting from the front view of these lines.

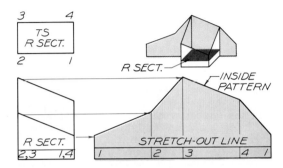

TS R SECT.

R SECT.

INSIDE PATTERN

R SECT.

STRETCH-OUT LINE

FIG. 32.39 The development of a rectangular prism with a beveled end to give an inside pattern. The stretch-out line is parallel to the right section.

32.15
Development of oblique prisms

The prism in Fig. 32.40 is inclined to the horizontal plane, but its fold lines are true length in the front view. The right section is drawn as an edge perpendicular to these fold lines, and the stretch-out line is drawn (Step 1). A true-size view of the right section is found in the auxiliary view.

In Step 2, the distances between the fold lines are transferred from the true-size right section to the stretch-out line. The lengths of the fold lines are found by projecting from the front view.

In Step 3, the ends of the prism are found and attached to the pattern so they can be folded into position.

In Fig. 32.41, the fold lines are true length in the top view; this enables you to draw the edge view of the right section perpendicular to the fold lines in the top view. The stretch-out line is drawn parallel to the edge view of the section, and the true size of the right

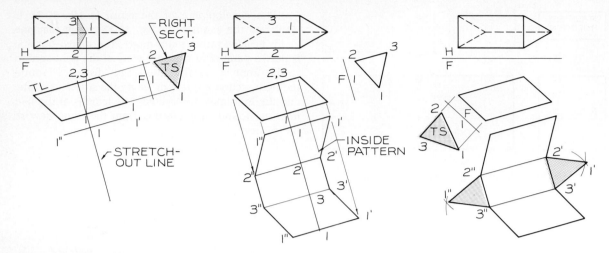

FIG. 32.40 Development of an oblique prism

Step 1 The edge view of the right section is perpendicular to the true-length axis of the prism in the front view. Determine the true-size view of the right section by an auxiliary view. Draw the stretch-out line parallel to the edge view of the right section. Line 1′–1″ is the first line of the development.

Step 2 Since the pattern is developed toward the right, from line 1′–1″, the next point is line 2′–2″ by referring to the auxiliary view. Transfer true-length lines 1–2, 2–3, and 3–1 from the right section to the stretch-out line to locate the elements. Determine the lengths of the bend lines by projection.

Step 3 Find the true-size views of the end pieces by projecting auxiliary views from the front view. Connect these surfaces to the development of the lateral sides to form the completed pattern. Fold lines are drawn with thin lines, and outside lines are drawn as object lines.

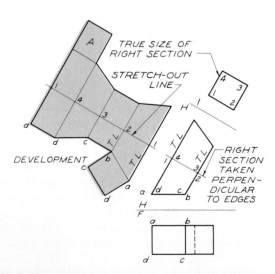

FIG. 32.41 The development of this oblique chute is found by locating the right section true size in the auxiliary view. The stretch-out line is drawn parallel to the right section.

section is found in an auxiliary view projected from the top view. The distances about the right section are transferred to the stretch-out line to locate the fold lines. The lengths of the fold lines are found by projecting from the top view. The end portions of the pattern are attached to the pattern to complete the construction.

A prism that does not project true length in either view can be developed as shown in Fig. 32.42. The fold lines are found true length in an auxiliary view projected from the front view. The right section will appear as an edge perpendicular to the fold lines in the auxiliary view. The true size of the right section is found in a secondary auxiliary view.

The stretch-out line is drawn parallel to the edge view on the right section. The fold lines are located on the stretch-out line by measuring around the right section in the secondary auxiliary view. The lengths of the fold lines are then projected to the development from the primary auxiliary view.

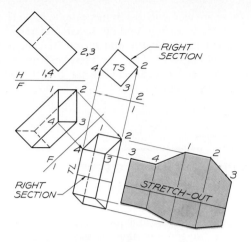

FIG. 32.42 The development of an oblique cylinder is found by locating an auxiliary view in which the fold lines are true length and a secondary auxiliary view in which the right section appears true size. The stretch-out line is drawn parallel to the right section.

32.16

Development of cylinders

The development of a cylinder is found in Fig. 32.43. Since the elements of the cylinder are true length in the front view, the right section will appear as an edge in this view and true size in the top view. The stretch-out line is drawn parallel to the edge view of the right section, and point 1 is chosen as the beginning point

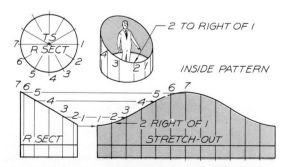

FIG. 32.43 The development of a right cylinder's inside pattern. The stretch-out line is parallel to the right section. Point 2 is to the right of point 1 for an inside pattern.

since it is the shortest element. To draw an inside pattern, assume you are standing on the inside looking at point 1 and you will see that point 2 is to the right of point 1; therefore the pattern is laid out with point 2 to the right of point 1.

The spacing between the elements in the top view can be conveniently done by drawing radial lines at 15° or 30° intervals. Using this technique, the elements will be equally spaced, making it convenient to lay them out along the stretch-out line. The lengths of the elements are found by projecting from the front view to complete the pattern.

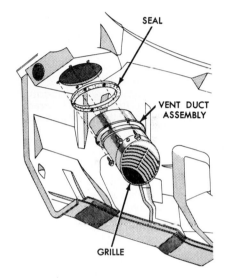

FIG. 32.44 This ventilator air duct was designed using development principles. (Courtesy of Ford Motor Co.)

An application of a developed cylinder with a beveled end is the air-conditioning duct from an automobile shown in Fig. 32.44. The development of a similar cylinder is shown in Fig. 32.45. The base of the front view is the edge view of the right section, which appears true size in the top view. Elements are located around the circumference of the right section in the top view. Two alternative methods are shown to illustrate how the elements are located at 30° intervals. One employs the 30°–60° triangle, and the other uses a compass with the radius equal to the radius of the right section.

The stretch-out line is drawn parallel to the right section in the front view. The total length is found mathematically by the formula $C = \pi D$. The stretch-out line is divided into the same number of divisions

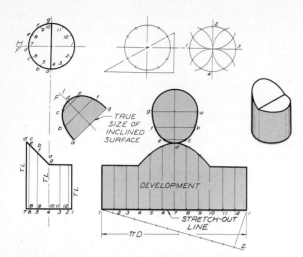

FIG. 32.45 The development of a right cylinder can be found by mathematically locating elements along the stretch-out line, which is equal in length to the circumference of the right section.

as there are elements, 12 in this case, to provide a high degree of accuracy in finding the circumference.

The pattern for the end of the cylinder is found by combining the partial top view and the auxiliary view. This end is connected to the overall pattern to complete the solution.

32.17
Development of oblique cylinders

The pattern for an oblique cylinder (Fig. 32.46) is found in the same manner as the previous examples but with the addition of one preliminary step: The right section must be found true-size in an auxiliary view. In Step 1, a series of equally spaced elements is located around the right section in the auxiliary view and is projected back to the true-length view. The

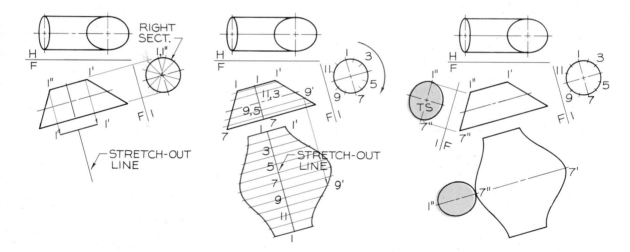

FIG. 32.46 Development of an oblique cylinder.

Step 1 The right section is an edge in the front view, in which it is perpendicular to the true-length axis. Construct an auxiliary view to determine the true size of the right section. Draw a stretch-out line parallel to the edge view of the right section. Locate element 1'–1". Divide the right section into equal divisions.

Step 2 Project these elements to the front view from the right section. Transfer measurements between the points in the auxiliary view to the stretch-out line to locate the elements in the development. Determine the lengths of the elements by projection.

Step 3 The development of the end pieces will require auxiliary views that project these surfaces as ellipses, as shown for the left end. Attach this true-size ellipse to the pattern. Note that the beginning line for the pattern was line 1'–1", the shortest element, for economy.

stretch-out line is drawn parallel to the edge view of the right section in the front view.

In Step 2, the spacing between the elements is laid out along the stretch-out line, and the elements are drawn through these points perpendicular to the stretch-out line. The lengths of the elements are found by projecting from the front view.

The ends of the cylinder are found in Step 3 to complete the pattern. Only one end pattern is shown as an example.

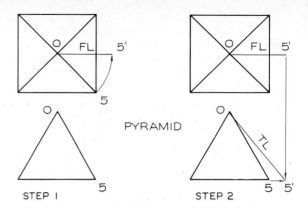

FIG. 32.48 True length by revolution.

Step 1   Corner 0–5 of a pyramid is found true length by revolving it into the frontal plane in the top view, 0–5'.

Step 2   Point 5' is projected to the front view where 0–5' is true length.

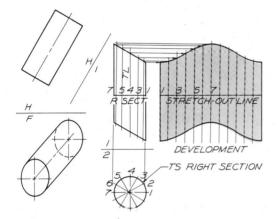

FIG. 32.47 The development of an oblique cylinder is found by constructing an auxiliary view in which the elements appear true length. The right section is found true size in a secondary auxiliary view.

A more general case is the oblique cylinder in Fig. 32.47, where the elements are not true length in the given views. A primary auxiliary view is used to find a view where the elements are true length, and a secondary auxiliary view is drawn to find the true-size view of the right section. The stretch-out line is drawn parallel to the edge view of the right section in the primary auxiliary view, and the elements are located along this line by transferring their distances apart from the true-size section.

The elements are drawn perpendicular to the stretch-out line. The length of each element is found by projecting from the primary auxiliary view. The endpoints are connected with a smooth curve to complete the pattern.

32.18
Development of pyramids

All lines used to draw a pattern must be true length. Pyramids have only a few lines that are true length in the given views; for this reason, the sloping corner lines must be found true length at the outset.

The corner lines of a pyramid can be found by revolution, as shown in Fig. 32.48. Line 0–5 is revolved in the frontal position of 0–5' in the top view (Step 1). Since 0–5' is a frontal line, it will be true length in the front view (Step 2).

The development of a pyramid is given in Fig. 32.49. Line 0–2 is revolved into the frontal plane in the top view to find its true length in the front view. Since this is a right pyramid, all bends are equal in length. Line 0–2' is used as a radius to construct the base circle for drawing the development. Distance 3–2 is transferred from the base in the top view to the development, where it forms a chord on the base circle. Lines 2–1, 1–4, and 4–3 are found in the same manner and in sequence. The bend lines are drawn as thin lines from the base to the apex, point 0.

A variation of this problem is given in Fig. 32.50, in which the pyramid has been truncated or cut at an angle to its axis. The development of the inside pattern is found in the same manner as the previous example;

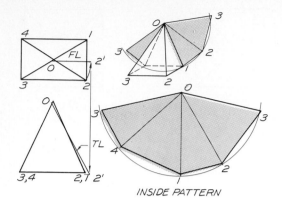

FIG. 32.49 Development of a right pyramid.

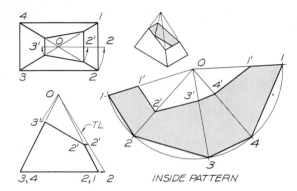

FIG. 32.50 The development of an inside pattern of a truncated pyramid. The corner lines are found true length by revolution.

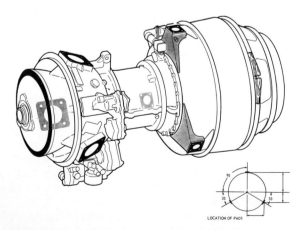

FIG. 32.51 Examples of pyramid shapes in the design of mounting pads for an engine. (Courtesy of Avco Lycoming.)

however, to establish the upper lines of the development, an additional step is required. The true-length lines from the apex to points 1′, 2′, 3′, and 4′ are found by revolution. These distances are located on their respective lines of the pattern to find the upper limits of the pattern.

The mounting pads in Fig. 32.51 are sections of pyramids that intersect an engine body.

The development of an oblique pyramid is shown in Fig. 32.52. The corner lines are found true length by revolution (Step 1). Using these true-length lines and those given in the principal views, the pattern is drawn by triangulation using a compass (Step 2). The upper limits of the pattern are found by measuring along the fold lines from point 0 (Step 3).

32.19
Development of cones

All elements of a right cone are equal in length, as illustrated in Fig. 32.53, where 0–6 is found true length by revolution. When revolved to 0–6′ position, it is a frontal line and is therefore true length in the front view where it is an outside element of the cone. Point 7′ is found by projecting horizontally to element 0–6′.

The right cone in Fig. 32.54 is developed by dividing the base into equally spaced elements in the top view and by projecting them to the base in the front view. These elements radiate to the apex at 0. The outside elements in the front view, 0–10 and 0–4, are true length.

Using element 0–10 as a radius, draw the base circle of the development. The elements are located along the base circle equal to the chordal distances between them around the base in the top view. You can see this is an inside pattern by inspecting the top view, where point 2 is to the right of point 1 when viewed from the inside.

The sheet metal conical vessel in Fig. 32.55 is an example of a large vessel that was designed using principles of development.

The development of a truncated cone is shown in Fig. 32.56. The pattern is found by laying out the total cone, ignoring the portion removed from it. The removed upper portion can be found by using true-length line 0–7′ as the radius in the pattern view. The hyperbolic sections through the front view of the cone

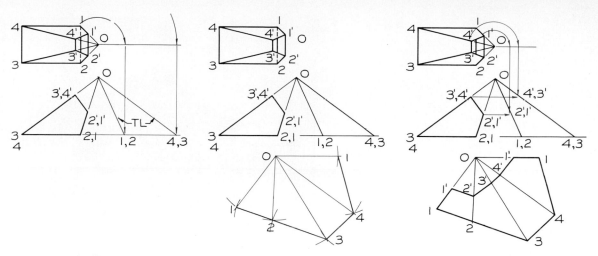

FIG. 32.52 Development of an oblique pyramid.

Step 1 Revolve each of the bend lines in the top view until they are parallel to the frontal plane. Project to the front view where the revolved lines are true length. Let point 0 remain stationary but project points 1, 2, 3, and 4 horizontally in the front view to the projectors from the top view.

Step 2 The base lines are true length in the top view. Using these true-length lines and the revolved lines in the front view, draw the development triangles. All triangles have one side and point 0 in common. This gives a development of the surface, excluding the truncated section.

Step 3 The true lengths of the lines from point 0 to points 1', 2', 3', and 4' are found by revolving them. These distances are laid off from point 0 along their respective lines to establish points along the upper edge of the pattern. The points are then connected by straight lines.

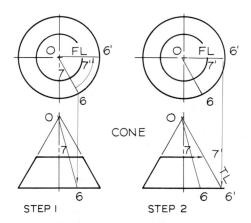

FIG. 32.53 True length by revolution.

Step 1 An element of a cone, 0–6, is revolved into a frontal plane in the top view.

Step 2 Point 6' is projected to the front view where it is the outside element of the cone and is true length. Line 0–7' is found TL by projecting to the outside element in the front view.

FIG. 32.54 The development of an inside pattern of a right cone. In this case, all elements are equal in length.

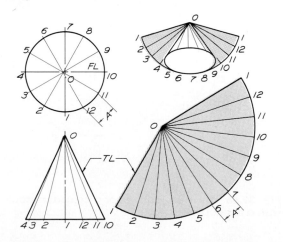

FIG. 32.55 An example of a conical shape formed from steel panels by applying principles of developments.

can be found on their respective elements in the top and front views. Lines 0–2' and 0–3' are projected horizontally to the true-length element 0–1 in the front view, where they will appear true length. These distances and others are measured off along their respective elements in the development to establish a smooth curve on the development.

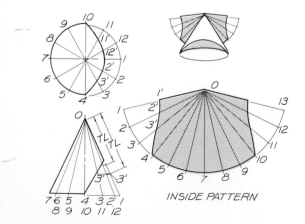

FIG. 32.56 The development of a conical surface with a side opening.

32.20
Development of oblique cones

The development of an oblique cone is found in Fig. 32.57. The elements of this cone are of varying lengths, but the pattern will be symmetrical about an axis.

Elements are located in the top view by dividing the base into equal arcs and drawing the elements to point 0, the apex. These elements are projected to the frontal view. Each element is found true length by revolving it into a frontal plane in the top view. When projected to the front view, the revolved lines are found true length in a true-length diagram.

The development is begun by constructing a series of adjacent triangles using the true-length elements and the chordal distances on the base in the top view. Line 0–1 is chosen for the line of separation since it is the shortest element. The base of the pattern is drawn as a smooth curve.

The distances from point 0 to the upper cut through the cone are found by projecting to their respective elements in the true-length diagram from the front view. Line 0–7' is shown as an example. These

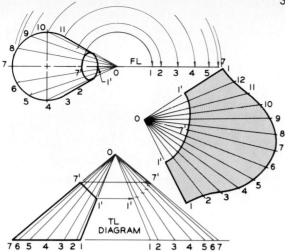

FIG. 32.57 The development of an oblique cone. All elements must be found true length by a true-length diagram before the pattern can be constructed.

is horizontal. However, an auxiliary view is necessary to find the true distance about the upper surface of the object.

The developed surface is found by triangulation using true-length lines from (1) the true-length diagram, (2) the horizontal base in the top view, and (3) the primary auxiliary view. To avoid confusion, each point should be labeled as it is laid out.

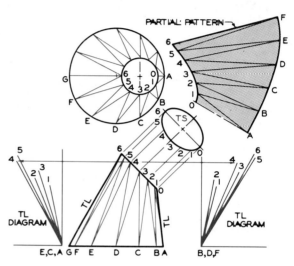

FIG. 32.58 The development of a partial pattern of a warped surface.

shorter elements are located on their respective elements in the pattern to give the upper limits of the pattern.

32.21
Development of warped surfaces

The geometric shape in Fig. 32.58 is an approximate cone with a warped surface; it is similar to the oblique cone in Fig. 32.57. The development of this surface will be an approximation, since a warped surface cannot be laid out on a flat surface.

The surface is divided into a series of triangles in the top and front views by dividing the upper and lower views into equal sectors. The true lengths of all lines are found in the true-length diagrams drawn on both sides of the front and by projecting horizontally from the ends of the lines. If necessary, review Fig. 32.57 to see how the true-length diagrams were found.

The chordal distances between the points on the base appear true-length in the top view since the base

32.22
Development of transition pieces

A transition piece is a form that transforms the section at one end to a different shape at the other (Fig. 32.59). Huge transition pieces can be seen in the industrial installation in Fig. 32.60.

The development of a transition piece is shown in Fig. 32.61. In Step 1, radial elements are drawn from each corner to the equally spaced points on the circular end of the piece. Each of these lines is found true length by revolution.

In Step 2, the true-length lines are used with the true-length chordal distance in the top view to lay out a series of adjacent triangles to form the pattern beginning with element *A*–2.

In Step 3, the triangles *A*–1–2 and *G*–3–4 are added at each end of the pattern to complete the development of half-pattern.

FIG. 32.60 Transition-piece developments are used to join a circular shape with a rectangular section. (Courtesy of Western Precipitation Group, Joy Manufacturing Co.)

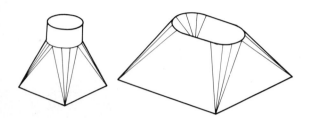

FIG. 32.59 Examples of transition pieces that join parts having different cross sections.

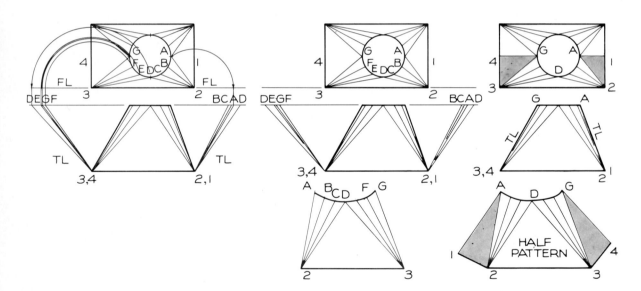

FIG. 32.61 Development of a transition piece.

Step 1   Divide the circular edge of the surface into equal parts in the top view. Connect these points with bend lines to the corner points, 2 and 3. Find the true length (TL) of these lines by revolving them into a frontal plane and projecting them to the front view. These lines represent elements on the surface of a cone.

Step 2   Using the TL lines found in the TL diagram and the lines on the circular edge in the top view, draw a series of triangles, which are joined together at common sides, to form the development. *Examples:* Arcs 2*D* and 2*C* are drawn from point 2. Point *C* is found by drawing arc *DC* from point *D*. Line *DC* is TL in the top view.

Step 3   Construct the remaining planes, *A*–1–2 and *G*–3–4, by triangulation to complete the inside half-pattern of the transition piece. Draw the fold lines as thin lines at the places where the surface is to be bent slightly. The line of departure for the pattern is chosen along *A*–1, the shortest possible line.

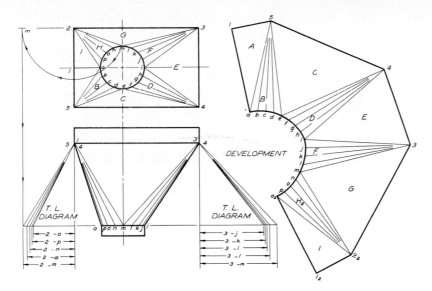

FIG. 32.62 The development of this transition piece is found by locating the conical elements true length in a true-length diagram. Line 2–*m* is found true length as an example. The pattern is found by a series of triangulations.

A similar transition piece is developed in Fig. 32.62 using the same techniques of construction. Elements are established at the corners of the given views and are found true length in the true-length diagrams by revolution. By triangulation, using true-length lines, the full pattern is drawn to complete the development.

32.23
Development of spheres— zone method

In Fig. 32.63, a development of a sphere is found by the zone method. A series of parallels, called *latitudes* in cartography, are drawn in the front view. Each is spaced an equal distance, *D*, apart along the surface in the front view. Distance *D* can be found mathematically to improve the accuracy of this step.

Cones are passed through the sphere's surface so that they pass through two parallels at the outer surface of the sphere. The largest cone with element *R*1 is found by extending it through where the equator and the next parallel intersect on the sphere's surface in the front view until *R*1 intersects the extended centerline of the sphere. Elements *R*2, *R*3, and *R*4 are found by repeating this process.

The development is begun by laying out the largest zone, using *R*1 as the radius, on the arc that rep-

resents the base of an imaginary cone. The breadth of the zone is found by laying off distance *D* from the front view to the development and drawing the upper portion of the zone with a radius equal to *R*1–*D* using the same center. No consideration is given to finding the arc lengths at this time.

The next zone is drawn using radius *R*2 with its center located on a line through the center of arc *R*1.

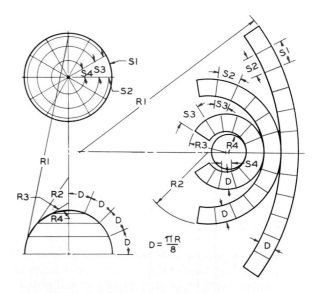

$$D = \frac{\pi R}{8}$$

FIG. 32.63 The zone method of finding the inside development of a spherical pattern.

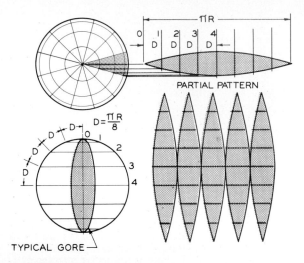

FIG. 32.64 The gore method of developing an inside pattern of a sphere.

The center of $R2$ is positioned along this line so that the arc drawn will be tangent to the preceding arc, which was drawn with radius $R1-D$. The upper arc of this second zone is drawn with a radius of $R2-D$. The remaining zones are constructed successively in this manner. The last zone will appear as a circle with $R4$ as its radius.

The lengths of the arcs can be established by dividing the top view with vertical cutting planes that radiate through the poles. These lines, which lie on the surface of the sphere, are called *longitudes* in cartography. Arc distances $S1$, $S2$, $S3$, and $S4$ are found on each parallel in the top view. These distances are measured off on the constructed arcs in the development. In this case, there are 12 divisions; smaller divisions would provide a more accurate measurement.

32.24
Development of spheres— gore method

Figure 32.64 illustrates an alternative method of developing a flat pattern for a sphere using a series of spherical elements called *gores*. Equally spaced vertical cutting planes are passed through the poles in the top view. Parallels are located in the front view by divid-

ing the surface into equal zones of dimension D. The gores are projected to the front view.

A true-size view of one of the gores is projected from the top view. Dimensions can be checked mathematically for all points. A series of these gores is laid out in sequence to complete the pattern.

32.25
Development of elbows

An elbow is a cylindrical shape that turns a 90° angle. An elbow that is also a transition piece is shown in Fig. 32.65.

FIG. 32.65 The radial bend of the cylinder was developed using the technique discussed in Fig. 32.66.

The method of constructing the pattern for an elbow is illustrated in Fig. 32.66. In Step 1, the 90° arc is drawn, and the cylinder is drawn to size about its centerline. Divide this arc into one division less than the number of pieces that will be used in the elbow. Since three pieces are to be used in this example, the arc is divided into two along line 0b. The two arcs are then bisected.

In Step 2, tangents to the arc are drawn at d, b, and 7. In Step 3, a half-circular view of the cylinder is drawn to locate equally spaced elements that are projected back to the given view parallel to the sides of the three pieces of the elbow. All elements will be true length in this view.

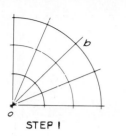
STEP 1

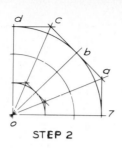
STEP 2

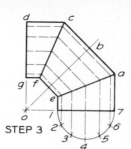
STEP 3

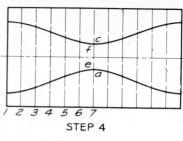
STEP 4

FIG. 32.66 Construction of an elbow.

Step 1 Divide the arc into one division less than the number of pieces desired in the elbow (three in this case) along 0b. Bisect the two arcs.

Step 2 Draw a tangent line through point b to intersect tangents dc and a–7.

Step 3 Draw a semicircle using 1–7 as the diameter. This is half of a right section that will be used to space elements along the stretch-out line.

Step 4 The three pieces can be cut from a rectangular piece of material whose height is equal to the sum of a–7, e–f, and cd. The curves for each cut are found by transferring the lengths of the elements to elements in the pattern.

In Step 4, the flat pattern is developed, with the patterns of the two short pieces of the elbow located at the top and bottom of the rectangular piece. The middle segment is located between these two smaller patterns so no material will be wasted and only two cuts will be necessary.

32.26
Development of straps

Figure 32.67 illustrates the steps of finding the development of a strap that has been bent to serve as a support bracket, between point A on a vertical surface and point B on horizontal surface. In Step 1, the strap is drawn in the side view where the strap appears as an edge using the specified radius to show the bend. The bend is divided into equal arcs and is developed into a flat piece in the vertical plane through point A. Point B' is located in this view.

In Step 2, the location of the hole at B' is found in the front view by projection. The true-size development of the strap is drawn in this view, and it is projected back to the side view to complete that view of the strap.

In Step 3, the projected view of the strap in its bent position in the front view can be found by using

projectors from the side view and the true pattern of the strap.

32.27
Intersections and developments in combination

Intersections between parts must be found before developments of each part can be completed. Figure 32.68 illustrates an example of this.

The intersection between the two prisms is found in the given top and front views. The pattern of the vertical prism is found to the left of the front view and the intersections are shown to indicate cuts that must be made in the pattern.

The pattern of the inclined prism cannot be found without the construction of an auxiliary view in which the fold lines appear true length. The true-size right section is found in a secondary auxiliary view. The fold lines of the pattern are laid off along a stretch-out line by transferring the distances around the right section to it.

The resulting two patterns can be cut out and folded to form shapes that will intersect as shown in the top and front views.

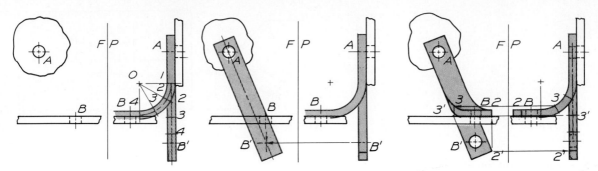

FIG. 32.67 Strap development.

Step 1 Construct the edge view of the strap in the side view. Locate points 1, 2, 3, and 4 on the neutral axis at the bend. Revolve this portion of the strap into the vertical plane and measure the distances along this view of the neutral axis. The hole is located at B' in this view.

Step 2 Construct the front view of B' by revolving point B parallel to the profile plane until it intersects the projector from B' in the side view. Draw the centerline of the true-size strap from A to B' in the front view. Add the outline of the strap around this centerline and around the holes at each end.

Step 3 Determine the projection of the strap in the front view by projecting points from the given views. Points 3 and 2 are shown in the views to illustrate the system of projection used. The ends of the strap are drawn in each view to form true projections.

FIG. 32.68 Two prisms intersect in this example. Their intersections are determined and the developments of each are found by using the principles covered in this chapter. The development of the inclined prism is found by using primary and secondary auxiliary views.

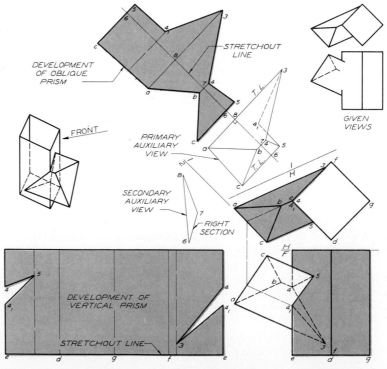

Problems

These problems are scaled to fit two problems per size A sheet when each grid is assigned a value of 0.20″ or 5 mm. When the grid is assigned a value of 0.40″ or 10 mm, one problem per size A sheet can be drawn.

Intersections

1–24.(Fig. 32.69) Lay out the problems assigned and find the intersections that are necessary to complete the views.

Developments

25–48.(Fig. 32.70) Lay out the problems assigned on a size A sheet. Place the long side of sheet in a horizontal direction to permit room at the right of the given views for the development to be drawn.

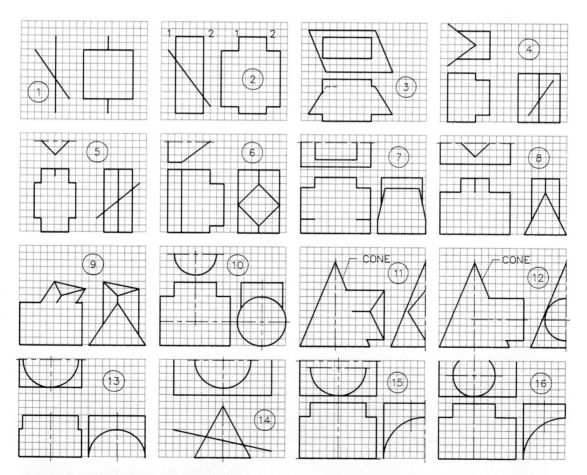

FIG. 32.69 Intersections. Problems 1–24.

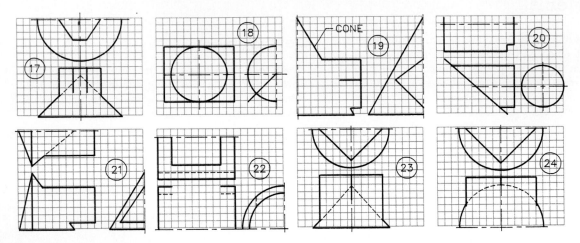

FIG. 32.69 (continued)

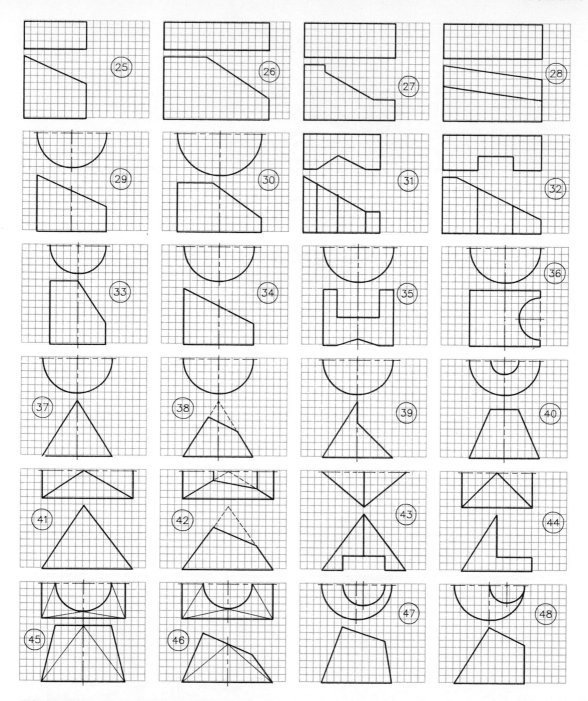

FIG. 32.70 Developments. Problems 25–48.

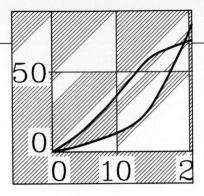

CHAPTER 33

Graphs

33.1
Introduction

Data and information expressed as numbers and words are often difficult to analyze or evaluate unless transcribed into graphical form. *Graphs* (*charts* is an acceptable term but is more appropriate when referring to maps, a specialized form of graphs) are a popular means of briefing other people on trends that might otherwise be difficult to communicate (Fig. 33.1). The trends of a plotted curve on a graph can be compared to the expressions on a person's face, which is a graph of sorts that reveals a person's feelings. For example, a flat curve shows no change, whereas an upwardly inclined curve indicates a positive increase. A downward curve, on the other hand, represents a negative result (Fig. 33.2).

This chapter will deal with the more commonly used graphs. The basic types are

1. Pie graphs
2. Bar graphs
3. Linear coordinate graphs
4. Logarithmic coordinate graphs
5. Semilogarithmic coordinate graphs
6. Polar graphs
7. Schematics and diagrams

FIG. 33.1 Graphs are helpful in presenting technical data to one's associates.

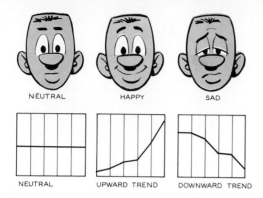

NEUTRAL   HAPPY   SAD

NEUTRAL   UPWARD TREND   DOWNWARD TREND

FIG. 33.2 Curves on a graph are similar to expressions on a face.

33.2
Size proportions of graphs

Graphs may be used to illustrate technical reports reproduced in quantity, and they may be used for projection by slide or overhead projectors. In all cases, the proportion of the graph must be determined so that it will match the proportion of the space or the format of the visual aid.

If a graph is to be photographed by a 35-mm camera (Fig. 33.3), the graph must conform to the standard size of the 35-mm film used. This proportion is approximately 2 × 3 (Fig. 33.4). The proportions of the area in which the graph is to be drawn can be enlarged or reduced by using the diagonal-line method.

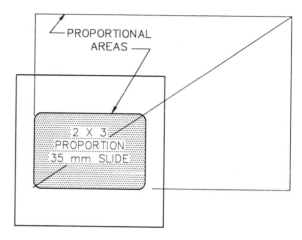

PROPORTIONAL AREAS

2 X 3 PROPORTION 35 mm SLIDE

FIG. 33.4 This diagonal-line method can be used for construction areas that are proportional to those of a 35-mm slide.

FIG. 33.3 When graphs are drawn to be photographed, they must be laid out at a proportion that will match the proportion of the film in the camera.

33.3
Pie graphs

Pie graphs compare the relationship of parts to a whole when there are only a few parts.

Figure 33.5 shows the distribution of skilled workers employed in industry.

The method of drawing a pie graph is shown in Fig. 33.6. The tabular data does not give as good an impression of the comparisons as does the pie graph, even though the data is quite simple.

To facilitate lettering within narrow spaces, the thin sectors should be placed as nearly horizontal as possible to provide more room for the label. The actual percentage should be given and it may be desirable to give the actual numbers or values as well in each sector.

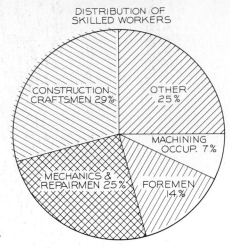

DISTRIBUTION OF SKILLED WORKERS

CONSTRUCTION CRAFTSMEN 29%

OTHER 25%

MACHINING OCCUP. 7%

MECHANICS & REPAIRMEN 25%

FOREMEN 14%

FIG. 33.5 A pie graph shows the relationship of the parts to a whole. It is effective when there are only a few parts.

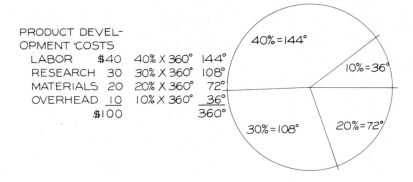

PRODUCT DEVEL-
OPMENT COSTS

LABOR	$40	40% X 360°	144°
RESEARCH	30	30% X 360°	108°
MATERIALS	20	20% X 360°	72°
OVERHEAD	10	10% X 360°	36°
	$100		360°

40%=144°

10%=36°

30%=108°

20%=72°

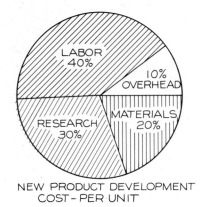

LABOR 40%

10% OVERHEAD

RESEARCH 30%

MATERIALS 20%

NEW PRODUCT DEVELOPMENT COST - PER UNIT

FIG. 33.6 Drawing pie graphs.

Step 1 The total sum of the parts is found and the percentage of each is found. Each percentage is multiplied by 360° to find the angle of each sector of the pie graph.

Step 2 The circle is drawn to represent the pie graph. Each sector is drawn using the degrees found in Step 1. The smaller sectors should be placed as nearly horizontal as possible.

Step 3 The sectors are labeled with their proper names and percentages. In some cases, it might be desirable to include the exact numbers in each sector as well.

33.4

Bar graphs

Since they are well understood by the general public, bar graphs are effective to compare values(Fig. 33.7). In this example, the bars show not only the overall production of timber (the total lengths of the bars) but also the portions of the total devoted to the three uses of the timber.

A bar graph can be composed of a single bar (Fig.33.8). The total length of the bar is 100%, and the

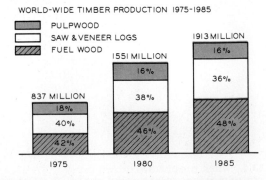

WORLD-WIDE TIMBER PRODUCTION 1975-1985

PULPWOOD
SAW & VENEER LOGS
FUEL WOOD

837 MILLION
18%
40%
42%

1551 MILLION
16%
38%
46%

1913 MILLION
16%
36%
48%

1975   1980   1985

FIG. 33.7 In this example, each bar represents 100% of the total amount and each bar represents different totals.

654

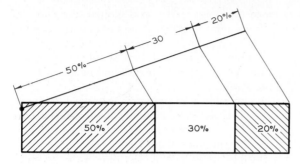

FIG. 33.8 The method of constructing a single bar where the sum of all the parts will be 100%.

bar is divided into lengths proportional to the percentages represented by each of the three parts of the bar.

The method of constructing a bar graph is given in Fig. 33.9. In this case, the title of the graph is placed inside the graph where space is available. Titles are often placed under or over the graph.

The data should be sorted by arranging the bars in ascending or descending order, since it is desirable to know how the data represented by the bars rank from category to category (Fig. 33.10B).

An alphabetical or numerical arrangement of bars results in a graph more difficult to evaluate (Fig. 33.10A).

If the data are sequential and involve time, such as sales per month, it would be less effective to rank the data in ascending order because it is important to see variations in the data over time.

Bars in a bar graph may be horizontal (Fig. 33.11) or vertical (Fig. 33.12). To show a true comparison of data, it is desirable that the bars begin at zero.

33.5
Linear coordinate graphs

A typical coordinate graph, with notes explaining its important features, is given in Fig. 33.13. The axes are linear if the divisions along the axes are equally spaced.

Points are plotted on the grid by using two measurements, called *coordinates*, made along each axis. The plotted points are indicated by using easy-to-draw symbols, such as circles, that can be drawn with a template.

The horizontal scale of the graph is called the *abscissa* or *X*-axis, and the vertical scale is called the *ordinate* or *Y*-axis.

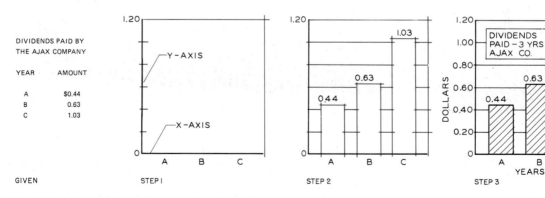

FIG. 33.9 Construction of a bar graph.

Given These data are to be plotted as a bar graph.

Step 1 Lay off the vertical and horizontal axes so that the data will fit on the grid. Make the bars begin at zero.

Step 2 Construct and label the bars. The width of the bars should be different from the space between them. Grid lines should not pass through them.

Step 3 Strengthen lines, title the graph, label the axes, and crosshatch the bars.

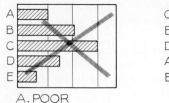
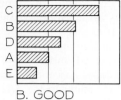

A. POOR   B. GOOD

FIG. 33.10 The bars at A are arranged alphabetically. The resulting graph is not as easy to evaluate as the one at B, where the bars have been arranged in descending order.

Once the points have been plotted, the curve is drawn from point to point. (The line drawn to represent the plotted points is called a curve whether it is a smooth or broken line.) The curve should not close up the plotted points; rather, they should be left as open circles or symbols.

The curve must be drawn as a heavy prominent line, since it is the most important part of the graph. In Fig. 33.13, there are two curves; therefore drawing them as different types of lines and labeling them with notes and leaders is helpful. The title of the graph is placed in a box inside the graph.

Units are given along the *X*- and *Y*-axis with labels that designate the units the graph is comparing.

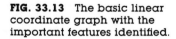
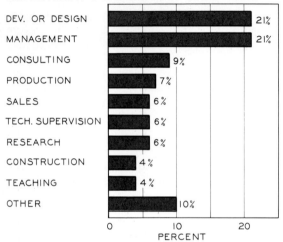

FIG. 33.11 A horizontal bar graph that is arranged in descending order to show where engineers are employed.

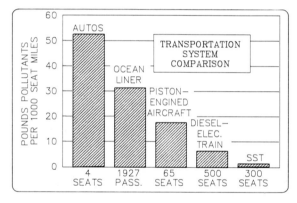

FIG. 33.12 This bar graph has the bars arranged in descending order to compare several sources of pollution.

FIG. 33.13 The basic linear coordinate graph with the important features identified.

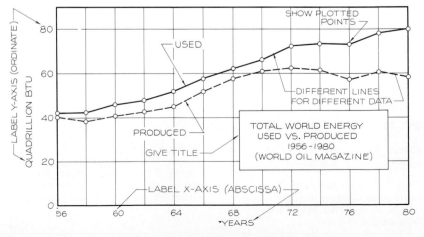

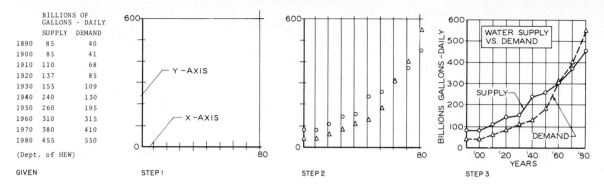

BILLIONS OF GALLONS - DAILY		
	SUPPLY	DEMAND
1890	85	40
1900	85	41
1910	110	68
1920	137	85
1930	155	109
1940	240	130
1950	260	195
1960	310	315
1970	380	410
1980	455	550

(Dept. of HEW)

GIVEN   STEP I   STEP 2   STEP 3

FIG. 33.14 Construction of a broken-line graph.

Given A record of water supply and water demand since 1890 plotted as a line graph.

Step 1 The vertical and horizontal axes are laid off to provide space for the largest values.

Step 2 The points are plotted over the respective years. Different symbols are used for each curve.

Step 3 The data points are connected with straight lines, the axes are labeled, the graph is titled, and the lines are strengthened.

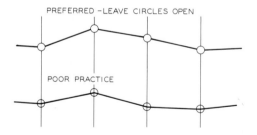

FIG. 33.15 The curve of a graph should be drawn from point to point, but it should not close up the symbols used to locate the plotted points.

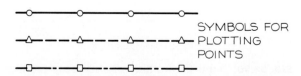

FIG. 33.16 Any of these symbols or lines can be effectively used to represent different curves on a single graph. The symbols are about $\frac{1}{8}$" (4 mm) in diameter.

Broken-line graphs

The steps involved in drawing a linear coordinate graph are shown in Fig. 33.14. Because the data points are ten years apart on the *X*-axis, it is impossible to assume the change in the data is a smooth, continual progression from point to point. Therefore the points are connected with a broken-line curve.

For the best appearance the plotted points should not be crossed by the curve or the grid lines of the graph (Fig. 33.15). Each circle or symbol used to plot points should be about $\frac{1}{8}$ in. (5 mm) in diameter. Several approved symbols and lines are shown in Fig. 33.16.

The title of a graph can be placed in any of the positions shown in Fig. 33.17. The title should never be one as meaningless as "graph" or "coordinate graph." Instead, it should explain the graph by giving the important information such as the company, date, source of data, and general comparisons being shown.

The proper calibration and labeling of axes is important to the appearance and usage of a graph. Figure 33.18A shows a properly executed axis. The axis in Fig. 33.18B has too many grid lines and too many units labeled along the axis. The units selected in Fig. 33.18C make it difficult to interpolate between the labeled values; for example, it is more difficult to locate a value such as 22 by eye on this scale than on the one in Fig. 33.18A.

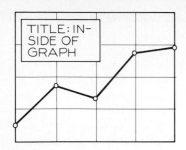
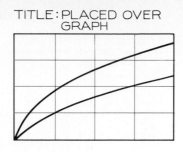
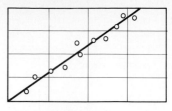

FIG. 33.17 Title placement on a graph.

A. The title of a graph can be lettered inside a box placed within the area of the graph. The perimeter lines of the box should not coincide with grid lines within the graph.

B. The title can be placed over the graph. The title should be drawn in $\frac{1}{8}$" letters or slightly larger.

C. The title can be placed under the graph. It is good practice to be consistent when a series of graphs is used in the same report.

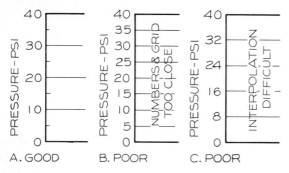

FIG. 33.18 The scale at A is the best. It has about the right number of grid lines and divisions, and the numbers are given in well-spaced, easy-to-interpolate form. The numbers at B are too close and there are too many grid lines. The units at C make interpolation difficult by eye.

Smooth-line graphs

The strength of cement as related to curing time has been plotted in Fig. 33.19. Since you know that the strength of cement changes gradually in relationship to curing time, the data points are connected with a smooth curve rather than a broken-line curve. Even if the data points do not lie on the curve, you can be reasonably certain the deviation is due to errors of measurement or the methods used in collecting the data.

Similarly, the strength of clay tile, as related to its absorption characteristics, is an example of data that yield a smooth curve (Fig. 33.20). Note that the plot-

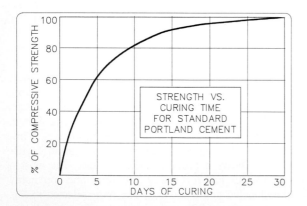

FIG. 33.19 When the process that is graphed involves gradual, continuous changes in relationships, the curve should be drawn as a smooth-line.

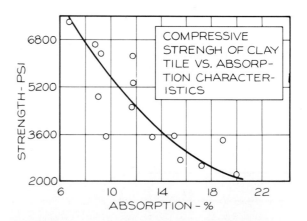

FIG. 33.20 If it is known that a relationship plotted in a graph should yield a smooth gradual curve, a smooth-line "best" curve is drawn to represent the average of the plotted points. You must use your judgment and knowledge of the data in cases of this type.

658

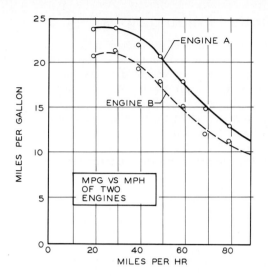

FIG. 33.21 These are "best" curves that approximate the data without necessarily passing through each data point. Inspection of the data tells you that this curve should be a smooth-line curve rather than a broken-line curve.

driven. In Fig. 33.21, two engines are compared with two smooth-line curves. The effect of speed on several automotive characteristics is compared in Fig. 33.22.

> When a smooth-line curve is used to connect data points, the implication is that you can *interpolate* between the plotted points to estimate other values. Points connected by a broken-line curve imply you cannot interpolate between the plotted points.

Straight-line graphs

Some graphs have neither broken-line curves nor smooth-line curves but straight-line curves as shown in Fig. 33.23. Using this graph, you can determine a third value from the two given values. For example, if you are driving 70 miles per hour and it takes 5 seconds to react and apply your brakes, you will have traveled 500 feet in this time.

ted data do not lie on the curve. Since you know this relationship should be represented by a smooth curve, the *best curve* is drawn to interpret the data to give an average representation of the points.

There is a smooth-line curve relationship between miles per gallon and the speed at which a car is

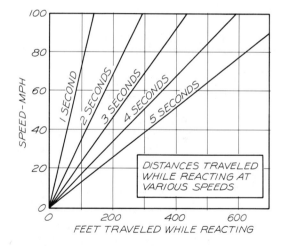

FIG. 33.23 A graph can be used to determine a third value when two variables are known. Taking this information from a graph is easier than computing each answer separately.

Two-scale coordinate graphs

Graphs can be drawn with different scales in combination, such as the one shown in Fig. 33.24. The vertical scale at the left is in units of pounds, and the one at the right is in degrees of temperature. Both curves

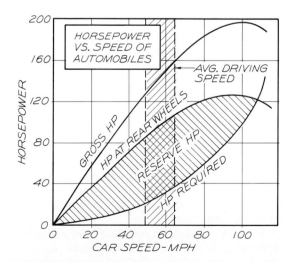

FIG. 33.22 A linear coordinate graph is used here to analyze data affecting the design of an automobile's power system.

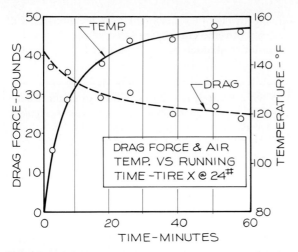

FIG. 33.24 This is a composite graph with different scales along each *Y*-axis. The curves are labeled so reference can be made to the applicable scale.

are drawn using their respective *Y*-axes, and each curve is labeled.

With graphs of this type, care must be taken to avoid confusing the reader. These graphs are effective when comparing related variables, such as the drag force and air temperature of a tire, as shown in this example.

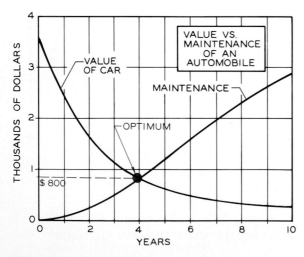

FIG. 33.25 This graph shows the optimum time to sell a car based on the intersection of two curves that represent the depreciation of the car's value and its increasing maintenance costs.

Optimization graphs

The optimization of the depreciation of an automobile and its increase in maintenance costs is shown in Fig. 33.25. These two sets of data are plotted, and the curves cross at an *X*-axis value of four years. At this time, the cost of maintenance is equal to the value of the car, indicating this might be a desirable time to exchange it for a new one.

Another optimization graph is constructed in Fig. 33.26. The manufacturing cost per unit is reduced as more units are made, but the warehousing cost increases. By adding the two curves, a third curve is found in Step 2. The "total" curve tells you that the optimum number to manufacture at a time is about 11,000 units. When more or fewer units are manufactured, the expense per unit is greater.

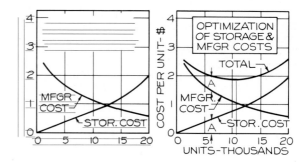

FIG. 33.26 Optimization graphs.

Step 1 Lay out the graph and plot the given curves.

Step 2 Add the two curves to find a third curve. Distance A is shown transferred to locate a point on the third curve. The lowest point of the "total" curve is the optimum point of 11,000 units.

Composite graphs

The graph in Fig. 33.27 is a composite (or combination) of an area graph and a coordinate graph. The lower curve is plotted first. The upper curve is found by adding the values to the lower curve so that the two areas represent the data. The upper curve is equal to the sum of the two *Y*-values.

Figure 33.28 shows a combination of a coordinate graph and a bar graph used to illustrate the Dow Jones industrial stock average. The bars represent the daily ranges in the index. The broken-line curve connects the points where the market closed for each day.

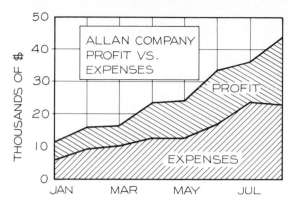

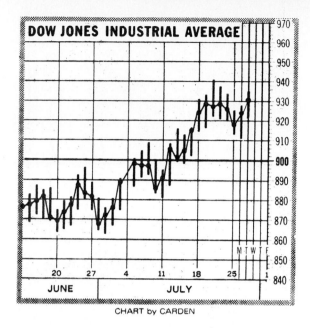

FIG. 33.27 This graph is a combination of a coordinate graph and an area graph. The upper curve represents the total of two values plotted, one above the other.

Break-even graphs

Break-even graphs are helpful in evaluating the marketing and manufacturing costs used to determine the selling cost of a product. The break-even graph in Fig. 33.29 reveals that 10,000 units must be sold at $3.50 each to cover the costs of manufacturing and development. Sales in excess of 10,000 result in profit.

FIG. 33.28 This graph is a combination of a coordinate graph and a bar graph. The bars represent the ranges of selling during a day, and the broken-line curve connects the points at which the market closed each day.

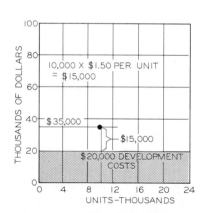

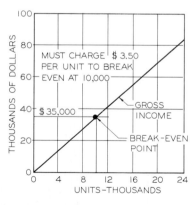

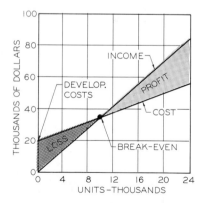

FIG. 33.29 Break-even graph.

Step 1 The graph is drawn to show the cost ($20,000 in this case) of developing the product. Each unit would cost $1.50 to manufacture. This is a total investment of $35,000 for 10,000 units.

Step 2 To break even at 10,000, the units must be sold for $3.50 each. Draw a line from zero through the break-even point for $35,000.

Step 3 The loss is $20,000 at zero units and becomes progressively less until the break-even point is reached. The profit is the difference between the cost and income to the right of the break-even point.

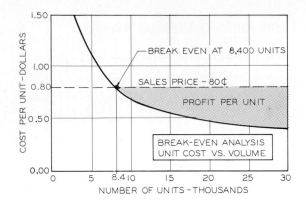

FIG. 33.30 The break-even point can be found on a graph that shows the relationship between the cost per unit, which includes the development cost, and the number of units produced. The sales price is a fixed price. The break-even point is reached when 8400 units have been sold at 80¢ each.

A second type of break-even graph (Fig. 33.30) uses the cost of manufacturing per unit versus the number of units produced. In this example, the development costs must be incorporated into the unit costs. The manufacturer can determine how many units

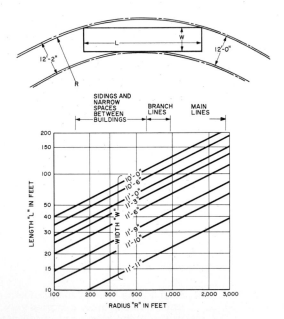

FIG. 33.31 This logarithmic graph shows the maximum load projection of 12 feet in relation to the length of a railroad car and the radius of the curve. (Courtesy of Plant Engineering.)

must be sold to break even at a given price or the price per unit if a given number is selected. In this example, a sales price of $0.80 requires that 8400 units be sold to break even.

33.6
Logarithmic coordinate graphs

Both scales of a logarithmic grid are calibrated into divisions equal to the logarithms of the units represented. Commercially printed logarithmic grid paper is available in many variations for graphing data.

The graph in Fig. 33.31 has a logarithmic grid and shows the geometry of standard railroad cars as they relate to the tracks to not exceed projection width of 12 feet around curves. Extremely large values can be shown on logarithmic grids since the lengths are considerably compressed.

33.7
Semilogarithmic coordinate graphs

Semilogarithmic graphs are referred to as *ratio graphs* because they give graphical representations of ratios.

One scale, usually the vertical scale, is logarithmic, and the other is linear (divided into equal divisions). Parallel curves on a semilogarithmic graph have equal percentage increases.

Figure 33.32 shows the same data plotted on a linear grid and on a semilogarithmic grid. The semilogarithmic graph reveals that the percent of change from 0 to 5 is greater for curve *B* than for curve *A*, since curve *B* is steeper. This comparison was not apparent in the plot on the linear grid.

The relationship between the linear scale and the logarithmic scale is shown in Fig. 33.33. Equal divisions along the linear scale have unequal ratios, and equal divisions along the log scale have equal ratios.

Log scales can be drawn to have one or many cycles. Each cycle increases by a factor of 10. For exam-

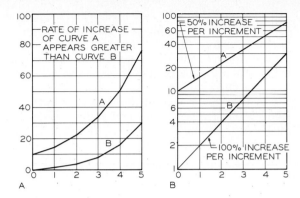

FIG. 33.32 When plotted on a standard grid, curve A appears to be increasing at a greater rate than curve B. However, the true rate of change can be seen when the same data are plotted on a semilogarithmic graph in part B.

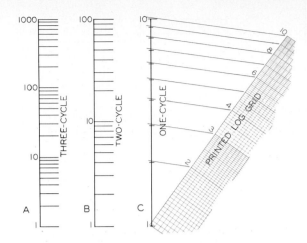

FIG. 33.34 Logarithmic paper can be purchased or drawn using several cycles. Three-, two-, and one-cycle scales are shown here. Calibrations can be drawn on a scale of any length by projecting from a printed scale as shown in part C.

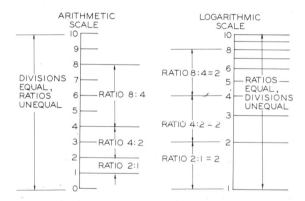

FIG. 33.33 The spacings on an arithmetic scale are equal with unequal ratios between points. The spacings on logarithmic scales are unequal, but equal spaces represent equal ratios.

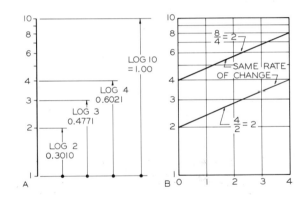

FIG. 33.35 **A.** A number's logarithm is used to locate its position on a log scale. **B.** This makes it possible to see the true rate of change at any location on a semilogarithmic graph.

ple, the scale in Fig. 33.34A is a three-cycle scale, and the one in Fig. 33.34B is a two-cycle scale. When these must be drawn to a special length, commercially printed log scales can be used to graphically transfer the calibrations to the scale being drawn (Fig. 33.34C).

In Fig. 33.35, the calibrations along the log scale are separated by the difference in their logarithms. The logarithms are laid off using a scale calibrated in decimal divisions. It can be seen in Fig. 33.35B that parallel straight-line curves yield equal ratios of increase. Figure 33.36 is an example of a semilogarithmic graph used to present industrial data.

Semilog graphs are sometimes misunderstood by people who do not realize they are different from linear coordinate graphs; also, zero values cannot be shown on log scales.

Percentage graphs

The percent that one number is of another, or the percent increase of one number that is greater than the other, can be determined by using a semilogarithmic graph (Fig. 33.37).

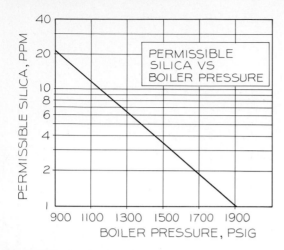

FIG. 33.36 A semilogarithmic graph is used to compare the permissible silica (parts per million) in relation to the boiler pressure.

Data plotted in Step 1 are used to find the percent that 30 is of 60, two points on the curve. The vertical distance between them is equal to the difference of their logarithms. This distance is subtracted from the log of 100 at the right of the graph to give a value of 50% as a direct reading.

In Step 2, the percent of increase between two points is transferred from the grid to the lower end of the log scale and measured upward since the increase is greater than zero. These methods can be used to find percent increases or decreases of any set of points on the grid.

33.8
Polar graphs

Polar graphs are drawn with a series of concentric circles with the origin at the center. Lines are drawn from the center toward the perimeter of the graph,

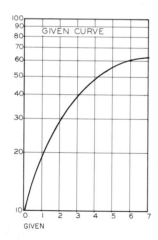
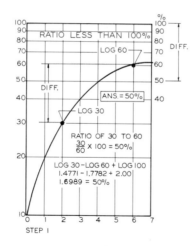
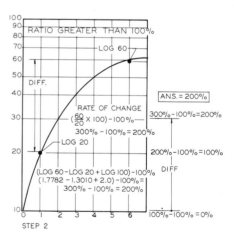

FIG. 33.37 Percentage graphs.

Given The data are plotted on a semilogarithmic graph to enable you to determine percentages and ratios in much the same manner that you use a slide rule.

Step 1 In finding the percent that a smaller number is of a larger number, you know that the percent will be less than 100%. The log of 30 is subtracted from the log of 60 with dividers and is transferred to the percent scale at the right, where 30 is found to be 50% of 60.

Step 2 To find the percent of increase, a smaller number is divided into a larger number to give a value greater than 100%. The difference between the logs of 60 and 20 is found with dividers and is measured upward from 100% at the right to find that the percent of increase is 200%.

where the data can be plotted through 360° by measuring values from the origin. For example, the illumination of a lamp is shown in Fig. 33.38, where the maximum lighting of the lamp is 550 lumens at 35° from the vertical.

This type of graph is used to plot the areas of illumination of all types of lighting fixtures and other applications. Polar graph paper is available commercially for drawing graphs of this type.

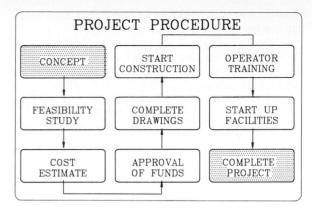

FIG. 33.39 This schematic shows a block diagram of the steps required to complete a project.

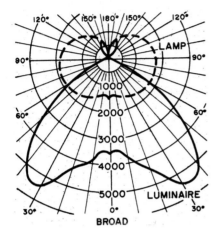

FIG. 33.38 A polar graph is used to show the illumination characteristics of luminaires.

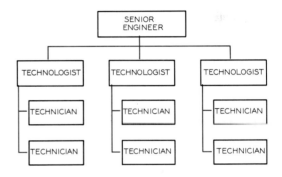

FIG. 33.40 This schematic shows the organization of a design team in a block diagram.

33.9
Schematics

The *block diagram* in Fig. 33.39 shows the progression of steps required to complete a construction project. Each step is blocked in and connected with arrows to explain the sequence of events.

The organization of a company or a group of people can be depicted in an *organizational chart* of the type shown in Fig. 33.40. The offices represented by the blocks in the lower part of the graph are responsible to the offices represented by the blocks above them. The lines of authority connecting the blocks suggest the routes for communication from one office to another in an upward or downward direction.

The drawing in Fig. 33.41 is not a graph, nor is it a true view of the apparatus; instead, it is a *schematic*

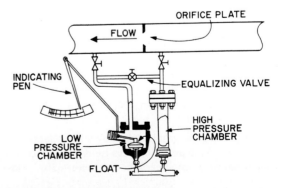

FIG. 33.41 A schematic showing the components of a gauge that measures the flow in a pipeline. (Courtesy of Plant Engineering.)

that effectively shows how the parts and their functions relate to one another.

Geographical graphs are used to combine maps and other relationships such as weather (Fig. 33.42). Different symbols represent the annual rainfall in various areas of the nation.

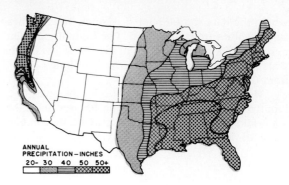

ANNUAL
PRECIPITATION–INCHES
20– 30 40 50 50+

FIG. 33.42 A map chart that shows the weather characteristics of various geographical areas. (Courtesy of the Structural Clay Products Institute.)

Problems

The problems below are to be drawn on Size A sheets in pencil or ink, as specified. Follow the techniques covered in this chapter and the examples given as you solve the problems.

Pie graphs

1. Draw a pie graph that compares the employment of male youth between the ages of 16 and 21: Operators–25%; Craftsmen–9%; Professionals, technicians, and managers–6%; Clerical and sales–17%; Service–11%; Farm workers–13%; and Laborers–19%.

2. Draw a pie graph that shows the relationship between the following members of the technological team: Engineers–985,000; Technicians–932,000; Scientists–410,000.

3. Construct a pie graph of the following percentages of the employment status of graduates of two-year technician programs one year after graduation: Employed–63%, Continuing full-time study–23%; Considering job offers–6%; Military–6%; Other–2%.

4. Construct a pie graph that shows the relationship between the types of degrees held by engineers in aeronautical engineering: Bachelor–65%; Master–29%; Ph.D.–6%.

5. Draw a pie graph for the following average annual expenditures of a state on public roads: Construction–$13,600,000; Maintenance–$7,100,000; Purchase and upkeep of equipment–$2,900,000; Bonds–$5,600,000; Engineering and administration–$1,600,000.

6. Draw a pie graph that shows the data given in Problem 10.

7. Draw a pie graph that shows the data given in Problem 11.

Bar graphs

8. Draw a bar graph that depicts the unemployment rate of high school graduates and dropouts in various age categories given in the following table:

Ages	Percent of Labor Force	
	Graduates	Dropouts
16–17	18	22
18–19	12.5	17.5
20–21	8	13
22–24	5	9

9. Draw a single bar that represents 100% of a die casting alloy. The proportional parts of the alloy are as follows: Tin–16%; Lead–24%; Zinc–38.8%; Aluminum–16.4%; Copper–4.8%.

10. Draw a bar graph that compares the number of skilled workers employed in various occupations. Arrange the graph for ease of interpretation and comparison of occupations. Use the following data: Carpenters–82,000; All-round machinists–310,000; Plumbers–350,000; Bricklayers–200,000; Appliance servicers–185,000; Automotive mechanics–760,000; Electricians–380,000; Painters–400,000.

11. Draw a bar graph that represents the flow of a river in cubic feet per second (cfs) as shown in the following table. Show bars that represent the data for ten days in the first month. Omit the second month.

Day of Month	Rate of Flow (in 100 cfs)	
	1st Month	2nd Month
1	19	19
2	130	70
3	228	79
4	129	33
5	65	19
6	32	14
7	17	15
8	13	11
9	22	19
10	32	27

12. Draw a bar graph that shows the airline distances in statute miles from New York to the cities listed in the table below. Arrange in ascending or descending order.

Berlin 3965

Buenos Aires 5300

Honolulu 4960

London. 3465

Manila 8510

Mexico City 2090

Moscow 4665

Paris. 3634

Tokyo 6740

13. Draw a bar graph that compares the corrosion resistance of the materials listed in the table below.

	Loss in Weight (%)	
	In Atmosphere	In Sea Water
Common Steel	100	100
10% Nickel Steel	70	80
25% Nickel Steel	20	55

14. Draw a bar graph using the data in Problem 1.

15. Draw a bar graph using the data in Problem 2.

16. Draw a bar graph using the data in Problem 3.

17. Construct a rectangular grid graph to show the accident experience of Company A. Plot the numbers of disabling accidents per million person-hours of work on the Y-axis. Years will be plotted on the X-axis. Data: 1973–0.63; 1974–0.76; 1975–0.99; 1976–0.95; 1977–0.55; 1978–0.76; 1979–0.68; 1980–0.55; 1981–0.73; 1982–0.52; 1983–0.46; 1984–0.53; 1985–0.49; 1986–0.55.

Linear coordinate graphs

18. Using the data given in Table 33.1, draw a linear coordinate graph that compares the supply and demand of water in the United States from 1890 to 1990 in billions of gallons of water per day.

19. Present the data in Table 33.2 in a linear coordinate graph to decide which lamps should be selected to provide economical lighting for an industrial plant. The table gives the candlepower directly under the lamps (0°) and at various an-

TABLE 33.1

	1890	1900	1910	1920	1930	1940	1950	1960	1970	1980	1990
Supply	80	90	110	135	155	240	270	315	380	450	460
Demand	35	35	60	80	110	125	200	320	410	550	570

TABLE 33.2

Angle with vertical	0°	10°	20°	30°	40°	50°	60°	70°	80°	90°
Candlepower (thous.) 2–400W	37	34	25	12	5.5	2.5	2	0.5	0.5	0.5
Candlepower (thous.) 1–1000W	22	21	19	16	12.3	7	3	2	0.5	0.5

gles from the vertical when the lamps are mounted at a height of 25 feet.

20. Construct a linear coordinate graph that shows the relationship in energy costs (mills per kilowatt-hour) and the percent capacity of two types of power plants. Plot energy costs along the Y-axis, and the capacity factor along the X-axis. The plotted curve will compare the costs of a nuclear plant with a gas- or oil-fired plant. Gas-fired plant data: 17 mills, 10%; 12 mills, 20%; 8 mills, 40%; 7 mills, 60%; 6 mills, 80%; 5.8 mills, 100%. Nuclear-plant data: 24 mills, 10%; 14 mills, 20%; 7 mills, 40%; 5 mills, 60%; 4.2 mills, 80%; 3.7 mills, 100%.

21. Plot the data from Problem 17 as a linear coordinate graph.

22. Construct a linear coordinate graph to show the relationship between the transverse resilience in inch-pounds (Y-axis) and the single-blow impact in foot-pounds (X-axis) of gray iron. Data: 21 fp, 375 ip; 22 fp, 350 ip; 23 fp, 380 ip; 30 fp, 400 ip; 32 fp, 420 ip; 33 fp, 410 ip; 38 fp, 510 ip; 45 fp, 615 ip; 50 fp, 585 ip; 60 fp, 785 ip; 70 fp, 900 ip; 75 fp, 920 ip.

23. Draw a linear coordinate graph to compare the two sets of data in the following table: capacity vs. diameter, and capacity vs. weight of a brine cooler. The horizontal scale is to be tons of capacity, and the vertical scales are to be outside diameter on the left, and weight (cwt) on the right.

Tons Refrigerating Capacity	Outside Diameter (inches)	Weight (cwt)
15	22	25
30	28	46
50	34	73
85	42	116
100	46	136
130	50	164
160	58	215
210	60	263

Use 20 × 20 graph paper 8½″ × 11″. Horizontal scale of 1″ = 40 tons. Vertical scales of 1″ = 10″ of outside DIA, and 1″ = 40 cwt (hundred weight).

24. Draw a linear coordinate graph that shows the voltage characteristics for a generator as given in the following table of values: Abscissa-armature current in amperes (I_a); ordinate-terminal voltage in volts (E_t).

I_a	E_t	I_a	E_t	I_a	E_t
0	288	31.1	181.8	41.5	68
5.4	275	35.4	156	40.5	42.5
11.8	257	39.7	108	39.5	26.5
15.6	247	40.5	97	37.8	16
22.2	224.5	40.7	90	13.0	0
26.2	217	41.4	77.5		

25. Draw a linear coordinate graph for the centrifugal pump test data in the table below. The units along the X-axis are to be gallons per minute. There will be four curves to represent the variables given.

Gallons Per Min.	Discharge Pressure	Water HP	Electric HP	Efficiency %
0	19.0	0.00	1.36	0.00
75	17.5	0.72	2.25	32.0
115	15.0	1.00	2.54	39.4
154	10.0	1.00	2.74	36.5
185	5.0	0.74	2.80	26.5
200	3.0	0.63	2.83	22.2

26. Draw a linear coordinate graph that compares two of the values shown in Table 33.3—ultimate strength and elastic limit—with degrees of temperature labeled along the X-axis.

27. Draw a linear coordinate graph that compares two of the values shown in the table in Problem 26—percent of elongation and percent of reduction of area of the cross section—with the degrees of temperature that will be represented along the X-axis.

TABLE 33.3

°F	Ultimate Strength	Elastic Limit	Elongation %	Reduction of Area %	Brinell Hardness No.
400	257,500	208,000	10.8	31.3	500
500	247,000	224,500	12.5	39.5	483
600	232,500	214,000	13.3	42.0	453
700	207,500	193,500	15.0	47.5	410
800	180,500	169,000	17.0	52.5	358
900	159,500	146,500	18.5	56.5	313
1000	142,500	128,500	20.3	59.2	285
1100	126,500	114,000	23.0	60.8	255
1200	114,500	96,500	26.3	67.8	230
1300	108,000	85,500	25.8	58.3	235

Break-even graphs

28. Construct a break-even graph that shows the earnings for a new product that has a development cost of $12,000. The first 8000 will cost $0.50 each to manufacture, and you wish to break even at this quantity. What would be the profit at volumes of 20,000 and 25,000?

29. Same as Problem 28 except that the development costs are $80,000, the manufacturing cost of the first 10,000 is $2.30 each, and the desired break-even point is 10,000. What would be the profit at volumes of 20,000 and 30,000?

30. A manufacturer has incorporated the manufacturing and development costs into a cost-per-unit estimate. He wishes to sell the product at $1.50 each. On the Y-axis, plot cost per unit in dollars; on the X-axis, plot number of units in thousands. Data: 1000, $2.55; 2000, $2.01; 3000, $1.55; 4000, $1.20; 5000, $0.98; 6000, $0.81; 7000, $0.80; 8000, $0.75; 9000, $0.73; 10,000, $0.70. How many must be sold to break even? What will be the total profit when 9000 are sold?

31. The cost per unit to produce a product by a manufacturing plant is given below. Construct a break-even graph with the cost per unit plotted on the Y-axis and the number of units on the X-axis. Data: 1000, $5.90; 2000, $4.50; 3000, $3.80; 4000, $3.20; 5000, $2.85; 6000, $2.55; 7000, $2.30; 8000, $2.17; 9000, $2.00; 10,000, $0.95.

Logarithmic graphs

32. Using the data given in Table 33.4, construct a logarithmic graph where the vibration amplitude (A) is plotted as the ordinate and the vibration frequency (F) as the abscissa. The data for curve 1 represent the maximum limits of machinery in good condition with no danger from vibration. The data for curve 2 are the lower limits of machinery that is being vibrated excessively to the danger point. The vertical scale should be three cycles and the horizontal scale two cycles.

33. Plot the data below on a two-cycle log graph to show the current in amperes (Y-axis) versus the voltage in volts (X-axis) of precision temperature-sensing resistors. Data: 1 volt, 1.9 amps; 2 volts, 4 amps; 4 volts, 8 amps; 8 volts, 17 amps; 10 volts, 20 amps; 20 volts, 30 amps; 40 volts, 36 amps; 80 volts, 31 amps; 100 volts, 30 amps.

TABLE 33.4

F	100	200	500	1000	2000	5000	10,000
A(1)	0.0028	0.002	0.0015	0.001	0.0006	0.0003	0.00013
A(2)	0.06	0.05	0.04	0.03	0.018	0.005	0.001

34. Plot the data from Problem 18 as a logarithmic graph.

35. Plot the data from Problem 24 as a logarithmic graph.

Semilogarithmic graphs

36. Construct a semilogarithmic graph with the Y-axis a two-cycle log scale from 1 to 100 and the X-axis a linear scale from 1 to 7. Plot the data below to show the survivability of a shelter at varying distances from a one-megaton bomb exploding in air. The data consists of overpressure in psi along the Y-axis, and distance from ground zero in miles along the X-axis. The data points represent an 80% chance of survival of the shelter. Data: 1 mile, 55 psi; 2 miles, 11 psi; 3 miles, 4.5 psi; 4 miles, 2.5 psi; 5 miles, 2.0 psi; 6 miles, 1.3 psi.

37. The growth of two divisions of a company, Division A and Division B, is given in the data below. Plot the data on a rectilinear graph and on a semilog graph. The semilog graph should have a one-cycle log scale on the Y-axis for sales in thousands of dollars, and a linear scale on the X-axis showing years for a six-year period. Data in dollars: 1 yr, A = \$11,700 and B = \$44,000; 2 yr, A = \$19,500 and B = \$50,000; 3 yr, A = \$25,000 and B = \$55,000; 4 yr, A = \$32,000 and B = \$64,000; 5 yr, A = \$42,000 and B = \$66,000; 6 yr, A = \$48,000 and B = \$75,000. Which division has the better growth rate?

38. Draw a semilog chart showing probable engineering progress. Use the following indices: 40,000 B.C. = 21; 30,000 B.C. = 21.5; 20,000 B.C. = 22; 16,000 B.C. = 23; 10,000 B.C. = 27; 6000 B.C. = 34; 4000 B.C. = 39; 2000 B.C. = 49; 500 B.C. = 60; A.D. 1900 = 100. Horizontal scale 1″ = 10,000 years. Height of cycle = about 5″. Two-cycle printed paper may be used if available.

39. Plot the data from Problem 24 as a semilogarithmic graph.

40. Plot the data from Problem 26 as a semilogarithmic graph.

Percentage graphs

41. Plot the data given in Problem 18 on a semilog graph to determine the percentages and ratios of the data. What is the percent of increase in the demand for water from 1890 to 1920? What percent of demand is the supply for 1900, 1930, and 1970?

42. Using the graph plotted in Problem 37, determine the percent of increase of Division A and Division B from Year 1 to Year 4. What percent of sales of Division A are the sales of Division B at the end of Year 2? At the end of Year 6?

43. Plot two values from Problem 26—water horsepower and electric horsepower—on semilog paper compared with gallons per minute along the x-axis. What percent of electric horsepower is water horsepower when 75 gallons per minute are being pumped? What is the percent increase of the electric horsepower from 0 to 185 gallons per minute?

Organizational charts

44. Draw an organization chart for a city government organized as follows: The electorate elects school board, city council, and municipal court officers; the city council is responsible for the civil service commission, city manager, and city planning board; the city manager's duties cover finance, public safety, public works, health and welfare, and law.

45. Draw an organization chart for a manufacturing plant. The sales manager, chief engineer, treasurer, and general manager are responsible to the president. The general manager has three department heads: master mechanic, plant superintendent, and purchasing agent. The plant superintendent has charge of the shop foremen, under whom are the working forces, as well as having direct charge of the shipping, tool and die, inspection, order, and stores and supplies departments.

Polar graphs

46. Construct a polar graph of the data given in Problem 19.

47. Construct a polar graph of the following illumination, in lumens at various angles, emitted from a luminaire. The zero-degrees position is vertically under the overhead lamp. Data: 0°, 12,000; 10°, 15,000; 20°, 10,000; 30°, 8000; 40°, 4200; 50°, 2500; 60°, 1000; 70°, 0. The illumination is symmetrical about the vertical.

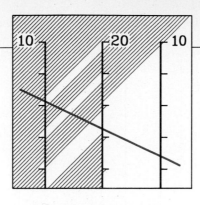

CHAPTER 34

Nomography

34.1
Nomography

An additional aid in analyzing data is a graphical computer called a *nomogram* or *nomograph*. Basically, a nomogram, or "number chart," is any graphical arrangement of calibrated scales and lines that may be used to facilitate calculations, usually those of a repetitive nature.

The term "nomogram" frequently denotes a specific type of scale arrangement called an *alignment chart*. Typical examples of alignment charts are shown in Fig. 34.1. Many other types are also used that have curved scales or other scale arrangements for more complex problems. The discussion of nomograms in this chapter will be limited to the simpler conversion, parallel-scale, and N-type graphs and their variations.

Using an alignment graph

An alignment graph is usually constructed to solve for one or more unknowns in a formula or empirical relationship between two or more quantities; for example, it can be used to convert degrees Celsius to degrees Fahrenheit or to find the size of a structural member to sustain a certain load. An alignment chart is read by placing a straightedge, or by drawing a line called an *isopleth*, across the scales of the chart and reading corresponding values from the scale on this line. The example in Fig. 34.2 shows readings for the formula $U + V = W$.

34.2
Alignment-graph scales

To construct any alignment graph, you must first determine the graduations of the scales that will give the desired relationships. Alignment-graph scales are called *functional scales*. A functional scale is graduated according to values of some function of a variable. A functional scale for $F(U) = U^2$ is illustrated in Fig. 34.3. It can be seen in this example that if a value of $U = 2$ was substituted into the equation, the position of U on the functional scale would be 4 units from zero, or $2^2 = 4$. This procedure can be repeated with all values of U by substitution.

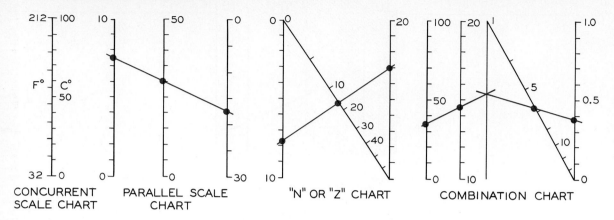

CONCURRENT SCALE CHART   PARALLEL SCALE CHART   "N" OR "Z" CHART   COMBINATION CHART

FIG. 34.1 Typical examples of types of alignment charts.

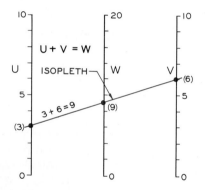

FIG. 34.2 Use of an isopleth to solve graphically for unknowns in the given equation.

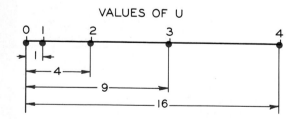

FIG. 34.3 Functional scale for units of measurement proportional to $F(U) = U^2$.

portionality is called the *scale modulus,* and it is given by the equation

$$m = \frac{L}{F(U_2) - F(U_1)},\qquad (1)$$

where

m = scale modulus, in inches per functional unit,

L = desired length of the scale, in inches,

$F(U_2)$ = functional value at the end of the scale,

$F(U_1)$ = functional value at the start of the scale.

For example, suppose we are to construct a functional scale for $F(U) = \sin U$ with $0° \le U \le 45°$ and a scale 6" in length. Thus, $L = 6"$, $F(U_2) = \sin 45° = 0.707$, $F(U_1) = \sin 0° = 0$. Therefore Eq. (1) can be written in the following form by substitution:

$$m = \frac{6}{0.707 - 0} = 8.49 \text{ inches per (sine) unit.}$$

The scale modulus

Since the graduations on a functional scale are spaced in proportion to values of the function, a proportionality, or scaling factor, is needed. This constant of pro-

The scale equation

Graduation and calibration of a functional scale are made possible by a *scale equation.* The general form of this equation may be written as a variation of Eq. (1)

in the following form:

$$X = m[F(U) - F(U_1)], \tag{2}$$

where

X = distance from the measuring point of the scale to any graduation point,

m = scale modulus,

$F(U)$ = functional value at the graduation point,

$F(U_1)$ = functional value at the measuring point of the scale.

For example, a functional scale is constructed for the previous equation, $F(U) = \sin U$ ($0° \leq U \leq 45°$). It has been determined that $m = 8.49$, $F(U) = \sin U$, and $F(U_1) = \sin 0° = 0$. Thus, by substitution (2) becomes

$$X = 8.49 (\sin U - 0) = 8.49 \sin U.$$

Using the equation, we can substitute values of U and construct a table of positions. In this case, the scale is calibrated at 5° intervals, as reflected in Table 34.1.

The values of X from the table give the positions, in inches, for the corresponding graduations, measured from the start of the scale ($U = 0°$); see Fig. 34.4. Note that the measuring point does *not* need to be at one end of the scale, but it is usually the most

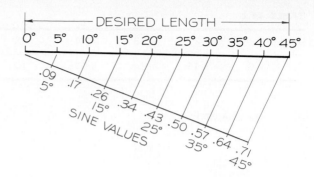

FIG. 34.5 A functional scale that shows the sine of the angles from 0° to 45° can be drawn graphically by the proportional-line method. The scale is drawn to a desired length and the sine values of angles at 5° intervals are laid off along a construction line that passes through the 0° end of the scale. These values are projected back to the scale.

convenient point, especially if the functional value is zero at that point.

A graphical method of locating the functional values along a scale can be found as shown in Fig. 34.5 by the proportional-line method. The sine functions are measured off along a line at 5° intervals, with the end of the line passing through the 0° end of the scale. The functions are transferred from the inclined line with parallel lines back to the scale where the functions are represented and labeled.

34.3
Concurrent scales

Concurrent scales are useful in the rapid conversion of one value into terms of a second system of measurement. Formulas of the type $F_1 - F_2$, which relate two variables, can be adapted to the concurrent-scale format. Typical examples might be the Fahrenheit–Celsius temperature relation,

$$°F = \frac{9}{5} °C + 32,$$

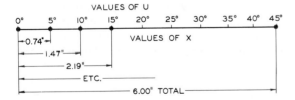

FIG. 34.4 Construction of a functional scale using values from Table 34.1, which were derived from the scale equation.

TABLE 34.1

U	0°	5°	10°	15°	20°	25°	30°	35°	40°	45°
X	0	0.74	1.47	2.19	2.90	3.58	4.24	4.86	5.45	6.00

TABLE 34.2

r	1	2	3	4	5	6	7	8	9	10
X_r	0	0.15	0.40	0.76	1.21	1.77	2.42	3.18	4.04	5.00

or the area of a circle,

$$A = \pi r^2.$$

Design of a concurrent-scale chart involves the construction of a functional scale for each side of the mathematical formula in such a manner that the *position* and *lengths* of each scale coincide. For example, to design a conversion chart 5 in. long that will give the areas of circles whose radii range from 1 to 10, we first write $F_1(A) = A$ $F_2(r) = \pi r^2$, and $r_1 = 1$, $r_2 = 10$. The scale modulus for r is

$$m_r = \frac{L}{F_2(r_2) - F_2(r_1)}$$
$$= \frac{5}{\pi(10)^2 - \pi(1)^2} = 0.0161.$$

Thus, the scale equation for r becomes

$$X_r = m_r\,[F_2(r) - F_2(r_1)]$$
$$= 0.0161[\pi r^2 - \pi(1)^2]$$
$$= 0.0161\pi(r^2 - 1)$$
$$= 0.0505(r^2 - 1).$$

A table of values for X_r and r may be completed as shown in Table 34.2. The r-scale can be drawn from this table, as shown in Fig. 34.6. From the original formula, $A = \pi r^2$, the limits of A are found to be $A_1 = \pi = 3.14$, and $A_2 = 100\pi = 314$. The scale modulus for concurrent scales is always the same for equal-length scales; therefore $m_A = m_r = 0.0161$, and the scale equation for A becomes

$$X_A = m_A[F_1(A) - F_1(A_1)]$$
$$= 0.0161\,(A - 3.14).$$

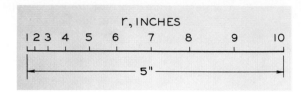

FIG. 34.6 Calibration of one scale of a concurrent scale chart using values from Table 34.2.

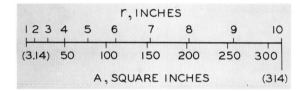

FIG. 34.7 The completed concurrent scale chart for the formula $A = \pi r^2$. Values for the A-scale are taken from Table 34.3.

The corresponding table of values is then computed for selected values of A, as shown in Table 34.3.

The A-scale is now superimposed on the r-scale; its calibrations have been placed on the other side of the line to facilitate reading (Fig. 34.7). If it is desired to expand or contract one of the scales, an alternative arrangement may be used, as shown in Fig. 34.8. The two scales are drawn parallel at any convenient distance, and calibrated in *opposite* directions. A different scale modulus and corresponding scale equation must be calculated for each scale if they are *not* the same length.

TABLE 34.3

A	(3.14)	50	100	150	200	250	300	(314)
X_A	0	0.76	1.56	2.36	3.16	3.96	4.76	5.00

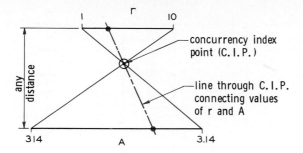

Fig. 34.8 Concurrent scale chart with unequal scales.

A graphical method can be used to construct concurrent scales, as shown in Fig. 34.9, by using the proportional-line method. Since there are 101.6 mm in 4 inches, the units of millimeters can be located on the upper side of the inch scale by projecting to the scale with a series of parallel projectors.

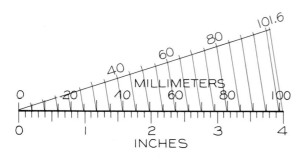

FIG. 34.9 The proportional-line method can be used to construct an alignment graph that converts inches to millimeters. This requires that the units at each end of the scales be known. For example, there are 101.6 millimeters in 4 inches.

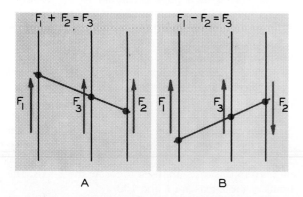

FIG. 34.10 Two common forms of parallel-scale alignment nomographs.

34.4
Construction of alignment graphs with three variables

For a formula of three functions (of one variable each), the general approach is to select the lengths and positions of *two* scales according to the range of variables and size of the chart desired. These are then calibrated by means of the scale equations, as shown in the preceding section. The position and calibration of the third scale will then depend on these initial constructions. Although definite mathematical relationships exist that may be used to locate the third scale, graphical constructions are simpler and usually less subject to error. Examples of the various forms are presented in the following sections.

34.5
Parallel-scale nomographs

Any formula of the type $F_1 + F_2 = F_3$ may be represented as a parallel-scale alignment chart, as shown in Fig. 34.10A. Note that all scales increase (functionally) in the same direction and that the function of the middle scale represents the *sum* of the other two. Reversing the direction of any scale changes the sign of its function in the formula, as for $F_1 - F_2 = F_3$ in Fig. 34.10B.

To illustrate this type of alignment graph, we will use the formula $Z = X + Y$, as shown in Fig. 34.11. The outer scales for X and Y are drawn and calibrated. They can be drawn to any length and positioned any distance apart, as shown in Fig. 34.12. Two sets of data that yield a Z of 8 in Fig. 34.11 (Step 1) are used to locate the parallel Z-scale. The Z-scale is drawn and divided into 16 equal units (Step 2). The finished nomograph (Step 3) can be used to add various values of X and Y to find their sums along the Z-scale.

A more complex alignment graph is illustrated in Fig. 34.13, where the formula $U + 2V = 3W$ is expressed in the form of a nomograph. First it is necessary to determine and calibrate the two outer scales for U and V; we can make them any convenient length and position them any convenient distance apart, as shown in Fig. 34.12. These scales are the basis for the construction shown in Fig. 34.13.

The limits of calibration for the middle scale are found by connecting the endpoints of the outer scales and substituting these values into the formula. Here,

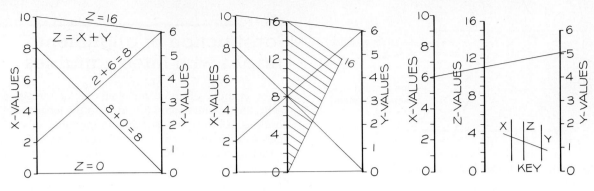

FIG. 34.11   Parallel-scale nomogram (linear).

Step 1   Two parallel scales are drawn at any length and calibrated. The location of the parallel Z-scale is found by selecting two sets of values that will give the same value of Z, 8 in this example. The ends of the Z-scale will have values of 0 and 16, the sum of the end values of X and Y.

Step 2   The Z-scale is drawn through the point located in Step 1 parallel to the other scales. The scale is calibrated from 0 to 16 by using the proportional-line method. Note that the two sets of X- and Y-values cross at 8, the sum of each set.

Step 3   The Z-scale is labeled and calibrated with easy-to-read units. A key is drawn to show how the nomograph is used. If the Y-scale were calibrated with 0 at the upper end instead of the bottom, a different Z-scale could be computed and the nomograph could be used for Z = X − Y.

W is found to be 0 and 10 at the extreme ends (Step 1). Two pairs of corresponding values of U and V are selected that will give the same value of W. For example, values of U = 0 and V = 7.5 give a value of 5 for W. We also find that W = 5 when U = 14 and V = 0.5. To verify this, we connect these corresponding pairs of values with isopleths to locate their intersection, which establishes the position of the W-scale.

Since the W-scale is linear (3W is a linear function), it may be subdivided into uniform intervals by

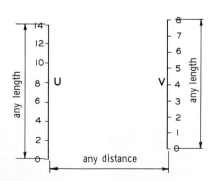

FIG. 34.12   Calibration of the outer scales for the formula $U + 2V = 3W$, where $0 \leq U \leq 14$ and $0 \leq V \leq 8$.

equal parts (Step 2). For a nonlinear scale, the scale modulus (and the scale equation) may be found in Step 2 by substituting length and its two end values into Eq. (1). The scales can be used to determine an infinite number of problem solutions (Step 3).

Parallel-scale graph with logarithmic scales

Problems involving formulas of the type $F_1 \times F_2 = F_3$ can be solved in a manner similar to the example given in Fig. 34.1 when logarithmic scales are used.

The first step in drawing a nomograph with logarithmic scales is learning how to transfer logarithmic functions to the scale. The graphical method is shown in Fig. 34.14, where units along the scale are found by projecting from a printed logarithmic scale with parallel lines.

The formula $Z = XY$ is converted into a nomograph in Fig. 34.15. The X- and Y-scales are drawn within the desired limits from 1 to 10 on each (Step 1). Sets of values of X and Y that yield the same value of Z, 10 in this case, are used to locate the Z-axis. The limits of the Z-axis are 1 and 100.

The Z-axis is drawn and calibrated as a two-cycle log scale (Step 2). A key is drawn to explain how an

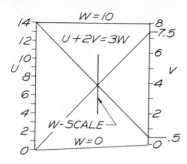
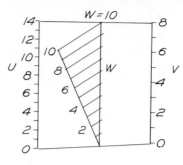
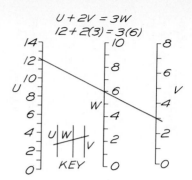

FIG. 34.13 Parallel-scale nomograph (linear).

Step 1 Substitute the end values of the U- and V-scales into the formula to find the end values of the W-scale: $W = 10$ and $W = 0$. Select two sets of U and V that will give the same value of W. *Example:* When $U = 0$ and $V = 7.5$, W will equal 5, and when $U = 14$ and $V = 0.5$, W will equal 5. Connect these sets of values; the intersection of their lines locates the W-scale.

Step 2 Draw the W-scale parallel to the outer scales; its length is controlled by the previously established lines of $W = 10$ and $W = 0$. Since this scale is 10 linear divisions long, divide it graphically into ten units as shown. This will be a linear scale constructed as shown in Fig. 34.11.

Step 3 The nomogram can be used as illustrated by selecting any two known variables and connecting them with an isopleth to determine the third unknown. A key is included to illustrate how the nomogram is to be used. An example of $U = 12$ and $V = 3$ is shown to verify the accuracy of the graph.

isopleth is used to add the logarithms of X and Y to give the log of Z (Step 3). When logarithms are added the result is multiplication. Had the Y-axis been calibrated in the opposite direction with 1 at the upper end and 10 at the lower end, a new Z-axis could have been calibrated, and the nomograph used for the formula, $Z = Y/X$, since it would be subtracting logarithms.

A more advanced example of this type of problem is the formula $R = S\sqrt{T}$, for $0.1 \le S \le 1.0$ and $1 \le$

$T \le 100$. Assume the scales to be 6 in. long. These scales need not be equal except for convenience. This formula may be converted into the required form by taking common logarithms of both sides, which gives

$$\log R = \log S + \frac{1}{2}\log T.$$

Thus we have

$$F_1(S) + F_2(T) = F_3(R), \qquad (3)$$

where $F_1(S) = \log S$, $F_2(T) = \frac{1}{2}\log T$, and $F_3(R) = \log R$. The scale modulus for $F_1(S)$ is from Eq. (3).

$$m_S = \frac{6}{\log 1.0 - \log 0.1} = \frac{6}{0 - (-1)} = 6 \qquad (4)$$

Choosing the scale measuring point from $S = 0.1$, we find from Eq. (4) that the scale equation for $F_1(S)$ is

$$X_S = 6(\log S - \log 0.1) = 6(\log S + 1). \qquad (5)$$

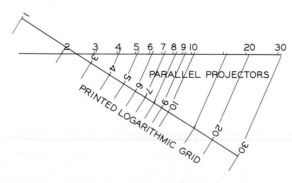

FIG. 34.14 Graphical calibration of a scale using logarithmic paper.

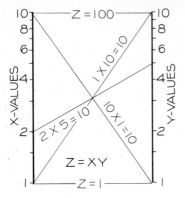
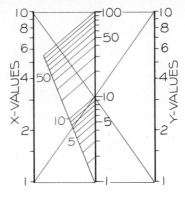
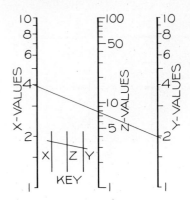

FIG. 34.15 Parallel-scale nomogram (log014thmic).

Step 1 For the equation $Z = XY$, parallel log scales are drawn that give the ranges of values for X and Y. Sets of X and Y points that yield the same value of Z, 10 in this example, are drawn. Their intersection locates the Z-scale with end values of 1 and 100.

Step 2 The Z-axis is graphically calibrated as a two-cycle logarithmic scale from 1 to 100. This scale is parallel to the X- and Y-scales.

Step 3 A key is drawn. An isopleth is drawn to show that 4 × 2 = 8. By reversing the Y-value scale from 1 downward to 10 and computing a different Z-scale, the nomogram could be used for the equation $Z = X/Y$.

Similarly, the scale modulus for $F_2(T)$ is

$$m_T = \frac{6}{\frac{1}{2}\log 100 - 1/2 \log 1}$$

$$= \frac{6}{\frac{1}{2}(2) - 1/2(0)} = 6. \quad (6)$$

Thus, the scale equation, measuring from $T = 1$, is

$$X_T = 6\left(\frac{1}{2}\log T - \frac{1}{2}\log 1\right) = 3 \log T. \quad (7)$$

The corresponding tables for the two-scale equations may be computed as shown in Tables 34.4 and 34.5. We shall position the two scales 5 in. apart, as shown in Fig. 34.16. The logarithmic scales are graduated using the values in Tables 34.4 and 34.5. The step-by-step procedure for constructing the remainder of the nomogram is given in Fig. 34.17 using the two outer scales determined here.

The end values of the middle (R) scale are found from the formula $R = S\sqrt{T}$ to be $R = 1.0\sqrt{100} = 10$ and $R = 0.1\sqrt{1} = 0.1$. Choosing a value of $R = 1.0$, we find that corresponding value pairs of S and T might be $S = 0.1$, $T = 100$, and $S = 1.0$, $T = 1.0$. We connect these pairs with isopleths (Step 1) and po-

TABLE 34.4

S	0.1	0.2	0.3	0.4	0.5	0.6	0.7	0.8	0.9	1.0
X_S	0	1.80	2.88	3.61	4.19	4.67	5.07	5.42	5.72	6.00

TABLE 34.5

T	1	2	4	6	8	10	20	40	60	80	100
X_T	0	0.91	1.80	2.33	2.71	3.00	3.91	4.81	5.33	5.77	6.00

TABLE 34.6

R	0.1	0.2	0.4	0.6	0.8	1.0	2.0	4.0	6.0	8.0	10.0
X_R	0	0.91	1.80	2.33	2.71	3.00	3.91	4.81	5.33	5.71	6.00

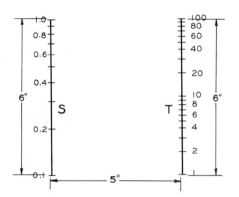

FIG. 34.16 Calibration of the outer scales for the formula $R = S\sqrt{T}$, where $0.1 \le S \le 1.0$ and $1 \le T \le 100$.

sition the middle scale at the intersection of the lines connecting the corresponding values. The R-scale is drawn parallel to the outer scales and is calibrated by deriving its scale modulus:

$$m_R = \frac{6}{\log 10 - \log 0.1} = \frac{6}{1 - (-1)} = 3.$$

Thus, its scale equation (measuring from $R = 0.1$) is

$$X_R = 3(\log R - \log 0.1) = 3(\log R + 1.0).$$

Table 34.6 is computed to give the values for the scale that are applied to the R-scale (Step 2). The finished nomogram can be used to compute the unknown variable when two variables are given (Step 3).

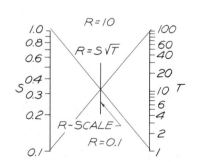
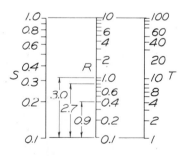
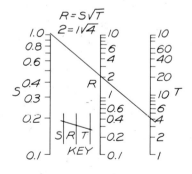

FIG. 34.17 Parallel-scale chart (logarithmic).

Step 1 Connect the end values of the outer scales to determine the extreme values of the R-scale, $R = 10$ and $R = 0.1$. Select corresponding values of S and T that will give the same value of R. Values of $S = 0.1$, $T = 100$ and $S = 1.0$, $T = 1.0$ give a value of $T = 1.0$. Connect the pairs to locate the R-scale.

Step 2 Draw the R-scale to extend from 0.1 to 10. Calibrate it by substituting values determined from its scale equation. These values have been computed and tabulated in Table 34.6. The resulting tabulation is a logarithmic, two-cycle scale.

Step 3 Add labels to the finished nomogram and draw a key to indicate how it is to be used. An isopleth has been used to determine R when $S = 1.0$ and $T = 4$. The result of 2 is the same as that obtained mathematically, thus verifying the accuracy of the chart.

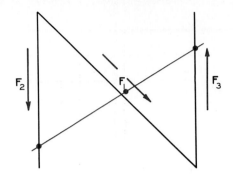

FIG. 34.18 An N-graph for solving an equation of the form $F_1 = F_2/F_3$.

34.6
N- or Z-graphs

Whenever F_2 and F_3 are linear functions, we can partially avoid using logarithmic scales for formulas of the type

$$F_1 = \frac{F_2}{F_3}.$$

Instead, we use an N-graph, as shown in Fig. 34.18. The outer scales, or "legs," of the N are functional scales and will therefore be linear if F_2 and F_3 are linear, whereas if the same formula were drawn as a parallel-scale graph, all scales would have to be logarithmic.

Some main features of the N-chart are

1. The outer scales are parallel functional scales of F_2 and F_3.

2. They increase (functionally) in *opposite* directions.

3. The diagonal scale connects the (functional) *zeros* of the outer scale.

4. In general, the diagonal scale is not a functional scale for the function F_1 and is nonlinear.

Construction of an N-graph is simplified because locating the middle (diagonal) scale is usually less of a problem than it is for a parallel-scale graph. Calibration of the diagonal scale is most easily accomplished by graphical methods.

The steps in constructing a basic N-graph of the equation $Z = Y/X$ are shown in Fig. 34.19. The diagonal is drawn to connect the zero ends of the scales (Step 1). Whole values are located along the diagonal by using combinations of X- and Y-values. It is important that the units located along the diagonal are whole values that are easy to interpolate between (Step 2). The diagonal is labeled, and a key is given explaining how to use the nomograph (Step 3). A

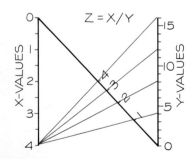
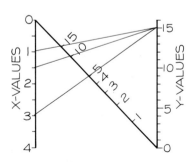
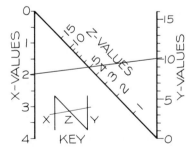

FIG. 34.19 An N-chart nomograph.

Step 1 A Z-nomograph for the equation of $Z = Y/X$ can be drawn with two parallel scales. The zero ends of each scale are connected with a diagonal scale. Isopleths are drawn to locate whole units along the diagonal.

Step 2 Additional isopleths are drawn to locate other whole units along the diagonal. It is important that the units labeled on the diagonal be whole units that are easy to interpolate between.

Step 3 The diagonal scale is labeled and a key is drawn. An example isopleth is drawn to show that $10/2 = 5$. The accuracy of the N-chart is greater at the 0 end of the diagonal. It approaches infinity at the other end.

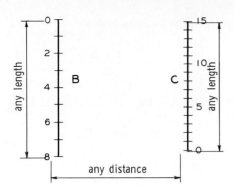

FIG. 34.20 Calibration of the outer scales of an N-graph for the equation $A = (B + 2)(C + 5)$.

follows the form of

$$F_1 = \frac{F_2}{F_3},$$

where $F_1(A) = A$, $F_2(B) = B + 2$, and $F_3(C) = C + 5$. Thus the outer scales will be for $B + 2$ and $C + 5$, and the diagonal scale will be for A.

The construction is begun in the same manner as for a parallel-scale graph by selecting the layout of the outer scales (Fig. 34.20). As before, the limits of the diagonal scale are determined by connecting the end-points on the outer scales, giving $A = 0.1$ for $B = 0$, $C = 15$ and $A = 2.0$ for $B = 8$, and $C = 0$, as shown in Fig. 34.21, which also gives the remainder of the construction.

The diagonal scale is located by finding the *function zeros* of the outer scales (i.e., the points where $B + 2 = 0$ or $B = -2$, and $C + 5 = 0$ or $C = -5$). The diagonal scale may then be drawn by connecting these points (Step 1). Calibration of the diagonal scale is most easily accomplished by substituting into the formula. Select the upper limit of an outer scale, for example, $B = 8$. This gives the formula

$$A = \frac{10}{C + 5}.$$

sample isopleth is also given verifying the correctness of the graphical relationship between the scales.

A more advanced N-graph is constructed for the equation

$$A = \frac{B + 2}{C + 5},$$

where $0 \leqslant B \leqslant 8$ and $0 \leqslant C \leqslant 15$. This equation

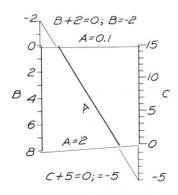

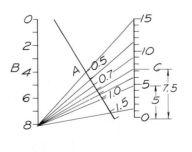

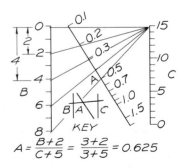

FIG. 34.21 Construction of an N-graph.

Step 1 Locate the diagonal scale by finding the functional zeros of the outer scales. This is done by setting $B + 2 = 0$ and $C + 5 = 0$, which gives a zero value for A.

Step 2 Select the upper limit of one of the outer scales, $B = 8$ in this case, and substitute it into the given equation to find a series of values of C for the desired values of A, as shown in Table 34.7. Draw isopleths from $B = 8$ to the values of C to calibrate the A-scale.

Step 3 Calibrate the remainder of the A-scale in the same manner by substituting $C = 15$ into the equation to determine a series of values on the B-scale for desired values on the A-scale, as listed in Table 34.8. Draw isopleths from $C = 15$ to calibrate the A-scale as shown.

TABLE 34.7

A	2.0	1.5	1.0	0.9	0.8	0.7	0.6	0.5
C	0	1.67	5.0	6.11	7.50	9.28	11.7	15.0

Solve this equation for the other outer scale variable,

$$C = \frac{10}{A} - 5.$$

Using this as a "scale equation," make a table of values for the desired values of A and corresponding values of C (up to the limit of C in the chart), as shown in Table 34.7. Connect isopleths from $B = 8$ to the tabulated values of C. Their intersections with the diagonal scale give the required calibrations for approximately half the diagonal scale (Step 2).

The remainder of the diagonal scale is calibrated by substituting the end value of the other outer scale

$(C = 15)$ into the formula, giving

$$A = \frac{B + 2}{20}.$$

Solving this for B yields

$$B = 20A - 2.$$

A table for the desired values of A can be constructed as shown in Table 34.8. Isopleths connecting $C = 15$ with the tabulated values of B will locate the remaining calibrations on the A-scale (Step 3).

TABLE 34.8

A	0.5	0.4	0.3	0.2	0.1
B	8.0	6.0	4.0	2.0	0

Problems

The following problems are to be solved on Size A sheets using the principles covered in this chapter. Problems involving geometric construction and mathematical calculations should show the construction and calculations as part of the solutions. If the calculations are extensive, include them on a separate sheet.

Concurrent scales

Construct concurrent scales for converting the following relationships of one type of unit to another. The range of units for the scales is given for each relationship.

1. Kilometers and miles:

   1.609 km = 1 mile, from 10 to 100 miles.

2. Liters and U.S. gallons:

   1 liter = 0.2692 U.S. gallons, from 1 to 10 liters.

3. Knots and miles per hour:

   1 knot = 1.15 miles per hour, from 0 to 45 knots.

4. Horsepower and British thermal units:

   1 horsepower = 42.4 Btu, from 0 to 1200 hp.

5. Centigrade and Fahrenheit:

   $°F = \frac{9}{5}°C + 32$, from 32°F to 212°F.

6. Radius and area of a circle:

   Area = πr^2, from $r = 0$ to 10.

7. Inches and millimeters:

   1 inch = 25.4 millimeters, from 0 to 5 inches.

8. Numbers and their logarithms:

   (use logarithm tables), numbers from 1 to 10.

Addition and subtraction nomographs

Construct parallel-scale nomographs to solve the following addition and subtraction problems.

9. $A = B + C$, where $B = 0$ to 10, and $C = 0$ to 5.

10. $Z = X + Y$, where $X = 0$ to 8, and $Y = 0$ to 12.

11. $Z = Y - X$, where $X = 0$ to 6, and $Y = 0$ to 24.

12. $A = C - B$, where $C = 0$ to 30, and $B = 0$ to 6.

13. $W = 2V + U$, where $U = 0$ to 12, and $V = 0$ to 9.

14. $W = 3U + V$, where $U = 0$ to 10, and $V = 0$ to 10.

15. Electrical current at a circuit junction:

$$I = I_1 + I_2$$

where

I = current entering the junction in amperes,
I_1 = current leaving the junction, varying from 2 to 15 amps,
I_2 = current leaving junction, varying from 7 to 36 amps.

16. Presssure change in fluid flowing in a pipe:

$$\Delta P = P_2 - P_1,$$

where

ΔP = pressure change between two points in pounds per square inch,
P_1 = pressure upstream, varying from 3 psi to 12 psi,
P_2 = pressure downstream, varying from 10 psi to 15 psi.

Multiplication and division— parallel scales

Construct parallel-scale nomographs with logarithmic scales that will perform the following multiplication and division operations.

17. Area of a rectangle: Area = Height × Width, where $H = 1$ to 10, and $W = 1$ to 12.

18. Area of a triangle: A = Base × Height/2, where $B = 1$ to 10, and $H = 1$ to 5.

19. Electrical potential between terminals of a conductor:

$$E = IR,$$

where

E = electrical potential in volts,
I = current, varying from 1 to 10 amperes,
R = resistance, varying from 5 to 30 ohms.

20. Pythagorean theorem:

$$C^2 = A^2 + B^2,$$

where

C = hypotenuse of a right triangle in centimeters,
A = one leg of the right triangle, varying from 5 to 50 cm,
B = second leg of the right triangle, varying from 20 to 80 cm.

21. Allowable presssure on a shaft bearing:

$$P = \frac{ZN}{100},$$

where

P = pressure in pounds per square inch,
Z = viscosity of lubricant from 15 to 50 cp (centipoises),
N = angular velocity of shaft from 10 to 1000 rpm.

22. Miles per gallon an automobile travels: mpg = miles/gallon. Miles vary from 1 to 500, and gallons vary from 1 to 24.

23. Cost per mile (cpm) of an automobile: cpm = cost/miles. Miles vary from 1 to 500, and cost varies from $1 to $28.

24. $R = S\sqrt{T}$, where S varies from 1 to 10, and T varies from 1 to 10.

25. Angular velocity of a rotating body:

$$W = \frac{V}{R},$$

where

W = angular velocity, in radians per second,
V = peripheral velocity, varying from 1 to 100 meters per second,
R = radius, varying from 0.1 to 1 meter.

N-graphs

Construct N-graphs that will solve the following equations.

26. Stress = P/A, where P varies from 0 to 1000 psi, and A varies from 0 to 15 square inches.

27. Volume of a cylinder:

$$V = \pi r^2 h,$$

where

V = volume in cubic inches,
r = radius, varying from 5 to 10 feet,
h = height, varying from 2 to 20 inches.

28. Same as Problem 17.   **33.** Same as Problem 22.

29. Same as Problem 18.   **34.** Same as Problem 23.

30. Some as Problem 19.   **35.** Same as Problem 24.

31. Same as Problem 20.   **36.** Same as Problem 25.

32. Same as Problem 21.

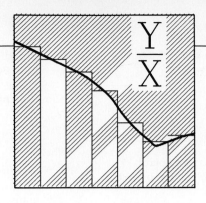

$$\frac{Y}{X}$$

CHAPTER 35

Empirical Equations and Calculus

35.1
Empirical data

Data gathered from laboratory experiments and tests or from actual field tests are called *empirical data*. Often empirical data can be transformed to equation form by means of one of three types of equations to be covered in this chapter.

    The analysis of empirical data begins by plotting the data on rectangular grids, logarithmic grids, or semilogarithmic grids. Curves are then drawn through each point to determine which of the grids renders a straight-line relationship (Fig. 35.1). When the data plots as a straight line, its equation may be determined.

35.2
Selection of points on a curve

Two methods of finding the equation of a curve are (1) the selected-points method and (2) the slope-in-

tercept method. These are compared on a linear graph in Fig. 35.2.

Selected-points method

Two widely separated points, such as (1, 30) and 4, 60), can be selected on the curve. These points are substituted in the equation below.

$$\frac{Y - 30}{X - 1} = \frac{60 - 30}{4 - 1}$$

The resulting data for the equation is

$$Y = 10X + 20.$$

Slope-intercept method

To apply the slope-intercept method, the intercept on the Y-axis where $X = 0$, must be known. If the X-axis is logarithmic, the log of $X = 1$ is 0, and the intercept must be found above the value of $X = 1$.

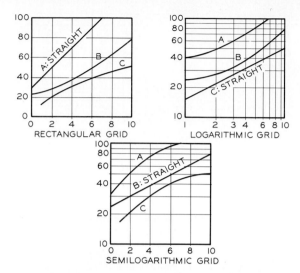

FIG. 35.1 Empirical data are plotted on each of these types of grids to determine which will render a straight-line plot. If the data can be plotted as a straight line on one of these grids, their equation can be found.

35.3

The linear equation: $Y = MX + B$

The curve fitting the experimental data plotted in Fig. 35.3 is a straight line; therefore these data are linear, meaning each measurement along the Y-axis is directly proportional to the X-axis units. We may use either the slope-intercept method or the selected-points method to find the equation for the data.

In the slope-intercept method, two known points are selected along the curve. The vertical and horizontal differences between the coordinates of each of these points are determined to establish the right triangle shown in Fig. 35.3A. In the slope-intercept equation, $Y = MX + B$, M is the tangent of the angle between the curve and the horizontal, B is the intercept of the curve with the Y-axis where $X = 0$, and X and Y are variables. In this example, $M = \frac{30}{5} = 6$, and the intercept is 20.

In Fig. 35.2, the data do not intercept the Y-axis; therefore the curve must be extended to find the intercept $B = 20$. The slope of the curve is found $(\Delta Y/\Delta X)$, and it is substituted into the slope-intercept form to give the equation $Y = 10X + 20$.

Other methods of converting data to equations are also used, but the two methods illustrated here make the best use of the graphical process and are the most direct methods of introducing these concepts.

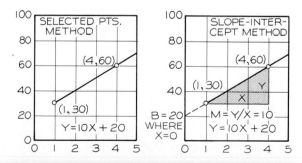

FIG. 35.2 The equation of a straight line on a grid can be determined by selecting any two points on the line. The slope-intercept method requires that the intercept be found where $X = 0$ on a semilog grid. This requires the extension of the curve to the Y-axis.

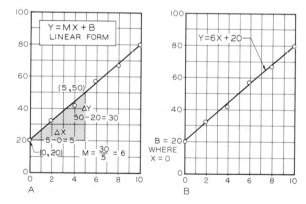

FIG. 35.3 A. A straight line on an arithmetic grid will have an equation in the form $Y = MX + B$. The slope, M, is found to be 6. **B.** The intercept, B, is found to be 20. The equation is written as $Y = 6X + 20$.

If the curve had sloped downward to the right, the slope would have been negative. By substituting this information into the slope-intercept equation, we obtain $Y = 6X + 20$, from which we can determine values of Y by substituting any value of X into the equation.

The selected-points method could also have been used to arrive at the same equation if the intercept

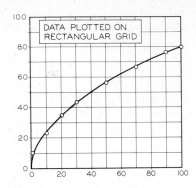
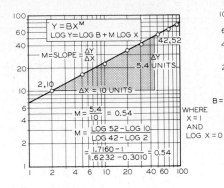
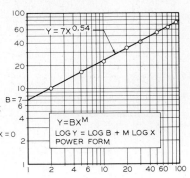

FIG. 35.4 The power equation: $Y = BX^M$.

Given The data plotted on the rectangular grid give an approximation of a parabola. Since the data do not form a straight line on the rectangular grid, the equation will not be linear.

Step 1 The data forms a straight line on a logarithmic grid. The slope, M, can be found graphically with an engineer's scale, setting dX at 10 units and measuring the slope (dY) using the same scale.

Step 2 The intercept $B = 7$ is found where $X = 1$. The slope and intercept are substituted into the equation, which then becomes $Y = 7X^{0.54}$.

were not known. By selecting two widely separated points, such as (2, 32) and 10, 80), one can write the equation in this form:

$$\frac{Y - 32}{X - 2} = \frac{80 - 32}{10 - 2}, \qquad \therefore Y = 6X + 20,$$

which results in the same equation as was found by the slope-intercept method ($Y = MX + B$).

35.4

The power equation: $Y = BX^M$

Data plotted on a logarithmic grid, as in (Fig. 35.4), are found to form a straight line (Step 1). Therefore we express the data in the form of a power equation in which Y is a function of X raised to a given power, or $Y = BX^M$. The equation of the data is obtained by using the point where the curve intersects the Y-axis where $X = 0$, and letting M equal the slope of the curve. Two known points are selected on the curve to form the slope triangle. The engineers' scale can then be used, when the cycles along the X- and Y-axes are equal, to measure the slope between the coordinates of the two points.

If the horizontal distance of the right triangle is drawn to be 1 or a multiple of 10, the vertical distance can be read directly. The slope M (tangent of the angle) is found to be 0.54 (Step 2). The intercept B is 7; thus, the equation is $Y = 7X^{0.54}$, which can be evalu-

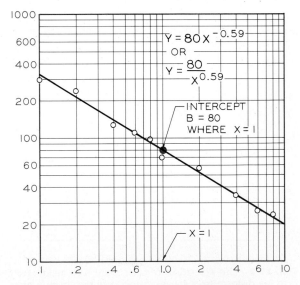

FIG. 35.5 When the slope-intercept equation is used, the intercept can be found only where $X = 1$. Therefore in this example, the intercept is found at the middle of the graph.

ated for each value of X by converting this power equation into the logarithmic form of log Y:

$$\log Y = \log B + M \log X,$$
$$\log Y = \log 7 + 0.54 \log X.$$

When the slope-intercept method is used, the intercept is found on the Y-axis where $X = 1$. In Fig. 35.5, the Y-axis at the left of the graph has an X value of 0.1; thus, the intercept is located midway across the graph where $X = 1$. This is analogous to the linear form of the equation, since the log of 1 is 0. The curve slopes downward to the right; thus the slope, M, is negative.

Base-10 logarithms are used in these examples, but natural logs could be used with e (2.718) as the base.

35.5
The exponential equation: $Y = BM^X$

When data plotted on a semilogarithmic grid (Fig. 35.6, Step 1) approximate a straight line, we can write the equation $Y - BM^X$, where B is the Y-intercept of

the curve, and M is the slope of the curve. To derive the equation, two points are selected along the curve so that a right triangle can be drawn to represent the differences between the coordinates of the points selected (Step 2). The slope of the curve is found to be

$$\log M = \frac{\log 40 - \log 6}{8 - 3} = 0.1648,$$

or

$$M = (10)^{0.1648} = 1.46.$$

The value of M can be substituted in the equation in following manner:

$$Y = BM^X \quad \text{or} \quad Y = 2(1.46)^X,$$
$$Y = B(10)^{MX} \quad \text{or} \quad Y = 2(10)^{0.1648X},$$

where X is a variable that can be substituted into the equation to give an infinite number of values for Y. We can write this equation in its logarithmic form, which enables us to solve it for the unknown value of Y for any given value of X. The equation can be written as

$$\log Y = \log B + X \log M,$$

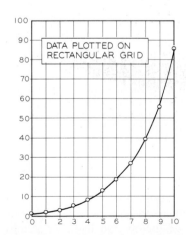
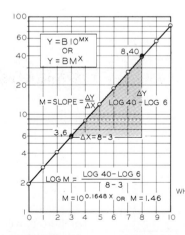
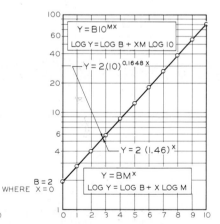

FIG. 35.6   The exponential equation: $Y = BM^X$.

Given   These data give a straight line on a semilogarithmic grid. Therefore the data fit the equation form, $Y = BM^X$.

Step 1   The slope must be found by mathematical calculations; it cannot be found graphically. The slope may be written in either of the forms shown here.

Step 2   The intercept $B = 2$ is found where $X = 0$. The slopes, M, and B, are substituted into the equation to give $Y = 2(10)^{0.1648X}$ or $Y = 2(1.46)^X$.

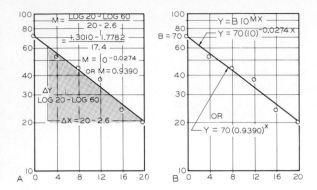

FIG. 35.7 A. When a curve slopes downward to the right, its slope is negative, as calculated here.
B. Two forms of a final equation are shown here by substitution.

35.6

Applications of empirical graphs

Figure 35.8 is an example of how empirical data can be plotted to compare the transverse strength and impact resistance of gray iron. Although the data are somewhat scattered, the best curve is drawn. Since the curve is a straight line on a linear graph, the equation of these data can be found by the equation

$$Y = MX + B.$$

or

$$\log Y = \log 2 + X \log 1.46.$$

To find the slope of a curve with a negative slope, the same methods are used. The curve of the data in Fig. 35.7 slopes downward to the right; therefore the slope is negative. Two points are selected to find the slope, M, which is the antilog of -0.0274. The intercept, 70, can be combined with the slope, M, to find the final equations, (Fig. 35.7B).

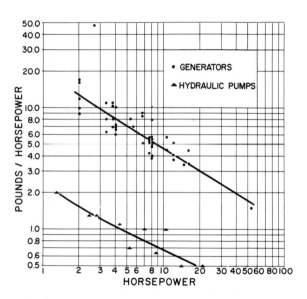

FIG. 35.9 Empirical data plotted on a logarithmic grid, showing the specific weight versus horsepower of electric generators and hydraulic pumps. The curve is the average of points plotted. (Courtesy of General Motors Corp., *Engineering Journal*.)

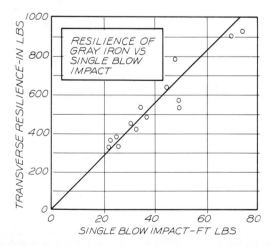

FIG. 35.8 The relationship between the transverse strength of gray iron and impact resistance results in a straight line with an equation of the form $Y = MX + B$.

Figure 35.9 is an example of how empirical data can be plotted to compare the specific weight (pounds per horsepower) of generators and hydraulic pumps versus horsepower. Note that the weight of these units decreases linearly as the horsepower increases. Therefore these data can be written in the form of the power equation

$$Y = BX^M.$$

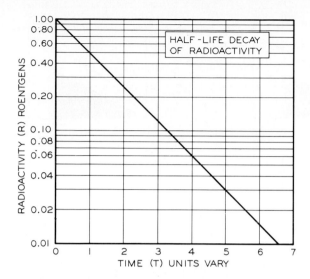

FIG. 35.10 The relative decay of radioactivity is plotted as a straight line on this semilog graph, making it possible for its equation to be found in the form $Y = BM^X$.

The half-life decay of radioactivity is plotted in Fig. 35.10 to show the relationship of decay to time. Since the half life of different isotopes varies, different units would have to be assigned to time along the X-axis; however, the curve would be a straight line for all isotopes. The exponential form of the equation discussed in Section 35.5 can be applied to find the equation for these data in the form of

$$Y = BM^X.$$

35.7
Introduction to graphical calculus

If the equation of the curve is known, traditional methods of calculus will solve the problem. However, many experimental data cannot be converted to standard equations. In these cases, it is desirable to use the graphical method of calculus, which provides relatively accurate solutions to irregular problems.

The two basic forms of calculus are (1) *differential calculus* and (2) *integral calculus*. Differential calculus is used to determine the rate of change of one variable with respect to another. For example, the curve plot-

ted in Fig. 35.11A represents the relationship between two variables. The rate of change at any instant along the curve is the slope of a line tangent to the curve at that particular point. This exact slope can be approximated by constructing a chord at a given interval (Fig. 35.11A). The slope of this chord can be measured by finding the tangent of $\Delta Y/\Delta X$. This slope can represent miles per hour, weight versus length, or a number of other rates important to the analysis of data.

Integral calculus is the reverse of differential calculus. Integration is the process of finding the area under a given curve, which can be thought of generally as the product of the two variables plotted on the X- and Y-axes. The area under a curve is approximated by dividing one of the variables into a number of very small intervals, which become small rectangular areas at a particular zone under the curve (Fig. 35.11B). The bars are extended so that as much of the square end of the bar is below the curve as above it and the average height of the bar is, therefore, near its midpoint.

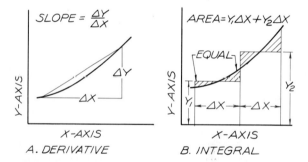

FIG. 35.11 A. The derivative of a curve is the change at any point that is the slope of the curve, Y/X. **B.** The integral of a curve is the cumulative area enclosed by the curve, which is the summation of the products of the areas.

35.8
Graphical differentiation

Graphical differentiation is the determination of the rate of change of two variables with respect to each other at any given point. Figure 35.12 illustrates the preliminary construction of the derivative scale that would be used to plot a continuous derivative.

STEP 1 The original data are plotted graphically, and the axes are labeled with the proper units of measure-

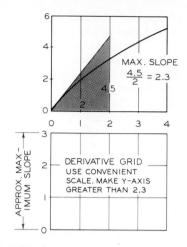
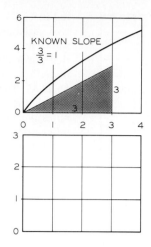
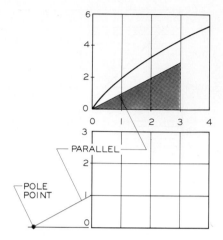

FIG. 35.12 Scales for graphical differentiation.

Step 1 The maximum slope, 2.3, of a curve is found by constructing a line tangent to the curve where it is steepest. The derivative grid is laid off with a maximum ordinate of 3.0 to accommodate the maximum value.

Step 2 A known slope is found on the given grid; this value is 1 in this example. The known slope has no relationship to the curve at this point.

Step 3 Construct a line from 1 on the Y-axis of the derivative grid parallel to the slope of the triangle constructed in the given grid. This line locates the pole point on the extension of the X-axis.

ment. A chord can be constructed to estimate the maximum slope by inspection. The maximum slope is estimated to be 2.3. A vertical scale is constructed in excess of this value to provide for the plotting of slopes that may exceed the estimate. The ordinate scale is drawn to a convenient scale to facilitate measurement.

STEP 2 A known slope is plotted on the given data grid. This slope need not be related to the curve in any way. In this case, the slope can be read directly as 1.

STEP 3 The pole can be found by drawing a line from the ordinate of 1 (the known slope) on the derivative scale parallel to the slope line. These similar triangles are used to obtain the pole.

The steps in completing the graphical differentiation are given in Fig. 35.13. Note that the same horizontal intervals used in the given curve are projected directly beneath on the derivative scale. The maximum slope of the data curve is estimated to be slightly less than 12. A scale is selected that will provide an ordinate that will accommodate the maximum slope. A line is drawn from point 12 on the ordinate axis of the derivative grid parallel to the known slope on the

given curve grid. The point of intersection of this line and the extension of the X-axis is the pole point.

A series of chords is constructed on the given curve. Lines are constructed parallel to these chords through point P and extended to the Y-axis of the derivative grid to locate bars at each interval. The interval between 0 and 1, where the curve is sharpest, was divided in half to provide a more accurate plot. A smooth curve is constructed through the top of these bars so that the area above the horizontal top of the bar is the same as that below it. The rate of change, $\Delta Y/\Delta X$, can be found at any interval of the variable X by reading directly from the derivative graph at the value of X in question.

35.9
Applications of graphical differentiation

The mechanical handling shuttle shown in Fig. 35.14 is used to convert rotational motion into linear motion. A scale drawing of the linkage components is

FIG. 35.13 Graphical differentiation.

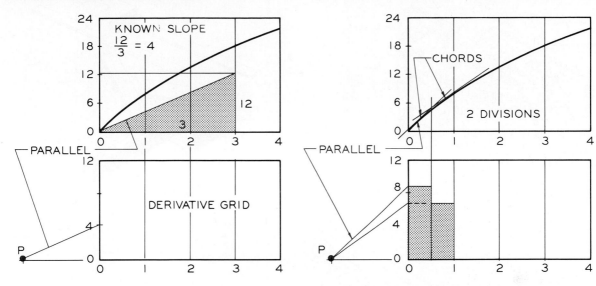

Required Find the derivative curve of the given data.

Step 1 Find the derivative grid and the pole point using the construction illustrated in Fig. 35.12.

Step 2 Construct chords between intervals on the given curve and draw lines parallel to them through point *P* on the derivative grid. These lines locate the heights of bars in their respective intervals. The first interval is divided in half since the curve is changing sharply in this interval.

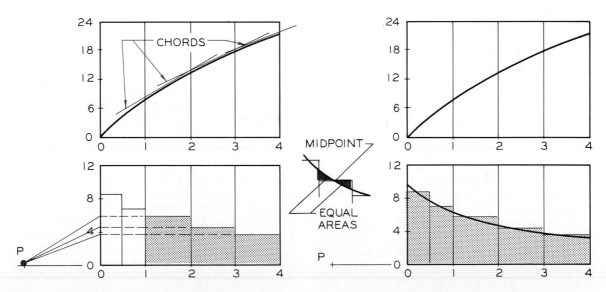

Step 3 Additional chords are drawn in the last two intervals. Lines parallel to these chords are drawn from the pole point to the *Y*-axis to find additional bars in their respective intervals.

Step 4 The vertical bars represent the slopes of the curve at different intervals. The derivative curve is drawn through the midpoints of the bars so that the areas under and above the bars are approximately equal.

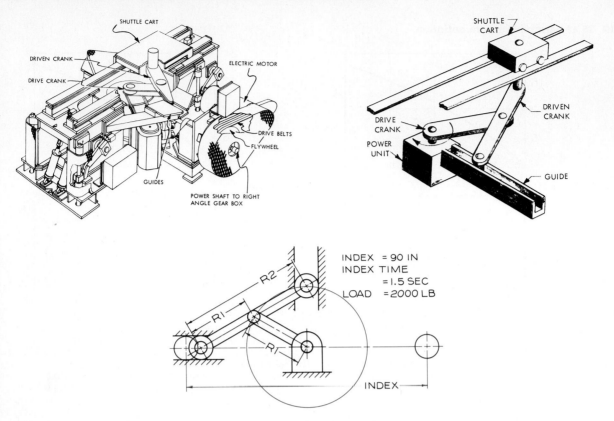

INDEX = 90 IN
INDEX TIME
= 1.5 SEC
LOAD = 2000 LB

FIG. 35.14 A pictorial and scale drawing of an electrically powered mechanical handling shuttle used to move automobile parts on an assembly line. (Courtesy of General Motors Corp.)

given so that graphical analysis can be applied to determine the resulting motion.

The linkage is drawn to show the end positions of point P, which will be used as the zero point for plotting the travel versus the degrees of revolution. Since rotation is constant at one revolution per three seconds, the degrees of revolution can be converted to time, as shown in the data curve at the top of Fig. 35.15. The drive crank, R_1, is revolved at 30° intervals, and the distance that point P travels from its end position is plotted on the graph to give the distance-versus-time relationship.

We determine the ordinate scale of the derivative grid by estimating the maximum slope of the given data curve, which is found to be a little less than 100 in./sec. A convenient scale is chosen that will have a maximum limit is 100 units. A slope of 40 is drawn on the given data curve to be used in determining the

location of pole P in the derivative grid. From point 40 on the derivative ordinate scale, we draw a line parallel to the known slope, which is found on the given grid. Point P is the point where this line intersects the extension of the X-axis.

A series of chords are drawn on the given curve to approximate the slope at various points. Lines are constructed through point P of the derivative scale parallel to the chord lines and extended to the ordinate scale. The points thus obtained are then projected across to their respective intervals to form vertical bars. A smooth curve is drawn through the tops of the bars to give an average of the bars. This curve can be used to find the velocity of the shuttle in inches per second at any time interval.

The construction of the second derivative curve, the acceleration, is similar to that of the first derivative. We estimate the maximum slope to be 200 in./sec/sec by inspecting the first derivative. An easily measured scale is established for the ordinate scale of the second derivative curve. Point P is found in the same manner as the first derivative.

Chords are drawn at intervals on the first derivative curve. Lines are drawn parallel to these chords from point P in the second derivative curve to the Y-axis, where they are projected horizontally to their respective intervals to form a series of bars. A smooth curve is drawn through the tops of the bars to give a close approximation of the average areas of the bars. A minus scale is given to indicate deceleration.

The maximum acceleration is found to be at the extreme endpoints, and the minimum acceleration is at 90°, where the velocity is the maximum. From the velocity and acceleration plots, it can be seen that the parts being handled by the shuttle are accelerated at a rapid rate until the maximum velocity is attained at 90°, at which time deceleration begins and continues until the parts come to rest.

35.10
Graphical integration

Integration is the process of determining the area (product of two variables) under a given curve. For example, if the Y-axis were pounds and the X-axis were feet, the integral curve would give the product of the variables, foot-pounds, at any interval of feet along the X-axis. Figure 35.16 depicts the method of constructing scales for graphical integration.

STEP 1 It is customary to locate the integral curve above the given data curve, since the integral will be an equation raised to a higher power. A line is drawn through the given data curve to approximate the total area under the curve, which is 4 × 5, or 20, square units of area. The ordinate scale is drawn on the integral curve in excess of 20 units to provide for any overage. The horizontal scale intervals are projected from the given curve to the integral grid.

STEP 2 The ordinate value on the integral scale represents the area under the curve as measured from the origin to that point on the given data curve. The ordinate at point 2 on the X-axis directly above the rectangle must be equal to its area of 8. A slope is drawn from the origin to the ordinate of 8.

STEP 3 Point P is found by drawing a line from point 4 on the given grid parallel to the slope established in the integral grid. This line intersects the extension of the X-axis at point P, the pole point.

This technique is used in Fig. 35.17 to integrate the equation of the given curve, $Y = 2X^2$. The total area under the curve can be estimated to be less than 40 units. This value becomes the maximum height of the Y-axis on the integral curve. A convenient scale is selected for the ordinate, and the pole point, P, is found.

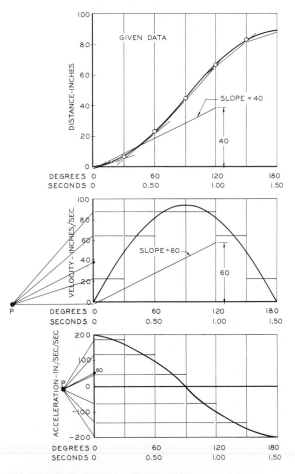

FIG. 35.15 Graphical determination of velocity and acceleration of the mechanical handling shuttle by differential calculus.

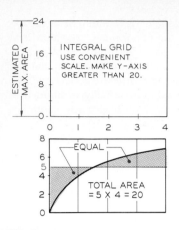
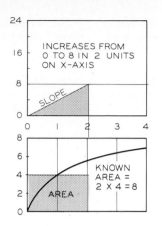
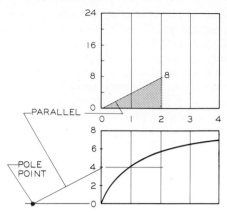

FIG. 35.16   Scales for graphical integration.

Step 1   To determine the maximum value on the *Y*-axis, a line is drawn to approximate the area under the given grid. This is found to be 20 and the *Y*-axis of the grid is constructed with 24 as the maximum value.

Step 2   A known area, 8 in this case, is found in the given grid. A slope line from 0 to 8 is constructed in the integral grid directly above the known area, which establishes the integral for this model.

Step 3   A line is drawn from 4 on the *Y*-axis of the given grid parallel to the slope line in the integral grid. This line from 4 crosses the extension of the *X*-axis to locate the pole point.

A series of vertical bars is constructed to approximate the areas under the curve at these intervals. The narrower the bars, the more accurate will be the resulting calculations. The interval between 1 and 2 was divided in half to provide a more accurate plot. The top lines of the bars are extended horizontally to the *Y*-axis, where the points are then connected by lines to point *P*. Lines are drawn parallel to *AP, BP, CP, DP,* and *EP* in the integral grid to correspond to the respective intervals in the given grid. The intersection points of the chords are connected by a smooth curve—the integral curve to give the cumulative product of the *X*- and *Y*-variables along the *X*-axis. For example, the area under the curve at $X = 3$ can be read as 18.

Mathematical integration gives the following result for the area under the curve from 0 to 3:

$$\text{Area} = \int_0^3 Y dX, \quad \text{where } Y = 2X^2;$$

$$A = \int_0^3 2X^2 dX = \left. \frac{2}{3X^3} \right|_0^3 = 18.$$

35.11
Applications of graphical integration

An example is shown in Fig. 35.18, where a truck exerts a total force of 36,000 lb on a beam of a bridge. From the load diagram shown in Fig. 35.19 we can, by integration, find the shear diagram, which indicates the points in the beam where failure due to crushing is most critical. In the shear diagram, the left-end resultant of 15.9 kips is drawn to scale from the axis. The first load of 4 kips acting in a downward direction is subtracted from this value directly over its point of application. The second load of 16 kips exerts a downward force and is subtracted from the 11.9 kips (15.9 − 4). The third load of 16 kips is also subtracted, and the right-end resultant will bring the shear diagram back to the *X*-axis.

The moment diagram is used to evaluate the bending characteristics of the applied loads in foot-pounds at any interval along the beam. The ordinate

FIG. 35.17 Graphical integration.

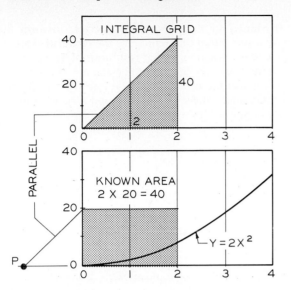

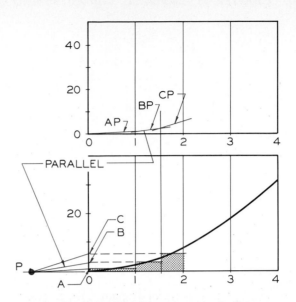

Required Plot the integral curve of the given data.

Step 1 Find the pole point, *P*, using the technique illustrated in Fig. 35.16.

Step 2 Construct bars to approximate the areas under the curve. The interval from 1 to 2 was divided in half to improve the approximation. The heights of the bars are projected to the *Y*-axis, and lines are drawn to the pole point. Sloping lines *AP*, *BP*, and *CP* are drawn in their respective intervals parallel to the lines drawn to the pole, *P*.

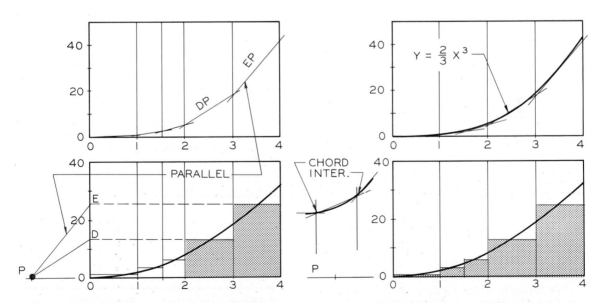

Step 3 Additional bars are drawn from 2 to 4 on the *X*-axis. The heights of the bars are projected to the *Y*-axis, and rays are drawn to the pole point, *P*. Lines *DP* and *DE* are drawn in their respective intervals and parallel to their rays in the integral grid.

Step 4 The straight lines connected in the integral grid represent chords of the integral curve. Construct the integral curve to pass through the points where the chords intersect. Ordinate value on the integral curve represents the cumulative area under the given curve from zero to that point on the *X*-axis.

695

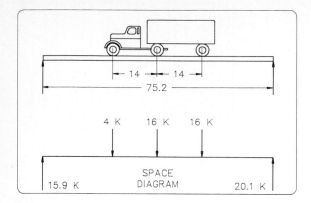

FIG. 35.18 The forces on a beam of a bridge loaded with a truck.

then drawn in their respective intervals to form a straight-line curve that represents the cumulative area of the shear diagram, which is in units of ft-kips. The maximum bending is scaled to be about 560 ft-kips at the center of the beam. This beam must be capable of withstanding a shear of 20.1 kips and a bending moment of 560 ft-kips.

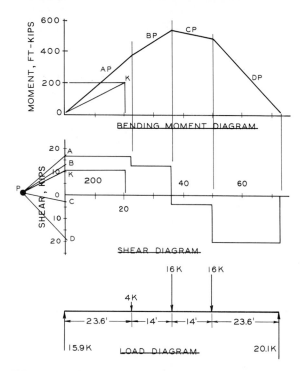

FIG. 35.19 Determination of shear and bending moment by graphical integration.

of any X-value in the moment diagram must represent the cumulative foot-pounds in the shear diagram as measured from either end of the beam.

Pole point P is located in the shear diagram by applying the method described in Fig. 35.18. A rectangular area of 200 ft-kips is found in the shear diagram. We estimate the total area to be less than 600 ft-kips, so we select a convenient scale that will allow an ordinate scale of 600 units for the moment diagram. We draw a known area of 200 (10 × 20) on the shear diagram. A diagonal line in the moment diagram is drawn that slopes upward from 0 to 200, where $X = 20$. The diagonal, OK, is transferred to the shear diagram, where it is drawn from the ordinate of the given rectangle to point P on the extension of the X-axis. Rays AP, BP, CP, and DP are found in the shear diagram by projecting horizontally from the various values of shear. In the moment diagram, these rays are

---

Problems

The following problems are to be solved on Size A sheets. Solutions involving mathematical calculations should show these calculations on separate sheets if space is not available on the sheet where the graphical solution is drawn.

Empirical equations—logarithmic

1. Find the equation of the data shown in the following table. The empirical data compare input

voltage *(V)* with input current amperes *(I)* to a heat pump.

Y-axis	V	0.8	1.3	1.75	1.85
X-axis	I	20	30	40	45

2. Find the equation of the data in the following table. The empirical data give the relationship be-

tween peak allowable current in amperes (I) with the overload operating time in cycles at 60 cycles per second (C). Place I on the Y-axis.

Y-axis	I	2000	1840	1640	1480	1300	1200	1000
X-axis	C	1	2	5	10	20	50	100

3. Find the equation of the data in the following table. The empirical data for a low-voltage circuit breaker used on a welding machine give the maximum loading during weld in amperes (rms) for the percent of duty (pdc). Place rms along the Y-axis and pdc along the X-axis.

Y-axis	rms	7500	5200	4400	3400	2300	1700
X-axis	pdc	3	6	9	15	30	60

4. Construct a three-cycle × three-cycle logarithmic graph to find the equation of a machine's vibration displacement in mills along the Y-axis and vibration frequency in cycles per minute (cpm) along the X-axis. Data: 100 cpm, 0.80 mills; 400 cpm, 0.22 mills; 1000 cpm, 0.09 mills; 10,000 cpm, 0.009 mills; 50,000 cpm, 0.0017 mills.

5. Find the equation of the data in the following table that compares the velocities of air moving over a plane surface in feet per second (v) at different heights in inches (y) above the surface. Plot Y-values on the Y-axis.

Y	V
0.1	18.8
0.2	21.0
0.3	22.6
0.4	24.1
0.6	26.0
0.8	27.3
1.2	29.2
1.6	30.6
2.4	32.4
3.2	33.7

6. Find the equation of the data in the following table that shows the distance traveled in feet (s) at various times in seconds (t) of a test vehicle. Plot s and the Y-axis and then t on the X-axis.

t	s
1	15.8
2	63.3
3	146.0
4	264.0
5	420.0
6	580.0

Empirical equations—linear

7. Construct a linear graph to determine the equation for the yearly cost of a compressor in relationship to the compressor's size in horsepower. The yearly cost should be plotted on the Y-axis and the compressor's size in horsepower on the X-axis. Data: 0 hp, $0; 50 hp, $2100; 100 hp, $4500; 150 hp, $6700; 200 hp, $9000; 250 hp, $11,400. What is the equation of these data?

8. Construct a linear graph to determine the equation for the cost of soil investigation by boring to determine the proper foundation design for varying sizes of buildings. Plot the cost of borings in dollars along the Y-axis and the building area in sq ft along the X-axis. Data 0 sq ft, $0; 25,000 sq ft, $35,000; 50,000 sq ft, $70,000; 750,000 sq ft, $100,000; 1,000,000 sq ft, $130,000.

9. Find the equation of the empirical data plotted in Fig. 35.8.

10. Plot the data in the table below on a linear graph and determine its equation. The empirical data show the deflection in centimeters of a spring (d) when it is loaded with different weights in kilograms (W). Plot W along the X-axis and d along the Y-axis.

w	d
0	0.45
1	1.10
2	1.45
3	2.03
4	2.38
5	3.09

11. Plot the data in the table below on a linear graph and determine its equation. The empirical data show the temperatures read from a Fahrenheit thermometer at B and a centigrade thermometer at A. Plot the A-values along the X-axis and the B-values along the Y-axis.

°A	°B
−6.8	20.0
6.0	43.0
16.0	60.8
32.2	90.0
52.0	125.8
76.0	169.0

Empirical equations—semilogarithmic

12. Construct a semilog graph of the following data to determine their equation. The Y-axis should be a two-cycle log scale and the X-axis a 10-unit linear scale. Plot the voltage (E) along the Y-axis and time (T) in sixteenths of a second along the X-axis to represent resistor voltage during capacitor charging. Data: 0, 10 volts; 2, 6 volts; 4, 3.6 volts; 6, 2.2 volts; 8, 1.4 volts; 10, 0.8 volts.

13. Find the equation of the data plotted in Fig. 35.10.

14. Construct a semilog graph of the following data to determine their equation. The Y-axis should be a three-cycle log scale and the X-axis a linear scale from 0 to 250. These data give a comparison of the reduction factor, R (Y-axis), with the mass thickness per square foot (X-axis), of a nuclear protection barrier. Data: 0, 1.0R; 100, 0.9R; 150, 0.028R; 200, 0.009R; 300, 0.0011R.

15. An engineering firm is considering its expansion by reviewing its past sales as shown in the table below. Their years of operation are represented by x, and N is their annual income in tens of thousands. Plot x along the X-axis and N along the Y-axis and determine the equation of their progress.

x	N
1	0.05
2	0.08
3	0.12
4	0.20
5	0.32
6	0.51
7	0.80
8	1.30
9	2.05
10	3.25

TABLE 35.1

A	X	0	40	80	120	160	200	240	280		
	Y	4.0	7.0	9.8	12.5	15.3	17.2	21.0	24.0		
B	X	1	2	5	10	20	50	100	200	500	1000
	Y	1.5	2.4	3.3	6.0	9.0	15.0	23.0	24.0	60.0	85.0
C	X	1	5	10	50	100	500	1000			
	Y	3	10	19	70	110	400	700			
D	X	2	4	6	8	10	12	14			
	Y	6.5	14.0	32.0	75.0	115.0	320.0	710.0			
E	X	0	2	4	6	8	10	12	14		
	Y	20	34	53	96	115	270	430	730		
F	X	0	1	2	3	4	5	6	7	8	9	10
	Y	1.8	2.1	2.2	2.5	2.7	3.0	3.4	3.7	4.1	4.5	5.0

Empirical equations—general types

16–21. Plot the experimental data shown in Table 35.1 on the grid where the data will appear as straight-line curves. Determine the equations of the data.

Calculus—differentiation

22. Plot the equation $Y = X^3/6$ as a rectangular graph. Graphically differentiate the curve to determine the first and second derivatives.

23. Plot the following data on a graph, and find the derivative curve of the data on a graph placed below the first: $Y = 2X^2$.

24. Plot the following equation on a graph, and find the derivative curve of the data on a graph placed below the first: $4Y = 8 - X^2$.

25. Plot the following data on a graph, and find the derivative curve of the data on a graph placed below the first: $3Y = X^2 + 16$.

26. Plot the following data on a graph, and find the derivative curve of the data on a graph placed below the first: $X = 3Y^2 - 5$.

Calculus—integration

27. Plot the following equation on a graph, and find the integral curve of the data on a graph placed above the first: $Y = X^2$.

28. Plot the following equation on a graph, and find the integral curve of the data on a graph placed above the first: $Y = 9 - X^2$.

29. Plot the following equation on a graph, and find the integral curve of the data on a graph placed above the first: $Y = X$.

30. Using graphical calculus, analyze a vertical strip 12 in. wide on the inside face of the dam in Fig. 35.20. The force on this strip will be 52.0 lb/in. at the bottom of the dam. The first graph will be pounds per inch (ordinate) versus height in inches (abscissa). The second graph will be the integral of the first to give shear in pounds (ordinate) versus height in inches (abscissa). The third

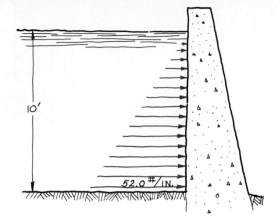

FIG. 35.20 Problem 30. Pressure on a 12-in.-wide section of a dam.

will be the integral of the second graph to give the moment in inch-pounds (ordinate) versus height in inches (abscissa). Convert these scales to give feet instead of inches.

31. A plot plan shows that a tract of land is bounded by a lake front (Fig. 35.21). By graphical integration, determine a graph that will represent the cumulative area of the land from points A to E. What is the total area? What is the area of each lot?

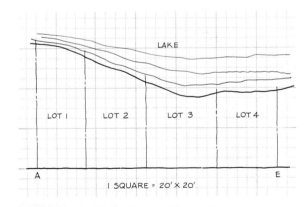

FIG. 35.21 Problem 31. Plot plan of a tract bounded by a lake front.

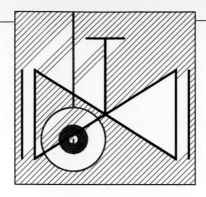

CHAPTER 36

Pipe Drafting

36.1
Introduction

An understanding of pipe drafting begins with a familiarity with the types of pipe available. The commonly used types of pipe are (1) steel pipe, (2) cast-iron pipe, (3) copper, brass, and bronze pipe and tubing, and (4) plastic pipe.

To ensure uniformity of size and strength of interchangeable components, the standards for the grades and weights of pipe and pipe fittings are specified by several organizations, including the American National Standards Institute (ANSI), the American Society for Test Materials (ASTM), the American Petroleum Institute (API), and the Manufacturers Standardization Society (MSS).

36.2
Welded and seamless steel pipe

Traditionally, steel pipe has been specified in three weights *standard* (STD), *extra strong* (XS), and *double extra strong* (XXS). These designations and their specifications are listed in the ANSI B 36.10–1979 standards. However, additional pipe designations, called *schedules*, have been introduced to provide the pipe designer with a wider selection of pipe.

The ten schedules are Schedule 10, Schedule 20, Schedule 30, Schedule 40, Schedule 60, Schedule 80, Schedule 100, Schedule 120, Schedule 140, and Schedule 160. The wall thicknesses of the pipes vary from the thinnest, in Schedule 10, to the thickest, in Schedule 160. The outside diameters are of a constant size for pipes of the same nominal size in all schedules.

Schedule designations correspond to STD, XS, and XXS specifications in some cases, as is shown partially in Table 36.1. This table has been abbreviated from the ANSI B 36.10–1979 tables by omitting a number of the pipe sizes and schedules. The most often used schedules are 40, 80, and 120.

Pipes from the smallest size up to and including 12-in. pipes are specified by their inside diameter (ID), which means the outside diameter (OD) is larger than the specified size. The inside diameters are the same size as the nominal sizes of the pipe for STD weight pipe. For XS and XXS pipe, the size of the inside diameters is slightly different from the nominal size. Be-

ginning with the 14-in. diameter pipes, the nominal sizes represent the outside diameters of the pipe.

The standard lengths for steel pipe are 20 ft and 40 ft. Seamless steel (SMLS STL) pipe is a smooth pipe with no weld seams along its length. Welded pipe is formed into a cylinder and is butt-welded (BW) at the seam, or it is joined with an electric resistance weld (ERW).

36.3
Cast-iron pipe

Cast-iron pipe is used for the transportation of liquids, water, gas, and sewage. When used as a sewerage pipe, cast-iron pipe is referred to as "soil

TABLE 36.1
DIMENSIONS AND WEIGHTS OF WELDED AND SEAMLESS STEEL PIPE (ANSI B36.10–1979)

	Inch Units			Identification			SI Units		
Inch Nominal Size	O.D. (in.)	Wall Thk. (in.)	Weight (lbs/ft)	*STD XS XXS	Sch. No.	O.D. (mm)	Wall Thk. (mm)	Weight (kg/m)
½	0.84	0.11	0.85	STD	40	21.3	2.8	1.3
1	1.32	0.13	1.68	STD	40	33.4	3.4	2.5
1	1.3	0.18	2.17	XS	80	33.4	4.6	3.2
1	1.3	0.36	3.66	XXS		33.4	9.1	5.5
2	2.38	0.22	3.65	STD	40	60.3	3.9	5.4
2	2.38	0.22	5.02	XS	80	60.3	5.5	7.5
2	2.38	0.44	9.03	XXS		60.3	11.1	13.4
4	4.50	0.23	10.79	STD	40	114.3	6.0	16.1
4	4.50	0.34	14.98	XS	80	114.3	8.6	42.6
4	4.50	0.67	27.54	XXS		114.3	17.1	41.0
8	8.63	0.32	28.55	STD	40	219.1	8.2	42.6
8	8.63	0.50	43.39	XS	80	219.1	12.7	64.6
8	8.63	0.88	74.40	XXS		219.1	22.2	107.9
12	12.75	0.38	49.56	STD		323.0	9.5	67.9
12	12.75	0.50	65.42	XS		323.0	12.7	97.5
12	12.75	1.00	125.4	XXS	120	133.9	25.4	187.0
14	†14.00	0.38	54.57	STD	30	355.6	9.5	87.3
14	14.00	0.50	72.08	XS		355.6	12.7	107.4
18	18.00	0.38	70.59	STD		457	9.5	106.2
18	18.00	0.50	93.45	XS		457	12.7	139.2
24	24.00	0.38	94.62	STD	20	610	9.5	141.1
24	24.00	0.50	125.49	XS		610	12.7	187.1
30	30.00	0.38	118.65	STD		762	9.5	176.8
30	30.00	0.50	157.53	XS	20	762	12.7	234.7
40	40.00	0.38	158.70	STD		1016	9.5	236.5
40	40.00	0.50	210.90	XS		1016	12.7	314.2

*Standard (STD)
X-strong (XS)
XX-strong (XXS)
†Beginning with 14-in. DIA pipe, the nominal size represents the outside diameter (O.D.).
This table has been compressed by omitting many of the available pipe sizes. The nominal sizes of pipes listed in the complete table are ⅛″, ¼″, ⅜″, ½″, ¾″, 1″, 1¼″, 1½″, 2″, 2½″, 3″, 3½″, 4″, 5″, 6″, 8″, 10″, 12″, 14″, 16″, 18″, 20″, 22″, ... (at 2″ increments up to 60″).

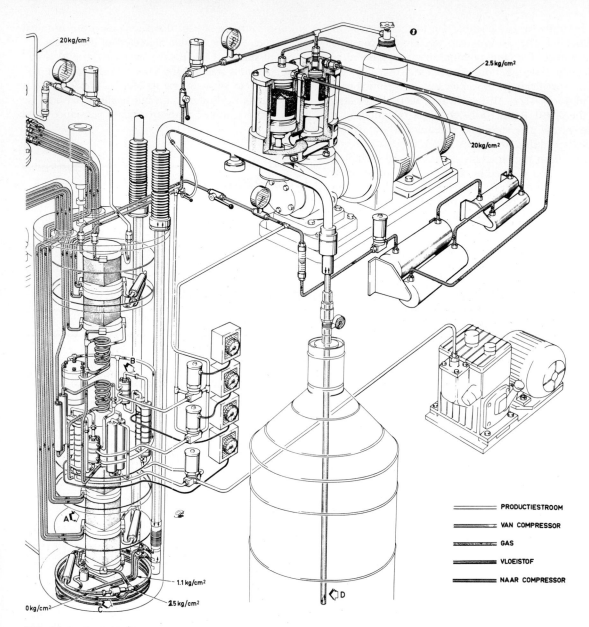

20kg/cm²

2.5kg/cm²

20kg/cm²

B

A

1.1 kg/cm²

0kg/cm²

2.5 kg/cm²

C

D

PRODUCTIESTROOM

VAN COMPRESSOR

GAS

VLOEISTOF

NAAR COMPRESSOR

FIG. 36.1 The small pipes in this illustration are called tubes because they are less than 2 in. in diameter and can be bent to form angles.

pipe." Cast-iron pipe is available in diameters of from 3 in. to 60 in.

The standard lengths of cast-iron pipe are 5 ft and 10 ft. Cast iron is more brittle and more subject to cracking when it is loaded than is steel pipe, so it should not be used where high pressures or weights will be applied to it.

36.4
Copper, brass, and bronze pipe

Copper, brass, and bronze are used to manufacture piping and tubing for applications where there must be a high resistance to corrosive elements, such as acidic soils and chemicals, that are transmitted through the pipes. Copper pipe is used when the pipes are placed within or under concrete slab foundations of buildings. These nonferrous materials ensure the pipes will not have to be replaced because of corrosion. The standard length of these pipes is 12 ft.

Tubing is a smaller-size pipe that can be easily bent when it is made of copper, brass, or bronze. An example of a system of tubing is shown in Fig. 36.1. The term piping applies to rigid pipes that are usually larger than 2 in. in diameter.

36.5
Miscellaneous pipes

Other materials used to manufacture pipes are aluminum, asbestos-cement, concrete, polyvinyl chloride (PVC), and various other plastics. Each of these materials has its special characteristics that make it desirable or economical for certain applications. The method the pipedrafter uses to design and detail piping systems is essentially the same regardless of the piping material used.

36.6
Pipe joints

The basic connection in a pipe system is the joint where two straight sections of pipe fit together. Figure 36.2 illustrates three types of joints: **screwed, welded,** and **flanged.**

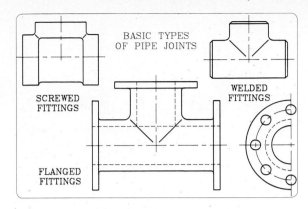

FIG. 36.2 The three basic types of joints are screwed, welded, and flanged joints.

SCREWED JOINTS are joined by pipe threads of the type covered in Chapter 18 and Appendix 10. Pipe threads are tapered at a ratio of 1 to 16 along the outside diameter (Fig. 36.3). As the pipes are screwed together, the threads bind to form a snug, locking fit. To improve the seal, a cementing compound is applied to the threads before joining.

WELDED JOINTS are joined by welded seams around the perimeter of the pipe to form butt welds. Welded joints are used extensively in "big inch" pipelines used in cross-country transporting of petroleum products.

FLANGED JOINTS shown in Fig. 36.4, are welded to the straight sections of pipe, which are then

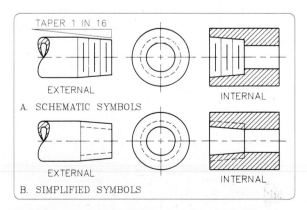

FIG. 36.3 A screwed pipe uses a pipe thread that binds tightly as pipes are screwed together.

FIG. 36.4 Types of flanged joints and the methods of attaching the flanges to the pipes. (Courtesy of Vogt Machine Co.)

bolted together around the perimeter of the flanges. Flanged joints form strong, rigid joints that can withstand high pressure and permit disassembly when needed. Several types of flange faces are shown in Fig. 36.5 and in Appendix 13.

BELL AND SPIGOT (B&S) JOINTS are used to join cast-iron pipes (Fig. 36.6). The spigot is placed inside the bell and the two are sealed with molten lead or a sealing ring that snaps into position to form a sealed joint.

SOLDERING is used to connect smaller pipes and tubular connections. Soldering is usually limited to nonferrous tubing. However, screwed fittings are also available to connect tubing (Fig. 36.7).

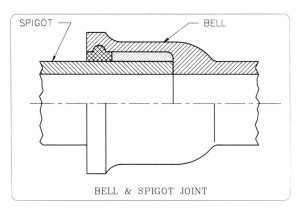

FIG. 36.6 A bell and spigot joint (B&S) is used to connect cast-iron pipes.

36.7
Screwed fittings

A number of standard fittings are shown in Fig. 36.8. The two types of graphical symbols used to represent fittings and pipe are *double-line symbols* and *single-line symbols*.

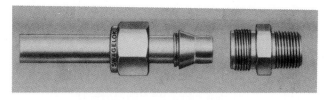

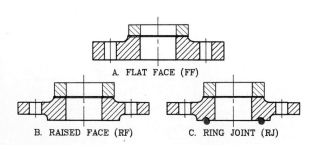

FIG. 36.5 Three types of flange faces are the raised face (RF), flat face (FF), and ring joint (RJ).

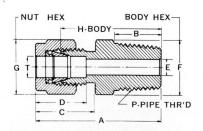

FIG. 36.7 This fitting is used to attach small tubing. (Courtesy of Crawford Fitting Co.)

| REDUCER | HALF COUPLING | PIPE CAP | SQUARE HEAD PLUG | HEX. HD. PLUG | ROUND HEAD PLUG | HEXAGON BUSHING | FLUSH BUSHING |

 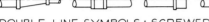 

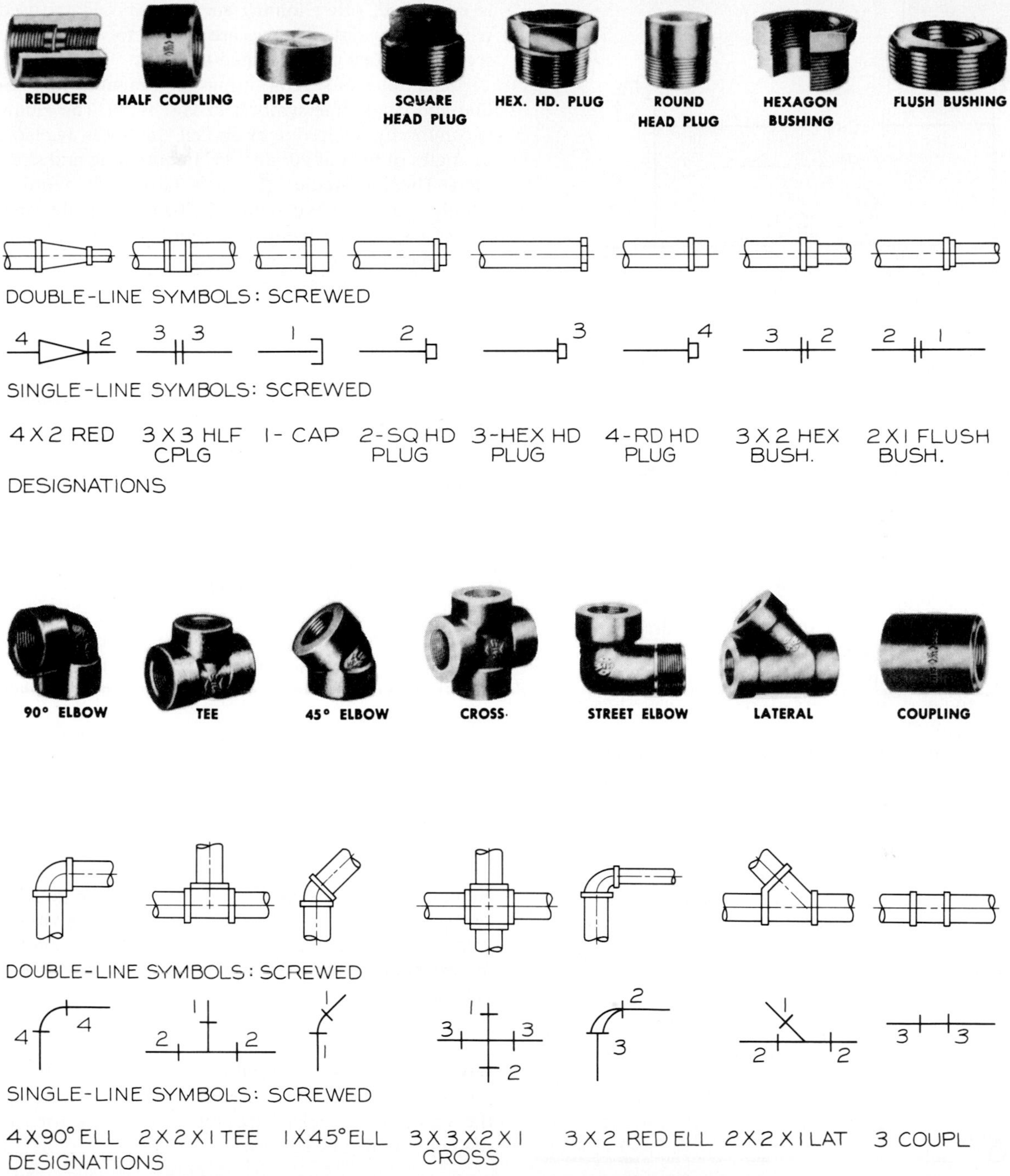

DOUBLE-LINE SYMBOLS: SCREWED

SINGLE-LINE SYMBOLS: SCREWED

4 X 2 RED   3 X 3 HLF CPLG   1-CAP   2-SQ HD PLUG   3-HEX HD PLUG   4-RD HD PLUG   3 X 2 HEX BUSH.   2 X 1 FLUSH BUSH.

DESIGNATIONS

| 90° ELBOW | TEE | 45° ELBOW | CROSS | STREET ELBOW | LATERAL | COUPLING |

     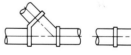 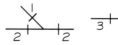

DOUBLE-LINE SYMBOLS: SCREWED

 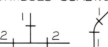

SINGLE-LINE SYMBOLS: SCREWED

4 X 90° ELL   2 X 2 X 1 TEE   1 X 45° ELL   3 X 3 X 2 X 1 CROSS   3 X 2 RED ELL   2 X 2 X 1 LAT   3 COUPL

DESIGNATIONS

FIG. 36.8 Examples of standard screwed fittings along with the single-line and double-line symbols that represent them. Nominal pipe sizes can be indicated by numbers placed near the joints. The major flow direction is labeled first and the branches labeled second. The large openings are labeled to precede the smaller openings.

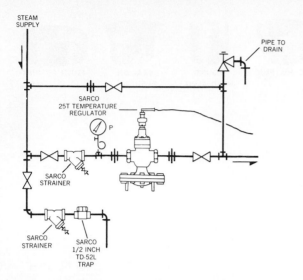

FIG. 36.9 A single-line piping system with the major valves represented as double-line symbols. (Courtesy of Sarco, Inc.)

Double-line symbols are more representative of the fittings and pipes since they are drawn to scale with double lines. Single-line symbols are more symbolic since the size of the pipe is drawn with a single line, and the fittings are drawn as single-line schematic symbols.

Fittings are available in three weights: standard (STD), extra strong (XS), and double extra strong (XXS). These weights match the standard weights of the pipes with which they will be connected. Other weights of fittings are available, but these three are stocked by practically all suppliers.

A piping system of screwed fittings is shown in Fig. 36.9, with double-line symbols in a single-line system to call attention to them. These could just as well have been drawn using the single-line symbols.

The most common symbols for representing fittings are shown in Table 36.2 on pages 708-709; these have been extracted from the ANSI Z 32.2.3 standards.

36.8
Flanged fittings

Flanges are used to connect fittings into a piping system when heavy loads are supported in large pipes and where pressures are great. Because using flanges

is expensive, other joining methods should be used whenever possible. Flanges are welded to straight pipe sections so they can be bolted together.

Examples of several fittings are drawn as double-line and single-line symbols in Fig. 36.10. The elbow is commonly referred to as an "ell," and it is available in angles of turn of 90° and 45° in both long and short radii. The long-radius (LR) ells have radii approximately 1.5 times the nominal diameter of the large end of the ell. The radius of a short-radius ell is equal to the diameter of the larger end.

A table of dimensions for 125-lb and 250-lb cast-iron fittings is given in Appendixes 11–13.

36.9
Welded fittings

Welding is a method of joining pipes and fittings for permanent, pressure-resistant joints. Examples of double-line and single-line fittings connected by welding are shown in Fig. 36.11. Fittings are available with beveled edges prepared for welding.

A piping layout in Fig. 36.12 illustrates a series of welded joints with a double-line drawing. The location of the welded joints has been dimensioned. So that flanges can be used, several flanged fittings have been welded into the system.

36.10
Valves

Valves are used to regulate the flow within a pipeline or to turn off the flow completely. Types of valves are *gate, globe, angle, check, safety, diaphragm, float,* and *relief,* to name a few. The four basic types of valves—gate, globe, angle, and check—are shown in Fig. 36.13 using single-line symbols.

GATE VALVES are used to turn the flow within a pipe on or off with the least restriction of flow through the valve. These valves are not meant to be used to regulate the degree of flow.

GLOBE VALVES are used to turn the flow on and off and to regulate the flow to a desired level.

ANGLE VALVES are types of globe valves that turn at 90° angles at bends in the piping system. They

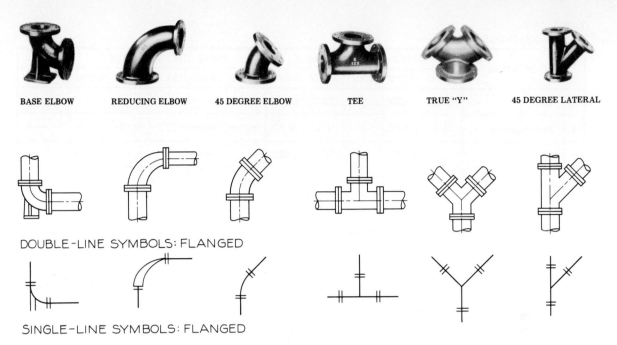

FIG. 36.10 Examples of standard flanged fittings along with the single-line and double-line symbols that represent them.

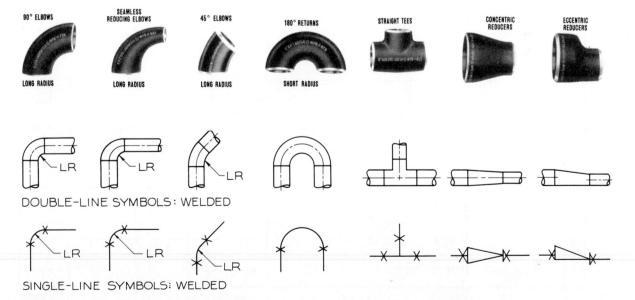

FIG. 36.11 Examples of standard welded fittings along with the single-line and double-line symbols that represent them.

TABLE 36.2
PIPING SYMBOLS

TYPE OF FITTING		DOUBLE LINE CONVENTION					SINGLE LINE CONVENTION					FLOW DIAGRAM
		FLANGED	SCREWED	B & S	WELDED	SOLDERED	FLANGED	SCREWED	B & S	WELDED	SOLDERED	
1	Joint											
2	Joint - Expansion											
3	Union											
4	Sleeve											
5	Reducer											
6	Reducer - Eccentric											
7	Reducing Flange											
8	Bushing											
9	Elbow - 45°											
10	Elbow - 90°											
11	Elbow - Long radius											
12	Elbow - (turned up)											
13	Elbow - (turned down)											
14	Elbow - Side outlet (outlet up)											
15	Elbow - Side outlet (outlet down)											
16	Elbow - Base											
17	Elbow - Double branch											
18	Elbow - Reducing											
19	Lateral											
20	Tee											
21	Tee - Single sweep											

Cont.

TYPE OF FITTING		DOUBLE LINE CONVENTION					SINGLE LINE CONVENTION					FLOW DIAGRAM
		FLANGED	SCREWED	B & S	WELDED	SOLDERED	FLANGED	SCREWED	B & S	WELDED	SOLDERED	
22	Tee-Double sweep											
23	Tee-(outlet up)											
24	Tee-(outlet down)											
25	Tee-Side outlet (outlet up)											
26	Tee-Side outlet (outlet down)											
27	Cross											
28	Valve - Globe											
29	Valve - Angle											
30	Valve - Motor operated globe											Motor operated.
31	Valve - Gate											
32	Valve - Angle gate											
33	Valve - Motor operated gate											Motor operated
34	Valve - Check											
35	Valve - Angle check											
36	Valve - Safety											
37	Valve - Angle safety											
38	Valve - Quick opening											
39	Valve - Float operating											
40	Stop Cock											

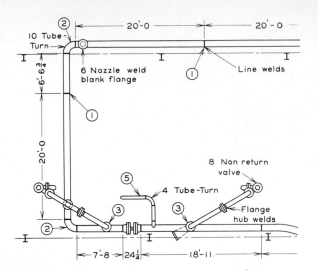

FIG. 36.12 A piping layout that uses a series of welded and flanged joints. This system is drawn using double-line symbols.

have the same controlling features as the straight globe valves.

CHECK VALVES restrict the flow in the pipe to only one direction. A backward flow is checked by either a movable piston or a swinging washer that will be activated by a change of flow.

The symbols for the other types of valves are shown in Fig. 36.14.

36.11
Fittings in orthographic views

Fittings and valves must be shown from any view in orthographic projection. Orthographic views of typical fittings and valves are shown in Fig. 36.15. These fittings are drawn as single-line screwed fittings, but the same general principles can be used to represent other types of joints as double-line drawings. Observe the

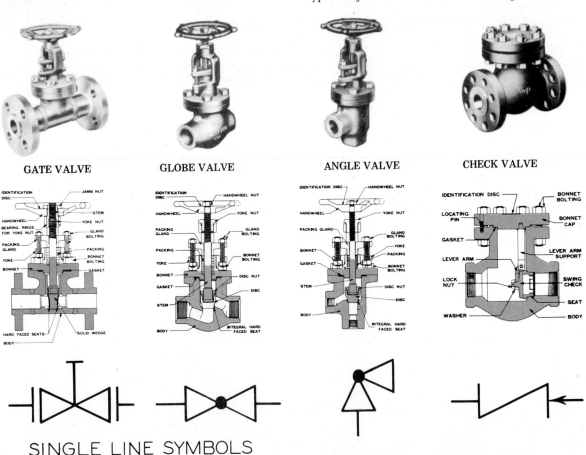

GATE VALVE   GLOBE VALVE   ANGLE VALVE   CHECK VALVE

SINGLE LINE SYMBOLS

FIG. 36.13 The four basic types of valves are gate, globe, angle, and check valves. (Courtesy of Vogt Machine Co.)

710

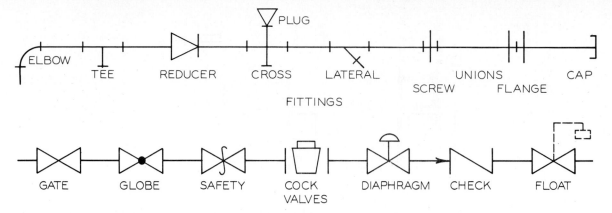

FITTINGS

FIG. 36.14 Examples of types of valves and fittings.

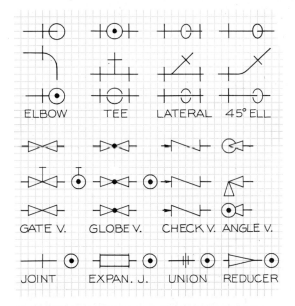

FIG. 36.15 Fittings and valves drawn with single-line symbols are drawn differently in these various orthographic views.

various views of the fittings and notice how the direction of an elbow can be shown by a slight variation in the different views.

A piping system is shown in a single orthographic view in Fig. 36.16, where a combination of double-line and single-line symbols are drawn. Arrows are used to give the direction of flow in the system. Different joints are screwed, welded, and flanged. Horizontal elevation lines give the heights of each horizon-

tal pipe. Station $5 + 12 - 0\text{-}\frac{1}{4}''$ represents a distance of 500 feet plus $12' - 0\text{-}\frac{1}{4}''$, or $512' - 0\text{-}\frac{1}{4}''$ from the beginning station point of $0 + 00$.

The dimensions in Fig. 36.16 are measured from the centerlines of the pipes, indicated by the CL symbols. In some cases, the elevations of the pipes are dimensioned to the bottom of the pipe, which is abbreviated BOP (Fig. 36.21).

36.12
Piping systems in pictorial

Isometric and axonometric drawings of piping systems are called *spool drawings;* they can be drawn using either single-line or double-line symbols.

A three-dimensional piping system is sketched orthographically in Fig. 36.17 with top and front views. Although this is a relatively simple three-dimensional system, a thorough understanding of orthographic projection is required to read the drawing.

In Fig. 36.18, the piping system is drawn with all the pipes revolved into the same horizontal plane. The vertical pipes and their fittings are drawn true size in the top view. This is called a *developed* pipe drawing. The fittings and pipe sizes are noted on this preliminary sketch from which the finished drawing will be made in Fig. 36.19.

An axonometric schematic, drawn in Fig. 36.20, explains the three-dimensional relationship of the parts of the system. The rounded bends in the elbows in an isometric drawing can be constructed with ellipses using the isometric ellipse template, or the cor-

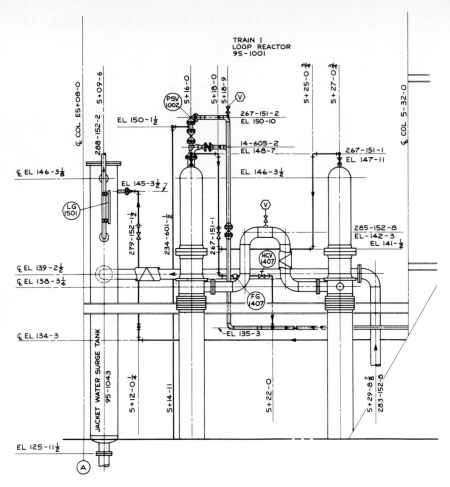

FIG. 36.16 This piping system is drawn using a combination of single-line and double-line symbols. The connections are shown as screwed, welded, and flanged. (Courtesy of Bechtel Corp.)

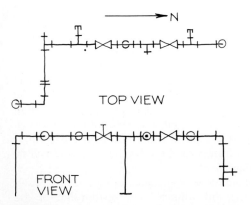

FIG. 36.17 Top and front views are used to represent this three-dimensional piping system with single-line symbols.

FIG. 36.18 The vertical pipes shown in Fig. 36.17 are revolved into the horizontal plane to form a developed drawing. The fittings and valves are noted on the sketch.

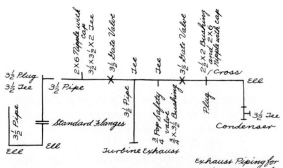

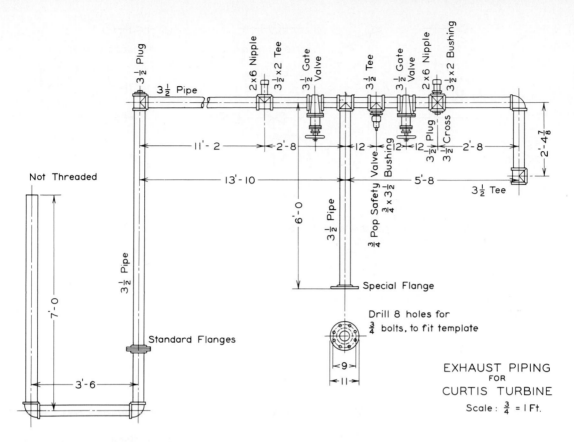

FIG. 36.19 A finished developed drawing that shows all the components in the system true size with double-line symbols.

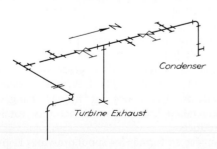

FIG. 36.20 An axonometric pictorial of the pipe system shown in Fig. 36.19 is drawn to give a three-dimensional picture of the system.

ners can be drawn square to reduce the effort and time required.

Piping layouts are often drawn as isometrics to clarify complex systems that are hard to interpret when shown in orthographic views. Some of the more often used piping connectors are shown in Fig. 36.21 as both orthographic and isometric views. Ellipse templates are used to represent circular arcs in isometric. These are single-line representations since the pipes are represented by single, heavy lines.

A portion of a pipe system is shown in Fig. 36.22 as an isometric drawing.

A north arrow is drawn on the plan view of the piping system in Fig. 36.17 to orient the isometric pictorial. This north direction is not necessarily related to compass north; rather, it is a direction parallel to a major set of pipes within the system.

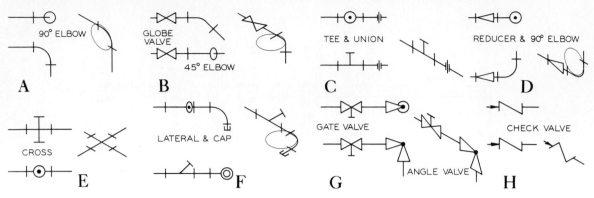

FIG. 36.21 A comparison of orthographic views and isometric pictorials of single-line representations of piping symbols. Note that the isometric template is used for constructing rounded corners in the isometric drawings.

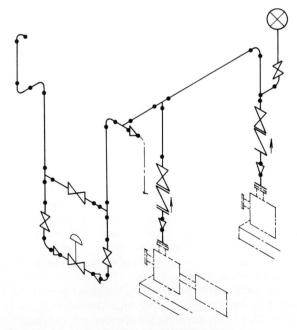

FIG. 36.22 A typical piping system drawn in isometric using the appropriate symbols.

36.13
Dimensioned isometrics

An isometric drawing can be drawn as a fully dimensioned and specified drawing from which a piping system can be constructed. The spool drawing in Fig. 36.23 is an example where the specifications for the pipe, fittings, flanges, and valves are noted on the drawing and are itemized in the bill of materials.

A number of abbreviations are used to specify piping components and fittings. Part number 1, for example, is an 8-in. diameter pipe of a Schedule 40 weight that is made of seamless steel by the open hearth (OH) process. Instead of OH, EF may be used, which is the abbreviation for electric furnace. Standard abbreviations associated with pipe drawings and specifications are given in Table 36.3.

Under the column "mat" (materials), you will notice a code beginning with the letter A, such as A–53. The letter A represents a grade of carbon steel listed in Table A of ANSI B31.3, *Petroleum Refinery Piping Standards*. The codes for fittings, flanges, and valves are taken from the manufacturers' catalogs of these products.

A suggested format for spool drawings is given in Fig. 36.24. This format is used by the Bechtel Corporation, a major construction company, in designing and constructing pipelines and refineries.

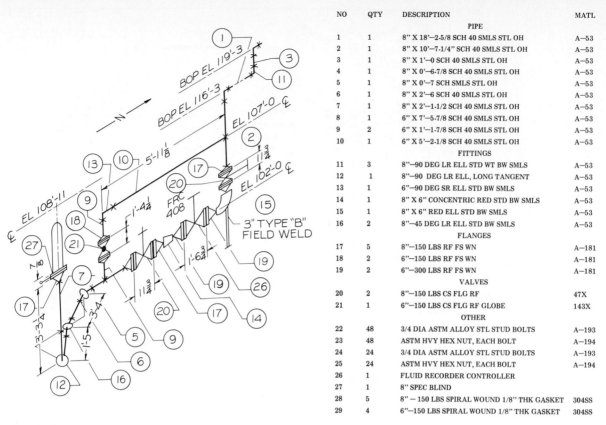

NO	QTY	DESCRIPTION	MATL
		PIPE	
1	1	8" X 18'–2-5/8 SCH 40 SMLS STL OH	A–53
2	1	8" X 10'–7-1/4" SCH 40 SMLS STL OH	A–53
3	1	8" X 1'–0 SCH 40 SMLS STL OH	A–53
4	1	8" X 0'–6-7/8 SCH 40 SMLS STL OH	A–53
5	1	8" X 0'–7 SCH SMLS STL OH	A–53
6	1	8" X 2'–6 SCH 40 SMLS STL OH	A–53
7	1	8" X 2'–1-1/2 SCH 40 SMLS STL OH	A–53
8	1	6" X 7'–5-7/8 SCH 40 SMLS STL OH	A–53
9	2	6" X 1'–1-7/8 SCH 40 SMLS STL OH	A–53
10	1	6" X 5'–2-1/8 SCH 40 SMLS STL OH	A–53
		FITTINGS	
11	3	8"–90 DEG LR ELL STD WT BW SMLS	A–53
12	1	8"–90 DEG LR ELL, LONG TANGENT	A–53
13	1	6"–90 DEG SR ELL STD BW SMLS	A–53
14	1	8" X 6" CONCENTRIC RED STD BW SMLS	A–53
15	1	8" X 6" RED ELL STD BW SMLS	A–53
16	2	8"–45 DEG LR ELL STD BW SMLS	A–53
		FLANGES	
17	5	8"–150 LBS RF FS WN	A–181
18	2	6"–150 LBS RF FS WN	A–181
19	2	6"–300 LBS RF FS WN	A–181
		VALVES	
20	2	8"–150 LBS CS FLG RF	47X
21	1	6"–150 LBS CS FLG RF GLOBE	143X
		OTHER	
22	48	3/4 DIA ASTM ALLOY STL STUD BOLTS	A–193
23	48	ASTM HVY HEX NUT, EACH BOLT	A–194
24	24	3/4 DIA ASTM ALLOY STL STUD BOLTS	A–193
25	24	ASTM HVY HEX NUT, EACH BOLT	A–194
26	1	FLUID RECORDER CONTROLLER	
27	1	8" SPEC BLIND	
28	5	8" – 150 LBS SPIRAL WOUND 1/8" THK GASKET	304SS
29	4	6"–150 LBS SPIRAL WOUND 1/8" THK GASKET	304SS

FIG. 36.23 This dimensioned isometric pictorial is called a "spool drawing." Because it is sufficiently complete, it can serve as a working drawing when used with the bill of materials.

TABLE 36.3
STANDARD ABBREVIATIONS ASSOCIATED WITH PIPE SPECIFICATIONS

AVG	average	FS	forged steel	SPEC	specification
BC	bolt circle	FSS	forged stainless steel	SR	short radius
BE	beveled ends	FW	field weld	SS	stainless steel
BF	blind flange	GALV	galvanized	STD	standard
BM	bill of materials	GR	grade	STL	steel
BOP	bottom of pipe	ID	inside diameter	STM	steam
B&S	bell & spigot	INS	insulate	SW	socketweld
BWG	Birmingham wire gauge	IPS	iron pipe size	SWP	standard working pressure
CAS	cast alloy steel	LR	long radius	TC	Test connection
CI	cast iron	LW	lap weld	TE	threaded end
CO	clean out	MI	malleable iron	TEMP	temperature
CONC	concentric	MFG	manufacture	T&G	tongue & groove
CPLG	cloupling	OD	outside diameter	TOS	top of steel
CS	carbon steel, cast steel	OH	open hearth	TYP	typical
DWG	drawing	PE	plain end—not beveled	VC	vitrified clay
ECC	eccentric	PR	pair	WE	weld end
EF	electric furnace	RED	reducer	WN	weld neck
EFW	electric fusion weld	RF	raised face	WB	welded bonnet
ELEV	elevation	RTG or RJ	right type joint	WT	weight
ERW	electric resistance weld	SCH	schedule	XS	extra strong
FF	flat face	SCRD	screwed	XXS	double extra strong
FLG	flange	SMLS	seamless		
FOB	flat on bottom	SO	slip-on		

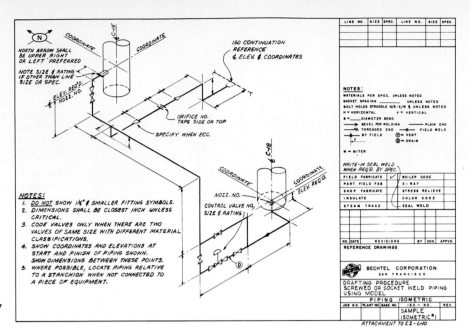

FIG. 36.24 A suggested format for preparing spool drawings. (Courtesy of the Bechtel Corp.)

Problems

1. On a Size A sheet, draw five orthographic views of the fittings listed below. The views should include the front view, the top view, the bottom view, and the left and right views. Draw two fittings per page. Refer to Fig. 36.15 and Table 36.2 as guides in making these drawings. Use single-line symbols to draw the following screwed fittings: 90° ell, 45° ell, tee, lateral, cap reducing ell, cross, concentric reducer, check valve, union, globe valve, gate valve, and bushing.

2. Same as Problem 1, but draw the fittings as flanged fittings.

3. Same as Problem 1, but draw the fittings as welded fittings.

4. Same as Problem 1, but draw the fittings as double-line screwed fittings.

5. Same as Problem 1, but draw the fittings as double-line flanged fittings.

6. Same as Problem 1, but draw the fittings as double-line welded fittings.

7. Convert the single-line sketch in Fig. 36.25 into a double-line system that will fit on a Size A sheet.

8. Convert the pipe system shown in isometric in Fig. 36.18 into a double-line isometric drawing that will fit on a Size B sheet.

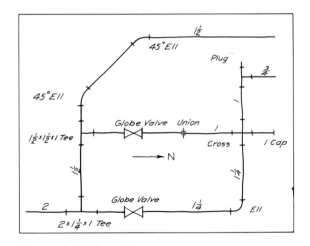

FIG. 36.25 Problem 7.

9. Convert the pipe system given in Fig. 36.18 into a two-view orthographic drawing using single-line symbols.

10. Convert the isometric drawing of the pipe system in Fig. 36.24 into a two-view orthographic drawing that will fit on a Size B sheet. Take the measurements from the given drawing, and select a convenient scale.

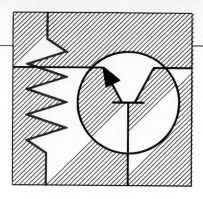

CHAPTER 37

Electric/Electronics Drafting

37.1
Introduction

Electric/electronics drafting is a specialty area of the field of drafting technology. *Electrical drafting* is related to the transmission of electrical power that is used in large quantities in homes and industry for lighting, heating, and equipment operation. *Electronics drafting* deals with circuits in which integrated circuits or transistors are used, where power is used in small quantities; examples of electronic equipment are radios, televisions, computers, and video cassette recorders.

Electronics drafters are responsible for the preparation of drawings that will be used in fabricating the circuit, and thereby bringing the product into being. They work from sketches and specifications developed by the engineer or electronics technologist. This chapter will review the drafting principles necessary for the preparation of electronic diagrams.

A major portion of the text of this chapter has been extracted from ANSI Y14.15, *Electrical and Electronics Diagrams*, the standards that regulate the draft-

ing techniques used in this area. The symbols used were taken from ANSI Y32.2, *Graphic Symbols for Electrical and Electronics Diagrams*.

37.2
Types of diagrams

Electronic circuits are classified and drawn in the format of one of the following types of diagrams:

1. single-line diagrams,
2. schematic diagrams, or
3. connection diagrams.

The suggested line weights for drawing these diagrams are shown in Fig. 37.1.

Single-line diagrams

Single-line diagrams use single lines and graphic symbols to show an electric circuit or system of circuits

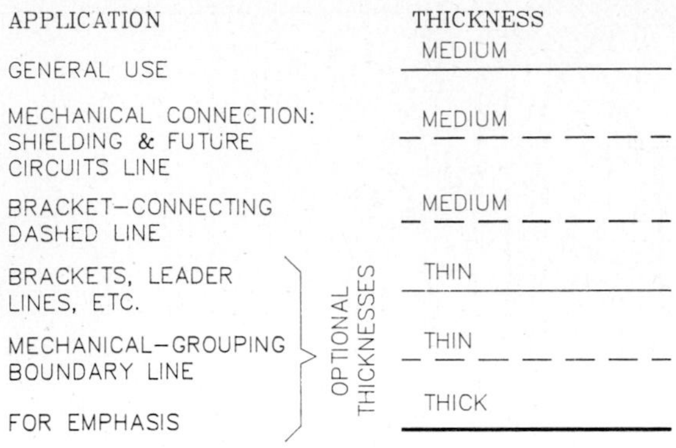

APPLICATION	THICKNESS
GENERAL USE	MEDIUM
MECHANICAL CONNECTION: SHIELDING & FUTURE CIRCUITS LINE	MEDIUM
BRACKET—CONNECTING DASHED LINE	MEDIUM
BRACKETS, LEADER LINES, ETC.	THIN
MECHANICAL—GROUPING BOUNDARY LINE	THIN
FOR EMPHASIS	THICK

OPTIONAL THICKNESSES

FIG. 37.1 The recommended line weights for drawing electronics diagrams.

and the parts and devices within it. A single line is used to represent both AC and DC systems, as illustrated in Fig. 37.2. An example of a single-line diagram of an audio system is shown in Fig. 37.3. Primary circuits are indicated by thick connecting lines, and medium lines represent connections to the current and potential sources.

Single-line diagrams show the connections of meters, major equipment, and instruments. In addition, ratings are often given to provide such information as kilowatts, voltages, cycles and revolutions per minute, and generator ratings.

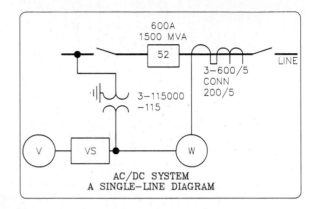

FIG. 37.2 A portion of a single-line diagram where heavy lines represent the primary circuits, and medium lines represent the connections to the current and potential sources.

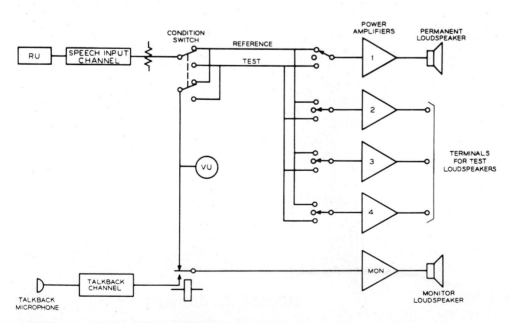

FIG. 37.3 A typical single-line diagram for illustrating electronics and communications circuits. (Courtesy of ANSI.)

Schematic diagrams

Schematic diagrams use graphic symbols to show the electrical connections and functions of a specific circuit arrangement. Although the schematic diagram enables one to trace the circuit and its functions, the physical size, shapes, and locations at various components are not given. A schematic diagram is illustrated in Fig. 37.4 on pages 720-721.

Connection diagrams

Connection diagrams show the connections and installations of the parts and devices of the system. In addition to showing the internal connections, external connections, or both, they show the physical arrangement of the parts. A three-dimensional connection diagram is shown in Fig. 37.5.

37.3
Schematic diagram connecting symbols

The most basic symbols of a circuit are those used to represent connections of parts within the circuit. Using dots to show connections is optional; it is preferable

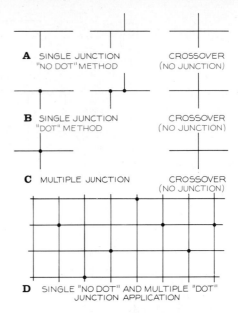

A SINGLE JUNCTION "NO DOT" METHOD   CROSSOVER (NO JUNCTION)

B SINGLE JUNCTION "DOT" METHOD   CROSSOVER (NO JUNCTION)

C MULTIPLE JUNCTION   CROSSOVER (NO JUNCTION)

D SINGLE "NO DOT" AND MULTIPLE "DOT" JUNCTION APPLICATION

FIG. 37.6 A. Connections should be shown with single-point junctions. **B.** Dots may be used to call attention to connectors. **C.** Dots *must* be used when there are multiples of the type shown here.

to omit them if clarity is not sacrificed by so doing. Connections, or junctions, are indicated by using small black dots (Fig. 37.6A). The dots distinguish between connecting lines and those that simply pass over each other (Fig. 37.6B). It is preferred that connecting wires have single junctions wherever possible. When the layout of a circuit does not permit the use of single junctions, and lines within the circuit must cross, dots must be used to distinguish between crossing and connecting lines (Figs. 36.6A and B).

INTERRUPTED PATHS are breaks in lines within a schematic diagram; they are used to conserve space when this can be done without confusion. For example, the circuit in Fig. 37.7 has been interrupted; instead, of connecting the left and right sides of the circuit, the lines are labeled to correspond to the matching notes at the other side of the interrupted circuit.

Occasionally sets of lines in a horizontal or vertical direction will be interrupted (Fig. 37.8). Brackets will be used to interrupt the circuit, and notes will be placed outside the brackets to indicate the destinations of the wires or their connections.

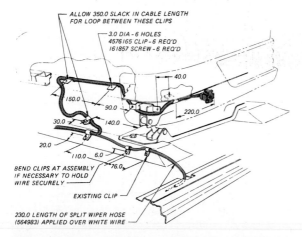

FIG. 37.5 This is a three-dimensional connection diagram that shows the circuit and its components with the dimensions necessary to explain how it is connected or installed. (Courtesy of the General Motors Corp.)

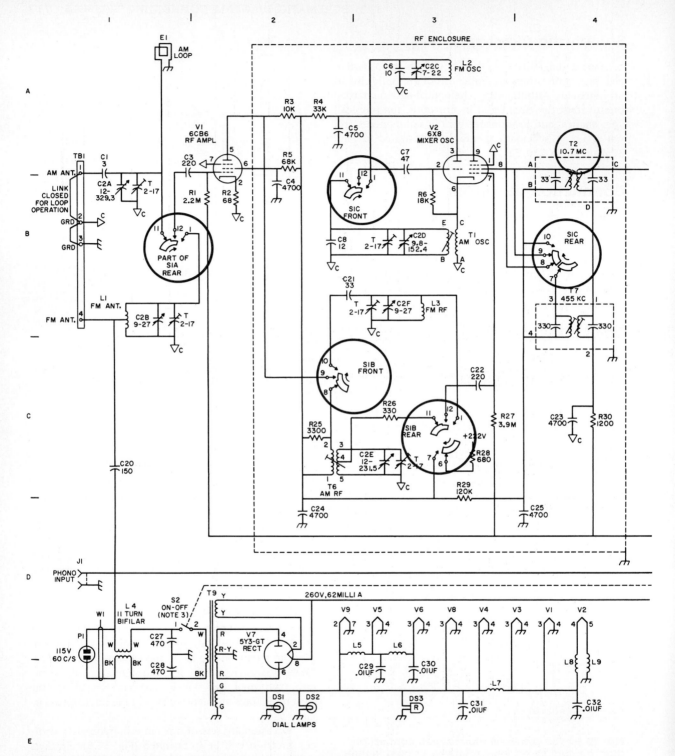

FIG. 37.4 A typical schematic diagram of an AM/FM radio circuit with all parts of the system labeled. The title of a drawing of this type should specify it is a schematic diagram as well as the type of circuit. (Courtesy of ANSI.)

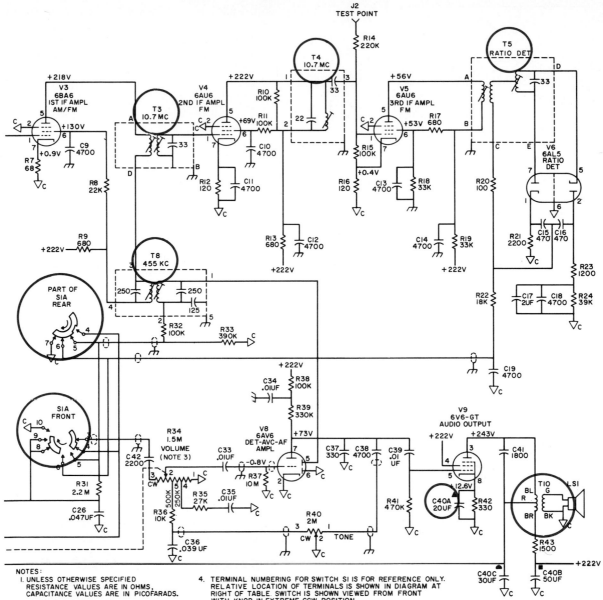

NOTES:
1. UNLESS OTHERWISE SPECIFIED RESISTANCE VALUES ARE IN OHMS, CAPACITANCE VALUES ARE IN PICOFARADS.

2. VOLTAGES MEASURED TO COMMON WIRING WITH VTVM.

3. SWITCH S2 IS PART OF POTENTIOMETER R34. TERMINAL NUMBERING IS FOR REFERENCE ONLY. RELATIVE LOCATION OF TERMINALS IS SHOWN IN DIAGRAM AT RIGHT.

4. TERMINAL NUMBERING FOR SWITCH S1 IS FOR REFERENCE ONLY. RELATIVE LOCATION OF TERMINALS IS SHOWN IN DIAGRAM AT RIGHT OF TABLE. SWITCH IS SHOWN VIEWED FROM FRONT WITH KNOB IN EXTREME CCW POSITION.

R34 TERMINAL SIDE

POS	FUNCTION	SWITCH SECTIONS AND TERMINALS CONNECTED					
		SIA		SIB		SIC	
		FRONT	REAR	FRONT	REAR	FRONT	REAR
I	PHONO (SHOWN)	3-4	5-6-7	—	—	—	8-9
2	AM RADIO	3-5, 8-9	4-6-7, 11-12	8-9	6-7, 11-12	11-12	7-8, 9-10
3	FM RADIO	3-6, 9-10	4-5-7, 12-1	9-10	6-7, 12-1	12-1	8-9

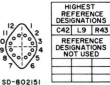

HIGHEST REFERENCE DESIGNATIONS		
C42	L9	R43
REFERENCE DESIGNATIONS NOT USED		

SD-802151

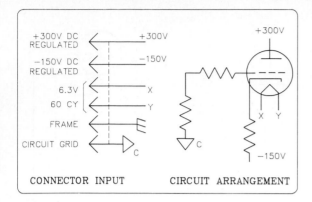

FIG. 37.7 Circuits may be interrupted and connections not shown by lines if they are properly labeled to clarify their relationship to the removed part of the circuit. The connections above are labeled to match those on the left and right sides of the illustration.

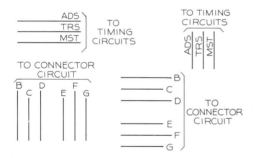

FIG. 37.8 Brackets and notes may be used to specify the destinations of interrupted circuits.

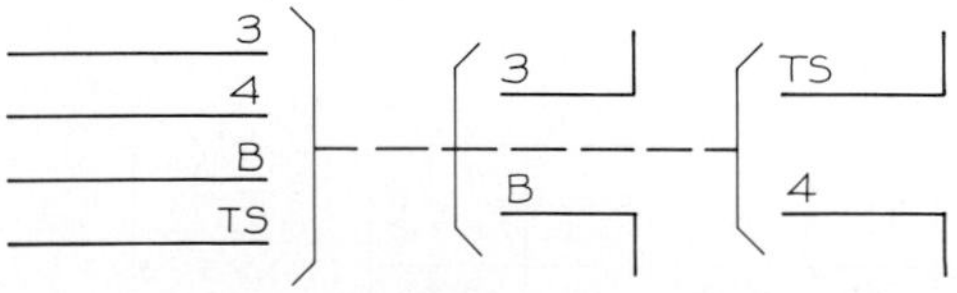

FIG. 37.9 The connections of interrupted circuits can be indicated by using brackets and a dashed line in addition to labeling the lines. The dashed line should not be drawn to appear as an extension of one of the lines in the circuit.

In some cases, a dashed line is used to connect brackets that interrupt circuits (Fig. 37.9). The dashed line should be drawn so that it will not be mistaken as a continuation of one of the lines within the bracket.

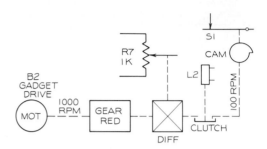

FIG. 37.10 If mechanical functions are closely related to electrical functions, it may be desirable to link the mechanical components within the schematic diagram.

MECHANICAL LINKAGES that are closely related to electronic functions may be shown as part of a schematic diagram (Fig. 37.10). This type of arrangement helps clarify the relationship of the electronics circuit to the mechanical components.

37.4
Graphic symbols

The electronics drafter must be familiar with the basic graphic symbols used to represent the parts and devices within electrical and electronics circuits.

The symbols covered in this article are extracted from the ANSI Y32.2 standards and are adequate for practically all diagrams. However, when a highly specialized part needs to be shown and a symbol for it is not provided in these standards, it is permissible for the drafter to develop a symbol provided it is properly labeled and its meaning clearly conveyed.

The symbols presented in Figs. 37.11–37.15 are drawn on a grid of 5 mm (0.20 in.) squares that has been reduced. The actual dimensions of the symbols can be approximated by using the grid and equating each square to its full-size measurement of 5 mm. Symbols may be drawn larger or smaller to fit the size

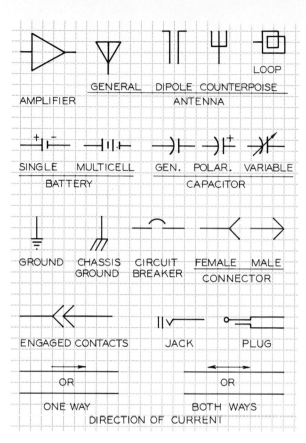

FIG. 37.11 These graphic symbols parts within a schematic diagram are drawn on a 5 mm (0.20 in.) grid. The suggested sizes of the symbols can be found by taking the dimensions from the grid to draw the symbols full size.

FIG. 37.12 The upper six symbols are used to represent often used types of electron tubes. The interpretation of the parts that make up each symbol is given in the lower half of the figure.

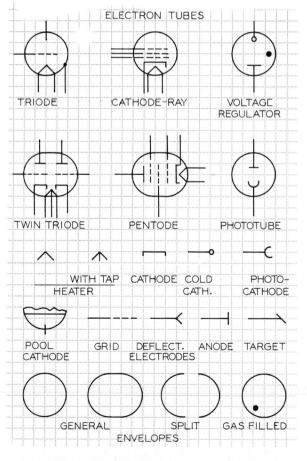

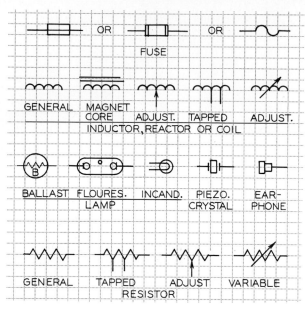

FIG. 37.13 Graphics symbols of standard circuit components.

FIG. 37.14 Graphics symbols for semiconductors. The arrows in the middle of the figure illustrate the meanings of the arrows used in the transistor symbols.

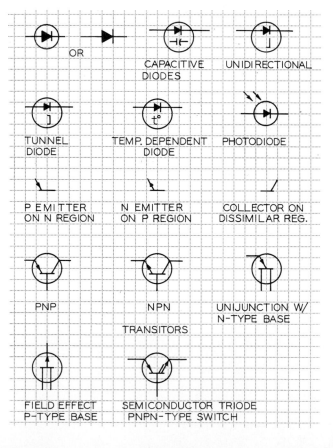

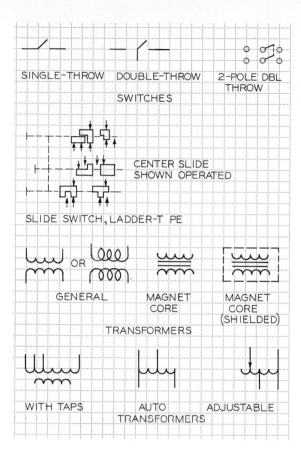

FIG. 37.15 Graphics symbols for representing switches and transformers.

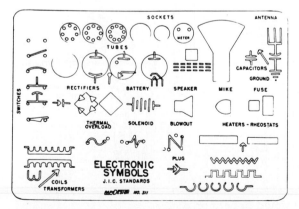

FIG. 37.16 Many types of templates are available for drawing the graphics symbols in schematic drawings. (Courtesy of Frederick Post Co.)

of your layout provided the relative proportions of the symbols are kept about the same. The symbols in these figures are but a few of the more commonly used symbols. Nearly 600 different variations of the basic electrical/electronics symbols are listed in the ANSI standards.

Electronics symbols can be drawn with conventional drawing equipment and instruments by approximating the proportions of the symbols; drafting templates of electronics symbols are also available (Fig. 37.16).

Some of the symbols provide designations of their sizes or ratings. The need for this information depends on the requirements and the usage of the schematic diagrams.

37.5
Terminals

Terminals are the ends of devices that are attached in a circuit with connecting wires. Examples of devices with terminals are switches, relays, and transformers. The graphic symbol for a terminal is an open circle; it is the same size as the solid circle used to indicate a connection.

SWITCHES are used to turn a circuit on or off, or to actuate a certain part of it while turning another part off. Examples of labeling switches are shown in the schematic diagram in Fig. 37.4. In this case, a table clarifies the switching connections of the terminals.

When a group of parts is enclosed or shielded (drawn enclosed with dashed lines) and the terminal circles have been omitted, the terminal markings should be placed immediately outside the enclosure, as shown in Fig. 37.4 at T2, T3, T4, T5, and T8. Terminal identifications should be added to the graphic symbols that correspond to the actual physical markings that appear on or near the terminals of the part, such as 10.7 MC for transformer T3 in Fig. 37.4. Several examples of notes and symbols that explain the parts of a diagram are shown in Figs. 37.17, 37.18, and 37.19.

Colored wires or symbols are often used to identify the various leads that connect to terminals. When colored wires need to be identified on a diagram, the colors are lettered on the drawing, as was done for transformer T10 in Fig. 37.4. An example of the iden-

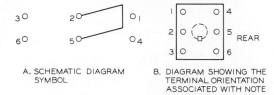

FIG. 37.17 A. An example of a method of labeling the terminals of a toggle switch on a schematic diagram. **B.** A diagram that illustrates the toggle switch when viewed from its rear.

A. SCHEMATIC DIAGRAM SYMBOL

B. DIAGRAM SHOWING THE TERMINAL ORIENTATION ASSOCIATED WITH NOTE

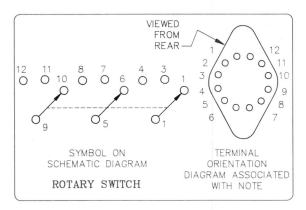

FIG. 37.18 An example of a rotary switch as it would appear on a schematic diagram, and a diagram that shows the numbered terminals of the switch when viewed from its rear.

tification of a capacitor, C40, marked with geometric symbols is shown in Fig. 37.4.

ROTARY TERMINALS are used to regulate the resistance in some circuits, and the direction of rotation of the dial is indicated on the schematic diagram. The abbreviations CW (clockwise) or CCW (counterclockwise) are placed adjacent to the movable contact when it is in its extreme clockwise position, as shown in Fig. 37.20A. The movable contact has an arrow at its end.

If the device terminals are not marked, numbers may be used with the resistor symbol and the number 2 assigned to the adjustable contact (Fig. 37.20B). Other fixed types may be sequentially numbered and added (Fig. 37.20C).

The position of a switch as it relates to the function of a circuit should be indicated on a schematic

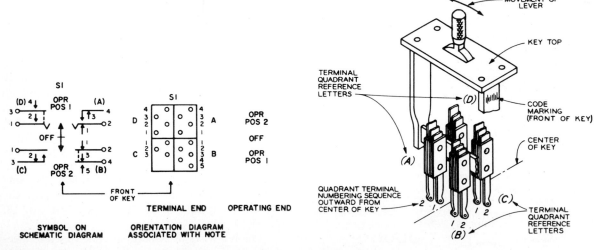

FIG. 37.19 A. An example of a typical lever switch as it would appear on a schematic diagram, and' an orientation diagram that shows the numbered terminals of the switch when viewed from its operating end. **B.** A pictorial of the lever switch and its four quadrants. (Courtesy of ANSI.)

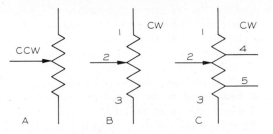

FIG. 37.20 **A.** To indicate the direction of rotation of rotary switches on a schematic diagram, the abbreviations CW (clockwise) and CCW (counterclockwise) are placed near the movable contact. **B.** If the device terminals are not marked, numbers may be used with the resistor symbols and the number 2 assigned to the adjustable contact. **C.** Additional contacts may be labeled as shown.

COMPLEX SWITCHES

SI VOLTAGE TEST	
FUNCTION	TERM
OFF	1–2
+100V REG	1–3
+150V REG	1–4
+300 V UNREG	1–5
+450V UNREG	1–6

FUNCTIONS SHOWN IN TABULAR FORM

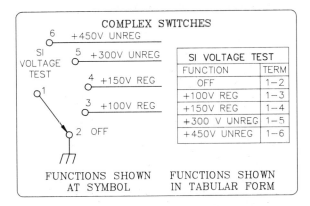

FIG. 37.21 For more complex switches, position-to-position function relations may be shown using symbols on the schematic diagram, or they may be shown by a table of values located elsewhere on the diagram.

diagram. A method of showing functions of a variable switch is shown in Fig. 37.21. The arrow represents the movable end of the switch that can be positioned to connect with several circuits. The different functional positions of the rotary switch are shown both by symbol and by table.

Another method of representing a rotary switch is shown in Fig. 37.22 by symbol and by table. Because of the complexity of this particular switch, the tabular form is preferred. The dashes between the numbers in the table indicate the numbers have been connected.

S1 (REAR)		
POS	FUNCTION	TERM.
1	OFF (SHOWN)	1–2, 5–6, 9–10
2	STANDBY	1–3, 5–7, 9–11
3	OPERATE	1–4, 5–8, 9–12

FUNCTIONS SHOWN IN TABULAR FORM

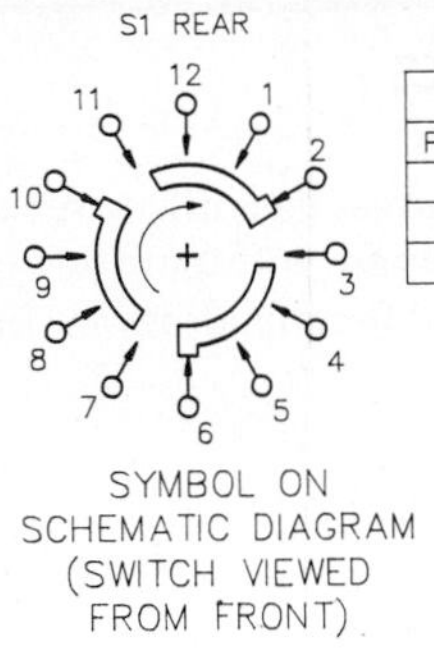

FIG. 37.22 A rotary switch may be shown on a schematic diagram with its terminals labeled as shown at the left, or its functions can be given in a table placed elsewhere on the drawing as shown at the right. Dashes are used to indicate the linkage of the numbered terminals. For example, 1–2 means terminals 1 and 2 are connected in the "off" position.

For example, when the switch is in position 2 the following terminals are connected: 1 and 3, 5 and 7, and 9 and 11. A table of this type is used at the bottom of Fig. 37.4.

Electron tubes have pins that fit into sockets, which have terminals that connect with circuits. Pins are labeled with numbers placed outside the symbol used to represent the tube (Fig. 37.23) and with the tube viewed from its bottom, are numbered in a clockwise direction.

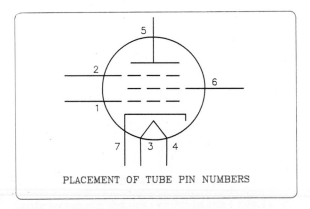

FIG. 37.23 Tube pin numbers should be placed outside the tube envelope and adjacent to the connecting lines.

37.6

Separation of parts

In complex circuits, it is often advantageous to separate elements of a multielement part with portions of the graphic symbols drawn in different locations on

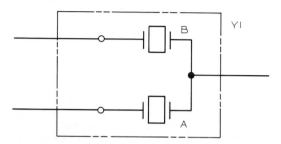

FIG. 37.24 As subdivisions within the complete part, crystals A and B are referred to as Y1A and Y1B.

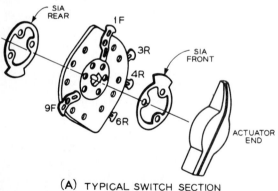

(A) TYPICAL SWITCH SECTION

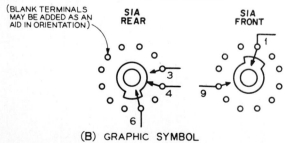

(B) GRAPHIC SYMBOL

FIG. 37.25 Parts of rotary switches are designated with suffix letters A, B, C, etc. and are referred to as S1A, S1B, S1C, etc. The words FRONT and REAR are added to these designations when both sides of the switch are used. (Courtesy of ANSI.)

the drawing. An example of this method of separation is the switch labeled S1A, S1B, S1C, and so on in Fig. 37.4. The switch is labeled S1, and the letters that follow, called suffixes, designate different parts of the switch. Suffix letters may also be used to label subdivisions of an enclosed unit made up of a series of internal parts, such as the crystal unit shown in Fig. 37.24. These crystals are referred to as Y1A and Y1B.

Rotary switches of the type shown in Fig. 37.25 are designated S1A, S1B, and so on. The suffix letters are labeled in sequence beginning with the knob and working away from it. Each end of the various sections of the switch should be viewed from the same end. When the rear and front of the switches need to be used, the words FRONT and REAR are added to the designations.

Portions of items such as terminal boards, connectors, or rotary switches may be separated on a diagram. The words PART OF may precede the identification of the portion of the circuit of which it is a part, as shown in Fig. 37.26A. A second method of showing a part of a system is by using conventional break lines, which makes the note, PART OF, unnecessary.

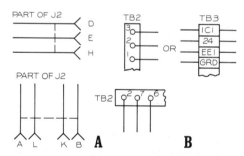

FIG. 37.26 The portions of connectors or terminal boards are functionally separated on a diagram; the words PART OF may precede the reference designation of the entire portion. Or, conventional breaks can be used to indicate graphically that the part drawn is only a portion of the whole.

37.7

Reference designations

A combination of letters and numbers that identify items on a schematic diagram are called reference designations. These designations identify the components not only on the drawing but also in the documents referring to them. Reference designations should be placed close to the symbols that represent the replaceable items of a circuit on a drawing. Items not separately replaceable may be identified if this is thought necessary. Mounting devices for electron tubes, lamps, fuses, and so forth are seldom identified on schematic diagrams.

Standard practice is to begin each reference designation with an uppercase letter that may be followed by a numeral with no hyphen between them. The number usually represents a portion of the part. The lowest number of a designation should be assigned to begin at the upper left of the schematic diagram and proceed consecutively from left to right and top to bottom throughout the drawing.

Some of the standard abbreviations that designate parts of an assembly are amplifier–A, battery–BT, capacitor–C, connector–J, piezoelectric crystal–Y, fuse–F, electron tube–V, generator–G, rectifier–CR, resistor–R, transformer–T, and transistor–Q.

As the circuit is being designed, some of the numbered elements may be deleted from the drawing. The remaining numbered elements should not be renumbered even though there is a missing element within the sequence of numbers. Instead, a table like the one shown in Fig. 37.27 can be used to list the parts omitted from the circuit. The highest designations are also given in the table to be sure that all parts were considered.

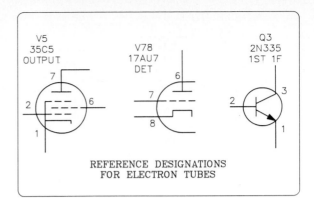

REFERENCE DESIGNATIONS
FOR ELECTRON TUBES

FIG. 37.28 Three lines of notes can be used with electron tubes to specify reference designations, type designations, and function. This information should be located adjacent to the symbol and preferably above it.

Electron tubes are labeled not only with reference designations but with type designation and circuit function also, as shown in Fig. 37.28. This information is labeled in three lines, such as V5/35C5/OUTPUT, which are located adjacent to the symbol.

37.8
Numerical units of function

Functional units such as the values of resistance, capacitance, inductance, and voltage should be specified with the fewest number of zeros by using the multipliers in Fig. 37.29A as prefixes. Examples using this method of expression are shown in Figs. 37.29B and C, where units of resistance and capacitance are given. When four-digit numbers are given, the commas should be omitted (e.g., one thousand should be written 1000, not 1,000). You should recognize and use the lowercase or uppercase prefixes indicated in Fig. 37.29A.

Where certain units are repeated, general notes can be used to reduce time and effort:

UNLESS OTHERWISE SPECIFIED: RESISTANCE
    VALUES ARE IN OHMS
CAPACITANCE VALUES ARE IN
    MICROFARADS.

CAPACITANCE VALUES ARE IN PICOFARADS.

HIGHEST REFERENCE DESIGNATIONS	
R72	C40
REFERENCE DESIGNATIONS NOT USED	
R8, R10, R61	C12, C15, C17
R64, R70	C20, C22

FIG. 37.27 Reference designations are used to identify parts of a circuit. They are labeled in a numerical sequence from left to right beginning at the upper left of the diagram. If parts are later deleted from the system, the ones deleted should be listed in a table along with the highest reference number designations.

A. MULTIPLIERS

MULTIPLIER	PREFIX	SYMBOL	
		METHOD 1	METHOD 2
10^{12}	TERA	T	T
10^9	GIGA	G	G
10^6 (1,000,000)	MEGA	M	M
10^3 (1,000)	KILO	k	K
10^{-3} (0.001)	MILLI	m	MILLI
10^{-6} (0.000,001)	MICRO	μ	U
10^{-9}	NANO	n	N
10^{-12}	PICO	p	P
10^{-13}	FEMTO	f	F
10^{-16}	ATTO	a	A

B. RESISTANCE

RANGE IN OHMS	EXPRESS AS	EXAMPLE
LESS THAN 1,000	OHMS	0.031 470
1,000 TO 99,999	OHMS OR KILOHMS	1800 15,853 10k 82k
100,000 to 999,999	KILOHMS OR MEGOHMS	220k 0.22M
1,000,000 OR MORE	MEGOHMS	3.3M

C. CAPACITANCE

RANGE IN PICOFARADS	EXPRESS AS	EXAMPLE
LESS THAN 10,000	PICOFARADS	152.4pF 4700pF
10,000 OR MORE	MICROFARADS	0.015μF 30μF

FIG. 37.29 A. Multipliers should be used to reduce the number of zeros in a number. **B.** and **C.** Examples of expressing units of capacitance and resistance.

A note for specifying capacitance values is

CAPACITANCE VALUES SHOWN AS NUMBERS EQUAL TO OR GREATER THAN UNITS ARE IN pF AND NUMBERS LESS THAN UNITY ARE IN μF.

Examples of the placement of the reference designations and the numerical values of resistors are shown in Fig. 37.30.

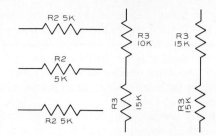

FIG. 37.30 Methods of labeling the units of resistance on a schematic diagram.

37.9
Functional identification of parts

The readability of a circuit is improved if parts are labeled to indicate their functions. Test points are labeled on drawings with the letters Tp and their suffix numbers. The sequence of the suffix numbers should be the same as the sequence of troubleshooting the circuit when it is defective. As an alternative, the test function can be indicated on the diagram below the reference designation.

Additional information may be included on the schematic diagram to aid in the maintenance of the

FIG. 37.31 The result of the miniaturization of electronic devices can be seen here where a solid-state logic technology (SLT) chip the size of a dot gives the same function as its predecessors, the vacuum tube and the transistor. (Courtesy of IBM.)

system:

- DC resistance of windings and coils
- Critical input and output impedance values
- Wave shapes (voltage or current) at significant points
- Wiring requirements for critical ground points, shielding, pairing, etc.
- Power or voltage ratings of parts
- Caution notation for electrical hazards at maintenance points
- Circuit voltage values at significant points (tube pins, test points, terminal boards, etc.)
- Zones (grid system) on complex schematics
- Signal flow direction in main signal paths.

37.10
Printed circuits

Printed circuits are universally used for miniature electronic components and computer systems. Some sense of the degree of miniaturization that has occurred in the electronics industry can be gathered from Fig. 37.31, where the function of a vacuum tube has been replaced by a silicon chip transistor the size of a dot.

FIG. 37.32 This is a magnified view of a printed circuit board where the circuit has been printed and etched on the board and the devices have been soldered into position. (Courtesy of Bishop Industries Corp.)

The drawings of printed circuits are drawn four or more times the size of the circuit ultimately printed. The drawings are usually drawn in black India ink on acetate film and are then photographically reduced to the desired size. The circuit is "printed" onto an insulated board made of plastic or ceramics, and the devices within the circuit are connected (Fig. 37.32).

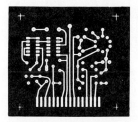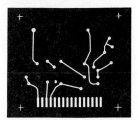

FIG. 37.33 A printed circuit attached to two sides of the circuit board requires two circuit drawings, one for each side. The drawings are photographically converted to reduced negatives for printing. (Courtesy of Bishop Industries Corp.)

Some printed circuits are printed on both sides of the circuit board, which requires two photographic negatives, as shown in Fig. 37.33, made from positive drawings (black lines on a white background). Each drawing for each side can be made on separate sheets of acetate that are laid over each other when the second diagram is drawn. However, a more efficient method of making a single drawing, from which two negatives are photographically made, uses red and blue tape (Fig. 37.34). A filter on the process camera drops out the red for one negative, and a different filter drops out the blue for the second negative.

37.11
Shortcut symbols

Several manufacturers produce preprinted symbols that can be used for "drawing" electronic and printed circuits. The symbols are available on sheets or tapes that can be burnished onto the surface of the drawing to form a permanent schematic diagram (Fig. 37.35). The symbols can be connected with matching tape to represent wires between them instead of drawing the

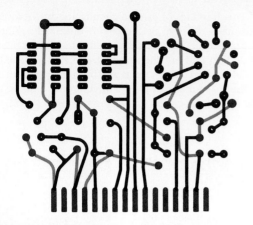

FIG. 37.34 By using two colors, such as blue and red, one circuit drawing can be made. And from the same drawing, two negatives can be made by using camera filters that screen out one of the colors with each shot. The circuits are then printed on each side of the board. (Courtesy of Bishop Industries Corp.)

lines. Schematic diagrams made with these materials are of high quality and hold up well when reduced for publication in specifications or technical journals.

37.12
Installation drawings

Many types of electric/electronics drawings are used to produce the finished installation, from the designer who visualized the system to the contractor who builds it. Drawings are used to design the circuit, detail its parts for fabrication, specify the arrangement of the devices within the system, and instruct the contractor how to install the project.

A combination arrangement and wiring diagram drawing is shown in Fig. 37.36, where the system is shown in a front and right-side view. The wiring diagram explains how the wires and components within the system are connected for the metal-encased switchgear. Busses are conductors for the primary circuits.

FIG. 37.35 Stick-on symbols are available for laying out schematic diagrams. The resulting layouts have a contrast and sharpness better than that found in drawn schematic diagrams. (Courtesy of Bishop Industries Corp.)

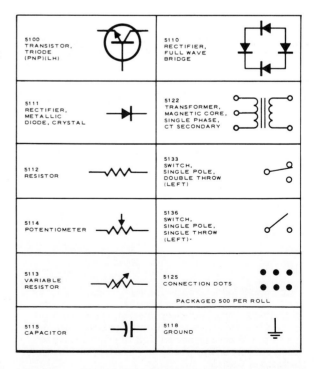

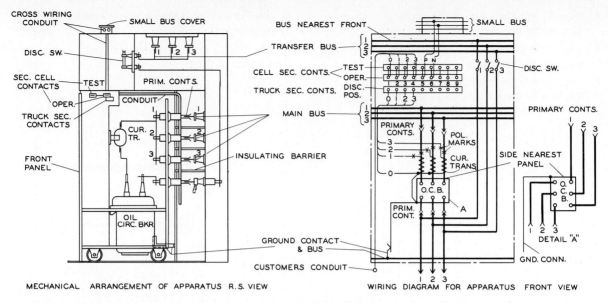

FIG. 37.36 This drawing shows views of a metal-enclosed switchgear to describe the arrangement of the apparatus; it also gives the wiring diagram for the unit.

Problems

1. (Fig. 37.37) On a size A sheet, make a schematic diagram of the circuit.

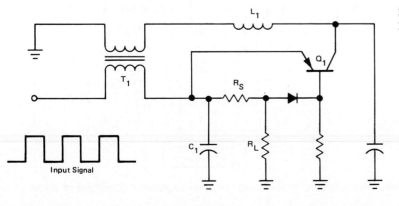

FIG. 37.37 Problem 1. A low-pass inductive-input filter. (Courtesy of NASA.)

2. (Fig. 37.38) On a Size A sheet, make a schematic diagram of the circuit.

3. (Fig. 37.39) On a Size A sheet, make a schematic diagram of the circuit.

4. (Fig. 37.40) On a Size A sheet, make a schematic diagram of the circuit.

5. (Fig. 37.41) On a Size A sheet, make a schematic diagram of the circuit.

6. (Fig. 37.42) On a Size B sheet, make a schematic diagram of the circuit.

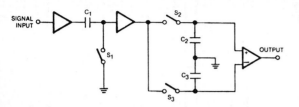

FIG. 37.38 Problem 2. A quadruple-sampling processor. (Courtesy of NASA.)

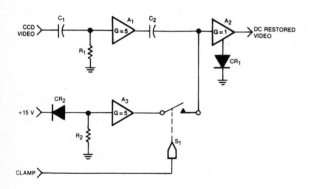

FIG. 37.39 Problem 3. A temperature-compensating DC restorer circuit designed to condition the video circuit of a Fairchild area-image sensor. (Courtesy of NASA.)

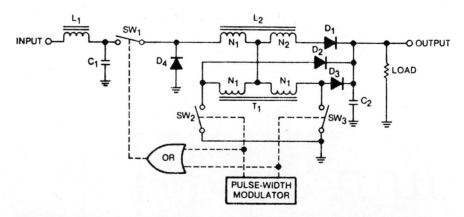

FIG. 37.40 Problem 4. A "buck/boost" voltage regulator. (Courtesy of NASA.)

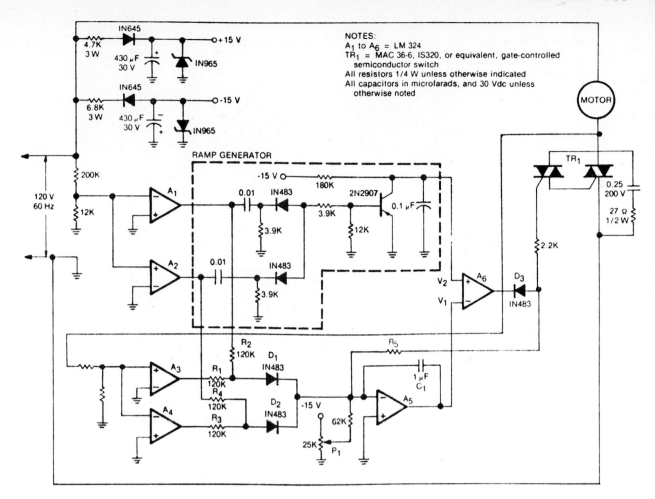

FIG. 37.41 Problem 5. An improved power-factor controller. (Courtesy of NASA.)

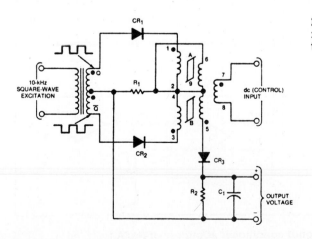

FIG. 37.42 Problem 6. A magnetic-amplifier DC transducer. (Courtesy of NASA.)

7. (Fig. 37.43) On a Size C sheet, make a schematic diagram of the circuit.

8. (Fig. 37.44) On a Size C sheet, make a schematic diagram of the circuit, and give the parts list.

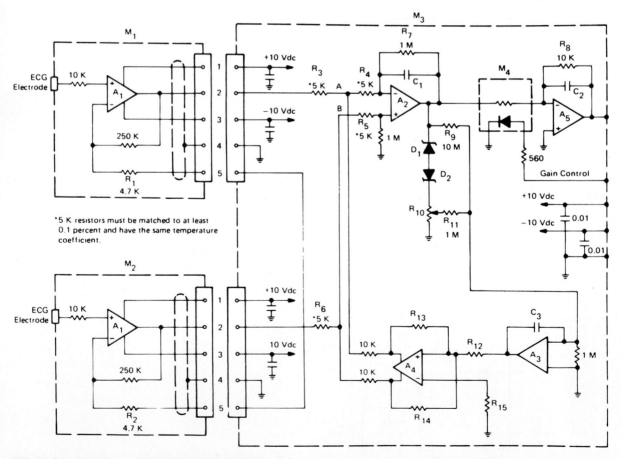

FIG. 37.43 Problem 7. An electrocardiograph (EKG) signal conditioner. (Courtesy of NASA.)

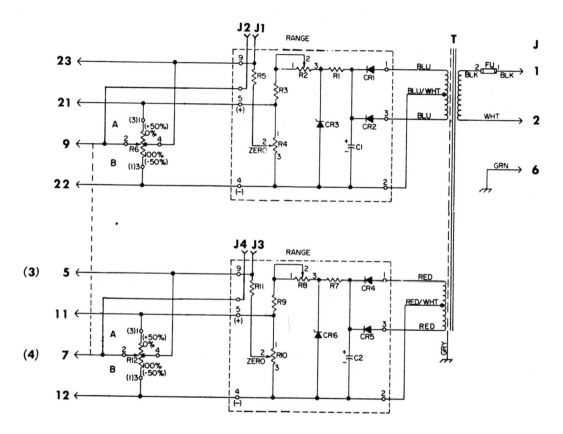

COMPONENT	DESCRIPTION	
CR1, CR2, CR3, CR4	DIODE	1N3193
CR3 & CR6	ZENER DIODE	24V 3W ±.5V TOL
R1 & R7	RESISTOR	560 Ω 5W 5%
R3 & R9	RESISTOR	56 Ω
R5 & R11	RESISTOR	2.7K 1/2W 5%
R2 & R8	TRIM POT	50 Ω
R4 & R10	TRIM POT	5K
R6 & R12	POT	500 Ω 2W DUAL
C1 & C2	CAPACITOR	60 MFD 60V
T	TRANSFORMER	
FU	FUSE 1 AMP	
M	EDGEWISE METER	
J	CONNECTOR	
J1, J2, J3, J4	TEST JACK	

TYPE	R6 SCALE		M SCALE	
	A	B	BOTTOM	TOP
RS1100C	0%	100%	0%	100%
RS1110C	0%	100%	PER ENGINEERING DATA	
RS3100C	+50%	-50%	-50%	+50%
RS4100C	0%	100%	0%	100%
RS2120C	100%	0%	OMIT	
RS2100C	100%	0%	0%	100%

NOTE: OUTPUT WIRED TO TERMINALS 3 AND 4 ON
TYPES RS1100C, RS1110C, AND RS3100C.

OUTPUT WIRED TO TERMINALS 5 AND 7 ON
TYPES RS2100C, RS2120C, AND RS4100C.

FIG. 37.44 Problem 8. A schematic diagram of a dual output manual station and its parts list. (Courtesy of NASA.)

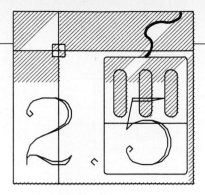

CHAPTER 38

AutoCAD Update

38.1
Introduction

Version 2.5 of AutoCAD was released in late summer of 1986 after this textbook was in the final proofing stages. Rather than issue the book without covering the new features, this chapter was added. The examples of drawings made by AutoCAD 2.15 (see Chapter 10) will be changed only slightly when version 2.5 is used. Also, the ATTRIBUTE, SKETCH, and SCRIPT commands have been included.

Version 2.5 operates on a variety of computers and runs under the PC-DOS, MS-DOS, or UNIX operating systems. Machines with 640K of RAM and a 10Mb hard disk are recommended for use with this updated version. The IBM XT and AT machines (and their compatibles) are well suited.

38.2
The root menu

The root menu displayed when AutoCAD 2.5. is used includes more commands than the earlier version (Fig. 38.1). By placing the cursor on "AUTOCAD," the root menu can always be recalled. The four stars, "****," can be selected to activate the OSNAP command.

Several submenus are longer than will fit on the screen at one time, so the NEXT command is used to obtain them.

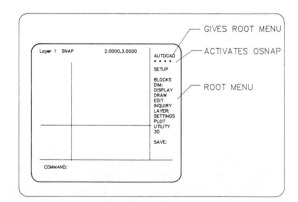

FIG. 38.1 The menu of AutoCAD version 2.5 is different from previous versions. By selecting "AutoCAD," the root menu is shown. When "****" is selected, the OSNAP mode is activated.

PDMODE POINTER ENTITIES
PDSIZE SETS THE SIZE
OF ENTITIES

·   0   A DOT AT THE POINT (DEFAULT)

    1   NO MARKS

+   2   A CROSS AT THE POINT

×   3   AN X AT THE POINT

| 4   A VERTICAL UPWARD FROM THE POINT

FIG. 38.2 The pointer entities (markers) used to locate points on a drawing can be selected and sized using the PDMODE and PDSIZE commands.

38.3
POINT command

The POINT command is used to locate points on a drawing. The point markers can be any of those shown in Fig. 38.2. The marker is selected by using two commands under the SETVAR command: PDMODE and PDSIZE. The basic markers can be set in the following manner:

```
Command: SETVAR Variable name
or ?:PDMODE
New value for PDMODE <0>: 3
(Marker = X)
```

PDMODE FIGURES	.		+	×		
BASIC VALUES	0	1	2	3	4
BASIC VALUES PLUS 32	⊙	○	⊕	⊗	☉
	32	33	34	35	36
BASIC VALUES PLUS 64	□	□	⊞	⊠	▯
	64	65	66	67	68
BASIC VALUES PLUS 96	⊡	◻	⊕	⊠	▯
	96	97	98	99	100

FIG. 38.3 The basic pointer symbols can be changed by adding 32, 64, and 96 to the basic symbols, 1 through 4.

The PDSIZE command is used to assign a size to the marker in the following way:

```
Command: SETVAR Variable name
or ?: PDSIZE
New value for PDSIZE <0.000>:
0.1 (Size of marker)
```

The PDSIZE variable gives the size of the marker on the drawing. If this value is a negative number, the marker will be sized as a percentage of the screen and will appear the same size regardless of the zooming that may occur. When entered as a positive number, the size of the marker will vary with each zoom.

Additional styles of markers can be obtained by adding 32, 64, and 96 to the basic values of 0 through 4 (Fig. 38.3).

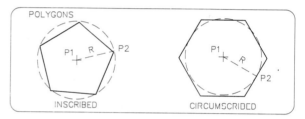

FIG. 38.4 The POLYGON command can be used to draw polygons inscribed in or circumscribed about a circle.

38.4
POLYGON command

A many-sided figure composed of equal sides can be drawn using the POLYGON command. The center is located, the number of sides given, and inscribing or circumscribing the polygon entered (Fig. 38.4).

```
Command: POLYGON
Number of sides: 5
Edge <Center of polygon>: (Lo-
cate center)
Inscribed in circle\Circum-
scribed about circle (IC): I
Radius of circle: (Type length or se-
lect with pointer)
```

When the EDGE option is used, the next prompt asks:

First endpoint of edge: (Select point)
Second endpoint of edge: (Select point)

The polygon will then be drawn in a counterclockwise direction about the center point. A maximum of 1024 sides can be drawn with this command.

38.5
Tangent options of the CIRCLE command

By using the CIRCLE command, a circle can be drawn tangent to a circle and a line, three lines, three circles, or two lines and a circle. A circle drawn tangent to a line and a circle is shown in Fig. 38.5. The TTR option is chosen, the radius is given, and points on the circle and line are selected.

An arc is drawn tangent to three lines in Fig. 38.6 by using the 3P option of the CIRCLE command. In this case, the radius cannot be given since it must be computed.

A circle can be drawn tangent to two lines and a circle or tangent to three circles, as shown in Fig. 38.7. These circles are also found by using the 3P option.

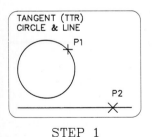

FIG. 38.5 TANGENT command (TTR)

Step 1: Command: <u>CIRCLE</u> 3P/TTR/<Center Pt>: <u>TTR</u>
Enter Tangent spec: (Select point on circle.
Enter second Tangent spec: (Select point on line.)

Step 2: Radius: <u>2</u> (CR)
(The circle with radius = 2 is drawn tangent to the line and circle.)

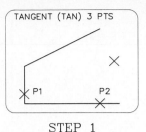 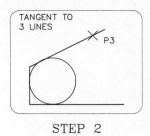

FIG. 38.6 Tangent to three lines

Step 1: Command: <u>CIRCLE</u> 3P/2/TTR/ (Center point): <u>3P</u>
First point: <u>TAN</u>
to (Select point on line.)
Second point: <u>TAN</u>
to (Select point on second line.)

Step 2: Third point: <u>TAN</u>
to (Select pt on third line.)
(The circle is drawn tangent to 3 lines.)

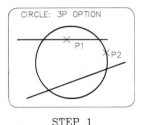 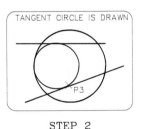

FIG. 38.7 Tangent to lines arcs

Step 1 Command: <u>CIRCLE</u>
3P/2P/TTR/<Center point>: <u>3P</u>
First point: <u>TAN</u>
to (Select *P1* on line)
Second point: <u>TAN</u>
to (Select *P2* on circle)

Step 2: Third point: <u>TAN</u> to (Select P3 on line)
(A circle is drawn tangent to the lines and circle.)

38.6
FILLET command

Arcs can be drawn tangent to circles or arcs, as shown in Fig. 38.8. The radius must be specified and points on two circles selected. AutoCAD will draw fillet arcs

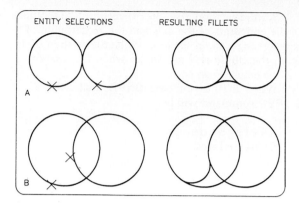

FIG. 38.8 Tangent arcs

Step 1: Command: <u>FILLET</u>
Polyline/Radius/<Select two objects>: <u>R</u>
Enter fillet radius <0.0000>: <u>1.2</u> (example) (CR)

Step 2: Command: (CR)
FILLET Polyline/Radius/<Select two objects>: (Select a pt on each circle and the fillet arc is drawn.)

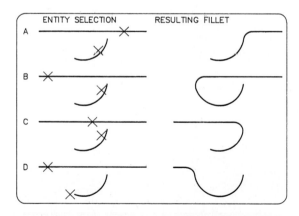

FIG. 38.9 When a fillet radius has been selected, two points on each line or arc can be selected for filleting, as shown here. Each fillet is determined by the location of the points selected.

that most nearly approximate the locations selected on the circles.

Examples of fillets that connect lines and arcs are shown in Fig. 38.9. The location of the selected points on the two entities determines the position of the fillet.

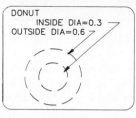
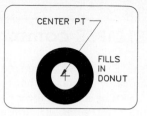

FIG. 38.10 The open donut

Step 1: Command: <u>DONUT</u>
Inside diameter <0.0000>: <u>.30</u> (CR)
Outside diameter <0.0000>: <u>.60</u> (CR)

Step 2: Center of donut: (Select point and donut is drawn.)

38.7
DONUT command

A "donut" can be drawn by using the DONUT command in which the inside diameter, outside diameter, and center point are given (Fig. 38.10). By setting the inside diameter to zero, DONUT will draw a solid circle (Fig. 38.11).

FIG. 38.11 The solid donut

Step 1: Command: <u>DONUT</u>
Inside diameter <0.0000>: <u>0</u> (CR)
Outside diameter <0.0000>: <u>.6</u> (CR)

Step 2: Center of donut: (Select point and donut is drawn without a center hole.)

38.8
ELLIPSE command

The ELLIPSE command can be used to draw ellipses by several methods. In Fig. 38.12, an ellipse is drawn by selecting the endpoints of the major axis and the third point, P3, through which the ellipse will pass.

A second method (Fig. 38.13) is begun by locating the center of the ellipse, P1, then the axis endpoint, P2, and the endpoint of the other axis, P3. The ellipse will be drawn to pass through points P2 and P3.

A third method (Fig. 38.14) requires the location of the endpoints of the axis and the specification of a rotation angle about the axis. When the rotation angle is 0°, the ellipse is a full circle; when the rotation is 90° the ellipse is an edge.

When in the isometric SNAP mode, the EL-LIPSE command will be:

```
<Axis endpoint 1>
CenterIsocircle: I
```

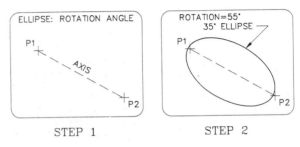

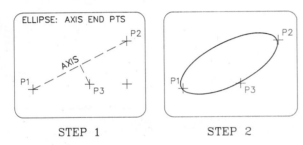

FIG. 38.12 Ellipse—endpoints

Step 1: Command: <u>ELLIPSE</u>
<Axis endpoint 1>/ Center: <u>P1</u>
Axis endpoint 2: <u>P2</u>
<Other axis distance>/Rotation: <u>P3</u>

Step 2: The ellipse is drawn throug *P1* and *P2* point a specified distance perpendicular to *P1-P2*.

FIG. 38.14 Ellipse—rotation angle

Step 1: Command: <u>ELLIPSE</u>
<Axis endpoint>/Center: <u>P1</u>
Axis endpoint 2: <u>P2</u>

Step 2: <Other axis distance>/Rotation: <u>R</u> (CR)
Rotation around major axis: <u>55</u> (CR)
(The ellipse is drawn.)

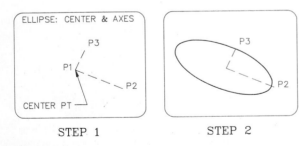

FIG. 38.13 Ellipse—center and axes

Step 1: Command: <u>ELLIPSE</u>
(Axis endpoint 1)/Center: <u>C</u> (CR)
Center of ellipse: <u>P1</u>
Axis endpoint: <u>P2</u>

Step 2: <Other axis distance>/Rotation: <u>P3</u>
(The ellipse is drawn.)

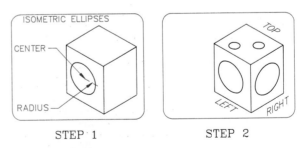

FIG. 38.15 Ellipse—isometric mode

Step 1: When in the isometric SNAP mode:
Command: <u>ELLIPSE</u>
<Axis endpoint 1>/Center/Isocircle: <u>I</u> (CR)
Center of circle: (Select center.)
<Circle radius>/Diameter: (Select radius.)

Step 2: The isometric ellipse is drawn on the current ISOPLANE

By selecting the ISOCIRCLE option, ellipses can be automatically drawn in the correct orientation on each of the three ISOPLANEs. The isometric ellipses can be selected using the previous techniques (Fig. 38.15).

Portions of ellipses drawn with these new commands can be removed by the BREAK command, unlike in version 2.15.

38.9
TEXT command

Two new TEXT modes are available in version 2.5: the FIT and MIDDLE options (Fig. 38.16). When using the FIT option, the height of the text does not change, but its width is stretched or compressed to fit between the two selected points.

The MIDDLE option centers a word or a line about a point in the center of the height of the letters used and centered about the width of the word or line.

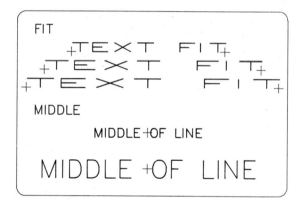

FIG. 38.16 Two new TEXT options are FIT and MIDDLE. When using FIT, a string of text is stretched to fit between two points without changing its height. Using MIDDLE places the center of the string and letters at the point selected.

38.10
SCALE command

The SCALE command permits you to make drawn objects larger or smaller. For example, the desk in Fig. 38.17 is enlarged by windowing the desk, selecting a

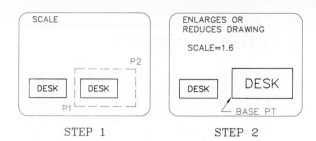

FIG. 38.17 Scaling—scale factor

Step 1: Command: <u>SCALE</u> (CR)
Select objects: (Window DESK with <u>P1</u> and <u>P2</u>.)

Step 2: Base point: (Select base point.)
<Scale factor>/Reference: <u>1.6</u> (CR)
(The desk is drawn 60% larger.)

base point, and giving it a scale factor of 1.6. The text and the drawing are enlarged in the X- and Y-directions.

A second option of the SCALE command enables you to select a length of a given figure (Fig. 38.18), specify its present length, and then specify the new length, which is a ratio of the first specified dimension. The lengths can be given by using the cursor or by typing them in at the keyboard as numeric values.

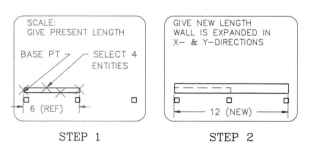

FIG. 38.18 Scaling—reference dimensions

Step 1: Command: <u>SCALE</u> (CR)
Select objects: (Select points on each line.)
Base point: (Select point.)
<Scale factor>/Reference: <u>R</u> (CR)
Reference length <1>: <u>6</u> (CR)

Step 2: New length: <u>12</u> (CR)
(The drawing is enlarged in all directions.)

38.11
STRETCH command

The STRETCH command is used to move a portion of a drawing while retaining the connections it has with other lines and entities. The window in a floor plan in Fig. 38.19 is moved to a new position by the STRETCH command while remaining attached to the line of the wall. A CROSSING window is used to select the line that will be stretched.

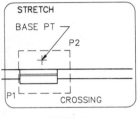

STEP 1   STEP 2

FIG. 38.19   STRETCH command

Step 1: Command: STRETCH
Select objects to stretch by window . . .
Select objects: C (Crossing window)
First corner: P1
Other corner: P2
Base point: (Select base point.)

Step 2: New point: (Select new point.)
(The window is repositioned.)

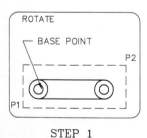

STEP 1   STEP 2

FIG. 38.20   ROTATE

Step 1: Command: ROTATE
Select objects: (Window with *P1* and *P2*.)

Step 2: Base point: (Select point.)
<Rotation angle>/Reference: 45 (CR)
(Object is rotated 45° CCW.)

38.12
ROTATE command

An object drawn on the screen can be rotated about a base point by using the ROTATE command (Fig. 38.20). The object, which may comprise a number of lines, is windowed, a base point is selected, and a rotation angle is typed in at the keyboard, or the cursor is used to show the angle of rotation on the screen. Lines from several layers can be rotated if they are included within the window used to select the object.

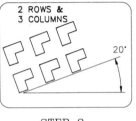

STEP 1   STEP 2

FIG. 38.21   Array at an angle

Step 1: The grid must be rotated using the SNAP mode at the desired angle. Draw the object in the lower left corner of the array.
Command: ARRAY
Select objects: (Window the object.)
Rectangular or Polar array <R/P>: R (CR)

Step 2: Number of rows(---) <1>: 2 (CR)
Number of columns (|||) <1>: 3 (CR)
Unit cell or distance between rows (---): 2 (CR)
Distance between columns (|||): 3 (CR)

38.13
Rotated rectangular arrays

Rectangular arrays may be drawn at angles (Fig. 38.21) if the SNAP mode has been rotated to some angle other than zero. The first object is then drawn with the DRAW commands in the lower left corner of the ARRAY. The ARRAY command is activated, and the number of rows, number of columns, and cell distances are given in response to the command prompts. The array will be drawn at the angle of the SNAP mode.

38.14
Mirrored text (MIRRTEXT)

MIRRTEXT is a system variable that is a subcommand of the SETVAR command, which is used for mirroring a drawing that has text. By setting MIRRTEXT to 0, the text will not be mirrored, but the drawing will be (Fig. 38.22). If MIRRTEXT is set to 1, the text will be mirrored along with the drawing.

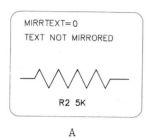
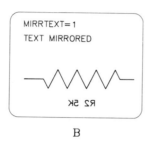

A   B

FIG. 38.22 Mirrored text
When the variable MIRRTEXT is given a value of 0, the text will not be mirrored. When MIRRTEXT is given a value of 1, the text will be mirrored.

38.15
BREAK command

A portion of a line may be broken away by using the BREAK command (Fig. 38.23). The line to be broken is windowed, two points are selected to indicate the ends of the break, and that portion of the line is removed.

38.16
TRIM command

By using the TRIM command, edges of entities can be used as cutting edges to trim a line, as shown in Fig. 38.24. You are prompted to select the entities that will serve as cutting edges, and the objects between the cutting edges are selected. The cutting edges trim the lines, giving perfect intersections at the crossing points.

Using the CROSSING option of the TRIM command, a window is placed around a set of intersecting

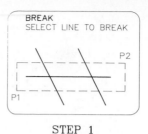
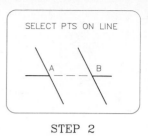

STEP 1   STEP 2

FIG. 38.23 BREAK

Step 1: Command: BREAK
Select object: (Window with P1 and P2.)

Step 2: Enter first point: A
Enter second point: B
(Line AB is removed. The OSNAP command can be used to select intersection points, if needed.)

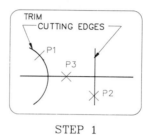
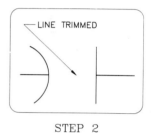

STEP 1   STEP 2

FIG. 38.24 TRIM EDGES

Step 1: Command: TRIM
Select cutting edge(s) . . .
Select objects: P1
Select objects: P2

Step 2: Select object to trim: P3
(Line between cutting edges is removed.)

lines (Fig. 38.25), and lines selected to be the cutting edges are crossed. The portions of the lines between the cutting edges are selected one at a time and are then removed, or trimmed.

38.17
EXTEND command

Lines and arcs can be lengthened to intersect a selected entity by using the EXTEND command (Fig. 38.26). You are first prompted to select the boundary

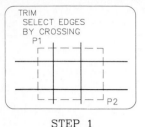
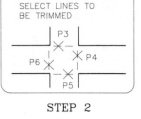

STEP 1   STEP 2

FIG. 38.25 Trim by crossing

Step 1: Command: <u>TRIM</u>
Select cutting edge(s) . . .
Select objects: <u>Crossing</u>
First corner: <u>P1</u>
Other corner: <u>P2</u>

Step 2: Select object to trim: (Select <u>P3, P4, P5, P6</u>.)
(The lines are trimmed one at a time.)

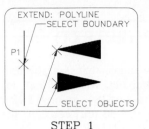
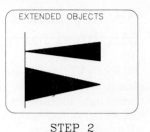

STEP 1   STEP 2

FIG 38.27 EXTEND a polyline

Step 1: Command: <u>EXTEND</u>
Select boundary edge(s) . . .
Select objects: <u>P1</u>.

Step 2: Select object to extend: (Select ends of
PLINES.)
(Both PLINES are extended to the border.)

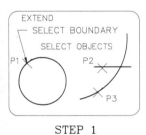
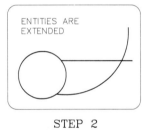

STEP 1   STEP 2

FIG. 38.26 EXTEND

Step 1: Command: <u>EXTEND</u>
Select boundary edge(s) . . .
Select objects: <u>P1</u>

Step2: Select object to extend: <u>P2</u> and <u>P3</u>
(The line and arc are extended to the boundary.)

entity and then the line or arc to be extended. More
than one entity can be extended at a time to join the
previously selected boundary entity.

A POLYLINE can be extended with the EX-
TEND command to a selected boundary, as shown in
Fig. 38.27. The boundary is selected, the ends of the
PLINES are selected, and both PLINES are ex-
tended. The EXTEND command will not work on
"closed" PLINES, such as a polygon.

38.18
DIVIDE command

An entity can be divided into a specified number of
equal segments by using the DIVIDE command (Fig.
38.28). The entity is selected by locating a point on it
with the cursor. You will be prompted for the number
of segments into which it is to be divided, and markers

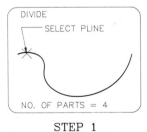
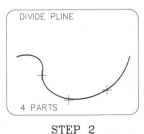

STEP 1   STEP 2

Fig. 38.28 DIVIDE a line

Step 1: Command: <u>DIVIDE</u>
Select object to divide: (Select PLINE.)

Step 2: <Number of segments>/Block: <u>4</u> (CR)
(PDMODE symbols are placed along the line,
dividing it.)

will be placed on the line at the ends of the segments. The markers will be of the type and size currently set by the PDMODE and PDSIZE variables under the SETVAR command.

The BLOCK option under the DIVIDE command allows you to select a previously saved block to mark the ends of the segments of a divided line (Fig. 38.29). The blocks can be either ALIGNED or NOT ALIGNED as shown. In this example, the BLOCKS are rectangles, but they could have been drawn in any shape.

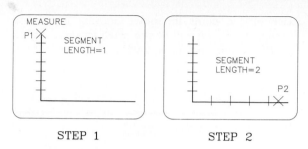

FIG. 38.30 MEASURE

Step 1: Command: <u>MEASURE</u>
Select object to measure: <u>P1</u>
<Segment length>/Block: <u>0.1</u> (CR)
(The line is divided into 0.1 divisions starting at the end nearest *P1*.)

Step 2: Command: <u>MEASURE</u>
Select object to measure: <u>P2</u>
<Segment length>/Block: <u>0.2</u> (CR)
(The line is divided into 0.2 divisions.)

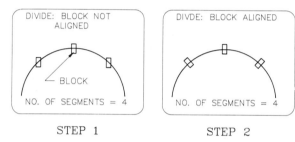

Fig 38.29 DIVIDE an arc

Step 1: Command: <u>DIVIDE</u>
Select object to divide: (Select arc.)
<Number of segments>/Block: <u>B</u> (CR)
Block name to insert: <u>RECT</u>
Align block with object? <Y> <u>N</u> (CR)
Number of segments: <u>4</u> (CR)

Step 2: By responding:
Align block with object? <Y> (CR)
the blocks will radiate from the arc's center.

38.20
OFFSET command

An entity can be drawn parallel to and offset from another entity by the OFFSET command (Fig. 38.31). In this example, a polyline is drawn offset from a given polyline. You will be prompted for the offset dis-

38.19
MEASURE command

Markers can be placed along an arc, circle, polyline, or line at a specified distance apart (Fig. 38.30) by using the MEASURE command. The segment length option asks you for the entity to be segmented, which you must select; then it asks for the segment length. Markers will be placed along the line (or entity) beginning with the end nearest to the location of the point selected. The last segment probably will not be equal to the specified segment length.

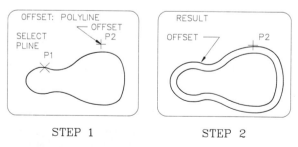

FIG. 38.31 OFFSET

Step 1: Command: <u>OFFSET</u>
Offset distance or Through <Through>: <u>T</u> (CR)
Select object to offset: <u>P1</u>
Through point: <u>P2</u>

Step 2: An enlarged PLINE is drawn that passes through *P2*.

tance or the point the offset drawing is to pass through. Next, you will be prompted for the entity (a PLINE, in this case) to be used as the pattern for the offset PLINE.

38.21
SETVAR command

Many modes, sizes, limits, and variables remaining in AutoCAD's memory under the SETVAR command are given in Table 38.1 on pages 750-751. These variables can be inspected and changed by means of the SETVAR command, unless they are read-only commands. Some variables are saved in AutoCAD's configuration file, and others are saved in drawings that are made.

To review system variables, use the SETVAR command:

```
Command: SETVAR
Variable name or ?: ?
```

A complete listing of the current variables will be given on the screen. To change one or more variables:

```
Command: SETVAR
Variable name or ?: TEXTSIZE
<0.20>: .125
```

The default value, given in brackets, can be changed by typing a new value, 0.125. By entering the SETVAR command with an apostrophe in front of it, you can use it transparently while another command is in progress.

38.22
Dimensioning variables

Version 2.5 introduces several new DIM VARS (dimensioning variables): DIMRND, DIMDLE, DIMZIN, DIMALT, DIMALTF, DIMALTD, DIMLFAC, and DIMBLK.

DIMRND rounds all dimensions to within a specified value. If DIMRND is set at 0.50, all dimensions will be rounded to the nearest 0.50 unit. This variable does not apply to angular dimensions.

DIMDLE (dimension line extension) extends the dimension line a specified distance beyond the dimen-

sion line when ticks are being used instead of arrows. For example, a value of 0.12 would extend the dimension line 0.12 beyond the extension line.

DIMZIN (zero-inch editing) controls the display of inches when architectural units of feet and inches are being used. By setting this variable "on," 3 feet would be given as 3'-0"; if set to "off," the dimension would appear as 3'.

DIMALT, DIMALTF, and DIMALTD obtain dual dimensions on a drawing—show inches and millimeters, for example. First, set DIMALT (alternate units) "on" and set the values of DIMALTF (alternate units scale factor); then set DIMALTD (alternate units decimal places) to the desired number of decimal places. If DIMALT is "on," DIMALTF is set to its default value of 25.4, and DIMALTD is set to 1 decimal place, a linear dimension of 24 inches would be given as 24 [609.6]. The number in brackets, the alternate value, represents the equivalent of 24 inches in millimeters.

DIMLFAC (length factor), a global scale factor, scales all linear dimensions, radii, and diameters. If applied before DIMALT, DIMLFAC will affect both dimension values.

DIMBLK (dimensioning block) is the name you assign to a block that you have created for specially designed arrows at the ends of dimension lines. If you set DIMBLK to DOT, you will get a dimension line with a dot located at the intersection of the dimension and extension lines. To create your own arrow, draw a right arrow with a segment of the dimension line attached. The total length (including the dimension line segment) should be one drawing unit long. Make a BLOCK of the arrowhead and dimension line, set the insertion point at the tip of the arrowhead where it will join the extension line, enter the dimensioning command (DIM), and type DIMBLK. When prompted, assign the arrow BLOCK to DIMBLK. Now, when you dimension a linear distance, your special arrowheads will appear. To stop using the special arrowhead, change the DIMBLK name to a period (".").

38.23
Attributes

ATTRIBUTES are combinations of drawings, saved as BLOCKS and text. A drawing can be made with attributes defined using the ATTDEF command, saved as a BLOCK, and then inserted repetitively.

With each insertion, you will be prompted for attribute values that will be written on the drawing or left on file as an invisible value for listing in an attribute report.

An attribute drawing is shown in Fig. 38.32, where a title block is drawn using the DRAW command. To define the attributes, use the ATTDEF command:

```
Command: ATTDEF
Attribute  modes—Invisible:  N
Constant:N Verify:N
Enter (ICV) to change. RETURN
when done: V
```

By entering *I, C,* or *V,* one at a time, you can change the settings of invisible, constant, and verify modes of the attributes. If the invisible mode is on (*Y*), the at-tribute values will not be shown on the drawing to prevent clutter. If the constant mode is on (*Y*), a fixed attribute value will be assigned to each BLOCK insertion. If the verify mode is on (*Y*), you will be able to verify the attribute values at insertion time.

The next prompts are:

```
Attribute tag: DATE (On blanks al-
lowed)
Attribute prompt: DATE?
Default Attribute value: (Blank if
none)
Attribute value: (Blank)
Start point or Align/Center/Fit/
Middle/Right/Style: C
Height <0.100>: .125
Rotation angle <0>: (CR)
```
(The attribute tag appears on the drawing.)

Multiple attributes can be added to each drawing. Once all attributes have been added, the drawing is windowed as a BLOCK in the usual manner.

When the BLOCK is inserted, the attribute prompts you, and default values will be shown on the screen:

```
Enter attribute values
Name: <NONE>: BILLY BOB
Scale? <full size>: (CR) (Accepts
full size default)
Date: <NONE>: 9-26-87
```

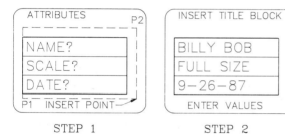

FIG. 38.32 Attribute definitions

STEP 1   STEP 2

Step 1: Draw the title block
Command: ATTDEF
Attribute Modes—Invisible:N Constant:N Verify:N
Enter (ICV) to change, RETURN when done: V (CR)
Attribute tag: NAME?
Attribute prompt: NAME?
Default attribute value: (Blank)
Start point or Align/Center/Fit/Middle/Right/Style:
(Locate tag on title block.)
Height <0.0000>: .125 (CR)
Insertion angle <0>: (CR)
(Repeat this procedure for other tags: SCALE and DATE.)

Step 2: BLOCK the title block with a *P1 and P2* window.
Command: INSERT
Block name <or ?>: TITLE
Insertion point: (Select point.)
X scale factor <1>/Corner/XYZ: (CR)
Y scale factor <default=X>: (CR)
Rotation angle <0>: (CR)
(Respond to the attribute prompts to complete the title block.)

A listing of these attribute values will be given as you carriage return (CR) through the list, enabling you to verify the correctness of your entries. When the end of the list is reached, the values will be plotted on the drawing in the places previously assigned (Fig. 38.33).

The visibility of the attributes can be changed by the ATTDISP command, which may be different from the ATTDEF command specifications:

```
Command: ATTDISP
Normal/On/Off <current value>: N
```

By entering *N,* the attributes specified either hidden or visible will appear that way on the drawing. The ON command will display all values on the drawing, whether hidden or visible. The OFF command does not show any of the values on the drawing. For the values to be redisplayed, the drawing must be REGENerated.

TABLE 38.1
VARIABLES AVAILABLE UNDER THE SETVAR COMMAND

Variable Name	ADE	Type	Saved In	Meaning
AFLAGS	2	integer		Attribute flags bit-code for ATTDEF command (sum of the following): 1 = Invisible 2 = Constant 4 = Verify
ANGBASE	1	real	drawing	Angle 0 direction
ANGDIR	1	integer	drawing	1 = clockwise angles, 0 = counterclockwise
APERTURE	2	integer	config	Object snap target height, in pixels
AREA		real		Area computed by AREA, LIST, or DBLIST (read-only)
ATTMODE	2	integer	drawing	Attribute display mode (0 = off, 1 = normal, 2 = on)
AUNITS	1	integer	drawing	Angular units mode (0 = decimal degrees, 1 = degrees/minutes/seconds, 2 = grads, 2 = radians, 4 = surveyor's units)
AUPREC	1	integer	drawing	Angular units decimal places
AXISMODE	1	integer	drawing	Axis on if 1, off if 0
AXISUNIT	1	point	drawing	Axis spacing, X and Y
BLIPMODE		integer	drawing	Marker blips on if 1, off is 0
CDATE		real		Calendar date/time (read-only) (special format; see below)
CECOLOR		string	drawing	Current entity color (read-only)
CELTYPE		string	drawing	Current entity linetype (read-only)
CHAMFERA	1	real	drawing	First chamfer distance
CHAMFERB	1	real	drawing	Second chamfer distance
CLAYER		string	drawing	Current layer (read-only)
CMDECHO	3	integer		When AutoLISP's (command) function is used, prompts and input are echoed if this variable is 1, but not if it is 0
COORDS	1	integer	drawing	If zero, coordinate display is updated on point picks only. If 1, display of absolute coordinates is continuously updated. If 2, distance and angle from last point are displayed when a distance or angle is requested.
DATE		real		Julian date/time (read-only)
DISTANCE		real		Distance computed by DIST, LIST, or DBLIST (read-only)
DRAGMODE	2	integer	drawing	0 = no dragging, 1 = on if requested, 2 = auto
DRAGP1	2	integer	config	Regen-drag input sampling rate
DRAGP2	2	integer	config	Fast-drag input sampling rate
DWGNAME		string		Drawing name (read-only)
DWGPREFIX		string		Drive/directory prefix for drawing (read-only)
ELEVATION	3	real	drawing	Current 3D elevation
EXPERT		integer		If 1, suppresses certain "are you sure?" prompts, such as "About to regen, proceed?" and "Do you really want to turn the current layer off?". If 0, these prompts are issued normally.
EXTMAX		point	drawing	Upper right "drawing uses" extents (read-only)
EXTMIN		point	drawing	Lower left "drawing uses" extents (read-only)
FILLETRAD	1	real	drawing	Fillet radius
FILLMODE		integer	drawing	Fill mode on if 1, off if 0
GRIDMODE		integer	drawing	Grid on if 1, off if 0
GRIDUNIT		point	drawing	Grid spacing, X and Y
HIGHLIGHT	3	integer		Object selection highlighting on if 1, off if 0
INSBASE		point	drawing	Insertion base point (set by BASE command)
LASTANGLE		real		The end angle of the last arc entered (read-only)
LASTPOINT		point		Referenced by "@" during keyboard point entry

LIMCHECK		integer	drawing	Limits checking on if 1, off if 0
LIMMAX		point	drawing	Upper right drawing limits
LIMMIN		point	drawing	Lower left drawing limits
LTSCALE		real	drawing	Global linetype scale factor
LUNITS	1	integer	drawing	Linear units mode (1 = scientific, 2 = decimal, 3 = engineering, 4 = architectural)
LUPREC	1	integer	drawing	Linear units decimal places or denominator
MENUECHO		integer		0 = menu input is echoed and prompts are displayed normally. 1 = menu input is not echoed (but ^P in a menu item toggles echoing on). 2 = prompts are suppressed if input is coming from a menu item. 3 = menu echoing and prompts are both suppressed.
MIRRTEXT	2	integer	drawing	MIRROR reflects text if nonzero, retains text direction if zero.
ORTHOMODE		integer	drawing	Ortho mode on if 1, off 0
OSMODE	2	integer	drawing	Object snap modes bit-code (sum of the following): 1 = Endpoint 32 = Intersection 2 = Midpoint 64 = Insert 4 = Center 128 = Perpendicular 8 = Node 256 = Nearest 16 = Quadrant 512 = Quick
PDMODE		integer	drawing	Point display mode
PDSIZE		real	drawing	Point display size
PERIMETER		real		Perimeter computed by AREA, LIST, or DBLIST (read-only)
PICKBOX		integer	config	Object selection target height, in pixels
QTEXTMODE		integer	drawing	Quick text mode on if 1, off if 0
REGENMODE		integer	drawing	REGENAUTO on if 1, off if 0
SCREENSIZE		point		Graphics screen size in pixels, X and Y (read-only)
SKETCHINC	1	real	drawing	Sketch record increment
SKPOLY	3	integer	drawing	SKETCH generates lines if 0, Polylines if 1
SNAPANG	2	real	drawing	Snap / grid rotation angle
SNAPBASE	2	point	drawing	Snap / grid origin point
SNAPISOPAIR	2	integer	drawing	Current isometric plane (0 = left, 1 = top, 2 = right)
SNAPMODE		integer	drawing	Snap mode on if 1, off if 0
SNAPSTYL	2	integer	drawing	Snap style (0 = stand, 1 = isometric)
SNAPUNIT		point	drawing	Snap spacing, X and Y
TDCREATE		real	drawing	Time and date of drawing creation (read-only)
TDINDWG		real	drawing	Total editing time (read-only)
TDUPDATE		real	drawing	Time and date of last update/save (read-only)
TDUSRTIMER		real	drawing	User elapsed timer (read-only)
TEXTSIZE		real	drawing	Default text height
THICKNESS	3	real	drawing	Current 3D thickness
TRACEWID		real	drawing	Default trace width
VIEWCTR		point	drawing	Center of current view (read-only)
VIEWSIZE		real	drawing	Current view height, in drawing units (read-only)
VPOINTX	3	real	drawing	X component of current 3D view point (read-only)
VPOINTY	3	real	drawing	Y component of current 3D view point (read-only)
VPOINTZ	3	real	drawing	Z component of current 3D view point (read-only)

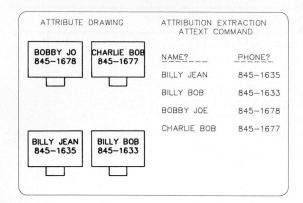

FIG. 38.33 A listing of the attributes on a drawing can be made using the ATTEXT command in conjunction with BASIC. More elaborate listings can be obtained when attributes are used with a database program.

The ATTEDIT command is used to edit attributes one at a time or globally, where all are changed at one time. The command is used as follows:

```
Command: ATTEDIT
Edit attributes one by one?
<Y> Y
```

By responding yes (*Y*), you may edit any of the attributes currently visible on the screen. By responding no (*N*), you may edit all attributes with global editing. You can select the attributes to edit by responding the following prompts:

```
Block name specification <*>:
Attribute tag specification
<*>:
Attribute value specification
<*>:
```

If you enter a CR after the BLOCK name specification, attributes of BLOCKS with all names will be selected for editing. If you specified the BLOCK name but entered a CR after the tag specification, attributes for all tags will selected. However, if you responded with *NAME* for the tag specification, only the attributes of the tag *NAME* will be selected. By entering a \, you will be able to edit null-value attributes.

You will now be prompted:

```
Select ATTRIBUTES:
```

Select the attributes by pointing to them on the drawing or by windowing the group to be edited. Each attribute will be marked with an "X" on the drawing to indicate those eligible for editing. The next prompt is:

```
Value/Position/Height/Angle/
Style/Layer/Color/Next <N>:
```

By selecting the first letter of these commands, you may change the attributes in many ways by responding to autoCAD's prompts.

Global editing may be done by using the following sequence of commands:

```
Command: ATTEDIT
Edit attributes one by one?
<Y> N  (Global option)
Global edit of Attribute val-
ues.
Edit only Attributes visible
on screen? <Y> N
```
(Screen goes to Text mode if N)
(Drawing must be regenerated afterwards.)

The next steps involve selecting the attributes in the same manner as when they were done one at a time. In this case, changes are applied uniformly to all selected blocks and attributes.

38.24
Attribute listing (ATTEXT)

The ATTEXT command can be used for producing tabular listings—such as parts lists, inventories, and bills of materials—when used with a text editor or database program. An elementary application of obtaining a printed report is given below.

A BLOCK with attributes is INSERTed four times, as shown in Fig. 38.33, and the drawing is saved under the name OFFICE. Begin by loading AutoCAD, and the OFFICE file, and then enter the ATTEXT command:

```
Command: ATTEXT
CDF, SDF or DXF Attribute ex-
tract (or Entities)?<C> D
(This file can be read by other programs.)
Extract file names <OFFICE>:
(CR) Accept name, OFFICE.)
Command: END
```

Leave AutoCad, go to the directory where BASIC is stored and load BASIC, then load the ATTEXT command:

```
Load "C:\ACAD\OFFICE"
Type RUN and RETURN
Extract file name: OFFICE
```

Your BASIC program will print a bill of materials, as shown in Fig. 38.33. Return to DOS by typing SYSTEM.

38.25
Digitizing with the tablet

Drawings on paper can be digitized into the computer point by point when a tablet is available. Tablets vary in size from 11″ × 8.5″ to several square feet.

To calibrate a drawing and tablet for digitizing, tape the drawing to the tablet and use the following steps:

```
Command: TABLET
Option (ON/OFF/CAL/CFG): CAL
Calibrate tablet for use
Digitize first known point:
(Digitize pt.)
Enter coordinates for first
point: 1,1
Digitize second known point:
(Digitize pt.)
Enter coordinates for second
point: 10,1
```

You should digitize points from left to right or from bottom to top of the drawing. Furthermore, your limits must be large enough to contain the limits on the tablet. You may now use the AutoCAD root menu for copying the drawing. For example, activate the LINE command, select a beginning point on the tablet with your pointer, and select other points in sequence.

Use the ON or OFF commands to turn the tablet mode on or off. When turned off, your pointer can be used to access the root menu at the right of the tablet. Function key F10 on the IBM XT can also be used for this purpose.

38.26
SKETCH command

The SKETCH command can be used with the tablet for tracing drawings composed of irregular lines (Fig. 38.34). Begin by attaching the drawing on the tablet, and calibrate the tablet as discussed above. Enter the SKETCH command:

```
Command: SKETCH
Record increment <0.1>: 0.01
Sketch. Pen eXit Quit Record
Erase Connect.
```

The record increment specifies the distances between the endpoints of the connecting lines that represent the irregular lines you sketch. The other command options are defined below:

Pen	Raises or lowers pen
eXit	Records lines and exits
Quit	Discards temporary lines and exits
Record	Records temporary lines
Erase	Erases selected lines
Connect	Connects current line to last endpoint
.(period)	Line from endpoint of last line to the current location of the pointer

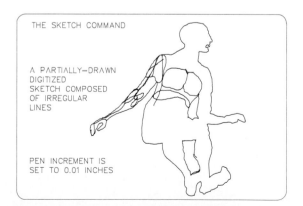

FIG. 38.34 This drawing was made using the SKETCH command at a tablet instead of a mouse. An original drawing was taped to the tablet, and the cursor was used to trace over it using increments of 0.01″.

Begin sketching by moving your pointer to the first point with the pen up, lower the pen (*P*), and trace over the line to be traced with the pointer. An irregular line is drawn on the screen. Erase by raising the pen (*P*), enter Erase (*E*), move the pointer backwards from the current point, and select the point where you want the erasure to stop. All drawn lines are temporary until you select Record (*R*) or eXit (*X*). After recording the lines, you can begin new lines by repeating these steps. The mouse is unsatisfactory for sketching; for this type of digitizing only a tablet can be effectively used.

38.27
Slide shows

Several commands can be used to create a slide show on the computer screen: SCRIPT, MSLIDE, RSCRIPT, DELAY, and RESUME. MSLIDE is the first to become familiar with because it is used to make a slide.

Enter the drawing editor, and make a drawing that you wish to use as a slide. Enter the MSLIDE command and respond to the slide file prompt with a name of your choice, *SLIDE1*, for example. Erase the drawing and make a second drawing, enter MSLIDE, and name it *SLIDE2*. You may make as many slides as you wish by following these steps. When finished, erase the last drawing from the screen.

Enter the command VSLIDE and type *SLIDE1*, and this image will be recalled and displayed on the screen. Type REDRAW to remove *SLIDE1* from the screen. Recall *SLIDE2* in the same manner. Now that you have checked your slides, QUIT, and EXIT AutoCAD (or use the SHELL command).

To create a SCRIPT file, obtain the DOS prompt, C>, and start a script file that will be called

DISPLAY.SCR in the following manner:

```
C> COPY CON:C:\ACAD
\DISPLAY.SCR (CR)
```

Type the following script file. Notice that the drive and directory, *C:\ACAD*, is typed in front of each slide name.

```
VSLIDE C:\ACAD\SLIDE1 (CR)
DELAY 2000 (CR)
VSLIDE C:\ACAD\SLIDE2 (CR)
DELAY 1000 (CR)
RSCRIPT (CR)
RESUME (CR)
```

(Press function key F6 and CR.)

The DELAY command is specified in milliseconds; this value will be an approximation because this speed varies with different types of computers.

To show the slides in accordance with the SCRIPT, load AutoCAD's drawing editor by calling up any of the new drawings. Type SCRIPT and *DISPLAY* (the name of SCRIPT) in response to the file name prompt. The slides will be shown in sequence with the delays between each slide; then the sequence will recycle, beginning with the first slide. This sequence can be stopped by pressing Control C or the back space key. If the RSCRIPT command had been omitted from the SCRIPT file, the two slides would have been shown, and the sequence would have stopped. To repeat the show, you would then type in RESUME. When through, QUIT to return to the main menu.

Many more advanced slides can be developed using methods covered in AutoCAD's manual. For example, the SCRIPT file can include such variables as LIMITS, SNAP, GRID, UNITS, TEXT, which are activated by SCRIPT each time it is called up.

Appendix Contents

APPENDIX 1
ABBREVIATIONS (ANSI Z 32.13)

Word	Abbreviation	Word	Abbreviation	Word	Abbreviation
Abbreviate	ABBR	Bench mark	BM	Centigrade	C
Absolute	ABS	Between	BET.	Centigram	CG
Account	ACCT	Between centers	BC	Centimeter	cm
Actual	ACT.	Between		Chain	CH
Adapter	ADPT	perpendiculars	BP	Chamfer	CHAM
Addendum	ADD.	Bevel	BEV	Change notice	CN
Adjust	ADJ	Bill of material	B/M	Change order	CO
Advance	ADV	Birmingham wire gage	BWG	Channel	CHAN
After	AFT.	Blueprint	BP	Check	CHK
Aggregate	AGGR	Board	BD	Check valve	CV
Air condition	AIR COND	Boiler	BLR	Chemical	CHEM
Airport	AP	Bolt circle	BC	Chord	CHD
Airplane	APL	Both sides	BS	Circle	CIR
Allowance	ALLOW	Bottom	BOT	Circuit	CKT
Alloy	ALY	Bottom chord	BC	Circular	CIR
Alteration	ALT	Boundary	BDY	Circular pitch	CP
Alternate	ALT	Bracket	BRKT	Circumference	CIRC
Alternating current	AC	Brake horsepower	BHP	Clockwise	CW
Altitude	ALT	Brass	BRS	Coated	CTD
Aluminum	AL	Brazing	BRZG	Cold drawn	CD
American National		Break	BRK	Cold drawn steel	CDS
Standard	AMER NATL STD	Breaker	BKR	Cold finish	CF
American wire gage	AWG	Bridge	BRDG	Cold punched	CP
Ammeter	AM	Brinnell hardness	BH	Cold rolled	CR
Amount	AMT	British Standard	BR STD	Cold rolled steel	CRS
Ampere	AMP	British Thermal Units	BTU	Column	COL
Anneal	ANL	Broach	BRO	Combination	COMB.
Antenna	ANT.	Bronze	BRZ	Combustion	COMB
Apparatus	APP	Brown & Sharp (Wire gage,		Commutator	COMM
Appendix	APPX	same as AWG)	B&S	Company	CO
Approved	APPD	Building	BLDG	Concentric	CONC
Approximate	APPROX	Bulkhead	BHD	Concrete	CONC
Arc weld	ARC/W	Bureau	BU	Condition	COND
Area	A	Bureau of Standards	BU STD	Connect	CONN
Armature	ARM.	Bushing	BUSH.	Contact	CONT
Asbestos	ASB	Button	BUT.	Cord	CD
Asphalt	ASPH	Buzzer	BUZ	Corporation	CORP
Assembly	ASSY	By-pass	BYP	Corrugate	CORR
Association	ASSN			Cotter	COT
Atomic	AT	Cabinet	CAB.	Counterclockwise	CCW
Authorized	AUTH	Cadmium plate	CD PL	Counterbore	CBORE
Auxiliary	AUX	Calculate	CALC	Counterdrill	CDRILL
Avenue	AVE	Calibrate	CAL	Counterpunch	CPUNCH
Average	AVG	Calorie	CAL	Countersink	CSK
Avoirdupois	AVDP	Capacitor	CAP	Coupling	CPLG
Azimuth	AZ	Cap screw	CAP SCR	Crank	CRK
		Case harden	CH	Cross section	XSECT
Babbitt	BAB	Cast iron	CI	Cubic	CU
Back pressure	BP	Cast steel	CS	Cubic centimeter	cc
Balance	BAL	Casting	CSTG	Cubic feet per minute	CFM
Ball bearing	BB	Castle nut	CAS NUT	Cubic feet per second	CFS
Barometer	BAR	Catalogue	CAT.	Cubic foot	CU FT
Barrel	BBL	Cement	CEM	Cubic inch	CU IN.
Base line	BL	Center	CTR	Cubic meter	CU M
Base plate	BP	Centerline	CL	Cubic yard	CU YD
Battery	BAT.	Center of gravity	CG	Current	CUR
Bearing	BRG	Center to center	C to C	Cylinder	CYL

Cont.

APPENDIX 1
ABBREVIATIONS (ANSI Z 32.13) (Cont.)

Word	Abbreviation	Word	Abbreviation	Word	Abbreviation
Decimal	DEC	Fillister	FIL	Illustrate	ILLUS
Dedendum	DED	Filter	FLT	Inch	(") IN.
Degree	(°) DEG	Finish	FIN.	Inches per second	IPS
Department	DEPT	Finish all over	FAO	Include	INCL
Design	DSGN	Flange	FLG	Industrial	IND
Detail	DET	Flat head	FH	Information	INFO
Develop	DEV	Fluid	FL	Inside diameter	ID
Diagonal	DIAG	Focus	FOC	Instrument	INST
Diagram	DIAG	Foot	(') FT	Insulate	INS
Diameter	DIA	Forging	FORG	Interior	INT
Diametrical pitch	DP	Forward	FWD	Internal	INT
Dimension	DIM.	Foundation	FDN	Intersect	INT
Direct current	DC	Foundry	FDRY	Iron	I
Discharge	DISCH	Frequency	FREQ	Irregular	IRREG
Distance	DIST	Front	FR		
District	DIST			Jack	J
Ditto	DO.			Joint	JT
Dovetail	DVTL	Gage	GA	Junction	JCT
Dowel	DWL	Gallon	GAL	Junction box	JB
Down	DN	Galvanize	GALV		
Dozen	DOZ	Galvanized iron	GI	Key	K
Drafting	DFTG	Galvanized steel	GS	Keyseat	KST
Draftsman	DFTSMN	Gasket	GSKT	Keyway	KWY
Drawing	DWG	General	GEN	Kiln-dried	KD
Drill	DR	Government	GOVT	Kip (1000 lb)	K
Drive	DR	Governor	GOV	Knots	KN
Drop forge	DF	Grade	GR		
		Grade line	GL	Laboratory	LAB
		Gram	G	Lateral	LAT
Each	EA	Gravity	G	Latitude	LAT
East	E	Grind	GRD	Left	L
Eccentric	ECC	Groove	GRV	Left hand	LH
Effective	EFF	Ground	GRD	Length	LG
Elbow	ELL	Gypsum	GYP	Letter	LTR
Electric	ELEC			Light	LT
Elevation	ELEV			Line	L
Engineer	ENGR	Half-round	½ RD	Logarithm	LOG.
Equal	EQ	Handle	HDL	Lubricate	LUB
Equipment	EQUIP.	Hanger	HGR	Lumber	LBR
Equivalent	EQUIV	Hard	H		
Estimate	EST	Hard-drawn	HD	Machine	MACH
Exterior	EXT	Harden	HDN	Malleable	MALL
Extra heavy	X HVY	Hardware	HDW	Malleable iron	MI
Extra strong	X STR	Head	HD	Manhole	MH
		Headless	HDLS	Manual	MAN.
		Headquarters	HQ	Manufacture	MFR
Fabricate	FAB	Heat	HT	Material	MATL
Face to face	F to F	Heat treat	HT TR	Maximum	MAX
Fahrenheit	F	Hexagon	HEX	Mechanical	MECH
Fairing	FAIR.	High-pressure	HP	Mechanism	MECH
Far side	FS	High-speed	HS	Median	MED
Federal	FED.	Horizontal	HOR	Metal	MET.
Feet	(') FT	Horsepower	HP	Meter (Instrument or	
Feet per minute	FPM	Hot rolled	HR	measure of length)	M
Feet per second	FPS	Hot rolled steel	HRS	Miles	MI
Field	FLD	Hour	HR	Miles per gallon	MPG
Figure	FIG.	Hundredweight	CWT	Miles per hour	MPH
Fillet	FIL	Hydraulic	HYD		

APPENDIX 1
ABBREVIATIONS (ANSI Z 32.13) (Cont.)

Word	Abbreviation	Word	Abbreviation	Word	Abbreviation
Millimeter	MM	Precast	PRCST	Shaft	SFT
Minimum	MIN	Prefabricated	PREFAB	Sketch	SK
Minute	(') MIN	Preferred	PFD	Sleeve	SLV
Miscellaneous	MISC	Prepare	PREP	Slide	SL
Mixture	MIX.	Pressure	PRESS.	Slotted	SLOT.
Model	MOD	Pressure angle	PA.	Socket	SOC
Month	MO	Process	PROC	Solder	SLD
Morse taper	MOR T	Production	PROD	South	S
Multiple	MULT	Profile	PF	Space	SP
		Project	PROJ	Special	SPL
National	NATL	Proof	PRF	Specific gravity	SP GR
Near side	NS			Spherical	SPHER
Negative	NEG	Quadrant	QUAD	Spot faced	SF
Neutral	NEUT	Quart	QT	Spring	SPG
Nipple	NIP.	Quarter	QTR	Square	SQ
Nominal	NOM	Quarter-round	¼ RD	Stainless	STN
Normal	NOR			Stainless steel	SST
North	N	Radial	RAD	Standard	STD
Not to scale	NTS	Radius	R	Station	STA
Number	NO.	Railroad	RR	Steel	STL
		Ream	RM	Stock	STK
Obsolete	OBS	Received	RECD	Straight	STR
Octagon	OCT	Record	REC	Structural	STR
Ohm	Ω	Rectangle	RECT	Substitute	SUB
On center	OC	Reference	REF	Summary	SUM.
Opposite	OPP	Reference line	REF L	Supply	SUP
Optical	OPT	Relief	REL	Surface	SUR
Original	ORIG	Remove	REM	Symbol	SYM
Ounce	OZ	Require	REQ	Symmetrical	SYM
Outlet	OUT.	Required	REQD	System	SYS
Outside diameter	OD	Return	RET.		
Outside face	OF	Revise	REV	Tangent	TAN.
Outside radius	OR	Revolution	REV	Taper	TPR
Overall	OA	Revolutions per minute	RPM	Technical	TECH
		Rheostat	RHEO	Temperature	TEMP
Pack	PK	Right	R	Template	TEMP
Packing	PKG	Right hand	RH	Tensile strength	TS
Parallel	PAR.	Rivet	RIV	Tension	TENS.
Part	PT	Rockwell hardness	RH	Thick	THK
Patent	PAT.	Roller bearing	RB	Thousand	M
Permanent	PERM	Room	RM	Thousand pound	KIP
Perpendicular	PERP	Root diameter	RD	Thread	THD
Photograph	PHOTO	Root mean square	RMS	Tolerance	TOL
Piece	PC	Rough	RGH	Tongue & groove	T&G
Pint	PT	Round	RD	Tool steel	TS
Pitch	P	Rubber	RUB.	Tooth	T
Pitch circle	PC			Total	TOT
Pitch diameter	PD	Safety	SAF	Transfer	TRANS
Plastic	PLSTC	Sand blast	SD BL	Typical	TYP
Plate	PL	Schedule	SCH		
Point	PT	Screen	SCRN	Ultimate	ULT
Polish	POL	Screw	SCR	Unit	U
Position	POS	Sea level	SL	Universal	UNIV
Positive	POS	Second	SEC		
Pound	LB	Section	SECT	Vacuum	VAC
Pounds per square inch	PSI	Separate	SEP	Valve	V
Power	PWR	Set screw	SS	Variable	VAR

Cont.

APPENDIX 1
ABBREVIATIONS (ANSI Z 32.13) (Cont.)

Word	Abbreviation	Word	Abbreviation	ABBREVIATIONS FOR COLORS	
Vertical	VERT	West	W	Amber	AMB
Volt	V	Width	W	Black	BLK
Voltmeter	VM	Wood	WD	Blue	BLU
Volume	VOL	Woodruff	WDF	Brown	BRN
		Wrought iron	WI	Green	GRN
Washer	WASH.			Orange	ORN
Watt	W	Yard	YD	White	WHT
Weight	WT	Year	YR	Yellow	YEL

APPENDIX 2
CONVERSION TABLES

Length conversions		
Angstrom units	$\times\ 1 \times 10^{-10}$	= meters
	$\times\ 1 \times 10^{-4}$	= microns
	$\times\ 1.650\ 763\ 73 \times 10^{-4}$	= wavelengths of orange-red line of krypton 86
Cables	$\times\ 120$	= fathoms
	$\times\ 720$	= feet
	$\times\ 219.456$	= meters
Fathoms	$\times\ 6$	= feet
	$\times\ 1.828\ 8$	= meters
Feet	$\times\ 12$	= inches
	$\times\ 0.3048$	= meters
Furlongs	$\times\ 660$	= feet
	$\times\ 201.168$	= meters
	$\times\ 220$	= yards
Inches	$\times\ 2.54 \times 10^{8}$	= Angstroms
	$\times\ 25.4$	= millimeters
	$\times\ 8.333\ 33 \times 10^{-2}$	= feet
Kilometers	$\times\ 3.280\ 839 \times 10^{3}$	= feet
	$\times\ 0.62$	= miles
	$\times\ 0.539\ 956$	= nautical miles
	$\times\ 0.621\ 371$	= statute miles
	$\times\ 1.093\ 613 \times 10^{3}$	= yards
Light-years	$\times\ 9.460\ 55 \times 10^{12}$	= kilometers
	$\times\ 5.878\ 51 \times 10^{12}$	= statute miles
Meters	$\times\ 1 \times 10^{10}$	= Angstroms
	$\times\ 3.280\ 839\ 9$	= feet
	$\times\ 39.370\ 079$	= inches
	$\times\ 1.093\ 61$	= yards
Microns	$\times\ 10^{4}$	= Angstroms
	$\times\ 10^{-4}$	= centimeters
	$\times\ 10^{-6}$	= meters
Nautical Miles (International)	$\times\ 8.439\ 049$	= cables
	$\times\ 6.076\ 115\ 49 \times 10^{3}$	= feet
	$\times\ 1.852 \times 10^{3}$	= meters
	$\times\ 1.150\ 77$	= statute miles

Length conversions

Statute Miles	$\times$ 5.280 $\times$ 10^3	= feet
	$\times$ 8	= furlongs
	$\times$ 6.336 0 $\times$ 10^4	= inches
	$\times$ 1.609 34	= kilometers
	$\times$ 8.689 7 $\times$ 10^{-1}	= nautical miles
Miles	$\times$ 10^{-3}	= inches
	$\times$ 2.54 $\times$ 10^{-2}	= millimeters
	$\times$ 25.4	= micrometers
	$\times$ 0.61	= kilometers
Yards	$\times$ 3	= feet
	$\times$ 9.144 $\times$ 10^{-1}	= meters
Feet/hour	$\times$ 3.048 $\times$ 10^{-4}	= kilometers/hour
	$\times$ 1.645 788 $\times$ 10^{-4}	= knots
Feet/minute	$\times$ 0.3048	= meters/minute
	$\times$ 5.08 $\times$ 10^{-3}	= meters/second
Feet/second	$\times$ 1.097 28	= kilometers/hour
	$\times$ 18.288	= meters/minute
Kilometers/hour	$\times$ 3.280 839 $\times$ 10^3	= feet/hour
	$\times$ 54.680 66	= feet/minute
	$\times$ 0.277 777	= meters/second
	$\times$ 0.621 371	= miles/hour
Kilometers/minute	$\times$ 3.280 839 $\times$ 10^3	= feet/minute
	$\times$ 37.282 27	= miles/hour
Knots	$\times$ 6.076 115 $\times$ 10^3	= feet/hour
	$\times$ 101.268 5	= feet/minute
	$\times$ 1.687 809	= feet/second
	$\times$ 1.852	= kilometers/hour
	$\times$ 30.866	= meters/minute
	$\times$ 0.514 4	= meters/second
	$\times$ 1.150 77	= statute miles/hour
Meters/hour	$\times$ 3.280 839	= feet/hour
	$\times$ 88	= feet/minute
	$\times$ 1.466	= feet/second
	$\times$ 1 $\times$ 10^{-3}	= kilometers/hour
	$\times$ 1.667 $\times$ 10^{-2}	= meters/minute
Feet/second^2	$\times$ 1.097 28	= kilometers/hour/second
	$\times$ 0.304 8	= meters/second^2

Area conversions

Acres	$\times$ 4.046 85 $\times$ 10^{-3}	= square kilometers
	$\times$ 4.046 856 $\times$ 10^3	= square meters
	$\times$ 4.356 0 $\times$ 10^4	= square feet
Ares	$\times$ 2.471 053 8 $\times$ 10^{-2}	= acres
	$\times$ 1	= square dekameters
	$\times$ 10^2	= square meters
Barns	$\times$ 1 $\times$ 10^{-28}	= square meters
Circular mils	$\times$ 1 $\times$ 10^{-6}	= circular inches
	$\times$ 5.067 074 8 $\times$ 10^{-4}	= square millimeters
	$\times$ 0.785 398 1	= square mils

Cont.

Area conversions

Hectares	$\times$ 2.471 05	= acres
	$\times 10^2$	= ares
	$\times 10^4$	= square meters
Square feet	$\times$ 2.295 684 $\times 10^{-5}$	= acres
	$\times$ 9.290 3 $\times 10^{-4}$	= ares
	$\times$ 144	= square inches
	$\times$ 9.290 304 $\times 10^{-2}$	= square meters
Square inches	$\times$ 1.273 239 5 $\times 10^6$	= circular mils
	$\times$ 6.944 4 $\times 10^{-3}$	= square feet
	$\times$ 6.451 6 $\times 10^{-4}$	= square meters
Square kilometers	$\times$ 247.105 38	= acres
	$\times$ 1.076 391 0 $\times 10^7$	= square feet
	$\times$ 1.000	= cubic meters
	$\times$ 1.307 950 6	= cubic yards
	$\times$ 219.969	= imperial gallons

Volume conversions

Liters	$\times 10^3$	= cubic centimeters
	$\times$ 1.000 $\times 10^6$	= cubic millimeters
	$\times$ 1.000 $\times 10^{-3}$	= cubic meters
	$\times$ 61.023 74	= cubic inches
	$\times$ 3.531 5 $\times 10^{-2}$	= cubic feet
	$\times$ 1.307 95 $\times 10^{-3}$	= cubic yards
	$\times$ 0.22	= gallons
	$\times$ 0.219 969	= imperial gallons
	$\times$ 0.879 877	= imperial quarts
Imperial pints	$\times$ 0.125	= imperial gallons
	$\times$ 0.568 261	= liters
	$\times$ 20	= imperial fluid ounces
	$\times$ 0.5	= imperial quarts
	$\times$ 568.260 9	= cubic centimeters
Imperial quarts	$\times$ 1.136 52 $\times 10^3$	= cubic centimeters
	$\times$ 69.354 8	= cubic inches
	$\times$ 1.136 522 8	= liters

Power conversions

British Thermal Units/hour	$\times$ 2.928 7 $\times 10^{-4}$	= kilowatts
	$\times$ 0.292 875	= watts
BTU/minute	$\times$ 1.757 25 $\times 10^{-2}$	= kilowatts
BTU/pound	$\times$ 2.324 4	= joules/gram
BTU/second	$\times$ 1.413 91	= horsepower
	$\times$ 107.514	= kilogrammeters/second
	$\times$ 1.054 35	= kilowatts
	$\times$ 1.054 35 $\times 10^3$	= watts
Foot-pound-force/hour	$\times$ 5.050 $\times 10^{-7}$	= horsepower
	$\times$ 3.766 16 $\times 10^{-7}$	= kilowatts
Foot-pound-force/ minute	$\times$ 3.030 303 $\times 10^{-5}$	= horsepower
	$\times$ 2.259 70 $\times 10^{-2}$	= joules/second
	$\times$ 2.259 70 $\times 10^{-5}$	= kilowatts
Horsepower	$\times$ 42.435 6	= BTU/minute
	$\times$ 550	= footpounds/second
	$\times$ 0.746	= kilowatts
	$\times$ 746	= joules/second

CONVERSION TABLES (Cont.)

Power conversions

Kilogrammeters/second	× 9.806 65	= watts
Kilowatts	× 3.414 43 × 10^3	= BTU/hour
	× 2.655 22 × 10^6	= footpounds/hour
	× 4.425 37 × 10^4	= footpounds/minute
	× 737.562	= footpounds/second
	× 1.019 726 × 10^7	= gramcentimeters/second
	× 1.341 02	= horsepower
	× 3.6 × 10^6	= joules/hour
	× 10^3	= joules/second
	× 3.671 01 × 10^5	= kilogrammeters/hour
	× 999.835	= international watt
Watts	× 44.253 7	= footpounds/minute
	× 1.341 02 × 10^{-3}	= horsepower
	× 1	= joules/second

Time conversions

(No attempt has been made in this brief treatment to correlate solar, mean solar, sidereal, and mean sidereal days.)

Mean solar days	× 24	= mean solar hours
Mean solar hours	× 3.600 × 10^3	= mean solar seconds
	× 60	= mean solar minutes

Angle conversions

Degrees	× 60	= minutes
	× 1.745 329 3 × 10^{-2}	= radians
Degrees/foot	× 5.726 145 × 10^{-4}	= radians/centimeter
Degrees/minute	× 2.908 8 × 10^{-4}	= radians/second
	× 4.629 629 × 10^{-5}	= revolutions/second
Degrees/second	× 1.745 329 3 × 10^{-2}	= radians/second
	× 0.166	= revolutions/minute
	× 2.77 × 10^{-3}	= revolutions/second
Minutes	× 1.667 × 10^{-2}	= degrees
	× 2.908 8 × 10^{-4}	= radians
	× 60	= seconds
Radians	× 0.159 154	= circumferences
	× 57.295 77	= degrees
	× 3.437 746 × 10^3	= minutes
Seconds	× 2.777 × 10^{-4}	= degrees
	× 1.667 × 10^{-2}	= minutes
	× 4.848 136 8 × 10^{-6}	= radians
Steradians	× 0.159 154 9	= hemispheres
	× 7.957 74 × 10^{-2}	= spheres
	× 0.636 619 7	= spherical right angles

Mass conversions

Grains	× 6.479 8 × 10^{-2}	= grams
	× 2.285 71 × 10^{-3}	= avoirdupois ounces

Cont.

APPENDIX 2
CONVERSION TABLES (Cont.)

Mass conversions

Grams	$\times$ 15.432 358	= grains
	$\times$ 3.527 396 $\times$ 10^{-2}	= avoirdupois ounces
	$\times$ 2.204 62 $\times$ 10^{-3}	= avoirdupois pounds
Kilograms	$\times$ 564.383 4	= avoirdupois drams
	$\times$ 2.204 622 6	= avoirdupois pounds
	$\times$ 2.2	= pounds
	$\times$ 9.842 065 $\times$ 10^{-4}	= long tons
	$\times$ 10^{-3}	= metric tons
	$\times$ 1.102 31 $\times$ 10^{-3}	= short tons
Avoirdupois ounces	$\times$ 28.349 5	= grams
	$\times$ 6.25 $\times$ 10^{-2}	= avoirdupois pounds
	$\times$ 0.911 458	= troy ounces
Avoirdupois pounds	$\times$ 256	= drams
	$\times$ 4.535 923 7 $\times$ 10^2	= grams
	$\times$ 0.453 592 4	= kilograms
	$\times$ 16	= ounces
Long tons	$\times$ 2.24 $\times$ 10^3	= avoirdupois pounds
	$\times$ 1.106 046 9	= metric tons
	$\times$ 1.12	= short tons
Metric tons	$\times$ 10^3	= kilograms
	$\times$ 2.204 622 $\times$ 10^3	= avoirdupois pounds
Short tons	$\times$ 2 $\times$ 10^3	= avoirdupois pounds
	$\times$ 907.184 74	= kilograms

Force conversions

Dynes	$\times$ 10^{-5}	= newtons
Newtons	$\times$ 10^5	= dynes
	$\times$ 0.224 808	= pounds-force
Pounds	$\times$ 4.448 22	= newtons

Energy conversions

British Thermal Units (thermochemical)	$\times$ 1.054 35 $\times$ 10^3	= joules
	$\times$ 2.928 27 $\times$ 10^{-4}	= kilowatthours
	$\times$ 1.054 35 $\times$ 10^3	= wattseconds
Foot-pound-force	$\times$ 1.355 818 0	= joules
	$\times$ 0.138 255	= kilogramforce-meters
	$\times$ 3.766 16 $\times$ 10^{-7}	= kilowatthours
	$\times$ 1.355 818 0	= newtonmeters
Joules	$\times$ 9.484 5 $\times$ 10^{-4}	= British Thermal Units
	$\times$ 0.737 562	= foot-pounds-force
	$\times$ 0.101 971 6	= kilogramforce-meters
	$\times$ 2.777 7 $\times$ 10^{-7}	= kilowatthours
	$\times$ 1	= wattseconds
Kilogramforce-meters	$\times$ 9.287 7 $\times$ 10^{-3}	= British Thermal Units
	$\times$ 7.233 01	= foot-pounds-force
	$\times$ 9.806 65	= joules
	$\times$ 9.806 65	= newtonmeters
	$\times$ 2.724 0 $\times$ 10^{-3}	= watthours

APPENDIX 2
CONVERSION TABLES (Cont.)

Energy conversions

Kilowatthours	$\times$ 3.409 52 $\times$ 10^3	= British Thermal Units
	$\times$ 2.655 22 $\times$ 10^6	= foot-pounds-force
	$\times$ 1.341 02	= horsepowerhours
	$\times$ 3.6 $\times$ 10^6	= joules
	$\times$ 3.670 98 $\times$ 10^5	= kilogramforce-meters
Newtonmeters	$\times$ 0.101 971	= kilogramforce-meters
	$\times$ 0.737 562	= poundforce-feet
Watthours	$\times$ 3.414 43	= British Thermal Units
	$\times$ 2.655 22 $\times$ 10^3	= foot-pounds-force
	$\times$ 3.6 $\times$ 10^3	= joules
	$\times$ 3.670 98 $\times$ 10^2	= kilogramforce-meters

Pressure conversions

Atmospheres	$\times$ 1.013 25	= bars
	$\times$ 1.033 23 $\times$ 10^3	= grams/square centimeter
	$\times$ 1.033 23 $\times$ 10^7	= grams/square meter
	$\times$ 14.696 0	= pounds/square inch
	$\times$ 760	= torrs
	$\times$ 101	= kilopascals
Bars	$\times$ 0.986 923	= atmospheres
	$\times$ 10^6	= baryes
	$\times$ 1.019 716 $\times$ 10^7	= grams/square meter
	$\times$ 1.019 716 $\times$ 10^4	= kilogramsforce/square meter
	$\times$ 14.503 8	= poundsforce/square inch
Baryes	$\times$ 10^{-6}	= bars
Inches of mercury	$\times$ 3.386 4 $\times$ 10^{-2}	= bars
	$\times$ 345.316	= kilogramsforce/square meter
	$\times$ 70.726 2	= poundsforce/square foot
Pascal	$\times$ 1	= newton/square meter

APPENDIX 3

LOGARITHMS OF NUMBERS

N	0	1	2	3	4	5	6	7	8	9
1.0	.0000	.0043	.0086	.0128	.0170	.0212	.0253	.0294	.0334	.0374
1.1	.0414	.0453	.0492	.0531	.0569	.0607	.0645	.0682	.0719	.0755
1.2	.0792	.0828	.0864	.0899	.0934	.0969	.1004	.1038	.1072	.1106
1.3	.1139	.1173	.1206	.1239	.1271	.1303	.1335	.1367	.1399	.1430
1.4	.1461	.1492	.1523	.1553	.1584	.1614	.1644	.1673	.1703	.1732
1.5	.1761	.1790	.1818	.1847	.1875	.1903	.1931	.1959	.1987	.2014
1.6	.2041	.2068	.2095	.2122	.2148	.2175	.2201	.2227	.2253	.2279
1.7	.2304	.2330	.2355	.2380	.2405	.2430	.2455	.2480	.2504	.2529
1.8	.2553	.2577	.2601	.2625	.2648	.2672	.2695	.2718	.2742	.2765
1.9	.2788	.2810	.2833	.2856	.2878	.2900	.2923	.2945	.2967	.2989
2.0	.3010	.3032	.3054	.3075	.3096	.3118	.3139	.3160	.3181	.3201
2.1	.3222	.3243	.3263	.3284	.3304	.3324	.3345	.3365	.3385	.3404
2.2	.3424	.3444	.3464	.3483	.3502	.3522	.3541	.3560	.3579	.3598
2.3	.3617	.3636	.3655	.3674	.3692	.3711	.3729	.3747	.3766	.3784
2.4	.3802	.3820	.3838	.3856	.3874	.3892	.3909	.3927	.3945	.3962
2.5	.3979	.3997	.4014	.4031	.4048	.4065	.4082	.4099	.4116	.4133
2.6	.4150	.4166	.4183	.4200	.4216	.4232	.4249	.4265	.4281	.4298
2.7	.4314	.4330	.4346	.4362	.4378	.4393	.4409	.4425	.4440	.4456
2.8	.4472	.4487	.4502	.4518	.4533	.4548	.4564	.4579	.4594	.4609
2.9	.4624	.4639	.4654	.4669	.4683	.4698	.4713	.4728	.4742	.4757
3.0	.4771	.4786	.4800	.4814	.4829	.4843	.4857	.4871	.4886	.4900
3.1	.4914	.4928	.4942	.4955	.4969	.4983	.4997	.5011	.5024	.5038
3.2	.5051	.5065	.5079	.5092	.5105	.5119	.5132	.5145	.5159	.5172
3.3	.5185	.5198	.5211	.5224	.5237	.5250	.5263	.5276	.5289	.5302
3.4	.5315	.5328	.5340	.5353	.5366	.5378	.5391	.5403	.5416	.5428
3.5	.5441	.5453	.5465	.5478	.5490	.5502	.5514	.5527	.5539	.5551
3.6	.5563	.5575	.5587	.5599	.5611	.5623	.5635	.5647	.5658	.5670
3.7	.5682	.5694	.5705	.5717	.5729	.5740	.5752	.5763	.5775	.5786
3.8	.5798	.5809	.5821	.5832	.5843	.5855	.5866	.5877	.5888	.5899
3.9	.5911	.5922	.5933	.5944	.5955	.5966	.5977	.5988	.5999	.6010
4.0	.6021	.6031	.6042	.6053	.6064	.6075	.6085	.6096	.6107	.6117
4.1	.6128	.6138	.6149	.6160	.6170	.6180	.6191	.6201	.6212	.6222
4.2	.6232	.6243	.6253	.6263	.6274	.6284	.6294	.6304	.6314	.6325
4.3	.6335	.6345	.6355	.6365	.6375	.6385	.6395	.6405	.6415	.6425
4.4	.6435	.6444	.6454	.6464	.6474	.6484	.6493	.6503	.6513	.6522
4.5	.6532	.6542	.6551	.6561	.6571	.6580	.6590	.6599	.6609	.6618
4.6	.6628	.6637	.6646	.6656	.6665	.6675	.6684	.6693	.6702	.6712
4.7	.6721	.6730	.6739	.6749	.6758	.6767	.6776	.6785	.6794	.6803
4.8	.6812	.6821	.6830	.6839	.6848	.6857	.6866	.6875	.6884	.6893
4.9	.6902	.6911	.6920	.6928	.6937	.6946	.6955	.6964	.6972	.6981
5.0	.6990	.6998	.7007	.7016	.7024	.7033	.7042	.7050	.7059	.7067
5.1	.7076	.7084	.7093	.7101	.7110	.7118	.7126	.7135	.7143	.7152
5.2	.7160	.7168	.7177	.7185	.7193	.7202	.7210	.7218	.7226	.7235
5.3	.7243	.7251	.7259	.7267	.7275	.7284	.7292	.7300	.7308	7316
5.4	.7324	.7332	.7340	.7348	.7356	.7364	.7372	.7380	.7388	.7396
N	0	1	2	3	4	5	6	7	8	9

APPENDIX 3
LOGARITHMS OF NUMBERS (Cont.)

N	0	1	2	3	4	5	6	7	8	9
5.5	.7404	.7412	.7419	.7427	.7435	.7443	.7451	.7459	.7466	.7474
5.6	.7482	.7490	.7497	.7505	.7513	.7520	.7528	.7536	.7543	.7551
5.7	.7559	.7566	.7574	.7582	.7589	.7597	.7604	.7612	.7619	.7627
5.8	.7634	.7642	.7649	.7657	.7664	.7672	.7679	.7686	.7694	.7701
5.9	.7709	.7716	.7723	.7731	.7738	.7745	.7752	.7760	.7767	.7774
6.0	.7782	.7789	.7796	.7803	.7810	.7818	.7825	.7832	.7839	.7846
6.1	.7853	.7860	.7868	.7875	.7882	.7889	.7896	.7903	.7910	.7917
6.2	.7924	.7931	.7938	.7945	.7952	.7959	.7966	.7973	.7980	.7987
6.3	.7993	.8000	.8007	.8014	.8021	.8028	.8035	.8041	.8048	.8055
6.4	.8062	.8069	.8075	.8082	.8089	.8096	.8102	.8109	.8116	.8122
6.5	.8129	.8136	.8142	.8149	.8156	.8162	.8169	.8176	.8182	.8189
6.6	.8195	.8202	.8209	.8215	.8222	.8228	.8235	.8241	.8248	.8254
6.7	.8261	.8267	.8274	.8280	.8287	.8293	.8299	.8306	.8312	.8319
6.8	.8325	.8331	.8338	.8344	.8351	.8357	.8363	.8370	.8376	.8382
6.9	.8388	.8395	.8401	.8407	.8414	.8420	.8426	.8432	.8439	.8445
7.0	.8451	.8457	.8463	.8470	.8476	.8482	.8488	.8494	.8500	.8506
7.1	.8513	.8519	.8525	.8531	.8537	.8543	.8549	.8555	.8561	.8567
7.2	.8573	.8579	.8585	.8591	.8597	.8603	.8609	.8615	.8621	.8627
7.3	.8633	.8639	.8645	.8651	.8657	.8663	.8669	.8675	.8681	.8686
7.4	.8692	.8698	.8704	.8710	.8716	.8722	.8727	.8733	.8739	.8745
7.5	.8751	.8756	.8762	.8768	.8774	.8779	.8785	.8791	.8797	.8802
7.6	.8808	.8814	.8820	.8825	.8831	.8837	.8842	.8848	.8854	.8859
7.7	.8865	.8871	.8876	.8882	.8887	.8893	.8899	.8904	.8910	.8915
7.8	.8921	.8927	.8932	.8938	.8943	.8949	.8954	.8960	.8965	.8971
7.9	.8976	.8982	.8987	.8993	.8998	.9004	.9009	.9015	.9020	.9025
8.0	.9031	.9036	.9042	.9047	.9053	.9058	.9063	.9069	.9074	.9079
8.1	.9085	.9090	.9096	.9101	.9106	.9112	.9117	.9122	.9128	.9133
8.2	.9138	.9143	.9149	.9154	.9159	.9165	.9170	.9175	.9180	.9186
8.3	.9191	.9196	.9201	.9206	.9212	.9217	.9222	.9227	.9232	.9238
8.4	.9243	.9248	.9253	.9258	.9263	.9269	.9274	.9279	.9284	.9289
8.5	.9294	.9299	.9304	.9309	.9315	.9320	.9325	.9330	.9335	.9340
8.6	.9345	.9350	.9355	.9360	.9365	.9370	.9375	.9380	.9385	.9390
8.7	.9395	.9400	.9405	.9410	.9415	.9420	.9425	.9430	.9435	.9440
8.8	.9445	.9450	.9455	.9460	.9465	.9469	.9474	.9479	.9484	.9489
8.9	.9494	.9499	.9504	.9509	.9513	.9518	.9523	.9528	.9533	.9538
9.0	.9542	.9547	.9552	.9557	.9562	.9566	.9571	.9576	.9581	.9586
9.1	.9590	.9595	.9600	.9605	.9609	.9614	.9619	.9624	.9628	.9633
9.2	.9638	.9643	.9647	.9652	.9657	.9661	.9666	.9671	.9675	.9680
9.3	.9685	.9689	.9694	.9699	.9703	.9708	.9713	.9717	.9722	.9727
9.4	.9731	.9736	.9741	.9745	.9750	.9754	.9759	.9763	.9768	.9773
9.5	.9777	.9782	.9786	.9791	.9795	.9800	.9805	.9809	.9814	.9818
9.6	.9823	.9827	.9832	.9836	.9841	.9845	.9850	.9854	.9859	.9863
9.7	.9868	.9872	.9877	.9881	.9886	.9890	.9894	.9899	.9903	.9908
9.8	.9912	.9917	.9921	.9926	.9930	.9934	.9939	.9943	.9948	.9952
9.9	.9956	.9961	.9965	.9969	.9974	.9978	.9983	.9987	.9991	.9996
N	0	1	2	3	4	5	6	7	8	9

VALUES OF TRIGONOMETRIC FUNCTIONS

Degrees	Radians	Sine	Tangent	Cotangent	Cosine		
0° 00'	.0000	.0000	.0000		1.0000	1.5708	90° 00'
10'	.0029	.0029	.0029	343.77	1.0000	1.5679	50'
20'	.0058	.0058	.0058	171.89	1.0000	1.5650	40'
30'	.0087	.0087	.0087	114.59	1.0000	1.5621	30'
40'	.0116	.0116	.0116	85.940	.9999	1.5592	20'
50'	.0145	.0145	.0145	68.750	.9999	1.5563	10'
1° 00'	.0175	.0175	.0175	57.290	.9998	1.5533	89° 00'
10'	.0204	.0204	.0204	49.104	.9998	1.5504	50'
20'	.0233	.0233	.0233	42.964	.9997	1.5475	40'
30'	.0262	.0262	.0262	38.188	.9997	1.5446	30'
40'	.0291	.0291	.0291	34.368	.9996	1.5417	20'
50'	.0320	.0320	.0320	31.242	.9995	1.5388	10'
2° 00'	.0349	.0349	.0349	28.636	.9994	1.5359	88° 00'
10'	.0378	.0378	.0378	26.432	.9993	1.5330	50'
20'	.0407	.0407	.0407	24.542	.9992	1.5301	40'
30'	.0436	.0436	.0437	22.904	.9990	1.5272	30'
40'	.0465	.0465	.0466	21.470	.9989	1.5243	20'
50'	.0495	.0494	.0495	20.206	.9988	1.5213	10'
3° 00'	.0524	.0523	.0524	19.081	.9986	1.5184	87° 00'
10'	.0553	.0552	.0553	18.075	.9985	1.5155	50'
20'	.0582	.0581	.0582	17.169	.9983	1.5126	40'
30'	.0611	.0610	.0612	16.350	.9981	1.5097	30'
40'	.0640	.0640	.0641	15.605	.9980	1.5068	20'
50'	.0669	.0669	.0670	14.924	.9978	1.5039	10'
4° 00'	.0698	.0698	.0699	14.301	.9976	1.5010	86° 00'
10'	.0727	.0727	.0729	13.727	.9974	1.4981	50'
20'	.0756	.0756	.0758	13.197	.9971	1.4952	40'
30'	.0785	.0785	.0787	12.706	.9969	1.4923	30'
40'	.0814	.0814	.0816	12.251	.9967	1.4893	20'
50'	.0844	.0843	.0846	11.826	.9964	1.4864	10'
5° 00'	.0873	.0872	.0875	11.430	.9962	1.4835	85° 00'
10'	.0902	.0901	.0904	11.059	.9959	1.4806	50'
20'	.0931	.0929	.0934	10.712	.9957	1.4777	40'
30'	.0960	.0958	.0963	10.385	.9954	1.4748	30'
40'	.0989	.0987	.0992	10.078	.9951	1.4719	20'
50'	.1018	.1016	.1022	9.7882	.9948	1.4690	10'
6° 00'	.1047	.1045	.1051	9.5144	.9945	1.4661	84° 00'
10'	.1076	.1074	.1080	9.2553	.9942	1.4632	50'
20'	.1105	.1103	.1110	9.0098	.9939	1.4603	40'
30'	.1134	.1132	.1139	8.7769	.9936	1.4573	30'
40'	.1164	.1161	.1169	8.5555	.9932	1.4544	20'
50'	.1193	.1190	.1198	8.3450	.9929	1.4515	10'
7° 00'	.1222	.1219	.1228	8.1443	.9925	1.4486	83° 00'
10'	.1251	.1248	.1257	7.9530	.9922	1.4457	50'
20'	.1280	.1276	.1287	7.7704	.9918	1.4428	40'
30'	.1309	.1305	.1317	7.5958	.9914	1.4399	30'
40'	.1338	.1334	.1346	7.4287	.9911	1.4370	20'
50'	.1367	.1363	.1376	7.2687	.9907	1.4341	10'
8° 00'	.1396	.1392	.1405	7.1154	.9903	1.4312	82° 00'
10'	.1425	.1421	.1435	6.9682	.9899	1.4283	50'
20'	.1454	.1449	.1465	6.8269	.9894	1.4254	40'
30'	.1484	.1478	.1495	6.6912	.9890	1.4224	30'
40'	.1513	.1507	.1524	6.5606	.9886	1.4195	20'
50'	.1542	.1536	.1554	6.4348	.9881	1.4166	10'
9° 00'	.1571	.1564	.1584	6.3138	.9877	1.4137	81° 00'
		Cosine	Cotangent	Tangent	Sine	Radians	Degrees

APPENDIX 4
VALUES OF TRIGONOMETRIC FUNCTIONS (Cont.)

Degrees	Radians	Sine	Tangent	Cotangent	Cosine		
9° 00'	.1571	.1564	.1584	6.3138	.9877	1.4137	81° 00'
10'	.1600	.1593	.1614	6.1970	.9872	1.4108	50'
20'	.1629	.1622	.1644	6.0844	.9868	1.4079	40'
30'	.1658	.1650	.1673	5.9758	.9863	1.4050	30'
40'	.1687	.1679	.1703	5.8708	.9858	1.4021	20'
50'	.1716	.1708	.1733	5.7694	.9853	1.3992	10'
10° 00'	.1745	.1736	.1763	5.6713	.9848	1.3963	80° 00'
10'	.1774	.1765	.1793	5.5764	.9843	1.3934	50'
20'	.1804	.1794	.1823	5.4845	.9838	1.3904	40'
30'	.1833	.1822	.1853	5.3955	.9833	1.3875	30'
40'	.1862	.1851	.1883	5.3093	.9827	1.3846	20'
50'	.1891	.1880	.1914	5.2257	.9822	1.3817	10'
11° 00'	.1920	.1908	.1944	5.1446	.9816	1.3788	79° 00'
10'	.1949	.1937	.1974	5.0658	.9811	1.3759	50'
20'	.1978	.1965	.2004	4.9894	.9805	1.3730	40'
30'	.2007	.1994	.2035	4.9152	.9799	1.3701	30'
40'	.2036	.2022	.2065	4.8430	.9793	1.3672	20'
50'	.2065	.2051	.2095	4.7729	.9787	1.3643	10'
12° 00'	.2094	.2079	.2126	4.7046	.9781	1.3614	78° 00'
10'	.2123	.2108	.2156	4.6382	.9775	1.3584	50'
20'	.2153	.2136	.2186	4.5736	.9769	1.3555	40'
30'	.2182	.2164	.2217	4.5107	.9763	1.3526	30'
40'	.2211	.2193	.2247	4.4494	.9757	1.3497	20'
50'	.2240	.2221	.2278	4.3897	.9750	1.3468	10'
13° 00'	.2269	.2250	.2309	4.3315	.9744	1.3439	77° 00'
10'	.2298	.2278	.2339	4.2747	.9737	1.3410	50'
20'	.2327	.2306	.2370	4.2193	.9730	1.3381	40'
30'	.2356	.2334	.2401	4.1653	.9724	1.3352	30'
40'	.2385	.2363	.2432	4.1126	.9717	1.3323	20'
50'	.2414	.2391	.2462	4.0611	.9710	1.3294	10'
14° 00'	.2443	.2419	.2493	4.0108	.9703	1.3265	76° 00'
10'	.2473	.2447	.2524	3.9617	.9696	1.3235	50'
20'	.2502	.2476	.2555	3.9136	.9689	1.3206	40'
30'	.2531	.2504	.2586	3.8667	.9681	1.3177	30'
40'	.2560	.2532	.2617	3.8208	.9674	1.3148	20'
50'	.2589	.2560	.2648	3.7760	.9667	1.3119	10'
15° 00'	.2618	.2588	.2679	3.7321	.9659	1.3090	75° 00'
10'	.2647	.2616	.2711	3.6891	.9652	1.3061	50'
20'	.2676	.2644	.2742	3.6470	.9644	1.3032	40'
30'	.2705	.2672	.2773	3.6059	.9636	1.3003	30'
40'	.2734	.2700	.2805	3.5656	.9628	1.2974	20'
50'	.2763	.2728	.2836	3.5261	.9621	1.2945	10'
16° 00'	.2793	.2756	.2867	3.4874	.9613	1.2915	74° 00'
10'	.2822	.2784	.2899	3.4495	.9605	1.2886	50'
20'	.2851	.2812	.2931	3.4124	.9596	1.2857	40'
30'	.2880	.2840	.2962	3.3759	.9588	1.2828	30'
40'	.2909	.2868	.2994	3.3402	.9580	1.2799	20'
50'	.2938	.2896	.3026	3.3052	.9572	1.2770	10'
17° 00'	.2967	.2924	.3057	3.2709	.9563	1.2741	73° 00'
10'	.2996	.2952	.3089	3.2371	.9555	1.2712	50'
20'	.3025	.2979	.3121	3.2041	.9546	1.2683	40'
30'	.3054	.3007	.3153	3.1716	.9537	1.2654	30'
40'	.3083	.3035	.3185	3.1397	.9528	1.2625	20'
50'	.3113	.3062	.3217	3.1084	.9520	1.2595	10'
18° 00'	.3142	.3090	.3249	3.0777	.9511	1.2566	72° 00'
		Cosine	Cotangent	Tangent	Sine	Radians	Degrees

Cont.

APPENDIX 4
VALUES OF TRIGONOMETRIC FUNCTIONS (Cont.)

Degrees	Radians	Sine	Tangent	Cotangent	Cosine		
18° 00′	.3142	.3090	.3249	3.0777	.9511	1.2566	72° 00′
10′	.3171	.3118	.3281	3.0475	.9502	1.2537	50′
20′	.3200	.3145	.3314	3.0178	.9492	1.2508	40′
30′	.3229	.3173	.3346	2.9887	.9483	1.2479	30′
40′	.3258	.3201	.3378	2.9600	.9474	1.2450	20′
50′	.3287	.3228	.3411	2.9319	.9465	1.2421	10′
19° 00′	.3316	.3256	.3443	2.9042	.9455	1.2392	71° 00′
10′	.3345	.3283	.3476	2.8770	.9446	1.2363	50′
20′	.3374	.3311	.3508	2.8502	.9436	1.2334	40′
30′	.3403	.3338	.3541	2.8239	.9426	1.2305	30′
40′	.3432	.3365	.3574	2.7980	.9417	1.2275	20′
50′	.3462	.3393	.3607	2.7725	.9407	1.2246	10′
20° 00′	.3491	.3420	.3640	2.7475	.9397	1.2217	70° 00′
10′	.3520	.3448	.3673	2.7228	.9387	1.2188	50′
20′	.3549	.3475	.3706	2.6985	.9377	1.2159	40′
30′	.3578	.3502	.3739	2.6746	.9367	1.2130	30′
40′	.3607	.3529	.3772	2.6511	.9356	1.2101	20′
50′	.3636	.3557	.3805	2.6279	.9346	1.2072	10′
21° 00′	.3665	.3584	.3839	2.6051	.9336	1.2043	69° 00′
10′	.3694	.3611	.3872	2.5826	.9325	1.2014	50′
20′	.3723	.3638	.3906	2.5605	.9315	1.1985	40′
30′	.3752	.3665	.3939	2.5386	.9304	1.1956	30′
40′	.3782	.3692	.3973	2.5172	.9293	1.1926	20′
50′	.3811	.3719	.4006	2.4960	.9283	1.1897	10′
22° 00′	.3840	.3746	.4040	2.4751	.9272	1.1868	68° 00′
10′	.3869	.3773	.4074	2.4545	.9261	1.1839	50′
20′	.3898	.3800	.4108	2.4342	.9250	1.1810	40′
30′	.3927	.3827	.4142	2.4142	.9239	1.1781	30′
40′	.3956	.3854	.4176	2.3945	.9228	1.1752	20′
50′	.3985	.3881	.4210	2.3750	.9216	1.1723	10′
23° 00′	.4014	.3907	.4245	2.3559	.9205	1.1694	67° 00′
10′	.4043	.3934	.4279	2.3369	.9194	1.1665	50′
20′	.4072	.3961	.4314	2.3183	.9182	1.1636	40′
30′	.4102	.3987	.4348	2.2998	.9171	1.1606	30′
40′	.4131	.4014	.4383	2.2817	.9159	1.1577	20′
50′	.4160	.4041	.4417	2.2637	.9147	1.1548	10′
24° 00′	.4189	.4067	.4452	2.2460	.9135	1.1519	66° 00′
10′	.4218	.4094	.4487	2.2286	.9124	1.1490	50′
20′	.4247	.4120	.4522	2.2113	.9112	1.1461	40′
30′	.4276	.4147	.4557	2.1943	.9100	1.1432	30′
40′	.4305	.4173	.4592	2.1775	.9088	1.1403	20′
50′	.4334	.4200	.4628	2.1609	.9075	1.1374	10′
25° 00′	.4363	.4226	.4663	2.1445	.9063	1.1345	65° 00′
10′	.4392	.4253	.4699	2.1283	.9051	1.1316	50′
20′	.4422	.4279	.4734	2.1123	.9038	1.1286	40′
30′	.4451	.4305	.4770	2.0965	.9026	1.1257	30′
40′	.4480	.4331	.4806	2.0809	.9013	1.1228	20′
50′	.4509	.4358	.4841	2.0655	.9001	1.1199	10′
26° 00′	.4538	.4384	.4877	2.0503	.8988	1.1170	64° 00′
10′	.4567	.4410	.4913	2.0353	.8975	1.1141	50′
20′	.4596	.4436	.4950	2.0204	.8962	1.1112	40′
30′	4625	.4462	.4986	2.0057	.8949	1.1083	30′
40′	.4654	.4488	.5022	1.9912	.8936	1.1054	20′
50′	.4683	.4514	.5059	1.9768	.8923	1.1025	10′
27° 00′	.4712	.4540	.5095	1.9626	.8910	1.0996	63° 00′
		Cosine	Cotangent	Tangent	Sine	Radians	Degrees

APPENDIX 4
VALUES OF TRIGONOMETRIC FUNCTIONS (Cont.)

Degrees	Radians	Sine	Tangent	Cotangent	Cosine		
27° 00′	.4712	.4540	.5095	1.9626	.8910	1.0996	63° 00′
10′	.4741	.4566	.5132	1.9486	.8897	1.0966	50′
20′	.4771	.4592	.5169	1.9347	.8884	1.0937	40′
30′	.4800	.4617	.5206	1.9210	.8870	1.0908	30′
40′	.4829	.4643	.5243	1.9074	.8857	1.0879	20′
50′	.4858	.4669	.5280	1.8940	.8843	1.0850	10′
28° 00′	.4887	.4695	.5317	1.8807	.8829	1.0821	62° 00′
10′	.4916	.4720	.5354	1.8676	.8816	1.0792	50′
20′	.4945	.4746	.5392	1.8546	.8802	1.0763	40′
30′	.4974	.4772	.5430	1.8418	.8788	1.0734	30′
40′	.5003	.4797	.5467	1.8291	.8774	1.0705	20′
50′	.5032	.4823	.5505	1.8165	.8760	1.0676	10′
29° 00′	.5061	.4848	.5543	1.8040	.8746	1.0647	61° 00′
10′	.5091	.4874	.5581	1.7917	.8732	1.0617	50′
20′	.5120	.4899	.5619	1.7796	.8718	1.0588	40′
30′	.5149	.4924	.5658	1.7675	.8704	1.0559	30′
40′	.5178	.4950	.5696	1.7556	.8689	1.0530	20′
50′	.5207	.4975	.5735	1.7437	.8675	1.0501	10′
30° 00′	.5236	.5000	.5774	1.7321	.8660	1.0472	60° 00′
10′	.5265	.5025	.5812	1.7205	.8646	1.0443	50′
20′	.5294	.5050	.5851	1.7090	.8631	1.0414	40′
30′	.5323	.5075	.5890	1.6977	.8616	1.0385	30′
40′	.5352	.5100	.5930	1.6864	.8601	1.0356	20′
50′	.5381	.5125	.5969	1.6753	.8587	1.0327	10′
31° 00′	.5411	.5150	.6009	1.6643	.8572	1.0297	59° 00′
10′	.5440	.5175	.6048	1.6534	.8557	1.0268	50′
20′	.5469	.5200	.6088	1.6426	.8542	1.0239	40′
30′	.5498	.5225	.6128	1.6319	.8526	1.0210	30′
40′	.5527	.5250	.6168	1.6212	.8511	1.0181	20′
50′	.5556	.5275	.6208	1.6107	.8496	1.0152	10′
32° 00′	.5585	.5299	.6249	1.6003	.8480	1.0123	58° 00′
10′	.5614	.5324	.6289	1.5900	.8465	1.0094	50′
20′	.5643	.5348	.6330	1.5798	.8450	1.0065	40′
30′	.5672	.5373	.6371	1.5697	.8434	1.0036	30′
40′	.5701	.5398	.6412	1.5597	.8418	1.0007	20′
50′	.5730	.5422	.6453	1.5497	.8403	.9977	10′
33° 00′	.5760	.5446	.6494	1.5399	.8387	.9948	57° 00′
10′	.5789	.5471	.6536	1.5301	.8371	.9919	50′
20′	.5818	.5495	.6577	1.5204	.8355	.9890	40′
30′	.5847	.5519	.6619	1.5108	.8339	.9861	30′
40′	.5876	.5544	.6661	1.5013	.8323	.9832	20′
50′	.5905	.5568	.6703	1.4919	.8307	.9803	10′
34° 00′	.5934	.5592	.6745	1.4826	.8290	.9774	56° 00′
10′	.5963	.5616	.6787	1.4733	.8274	.9745	50′
20′	.5992	.5640	.6830	1.4641	.8258	.9716	40′
30′	.6021	.5664	.6873	1.4550	.8241	.9687	30′
40′	.6050	.5688	.6916	1.4460	.8225	.9657	20′
50′	.6080	.5712	.6959	1.4370	.8208	.9628	10′
35° 00′	.6109	.5736	.7002	1.4281	.8192	.9599	55° 00′
10′	.6138	.5760	.7046	1.4193	.8175	.9570	50′
20′	.6167	.5783	.7089	1.4106	.8158	.9541	40′
30′	.6196	.5807	.7133	1.4019	.8141	.9512	30′
40′	.6225	.5831	.7177	1.3934	.8124	.9483	20′
50′	.6254	.5854	.7221	1.3848	.8107	.9454	10′
36° 00′	.6283	.5878	.7265	1.3764	.8090	.9425	54° 00′
		Cosine	Cotangent	Tangent	Sine	Radians	Degrees

Cont.

APPENDIX 4
VALUES OF TRIGONOMETRIC FUNCTIONS (Cont.)

Degrees	Radians	Sine	Tangent	Cotangent	Cosine		
36° 00′	.6283	.5878	.7265	1.3764	.8090	.9425	54° 00′
10′	.6312	.5901	.7310	1.3680	.8073	.9396	50′
20′	.6341	.5925	.7355	1.3597	.8056	.9367	40′
30′	.6370	.5948	.7400	1.3514	.8039	.9338	30′
40′	.6400	.5972	.7445	1.3432	.8021	.9308	20′
50′	.6429	.5995	.7490	1.3351	.8004	.9279	10′
37° 00′	.6458	.6018	.7536	1.3270	.7986	.9250	53° 00′
10′	.6487	.6041	.7581	1.3190	.7969	.9221	50′
20′	.6516	.6065	.7627	1.3111	.7951	.9192	40′
30′	.6545	.6088	.7673	1.3032	.7934	.9163	30′
40′	.6574	.6111	.7720	1.2954	.7916	.9134	20′
50′	.6603	.6134	.7766	1.2876	.7898	.9105	10′
38° 00′	.6632	.6157	.7813	1.2799	.7880	.9076	52° 00′
10′	.6661	.6180	.7860	1.2723	.7862	.9047	50′
20′	.6690	.6202	.7907	1.2647	.7844	.9018	40′
30′	.6720	.6225	.7954	1.2572	.7826	.8988	30′
40′	.6749	.6248	.8002	1.2497	.7808	.8959	20′
50′	.6778	.6271	.8050	1.2423	.7790	.8930	10′
39° 00′	.6807	.6293	.8098	1.2349	.7771	.8901	51° 00′
10′	.6836	.6316	.8146	1.2276	.7753	.8872	50′
20′	.6865	.6338	.8195	1.2203	.7735	.8843	40′
30′	.6894	.6361	.8243	1.2131	.7716	.8814	30′
40′	.6923	.6383	.8292	1.2059	.7698	.8785	20′
50′	.6952	.6406	.8342	1.1988	.7679	.8756	10′
40° 00′	.6981	.6428	.8391	1.1918	.7660	.8727	50° 00′
10′	.7010	.6450	.8441	1.1847	.7642	.8698	50′
20′	.7039	.6472	.8491	1.1778	.7623	.8668	40′
30′	.7069	.6494	.8541	1.1708	.7604	.8639	30′
40′	.7098	.6517	.8591	1.1640	.7585	.8610	20′
50′	.7127	.6539	.8642	1.1571	.7566	.8581	10′
41° 00′	.7156	.6561	.8693	1.1504	.7547	.8552	49° 00′
10′	.7185	.6583	.8744	1.1436	.7528	.8523	50′
20′	.7214	.6604	.8796	1.1369	.7509	.8494	40′
30′	.7243	.6626	.8847	1.1303	.7490	.8465	30′
40′	.7272	.6648	.8899	1.1237	.7470	.8436	20′
50′	.7301	.6670	.8952	1.1171	.7451	.8407	10′
42° 00′	.7330	.6691	.9004	1.1106	.7431	.8378	48° 00′
10′	.7359	.6713	.9057	1.1041	.7412	.8348	50′
20′	.7389	.6734	.9110	1.0977	.7392	.8319	40′
30′	.7418	.6756	.9163	1.0913	.7373	.8290	30′
40′	.7447	.6777	.9217	1.0850	.7353	.8261	20′
50′	.7476	.6799	.9271	1.0786	.7333	.8232	10′
43° 00′	.7505	.6820	.9325	1.0724	.7314	.8203	47° 00′
10′	.7534	.6841	.9380	1.0661	.7294	.8174	50′
20′	.7563	.6862	.9435	1.0599	.7274	.8145	40′
30′	.7592	.6884	.9490	1.0538	.7254	.8116	30′
40′	.7621	.6905	.9545	1.0477	.7234	.8087	20′
50′	.7650	.6926	.9601	1.0416	.7214	.8058	10′
44° 00′	.7679	.6947	.9657	1.0355	.7193	.8029	46° 00′
10′	.7709	.6967	.9713	1.0295	.7173	.7999	50′
20′	.7738	.6988	.9770	1.0235	.7153	.7970	40′
30′	.7767	.7009	.9827	1.0176	.7133	.7941	30′
40′	.7796	.7030	.9884	1.0117	.7112	.7912	20′
50′	.7825	.7050	.9942	1.0058	.7092	.7883	10′
45° 00′	.7854	.7071	1.0000	1.0000	.7071	.7854	45° 00′
		Cosine	Cotangent	Tangent	Sine	Radians	Degrees

APPENDIX 5
WEIGHTS AND MEASURES

UNITED STATES SYSTEM

LINEAR MEASURE

Inches	Feet	Yards	Rods	Furlongs	Miles
1.0 =	.08333 =	.02778 =	.0050505 =	.00012626 =	.00001578
12.0 =	1.0 =	.33333 =	.0606061 =	.00151515 =	.00018939
36.0 =	3.0 =	1.0 =	.1818182 =	.00454545 =	.00056818
198.0 =	16.5 =	5.5 =	1.0 =	.025 =	.003125
7920.0 =	660.0 =	220.0 =	40.0 =	1.0 =	.125
63360.0 =	5280.0 =	1760.0 =	320.0 =	8.0 =	1.0

SQUARE AND LAND MEASURE

Sq. Inches	Square Feet	Square Yards	Sq. Rods	Acres	Sq. Miles
1.0 =	.006944 =	.000772			
144.0 =	1.0 =	.111111			
1296.0 =	9.0 =	1.0 =	.03306 =	.000207	
39204.0 =	272.25 =	30.25 =	1.0 =	.00625 =	.0000098
	43560.0 =	4840.0 =	160.0 =	1.0 =	.0015625
		3097600.0 =	102400.0 =	640.0 =	1.0

AVOIRDUPOIS WEIGHTS

Grains	Drams	Ounces	Pounds	Tons
1.0 =	.03657 =	.002286 =	.000143 =	.0000000714
27.34375 =	1.0 =	.0625 =	.003906 =	.00000195
437.5 =	16.0 =	1.0 =	.0625 =	.00003125
7000.0 =	256.0 =	16.0 =	1.0 =	.0005
14000000.0 =	512000.0 =	32000.0 =	2000.0 =	1.0

DRY MEASURE

Pints	Quarts	Pecks	Cubic Feet	Bushels
1.0 =	.5 =	.0625 =	.01945 =	.01563
2.0 =	1.0 =	.125 =	.03891 =	.03125
16.0 =	8.0 =	1.0 =	.31112 =	.25
51.42627 =	25.71314 =	3.21414 =	1.0 =	.80354
64.0 =	32.0 =	4.0 =	1.2445 =	1.0

LIQUID MEASURE

Gills	Pints	Quarts	U. S. Gallons	Cubic Feet
1.0 =	.25 =	.125 =	.03125 =	.00418
4.0 =	1.0 =	.5 =	.125 =	.01671
8.0 =	2.0 =	1.0 =	.250 =	.03342
32.0 =	8.0 =	4.0 =	1.0 =	.1337
			7.48052 =	1.0

METRIC SYSTEM

UNITS

Length—Meter   :   Mass—Gram   :   Capacity—Liter
for pure water at 4°C. (39.2°F.)
1 cubic decimeter or 1 liter = 1 kilogram

1000 Milli $\begin{cases} meters\ (mm) \\ grams\ (mg) \\ liters\ (ml) \end{cases}$ = 100 Centi $\begin{cases} meters\ (cm) \\ grams\ (cg) \\ liters\ (cl) \end{cases}$ = 10 Deci $\begin{cases} meters\ (dm) \\ grams\ (dg) \\ liters\ (dl) \end{cases}$ = 1 $\begin{cases} meter \\ gram \\ liter \end{cases}$

1000 $\begin{cases} meters \\ grams \\ liters \end{cases}$ = 100 Deka $\begin{cases} meters\ (dkm) \\ grams\ (dkg) \\ liters\ (dkl) \end{cases}$ = 10 Hecto $\begin{cases} meters\ (hm) \\ grams\ (hg) \\ liters\ (hl) \end{cases}$ = 1 Kilo $\begin{cases} meter\ (km) \\ gram\ (kg) \\ liter\ (kl) \end{cases}$

1 Metric Ton	= 1000 Kilograms
100 Square Meters	= 1 Are
100 Ares	= 1 Hectare
100 Hectares	= 1 Square Kilometer

APPENDIX 6
DECIMAL EQUIVALENTS AND TEMPERATURE CONVERSION

DECIMAL EQUIVALENTS — INCH-MILLIMETER CONVERSION TABLE

1/2	1/4	1/8	1/16	1/32	1/64	Decimals	Millimeters
				1	1	.015625	.396875
				1		.031250	.793750
					3	.046875	1.190625
			1			.062500	1.587500
					5	.078125	1.984375
				3		.093750	2.381250
					7	.109375	2.778125
		1				.125000	3.175000
					9	.140625	3.571875
				5		.156250	3.968750
					11	.171875	4.365625
			3			.187500	4.762500
					13	.203125	5.159375
				7		.218750	5.556250
					15	.234375	5.953125
	1					.250000	6.350000
					17	.265625	6.746875
				9		.281250	7.143750
					19	.296875	7.540625
			5			.312500	7.937500
					21	.328125	8.334375
				11		.343750	8.731250
					23	.359375	9.128125
		3				.375000	9.525000
					25	.390625	9.921875
				13		.406250	10.318750
					27	.421875	10.715625
			7			.437500	11.112500
					29	.453125	11.509375
				15		.468750	11.906250
					31	.484375	12.303125
1						.500000	12.700000
					33	.515625	13.096875
				17		.531250	13.493750
					35	.546875	13.890625
			9			.562500	14.287500
					37	.578125	14.684375
				19		.593750	15.081250
					39	.609375	15.478125
		5				.625000	15.875000
					41	.640625	16.271875
				21		.656250	16.668750
					43	.671875	17.065625
			11			.687500	17.462500
					45	.703125	17.859375
				23		.718750	18.256250
					47	.734375	18.653125
	3					.750000	19.050000
					49	.765625	19.446875
				25		.781250	19.843750
					51	.796875	20.240625
			13			.812500	20.637500
					53	.828125	21.034375
				27		.843750	21.431250
					55	.859375	21.828125
		7				.875000	22.225000
					57	.890625	22.621875
				29		.906250	23.018750
					59	.921875	23.415625
			15			.937500	23.812500
					61	.953125	24.209375
				31		.968750	24.606250
					63	.984375	25.003125
2	4	8	16	32	64	1.000000	25.400000

APPENDIX 6

DECIMAL EQUIVALENTS AND TEMPERATURE CONVERSION (Cont.)

TEMPERATURE CONVERSION

−210 to 0

C.	C. or F.	F.
−134	−210	−346
−129	−200	−328
−123	−190	−310
−118	−180	−292
−112	−170	−274
−107	−160	−256
−101	−150	−238
−95.6	−140	−220
−90.0	−130	−202
−84.4	−120	−184
−78.9	−110	−166
−73.3	−100	−148
−67.8	−90	−130
−62.2	−80	−112
−56.7	−70	−94
−51.1	−60	−76
−45.6	−50	−58
−40.0	−40	−40
−34.4	−30	−22
−28.9	−20	−4
−23.3	−10	14
−17.8	0	32

1 to 25

C.	C. or F.	F.
−17.2	1	33.8
−16.7	2	35.6
−16.1	3	37.4
−15.6	4	39.2
−15.0	5	41.0
−14.4	6	42.8
−13.9	7	44.6
−13.3	8	46.4
−12.8	9	48.2
−12.2	10	50.0
−11.7	11	51.8
−11.1	12	53.6
−10.6	13	55.4
−10.0	14	57.2
−9.44	15	59.0
−8.89	16	60.8
−8.33	17	62.6
−7.78	18	64.4
−7.22	19	66.2
−6.67	20	68.0
−6.11	21	69.8
−5.56	22	71.6
−5.00	23	73.4
−4.44	24	75.2
−3.89	25	77.0

26 to 50

C.	C. or F.	F.
−3.33	26	78.8
−2.78	27	80.6
−2.22	28	82.4
−1.67	29	84.2
−1.11	30	86.0
−0.56	31	87.8
0	32	89.6
0.56	33	91.4
1.11	34	93.2
1.67	35	95.0
2.22	36	96.8
2.78	37	98.6
3.33	38	100.4
3.89	39	102.2
4.44	40	104.0
5.00	41	105.8
5.56	42	107.6
6.11	43	109.4
6.67	44	111.2
7.22	45	113.0
7.78	46	114.8
8.33	47	116.6
8.89	48	118.4
9.44	49	120.2
10.0	50	122.0

51 to 75

C.	C. or F.	F.
10.6	51	123.8
11.1	52	125.6
11.7	53	127.4
12.2	54	129.2
12.8	55	131.0
13.3	56	132.8
13.9	57	134.6
14.4	58	136.4
15.0	59	138.2
15.6	60	140.0
16.1	61	141.8
16.7	62	143.6
17.2	63	145.4
17.8	64	147.2
18.3	65	149.0
18.9	66	150.8
19.4	67	152.6
20.0	68	154.4
20.6	69	156.2
21.1	70	158.0
21.7	71	159.8
22.2	72	161.6
22.8	73	163.4
23.3	74	165.2
23.9	75	167.0

76 to 100

C.	C. or F.	F.
24.4	76	168.8
25.0	77	170.6
25.6	78	172.4
26.1	79	174.2
26.7	80	176.0
27.2	81	177.8
27.8	82	179.6
28.3	83	181.4
28.9	84	183.2
29.4	85	185.0
30.0	86	186.8
30.6	87	188.6
31.1	88	190.4
31.7	89	192.2
32.2	90	194.0
32.8	91	195.8
33.3	92	197.6
33.9	93	199.4
34.4	94	201.2
35.0	95	203.0
35.6	96	204.8
36.1	97	206.6
36.7	98	208.4
37.2	99	210.2
37.8	100	212.0

101 to 340

C.	C. or F.	F.
43	110	230
49	120	248
54	130	266
60	140	284
66	150	302
71	160	320
77	170	338
82	180	356
88	190	374
93	200	392
99	210	410
100	212	413
104	220	428
110	230	446
116	240	464
121	250	482
127	260	500
132	270	518
138	280	536
143	290	554
149	300	572
154	310	590
160	320	608
166	330	626
171	340	644

341 to 490

C.	C. or F.	F.
177	350	662
182	360	680
188	370	698
193	380	716
199	390	734
204	400	752
210	410	770
216	420	788
221	430	806
227	440	824
232	450	842
238	460	860
243	470	878
249	480	896
254	490	914

491 to 750

C.	C. or F.	F.
260	500	932
266	510	950
271	520	968
277	530	986
282	540	1004
288	550	1022
293	560	1040
299	570	1058
304	580	1076
310	590	1094
316	600	1112
321	610	1130
327	620	1148
332	630	1166
338	640	1184
343	650	1202
349	660	1220
354	670	1238
360	680	1256
366	690	1274
371	700	1292
377	710	1310
382	720	1328
388	730	1346
393	740	1364
399	750	1382

INTERPOLATION FACTORS

C.		F.	C.		F.
0.56	1	1.8	3.33	6	10.8
1.11	2	3.6	3.89	7	12.6
1.67	3	5.4	4.44	8	14.4
2.22	4	7.2	5.00	9	16.2
2.78	5	9.0	5.56	10	18.0

NOTE:—The numbers in bold face type refer to the temperature either in degrees Centigrade or Fahrenheit which it is desired to convert into the other scale. If converting from Fahrenheit degrees to Centigrade degrees the equivalent temperature will be found in the left column, while if converting from degrees Centigrade to degrees Fahrenheit, the answer will be found in the column on the right.

$$°F = \frac{9}{5}(°C) + 32$$

$$°C = \frac{5}{9}(°F - 32)$$

APPENDIX 7

WEIGHTS AND SPECIFIC GRAVITIES

Substance	Weight Lb. per Cu. Ft.	Specific Gravity	Substance	Weight Lb. per Cu. Ft.	Specific Gravity
METALS, ALLOYS, ORES			**TIMBER, U. S. SEASONED**		
Aluminum, cast, hammered	165	2.55-2.75	Moisture Content by Weight:		
Brass, cast, rolled	534	8.4-8.7	Seasoned timber 15 to 20%		
Bronze, 7.9 to 14% Sn.	509	7.4-8.9	Green timber up to 50%		
Bronze, aluminum	481	7.7	Ash, white, red	40	0.62-0.65
Copper, cast, rolled	556	8.8-9.0	Cedar, white, red	22	0.32-0.38
Copper ore, pyrites	262	4.1-4.3	Chestnut	41	0.66
Gold, cast, hammered	1205	19.25-19.3	Cypress	30	0.48
Iron, cast, pig	450	7.2	Fir, Douglas spruce	32	0.51
Iron, wrought	485	7.6-7.9	Fir, eastern	25	0.40
Iron, spiegel-eisen	468	7.5	Elm, white	45	0.72
Iron, ferro-silicon	437	6.7-7.3	Hemlock	29	0.42-0.52
Iron ore, hematite	325	5.2	Hickory	49	0.74-0.84
Iron ore, hematite in bank	160-180	....	Locust	46	0.73
Iron ore, hematite loose	130-160	....	Maple, hard	43	0.68
Iron ore, limonite	237	3.6-4.0	Maple, white	33	0.53
Iron ore, magnetite	315	4.9-5.2	Oak, chestnut	54	0.86
Iron slag	172	2.5-3.0	Oak, live	59	0.95
Lead	710	11.37	Oak, red, black	41	0.65
Lead ore, galena	465	7.3-7.6	Oak, white	46	0.74
Magnesium, alloys	112	1.74-1.83	Pine, Oregon	32	0.51
Manganese	475	7.2-8.0	Pine, red	30	0.48
Manganese ore, pyrolusite	259	3.7-4.6	Pine, white	26	0.41
Mercury	849	13.6	Pine, yellow, long-leaf	44	0.70
Monel Metal	556	8.8-9.0	Pine, yellow, short-leaf	38	0.61
Nickel	565	8.9-9.2	Poplar	30	0.48
Platinum, cast, hammered	1330	21.1-21.5	Redwood, California	26	0.42
Silver, cast, hammered	656	10.4-10.6	Spruce, white, black	27	0.40-0.46
Steel, rolled	490	7.85	Walnut, black	38	0.61
Tin, cast, hammered	459	7.2-7.5			
Tin ore, cassiterite	418	6.4-7.0			
Zinc, cast, rolled	440	6.9-7.2			
Zinc ore, blende	253	3.9-4.2	**VARIOUS LIQUIDS**		
			Alcohol, 100%	49	0.79
			Acids, muriatic 40%	75	1.20
			Acids, nitric 91%	94	1.50
VARIOUS SOLIDS			Acids, sulphuric 87%	112	1.80
			Lye, soda 66%	106	1.70
Cereals, oats .... bulk	32	....	Oils, vegetable	58	0.91-0.94
Cereals, barley .... bulk	39	....	Oils, mineral, lubricants	57	0.90-0.93
Cereals, corn, rye .... bulk	48	....	Water, 4°C. max. density	62.428	1.0
Cereals, wheat .... bulk	48	....	Water, 100°C.	59.830	0.9584
Hay and Straw .... bales	20	....	Water, ice	56	0.88-0.92
Cotton, Flax, Hemp	93	1.47-1.50	Water, snow, fresh fallen	8	.125
Fats	58	0.90-0.97	Water, sea water	64	1.02-1.03
Flour, loose	28	0.40-0.50			
Flour, pressed	47	0.70-0.80			
Glass, common	156	2.40-2.60			
Glass, plate or crown	161	2.45-2.72	**GASES**		
Glass, crystal	184	2.90-3.00			
Leather	59	0.86-1.02	Air, 0°C. 760 mm.	.08071	1.0
Paper	58	0.70-1.15	Ammonia	.0478	0.5920
Potatoes, piled	42	....	Carbon dioxide	.1234	1.5291
Rubber, caoutchouc	59	0.92-0.96	Carbon monoxide	.0781	0.9673
Rubber goods	94	1.0-2.0	Gas, illuminating	.028-.036	0.35-0.45
Salt, granulated, piled	48	....	Gas, natural	.038-.039	0.47-0.48
Saltpeter	67	....	Hydrogen	.00559	0.0693
Starch	96	1.53	Nitrogen	.0784	0.9714
Sulphur	125	1.93-2.07	Oxygen	.0892	1.1056
Wool.	82	1.32			

The specific gravities of solids and liquids refer to water at 4°C., those of gases to air at 0°C. and 760 mm. pressure. The weights per cubic foot are derived from average specific gravities, except where stated that weights are for bulk, heaped or loose material, etc.

(Courtesy of the American Institute of Steel Construction.)

APPENDIX 7

WEIGHTS AND SPECIFIC GRAVITIES (Cont.)

Substance	Weight Lb. per Cu. Ft.	Specific Gravity	Substance	Weight Lb. per Cu. Ft.	Specific Gravity
ASHLAR MASONRY			**MINERALS**		
Granite, syenite, gneiss......	165	2.3-3.0	Asbestos................................	153	2.1-2.8
Limestone, marble................	160	2.3-2.8	Barytes..................................	281	4.50
Sandstone, bluestone..........	140	2.1-2.4	Basalt....................................	184	2.7-3.2
			Bauxite..................................	159	2.55
MORTAR RUBBLE			Borax.....................................	109	1.7-1.8
MASONRY			Chalk.....................................	137	1.8-2.6
Granite, syenite, gneiss......	155	2.2-2.8	Clay, marl..............................	137	1.8-2.6
Limestone, marble................	150	2.2-2.6	Dolomite................................	181	2.9
Sandstone, bluestone..........	130	2.0-2.2	Feldspar, orthoclase.............	159	2.5-2.6
			Gneiss, serpentine................	159	2.4-2.7
DRY RUBBLE MASONRY			Granite, syenite....................	175	2.5-3.1
Granite, syenite, gneiss......	130	1.9-2.3	Greenstone, trap...................	187	2.8-3.2
Limestone, marble................	125	1.9-2.1	Gypsum, alabaster................	159	2.3-2.8
Sandstone, bluestone..........	110	1.8-1.9	Hornblende............................	187	3.0
			Limestone, marble................	165	2.5-2.8
BRICK MASONRY			Magnesite..............................	187	3.0
Pressed brick.......................	140	2.2-2.3	Phosphate rock, apatite........	200	3.2
Common brick.......................	120	1.8-2.0	Porphyry................................	172	2.6-2.9
Soft brick..............................	100	1.5-1.7	Pumice, natural.....................	40	0.37-0.90
			Quartz, flint...........................	165	2.5-2.8
CONCRETE MASONRY			Sandstone, bluestone..........	147	2.2-2.5
Cement, stone, sand............	144	2.2-2.4	Shale, slate...........................	175	2.7-2.9
Cement, slag, etc.................	130	1.9-2.3	Soapstone, talc.....................	169	2.6-2.8
Cement, cinder, etc..............	100	1.5-1.7			
VARIOUS BUILDING			**STONE, QUARRIED, PILED**		
MATERIALS			Basalt, granite, gneiss........	96	....
Ashes, cinders......................	40-45	....	Limestone, marble, quartz	95	....
Cement, portland, loose......	90	....	Sandstone..............................	82	....
Cement, portland, set..........	183	2.7-3.2	Shale......................................	92	....
Lime, gypsum, loose............	53-64	....	Greenstone, hornblende.......	107	....
Mortar, set.............................	103	1.4-1.9			
Slags, bank slag...................	67-72	....			
Slags, bank screenings.......	98-117	....			
Slags, machine slag..............	96	....			
Slags, slag sand...................	49-55	....	**BITUMINOUS SUBSTANCES**		
			Asphaltum.............................	81	1.1-1.5
EARTH, ETC., EXCAVATED			Coal, anthracite.....................	97	1.4-1.7
Clay, dry................................	63	....	Coal, bituminous...................	84	1.2-1.5
Clay, damp, plastic...............	110	....	Coal, lignite...........................	78	1.1-1.4
Clay and gravel, dry.............	100	....	Coal, peat, turf, dry..............	47	0.65-0.85
Earth, dry, loose...................	76	....	Coal, charcoal, pine..............	23	0.28-0.44
Earth, dry, packed................	95	....	Coal, charcoal, oak...............	33	0.47-0.57
Earth, moist, loose...............	78	....	Coal, coke.............................	75	1.0-1.4
Earth, moist, packed............	96	....	Graphite.................................	131	1.9-2.3
Earth, mud, flowing..............	108	....	Paraffine...............................	56	0.87-0.91
Earth, mud, packed..............	115	....	Petroleum..............................	54	0.87
Riprap, limestone.................	80-85	....	Petroleum, refined................	50	0.79-0.82
Riprap, sandstone................	90	....	Petroleum, benzine...............	46	0.73-0.75
Riprap, shale.........................	105	....	Petroleum, gasoline..............	42	0.66-0.69
Sand, gravel, dry, loose......	90-105	....	Pitch......................................	69	1.07-1.15
Sand, gravel, dry, packed....	100-120	....	Tar, bituminous.....................	75	1.20
Sand, gravel, dry, wet.........	118-120	....			
EXCAVATIONS IN WATER					
Sand or gravel......................	60	....	**COAL AND COKE, PILED**		
Sand or gravel and clay......	65	....	Coal, anthracite.....................	47-58	....
Clay..	80	....	Coal, bituminous, lignite......	40-54	....
River mud..............................	90	....	Coal, peat, turf......................	20-26	....
Soil...	70	....	Coal, charcoal.......................	10-14	....
Stone riprap..........................	65	....	Coal, coke.............................	23-32	....

The specific gravities of solids and liquids refer to water at 4°C., those of gases to air at 0°C. and 760 mm. pressure. The weights per cubic foot are derived from average specific gravities, except where stated that weights are for bulk, heaped or loose material, etc.

APPENDIX 8

WIRE AND SHEET METAL GAGES

WIRE AND SHEET METAL GAGES
IN DECIMALS OF AN INCH

Name of Gage	United States Standard Gage*		The United States Steel Wire Gage	American or Brown & Sharpe Wire Gage	New Birmingham Standard Sheet & Hoop Gage	British Imperial or English Legal Standard Wire Gage	Birmingham or Stubs Iron Wire Gage	Name of Gage
Principal Use	Uncoated Steel Sheets and Light Plates		Steel Wire except Music Wire	Non-Ferrous Sheets and Wire	Iron and Steel Sheets and Hoops	Wire	Strips, Bands, Hoops and Wire	Principal Use
Gage No.	Weight Oz. per Sq. Ft.	Approx. Thickness Inches	Thickness, Inches					Gage No.
7/0's			.4900		.6666	.500		7/0's
6/0's			.4615	.5800	.625	.464		6/0's
5/0's			.4305	.5165	.5883	.432	.500	5/0's
4/0's			.3938	.4600	.5416	.400	.454	4/0's
3/0's			.3625	.4096	.500	.372	.425	3/0's
2/0's			.3310	.3648	.4452	.348	.380	2/0's
0			.3065	.3249	.3964	.324	.340	0
1			.2830	.2893	.3532	.300	.300	1
2			.2625	.2576	.3147	.276	.284	2
3	160	.2391	.2437	.2294	.2804	.252	.259	3
4	150	.2242	.2253	.2043	.250	.232	.238	4
5	140	.2092	.2070	.1819	.2225	.212	.220	5
6	130	.1943	.1920	.1620	.1981	.192	.203	6
7	120	.1793	.1770	.1443	.1764	.176	.180	7
8	110	.1644	.1620	.1285	.1570	.160	.165	8
9	100	.1495	.1483	.1144	.1398	.144	.148	9
10	90	.1345	.1350	.1019	.1250	.128	.134	10
11	80	.1196	.1205	.0907	.1113	.116	.120	11
12	70	.1046	.1055	.0808	.0991	.104	.109	12
13	60	.0897	.0915	.0720	.0882	.092	.095	13
14	50	.0747	.0800	.0641	.0785	.080	.083	14
15	45	.0673	.0720	.0571	.0699	.072	.072	15
16	40	.0598	.0625	.0508	.0625	.064	.065	16
17	36	.0538	.0540	.0453	.0556	.056	.058	17
18	32	.0478	.0475	.0403	.0495	.048	.049	18
19	28	.0418	.0410	.0359	.0440	.040	.042	19
20	24	.0359	.0348	.0320	.0392	.036	.035	20
21	22	.0329	.0318	.0285	.0349	.032	.032	21
22	20	.0299	.0286	.0253	.0313	.028	.028	22
23	18	.0269	.0258	.0226	.0278	.024	.025	23
24	16	.0239	.0230	.0201	.0248	.022	.022	24
25	14	.0209	.0204	.0179	.0220	.020	.020	25
26	12	.0179	.0181	.0159	.0196	.018	.018	26
27	11	.0164	.0173	.0142	.0175	.0164	.016	27
28	10	.0149	.0162	.0126	.0156	.0148	.014	28
29	9	.0135	.0150	.0113	.0139	.0136	.013	29
30	8	.0120	.0140	.0100	.0123	.0124	.012	30
31	7	.0105	.0132	.0089	.0110	.0116	.010	31
32	6.5	.0097	.0128	.0080	.0098	.0108	.009	32
33	6	.0090	.0118	.0071	.0087	.0100	.008	33
34	5.5	.0082	.0104	.0063	.0077	.0092	.007	34
35	5	.0075	.0095	.0056	.0069	.0084	.005	35
36	4.5	.0067	.0090	.0050	.0061	.0076	.004	36
37	4.25	.0064	.0085	.0045	.0054	.0068		37
38	4	.0060	.0080	.0040	.0048	.0060		38
39			.0075	.0035	.0043	.0052		39
40			.0070	.0031	.0039	.0048		40

* U. S. Standard Gage is officially a weight gage, in oz. per sq. ft. as tabulated. The Approx. Thickness shown is the "Manufacturers' Standard" of the American Iron and Steel Institute, based on steel as weighing 501.81 lbs. per cu. ft. (489.6 true weight plus 2.5 percent for average over-run in area and thickness). The A.I.S.I. standard nomenclature for flat rolled carbon steel is as follows:

Widths, Inches	Thicknesses, Inch							
	0.2500 and thicker	0.2499 to 0.2031	0.2030 to 0.1875	0.1874 to 0.0568	0.0567 to 0.0344	0.0343 to 0.0255	0.0254 to 0.0142	0.0141 and thinner
To 3½ incl.	Bar	Bar	Strip	Strip	Strip	Strip	Sheet	Sheet
Over 3½ to 6 incl.	Bar	Bar	Strip	Strip	Strip	Sheet	Sheet	Sheet
" 6 to 12 "	Plate	Strip	Strip	Strip	Sheet	Sheet	Sheet	Sheet
" 12 to 32 "	Plate	Sheet	Sheet	Sheet	Sheet	Sheet	Sheet	Black Plate
" 32 to 48 "	Plate	Sheet	Sheet	Sheet	Sheet	Sheet	Sheet	Sheet
" 48	Plate	Plate	Plate	Sheet	Sheet	Sheet	Sheet	—

APPENDIX 9

AMERICAN STANDARD TAPER PIPE THREADS, NPT[1]

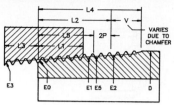

1	2	3	4	5	6	7	8	9	10	11
				Pitch Diameter at Beginning of External Thread E_0	Hand-Tight Engagement			Effective Thread, External		
Nominal Pipe Size	Outside Diameter of Pipe D	Threads per Inch n	Pitch of Thread p		Length[2] L_1		Dia E_1	Length L_2		Dia E_2
					In.	Thds		In.	Thds	In.
$\frac{1}{16}$	0.3125	27	0.03704	0.27118	0.160	4.32	0.28118	0.2611	7.05	0.28750
$\frac{1}{8}$	0.405	27	0.03704	0.36351	0.180	4.86	0.37476	0.2639	7.12	0.38000
$\frac{1}{4}$	0.540	18	0.05556	0.47739	0.200	3.60	0.48989	0.4018	7.23	0.50250
$\frac{3}{8}$	0.675	18	0.05556	0.61201	0.240	4.32	0.62701	0.4078	7.34	0.63750
$\frac{1}{2}$	0.840	14	0.07143	0.75843	0.320	4.48	0.77843	0.5337	7.47	0.79179
$\frac{3}{4}$	1.050	14	0.07143	0.96768	0.339	4.75	0.98887	0.5457	7.64	1.00179
1	1.315	$11\frac{1}{2}$	0.08696	1.21363	0.400	4.60	1.23863	0.6828	7.85	1.25630
$1\frac{1}{4}$	1.660	$11\frac{1}{2}$	0.08696	1.55713	0.420	4.83	1.58338	0.7068	8.13	1.60130
$1\frac{1}{2}$	1.900	$11\frac{1}{2}$	0.08696	1.79609	0.420	4.83	1.82234	0.7235	8.32	1.84130
2	2.375	$11\frac{1}{2}$	0.08696	2.26902	0.436	5.01	2.29627	0.7565	8.70	2.31630
$2\frac{1}{2}$	2.875	8	0.12500	2.71953	0.682	5.46	2.76216	1.1375	9.10	2.79062
3	3.500	8	0.12500	3.34062	0.766	6.13	3.38850	1.2000	9.60	3.41562
$3\frac{1}{2}$	4.000	8	0.12500	3.83750	0.821	6.57	3.88881	1.2500	10.00	3.91562
4	4.500	8	0.12500	4.33438	0.844	6.75	4.38712	1.3000	10.40	4.41562
5	5.563	8	0.12500	5.39073	0.937	7.50	5.44929	1.4063	11.25	5.47862
6	6.625	8	0.12500	6.44609	0.958	7.66	6.50597	1.5125	12.10	6.54062
8	8.625	8	0.12500	8.43359	1.063	8.50	8.50003	1.7125	13.70	8.54062
10	10.750	8	0.12500	10.54531	1.210	9.68	10.62094	1.9250	15.40	10.66562
12	12.750	8	0.12500	12.53281	1.360	10.88	12.61781	2.1250	17.00	12.66562
14 OD	14.000	8	0.12500	13.77500	1.562	12.50	13.87262	2.2500	18.90	13.91562
16 OD	16.000	8	0.12500	15.76250	1.812	14.50	15.87575	2.4500	19.60	15.91562
18 OD	18.000	8	0.12500	17.75000	2.000	16.00	17.87500	2.6500	21.20	17.91562
20 OD	20.000	8	0.12500	19.73750	2.125	17.00	19.87031	2.8500	22.80	19.91562
24 OD	24.000	8	0.12500	23.71250	2.375	19.00	23.86094	3.2500	26.00	23.91562

All dimensions are given in inches.

[1] The basic dimensions of the American Standard Taper Pipe Thread are given in inches to four or five decimal places. While this implies a greater degree of precision than is ordinarily attained, these dimensions are the basis of gage dimensions and are so expressed for the purpose of eliminating errors in computations.

[2] Also length of thin ring gage and length from gaging notch to small end of plug gage.

(Courtesy of ANSI; B2.1–1960.)

APPENDIX 10
AMERICAN STANDARD 250-LB CAST IRON FLANGED FITTINGS

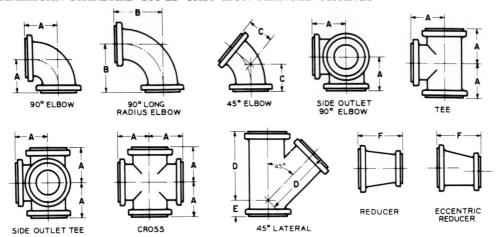

90° ELBOW 90° LONG RADIUS ELBOW 45° ELBOW SIDE OUTLET 90° ELBOW TEE

SIDE OUTLET TEE CROSS 45° LATERAL REDUCER ECCENTRIC REDUCER

Dimensions of 250-lb Cast Iron Flanged Fittings

Nominal Pipe Size	Flanges			Fittings		Straight					
	Dia of Flange	Thickness of Flange (Min)	Dia of Raised Face	Inside Dia of Fittings (Min)	Wall Thickness	Center to Face 90 Deg Elbow Tees, Crosses and True "Y"	Center to Face 90 Deg Long Radius Elbow	Center to Face 45 Deg Elbow	Center to Face Lateral	Short Center to Face True "Y" and Lateral	Face to Face Reducer
						A	B	C	D	E	F
1	4 7/8	11/16	2 11/16	1	7/16	4	5	2	6 1/2	2	....
1 1/4	5 1/4	3/4	3 1/16	1 1/4	7/16	4 1/4	5 1/2	2 1/2	7 1/4	2 1/4	....
1 1/2	6 1/8	13/16	3 9/16	1 1/2	7/16	4 1/2	6	2 3/4	8 1/2	2 1/2	....
2	6 1/2	7/8	4 3/16	2	7/16	5	6 1/2	3	9	2 1/2	5
2 1/2	7 1/2	1	4 15/16	2 1/2	1/2	5 1/2	7	3 1/2	10 1/2	2 1/2	5 1/2
3	8 1/4	1 1/8	5 11/16	3	9/16	6	7 3/4	3 1/2	11	3	6
3 1/2	9	1 3/16	6 5/16	3 1/2	9/16	6 1/2	8 1/2	4	12 1/2	3	6 1/2
4	10	1 1/4	6 15/16	4	5/8	7	9	4 1/2	13 1/2	3	7
5	11	1 3/8	8 5/16	5	11/16	8	10 1/4	5	15	3 1/2	8
6	12 1/2	1 7/16	9 11/16	6	3/4	8 1/2	11 1/2	5 1/2	17 1/2	4	9
8	15	1 5/8	11 15/16	8	13/16	10	14	6	20 1/2	5	11
10	17 1/2	1 7/8	14 1/16	10	15/16	11 1/2	16 1/2	7	24	5 1/2	12
12	20 1/2	2	16 7/16	12	1	13	19	8	27 1/2	6	14
14	23	2 1/8	18 15/16	13 1/4	1 1/8	15	21 1/2	8 1/2	31	6 1/2	16
16	25 1/2	2 1/4	21 1/16	15 1/4	1 1/4	16 1/2	24	9 1/2	34 1/2	7 1/2	18
18	28	2 3/8	23 5/16	17	1 3/8	18	26 1/2	10	37 1/2	8	19
20	30 1/2	2 1/2	25 9/16	19	1 1/2	19 1/2	29	10 1/2	40 1/2	8 1/2	20
24	36	2 3/4	30 5/16	23	1 5/8	22 1/2	34	12	47 1/2	10	24
30	43	3	37 3/16	29	2	27 1/2	41 1/2	15	....	....	30

All dimensions are given in inches.
(Courtesy of ANSI; B16.1–1967.)

APPENDIX 11

AMERICAN STANDARD 125-LB CAST IRON FLANGED FITTINGS

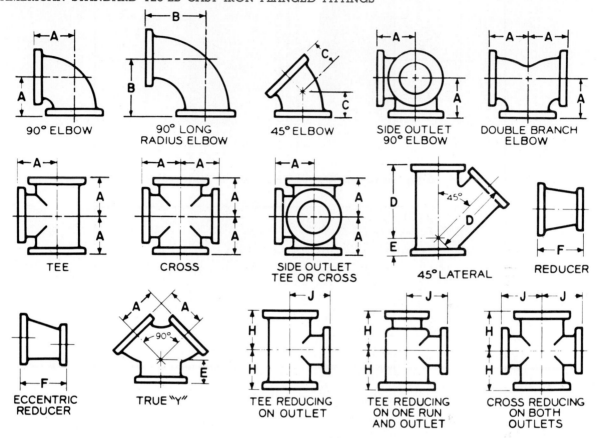

90° ELBOW

90° LONG
RADIUS ELBOW

45° ELBOW

SIDE OUTLET
90° ELBOW

DOUBLE BRANCH
ELBOW

TEE

CROSS

SIDE OUTLET
TEE OR CROSS

45° LATERAL

REDUCER

ECCENTRIC
REDUCER

TRUE "Y"

TEE REDUCING
ON OUTLET

TEE REDUCING
ON ONE RUN
AND OUTLET

CROSS REDUCING
ON BOTH
OUTLETS

Cont.

APPENDIX 11

AMERICAN STANDARD 125-LB CAST IRON FLANGED FITTINGS (Cont.)

	Flanges		General		Straight Fittings						Reducing Fittings (Short Body Patterns)		
											Tees and Crosses		
Nominal Pipe Size	Dia of Flange	Thickness of Flange (Min)	Inside Dia of Flange Fittings	Wall Thickness	Center to Face 90 deg Elbow Tees, Crosses True "Y" and Double Branch Elbow A	Center to Face 90 deg Long Radius El-bow B	Center to Face 45 deg Elbow C	Center to Face Lateral D	Short Center to Face True "Y" and Lateral E	Face to Face Reducer F	Size of Outlet and Smaller	Center to Face Run H	Center to Face Outlet or Side Outlet J
1	4¼	7/16	1	5/16	3½	5	1¾	5¾	1¾	. . .			
1¼	4⅝	½	1¼	5/16	3¾	5½	2	6¼	1¾	. . .			
1½	5	9/16	1½	5/16	4	6	2¼	7	2				
2	6	⅝	2	5/16	4½	6½	2½	8	2½	5			
2½	7	11/16	2½	5/16	5	7	3	9½	2½	5½			
3	7½	¾	3	⅜	5½	7¾	3	10	3	6			
3½	8½	13/16	3½	7/16	6	8½	3½	11½	3	6½	All reducing tees and crosses, sizes 16 in. and smaller, shall have same center to face dimensions as straight size fittings, corresponding to the size of the largest opening.		
4	9	15/16	4	½	6½	9	4	12	3	7			
5	10	15/16	5	½	7½	10¼	4½	13½	3½	8			
6	11	1	6	9/16	8	11½	5	14½	3½	9			
8	13½	1⅛	8	⅝	9	14	5½	17½	4½	11			
10	16	1 3/16	10	¾	11	16½	6½	20½	5	12			
12	19	1¼	12	13/16	12	19	7½	24½	5½	14			
14	21	1⅜	14	⅞	14	21½	7½	27	6	16			
16	23½	1 7/16	16	1	15	24	8	30	6½	18			
18	25	1 9/16	18	1 1/16	16½	26½	8½	32	7	19	12	13	15½
20	27½	1 11/16	20	1⅛	18	29	9½	35	8	20	14	14	17
24	32	1⅞	24	1¼	22	34	11	40½	9	24	16	15	19
30	38¾	2⅛	30	1 7/16	25	41½	15	49	10	30	20	18	23
36	46	2⅜	36	1⅝	28*	49	18	. . .	. . .	36	24	20	26
42	53	2⅝	42	1 13/16	31*	56½	21	. . .	. . .	42	24	23	30
48	59½	2¾	48	2	34*	64	24	. . .	. . .	48	30	26	34

All dimensions are given in inches.
(Courtesy of ANSI; B16.1–1967.)

APPENDIX 12

AMERICAN STANDARD 125-LB CAST IRON FLANGES*

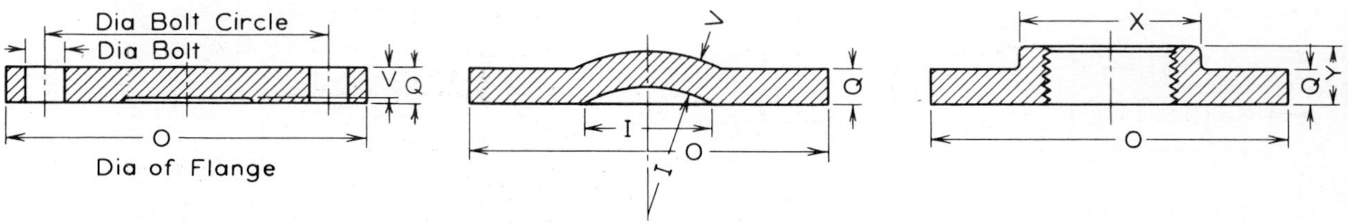

Size *I*	*O*	*Q*	*V*	*X*	*Y*	Dia. Bolt Circle	No. of Bolts	Dia. Bolts	Dia. Bolt Holes	Length of Bolts
1	4 1/4	7/16	—	1 15/16	0.68	3 1/8	4	1/2	5/8	1 3/4
1 1/4	4 5/8	1/2	—	2 5/16	0.76	3 1/2	4	1/2	5/8	2
1 1/2	5	9/16	—	2 9/16	0.87	3 7/8	4	1/2	5/8	2
2	6	5/8	—	3 1/16	1.00	4 3/4	4	5/8	3/4	2 1/4
2 1/2	7	11/16	—	3 9/16	1.14	5 1/2	4	5/8	3/4	2 1/2
3	7 1/2	3/4	—	4 1/4	1.20	6	4	5/8	3/4	2 1/2
3 1/2	8 1/2	13/16	—	4 13/16	1.25	7	8	5/8	3/4	2 3/4
4	9	15/16	—	5 5/16	1.30	7 1/2	8	5/8	3/4	3
5	10	15/16	—	6 7/16	1.41	8 1/2	8	3/4	7/8	3 1/4
6	11	1	—	7 9/16	1.51	9 1/2	8	3/4	7/8	3 1/4
8	13 1/2	1 1/8	—	9 11/16	1.71	11 3/4	8	3/4	7/8	3 1/2
10	16	1 3/16	—	11 15/16	1.93	14 1/4	12	7/8	1	3 3/4
12	19	1 1/4	12/16	14 1/16	2.13	17	12	7/8	1	3 3/4
14 O.D.	21	1 3/8	7/8	15 3/8	2.25	18 3/4	12	1	1 1/8	4 1/4
16 O.D.	23 1/2	1 7/16	1	17 1/2	2.45	21 1/4	16	1 1/8	1 1/8	4 1/2
18 O.D.	25	1 9/16	1 1/16	19 5/8	2.65	22 3/4	16	1 1/8	1 1/4	4 3/4

All dimensions in inches.

* Extracted from American Standards, "Cast-Iron Pipe Flanges and Flanged Fittings" (ANSI B16.1), with the permission of the publisher, The American Society of Mechanical Engineers.

APPENDIX 13
AMERICAN NATIONAL STANDARD 125-LB CAST IRON SCREWED FITTINGS*

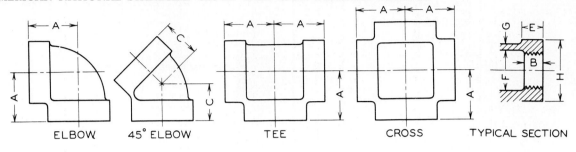

ELBOW   45° ELBOW   TEE   CROSS   TYPICAL SECTION

Nominal Pipe Size	A	C	B	E	F		G	H
			Min	Min	Min	Max	Min	Min
¼	0.81	0.73	0.32	0.38	0.540	0.584	0.110	0.93
⅜	0.95	0.80	0.36	0.44	0.675	0.719	0.120	1.12
½	1.12	0.88	0.43	0.50	0.840	0.897	0.130	1.34
¾	1.31	0.98	0.50	0.56	1.050	1.107	0.155	1.63
1	1.50	1.12	0.58	0.62	1.315	1.385	0.170	1.95
1¼	1.75	1.29	0.67	0.69	1.660	1.730	0.185	2.39
1½	1.94	1.43	0.70	0.75	1.900	1.970	0.200	2.68
2	2.25	1.68	0.75	0.84	2.375	2.445	0.220	3.28
2½	2.70	1.95	0.92	0.94	2.875	2.975	0.240	3.86
3	3.08	2.17	0.98	1.00	3.500	3.600	0.260	4.62
3½	3.42	2.39	1.03	1.06	4.000	4.100	0.280	5.20
4	3.79	2.61	1.08	1.12	4.500	4.600	0.310	5.79
5	4.50	3.05	1.18	1.18	5.563	5.663	0.380	7.05
6	5.13	3.46	1.28	1.28	6.625	6.725	0.430	8.28
8	6.56	4.28	1.47	1.47	8.625	8.725	0.550	10.63
10	8.08	5.16	1.68	1.68	10.750	10.850	0.690	13.12
12	9.50	5.97	1.88	1.88	12.750	12.850	0.800	15.47
14 O.D.	10.40	—	2.00	2.00	14.000	14.100	0.880	16.94
16 O.D.	11.82	—	2.20	2.20	16.000	16.100	1.000	19.30

All dimensions in inches.
* Extracted from American National Standards, "Cast-Iron Screwed Fittings, 125- and 250-lb" (ANSI B16.4), with the permission of the publisher, The American Society of Mechanical Engineers.

APPENDIX 14
AMERICAN NATIONAL STANDARD UNIFIED INCH SCREW THREADS (UN AND UNR THREAD FORM)*

Sizes		Basic Major Diameter	Series with Graded Pitches			Series with Constant Pitches								Sizes
Primary	Secondary		Coarse UNC	Fine UNF	Extra Fine UNEF	4UN	6UN	8UN	12UN	16UN	20UN	28UN	32UN	
0		0.0600	—	80	—	—	—	—	—	—	—	—	—	0
	1	0.0730	64	72	—	—	—	—	—	—	—	—	—	1
2		0.0860	56	64	—	—	—	—	—	—	—	—	—	2
	3	0.0990	48	56	—	—	—	—	—	—	—	—	—	3
4		0.1120	40	48	—	—	—	—	—	—	—	—	—	4
5		0.1250	40	44	—	—	—	—	—	—	—	—	—	5
6		0.1380	32	40	—	—	—	—	—	—	—	—	UNC	6
8		0.1640	32	36	—	—	—	—	—	—	—	—	UNC	8
10		0.1900	24	32	—	—	—	—	—	—	—	—	UNF	10
	12	0.2160	24	28	32	—	—	—	—	—	—	UNF	UNEF	12
$\frac{1}{4}$		0.2500	20	28	32	—	—	—	—	—	UNC	UNF	UNEF	$\frac{1}{4}$
$\frac{5}{16}$		0.3125	18	24	32	—	—	—	—	—	20	28	UNEF	$\frac{5}{16}$
$\frac{3}{8}$		0.3750	16	24	32	—	—	—	—	UNC	20	28	UNEF	$\frac{3}{8}$
$\frac{7}{16}$		0.4375	14	20	28	—	—	—	—	16	UNF	UNEF	32	$\frac{7}{16}$
$\frac{1}{2}$		0.5000	13	20	28	—	—	—	—	16	UNF	UNEF	32	$\frac{1}{2}$
$\frac{9}{16}$		0.5625	12	18	24	—	—	—	UNC	16	20	28	32	$\frac{9}{16}$
$\frac{5}{8}$		0.6250	11	18	24	—	—	—	12	16	20	28	32	$\frac{5}{8}$
	$\frac{11}{16}$	0.6875	—	—	24	—	—	—	12	16	20	28	32	$\frac{11}{16}$
$\frac{3}{4}$		0.7500	10	16	20	—	—	—	12	UNF	UNEF	28	32	$\frac{3}{4}$
	$\frac{13}{16}$	0.8125	—	—	20	—	—	—	12	16	UNEF	28	32	$\frac{13}{16}$
$\frac{7}{8}$		0.8750	9	14	20	—	—	—	12	16	UNEF	28	32	$\frac{7}{8}$
	$\frac{15}{16}$	0.9375	—	—	20	—	—	—	12	16	UNEF	28	32	$\frac{15}{16}$
1		1.0000	8	12	20	—	—	UNC	UNF	16	UNEF	28	32	1
	$1\frac{1}{16}$	1.0625	—	—	18	—	—	8	12	16	20	28	—	$1\frac{1}{16}$
$1\frac{1}{8}$		1.1250	7	12	18	—	—	8	UNF	16	20	28	—	$1\frac{1}{8}$
	$1\frac{3}{16}$	1.1875	—	—	18	—	—	8	12	16	20	28	—	$1\frac{3}{16}$
$1\frac{1}{4}$		1.2500	7	12	18	—	—	8	UNF	16	20	28	—	$1\frac{1}{4}$
	$1\frac{5}{16}$	1.3125	—	—	18	—	—	8	12	16	20	28	—	$1\frac{5}{16}$
$1\frac{3}{8}$		1.3750	6	12	18	—	UNC	8	UNF	16	20	28	—	$1\frac{3}{8}$
	$1\frac{7}{16}$	1.4375	—	—	18	—	6	8	12	16	20	28	—	$1\frac{7}{16}$
$1\frac{1}{2}$		1.5000	6	12	18	—	UNC	8	UNF	16	20	28	—	$1\frac{1}{2}$
	$1\frac{9}{16}$	1.5625	—	—	18	—	6	8	12	16	20	—	—	$1\frac{9}{16}$
$1\frac{5}{8}$		1.6250	—	—	18	—	6	8	12	16	20	—	—	$1\frac{5}{8}$
	$1\frac{11}{16}$	1.6875	—	—	18	—	6	8	12	16	20	—	—	$1\frac{11}{16}$
$1\frac{3}{4}$		1.7500	5	—	—	—	6	8	12	16	20	—	—	$1\frac{3}{4}$
	$1\frac{13}{16}$	1.8125	—	—	—	—	6	8	12	16	20	—	—	$1\frac{13}{16}$
$1\frac{7}{8}$		1.8750	—	—	—	—	6	8	12	16	20	—	—	$1\frac{7}{8}$
	$1\frac{15}{16}$	1.9375	—	—	—	—	6	8	12	16	20	—	—	$1\frac{15}{16}$
2		2.0000	$4\frac{1}{2}$	—	—	—	6	8	12	16	20	—	—	2
	$2\frac{1}{8}$	2.1250	—	—	—	—	6	8	12	16	20	—	—	$2\frac{1}{8}$
$2\frac{1}{4}$		2.2500	$4\frac{1}{2}$	—	—	—	6	8	12	16	20	—	—	$2\frac{1}{4}$
	$2\frac{3}{8}$	2.3750	—	—	—	—	6	8	12	16	20	—	—	$2\frac{3}{8}$
$2\frac{1}{2}$		2.5000	4	—	—	UNC	6	8	12	16	20	—	—	$2\frac{1}{2}$
	$2\frac{5}{8}$	2.6250	—	—	—	4	6	8	12	16	20	—	—	$2\frac{5}{8}$
$2\frac{3}{4}$		2.7500	4	—	—	UNC	6	8	12	16	20	—	—	$2\frac{3}{4}$
	$2\frac{7}{8}$	2.8750	—	—	—	4	6	8	12	16	20	—	—	$2\frac{7}{8}$

* Series designation shown indicates the UN thread form; however, the UNR thread form may be specified by substituting UNR in place of UN in all designations for external use only.

Cont.

APPENDIX 14
AMERICAN NATIONAL STANDARD UNIFIED INCH SCREW THREADS (UN AND UNR THREAD FORM)* (Cont.)

Sizes		Basic Major Diameter	Series with Graded Pitches			Series with Constant Pitches								Sizes
Primary	Secondary		Coarse UNC	Fine UNF	Extra Fine UNEF	4UN	6UN	8UN	12UN	16UN	20UN	28UN	32UN	
3		3.0000	4	—	—	UNC	6	8	12	16	20	—	—	3
	$3\frac{1}{8}$	3.1250	—	—	—	4	6	8	12	16	—	—	—	$3\frac{1}{8}$
$3\frac{1}{4}$		3.2500	4	—	—	UNC	6	8	12	16	—	—	—	$3\frac{1}{4}$
	$3\frac{3}{8}$	3.3750	—	—	—	4	6	8	12	16	—	—	—	$3\frac{3}{8}$
$3\frac{1}{2}$		3.5000	4	—	—	UNC	6	8	12	16	—	—	—	$3\frac{1}{2}$
	$3\frac{5}{8}$	3.6250	—	—	—	4	6	8	12	16	—	—	—	$3\frac{5}{8}$
$3\frac{3}{4}$		3.7500	4	—	—	UNC	6	8	12	16	—	—	—	$3\frac{3}{4}$
	$3\frac{7}{8}$	3.8750	—	—	—	4	6	8	12	16	—	—	—	$3\frac{7}{8}$
4		4.0000	4	—	—	UNC	6	8	12	16	—	—	—	4
	$4\frac{1}{8}$	4.1250	—	—	—	4	6	8	12	16	—	—	—	$4\frac{1}{8}$
$4\frac{1}{4}$		4.2500	—	—	—	4	6	8	12	16	—	—	—	$4\frac{1}{4}$
	$4\frac{3}{8}$	4.3750	—	—	—	4	6	8	12	16	—	—	—	$4\frac{3}{8}$
$4\frac{1}{2}$		4.5000	—	—	—	4	6	8	12	16	—	—	—	$4\frac{1}{2}$
	$4\frac{5}{8}$	4.6250	—	—	—	4	6	8	12	16	—	—	—	$4\frac{5}{8}$
$4\frac{3}{4}$		4.7500	—	—	—	4	6	8	12	16	—	—	—	$4\frac{3}{4}$
	$4\frac{7}{8}$	4.8750	—	—	—	4	6	8	12	16	—	—	—	$4\frac{7}{8}$
5		5.0000	—	—	—	4	6	8	12	16	—	—	—	5
	$5\frac{1}{8}$	5.1250	—	—	—	4	6	8	12	16	—	—	—	$5\frac{1}{8}$
$5\frac{1}{4}$		5.2500	—	—	—	4	6	8	12	16	—	—	—	$5\frac{1}{4}$
	$5\frac{3}{8}$	5.3750	—	—	—	4	6	8	12	16	—	—	—	$5\frac{3}{8}$
$5\frac{1}{2}$		5.5000	—	—	—	4	6	8	12	16	—	—	—	$5\frac{1}{2}$
	$5\frac{5}{8}$	5.6250	—	—	—	4	6	8	12	16	—	—	—	$5\frac{5}{8}$
$5\frac{3}{4}$		5.7500	—	—	—	4	6	8	12	16	—	—	—	$5\frac{3}{4}$
	$5\frac{7}{8}$	5.8750	—	—	—	4	6	8	12	16	—	—	—	$5\frac{7}{8}$
6		6.0000	—	—	—	4	6	8	12	16	—	—	—	6

(Courtesy of ANSI; B1.1–1974.)

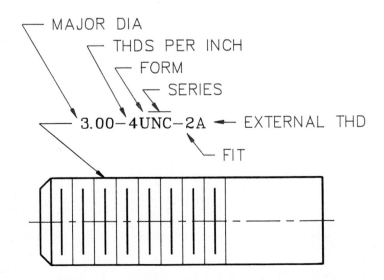

MAJOR DIA
THDS PER INCH
FORM
SERIES
3.00—4UNC—2A ← EXTERNAL THD
FIT

APPENDIX 15

TAP DRILL SIZES FOR AMERICAN NATIONAL AND UNIFIED COARSE AND FINE THREADS

$$p = \text{pitch} = \frac{1}{\text{No. thd per in}}$$

$$d = \text{depth} = p \times 0.650$$

$$f = \text{flat} = \frac{p}{8}$$

$$\text{pitch dia} = d - \frac{0.650}{N}$$

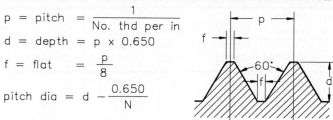

For nos. 575 and 585 screw thread micrometers

Size	Threads per inch NC UNC	NF UNF	Outside Diameter Inches	Pitch Diameter Inches	Root Diameter Inches	Tap Drill Approx. 75% Full Thread	Decimal Equiv. of Tap Drill
0	..	80	.0600	.0519	.0438	3/64	.0469
1	64	..	.0730	.0629	.0527	53	.0595
1	..	72	.0730	.0640	.0550	53	.0595
2	56	..	.0860	.0744	.0628	50	.0700
2	..	64	.0860	.0759	.0657	50	.0700
3	48	..	.0990	.0855	.0719	47	.0785
3	..	56	.0990	.0874	.0758	46	.0810
4	40	..	.1120	.0958	.0795	43	.0890
4	..	48	.1120	.0985	.0849	42	.0935
5	40	..	.1250	.1088	.0925	38	.1015
5	..	44	.1250	.1102	.0955	37	.1040
6	32	..	.1380	.1177	.0974	36	.1065
6	..	40	.1380	.1218	.1055	33	.1130
8	32	..	.1640	.1437	.1234	29	.1360
8	..	36	.1640	.1460	.1279	29	.1360
10	24	..	.1900	.1629	.1359	26	.1470
10	..	32	.1900	.1697	.1494	21	.1590
12	24	..	.2160	.1889	.1619	16	.1770
12	..	28	.2160	.1928	.1696	15	.1800
1/4	20	..	.2500	.2175	.1850	7	.2010
1/4	..	28	.2500	.2268	.2036	3	.2130
5/16	18	..	.3125	.2764	.2403	F	.2570
5/16	..	24	.3125	.2854	.2584	I	.2720
3/8	16	..	.3750	.3344	.2938	5/16	.3125
3/8	..	24	.3750	.3479	.3209	Q	.3320
7/16	14	..	.4375	.3911	.3447	U	.3680
7/16	..	20	.4375	.4050	.3726	25/64	.3906
1/2	13	..	.5000	.4500	.4001	27/64	.4219
1/2	..	20	.5000	.4675	.4351	29/64	.4531
9/16	12	..	.5625	.5084	.4542	31/64	.4844
9/16	..	18	.5625	.5264	.4903	33/64	.5156
5/8	11	..	.6250	.5660	.5069	17/32	.5312
5/8	..	18	.6250	.5889	.5528	37/64	.5781
3/4	10	..	.7500	.6850	.6201	21/32	.6562
3/4	..	16	.7500	.7094	.6688	11/16	.6875
7/8	9	..	.8750	.8028	.7307	49/64	.7656
7/8	..	14	.8750	.8286	.7822	13/16	.8125

Cont.

APPENDIX 15

TAP DRILL SIZES FOR AMERICAN NATIONAL AND UNIFIED COARSE AND FINE THREADS (Cont.)

Size	Threads per inch		Outside Diameter Inches	Pitch Diameter Inches	Root Diameter Inches	Tap Drill Approx. 75% Full Thread	Decimal Equiv. of Tap Drill
	NC UNC	NF UNF					
1	8	..	1.0000	.9188	.8376	$\frac{7}{8}$	.8750
1	..	12	1.0000	.9459	.8917	$\frac{59}{64}$	.9219
1⅛	7	..	1.1250	1.0322	.9394	$\frac{63}{64}$	.9844
1⅛	..	12	1.1250	1.0709	1.0168	$1\frac{3}{64}$	1.0469
1¼	7	..	1.2500	1.1572	1.0644	$1\frac{7}{64}$	1.1094
1¼	..	12	1.2500	1.1959	1.1418	$1\frac{11}{64}$	1.1719
1⅜	6	..	1.3750	1.2667	1.1585	$1\frac{7}{32}$	1.2187
1⅜	..	12	1.3750	1.3209	1.2668	$1\frac{19}{64}$	1.2969
1½	6	..	1.5000	1.3917	1.2835	$1\frac{11}{32}$	1.3437
1½	..	12	1.5000	1.4459	1.3918	$1\frac{27}{64}$	1.4219
1¾	5	..	1.7500	1.6201	1.4902	$1\frac{9}{16}$	1.5625
2	4½	..	2.0000	1.8557	1.7113	$1\frac{25}{32}$	1.7812
2¼	4½	..	2.2500	2.1057	1.9613	$2\frac{1}{32}$	2.0313
2½	4	..	2.5000	2.3376	2.1752	2¼	2.2500
2¾	4	..	2.7500	2.5876	2.4252	2½	2.5000
3	4	..	3.0000	3.8376	2.6752	2¾	2.7500
3¼	4	..	3.2500	3.0876	2.9252	3	3.0000
3½	4	..	3.5000	3.3376	3.1752	3¼	3.2500
3¾	4	..	3.7500	3.5876	3.4252	3½	3.5000
4	4	..	4.0000	3.3786	3.6752	3¾	3.7500

(Courtesy of the L. S. Starrett Company.)

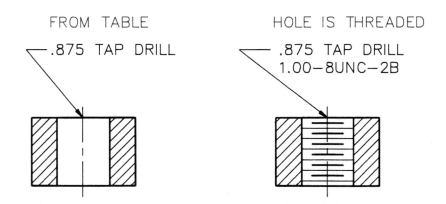

FROM TABLE — .875 TAP DRILL

HOLE IS THREADED — .875 TAP DRILL 1.00–8UNC–2B

APPENDIX 16

LENGTH OF THREAD ENGAGEMENT GROUPS

Nominal Size Diam. Over	To and Incl	Pitch P	Group S To and Incl	Over	Group N To and Incl	Group L Over
1.5	2.8	0.2	0.5	0.5	1.5	1.5
		0.25	0.6	0.6	1.9	1.9
		0.35	0.8	0.8	2.6	2.6
		0.4	1	1	3	3
		0.45	1.3	1.3	3.8	3.8
2.8	5.6	0.35	1	1	3	3
		0.5	1.5	1.5	4.5	4.5
		0.6	1.7	1.7	5	5
		0.7	2	2	6	6
		0.75	2.2	2.2	6.7	6.7
		0.8	2.5	2.5	7.5	7.5
5.6	11.2	0.75	2.4	2.4	7.1	7.1
		1	3	3	9	9
		1.25	4	4	12	12
		1.5	5	5	15	15
11.2	22.4	1	3.8	3.8	11	11
		1.25	4.5	4.5	13	13
		1.5	5.6	5.6	16	16
		1.75	6	6	18	18
		2	8	8	24	24
		2.5	10	10	30	30

Nominal Size Diam. Over	To and Incl	Pitch P	Group S To and Incl	Over	Group N To and Incl	Group L Over
22.4	45	1	4	4	12	12
		1.5	6.3	6.3	19	19
		2	8.5	8.5	25	25
		3	12	12	36	36
		3.5	15	15	45	45
		4	18	18	53	53
		4.5	21	21	63	63
45	90	1.5	7.5	7.5	22	22
		2	9.5	9.5	28	28
		3	15	15	45	45
		4	19	19	56	56
		5	24	24	71	71
		5.5	28	28	85	85
		6	32	32	95	95
90	180	2	12	12	36	36
		3	18	18	53	53
		4	24	24	71	71
		6	36	36	106	106
180	355	3	20	20	60	60
		4	26	26	80	80
		6	40	40	118	118

All dimensions are given in millimeters. (Courtesy of ISO Standards.)

APPENDIX 17
ISO METRIC SCREW THREAD STANDARD SERIES

Nominal Size Dia. (mm)			Pitches (mm)														Nominal Size Dia. (mm)
Column[a]			Series with Graded Pitches		Series with Constant Pitches												
1	2	3	Coarse	Fine	6	4	3	2	1.5	1.25	1	0.75	0.5	0.35	0.25	0.2	
0.25			0.075	—	—	—	—	—	—	—	—	—	—	—	—	—	0.25
0.3			0.08	—	—	—	—	—	—	—	—	—	—	—	—	—	0.3
	0.35		0.09	—	—	—	—	—	—	—	—	—	—	—	—	—	0.35
0.4			0.1	—	—	—	—	—	—	—	—	—	—	—	—	—	0.4
	0.45		0.1	—	—	—	—	—	—	—	—	—	—	—	—	—	0.45
0.5			0.125	—	—	—	—	—	—	—	—	—	—	—	—	—	0.5
	0.55		0.125	—	—	—	—	—	—	—	—	—	—	—	—	—	0.55
0.6			0.15	—	—	—	—	—	—	—	—	—	—	—	—	—	0.6
	0.7		0.175	—	—	—	—	—	—	—	—	—	—	—	—	—	0.7
0.8			0.2	—	—	—	—	—	—	—	—	—	—	—	—	—	0.8
	0.9		0.225	—	—	—	—	—	—	—	—	—	—	—	—	—	0.9
			0.25	—	—	—	—	—	—	—	—	—	—	—	—	0.2	1
	1.1		0.25	—	—	—	—	—	—	—	—	—	—	—	—	0.2	1.1
1.2			0.25	—	—	—	—	—	—	—	—	—	—	—	—	0.2	1.2
	1.4		0.3	—	—	—	—	—	—	—	—	—	—	—	—	0.2	1.4
1.6			0.35	—	—	—	—	—	—	—	—	—	—	—	—	0.2	1.6
	1.8		0.35	—	—	—	—	—	—	—	—	—	—	—	—	0.2	1.8
2			0.4	—	—	—	—	—	—	—	—	—	—	—	0.25	—	2
	2.2		0.45	—	—	—	—	—	—	—	—	—	—	—	0.25	—	2.2
2.5			0.45	—	—	—	—	—	—	—	—	—	—	0.35	—	—	2.5
3			0.5	—	—	—	—	—	—	—	—	—	—	0.35	—	—	3
	3.5		0.6	—	—	—	—	—	—	—	—	—	—	0.35	—	—	3.5
4			0.7	—	—	—	—	—	—	—	—	—	0.5	—	—	—	4
	4.5		0.75	—	—	—	—	—	—	—	—	—	0.5	—	—	—	4.5
5			0.8	—	—	—	—	—	—	—	—	—	0.5	—	—	—	5
		5.5	—	—	—	—	—	—	—	—	—	—	0.5	—	—	—	5.5
6			1	—	—	—	—	—	—	—	—	0.75	—	—	—	—	6
		7	1	—	—	—	—	—	—	—	—	0.75	—	—	—	—	7
8			1.25	1	—	—	—	—	—	—	1	0.75	—	—	—	—	8
		9	1.25	—	—	—	—	—	—	—	1	0.75	—	—	—	—	9
10			1.5	1.25	—	—	—	—	—	1.25	1	0.75	—	—	—	—	10
		11	1.5	—	—	—	—	—	—	—	1	0.75	—	—	—	—	11
12			1.75	1.25	—	—	—	—	1.5	1.25	1	—	—	—	—	—	12
	14		2	1.5	—	—	—	—	1.5	1.25[b]	1	—	—	—	—	—	14
		15	—	—	—	—	—	—	1.5	—	1	—	—	—	—	—	15
16			2	1.5	—	—	—	—	1.5	—	1	—	—	—	—	—	16
		17	—	—	—	—	—	—	1.5	—	1	—	—	—	—	—	17
	18		2.5	1.5	—	—	—	2	1.5	—	1	—	—	—	—	—	18
20			2.5	1.5	—	—	—	2	1.5	—	1	—	—	—	—	—	20
	22		2.5	1.5	—	—	—	2	1.5	—	1	—	—	—	—	—	22

[a] Thread diameter should be selected from columns 1, 2 or 3, with preference being in that order.

[b] Pitch 1.25 mm in combination with diameter 14 mm has been included for sparkplug applications.

[c] Diameter 35 mm has been included for bearing locknut applications.

The use of pitches shown in parentheses should be avoided wherever possible.

The pitches enclosed in the bold frame, together with the corresponding nominal diameters in columns 1 and 2, are those combinations which have been established by ISO Recommendations as a selected "coarse" and "fine" series for commercial fasteners.

APPENDIX 17
ISO METRIC SCREW THREAD STANDARD SERIES (Cont.)

Nominal Size Dia. (mm) Column[a]			Pitches (mm) Series with Graded Pitches		Series with Constant Pitches												Nominal Size Dia. (mm)
1	2	3	Coarse	Fine	6	4	3	2	1.5	1.25	1	0.75	0.5	0.35	0.25	0.2	
24			3	2	—	—	—	2	1.5	—	1	—	—	—	—	—	24
		25	—	—	—	—	—	2	1.5	—	1	—	—	—	—	—	25
		26	—	—	—	—	—	—	1.5	—	1	—	—	—	—	—	26
	27		3	2	—	—	—	2	1.5	—	1	—	—	—	—	—	27
		28	—	—	—	—	—	2	1.5	—	1	—	—	—	—	—	28
30			3.5	2	—	—	(3)	2	1.5	—	1	—	—	—	—	—	30
		32	—	—	—	—	—	2	1.5	—	—	—	—	—	—	—	32
	33		3.5	2	—	—	(3)	2	1.5	—	—	—	—	—	—	—	33
		35c	—	—	—	—	—	—	1.5	—	—	—	—	—	—	—	35c
36			4	3	—	—	—	2	1.5	—	—	—	—	—	—	—	36
		38	—	—	—	—	—	—	1.5	—	—	—	—	—	—	—	38
	39		4	3	—	—	—	2	1.5	—	—	—	—	—	—	—	39
		40	—	—	—	—	3	2	1.5	—	—	—	—	—	—	—	40
42			4.5	3	—	4	3	2	1.5	—	—	—	—	—	—	—	42
	45		4.5	3	—	4	3	2	1.5	—	—	—	—	—	—	—	45
48			5	3	—	4	3	2	1.5	—	—	—	—	—	—	—	48
		50	—	—	—	—	3	2	1.5	—	—	—	—	—	—	—	50
	52		5	3	—	4	3	2	1.5	—	—	—	—	—	—	—	52
		55	—	—	—	4	3	2	1.5	—	—	—	—	—	—	—	55
56			5.5	4	—	4	3	2	1.5	—	—	—	—	—	—	—	56
		58	—	—	—	4	3	2	1.5	—	—	—	—	—	—	—	58
	60		5.5	4	—	4	3	2	1.5	—	—	—	—	—	—	—	60
		62	—	—	—	4	3	2	1.5	—	—	—	—	—	—	—	62
64			6	4	—	4	3	2	1.5	—	—	—	—	—	—	—	64
		65	—	—	—	4	3	2	1.5	—	—	—	—	—	—	—	65
	68		6	4	—	4	3	2	1.5	—	—	—	—	—	—	—	68
		70	—	—	6	4	3	2	1.5	—	—	—	—	—	—	—	70
72			—	—	6	4	3	2	1.5	—	—	—	—	—	—	—	72
		75	—	—	—	4	3	2	1.5	—	—	—	—	—	—	—	75
	76		—	—	6	4	3	2	1.5	—	—	—	—	—	—	—	76
		78	—	—	—	—	—	2	—	—	—	—	—	—	—	—	78
80			—	—	6	4	3	2	1.5	—	—	—	—	—	—	—	80
		82	—	—	—	—	—	2	—	—	—	—	—	—	—	—	82
	85		—	—	6	4	3	2	—	—	—	—	—	—	—	—	85
90			—	—	6	4	3	2	—	—	—	—	—	—	—	—	90

Cont.

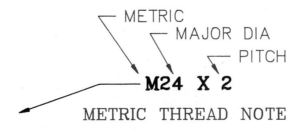

METRIC
MAJOR DIA
PITCH
M24 X 2
METRIC THREAD NOTE

APPENDIX 17
ISO METRIC SCREW THREAD STANDARD SERIES (Cont.)

Nominal Size Dia. (mm) Column[a]			Pitches (mm)													Nominal Size Dia. (mm)
			Series with Graded Pitches		Series with Constant Pitches												
1	2	3	Coarse	Fine	6	4	3	2	1.5	1.25	1	0.75	0.5	0.35	0.25	0.2	
	95		—	—	6	4	3	2	—	—	—	—	—	—	—	—	95
100			—	—	6	4	3	2	—	—	—	—	—	—	—	—	100
	105		—	—	6	4	3	2	—	—	—	—	—	—	—	—	105
110			—	—	6	4	3	2	—	—	—	—	—	—	—	—	110
	115		—	—	6	4	3	2	—	—	—	—	—	—	—	—	115
	120		—	—	6	4	3	2	—	—	—	—	—	—	—	—	120
125			—	—	6	4	3	2	—	—	—	—	—	—	—	—	125
	130		—	—	6	4	3	2	—	—	—	—	—	—	—	—	130
		135	—	—	6	4	3	2	—	—	—	—	—	—	—	—	135
140			—	—	6	4	3	2	—	—	—	—	—	—	—	—	140
		145	—	—	6	4	3	2	—	—	—	—	—	—	—	—	145
	150		—	—	6	4	3	2	—	—	—	—	—	—	—	—	150
		155	—	—	6	4	3	—	—	—	—	—	—	—	—	—	155
160			—	—	6	4	3	—	—	—	—	—	—	—	—	—	160
		165	—	—	6	4	3	—	—	—	—	—	—	—	—	—	165
	170		—	—	6	4	3	—	—	—	—	—	—	—	—	—	170
		175	—	—	6	4	3	—	—	—	—	—	—	—	—	—	175
180			—	—	6	4	3	—	—	—	—	—	—	—	—	—	180
		185	—	—	6	4	3	—	—	—	—	—	—	—	—	—	185
	190		—	—	6	4	3	—	—	—	—	—	—	—	—	—	190
		195	—	—	6	4	3	—	—	—	—	—	—	—	—	—	195
200			—	—	6	4	3	—	—	—	—	—	—	—	—	—	200
		205	—	—	6	4	3	—	—	—	—	—	—	—	—	—	205
	210		—	—	6	4	3	—	—	—	—	—	—	—	—	—	210
220			—	—	6	4	3	—	—	—	—	—	—	—	—	—	220
		225	—	—	6	4	3	—	—	—	—	—	—	—	—	—	225
		230	—	—	6	4	3	—	—	—	—	—	—	—	—	—	230
		235	—	—	6	4	3	—	—	—	—	—	—	—	—	—	235
	240		—	—	6	4	3	—	—	—	—	—	—	—	—	—	240
		245	—	—	6	4	3	—	—	—	—	—	—	—	—	—	245
250			—	—	6	4	3	—	—	—	—	—	—	—	—	—	250
		255	—	—	6	4	—	—	—	—	—	—	—	—	—	—	255
	260		—	—	6	4	—	—	—	—	—	—	—	—	—	—	260
		265	—	—	6	4	—	—	—	—	—	—	—	—	—	—	265
		270	—	—	6	4	—	—	—	—	—	—	—	—	—	—	270
		275	—	—	6	4	—	—	—	—	—	—	—	—	—	—	275
280			—	—	6	4	—	—	—	—	—	—	—	—	—	—	280
		285	—	—	6	4	—	—	—	—	—	—	—	—	—	—	285
		290	—	—	6	4	—	—	—	—	—	—	—	—	—	—	290
		295	—	—	6	4	—	—	—	—	—	—	—	—	—	—	295
	300		—	—	6	4	—	—	—	—	—	—	—	—	—	—	300

[1] Thread diameter should be selected from columns 1, 2, or 3; with preference being in that order.

APPENDIX 18
SQUARE AND ACME THREADS

Size	Threads per Inch	Size	Threads per Inch
$\frac{3}{8}$	12	2	$2\frac{1}{2}$
$\frac{7}{16}$	10	$2\frac{1}{4}$	2
$\frac{1}{2}$	10	$2\frac{1}{2}$	2
$\frac{9}{16}$	8	$2\frac{3}{4}$	2
$\frac{5}{8}$	8	3	$1\frac{1}{2}$
$\frac{3}{4}$	6	$3\frac{1}{4}$	$1\frac{1}{2}$
$\frac{7}{8}$	5	$3\frac{1}{2}$	$1\frac{1}{3}$
1	5	$3\frac{3}{4}$	$1\frac{1}{3}$
$1\frac{1}{8}$	4	4	$1\frac{1}{3}$
$1\frac{1}{4}$	4	$4\frac{1}{4}$	$1\frac{1}{3}$
$1\frac{1}{2}$	3	$4\frac{1}{2}$	1
$1\frac{3}{4}$	$2\frac{1}{2}$	over $4\frac{1}{2}$	1

2.00−2.5 SQUARE

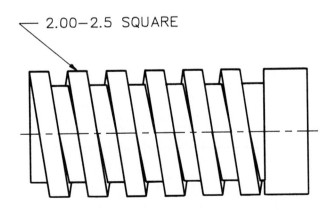

SQUARE THREAD NOTE

APPENDIX 19
AMERICAN STANDARD SQUARE BOLTS AND NUTS

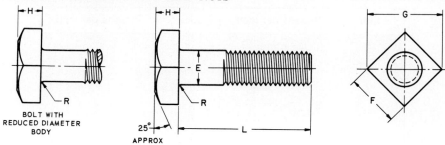

BOLT WITH
REDUCED DIAMETER
BODY

25°
APPROX

Dimensions of Square Bolts

Nominal Size or Basic Product Dia		Body Dia E	Width Across Flats F			Width Across Corners G		Height H			Radius of Fillet R
		Max	Basic	Max	Min	Max	Min	Basic	Max	Min	Max
1/4	0.2500	0.260	3/8	0.3750	0.362	0.530	0.498	11/64	0.188	0.156	0.031
5/16	0.3125	0.324	1/2	0.5000	0.484	0.707	0.665	13/64	0.220	0.186	0.031
3/8	0.3750	0.388	9/16	0.5625	0.544	0.795	0.747	1/4	0.268	0.232	0.031
7/16	0.4375	0.452	5/8	0.6250	0.603	0.884	0.828	19/64	0.316	0.278	0.031
1/2	0.5000	0.515	3/4	0.7500	0.725	1.061	0.995	21/64	0.348	0.308	0.031
5/8	0.6250	0.642	15/16	0.9375	0.906	1.326	1.244	27/64	0.444	0.400	0.062
3/4	0.7500	0.768	1 1/8	1.1250	1.088	1.591	1.494	1/2	0.524	0.476	0.062
7/8	0.8750	0.895	1 5/16	1.3125	1.269	1.856	1.742	19/32	0.620	0.568	0.062
1	1.0000	1.022	1 1/2	1.5000	1.450	2.121	1.991	21/32	0.684	0.628	0.093
1 1/8	1.1250	1.149	1 11/16	1.6875	1.631	2.386	2.239	3/4	0.780	0.720	0.093
1 1/4	1.2500	1.277	1 7/8	1.8750	1.812	2.652	2.489	27/32	0.876	0.812	0.093
1 3/8	1.3750	1.404	2 1/16	2.0625	1.994	2.917	2.738	29/32	0.940	0.872	0.093
1 1/2	1.5000	1.531	2 1/4	2.2500	2.175	3.182	2.986	1	1.036	0.964	0.093

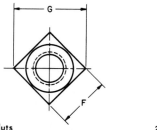

25°

Dimensions of Square Nuts

Nominal Size or Basic Major Dia of Thread		Width Across Flats F			Width Across Corners G		Thickness H		
		Basic	Max	Min	Max	Min	Basic	Max	Min
1/4	0.2500	7/16	0.4375	0.425	0.619	0.584	7/32	0.235	0.203
5/16	0.3125	9/16	0.5625	0.547	0.795	0.751	17/64	0.283	0.249
3/8	0.3750	5/8	0.6250	0.606	0.884	0.832	21/64	0.346	0.310
7/16	0.4375	3/4	0.7500	0.728	1.061	1.000	3/8	0.394	0.356
1/2	0.5000	13/16	0.8125	0.788	1.149	1.082	7/16	0.458	0.418
5/8	0.6250	1	1.0000	0.969	1.414	1.330	35/64	0.569	0.525
3/4	0.7500	1 1/8	1.1250	1.088	1.591	1.494	21/32	0.680	0.632
7/8	0.8750	1 5/16	1.3125	1.269	1.856	1.742	49/64	0.792	0.740
1	1.0000	1 1/2	1.5000	1.450	2.121	1.991	7/8	0.903	0.847
1 1/8	1.1250	1 11/16	1.6875	1.631	2.386	2.239	1	1.030	0.970
1 1/4	1.2500	1 7/8	1.8750	1.812	2.652	2.489	1 3/32	1.126	1.062
1 3/8	1.3750	2 1/16	2.0625	1.994	2.917	2.738	1 13/64	1.237	1.169
1 1/2	1.5000	2 1/4	2.2500	2.175	3.182	2.986	1 5/16	1.348	1.276

(Courtesy of ANSI; B18.2.1–1965 and ANSI; B18.2.2–1965.)

APPENDIX 20

AMERICAN STANDARD HEXAGON HEAD BOLTS AND NUTS

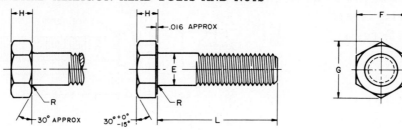

Dimensions of Hex Cap Screws (Finished Hex Bolts)

Nominal Size or Basic Product Dia		Body Dia E		Width Across Flats F			Width Across Corners G		Height H			Radius of Fillet R	
		Max	Min	Basic	Max	Min	Max	Min	Basic	Max	Min	Max	Min
1/4	0.2500	0.2500	0.2450	7/16	0.4375	0.428	0.505	0.488	5/32	0.163	0.150	0.025	0.015
5/16	0.3125	0.3125	0.3065	1/2	0.5000	0.489	0.577	0.557	13/64	0.211	0.195	0.025	0.015
3/8	0.3750	0.3750	0.3690	9/16	0.5625	0.551	0.650	0.628	15/64	0.243	0.226	0.025	0.015
7/16	0.4375	0.4375	0.4305	5/8	0.6250	0.612	0.722	0.698	9/32	0.291	0.272	0.025	0.015
1/2	0.5000	0.5000	0.4930	3/4	0.7500	0.736	0.866	0.840	5/16	0.323	0.302	0.025	0.015
9/16	0.5625	0.5625	0.5545	13/16	0.8125	0.798	0.938	0.910	23/64	0.371	0.348	0.045	0.020
5/8	0.6250	0.6250	0.6170	15/16	0.9375	0.922	1.083	1.051	25/64	0.403	0.378	0.045	0.020
3/4	0.7500	0.7500	0.7410	1 1/8	1.1250	1.100	1.299	1.254	15/32	0.483	0.455	0.045	0.020
7/8	0.8750	0.8750	0.8660	1 5/16	1.3125	1.285	1.516	1.465	35/64	0.563	0.531	0.065	0.040
1	1.0000	1.0000	0.9900	1 1/2	1.5000	1.469	1.732	1.675	39/64	0.627	0.591	0.095	0.060
1 1/8	1.1250	1.1250	1.1140	1 11/16	1.6875	1.631	1.949	1.859	11/16	0.718	0.658	0.095	0.060
1 1/4	1.2500	1.2500	1.2390	1 7/8	1.8750	1.812	2.165	2.066	25/32	0.813	0.749	0.095	0.060
1 3/8	1.3750	1.3750	1.3630	2 1/16	2.0625	1.994	2.382	2.273	27/32	0.878	0.810	0.095	0.060
1 1/2	1.5000	1.5000	1.4880	2 1/4	2.2500	2.175	2.598	2.480	15/16	0.974	0.902	0.095	0.060
1 3/4	1.7500	1.7500	1.7380	2 5/8	2.6250	2.538	3.031	2.893	1 3/32	1.134	1.054	0.095	0.060
2	2.0000	2.0000	1.9880	3	3.0000	2.900	3.464	3.306	1 7/32	1.263	1.175	0.095	0.060
2 1/4	2.2500	2.2500	2.2380	3 3/8	3.3750	3.262	3.897	3.719	1 3/8	1.423	1.327	0.095	0.060
2 1/2	2.5000	2.5000	2.4880	3 3/4	3.7500	3.625	4.330	4.133	1 17/32	1.583	1.479	0.095	0.060
2 3/4	2.7500	2.7500	2.7300	4 1/8	4.1250	3.988	4.763	4.546	1 11/16	1.744	1.632	0.095	0.060
3	3.0000	3.0000	2.9880	4 1/2	4.5000	4.350	5.196	4.959	1 7/8	1.935	1.815	0.095	0.060

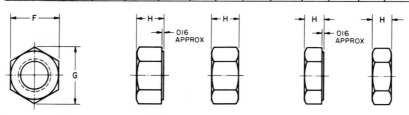

Dimensions of Hex Nuts and Hex Jam Nuts

Nominal Size or Basic Major Dia of Thread		Width Across Flats F			Width Across Corners G		Thickness Hex Nuts H			Thickness Hex Jam Nuts H		
		Basic	Max	Min	Max	Min	Basic	Max	Min	Basic	Max	Min
1/4	0.2500	7/16	0.4375	0.428	0.505	0.488	7/32	0.226	0.212	5/32	0.163	0.150
5/16	0.3125	1/2	0.5000	0.489	0.577	0.557	17/64	0.273	0.258	3/16	0.195	0.180
3/8	0.3750	9/16	0.5625	0.551	0.650	0.628	21/64	0.337	0.320	7/32	0.227	0.210
7/16	0.4375	11/16	0.6875	0.675	0.794	0.768	3/8	0.385	0.365	1/4	0.260	0.240
1/2	0.5000	3/4	0.7500	0.736	0.866	0.840	7/16	0.448	0.427	5/16	0.323	0.302
9/16	0.5625	7/8	0.8750	0.861	1.010	0.982	31/64	0.496	0.473	5/16	0.324	0.301
5/8	0.6250	15/16	0.9375	0.922	1.083	1.051	35/64	0.559	0.535	3/8	0.387	0.363
3/4	0.7500	1 1/8	1.1250	1.088	1.299	1.240	41/64	0.665	0.617	27/64	0.446	0.398
7/8	0.8750	1 5/16	1.3125	1.269	1.516	1.447	3/4	0.776	0.724	31/64	0.510	0.458
1	1.0000	1 1/2	1.5000	1.450	1.732	1.653	55/64	0.887	0.831	35/64	0.575	0.519
1 1/8	1.1250	1 11/16	1.6875	1.631	1.949	1.859	31/32	0.999	0.939	39/64	0.639	0.579
1 1/4	1.2500	1 7/8	1.8750	1.812	2.165	2.066	1 1/16	1.094	1.030	23/32	0.751	0.687
1 3/8	1.3750	2 1/16	2.0625	1.994	2.382	2.273	1 11/64	1.206	1.138	25/32	0.815	0.747
1 1/2	1.5000	2 1/4	2.2500	2.175	2.598	2.480	1 9/32	1.317	1.245	27/32	0.880	0.808

(Courtesy of ANSI; B18.2.1–1965 and ANSI; B18.2.2–1965.)

APPENDIX 21
FILLISTER HEAD AND ROUND HEAD CAP SCREWS

Fillister Head Cap Screws

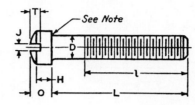

Nominal Size	D Body Diameter		A Head Diameter		H Height of Head		O Total Height of Head		J Width of Slot		T Depth of Slot	
	Max	Min	Max	Min	Max	Min	Max	Min	Max	Min	Max	Min
1/4	0.250	0.245	0.375	0.363	0.172	0.157	0.216	0.194	0.075	0.064	0.097	0.077
5/16	0.3125	0.307	0.437	0.424	0.203	0.186	0.253	0.230	0.084	0.072	0.115	0.090
3/8	0.375	0.369	0.562	0.547	0.250	0.229	0.314	0.284	0.094	0.081	0.142	0.112
7/16	0.4375	0.431	0.625	0.608	0.297	0.274	0.368	0.336	0.094	0.081	0.168	0.133
1/2	0.500	0.493	0.750	0.731	0.328	0.301	0.413	0.376	0.106	0.091	0.193	0.153
9/16	0.5625	0.555	0.812	0.792	0.375	0.346	0.467	0.427	0.118	0.102	0.213	0.168
5/8	0.625	0.617	0.875	0.853	0.422	0.391	0.521	0.478	0.133	0.116	0.239	0.189
3/4	0.750	0.742	1.000	0.976	0.500	0.466	0.612	0.566	0.149	0.131	0.283	0.223
7/8	0.875	0.866	1.125	1.098	0.594	0.556	0.720	0.668	0.167	0.147	0.334	0.264
1	1.000	0.990	1.312	1.282	0.656	0.612	0.803	0.743	0.188	0.166	0.371	0.291

All dimensions are given in inches.

The radius of the fillet at the base of the head:
  For sizes 1/4 to 3/8 in. incl. is 0.016 min and 0.031 max,
    7/16 to 9/16 in. incl. is 0.016 min and 0.047 max,
    5/8 to 1 in. incl. is 0.031 min and 0.062 max.

Round Head Cap Screws

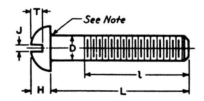

Nominal Size	D Body Diameter		A Head Diameter		H Height of Head		J Width of Slot		T Depth of Slot	
	Max	Min	Max	Min	Max	Min	Max	Min	Max	Min
1/4	0.250	0.245	0.437	0.418	0.191	0.175	0.075	0.064	0.117	0.097
5/16	0.3125	0.307	0.562	0.540	0.245	0.226	0.084	0.072	0.151	0.126
3/8	0.375	0.369	0.625	0.603	0.273	0.252	0.094	0.081	0.168	0.138
7/16	0.4375	0.431	0.750	0.725	0.328	0.302	0.094	0.081	0.202	0.167
1/2	0.500	0.493	0.812	0.786	0.354	0.327	0.106	0.091	0.218	0.178
9/16	0.5625	0.555	0.937	0.909	0.409	0.378	0.118	0.102	0.252	0.207
5/8	0.625	0.617	1.000	0.970	0.437	0.405	0.133	0.116	0.270	0.220
3/4	0.750	0.742	1.250	1.215	0.546	0.507	0.149	0.131	0.338	0.278

All dimensions are given in inches.

Radius of the fillet at the base of the head:
  For sizes 1/4 to 3/8 in. incl. is 0.016 min and 0.031 max,
    7/16 to 9/16 in. incl. is 0.016 min and 0.047 max,
    5/8 to 1 in. incl. is 0.031 min and 0.062 max.

(Courtesy of ANSI; B18.6.2–1956.)

APPENDIX 22

FLAT HEAD CAP SCREWS

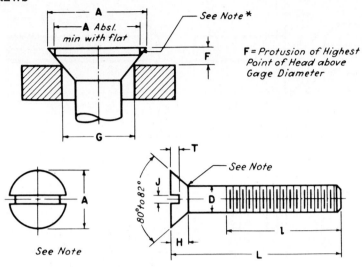

Nominal Size	D Body Diameter		A Head Diameter			G Gaging Diameter	H Height of Head	J Width of Slot		T Depth of Slot		F Protrusion Above Gaging Diameter	
	Max	Min	Max	Min	Absolute Min with Flat		Average	Max	Min	Max	Min	Max	Min
1/4	0.250	0.245	0.500	0.477	0.452	0.4245	0.140	0.075	0.064	0.068	0.045	0.0452	0.0307
5/16	0.3125	0.307	0.625	0.598	0.567	0.5376	0.177	0.084	0.072	0.086	0.057	0.0523	0.0354
3/8	0.375	0.369	0.750	0.720	0.682	0.6507	0.210	0.094	0.081	0.103	0.068	0.0594	0.0401
7/16	0.4375	0.431	0.8125	0.780	0.736	0.7229	0.210	0.094	0.081	0.103	0.068	0.0649	0.0448
1/2	0.500	0.493	0.875	0.841	0.791	0.7560	0.210	0.106	0.091	0.103	0.068	0.0705	0.0495
9/16	0.5625	0.555	1.000	0.962	0.906	0.8691	0.244	0.118	0.102	0.120	0.080	0.0775	0.0542
5/8	0.625	0.617	1.125	1.083	1.020	0.9822	0.281	0.133	0.116	0.137	0.091	0.0846	0.0588
3/4	0.750	0.742	1.375	1.326	1.251	1.2085	0.352	0.149	0.131	0.171	0.115	0.0987	0.0682
7/8	0.875	0.866	1.625	1.568	1.480	1.4347	0.423	0.167	0.147	0.206	0.138	0.1128	0.0776
1	1.000	0.990	1.875	1.811	1.711	1.6610	0.494	0.188	0.166	0.240	0.162	0.1270	0.0870
1 1/8	1.125	1.114	2.062	1.992	1.880	1.8262	0.529	0.196	0.178	0.257	0.173	0.1401	0.0964
1 1/4	1.250	1.239	2.312	2.235	2.110	2.0525	0.600	0.211	0.193	0.291	0.197	0.1542	0.1056
1 3/8	1.375	1.363	2.562	2.477	2.340	2.2787	0.665	0.226	0.208	0.326	0.220	0.1684	0.1151
1 1/2	1.500	1.488	2.812	2.720	2.570	2.5050	0.742	0.258	0.240	0.360	0.244	0.1825	0.1245

All dimensions are given in inches.

The maximum and minimum head diameters, A, are extended to the theoretical sharp corners.

The radius of the fillet at the base of the head shall not exceed 0.4 Max. D.

*Edge of head may be flat as shown or slightly rounded.

(Courtesy of ANSI; B18.6.2–1956.)

APPENDIX 23
MACHINE SCREWS

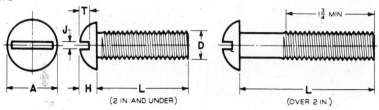

Dimensions of Slotted Round Head Machine Screws

Nominal Size	D Diameter of Screw	A Head Diameter		H Head Height		J Width of Slot		T Depth of Slot	
	Basic	Max	Min	Max	Min	Max	Min	Max	Min
0	0.0600	0.113	0.099	0.053	0.043	0.023	0.016	0.039	0.029
1	0.0730	0.138	0.122	0.061	0.051	0.026	0.019	0.044	0.033
2	0.0860	0.162	0.146	0.069	0.059	0.031	0.023	0.048	0.037
3	0.0990	0.187	0.169	0.078	0.067	0.035	0.027	0.053	0.040
4	0.1120	0.211	0.193	0.086	0.075	0.039	0.031	0.058	0.044
5	0.1250	0.236	0.217	0.095	0.083	0.043	0.035	0.063	0.047
6	0.1380	0.260	0.240	0.103	0.091	0.048	0.039	0.068	0.051
8	0.1640	0.309	0.287	0.120	0.107	0.054	0.045	0.077	0.058
10	0.1900	0.359	0.334	0.137	0.123	0.060	0.050	0.087	0.065
12	0.2160	0.408	0.382	0.153	0.139	0.067	0.056	0.096	0.073
1/4	0.2500	0.472	0.443	0.175	0.160	0.075	0.064	0.109	0.082
5/16	0.3125	0.590	0.557	0.216	0.198	0.084	0.072	0.132	0.099
3/8	0.3750	0.708	0.670	0.256	0.237	0.094	0.081	0.155	0.117
7/16	0.4375	0.750	0.707	0.328	0.307	0.094	0.081	0.196	0.148
1/2	0.5000	0.813	0.766	0.355	0.332	0.106	0.091	0.211	0.159
9/16	0.5625	0.938	0.887	0.410	0.385	0.118	0.102	0.242	0.183
5/8	0.6250	1.000	0.944	0.438	0.411	0.133	0.116	0.258	0.195
3/4	0.7500	1.250	1.185	0.547	0.516	0.149	0.131	0.320	0.242

All dimensions are given in inches.

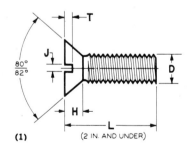

(1)

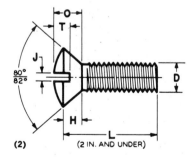

(2)

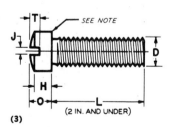

(3)

Three other common forms of machine screws are shown above: (1) flat head, (2) oval head, and (3) fillister head. Although dimension tables are not given for these three types of machine screws in this text, their general dimensions are closely related to those shown in the table above. Additional information on these screws can be obtained from ANSI; B18.6.3–1962.

(Courtesy of ANSI; B18.6.3–1962.)

APPENDIX 24

AMERICAN STANDARD MACHINE SCREWS

(The proportions of the screws can be found by multiplying the major diameter, D, by the factors given below.)

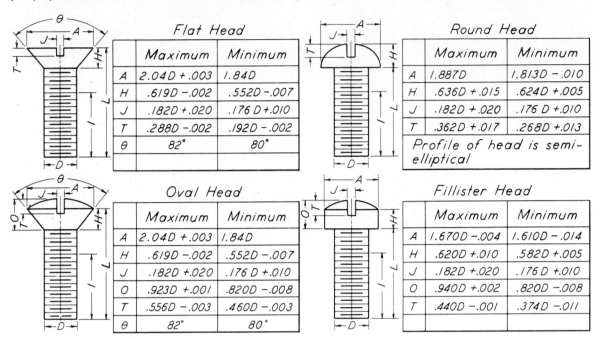

Flat Head

	Maximum	Minimum
A	2.04D + .003	1.84D
H	.619D − .002	.552D − .007
J	.182D + .020	.176 D + .010
T	.288D − .002	.192D − .002
θ	82°	80°

Round Head

	Maximum	Minimum
A	1.887D	1.813D − .010
H	.636D + .015	.624D + .005
J	.182D + .020	.176 D + .010
T	.362D + .017	.268D + .013

Profile of head is semi-elliptical

Oval Head

	Maximum	Minimum
A	2.04D + .003	1.84D
H	.619D − .002	.552D − .007
J	.182D + .020	.176 D + .010
O	.923D + .001	.820D − .008
T	.556D − .003	.460D − .003
θ	82°	80°

Fillister Head

	Maximum	Minimum
A	1.670D − .004	1.610D − .014
H	.620D + .010	.582D + .005
J	.182D + .020	.176 D + .010
O	.940D + .002	.820D − .008
T	.440D − .001	.374D − .011

APPENDIX 25

AMERICAN STANDARD MACHINE TAPERS*

No. of Taper	Taper per Foot (Basic)	Origin of Series	No. of Taper	Taper per Foot (Basic)	Origin of Series	No. of Taper	Taper per Foot (Basic)	Origin of Series	No. of Taper	Taper per Foot (Basic)	Origin of Series
0.239	0.50200	Brown & Sharpe	*	0.62326	Morse	250	0.750	¾ in. per ft.	600	0.750	¾ in. per ft.
.299	.50200	Brown & Sharpe	4½	.62400	Morse	300	.750	¾ in. per ft.	800	0.750	¾ in. per ft.
.375	.50200	Brown & Sharpe	5	.63151	Morse	350	.750	¾ in. per ft.	1000	0.750	¾ in. per ft.
1	.59858	Morse	6	.62565	Morse	400	.750	¾ in. per ft.	1200	0.750	¾ in. per ft.
2	.59941	Morse	7	.62400	Morse	450	.750	¾ in. per ft.			
3	.60235	Morse	200	.750	¾ in. per ft.	500	.750	¾ in. per ft.			

All dimensions in inches.
* Extracted from American Standards, "Machine Tapers, Self-Holding and Steep Taper Series" (ASA B5,10-1960), with the permission of the publisher, The American Society of Mechanical Engineers.

APPENDIX 26

AMERICAN NATIONAL STANDARD SQUARE HEAD SET SCREWS (ANSI B18.6.2)

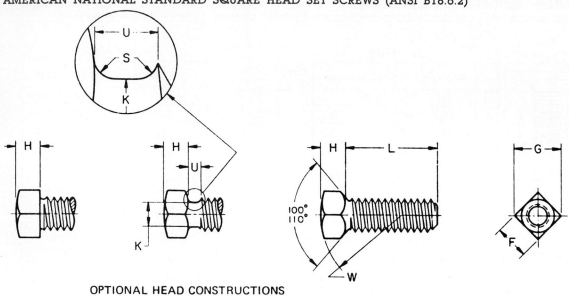

OPTIONAL HEAD CONSTRUCTIONS

Nominal Size[1] or Basic Screw Diameter		F Width Across Flats		G Width Across Corners		H Head Height		K Neck Relief Diameter		S Neck Relief Fillet Radius	U Neck Relief Width	W Head Radius
		Max	Min	Max	Min	Max	Min	Max	Min	Max	Min	Min
10	0.1900	0.188	0.180	0.265	0.247	0.148	0.134	0.145	0.140	0.027	0.083	0.48
1/4	0.2500	0.250	0.241	0.354	0.331	0.196	0.178	0.185	0.170	0.032	0.100	0.62
5/16	0.3125	0.312	0.302	0.442	0.415	0.245	0.224	0.240	0.225	0.036	0.111	0.78
3/8	0.3750	0.375	0.362	0.530	0.497	0.293	0.270	0.294	0.279	0.041	0.125	0.94
7/16	0.4375	0.438	0.423	0.619	0.581	0.341	0.315	0.345	0.330	0.046	0.143	1.09
1/2	0.5000	0.500	0.484	0.707	0.665	0.389	0.361	0.400	0.385	0.050	0.154	1.25
9/16	0.5625	0.562	0.545	0.795	0.748	0.437	0.407	0.454	0.439	0.054	0.167	1.41
5/8	0.6250	0.625	0.606	0.884	0.833	0.485	0.452	0.507	0.492	0.059	0.182	1.56
3/4	0.7500	0.750	0.729	1.060	1.001	0.582	0.544	0.620	0.605	0.065	0.200	1.88
7/8	0.8750	0.875	0.852	1.237	1.170	0.678	0.635	0.731	0.716	0.072	0.222	2.19
1	1.0000	1.000	0.974	1.414	1.337	0.774	0.726	0.838	0.823	0.081	0.250	2.50
1 1/8	1.1250	1.125	1.096	1.591	1.505	0.870	0.817	0.939	0.914	0.092	0.283	2.81
1 1/4	1.2500	1.250	1.219	1.768	1.674	0.966	0.908	1.064	1.039	0.092	0.283	3.12
1 3/8	1.3750	1.375	1.342	1.945	1.843	1.063	1.000	1.159	1.134	0.109	0.333	3.44
1 1/2	1.5000	1.500	1.464	2.121	2.010	1.159	1.091	1.284	1.259	0.109	0.333	3.75

[1] Where specifying nominal size in decimals, zeros preceding decimal and in the fourth decimal place shall be omitted.

APPENDIX 27
AMERICAN NATIONAL STANDARD POINTS FOR SQUARE HEAD SET SCREWS (ANSI B18.6.2)

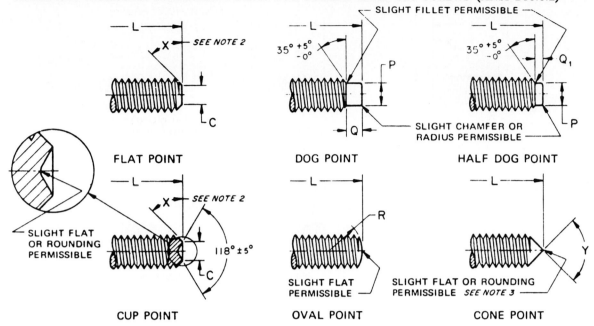

FLAT POINT · DOG POINT · HALF DOG POINT

CUP POINT · OVAL POINT · CONE POINT

Nominal Size[1] or Basic Screw Diameter		C — Cup and Flat Point Diameters		P — Dog and Half Dog Point Diameters		Q — Point Length				R — Oval Point Radius	Y — Cone Point Angle
						Dog		Half Dog		+0.031 −0.000	90° ±2° For These Nominal Lengths or Longer; 118° ±2° For Shorter Screws
		Max	Min	Max	Min	Max	Min	Max	Min		
10	0.1900	0.102	0.088	0.127	0.120	0.095	0.085	0.050	0.040	0.142	1/4
1/4	0.2500	0.132	0.118	0.156	0.149	0.130	0.120	0.068	0.058	0.188	5/16
5/16	0.3125	0.172	0.156	0.203	0.195	0.161	0.151	0.083	0.073	0.234	3/8
3/8	0.3750	0.212	0.194	0.250	0.241	0.193	0.183	0.099	0.089	0.281	7/16
7/16	0.4375	0.252	0.232	0.297	0.287	0.224	0.214	0.114	0.104	0.328	1/2
1/2	0.5000	0.291	0.270	0.344	0.334	0.255	0.245	0.130	0.120	0.375	9/16
9/16	0.5625	0.332	0.309	0.391	0.379	0.287	0.275	0.146	0.134	0.422	5/8
5/8	0.6250	0.371	0.347	0.469	0.456	0.321	0.305	0.164	0.148	0.469	3/4
3/4	0.7500	0.450	0.425	0.562	0.549	0.383	0.367	0.196	0.180	0.562	7/8
7/8	0.8750	0.530	0.502	0.656	0.642	0.446	0.430	0.227	0.211	0.656	1
1	1.0000	0.609	0.579	0.750	0.734	0.510	0.490	0.260	0.240	0.750	1 1/8
1 1/8	1.1250	0.689	0.655	0.844	0.826	0.572	0.552	0.291	0.271	0.844	1 1/4
1 1/4	1.2500	0.767	0.733	0.938	0.920	0.635	0.615	0.323	0.303	0.938	1 1/2
1 3/8	1.3750	0.848	0.808	1.031	1.011	0.698	0.678	0.354	0.334	1.031	1 5/8
1 1/2	1.5000	0.926	0.886	1.125	1.105	0.760	0.740	0.385	0.365	1.125	1 3/4

[1] Where specifying nominal size in decimals, zeros preceding decimal and in the fourth decimal place shall be omitted.
[2] Point angle X shall be 45° plus 5°, minus 0°, for screws of nominal lengths equal to or longer than those listed in Column Y, and 30° minimum for screws of shorter nominal lengths.
[3] The extent of rounding or flat at apex of cone point shall not exceed an amount equivalent to 10 per cent of the basic screw diameter.

APPENDIX 28
AMERICAN NATIONAL STANDARD SLOTTED HEADLESS SET SCREWS (ANSI B18.6.2)

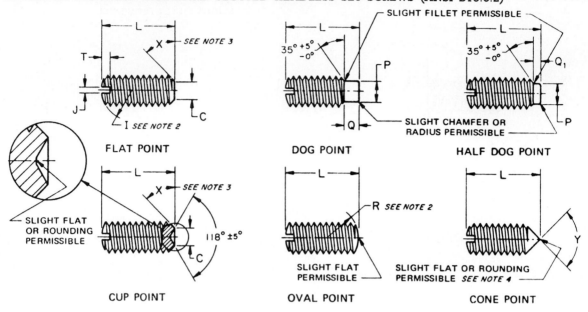

FLAT POINT   DOG POINT   HALF DOG POINT

CUP POINT   OVAL POINT   CONE POINT

Nominal Size[1] or Basic Screw Diameter	I[2] Crown Radius	J Slot Width		T Slot Depth		C Cup and Flat Point Diameters		P Dog Point Diameters		Point Length				R[2] Oval Point Radius	Y Cone Point Angle 90° ±2° For These Nominal Lengths or Longer; 118° ±2° For Shorter Screws
										Dog		Half Dog			
	Basic	Max	Min	Max	Min	Max	Min	Max	Min	Max	Min	Max	Min	Basic	
0 0.0600	0.060	0.014	0.010	0.020	0.016	0.033	0.027	0.040	0.037	0.032	0.028	0.017	0.013	0.045	5/64
1 0.0730	0.073	0.016	0.012	0.020	0.016	0.040	0.033	0.049	0.045	0.040	0.036	0.021	0.017	0.055	3/32
2 0.0860	0.086	0.018	0.014	0.025	0.019	0.047	0.039	0.057	0.053	0.046	0.042	0.024	0.020	0.064	7/64
3 0.0990	0.099	0.020	0.016	0.028	0.022	0.054	0.045	0.066	0.062	0.052	0.048	0.027	0.023	0.074	1/8
4 0.1120	0.112	0.024	0.018	0.031	0.025	0.061	0.051	0.075	0.070	0.058	0.054	0.030	0.026	0.084	5/32
5 0.1250	0.125	0.026	0.020	0.036	0.026	0.067	0.057	0.083	0.078	0.063	0.057	0.033	0.027	0.094	3/16
6 0.1380	0.138	0.028	0.022	0.040	0.030	0.074	0.064	0.092	0.087	0.073	0.067	0.038	0.032	0.104	3/16
8 0.1640	0.164	0.032	0.026	0.046	0.036	0.087	0.076	0.109	0.103	0.083	0.077	0.043	0.037	0.123	1/4
10 0.1900	0.190	0.035	0.029	0.053	0.043	0.102	0.088	0.127	0.120	0.095	0.085	0.050	0.040	0.142	1/4
12 0.2160	0.216	0.042	0.035	0.061	0.051	0.115	0.101	0.144	0.137	0.115	0.105	0.060	0.050	0.162	5/16
1/4 0.2500	0.250	0.049	0.041	0.068	0.058	0.132	0.118	0.156	0.149	0.130	0.120	0.068	0.058	0.188	5/16
5/16 0.3125	0.312	0.055	0.047	0.083	0.073	0.172	0.156	0.203	0.195	0.161	0.151	0.083	0.073	0.234	3/8
3/8 0.3750	0.375	0.068	0.060	0.099	0.089	0.212	0.194	0.250	0.241	0.193	0.183	0.099	0.089	0.281	7/16
7/16 0.4375	0.438	0.076	0.068	0.114	0.104	0.252	0.232	0.297	0.287	0.224	0.214	0.114	0.104	0.328	1/2
1/2 0.5000	0.500	0.086	0.076	0.130	0.120	0.291	0.270	0.344	0.334	0.255	0.245	0.130	0.120	0.375	9/16
9/16 0.5625	0.562	0.096	0.086	0.146	0.136	0.332	0.309	0.391	0.379	0.287	0.275	0.146	0.134	0.422	5/8
5/8 0.6250	0.625	0.107	0.097	0.161	0.151	0.371	0.347	0.469	0.456	0.321	0.305	0.164	0.148	0.469	3/4
3/4 0.7500	0.750	0.134	0.124	0.193	0.183	0.450	0.425	0.562	0.549	0.383	0.367	0.196	0.180	0.562	7/8

[1] Where specifying nominal size in decimals, zeros preceding decimal and in the fourth decimal place shall be omitted.

[2] Tolerance on radius for nominal sizes up to and including 5 (0.125 in.) shall be plus 0.015 in. and minus 0.000, and for larger sizes, plus 0.031 in. and minus 0.000. Slotted ends on screws may be flat at option of manufacturer.

[3] Point angle X shall be 45° plus 5°, minus 0°, for screws of nominal lengths equal to or longer than those listed in Column Y, and 30° minimum for screws of shorter nominal lengths.

[4] The extent of rounding or flat at apex of cone point shall not exceed an amount equivalent to 10 per cent of the basic screw diameter.

APPENDIX 29
TWIST DRILL SIZES
Number Size Drills

Size	Drill Diameter Inches	Drill Diameter mm	Size	Drill Diameter Inches	Drill Diameter mm	Size	Drill Diameter Inches	Drill Diameter mm	Size	Drill Diameter Inches	Drill Diameter mm
1	0.2280	5.7912	21	0.1590	4.0386	41	0.0960	2.4384	61	0.0390	0.9906
2	0.2210	5.6134	22	0.1570	3.9878	42	0.0935	2.3622	62	0.0380	0.9652
3	0.2130	5.4102	23	0.1540	3.9116	43	0.0890	2.2606	63	0.0370	0.9398
4	0.2090	5.3086	24	0.1520	3.8608	44	0.0860	2.1844	64	0.0360	0.9144
5	0.2055	5.2197	25	0.1495	3.7973	45	0.0820	2.0828	65	0.0350	0.8890
6	0.2040	5.1816	26	0.1470	3.7338	46	0.0810	2.0574	66	0.0330	0.8382
7	0.2010	5.1054	27	0.1440	3.6576	47	0.0785	1.9812	67	0.0320	0.8128
8	0.1990	5.0800	28	0.1405	3.5560	48	0.0760	1.9304	68	0.0310	0.7874
9	0.1960	4.9784	29	0.1360	3.4544	49	0.0730	1.8542	69	0.0292	0.7417
10	0.1935	4.9149	30	0.1285	3.2639	50	0.0700	1.7780	70	0.0280	0.7112
11	0.1910	4.8514	31	0.1200	3.0480	51	0.0670	1.7018	71	0.0260	0.6604
12	0.1890	4.8006	32	0.1160	2.9464	52	0.0635	1.6129	72	0.0250	0.6350
13	0.1850	4.6990	33	0.1130	2.8702	53	0.0595	1.5113	73	0.0240	0.6096
14	0.1820	4.6228	34	0.1110	2.8194	54	0.0550	1.3970	74	0.0225	0.5715
15	0.1800	4.5720	35	0.1100	2.7940	55	0.0520	1.3208	75	0.0210	0.5334
16	0.1770	4.4958	36	0.1065	2.7051	56	0.0465	1.1684	76	0.0200	0.5080
17	0.1730	4.3942	37	0.1040	2.6416	57	0.0430	1.0922	77	0.0180	0.4572
18	0.1695	4.3053	38	0.1015	2.5781	58	0.0420	1.0668	78	0.0160	0.4064
19	0.1660	4:2164	39	0.0995	2.5273	59	0.0410	1.0414	79	0.0145	0.3638
20	0.1610	4.0894	40	0.0980	2.4892	60	0.0400	1.0160	80	0.0135	0.3429

Metric Drill Sizes   Preferred sizes are in color type. Decimal-inch equivalents are for reference only.

Drill Diameter mm	in.	Drill Diameter mm	in.	Drill Diameter mm	in.	Drill Diameter mm	in.	Drill Diameter mm	in.	Drill Diameter mm	in.	Drill Diameter mm	in.
.40	.0157	1.03	.0406	2.20	.0866	5.00	.1969	10.00	.3937	21.50	.8465	48.00	1.8898
.42	.0165	1.05	.0413	2.30	.0906	5.20	.2047	10.30	.4055	22.00	.8661	50.00	1.9685
.45	.0177	1.08	.0425	2.40	.0945	5.30	.2087	10.50	.4134	23.00	.9055	51.50	2.0276
.48	.0189	1.10	.0433	2.50	.0984	5.40	.2126	10.80	.4252	24.00	.9449	53.00	2.0866
.50	.0197	1.15	.0453	2.60	.1024	5.60	.2205	11.00	.4331	25.00	.9843	54.00	2.1260
.52	.0205	1.20	.0472	2.70	.1063	5.80	.2283	11.50	.4528	26.00	1.0236	56.00	2.2047
.55	.0217	1.25	.0492	2.80	.1102	6.00	.2362	12.00	.4724	27.00	1.0630	58.00	2.2835
.58	.0228	1.30	.0512	2.90	.1142	6.20	.2441	12.50	.4921	28.00	1.1024	60.00	2.3622
.60	.0236	1.35	.0531	3.00	.1181	6.30	.2480	13.00	.5118	29.00	1.1417		
.62	.0244	1.40	.0551	3.10	.1220	6.50	.2559	13.50	.5315	30.00	1.1811		
.65	.0256	1.45	.0571	3.20	.1260	6.70	.2638	14.00	.5512	31.00	1.2205		
.68	.0268	1.50	.0591	3.30	.1299	6.80	.2677	14.50	.5709	32.00	1.2598		
.70	.0276	1.55	.0610	3.40	.1339	6.90	.2717	15.00	.5906	33.00	1.2992		
.72	.0283	1.60	.0630	3.50	.1378	7.10	.2795	15.50	.6102	34.00	1.3386		
.75	.0295	1.65	.0650	3.60	.1417	7.30	.2874	16.00	.6299	35.00	1.3780		
.78	.0307	1.70	.0669	3.70	.1457	7.50	.2953	16.50	.6496	36.00	1.4173		
.80	.0315	1.75	.0689	3.80	.1496	7.80	.3071	17.00	.6693	37.00	1.4567		
.82	.0323	1.80	.0709	3.90	.1535	8.00	.3150	17.50	.6890	38.00	1.4961		
.85	.0335	1.85	.0728	4.00	.1575	8.20	.3228	18.00	.7087	39.00	1.5354		
.88	.0346	1.90	.0748	4.10	.1614	8.50	.3346	18.50	.7283	40.00	1.5748		
.90	.0354	1.95	.0768	4.20	.1654	8.80	.3465	19.00	.7480	41.00	1.6142		
.92	.0362	2.00	.0787	4.40	.1732	9.00	.3543	19.50	.7677	42.00	1.6535		
.95	.0374	2.05	.0807	4.50	.1772	9.20	.3622	20.00	.7874	43.50	1.7126		
.98	.0386	2.10	.0827	4.60	.1811	9.50	.3740	20.50	.8071	45.00	1.7717		
1.00	.0394	2.15	.0846	4.80	.1890	9.80	.3858	21.00	.8268	46.50	1.8307		

Cont.

APPENDIX 29
TWIST DRILL SIZES (Cont.)
Letter Size Drills

Size	Drill Diameter		Size	Drill Diameter		Size	Drill Diameter		Size	Drill Diameter	
	Inches	mm		Inches	mm		Inches	mm		Inches	mm
A	0.234	5.944	H	0.266	6.756	O	0.316	8.026	V	0.377	9.576
B	0.238	6.045	I	0.272	6.909	P	0.323	8.204	W	0.386	9.804
C	0.242	6.147	J	0.277	7.036	Q	0.332	8.433	X	0.397	10.084
D	0.246	6.248	K	0.281	7.137	R	0.339	8.611	Y	0.404	10.262
E	0.250	6.350	L	0.290	7.366	S	0.348	8.839	Z	0.413	10.490
F	0.257	6.528	M	0.295	7.493	T	0.358	9.093			
G	0.261	6.629	N	0.302	7.601	U	0.368	9.347			

(Courtesy of General Motors Corporation.)

APPENDIX 30
STRAIGHT PINS

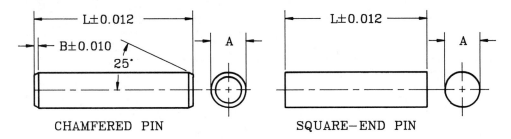

CHAMFERED PIN   SQUARE–END PIN

Nominal Diameter	Diameter A		Chamfer B
	Max	Min	
0.062	0.0625	0.0605	0.015
0.094	0.0937	0.0917	0.015
0.109	0.1094	0.1074	0.015
0.125	0.1250	0.1230	0.015
0.156	0.1562	0.1542	0.015
0.188	0.1875	0.1855	0.015
0.219	0.2187	0.2167	0.015
0.250	0.2500	0.2480	0.015
0.312	0.3125	0.3095	0.030
0.375	0.3750	0.3720	0.030
0.438	0.4375	0.4345	0.030
0.500	0.500	0.4970	0.030

All dimensions are given in inches.

These pins must be straight and free from burrs or any other defects that will affect their serviceability.

(Courtesy of ANSI; B5.20–1958.)

APPENDIX 31

STANDARD KEYS AND KEYWAYS

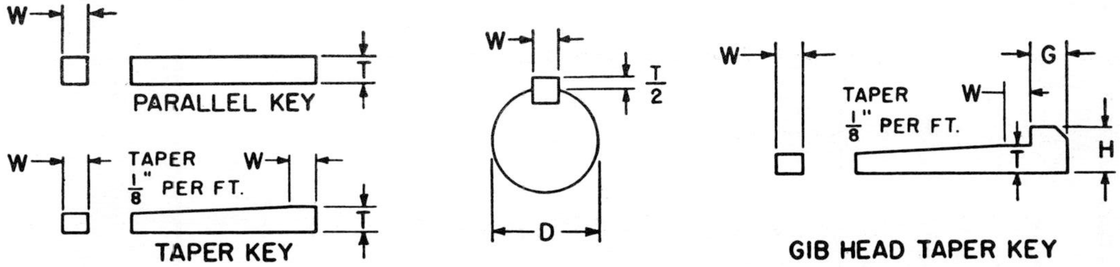
W
T
PARALLEL KEY
W
TAPER
1/8" PER FT.
W
T
TAPER KEY
W
T/2
D
W
TAPER
1/8" PER FT.
W
G
T
H
GIB HEAD TAPER KEY

Sprocket Bore (= Shaft Diam.)	Keyway Dimensions — Inches				Key Dimensions — Inches					Gib Head Dimensions — Inches				Key Tolerances Taper and Gib Head	
	For Square Key		For Flat Key		Square		Flat		Tolerance on W and T (−)	Square Key		Flat Key			
Inches D	Width W	Depth T/2	Width W	Depth T/2	Width W	Height T	Width W	Height T		H	G	H	G	W (−)	T (+)
1/2 — 9/16	1/8 x	1/16	1/8 x	3/64	1/8 x	1/8	1/8 x	3/32	0.002	1/4	7/32	3/16	1/8	0.002	0.002
5/8 — 7/8	3/16 x	3/32	3/16 x	1/16	3/16 x	3/16	3/16 x	1/8	0.002	5/16	9/32	1/4	3/16	0.002	0.002
15/16 — 1 1/4	1/4 x	1/8	1/4 x	3/32	1/4 x	1/4	1/4 x	3/16	0.002	7/16	11/32	5/16	1/4	0.002	0.002
1 5/16 — 1 3/8	5/16 x	5/32	5/16 x	1/8	5/16 x	5/16	5/16 x	1/4	0.002	9/16	13/32	3/8	5/16	0.002	0.002
1 7/16 — 1 3/4	3/8 x	3/16	3/8 x	1/8	3/8 x	3/8	3/8 x	1/4	0.002	11/16	15/32	7/16	3/8	0.002	0.002
1 13/16 — 2 1/4	1/2 x	1/4	1/2 x	3/16	1/2 x	1/2	1/2 x	3/8	0.0025	7/8	19/32	5/8	1/2	0.0025	0.0025
2 5/16 — 2 3/4	5/8 x	5/16	5/8 x	7/32	5/8 x	5/8	5/8 x	7/16	0.0025	1 1/16	23/32	3/4	5/8	0.0025	0.0025
2 7/8 — 3 1/4	3/4 x	3/8	3/4 x	1/4	3/4 x	3/4	3/4 x	1/2	0.0025	1 1/4	7/8	7/8	3/4	0.0025	0.0025
3 3/8 — 3 3/4	7/8 x	7/16	7/8 x	5/16	7/8 x	7/8	7/8 x	5/8	0.003	1 1/2	1	1 1/16	7/8	0.003	0.003
3 7/8 — 4 1/2	1 x	1/2	1 x	3/8	1 x	1	1 x	3/4	0.003	1 3/4	1 3/16	1 1/4	1	0.003	0.003
4 3/4 — 5 1/2	1 1/4 x	5/8	1 1/4 x	7/16	1 1/4 x	1 1/4	1 1/4 x	7/8	0.003	2	1 7/16	1 1/2	1 1/4	0.003	0.003
5 3/4 — 7 3/8	1 1/2 x	3/4	1 1/2 x	1/2	1 1/2 x	1 1/2	1 1/2 x	1	0.003	2 1/2	1 3/4	1 3/4	1 1/2	0.003	0.003
7 1/2 — 9 7/8	1 3/4 x	7/8	. .	. .	1 3/4 x	1 3/4	. .	. .	0.004	3	2	. .	. .	0.004	0.004
10 — 12 1/2	2 x	1	. .	. .	2 x	2	. .	. .	0.004	3 1/2	2 3/8	. .	. .	0.004	0.004

Standard Keyway Tolerances: Straight Keyway — Width (W) $^{+.005}_{-.000}$ Depth (T/2) $^{+.010}_{-.000}$

Taper Keyway — Width (W) $^{+.005}_{-.000}$ Depth (T/2) $^{+.000}_{-.010}$

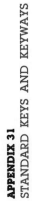

APPENDIX 32
WOODRUFF KEYS

USA STANDARD

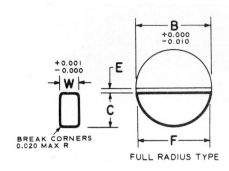

FULL RADIUS TYPE

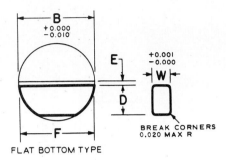

FLAT BOTTOM TYPE

Woodruff Keys

Key No.	Nominal Key Size W × B	Actual Length F +0.000−0.010	Height of Key				Distance Below Center E
			C		D		
			Max	Min	Max	Min	
202	1/16 × 1/4	0.248	0.109	0.104	0.109	0.104	1/64
202.5	1/16 × 5/16	0.311	0.140	0.135	0.140	0.135	1/64
302.5	3/32 × 5/16	0.311	0.140	0.135	0.140	0.135	1/64
203	1/16 × 3/8	0.374	0.172	0.167	0.172	0.167	1/64
303	3/32 × 3/8	0.374	0.172	0.167	0.172	0.167	1/64
403	1/8 × 3/8	0.374	0.172	0.167	0.172	0.167	1/64
204	1/16 × 1/2	0.491	0.203	0.198	0.194	0.188	3/64
304	3/32 × 1/2	0.491	0.203	0.198	0.194	0.188	3/64
404	1/8 × 1/2	0.491	0.203	0.198	0.194	0.188	3/64
305	3/32 × 5/8	0.612	0.250	0.245	0.240	0.234	1/16
405	1/8 × 5/8	0.612	0.250	0.245	0.240	0.234	1/16
505	5/32 × 5/8	0.612	0.250	0.245	0.240	0.234	1/16
605	3/16 × 5/8	0.612	0.250	0.245	0.240	0.234	1/16
406	1/8 × 3/4	0.740	0.313	0.308	0.303	0.297	1/16
506	5/32 × 3/4	0.740	0.313	0.308	0.303	0.297	1/16
606	3/16 × 3/4	0.740	0.313	0.308	0.303	0.297	1/16
806	1/4 × 3/4	0.740	0.313	0.308	0.303	0.297	1/16
507	5/32 × 7/8	0.866	0.375	0.370	0.365	0.359	1/16
607	3/16 × 7/8	0.866	0.375	0.370	0.365	0.359	1/16
707	7/32 × 7/8	0.866	0.375	0.370	0.365	0.359	1/16
807	1/4 × 7/8	0.866	0.375	0.370	0.365	0.359	1/16
608	3/16 × 1	0.992	0.438	0.433	0.428	0.422	1/16
708	7/32 × 1	0.992	0.438	0.433	0.428	0.422	1/16
808	1/4 × 1	0.992	0.438	0.433	0.428	0.422	1/16
1008	5/16 × 1	0.992	0.438	0.433	0.428	0.422	1/16
1208	3/8 × 1	0.992	0.438	0.433	0.428	0.422	1/16
609	3/16 × 1 1/8	1.114	0.484	0.479	0.475	0.469	5/64
709	7/32 × 1 1/8	1.114	0.484	0.479	0.475	0.469	5/64
809	1/4 × 1 1/8	1.114	0.484	0.479	0.475	0.469	5/64
1009	5/16 × 1 1/8	1.114	0.484	0.479	0.475	0.469	5/64

(Courtesy of ANSI; B17.2-1967.)

APPENDIX 33
WOODRUFF KEYSEATS

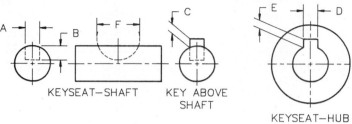

KEYSEAT—SHAFT   KEY ABOVE SHAFT   KEYSEAT—HUB

Keyseat Dimensions

Key Number	Nominal Size Key	Keyseat — Shaft					Key Above Shaft	Keyseat — Hub	
		Width A*		Depth B	Diameter F		Height C	Width D	Depth E
		Min	Max	+0.005 −0.000	Min	Max	+0.005 −0.005	+0.002 −0.000	+0.005 −0.000
202	1/16 × 1/4	0.0615	0.0630	0.0728	0.250	0.268	0.0312	0.0635	0.0372
202.5	1/16 × 5/16	0.0615	0.0630	0.1038	0.312	0.330	0.0312	0.0635	0.0372
302.5	3/32 × 5/16	0.0928	0.0943	0.0882	0.312	0.330	0.0469	0.0948	0.0529
203	1/16 × 3/8	0.0615	0.0630	0.1358	0.375	0.393	0.0312	0.0635	0.0372
303	3/32 × 3/8	0.0928	0.0943	0.1202	0.375	0.393	0.0469	0.0948	0.0529
403	1/8 × 3/8	0.1240	0.1255	0.1045	0.375	0.393	0.0625	0.1260	0.0685
204	1/16 × 1/2	0.0615	0.0630	0.1668	0.500	0.518	0.0312	0.0635	0.0372
304	3/32 × 1/2	0.0928	0.0943	0.1511	0.500	0.518	0.0469	0.0948	0.0529
404	1/8 × 1/2	0.1240	0.1255	0.1355	0.500	0.518	0.0625	0.1260	0.0685
305	3/32 × 5/8	0.0928	0.0943	0.1981	0.625	0.643	0.0469	0.0948	0.0529
405	1/8 × 5/8	0.1240	0.1255	0.1825	0.625	0.643	0.0625	0.1260	0.0685
505	5/32 × 5/8	0.1553	0.1568	0.1669	0.625	0.643	0.0781	0.1573	0.0841
605	3/16 × 5/8	0.1863	0.1880	0.1513	0.625	0.643	0.0937	0.1885	0.0997
406	1/8 × 3/4	0.1240	0.1255	0.2455	0.750	0.768	0.0625	0.1260	0.0685
506	5/32 × 3/4	0.1553	0.1568	0.2299	0.750	0.768	0.0781	0.1573	0.0841
606	3/16 × 3/4	0.1863	0.1880	0.2143	0.750	0.768	0.0937	0.1885	0.0997
806	1/4 × 3/4	0.2487	0.2505	0.1830	0.750	0.768	0.1250	0.2510	0.1310
507	5/32 × 7/8	0.1553	0.1568	0.2919	0.875	0.895	0.0781	0.1573	0.0841
607	3/16 × 7/8	0.1863	0.1880	0.2763	0.875	0.895	0.0937	0.1885	0.0997
707	7/32 × 7/8	0.2175	0.2193	0.2607	0.875	0.895	0.1093	0.2198	0.1153
807	1/4 × 7/8	0.2487	0.2505	0.2450	0.875	0.895	0.1250	0.2510	0.1310
608	3/16 × 1	0.1863	0.1880	0.3393	1.000	1.020	0.0937	0.1885	0.0997
708	7/32 × 1	0.2175	0.2193	0.3237	1.000	1.020	0.1093	0.2198	0.1153
808	1/4 × 1	0.2487	0.2505	0.3080	1.000	1.020	0.1250	0.2510	0.1310
1008	5/16 × 1	0.3111	0.3130	0.2768	1.000	1.020	0.1562	0.3135	0.1622
1208	3/8 × 1	0.3735	0.3755	0.2455	1.000	1.020	0.1875	0.3760	0.1935
609	3/16 × 1 1/8	0.1863	0.1880	0.3853	1.125	1.145	0.0937	0.1885	0.0997
709	7/32 × 1 1/8	0.2175	0.2193	0.3697	1.125	1.145	0.1093	0.2198	0.1153
809	1/4 × 1 1/8	0.2487	0.2505	0.3540	1.125	1.145	0.1250	0.2510	0.1310
1009	5/16 × 1 1/8	0.3111	0.3130	0.3228	1.125	1.145	0.1562	0.3135	0.1622

(Courtesy of ANSI; B17.2–1967.)

APPENDIX 34
TAPER PINS

Number	7/0	6/0	5/0	4/0	3/0	2/0	0	1	2	3	4	5	6	7	8	9	10
Size (large end)	0.0625	0.0780	0.0940	0.1090	0.1250	0.1410	0.1560	0.1720	0.1930	0.2190	0.2500	0.2890	0.3410	0.4090	0.4920	0.5910	0.7060
Length, L																	
0.375	X	X															
0.500	X	X	X	X	X	X	X										
0.625	X	X	X	X	X	X	X	X	X	X							
0.750		X	X	X	X	X	X	X	X	X							
0.875					X	X	X	X	X	X	X	X	X				
1.000			X	X		X	X	X	X	X	X	X	X				
1.250						X	X	X	X	X	X	X	X				
1.500							X	X	X	X	X	X	X				
1.750								X	X	X	X	X	X	X			
2.000									X	X	X	X	X	X			
2.250										X	X	X	X	X	X		
2.500										X	X	X	X	X	X		
2.750										X	X	X	X	X	X		
3.000													X	X	X		
3.250													X	X	X		
3.500													X	X	X	X	X
3.750													X	X	X	X	X
4.000														X	X	X	X
4.250														X	X	X	X
4.500															X	X	X
4.750															X	X	X
5.000																X	X
5.250																X	X
5.500																X	X
5.750																X	X
6.000																X	X

All dimensions are given in inches.

Standard reamers are available for pins given above the line.

Pins Nos. 11 (size 0.8600), 12 (size 1.032), 13 (size 1.241), and 14 (1.523) are special sizes—hence their lengths are special.

To find small diameter of pin, multiply the length by 0.2083 and subtract the result from the large diameter.

(Courtesy of ANSI; B5.20–1958.)

APPENDIX 35
PLAIN WASHERS

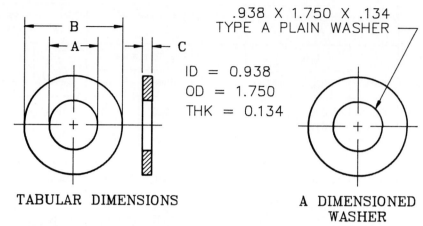

.938 X 1.750 X .134
TYPE A PLAIN WASHER

ID = 0.938
OD = 1.750
THK = 0.134

TABULAR DIMENSIONS

A DIMENSIONED WASHER

Dimensions of Preferred Sizes of Type A Plain Washers[a]

When specifying washers on drawings or in notes, give the inside diameter, outside diameter, and the thickness. *Example:* 0.938 × 1.750 × 0.134 TYPE A PLAIN WASHER.

Nominal Washer Size[b]			Inside Diameter A			Outside Diameter B			Thickness C		
			Basic	Plus	Minus	Basic	Plus	Minus	Basic	Max	Min
				Tolerance			Tolerance				
—	—		0.078	0.000	0.005	0.188	0.000	0.005	0.020	0.025	0.016
—	—		0.094	0.000	0.005	0.250	0.000	0.005	0.020	0.025	0.016
—	—		0.125	0.008	0.005	0.312	0.008	0.005	0.032	0.040	0.025
No. 6	0.138		0.156	0.008	0.005	0.375	0.015	0.005	0.049	0.065	0.036
No. 8	0.164		0.188	0.008	0.005	0.438	0.015	0.005	0.049	0.065	0.036
No. 10	0.190		0.219	0.008	0.005	0.500	0.015	0.005	0.049	0.065	0.036
$\frac{3}{16}$	0.188		0.250	0.015	0.005	0.562	0.015	0.005	0.049	0.065	0.036
No. 12	0.216		0.250	0.015	0.005	0.562	0.015	0.005	0.065	0.080	0.051
$\frac{1}{4}$	0.250	N	0.281	0.015	0.005	0.625	0.015	0.005	0.065	0.080	0.051
$\frac{1}{4}$	0.250	W	0.312	0.015	0.005	0.734[c]	0.015	0.007	0.065	0.080	0.051
$\frac{5}{16}$	0.312	N	0.344	0.015	0.005	0.688	0.015	0.007	0.065	0.080	0.051
$\frac{5}{16}$	0.312	W	0.375	0.015	0.005	0.875	0.030	0.007	0.083	0.104	0.064
$\frac{3}{8}$	0.375	N	0.406	0.015	0.005	0.812	0.015	0.007	0.065	0.080	0.051
$\frac{3}{8}$	0.375	W	0.438	0.015	0.005	1.000	0.030	0.007	0.083	0.104	0.064
$\frac{7}{16}$	0.438	N	0.469	0.015	0.005	0.922	0.015	0.007	0.065	0.080	0.051
$\frac{7}{16}$	0.438	W	0.500	0.015	0.005	1.250	0.030	0.007	0.083	0.104	0.064
$\frac{1}{2}$	0.500	N	0.531	0.015	0.005	1.062	0.030	0.007	0.095	0.121	0.074
$\frac{1}{2}$	0.500	W	0.562	0.015	0.005	1.375	0.030	0.007	0.109	0.132	0.086

[a] Preferred sizes are for the most part from series previously designated "Standard Plate" and "SAE." Where common sizes existed in the two series, the SAE size is designated "N" (narrow) and the Standard Plate "W" (wide). These sizes as well as all other sizes of Type A Plain Washers are to be ordered by ID, OD, and thickness dimensions.

[b] Nominal washer sizes are intended for use with comparable nominal screw or bolt sizes.

[c] The 0.734 in., 1.156 in., and 1.469 in. outside diameters avoid washers which could be used in coin-operated devices.

Cont.

APPENDIX 35
PLAIN WASHERS (Cont.)

Nominal Washer Size[b]			Inside Diameter A			Outside Diameter B			Thickness C		
				Tolerance			Tolerance				
			Basic	Plus	Minus	Basic	Plus	Minus	Basic	Max	Min
$\frac{9}{16}$	0.562	N	0.594	0.015	0.005	1.156[c]	0.030	0.007	0.095	0.121	0.074
$\frac{9}{16}$	0.562	W	0.625	0.015	0.005	1.469[c]	0.030	0.007	0.109	0.132	0.086
$\frac{5}{8}$	0.625	N	0.656	0.030	0.007	1.312	0.030	0.007	0.095	0.121	0.074
$\frac{5}{8}$	0.625	W	0.688	0.030	0.007	1.750	0.030	0.007	0.134	0.160	0.108
$\frac{3}{4}$	0.750	N	0.812	0.030	0.007	1.469	0.030	0.007	0.134	0.160	0.108
$\frac{3}{4}$	0.750	W	0.812	0.030	0.007	2.000	0.030	0.007	0.148	0.177	0.122
$\frac{7}{8}$	0.875	N	0.938	0.030	0.007	1.750	0.030	0.007	0.134	0.160	0.108
$\frac{7}{8}$	0.875	W	0.938	0.030	0.007	2.250	0.030	0.007	0.165	0.192	0.136
1	1.000	N	1.062	0.030	0.007	2.000	0.030	0.007	0.134	0.160	0.108
1	1.000	W	1.062	0.030	0.007	2.500	0.030	0.007	0.165	0.192	0.136
$1\frac{1}{8}$	1.125	N	1.250	0.030	0.007	2.250	0.030	0.007	0.134	0.160	0.108
$1\frac{1}{8}$	1.125	W	1.250	0.030	0.007	2.750	0.030	0.007	0.165	0.192	0.136
$1\frac{1}{4}$	1.250	N	1.375	0.030	0.007	2.500	0.030	0.007	0.165	0.192	0.136
$1\frac{1}{4}$	1.250	W	1.375	0.030	0.007	3.000	0.030	0.007	0.165	0.192	0.136
$1\frac{3}{8}$	1.375	N	1.500	0.030	0.007	2.750	0.030	0.007	0.165	0.192	0.136
$1\frac{3}{8}$	1.375	W	1.500	0.045	0.010	3.250	0.045	0.010	0.180	0.213	0.153
$1\frac{1}{2}$	1.500	N	1.625	0.030	0.007	3.000	0.030	0.007	0.165	0.192	0.136
$1\frac{1}{2}$	1.500	W	1.625	0.045	0.010	3.500	0.045	0.010	0.180	0.213	0.153
$1\frac{5}{8}$	1.625		1.750	0.045	0.010	3.750	0.045	0.010	0.180	0.213	0.153
$1\frac{3}{4}$	1.750		1.875	0.045	0.010	4.000	0.045	0.010	0.180	0.213	0.153
$1\frac{7}{8}$	1.875		2.000	0.045	0.010	4.250	0.045	0.010	0.180	0.213	0.153
2	2.000		2.125	0.045	0.010	4.500	0.045	0.010	0.180	0.213	0.153
$2\frac{1}{4}$	2.250		2.375	0.045	0.010	4.750	0.045	0.010	0.220	0.248	0.193
$2\frac{1}{2}$	2.500		2.625	0.045	0.010	5.000	0.045	0.010	0.238	0.280	0.210
$2\frac{3}{4}$	2.750		2.875	0.065	0.010	5.250	0.065	0.010	0.259	0.310	0.228
3	3.000		3.125	0.065	0.010	5.500	0.065	0.010	0.284	0.327	0.249

(Courtesy of ANSI; B27.2–1965.)

APPENDIX 36
LOCK WASHERS (ANSI B27.1)

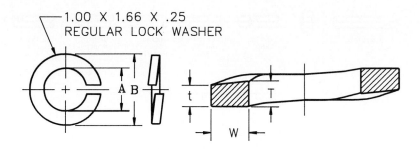

1.00 X 1.66 X .25
REGULAR LOCK WASHER

Dimensions of Regular* Helical Spring Lock Washers

Nominal Washer Size		Inside Diameter A		Outside Diameter B	Washer Section	
		Min	Max	Max**	Width W Min	Thickness $\frac{T+t}{2}$ Min
No. 2	0.086	0.088	0.094	0.172	0.035	0.020
No. 3	0.099	0.101	0.107	0.195	0.040	0.025
No. 4	0.112	0.115	0.121	0.209	0.040	0.025
No. 5	0.125	0.128	0.134	0.236	0.047	0.031
No. 6	0.138	0.141	0.148	0.250	0.047	0.031
No. 8	0.164	0.168	0.175	0.293	0.055	0.040
No. 10	0.190	0.194	0.202	0.334	0.062	0.047
No. 12	0.216	0.221	0.229	0.377	0.070	0.056
¼	0.250	0.255	0.263	0.489	0.109	0.062
⁵⁄₁₆	0.312	0.318	0.328	0.586	0.125	0.078
³⁄₈	0.375	0.382	0.393	0.683	0.141	0.094
⁷⁄₁₆	0.438	0.446	0.459	0.779	0.156	0.109
½	0.500	0.509	0.523	0.873	0.171	0.125
⁹⁄₁₆	0.562	0.572	0.587	0.971	0.188	0.141
⅝	0.625	0.636	0.653	1.079	0.203	0.156
¹¹⁄₁₆	0.688	0.700	0.718	1.176	0.219	0.172
¾	0.750	0.763	0.783	1.271	0.234	0.188
¹³⁄₁₆	0.812	0.826	0.847	1.367	0.250	0.203
⅞	0.875	0.890	0.912	1.464	0.266	0.219
¹⁵⁄₁₆	0.938	0.954	0.978	1.560	0.281	0.234
1	1.000	1.017	1.042	1.661	0.297	0.250
1¹⁄₁₆	1.062	1.080	1.107	1.756	0.312	0.266
1⅛	1.125	1.144	1.172	1.853	0.328	0.281
1³⁄₁₆	1.188	1.208	1.237	1.950	0.344	0.297
1¼	1.250	1.271	1.302	2.045	0.359	0.312
1⁵⁄₁₆	1.312	1.334	1.366	2.141	0.375	0.328
1⅜	1.375	1.398	1.432	2.239	0.391	0.344
1⁷⁄₁₆	1.438	1.462	1.497	2.334	0.406	0.359
1½	1.500	1.525	1.561	2.430	0.422	0.375

*Formerly designated Medium Helical Spring Lock Washers.

**The maximum outside diameters specified allow for the commercial tolerances on cold drawn wire.

APPENDIX 37
COTTER PINS

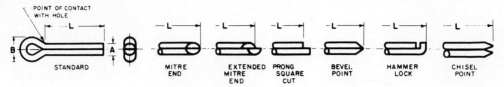

Nominal Diameter	Diameter A		Outside Eye Diameter B Min	Hole Sizes Recommended
	Max	Min		
0.031	0.032	0.028	$\frac{1}{16}$	$\frac{3}{64}$
0.047	0.048	0.044	$\frac{3}{32}$	$\frac{1}{16}$
0.062	0.060	0.056	$\frac{1}{8}$	$\frac{5}{64}$
0.078	0.076	0.072	$\frac{5}{32}$	$\frac{3}{32}$
0.094	0.090	0.086	$\frac{3}{16}$	$\frac{7}{64}$
0.109	0.104	0.100	$\frac{7}{32}$	$\frac{1}{8}$
0.125	0.120	0.116	$\frac{1}{4}$	$\frac{9}{64}$
0.141	0.134	0.130	$\frac{9}{32}$	$\frac{5}{32}$
0.156	0.150	0.146	$\frac{5}{16}$	$\frac{11}{64}$
0.188	0.176	0.172	$\frac{3}{8}$	$\frac{13}{64}$
0.219	0.207	0.202	$\frac{7}{16}$	$\frac{15}{64}$
0.250	0.225	0.220	$\frac{1}{2}$	$\frac{17}{64}$
0.312	0.280	0.275	$\frac{5}{8}$	$\frac{5}{16}$
0.375	0.335	0.329	$\frac{3}{4}$	$\frac{3}{8}$
0.438	0.406	0.400	$\frac{7}{8}$	$\frac{7}{16}$
0.500	0.473	0.467	1	$\frac{1}{2}$
0.625	0.598	0.590	$1\frac{1}{4}$	$\frac{5}{8}$
0.750	0.723	0.715	$1\frac{1}{2}$	$\frac{3}{4}$

All dimensions are given in inches.

A certain amount of leeway is permitted in the design of the head; however, the outside diameters given should be adhered to.

Prongs are to be parallel; ends shall not be open.

Points may be blunt, bevel, extended prong, mitre, etc., and purchaser may specify type required.

Lengths shall be measured as shown on the above illustration (L-dimension).

Cotter pins shall be free from burrs or any defects that will affect their serviceability.

(Courtesy of ANSI; B5.20–1958.)

APPENDIX 38
AMERICAN STANDARD RUNNING AND SLIDING FITS

Limits are in thousandths of an inch.

Limits for hole and shaft are applied algebraically to the basic size to obtain the limits of size for the parts.

Data in bold face are in accordance with ABC agreements.

Symbols H5, g5, etc., are Hole and Shaft designations used in ABC System.

Nominal Size Range Inches Over — To	Class RC 1 Limits of Clearance	Class RC 1 Standard Limits Hole H5	Class RC 1 Standard Limits Shaft g4	Class RC 2 Limits of Clearance	Class RC 2 Standard Limits Hole H6	Class RC 2 Standard Limits Shaft g5	Class RC 3 Limits of Clearance	Class RC 3 Standard Limits Hole H7	Class RC 3 Standard Limits Shaft f6	Class RC 4 Limits of Clearance	Class RC 4 Standard Limits Hole H8	Class RC 4 Standard Limits Shaft f7
0 — 0.12	0.1 0.45	+ 0.2 0	− 0.1 − 0.25	0.1 0.55	+ 0.25 0	− 0.1 − 0.3	0.3 0.95	+ 0.4 0	− 0.3 − 0.55	0.3 1.3	+ 0.6 0	− 0.3 − 0.7
0.12 — 0.24	0.15 0.5	+ 0.2 0	− 0.15 − 0.3	0.15 0.65	+ 0.3 0	− 0.15 − 0.35	0.4 1.12	+ 0.5 0	− 0.4 − 0.7	0.4 1.6	+ 0.7 0	− 0.4 − 0.9
0.24 — 0.40	0.2 0.6	0.25 0	− 0.2 − 0.35	0.2 0.85	+ 0.4 0	− 0.2 − 0.45	0.5 1.5	+ 0.6 0	− 0.5 − 0.9	0.5 2.0	+ 0.9 0	− 0.5 − 1.1
0.40 — 0.71	0.25 0.75	+ 0.3 0	− 0.25 − 0.45	0.25 0.95	+ 0.4 0	− 0.25 − 0.55	0.6 1.7	+ 0.7 0	− 0.6 − 1.0	0.6 2.3	+ 1.0 0	− 0.6 − 1.3
0.71 — 1.19	0.3 0.95	+ 0.4 0	− 0.3 − 0.55	0.3 1.2	+ 0.5 0	− 0.3 − 0.7	0.8 2.1	+ 0.8 0	− 0.8 − 1.3	0.8 2.8	+ 1.2 0	− 0.8 − 1.6
1.19 — 1.97	0.4 1.1	+ 0.4 0	− 0.4 − 0.7	0.4 1.4	+ 0.6 0	− 0.4 − 0.8	1.0 2.6	+ 1.0 0	− 1.0 − 1.6	1.0 3.6	+ 1.6 0	− 1.0 − 2.0
1.97 — 3.15	0.4 1.2	+ 0.5 0	− 0.4 − 0.7	0.4 1.6	+ 0.7 0	− 0.4 − 0.9	1.2 3.1	+ 1.2 0	− 1.2 − 1.9	1.2 4.2	+ 1.8 0	− 1.2 − 2.4
3.15 — 4.73	0.5 1.5	+ 0.6 0	− 0.5 − 0.9	0.5 2.0	+ 0.9 0	− 0.5 − 1.1	1.4 3.7	+ 1.4 0	− 1.4 − 2.3	1.4 5.0	+ 2.2 0	− 1.4 − 2.8
4.73 — 7.09	0.6 1.8	+ 0.7 0	− 0.6 − 1.1	0.6 2.3	+ 1.0 0	− 0.6 − 1.3	1.6 4.2	+ 1.6 0	− 1.6 − 2.6	1.6 5.7	+ 2.5 0	− 1.6 − 3.2
7.09 — 9.85	0.6 2.0	+ 0.8 0	− 0.6 − 1.2	0.6 2.6	+ 1.2 0	− 0.6 − 1.4	2.0 5.0	+ 1.8 0	− 2.0 − 3.2	2.0 6.6	+ 2.8 0	− 2.0 − 3.8
9.85 — 12.41	0.8 2.3	+ 0.9 0	− 0.8 − 1.4	0.8 2.9	+ 1.2 0	− 0.8 − 1.7	2.5 5.7	+ 2.0 0	− 2.5 − 3.7	2.5 7.5	+ 3.0 0	− 2.5 − 4.5
12.41 — 15.75	1.0 2.7	+ 1.0 0	− 1.0 − 1.7	1.0 3.4	+ 1.4 0	− 1.0 − 2.0	3.0 6.6	+ 0	− 3.0 − 4.4	3.0 8.7	+ 3.5 0	− 3.0 − 5.2
15.75 — 19.69	1.2 3.0	+ 1.0 0	− 1.2 − 2.0	1.2 3.8	+ 1.6 0	− 1.2 − 2.2	4.0 8.1	+ 1.6 0	− 4.0 − 5.6	4.0 10.5	+ 4.0 0	− 4.0 − 6.5
19.69 — 30.09	1.6 3.7	+ 1.2 0	− 1.6 − 2.5	1.6 4.8	+ 2.0 0	− 1.6 − 2.8	5.0 10.0	+ 3.0 0	− 5.0 − 7.0	5.0 13.0	+ 5.0 0	− 5.0 − 8.0
30.09 — 41.49	2.0 4.6	+ 1.6 0	− 2.0 − 3.0	2.0 6.1	+ 2.5 0	− 2.0 − 3.6	6.0 12.5	+ 4.0 0	− 6.0 − 8.5	6.0 16.0	+ 6.0 0	− 6.0 − 10.0
41.49 — 56.19	2.5 5.7	+ 2.0 0	− 2.5 − 3.7	2.5 7.5	+ 3.0 0	− 2.5 − 4.5	8.0 16.0	+ 5.0 0	− 8.0 − 11.0	8.0 21.0	+ 8.0 0	− 8.0 − 13.0
56.19 — 76.39	3.0 7.1	+ 2.5 0	− 3.0 − 4.6	3.0 9.5	+ 4.0 0	− 3.0 − 5.5	10.0 20.0	+ 6.0 0	− 10.0 − 14.0	10.0 26.0	+10.0 0	− 10.0 − 16.0
76.39 — 100.9	4.0 9.0	+ 3.0 0	− 4.0 − 6.0	4.0 12.0	+ 5.0 0	− 4.0 − 7.0	12.0 25.0	+ 8.0 0	− 12.0 − 17.0	12.0 32.0	+12.0 0	− 12.0 − 20.0
100.9 — 131.9	5.0 11.5	+ 4.0 0	− 5.0 − 7.5	5.0 15.0	+ 6.0 0	− 5.0 − 9.0	16.0 32.0	+10.0 0	− 16.0 − 22.0	16.0 36.0	+16.0 0	− 16.0 − 26.0
131.9 — 171.9	6.0 14.0	+ 5.0 0	− 6.0 − 9.0	6.0 19.0	+ 8.0 0	− 6.0 − 11.0	18.0 38.0	+ 8.0 0	− 18.0 − 26.0	18.0 50.0	+20.0 0	− 18.0 − 30.0
171.9 — 200	8.0 18.0	+ 6.0 0	− 8.0 − 12.0	8.0 22.0	+10.0 0	− 8.0 − 12.0	22.0 48.0	+16.0 0	− 22.0 − 32.0	22.0 63.0	+25.0 0	− 22.0 − 38.0

(Courtesy of USASI; B4.1–1955.)

Cont.

AMERICAN STANDARD RUNNING AND SLIDING FITS (Cont.)

Class RC 5 Limits of Clearance	RC 5 Hole H8	RC 5 Shaft e7	Class RC 6 Limits of Clearance	RC 6 Hole H9	RC 6 Shaft e8	Class RC 7 Limits of Clearance	RC 7 Hole H9	RC 7 Shaft d8	Class RC 8 Limits of Clearance	RC 8 Hole H10	RC 8 Shaft c9	Class RC 9 Limits of Clearance	RC 9 Hole H11	RC 9 Shaft	Nominal Size Range Inches Over	To
0.6 / 1.6	+0.6 / −0	−0.6 / −1.0	0.6 / 2.2	+1.0 / −0	−0.6 / −1.2	1.0 / 2.6	+1.0 / 0	−1.0 / −1.6	2.5 / 5.1	+1.6 / 0	−2.5 / −3.5	4.0 / 8.1	+2.5 / 0	−4.0 / −5.6	0	0.12
0.8 / 2.0	+0.7 / −0	−0.8 / −1.3	0.8 / 2.7	+1.2 / −0	−0.8 / −1.5	1.2 / 3.1	+1.2 / 0	−1.2 / −1.9	2.8 / 5.8	+1.8 / 0	−2.8 / −4.0	4.5 / 9.0	+3.0 / 0	−4.5 / −6.0	0.12	0.24
1.0 / 2.5	+0.9 / −0	−1.0 / −1.6	1.0 / 3.3	+1.4 / −0	−1.0 / −1.9	1.6 / 3.9	+1.4 / 0	−1.6 / −2.5	3.0 / 6.6	+2.2 / 0	−3.0 / −4.4	5.0 / 10.7	+3.5 / 0	−5.0 / −7.2	0.24	0.40
1.2 / 2.9	+1.0 / −0	−1.2 / −1.9	1.2 / 3.8	+1.6 / −0	−1.2 / −2.2	2.0 / 4.6	+1.6 / 0	−2.0 / −3.0	3.5 / 7.9	+2.8 / 0	−3.5 / −5.1	6.0 / 12.8	+4.0 / −0	−6.0 / −8.8	0.40	0.71
1.6 / 3.6	+1.2 / −0	−1.6 / −2.4	1.6 / 4.8	+2.0 / −0	−1.6 / −2.8	2.5 / 5.7	+2.0 / 0	−2.5 / −3.7	4.5 / 10.0	+3.5 / 0	−4.5 / −6.5	7.0 / 15.5	+5.0 / 0	−7.0 / −10.5	0.71	1.19
2.0 / 4.6	+1.6 / −0	−2.0 / −3.0	2.0 / 6.1	+2.5 / −0	−2.0 / −3.6	3.0 / 7.1	+2.5 / 0	−3.0 / −4.6	5.0 / 11.5	+4.0 / 0	−5.0 / −7.5	8.0 / 18.0	+6.0 / 0	−8.0 / −12.0	1.19	1.97
2.5 / 5.5	+1.8 / −0	−2.5 / −3.7	2.5 / 7.3	+3.0 / −0	−2.5 / −4.3	4.0 / 8.8	+3.0 / 0	−4.0 / −5.8	6.0 / 13.5	+4.5 / 0	−6.0 / −9.0	9.0 / 20.5	+7.0 / 0	−9.0 / −13.5	1.97	3.15
3.0 / 6.6	+2.2 / −0	−3.0 / −4.4	3.0 / 8.7	+3.5 / −0	−3.0 / −5.2	5.0 / 10.7	+3.5 / 0	−5.0 / −7.2	7.0 / 15.5	+5.0 / 0	−7.0 / −10.5	10.0 / 24.0	+9.0 / 0	−10.0 / −15.0	3.15	4.73
3.5 / 7.6	+2.5 / −0	−3.5 / −5.1	3.5 / 10.0	+4.0 / −0	−3.5 / −6.0	6.0 / 12.5	+4.0 / 0	−6.0 / −8.5	8.0 / 18.0	+6.0 / 0	−8.0 / −12.0	12.0 / 28.0	+10.0 / 0	−12.0 / −18.0	4.73	7.09
4.0 / 8.6	+2.8 / −0	−4.0 / −5.8	4.0 / 11.3	+4.5 / 0	−4.0 / −6.8	7.0 / 14.3	+4.5 / 0	−7.0 / −9.8	10.0 / 21.5	+7.0 / 0	−10.0 / −14.5	15.0 / 34.0	+12.0 / 0	−15.0 / −22.0	7.09	9.85
5.0 / 10.0	+3.0 / 0	−5.0 / −7.0	5.0 / 13.0	+5.0 / 0	−5.0 / −8.0	8.0 / 16.0	+5.0 / 0	−8.0 / −11.0	12.0 / 25.0	+8.0 / 0	−12.0 / −17.0	18.0 / 38.0	+12.0 / 0	−18.0 / −26.0	9.85	12.41
6.0 / 11.7	+3.5 / 0	−6.0 / −8.2	6.0 / 15.5	+6.0 / 0	−6.0 / −9.5	10.0 / 19.5	+6.0 / 0	−10.0 / −13.5	14.0 / 29.0	+9.0 / 0	−14.0 / −20.0	22.0 / 45.0	+14.0 / 0	−22.0 / −31.0	12.41	15.75
8.0 / 14.5	+4.0 / 0	−8.0 / −10.5	8.0 / 18.0	+6.0 / 0	−8.0 / −12.0	12.0 / 22.0	+6.0 / 0	−12.0 / −16.0	16.0 / 32.0	+10.0 / 0	−16.0 / −22.0	25.0 / 51.0	+16.0 / 0	−25.0 / −35.0	15.75	19.69
10.0 / 18.0	+5.0 / 0	−10.0 / −13.0	10.0 / 23.0	+8.0 / 0	−10.0 / −15.0	16.0 / 29.0	+8.0 / 0	−16.0 / −21.0	20.0 / 40.0	+12.0 / 0	−20.0 / −28.0	30.0 / 62.0	+20.0 / 0	−30.0 / −42.0	19.69	30.09
12.0 / 22.0	+6.0 / 0	−12.0 / −16.0	12.0 / 28.0	+10.0 / 0	−12.0 / −18.0	20.0 / 36.0	+10.0 / 0	−20.0 / −26.0	25.0 / 51.0	+16.0 / 0	−25.0 / −35.0	40.0 / 81.0	+25.0 / 0	−40.0 / −56.0	30.09	41.49
16.0 / 29.0	+8.0 / 0	−16.0 / −21.0	16.0 / 36.0	+12.0 / 0	−16.0 / −24.0	25.0 / 45.0	+12.0 / 0	−25.0 / −33.0	30.0 / 62.0	+20.0 / 0	−30.0 / −42.0	50.0 / 100	+30.0 / 0	−50.0 / −70.0	41.49	56.19
20.0 / 36.0	+10.0 / 0	−20.0 / −26.0	20.0 / 46.0	+16.0 / 0	−20.0 / −30.0	30.0 / 56.0	+16.0 / 0	−30.0 / −40.0	40.0 / 81.0	+25.0 / 0	−40.0 / −56.0	60.0 / 125	+40.0 / 0	−60.0 / −85.0	56.19	76.39
25.0 / 45.0	+12.0 / 0	−25.0 / −33.0	25.0 / 57.0	+20.0 / 0	−25.0 / −37.0	40.0 / 72.0	+20.0 / 0	−40.0 / −52.0	50.0 / 100	+30.0 / 0	−50.0 / −70.0	80.0 / 160	+50.0 / 0	−80.0 / −110	76.39	100.9
30.0 / 56.0	+16.0 / 0	−30.0 / −40.0	30.0 / 71.0	+25.0 / 0	−30.0 / −46.0	50.0 / 91.0	+25.0 / 0	−50.0 / −66.0	60.0 / 125	+40.0 / 0	−60.0 / −85.0	100 / 200	+60.0 / 0	−100 / −140	100.9	131.9
35.0 / 67.0	+20.0 / 0	−35.0 / −47.0	35.0 / 85.0	+30.0 / 0	−35.0 / −55.0	60.0 / 110	+30.0 / 0	−60.0 / −80.0	80.0 / 160	+50.0 / 0	−80.0 / −110	130 / 260	+80.0 / 0	−130 / −180	131.9	171.9
45.0 / 86.0	+25.0 / 0	−45.0 / −61.0	45.0 / 110	+40.0 / 0	−45.0 / −70.0	80.0 / 145.0	+40.0 / 0	−80.0 / −105.0	100 / 200	+60.0 / 0	−100 / −140	150 / 310	+100 / 0	−150 / −210	171.9	200

(Courtesy of ANSI; B4.1–1955.)

APPENDIX 39

AMERICAN STANDARD CLEARANCE LOCATIONAL FITS

Limits are in thousandths of an inch.
Limits for hole and shaft are applied algebraically to the basic size to obtain the limits of size for the parts.
Data in bold face are in accordance with ABC agreements.
Symbols H9,f8, etc., are Hole and Shaft designations used in ABC System.

Nominal Size Range Inches (Over – To)	Class LC 1 Limits of Clearance	Class LC 1 Hole H6	Class LC 1 Shaft h5	Class LC 2 Limits of Clearance	Class LC 2 Hole H7	Class LC 2 Shaft h6	Class LC 3 Limits of Clearance	Class LC 3 Hole H8	Class LC 3 Shaft h7	Class LC 4 Limits of Clearance	Class LC 4 Hole H10	Class LC 4 Shaft h9	Class LC 5 Limits of Clearance	Class LC 5 Hole H7	Class LC 5 Shaft g6
0 – 0.12	0 / 0.45	+0.25 / −0	+0 / −0.2	0 / 0.65	+0.4 / −0	+0 / −0.25	0 / 1	+0.6 / −0	+0 / −0.4	0 / 2.6	+1.6 / −0	+0 / −1.0	0.1 / 0.75	+0.4 / −0	−0.1 / −0.35
0.12 – 0.24	0 / 0.5	+0.3 / −0	+0 / −0.2	0 / 0.8	+0.5 / −0	+0 / −0.3	0 / 1.2	+0.7 / −0	+0 / −0.5	0 / 3.0	+1.8 / −0	+0 / −1.2	0.15 / 0.95	+0.5 / −0	−0.15 / −0.45
0.24 – 0.40	0 / 0.65	+0.4 / −0	+0 / −0.25	0 / 1.0	+0.6 / −0	+0 / −0.4	0 / 1.5	+0.9 / −0	+0 / −0.6	0 / 3.6	+2.2 / −0	+0 / −1.4	0.2 / 1.2	+0.6 / −0	−0.2 / −0.6
0.40 – 0.71	0 / 0.7	+0.4 / −0	+0 / −0.3	0 / 1.1	+0.7 / −0	+0 / −0.4	0 / 1.7	+1.0 / −0	+0 / −0.7	0 / 4.4	+2.8 / −0	+0 / −1.6	0.25 / 1.35	+0.7 / −0	−0.25 / −0.65
0.71 – 1.19	0 / 0.9	+0.5 / −0	+0 / −0.4	0 / 1.3	+0.8 / −0	+0 / −0.5	0 / 2	+1.2 / −0	+0 / −0.8	0 / 5.5	+3.5 / −0	+0 / −2.0	0.3 / 1.6	+0.8 / −0	−0.3 / −0.8
1.19 – 1.97	0 / 1.0	+0.6 / −0	+0 / −0.4	0 / 1.6	+1.0 / −0	+0 / −0.6	0 / 2.6	+1.6 / −0	+0 / −1	0 / 6.5	+4.0 / −0	+0 / −2.5	0.4 / 2.0	+1.0 / −0	−0.4 / −1.0
1.97 – 3.15	0 / 1.2	+0.7 / −0	+0 / −0.5	0 / 1.9	+1.2 / −0	+0 / −0.7	0 / 3	+1.8 / −0	+0 / −1.2	0 / 7.5	+4.5 / −0	+0 / −3	0.4 / 2.3	+1.2 / −0	−0.4 / −1.1
3.15 – 4.73	0 / 1.5	+0.9 / −0	+0 / −0.6	0 / 2.3	+1.4 / −0	+0 / −0.9	0 / 3.6	+2.2 / −0	+0 / −1.4	0 / 8.5	+5.0 / −0	+0 / −3.5	0.5 / 2.8	+1.4 / −0	−0.5 / −1.4
4.73 – 7.09	0 / 1.7	+1.0 / −0	+0 / −0.7	0 / 2.6	+1.6 / −0	+0 / −1.0	0 / 4.1	+2.5 / −0	+0 / −1.6	0 / 10	+6.0 / −0	+0 / −4	0.6 / 3.2	+1.6 / −0	−0.6 / −1.6
7.09 – 9.85	0 / 2.0	+1.2 / −0	+0 / −0.8	0 / 3.0	+1.8 / −0	+0 / −1.2	0 / 4.6	+2.8 / −0	+0 / −1.8	0 / 11.5	+7.0 / −0	+0 / −4.5	0.6 / 3.6	+1.8 / −0	−0.6 / −1.8
9.85 – 12.41	0 / 2.1	+1.2 / −0	+0 / −0.9	0 / 3.2	+2.0 / −0	+0 / −1.2	0 / 5	+3.0 / −0	+0 / −2.0	0 / 13	+8.0 / −0	+0 / −5	0.7 / 3.9	+2.0 / −0	−0.7 / −1.9
12.41 – 15.75	0 / 2.4	+1.4 / −0	+0 / −1.0	0 / 3.6	+2.2 / −0	+0 / −1.4	0 / 5.7	+3.5 / −0	+0 / −2.2	0 / 15	+9.0 / −0	+0 / −6	0.7 / 4.3	+2.2 / −0	−0.7 / −2.1
15.75 – 19.69	0 / 2.6	+1.6 / −0	+0 / −1.0	0 / 4.1	+2.5 / −0	+0 / −1.6	0 / 6.5	+4 / −0	+0 / −2.5	0 / 16	+10.0 / −0	+0 / −6	0.8 / 4.9	+2.5 / −0	−0.8 / −2.4
19.69 – 30.09	0 / 3.2	+2.0 / −0	+0 / −1.2	0 / 5.0	+3 / −0	+0 / −2	0 / 8	+5 / −0	+0 / −3	0 / 20	+12.0 / −0	+0 / −8	0.9 / 5.9	+3.0 / −0	−0.9 / −2.9
30.09 – 41.49	0 / 4.1	+2.5 / −0	+0 / −1.6	0 / 6.5	+4 / −0	+0 / −2.5	0 / 10	+6 / −0	+0 / −4	0 / 26	+16.0 / −0	+0 / −10	1.0 / 7.5	+4.0 / −0	−1.0 / −3.5
41.49 – 56.19	0 / 5.0	+3.0 / −0	+0 / −2.0	0 / 8.0	+5 / −0	+0 / −3	0 / 13	+8 / −0	+0 / −5	0 / 32	+20.0 / −0	+0 / −12	1.2 / 9.2	+5.0 / −0	−1.2 / −4.2
56.19 – 76.39	0 / 6.5	+4.0 / −0	+0 / −2.5	0 / 10	+6 / −0	+0 / −4	0 / 16	+10 / −0	+0 / −6	0 / 41	+25.0 / −0	+0 / −16	1.2 / 11.2	+6.0 / −0	−1.2 / −5.2
76.39 – 100.9	0 / 8.0	+5.0 / −0	+0 / −3.0	0 / 13	+8 / −0	+0 / −5	0 / 20	+12 / −0	+0 / −8	0 / 50	+30.0 / −0	+0 / −20	1.4 / 14.4	+8.0 / −0	−1.4 / −6.4
100.9 – 131.9	0 / 10.0	+6.0 / −0	+0 / −4.0	0 / 16	+10 / −0	+0 / −6	0 / 26	+16 / −0	+0 / −10	0 / 65	+40.0 / −0	+0 / −25	1.6 / 17.6	+10.0 / −0	−1.6 / −7.6
131.9 – 171.9	0 / 13.0	+8.0 / −0	+0 / −5.0	0 / 20	+12 / −0	+0 / −8	0 / 32	+20 / −0	+0 / −12	0 / 8	+50.0 / −0	+0 / −30	1.8 / 21.8	+12.0 / −0	−1.8 / −9.8
171.9 – 200	0 / 16.0	+10.0 / −0	+0 / −6.0	0 / 26	+16 / −0	+0 / −10	0 / 41	+25 / −0	+0 / −16	0 / 100	+60.0 / −0	+0 / −40	1.8 / 27.8	+16.0 / −0	−1.8 / −11.8

(Courtesy of USASI; B4.1–1955.)

Cont.

APPENDIX 39
AMERICAN STANDARD CLEARANCE LOCATIONAL FITS (Cont.)

Class LC 6			Class LC 7			Class LC 8			Class LC 9			Class LC 10			Class LC 11			Nominal Size Range Inches	
Limits of Clearance	Hole H9	Shaft f8	Limits of Clearance	Hole H10	Shaft e9	Limits of Clearance	Hole H10	Shaft d9	Limits of Clearance	Hole H11	Shaft c10	Limits of Clearance	Hole H12	Shaft	Limits of Clearance	Hole H13	Shaft	Over	To
0.3 / 1.9	+1.0 / 0	−0.3 / −0.9	0.6 / 3.2	+1.6 / 0	−0.6 / −1.6	1.0 / 3.6	+1.6 / −0	−1.0 / −2.0	2.5 / 6.6	+2.5 / −0	−2.5 / −4.1	4 / 12	+4 / −0	−4 / −8	5 / 17	+6 / −0	−5 / −11	0	0.12
0.4 / 2.3	+1.2 / 0	−0.4 / −1.1	0.8 / 3.8	+1.8 / 0	−0.8 / −2.0	1.2 / 4.2	+1.8 / −0	−1.2 / −2.4	2.8 / 7.6	+3.0 / −0	−2.8 / −4.6	4.5 / 14.5	+5 / −0	−4.5 / −9.5	6 / 20	+7 / −0	−6 / −13	0.12	0.24
0.5 / 2.8	+1.4 / 0	−0.5 / −1.4	1.0 / 4.6	+2.2 / 0	−1.0 / −2.4	1.6 / 5.2	+2.2 / −0	−1.6 / −3.0	3.0 / 8.7	+3.5 / −0	−3.0 / −5.2	5 / 17	+6 / −0	−5 / −11	7 / 25	+9 / −0	−7 / −16	0.24	0.40
0.6 / 3.2	+1.6 / 0	−0.6 / −1.6	1.2 / 5.6	+2.8 / 0	−1.2 / −2.8	2.0 / 6.4	+2.8 / −0	−2.0 / −3.6	3.5 / 10.3	+4.0 / −0	−3.5 / −6.3	6 / 20	+7 / −0	−6 / −13	8 / 28	+10 / −0	−8 / −18	0.40	0.71
0.8 / 4.0	+2.0 / 0	−0.8 / −2.0	1.6 / 7.1	+3.5 / 0	−1.6 / −3.6	2.5 / 8.0	+3.5 / −0	−2.5 / −4.5	4.5 / 13.0	+5.0 / −0	−4.5 / −8.0	7 / 23	+8 / −0	−7 / −15	10 / 34	+12 / −0	−10 / −22	0.71	1.19
1.0 / 5.1	+2.5 / 0	−1.0 / −2.6	2.0 / 8.5	+4.0 / 0	−2.0 / −4.5	3.0 / 9.5	+4.0 / −0	−3.0 / −5.5	5 / 15	+6 / −0	−5 / −9	8 / 28	+10 / −0	−8 / −18	12 / 44	+16 / −0	−12 / −28	1.19	1.97
1.2 / 6.0	+3.0 / 0	−1.2 / −3.0	2.5 / 10.0	+4.5 / 0	−2.5 / −5.5	4.0 / 11.5	+4.5 / −0	−4.0 / −7.0	6 / 17.5	+7 / −0	−6 / −10.5	10 / 34	+12 / −0	−10 / −22	14 / 50	+18 / −0	−14 / −32	1.97	3.15
1.4 / 7.1	+3.5 / 0	−1.4 / −3.6	3.0 / 11.5	+5.0 / 0	−3.0 / −6.5	5.0 / 13.5	+5.0 / −0	−5.0 / −8.5	7 / 21	+9 / −0	−7 / −12	11 / 39	+14 / −0	−11 / −25	16 / 60	+22 / −0	−16 / −38	3.15	4.73
1.6 / 8.1	+4.0 / 0	−1.6 / −4.1	3.5 / 13.5	+6.0 / 0	−3.5 / −7.5	6 / 16	+6 / −0	−6 / −10	8 / 24	+10 / −0	−8 / −14	12 / 44	+16 / −0	−12 / −28	18 / 68	+25 / −0	−18 / −43	4.73	7.09
2.0 / 9.3	+4.5 / 0	−2.0 / −4.8	4.0 / 15.5	+7.0 / 0	−4.0 / −8.5	7 / 18.5	+7 / −0	−7 / −11.5	10 / 29	+12 / −0	−10 / −17	16 / 52	+18 / −0	−16 / −34	22 / 78	+28 / −0	−22 / −50	7.09	9.85
2.2 / 10.2	+5.0 / 0	−2.2 / −5.2	4.5 / 17.5	+8.0 / 0	−4.5 / −9.5	7 / 20	+8 / −0	−7 / −12	12 / 32	+12 / −0	−12 / −20	20 / 60	+20 / −0	−20 / −40	28 / 88	+30 / −0	−28 / −58	9.85	12.41
2.5 / 12.0	+6.0 / 0	−2.5 / −6.0	5.0 / 20.0	+9.0 / 0	−5 / −11	8 / 23	+9 / −0	−8 / −14	14 / 37	+14 / −0	−14 / −23	22 / 66	+22 / −0	−22 / −44	30 / 100	+35 / −0	−30 / −65	12.41	15.75
2.8 / 12.8	+6.0 / 0	−2.8 / −6.8	5.0 / 21.0	+10.0 / 0	−5 / −11	9 / 25	+10 / −0	−9 / −15	16 / 42	+16 / −0	−16 / −26	25 / 75	+25 / −0	−25 / −50	35 / 115	+40 / −0	−35 / −75	15.75	19.69
3.0 / 16.0	+8.0 / 0	−3.0 / −8.0	6.0 / 26.0	+12.0 / −0	−6 / −14	10 / 30	+12 / −0	−10 / −18	18 / 50	+20 / −0	−18 / −30	28 / 88	+30 / −0	−28 / −58	40 / 140	+50 / −0	−40 / −90	19.69	30.09
3.5 / 19.5	+10.0 / 0	−3.5 / −9.5	7.0 / 33.0	+16.0 / −0	−7 / −17	12 / 38	+16 / −0	−12 / −22	20 / 61	+25 / −0	−20 / −36	30 / 110	+40 / −0	−30 / −70	45 / 165	+60 / −0	−45 / −105	30.09	41.49
4.0 / 24.0	+12.0 / 0	−4.0 / −12.0	8.0 / 40.0	+20.0 / −0	−8 / −20	14 / 46	+20 / −0	−14 / −26	25 / 75	+30 / −0	−25 / −45	40 / 140	+50 / −0	−40 / −90	60 / 220	+80 / −0	−60 / −140	41.49	56.19
4.5 / 30.5	+16.0 / 0	−4.5 / −14.5	9.0 / 50.0	+25.0 / −0	−9 / −25	16 / 57	+25 / −0	−16 / −32	30 / 95	+40 / −0	−30 / −55	50 / 170	+60 / −0	−50 / 110	70 / 270	+100 / −0	−70 / −170	56.19	76.39
5.0 / 37.0	+20.0 / 0	−5 / −17	10.0 / 60.0	+30.0 / −0	−10 / −30	18 / 68	+30 / −0	−18 / −38	35 / 115	+50 / −0	−35 / −65	50 / 210	+80 / −0	−50 / −130	80 / 330	+125 / −0	−80 / −205	76.39	100.9
6.0 / 47.0	+25.0 / 0	−6 / −22	12.0 / 67.0	+40.0 / −0	−12 / −27	20 / 85	+40 / −0	−20 / −45	40 / 140	+60 / −0	−40 / −80	60 / 260	+100 / −0	−60 / −160	90 / 410	+160 / −0	−90 / −250	100.9	131.9
7.0 / 57.0	+30.0 / 0	−7 / −27	14.0 / 94.0	+50.0 / −0	−14 / −44	25 / 105	+50 / −0	−25 / −55	50 / 180	+80 / −0	−50 / −100	80 / 330	+125 / −0	−80 / −205	100 / 500	+200 / −0	−100 / −300	131.9	171.9
7.0 / 72.0	+40.0 / 0	−7 / −32	14.0 / 114.0	+60.0 / −0	−14 / −54	25 / 125	+60 / −0	−25 / −65	50 / 210	+100 / −0	−50 / −110	90 / 410	+160 / −0	−90 / −250	125 / 625	+250 / −0	−125 / −375	171.9	200

(Courtesy of ANSI; B4.1–1955.)

APPENDIX 40
AMERICAN STANDARD TRANSITION LOCATIONAL FITS

Limits are in thousandths of an inch.

Limits for hole and shaft are applied algebraically to the basic size to obtain the limits of size for the mating parts.

Data in bold face are in accordance with ABC agreements.

"Fit" represents the maximum interference (minus values) and the maximum clearance (plus values).

Symbols H7, js6, etc., are Hole and Shaft designations used in ABC System.

Nominal Size Range Inches Over	To	Class LT 1 Fit	Hole H7	Shaft js6	Class LT 2 Fit	Hole H8	Shaft js7	Class LT 3 Fit	Hole H7	Shaft k6	Class LT 4 Fit	Hole H8	Shaft k7	Class LT 5 Fit	Hole H7	Shaft n6	Class LT 6 Fit	Hole H7	Shaft n7
0 –	0.12	−0.10 / +0.50	+0.4 / −0	+0.10 / −0.10	−0.2 / +0.8	+0.6 / −0	+0.2 / −0.2							−0.5 / +0.15	+0.4 / −0	+0.5 / +0.25	−0.65 / +0.15	+0.4 / −0	+0.65 / +0.25
0.12 –	0.24	−0.15 / +0.65	+0.5 / −0	+0.15 / −0.15	−0.25 / +0.95	+0.7 / −0	+0.25 / −0.25							−0.6 / +0.2	+0.5 / −0	+0.6 / +0.3	−0.8 / +0.2	+0.5 / −0	+0.8 / +0.3
0.24 –	0.40	−0.2 / +0.8	+0.6 / −0	+0.2 / −0.2	−0.3 / +1.2	+0.9 / −0	+0.3 / −0.3	−0.5 / +0.5	+0.6 / −0	+0.5 / +0.1	−0.7 / +0.8	+0.9 / −0	+0.7 / +0.1	−0.8 / +0.2	+0.6 / −0	+0.8 / +0.4	−1.0 / +0.2	+0.6 / −0	+1.0 / +0.4
0.40 –	0.71	−0.2 / +0.9	+0.7 / −0	+0.2 / −0.2	−0.35 / +1.35	+1.0 / −0	+0.35 / −0.35	−0.5 / +0.6	+0.7 / −0	+0.5 / +0.1	−0.8 / +0.9	+1.0 / −0	+0.8 / +0.1	−0.9 / +0.2	+0.7 / −0	+0.9 / +0.5	−1.2 / +0.2	+0.7 / −0	+1.2 / +0.5
0.71 –	1.19	−0.25 / +1.05	+0.8 / −0	+0.25 / −0.25	−0.4 / +1.6	+1.2 / −0	+0.4 / −0.4	−0.6 / +0.7	+0.8 / −0	+0.6 / +0.1	−0.9 / +1.1	+1.2 / −0	+0.9 / +0.1	−1.1 / +0.2	+0.8 / −0	+1.1 / +0.6	−1.4 / +0.2	+0.8 / −0	+1.4 / +0.6
1.19 –	1.97	−0.3 / +1.3	+1.0 / −0	+0.3 / −0.3	−0.5 / +2.1	+1.6 / −0	+0.5 / −0.5	−0.7 / +0.9	+1.0 / −0	+0.7 / +0.1	−1.1 / +1.5	+1.6 / −0	+1.1 / +0.1	−1.3 / +0.3	+1.0 / −0	+1.3 / +0.7	−1.7 / +0.3	+1.0 / −0	+1.7 / +0.7
1.97 –	3.15	−0.3 / +1.5	+1.2 / −0	+0.3 / −0.3	−0.6 / +2.4	+1.8 / −0	+0.6 / −0.6	−0.8 / +1.1	+1.2 / −0	+0.8 / +0.1	−1.3 / +1.7	+1.8 / −0	+1.3 / +0.1	−1.5 / +0.4	+1.2 / −0	+1.5 / +0.8	−2.0 / +0.4	+1.2 / −0	+2.0 / +0.8
3.15 –	4.73	−0.4 / +1.8	+1.4 / −0	+0.4 / −0.4	−0.7 / +2.9	+2.2 / −0	+0.7 / −0.7	−1.0 / +1.3	+1.4 / −0	+1.0 / +0.1	−1.5 / +2.1	+2.2 / −0	+1.5 / +0.1	−1.9 / +0.4	+1.4 / −0	+1.9 / +1.0	−2.4 / +0.4	+1.4 / −0	+2.4 / +1.0
4.73 –	7.09	−0.5 / +2.1	+1.6 / −0	+0.5 / −0.5	−0.8 / +3.3	+2.5 / −0	+0.8 / −0.8	−1.1 / +1.5	+1.6 / −0	+1.1 / +0.1	−1.7 / +2.4	+2.5 / −0	+1.7 / +0.1	−2.2 / +0.4	+1.6 / −0	+2.2 / +1.2	−2.8 / +0.4	+1.6 / −0	+2.8 / +1.2
7.09 –	9.85	−0.6 / +2.4	+1.8 / −0	+0.6 / −0.6	−0.9 / +3.7	+2.8 / −0	+0.9 / −0.9	−1.4 / +1.6	+1.8 / −0	+1.4 / +0.2	−2.0 / +2.6	+2.8 / −0	+2.0 / +0.2	−2.6 / +0.4	+1.8 / −0	+2.6 / +1.4	−3.2 / +0.4	+1.8 / −0	+3.2 / +1.4
9.85 –	12.41	−0.6 / +2.6	+2.0 / −0	+0.6 / −0.6	−1.0 / +4.0	+3.0 / −0	+1.0 / −1.0	−1.4 / +1.8	+2.0 / −0	+1.4 / +0.2	−2.2 / +2.8	+3.0 / −0	+2.2 / +0.2	−2.6 / +0.6	+2.0 / −0	+2.6 / +1.4	−3.4 / +0.6	+2.0 / −0	+3.4 / +1.4
12.41 –	15.75	−0.7 / +2.9	+2.2 / −0	+0.7 / −0.7	−1.0 / +4.5	+3.5 / −0	+1.0 / −1.0	−1.6 / +2.0	+2.2 / −0	+1.6 / +0.2	−2.4 / +3.3	+3.5 / −0	+2.4 / +0.2	−3.0 / +0.6	+2.2 / −0	+3.0 / +1.6	−3.8 / +0.6	+2.2 / −0	+3.8 / +1.6
15.75 –	19.69	−0.8 / +3.3	+2.5 / −0	+0.8 / −0.8	−1.2 / +5.2	+4.0 / −0	+1.2 / −1.2	−1.8 / +2.3	+2.5 / −0	+1.8 / +0.2	−2.7 / +3.8	+4.0 / −0	+2.7 / +0.2	−3.4 / +0.7	+2.5 / −0	+3.4 / +1.8	−4.3 / +0.7	+2.5 / −0	+4.3 / +1.8

(Courtesy of ANSI; B4.1–1955.)

APPENDIX 41
AMERICAN STANDARD INTERFERENCE LOCATIONAL FITS

Limits are in thousandths of an inch.
Limits for hole and shaft are applied algebraically to the
basic size to obtain the limits of size for the parts.
Data in bold face are in accordance with ABC agreements.
Symbols H7, p6, etc., are Hole and Shaft designations
used in ABC System.

Nominal Size Range Inches Over	To	Class LN 1 Limits of Interference	Class LN 1 Standard Limits Hole H6	Class LN 1 Standard Limits Shaft n5	Class LN 2 Limits of Interference	Class LN 2 Standard Limits Hole H7	Class LN 2 Standard Limits Shaft p6	Class LN 3 Limits of Interference	Class LN 3 Standard Limits Hole H7	Class LN 3 Standard Limits Shaft r6
0	0.12	0 0.45	+ 0.25 − 0	+0.45 +0.25	0 0.65	+ 0.4 − 0	+ 0.65 + 0.4	0.1 0.75	+ 0.4 − 0	+ 0.75 + 0.5
0.12	0.24	0 0.5	+ 0.3 − 0	+0.5 +0.3	0 0.8	+ 0.5 − 0	+ 0.8 + 0.5	0.1 0.9	+ 0.5 0	+ 0.9 + 0.6
0.24	0.40	0 0.65	+ 0.4 − 0	+0.65 +0.4	0 1.0	+ 0.6 − 0	+ 1.0 + 0.6	0.2 1.2	+ 0.6 − 0	+ 1.2 + 0.8
0.40	0.71	0 0.8	+ 0.4 − 0	+0.8 +0.4	0 1.1	+ 0.7 − 0	+ 1.1 + 0.7	0.3 1.4	+ 0.7 − 0	+ 1.4 + 1.0
0.71	1.19	0 1.0	+ 0.5 − 0	+1.0 +0.5	0 1.3	+ 0.8 − 0	+ 1.3 + 0.8	0.4 1.7	+ 0.8 − 0	+ 1.7 + 1.2
1.19	1.97	0 1.1	+ 0.6 − 0	+1.1 +0.6	0 1.6	+ 1.0 − 0	+ 1.6 + 1.0	0.4 2.0	+ 1.0 − 0	+ 2.0 + 1.4
1.97	3.15	0.1 1.3	+ 0.7 − 0	+1.3 +0.7	0.2 2.1	+ 1.2 − 0	+ 2.1 + 1.4	0.4 2.3	+ 1.2 − 0	+ 2.3 + 1.6
3.15	4.73	0.1 1.6	+ 0.9 − 0	+1.6 +1.0	0.2 2.5	+ 1.4 − 0	+ 2.5 + 1.6	0.6 2.9	+ 1.4 − 0	+ 2.9 + 2.0
4.73	7.09	0.2 1.9	+ 1.0 − 0	+1.9 +1.2	0.2 2.8	+ 1.6 − 0	+ 2.8 + 1.8	0.9 3.5	+ 1.6 − 0	+ 3.5 + 2.5
7.09	9.85	0.2 2.2	+ 1.2 − 0	+2.2 +1.4	0.2 3.2	+ 1.8 − 0	+ 3.2 + 2.0	1.2 4.2	+ 1.8 − 0	+ 4.2 + 3.0
9.85	12.41	0.2 2.3	+ 1.2 − 0	+2.3 +1.4	0.2 3.4	+ 2.0 − 0	+ 3.4 + 2.2	1.5 4.7	+ 2.0 − 0	+ 4.7 + 3.5
12.41	15.75	0.2 2.6	+ 1.4 − 0	+2.6 +1.6	0.3 3.9	+ 2.2 − 0	+ 3.9 + 2.5	2.3 5.9	+ 2.2 − 0	+ 5.9 + 4.5
15.75	19.69	0.2 2.8	+ 1.6 − 0	+2.8 +1.8	0.3 4.4	+ 2.5 − 0	+ 4.4 + 2.8	2.5 6.6	+ 2.5 − 0	+ 6.6 + 5.0
19.69	30.09		+ 2.0 − 0		0.5 5.5	+ 3 − 0	+ 5.5 + 3.5	4 9	+ 3 − 0	+ 9 + 7
30.09	41.49		+ 2.5 − 0		0.5 7.0	+ 4 − 0	+ 7.0 + 4.5	5 11.5	+ 4 − 0	+11.5 + 9
41.49	56.19		+ 3.0 − 0		1 9	+ 5 − 0	+ 9 + 6	7 15	+ 5 − 0	+15 +12
56.19	76.39		+ 4.0 − 0		1 11	+ 6 − 0	+11 + 7	10 20	+ 6 − 0	+20 +16
76.39	100.9		+ 5.0 − 0		1 14	+ 8 − 0	+14 + 9	12 25	+ 8 − 0	+25 +20
100.9	131.9		+ 6.0 − 0		2 18	+10 − 0	+18 +12	15 31	+10 − 0	+31 +25
131.9	171.9		+ 8.0 − 0		4 24	+12 − 0	+24 +16	18 38	+12 − 0	+38 +30
171.9	200		+10.0 − 0		4 30	+16 − 0	+30 +20	24 50	+16 − 0	+50 +40

(Courtesy of ANSI; B4.1–1955.)

APPENDIX 42
AMERICAN STANDARD FORCE AND SHRINK FITS

Limits are in thousandths of an inch.
Limits for hole and shaft are applied algebraically to the basic size to obtain the limits of size for the parts.
Data in bold face are in accordance with ABC agreements.
Symbols H7, s6, etc., are Hole and Shaft designations used in ABC System.

Nominal Size Range Inches Over — To	Class FN 1 Limits of Interference	Standard Limits Hole H6	Shaft	Class FN 2 Limits of Interference	Standard Limits Hole H7	Shaft s6	Class FN 3 Limits of Interference	Standard Limits Hole H7	Shaft t6	Class FN 4 Limits of Interference	Standard Limits Hole H7	Shaft u6	Class FN 5 Limits of Interference	Standard Limits Hole H8	Shaft x7
0 — 0.12	0.05 0.5	+0.25 −0	+0.5 +0.3	0.2 0.85	+0.4 −0	+0.85 +0.6				0.3 0.95	+0.4 −0	+0.95 +0.7	0.3 1.3	+0.6 −0	+1.3 +0.9
0.12 — 0.24	0.1 0.6	+0.3 −0	+0.6 +0.4	0.2 1.0	+0.5 −0	+1.0 +0.7				0.4 1.2	+0.5 −0	+1.2 +0.9	0.5 1.7	+0.7 −0	+1.7 +1.2
0.24 — 0.40	0.1 0.75	+0.4 −0	+0.75 +0.5	0.4 1.4	+0.6 −0	+1.4 +1.0				0.6 1.6	+0.6 −0	+1.6 +1.2	0.5 2.0	+0.9 −0	+2.0 +1.4
0.40 — 0.56	0.1 0.8	−0.4 −0	+0.8 +0.5	0.5 1.6	+0.7 −0	+1.6 +1.2				0.7 1.8	+0.7 −0	+1.8 +1.4	0.6 2.3	+1.0 −0	+2.3 +1.6
0.56 — 0.71	0:2 0.9	+0.4 −0	+0.9 +0.6	0.5 1.6	+0.7 −0	+1.6 +1.2				0.7 1.8	+0.7 −0	+1.8 +1.4	0.8 2.5	+1.0 −0	+2.5 +1.8
0.71 — 0.95	0.2 1.1	+0.5 −0	+1.1 +0.7	0.6 1.9	+0.8 −0	+1.9 +1.4				0.8 2.1	+0.8 −0	+2.1 +1.6	1.0 3.0	+1.2 −0	+3.0 +2.2
0.95 — 1.19	0.3 1.2	+0.5 −0	+1.2 +0.8	0.6 1.9	+0.8 −0	+1.9 +1.4	0.8 2.1	+0.8 −0	+2.1 +1.6	1.0 2.3	+0·8 −0	+2.3 +1.8	1.3 3.3	+1.2 −0	+3.3 +2.5
1.19 — 1.58	0.3 1.3	+0.6 −0	+1.3 +0.9	0.8 2.4	+1.0 −0	+2.4 +1.8	1.0 2.6	+1.0 −0	+2.6 +2.0	1.5 3.1	+1.0 −0	+3.1 +2.5	1.4 4.0	+1.6 −0	+4.0 +3.0
1.58 — 1.97	0.4 1.4	+0.6 −0	+1.4 +1.0	0.8 2.4	+1.0 −0	+2.4 +1.8	1.2 2.8	+1.0 −0	+2.8 +2.2	1.8 3.4	+1.0 −0	+3.4 +2.8	2.4 5.0	+1.6 −0	+5.0 +4.0
1.97 — 2.56	0.6 1.8	+0.7 −0	+1.8 +1.3	0.8 2.7	+1.2 −0	+2.7 +2.0	1.3 3.2	+1.2 −0	+3.2 +2.5	2.3 4.2	+1.2 −0	+4.2 +3.5	3.2 6.2	+1.8 −0	+6.2 +5.0
2.56 — 3.15	0.7 1.9	+0.7 −0	+1.9 +1.4	1.0 2.9	+1.2 −0	+2.9 +2.2	1.8 3.7	+1.2 −0	+3.7 +3.0	2.8 4.7	+1.2 −0	+4.7 +4.0	4.2 7.2	+1.8 −0	+7.2 +6.0
3.15 — 3.94	0.9 2.4	+0.9 −0	+2.4 +1.8	1.4 3.7	+1.4 −0	+3.7 +2.8	2.1 4.4	+1.4 −0	+4.4 +3.5	3.6 5.9	+1.4 −0	+5.9 +5.0	4.8 8.4	+2.2 −0	+8.4 +7.0
3.94 — 4.73	1.1 2.6	+0.9 −0	+2.6 +2.0	1.6 3.9	+1.4 −0	+3.9 +3.0	2.6 4.9	+1.4 −0	+4.9 +4.0	4.6 6.9	+1.4 −0	+6.9 +6.0	5.8 9.4	+2.2 −0	+9.4 +8.0
4.73 — 5.52	1.2 2.9	+1.0 −0	+2.9 +2.2	1.9 4.5	+1.6 −0	+4.5 +3.5	3.4 6.0	+1.6 −0	+6.0 +5.0	5.4 8.0	+1.6 −0	+8.0 +7.0	7.5 11.6	+2.5 −0	+11.6 +10.0
5.52 — 6.30	1.5 3.2	+1.0 −0	+3.2 +2.5	2.4 5.0	+1.6 −0	+5.0 +4.0	3.4 6.0	+1.6 −0	+6.0 +5.0	5.4 8.0	+1.6 −0	+8.0 +7.0	9.5 13.6	+2.5 −0	+13.6 +12.0
6.30 — 7.09	1.8 3.5	+1.0 −0	+3.5 +2.8	2.9 5.5	+1.6 −0	+5.5 +4.5	4.4 7.0	+1.6 −0	+7.0 +6.0	6.4 9.0	+1.6 −0	+9.0 +8.0	9.5 13.6	+2.5 −0	+13.6 +12.0
7.09 — 7.88	1.8 3.8	+1.2 −0	+3.8 +3.0	3.2 6.2	+1.8 −0	+6.2 +5.0	5.2 8.2	+1.8 −0	+8.2 +7.0	7.2 10.2	+1.8 −0	+10.2 +9.0	11.2 15.8	+2.8 −0	+15.8 +14.0
7.88 — 8.86	2.3 4.3	+1.2 −0	+4.3 +3.5	3.2 6.2	+1.8 −0	+6.2 +5.0	5.2 8.2	+1.8 −0	+8.2 +7.0	8.2 11.2	+1.8 −0	+11.2 +10.0	13.2 17.8	+2.8 −0	+17.8 +16.0
8.86 — 9.85	2.3 4.3	+1.2 −0	+4.3 +3.5	4.2 7.2	+1.8 −0	+7.2 +6.0	6.2 9.2	+1.8 −0	+9.2 +8.0	10.2 13.2	+1.8 −0	+13.2 +12.0	13.2 17.8	+2.8 −0	+17.8 +16.0
9.85 — 11.03	2.8 4.9	+1.2 −0	+4.9 +4.0	4.0 7.2	+2.0 −0	+7.2 +6.0	7.0 10.2	+2.0 −0	+10.2 +9.0	10.0 13.2	+2.0 −0	+13.2 +12.0	15.0 20.0	+3.0 −0	+20.0 +18.0
11.03 — 12.41	2.8 4.9	+1.2 −0	+4.9 +4.0	5.0 8.2	+2.0 −0	+8.2 +7.0	7.0 10.2	+2.0 −0	+10.2 +9.0	12.0 15.2	+2.0 −0	+15.2 +14.0	17.0 22.0	+3.0 −0	+22.0 +20.0
12.41 — 13.98	3.1 5.5	+1.4 −0	+5.5 +4.5	5.8 9.4	+2.2 −0	+9.4 +8.0	7.8 11.4	+2.2 −0	+11.4 +10.0	13.8 17.4	+2.2 −0	+17.4 +16.0	18.5 24.2	+3.5 +0	+24.2 +22.0
13.98 — 15.75	3.6 6.1	+1.4 −0	+6.1 +5.0	5.8 9.4	+2.2 −0	+9.4 +8.0	9.8 13.4	+2.2 −0	+13.4 +12.0	15.8 19.4	+2.2 −0	+19.4 +18.0	21.5 27.2	+3.5 −0	+27.2 +25.0
15.75 — 17.72	4.4 7.0	+1.6 −0	+7.0 +6.0	6.5 10.6	+2.5 −0	+10.6 +9.0	9.5 13.6	+2.5 −0	+13.6 +12.0	17.5 21.6	+2.5 −0	+21.6 +20.0	24.0 30.5	+4.0 −0	+30.5 +28.0
17.72 — 19.69	4.4 7.0	+1.6 −0	+7.0 +6.0	7.5 11.6	+2.5 −0	+11.6 +10.0	11.5 15.6	+2.5 −0	+15.6 +14.0	19.5 23.6	+2.5 −0	+23.6 +22.0	26.0 32.5	+4.0 −0	+32.5 +30.0

(Courtesy of ANSI; B4.1–1955.)

APPENDIX 43
THE INTERNATIONAL TOLERANCE GRADES (ANSI B4.2)

Dimensions are in mm.

Basic sizes Over	Up to and including	IT01	IT0	IT1	IT2	IT3	IT4	IT5	IT6	IT7	IT8	IT9	IT10	IT11	IT12	IT13	IT14	IT15	IT16
0	3	0.0003	0.0005	0.0008	0.0012	0.002	0.003	0.004	0.006	0.010	0.014	0.025	0.040	0.060	0.100	0.140	0.250	0.400	0.600
3	6	0.0004	0.0006	0.001	0.0015	0.0025	0.004	0.005	0.008	0.012	0.018	0.030	0.048	0.075	0.120	0.180	0.300	0.480	0.750
6	10	0.0004	0.0006	0.001	0.0015	0.0025	0.004	0.006	0.009	0.015	0.022	0.036	0.058	0.090	0.150	0.220	0.360	0.580	0.900
10	18	0.0005	0.0008	0.0012	0.002	0.003	0.005	0.008	0.011	0.018	0.027	0.043	0.070	0.110	0.180	0.270	0.430	0.700	1.100
18	30	0.0006	0.001	0.0015	0.0025	0.004	0.006	0.009	0.013	0.021	0.033	0.052	0.084	0.130	0.210	0.330	0.520	0.840	1.300
30	50	0.0006	0.001	0.0015	0.0025	0.004	0.007	0.011	0.016	0.025	0.039	0.062	0.100	0.160	0.250	0.390	0.620	1.000	1.600
50	80	0.0008	0.0012	0.002	0.003	0.005	0.008	0.013	0.019	0.030	0.046	0.074	0.120	0.190	0.300	0.460	0.740	1.200	1.900
80	120	0.001	0.0015	0.0025	0.004	0.006	0.010	0.015	0.022	0.035	0.054	0.087	0.140	0.220	0.350	0.540	0.870	1.400	2.200
120	180	0.0012	0.002	0.0035	0.005	0.008	0.012	0.018	0.025	0.040	0.063	0.100	0.160	0.250	0.400	0.630	1.000	1.600	2.500
180	250	0.002	0.003	0.0045	0.007	0.010	0.014	0.020	0.029	0.046	0.072	0.115	0.185	0.290	0.460	0.720	1.150	1.850	2.900
250	315	0.0025	0.004	0.006	0.008	0.012	0.016	0.023	0.032	0.052	0.081	0.130	0.210	0.320	0.520	0.810	1.300	2.100	3.200
315	400	0.003	0.005	0.007	0.009	0.013	0.018	0.025	0.036	0.057	0.089	0.140	0.230	0.360	0.570	0.890	1.400	2.300	3.600
400	500	0.004	0.006	0.008	0.010	0.015	0.020	0.027	0.040	0.063	0.097	0.155	0.250	0.400	0.630	0.970	1.550	2.500	4.000
500	630	0.0045	0.006	0.009	0.011	0.016	0.022	0.030	0.044	0.070	0.110	0.175	0.280	0.440	0.700	1.100	1.750	2.800	4.400
630	800	0.005	0.007	0.010	0.013	0.018	0.025	0.035	0.050	0.080	0.125	0.200	0.320	0.500	0.800	1.250	2.000	3.200	5.000
800	1000	0.0055	0.008	0.011	0.015	0.021	0.029	0.040	0.056	0.090	0.140	0.230	0.360	0.560	0.900	1.400	2.300	3.600	5.600
1000	1250	0.0065	0.009	0.013	0.018	0.024	0.034	0.046	0.066	0.105	0.165	0.260	0.420	0.660	1.050	1.650	2.600	4.200	6.600
1250	1600	0.008	0.011	0.015	0.021	0.029	0.040	0.054	0.078	0.125	0.195	0.310	0.500	0.780	1.250	1.950	3.100	5.000	7.800
1600	2000	0.009	0.013	0.018	0.025	0.035	0.048	0.065	0.092	0.150	0.230	0.370	0.600	0.920	1.500	2.300	3.700	6.000	9.200
2000	2500	0.011	0.015	0.022	0.030	0.041	0.057	0.077	0.110	0.175	0.280	0.440	0.700	1.100	1.750	2.800	4.400	7.000	11.000
2500	3150	0.013	0.018	0.026	0.036	0.050	0.069	0.093	0.135	0.210	0.330	0.540	0.860	1.350	2.100	3.300	5.400	8.600	13.500

Tolerance grades[3]

[3] IT Values for tolerance grades larger than IT16 can be calculated by using the following formulas:
IT17 = IT12 × 10; IT18 = IT13 × 10; etc.

APPENDIX 44

PREFERRED HOLE BASIS CLEARANCE FITS—CYLINDRICAL FITS (ANSI B4.2)

AMERICAN NATIONAL STANDARD
PREFERRED METRIC LIMITS AND FITS

ANSI B4.2-1978

Dimensions in mm.

BASIC SIZE		LOOSE RUNNING Hole H11	LOOSE RUNNING Shaft c11	LOOSE RUNNING Fit	FREE RUNNING Hole H9	FREE RUNNING Shaft d9	FREE RUNNING Fit	CLOSE RUNNING Hole H8	CLOSE RUNNING Shaft f7	CLOSE RUNNING Fit	SLIDING Hole H7	SLIDING Shaft g6	SLIDING Fit	LOCATIONAL CLEARANCE Hole H7	LOCATIONAL CLEARANCE Shaft h6	LOCATIONAL CLEARANCE Fit
1	MAX	1.060	0.940	0.180	1.025	0.980	0.070	1.014	0.994	0.030	1.010	0.998	0.018	1.010	1.000	0.016
	MIN	1.000	0.880	0.060	1.000	0.955	0.020	1.000	0.984	0.006	1.000	0.992	0.002	1.000	0.994	0.000
1.2	MAX	1.260	1.140	0.180	1.225	1.180	0.070	1.214	1.194	0.030	1.210	1.198	0.018	1.210	1.200	0.016
	MIN	1.200	1.080	0.060	1.200	1.155	0.020	1.200	1.184	0.006	1.200	1.192	0.002	1.200	1.194	0.000
1.6	MAX	1.660	1.540	0.180	1.625	1.580	0.070	1.614	1.594	0.030	1.610	1.598	0.018	1.610	1.600	0.016
	MIN	1.600	1.480	0.060	1.600	1.555	0.020	1.600	1.584	0.006	1.600	1.592	0.002	1.600	1.594	0.000
2	MAX	2.060	1.940	0.180	2.025	1.980	0.070	2.014	1.994	0.030	2.010	1.998	0.018	2.010	2.000	0.020
	MIN	2.000	1.880	0.060	2.000	1.955	0.020	2.000	1.984	0.006	2.000	1.992	0.002	2.000	1.994	0.000
2.5	MAX	2.560	2.440	0.180	2.525	2.480	0.070	2.514	2.494	0.030	2.510	2.498	0.018	2.510	2.500	0.020
	MIN	2.500	2.380	0.060	2.500	2.455	0.020	2.500	2.484	0.006	2.500	2.492	0.002	2.500	2.494	0.000
3	MAX	3.060	2.940	0.180	3.025	2.980	0.070	3.014	2.994	0.030	3.010	2.998	0.018	3.010	3.000	0.016
	MIN	3.000	2.880	0.060	3.000	2.955	0.020	3.000	2.984	0.006	3.000	2.992	0.002	3.000	2.994	0.000
4	MAX	4.075	3.930	0.220	4.030	3.970	0.090	4.018	3.990	0.040	4.012	3.996	0.024	4.012	4.000	0.020
	MIN	4.000	3.855	0.070	4.000	3.940	0.030	4.000	3.978	0.010	4.000	3.988	0.004	4.000	3.992	0.000
5	MAX	5.075	4.930	0.220	5.030	4.970	0.090	5.018	4.990	0.040	5.012	4.996	0.024	5.012	5.000	0.020
	MIN	5.000	4.855	0.070	5.000	4.940	0.030	5.000	4.978	0.010	5.000	4.988	0.004	5.000	4.992	0.000
6	MAX	6.075	5.930	0.220	6.030	5.970	0.090	6.018	5.990	0.040	6.012	5.996	0.024	6.012	6.000	0.020
	MIN	6.000	5.855	0.070	6.000	5.940	0.030	6.000	5.978	0.010	6.000	5.988	0.004	6.000	5.992	0.000
8	MAX	8.090	7.920	0.260	8.036	7.960	0.112	8.022	7.987	0.050	8.015	7.995	0.029	8.015	8.000	0.024
	MIN	8.000	7.830	0.080	8.000	7.924	0.040	8.000	7.972	0.013	8.000	7.986	0.005	8.000	7.991	0.000
10	MAX	10.090	9.920	0.260	10.036	9.960	0.112	10.022	9.987	0.050	10.015	9.995	0.029	10.015	10.000	0.024
	MIN	10.000	9.830	0.080	10.000	9.924	0.040	10.000	9.972	0.013	10.000	9.986	0.005	10.000	9.991	0.000
12	MAX	12.110	11.905	0.315	12.043	11.950	0.136	12.027	11.984	0.061	12.018	11.994	0.035	12.018	12.000	0.029
	MIN	12.000	11.795	0.095	12.000	11.907	0.050	12.000	11.966	0.016	12.000	11.983	0.006	12.000	11.989	0.000
16	MAX	16.110	15.905	0.315	16.043	15.950	0.136	16.027	15.984	0.061	16.018	15.994	0.035	16.018	16.000	0.029
	MIN	16.000	15.795	0.095	16.000	15.907	0.050	16.000	15.966	0.016	16.000	15.983	0.006	16.000	15.989	0.000
20	MAX	20.130	19.890	0.370	20.052	19.935	0.169	20.033	19.980	0.074	20.021	19.993	0.041	20.021	20.000	0.034
	MIN	20.000	19.760	0.110	20.000	19.883	0.065	20.000	19.959	0.020	20.000	19.980	0.007	20.000	19.987	0.000
25	MAX	25.130	24.890	0.370	25.052	24.935	0.169	25.033	24.980	0.074	25.021	24.993	0.041	25.021	25.000	0.034
	MIN	25.000	24.760	0.110	25.000	24.883	0.065	25.000	24.959	0.020	25.000	24.980	0.007	25.000	24.987	0.000
30	MAX	30.130	29.890	0.370	30.052	29.935	0.169	30.033	29.980	0.074	30.021	29.993	0.041	30.021	30.000	0.034
	MIN	30.000	29.760	0.110	30.000	29.883	0.065	30.000	29.959	0.020	30.000	29.980	0.007	30.000	29.987	0.000

Cont.

APPENDIX 44
PREFERRED HOLE BASIS CLEARANCE FITS—CYLINDRICAL FITS (Cont.)

AMERICAN NATIONAL STANDARD
PREFERRED METRIC LIMITS AND FITS

ANSI B4.2-1978

Dimensions in mm.

BASIC SIZE		LOOSE RUNNING			FREE RUNNING			CLOSE RUNNING			SLIDING			LOCATIONAL CLEARANCE		
		Hole H11	Shaft c11	Fit	Hole H9	Shaft d9	Fit	Hole H8	Shaft f7	Fit	Hole H7	Shaft g6	Fit	Hole H7	Shaft h6	Fit
40	MAX	40.160	39.880	0.440	40.062	39.920	0.204	40.039	39.975	0.089	40.025	39.991	0.050	40.025	40.000	0.041
	MIN	40.000	39.720	0.120	40.000	39.858	0.080	40.000	39.950	0.025	40.000	39.975	0.009	40.000	39.984	0.000
50	MAX	50.160	49.870	0.450	50.062	49.920	0.204	50.039	49.975	0.089	50.025	49.991	0.050	50.025	50.000	0.041
	MIN	50.000	49.710	0.130	50.000	49.858	0.080	50.000	49.950	0.025	50.000	49.975	0.009	50.000	49.984	0.000
60	MAX	60.190	59.860	0.520	60.074	59.900	0.248	60.046	59.970	0.106	60.030	59.990	0.059	60.030	60.000	0.049
	MIN	60.000	59.670	0.140	60.000	59.826	0.100	60.000	59.940	0.030	60.000	59.971	0.010	60.000	59.981	0.000
80	MAX	80.190	79.850	0.530	80.074	79.900	0.248	80.046	79.970	0.106	80.030	79.990	0.059	80.030	80.000	0.049
	MIN	80.000	79.660	0.150	80.000	79.826	0.100	80.000	79.940	0.030	80.000	79.971	0.010	80.000	79.981	0.000
100	MAX	100.220	99.830	0.610	100.087	99.880	0.294	100.054	99.964	0.125	100.035	99.988	0.069	100.035	100.000	0.057
	MIN	100.000	99.610	0.170	100.000	99.793	0.120	100.000	99.929	0.036	100.000	99.966	0.012	100.000	99.978	0.000
120	MAX	120.220	119.820	0.620	120.087	119.880	0.294	120.054	119.964	0.125	120.035	119.988	0.069	120.035	120.000	0.057
	MIN	120.000	119.600	0.180	120.000	119.793	0.120	120.000	119.929	0.036	120.000	119.966	0.012	120.000	119.978	0.000
160	MAX	160.250	159.790	0.710	160.100	159.855	0.345	160.063	159.957	0.146	160.040	159.986	0.079	160.040	160.000	0.065
	MIN	160.000	159.540	0.210	160.000	159.755	0.145	160.000	159.917	0.043	160.000	159.961	0.014	160.000	159.975	0.000
200	MAX	200.290	199.760	0.820	200.115	199.830	0.400	200.072	199.950	0.168	200.046	199.985	0.090	200.046	200.000	0.075
	MIN	200.000	199.470	0.240	200.000	199.715	0.170	200.000	199.904	0.050	200.000	199.956	0.015	200.000	199.971	0.000
250	MAX	250.290	249.720	0.860	250.115	249.830	0.400	250.072	249.950	0.168	250.046	249.985	0.090	250.046	250.000	0.075
	MIN	250.000	249.430	0.280	250.000	249.715	0.170	250.000	249.904	0.050	250.000	249.956	0.015	250.000	249.971	0.000
300	MAX	300.320	299.670	0.970	300.130	299.810	0.450	300.081	299.944	0.189	300.052	299.983	0.101	300.052	300.000	0.084
	MIN	300.000	299.350	0.330	300.000	299.680	0.190	300.000	299.892	0.056	300.000	299.951	0.017	300.000	299.968	0.000
400	MAX	400.360	399.600	1.120	400.140	399.790	0.490	400.089	399.938	0.208	400.057	399.982	0.111	400.057	400.000	0.093
	MIN	400.000	399.240	0.400	400.000	399.650	0.210	400.000	399.881	0.062	400.000	399.946	0.018	400.000	399.964	0.000
500	MAX	500.400	499.520	1.280	500.155	499.770	0.540	500.097	499.932	0.228	500.063	499.980	0.123	500.063	500.000	0.103
	MIN	500.000	499.120	0.480	500.000	499.615	0.230	500.000	499.869	0.068	500.000	499.940	0.020	500.000	499.960	0.000

APPENDIX 45

PREFERRED HOLE BASIS TRANSITION AND INTERFERENCE FITS—CYLINDRICAL FITS (ANSI B4.2)

AMERICAN NATIONAL STANDARD
PREFERRED METRIC LIMITS AND FITS

ANSI B4.2-1978

Dimensions in mm.

BASIC SIZE		LOCATIONAL TRANSN. Hole H7	Shaft k6	Fit	LOCATIONAL TRANSN. Hole H7	Shaft n6	Fit	LOCATIONAL INTERF. Hole H7	Shaft p6	Fit	MEDIUM DRIVE Hole H7	Shaft s6	Fit	FORCE Hole H7	Shaft u6	Fit
1	MAX	1.010	1.006	0.010	1.010	1.010	0.006	1.010	1.012	0.004	1.010	1.020	-0.004	1.010	1.024	-0.008
	MIN	1.000	1.000	-0.006	1.000	1.004	-0.010	1.000	1.006	-0.012	1.000	1.014	-0.020	1.000	1.018	-0.024
1.2	MAX	1.210	1.206	0.010	1.210	1.210	0.006	1.210	1.212	0.004	1.210	1.220	-0.004	1.210	1.224	-0.008
	MIN	1.200	1.200	-0.006	1.200	1.204	-0.010	1.200	1.206	-0.012	1.200	1.214	-0.020	1.200	1.218	-0.024
1.6	MAX	1.610	1.606	0.010	1.610	1.610	0.006	1.610	1.612	0.004	1.610	1.620	-0.004	1.610	1.624	-0.008
	MIN	1.600	1.600	-0.006	1.600	1.604	-0.010	1.600	1.606	-0.012	1.600	1.614	-0.020	1.600	1.618	-0.024
2	MAX	2.010	2.006	0.010	2.010	2.010	0.006	2.010	2.012	0.004	2.010	2.020	-0.004	2.010	2.024	-0.008
	MIN	2.000	2.000	-0.006	2.000	2.004	-0.010	2.000	2.006	-0.012	2.000	2.014	-0.020	2.000	2.018	-0.024
2.5	MAX	2.510	2.506	0.010	2.510	2.510	0.006	2.510	2.512	0.004	2.510	2.520	-0.004	2.510	2.524	-0.008
	MIN	2.500	2.500	-0.006	2.500	2.504	-0.010	2.500	2.506	-0.012	2.500	2.514	-0.020	2.500	2.518	-0.024
3	MAX	3.010	3.006	0.010	3.010	3.010	0.006	3.010	3.012	0.004	3.010	3.020	-0.004	3.010	3.024	-0.008
	MIN	3.000	3.000	-0.006	3.000	3.004	-0.010	3.000	3.006	-0.012	3.000	3.014	-0.020	3.000	3.018	-0.024
4	MAX	4.012	4.009	0.011	4.012	4.016	0.004	4.012	4.020	0.000	4.012	4.027	-0.007	4.012	4.031	-0.011
	MIN	4.000	4.001	-0.009	4.000	4.008	-0.016	4.000	4.012	-0.020	4.000	4.019	-0.027	4.000	4.023	-0.031
5	MAX	5.012	5.009	0.011	5.012	5.016	0.004	5.012	5.020	0.000	5.012	5.027	-0.007	5.012	5.031	-0.011
	MIN	5.000	5.001	-0.009	5.000	5.008	-0.016	5.000	5.012	-0.020	5.000	5.019	-0.027	5.000	5.023	-0.031
6	MAX	6.012	6.009	0.011	6.012	6.016	0.004	6.012	6.020	0.000	6.012	6.027	-0.007	6.012	6.031	-0.011
	MIN	6.000	6.001	-0.009	6.000	6.008	-0.016	6.000	6.012	-0.020	6.000	6.019	-0.027	6.000	6.023	-0.031
8	MAX	8.015	8.010	0.014	8.015	8.019	0.005	8.015	8.024	0.000	8.015	8.032	-0.008	8.015	8.037	-0.013
	MIN	8.000	8.001	-0.010	8.000	8.010	-0.019	8.000	8.015	-0.024	8.000	8.023	-0.032	8.000	8.028	-0.037
10	MAX	10.015	10.010	0.014	10.015	10.019	0.005	10.015	10.024	0.000	10.015	10.032	-0.008	10.015	10.037	-0.013
	MIN	10.000	10.001	-0.010	10.000	10.010	-0.019	10.000	10.015	-0.024	10.000	10.023	-0.032	10.000	10.028	-0.037
12	MAX	12.018	12.012	0.017	12.018	12.023	0.006	12.018	12.029	0.000	12.018	12.039	-0.010	12.018	12.044	-0.015
	MIN	12.000	12.001	-0.012	12.000	12.012	-0.023	12.000	12.018	-0.029	12.000	12.028	-0.039	12.000	12.033	-0.044
16	MAX	16.018	16.012	0.017	16.018	16.023	0.006	16.018	16.029	0.000	16.018	16.039	-0.010	16.018	16.044	-0.015
	MIN	16.000	16.001	-0.012	16.000	16.012	-0.023	16.000	16.018	-0.029	16.000	16.028	-0.039	16.000	16.033	-0.044
20	MAX	20.021	20.015	0.019	20.021	20.028	0.006	20.021	20.035	-0.001	20.021	20.048	-0.014	20.021	20.054	-0.020
	MIN	20.000	20.002	-0.015	20.000	20.015	-0.028	20.000	20.022	-0.035	20.000	20.035	-0.048	20.000	20.041	-0.054
25	MAX	25.021	25.015	0.019	25.021	25.028	0.006	25.021	25.035	-0.001	25.021	25.048	-0.014	25.021	25.061	-0.027
	MIN	25.000	25.002	-0.015	25.000	25.015	-0.028	25.000	25.022	-0.035	25.000	25.035	-0.048	25.000	25.048	-0.061
30	MAX	30.021	30.015	0.019	30.021	30.028	0.006	30.021	30.035	-0.001	30.021	30.048	-0.014	30.021	30.061	-0.027
	MIN	30.000	30.002	-0.015	30.000	30.015	-0.028	30.000	30.022	-0.035	30.000	30.035	-0.048	30.000	30.048	-0.061

Cont.

APPENDIX 45

PREFERRED HOLE BASIS TRANSITION AND INTERFERENCE FITS—CYLINDRICAL FITS (ANSI B4.2) (Cont.)

AMERICAN NATIONAL STANDARD
PREFERRED METRIC LIMITS AND FITS

ANSI B4.2-1978

Dimensions in mm.

BASIC SIZE		LOCATIONAL TRANSN.			LOCATIONAL TRANSN.			LOCATIONAL INTERF.			MEDIUM DRIVE			FORCE		
		Hole H7	Shaft k6	Fit	Hole H7	Shaft n6	Fit	Hole H7	Shaft p6	Fit	Hole H7	Shaft s6	Fit	Hole H7	Shaft u6	Fit
40	MAX	40.025	40.018	0.023	40.025	40.033	0.008	40.025	40.042	-0.001	40.025	40.059	-0.018	40.025	40.076	-0.035
	MIN	40.000	40.002	-0.018	40.000	40.017	-0.033	40.000	40.026	-0.042	40.000	40.043	-0.059	40.000	40.060	-0.076
50	MAX	50.025	50.018	0.023	50.025	50.033	0.008	50.025	50.042	-0.001	50.025	50.059	-0.018	50.025	50.086	-0.045
	MIN	50.000	50.002	-0.018	50.000	50.017	-0.033	50.000	50.026	-0.042	50.000	50.043	-0.059	50.000	50.070	-0.086
60	MAX	60.030	60.021	0.028	60.030	60.039	0.010	60.030	60.051	-0.002	60.030	60.072	-0.023	60.030	60.106	-0.057
	MIN	60.000	60.002	-0.021	60.000	60.020	-0.039	60.000	60.032	-0.051	60.000	60.053	-0.072	60.000	60.087	-0.106
80	MAX	80.030	80.021	0.028	80.030	80.039	0.010	80.030	80.051	-0.002	80.030	80.078	-0.029	80.030	80.121	-0.072
	MIN	80.000	80.002	-0.021	80.000	80.020	-0.039	80.000	80.032	-0.051	80.000	80.059	-0.078	80.000	80.102	-0.121
100	MAX	100.035	100.025	0.032	100.035	100.045	0.012	100.035	100.059	-0.002	100.035	100.093	-0.036	100.035	100.146	-0.089
	MIN	100.000	100.003	-0.025	100.000	100.023	-0.045	100.000	100.037	-0.059	100.000	100.071	-0.093	100.000	100.124	-0.146
120	MAX	120.035	120.025	0.032	120.035	120.045	0.012	120.035	120.059	-0.002	120.035	120.101	-0.044	120.035	120.166	-0.109
	MIN	120.000	120.003	-0.025	120.000	120.023	-0.045	120.000	120.037	-0.059	120.000	120.079	-0.101	120.000	120.144	-0.166
160	MAX	160.040	160.028	0.037	160.040	160.052	0.013	160.040	160.068	-0.003	160.040	160.125	-0.060	160.040	160.215	-0.150
	MIN	160.000	160.003	-0.028	160.000	160.027	-0.052	160.000	160.043	-0.068	160.000	160.100	-0.125	160.000	160.190	-0.215
200	MAX	200.046	200.033	0.042	200.046	200.060	0.015	200.046	200.079	-0.004	200.046	200.151	-0.076	200.046	200.265	-0.190
	MIN	200.000	200.004	-0.033	200.000	200.031	-0.060	200.000	200.050	-0.079	200.000	200.122	-0.151	200.000	200.236	-0.265
250	MAX	250.046	250.033	0.042	250.046	250.060	0.015	250.046	250.079	-0.004	250.046	250.169	-0.094	250.046	250.313	-0.238
	MIN	250.000	250.004	-0.033	250.000	250.031	-0.060	250.000	250.050	-0.079	250.000	250.140	-0.169	250.000	250.284	-0.313
300	MAX	300.052	300.036	0.048	300.052	300.066	0.018	300.052	300.088	-0.004	300.052	300.202	-0.118	300.052	300.382	-0.298
	MIN	300.000	300.004	-0.036	300.000	300.034	-0.066	300.000	300.056	-0.088	300.000	300.170	-0.202	300.000	300.350	-0.382
400	MAX	400.057	400.040	0.053	400.057	400.073	0.020	400.057	400.098	-0.005	400.057	400.244	-0.151	400.057	400.471	-0.378
	MIN	400.000	400.004	-0.040	400.000	400.037	-0.073	400.000	400.062	-0.098	400.000	400.208	-0.244	400.000	400.435	-0.471
500	MAX	500.063	500.045	0.058	500.063	500.080	0.023	500.063	500.108	-0.005	500.063	500.292	-0.189	500.063	500.580	-0.477
	MIN	500.000	500.005	-0.045	500.000	500.040	-0.080	500.000	500.068	-0.108	500.000	500.252	-0.292	500.000	500.540	-0.580

APPENDIX 46

PREFERRED SHAFT BASIS CLEARANCE FITS—CYLINDRICAL FITS (ANSI B4.2)

AMERICAN NATIONAL STANDARD
PREFERRED METRIC LIMITS AND FITS

ANSI B4.2-1978

Dimensions in mm.

BASIC SIZE		LOOSE RUNNING Hole C11	LOOSE RUNNING Shaft h11	LOOSE RUNNING Fit	FREE RUNNING Hole D9	FREE RUNNING Shaft h9	FREE RUNNING Fit	CLOSE RUNNING Hole F8	CLOSE RUNNING Shaft h7	CLOSE RUNNING Fit	SLIDING Hole G7	SLIDING Shaft h6	SLIDING Fit	LOCATIONAL CLEARANCE Hole H7	LOCATIONAL CLEARANCE Shaft h6	LOCATIONAL CLEARANCE Fit
1	MAX	1.120	1.000	0.180	1.045	1.000	0.070	1.020	1.000	0.030	1.012	1.000	0.018	1.010	1.000	0.016
	MIN	1.060	0.940	0.060	1.020	0.975	0.020	1.006	0.990	0.006	1.002	0.994	0.002	1.000	0.994	0.000
1.2	MAX	1.320	1.200	0.180	1.245	1.200	0.070	1.220	1.200	0.030	1.212	1.200	0.018	1.210	1.200	0.016
	MIN	1.260	1.140	0.060	1.220	1.175	0.020	1.206	1.190	0.006	1.202	1.194	0.002	1.200	1.194	0.000
1.6	MAX	1.720	1.600	0.180	1.645	1.600	0.070	1.620	1.600	0.030	1.612	1.600	0.018	1.610	1.600	0.016
	MIN	1.660	1.540	0.060	1.620	1.575	0.020	1.606	1.590	0.006	1.602	1.594	0.002	1.600	1.594	0.000
2	MAX	2.120	2.000	0.180	2.045	2.000	0.070	2.020	2.000	0.030	2.012	2.000	0.018	2.010	2.000	0.016
	MIN	2.060	1.940	0.060	2.020	1.975	0.020	2.006	1.990	0.006	2.002	1.994	0.002	2.000	1.994	0.000
2.5	MAX	2.620	2.500	0.180	2.545	2.500	0.070	2.520	2.500	0.030	2.512	2.500	0.018	2.510	2.500	0.016
	MIN	2.560	2.440	0.060	2.520	2.475	0.020	2.506	2.490	0.006	2.502	2.494	0.002	2.500	2.494	0.000
3	MAX	3.120	3.000	0.180	3.045	3.000	0.070	3.020	3.000	0.030	3.012	3.000	0.018	3.010	3.000	0.016
	MIN	3.060	2.940	0.060	3.020	2.975	0.020	3.006	2.990	0.006	3.002	2.994	0.002	3.000	2.994	0.000
4	MAX	4.145	4.000	0.220	4.060	4.000	0.090	4.028	4.000	0.040	4.016	4.000	0.024	4.012	4.000	0.020
	MIN	4.070	3.925	0.070	4.030	3.970	0.030	4.010	3.988	0.010	4.004	3.992	0.004	4.000	3.992	0.000
5	MAX	5.145	5.000	0.220	5.060	5.000	0.090	5.028	5.000	0.040	5.016	5.000	0.024	5.012	5.000	0.020
	MIN	5.070	4.925	0.070	5.030	4.970	0.030	5.010	4.988	0.010	5.004	4.992	0.004	5.000	4.992	0.000
6	MAX	6.145	6.000	0.220	6.060	6.000	0.090	6.028	6.000	0.040	6.016	6.000	0.024	6.012	6.000	0.020
	MIN	6.070	5.925	0.070	6.030	5.970	0.030	6.010	5.988	0.010	6.004	5.992	0.004	6.000	5.992	0.000
8	MAX	8.170	8.000	0.260	8.076	8.000	0.112	8.035	8.000	0.050	8.020	8.000	0.029	8.015	8.000	0.024
	MIN	8.080	7.910	0.080	8.040	7.964	0.040	8.013	7.985	0.013	8.005	7.991	0.005	8.000	7.991	0.000
10	MAX	10.170	10.000	0.260	10.076	10.000	0.112	10.035	10.000	0.050	10.020	10.000	0.029	10.015	10.000	0.024
	MIN	10.080	9.910	0.080	10.040	9.964	0.040	10.013	9.985	0.013	10.005	9.991	0.005	10.000	9.991	0.000
12	MAX	12.205	12.000	0.315	12.093	12.000	0.136	12.043	12.000	0.061	12.024	12.000	0.035	12.018	12.000	0.029
	MIN	12.095	11.890	0.095	12.050	11.957	0.050	12.016	11.982	0.016	12.006	11.989	0.006	12.000	11.989	0.000
16	MAX	16.205	16.000	0.315	16.093	16.000	0.136	16.043	16.000	0.061	16.024	16.000	0.035	16.018	16.000	0.029
	MIN	16.095	15.890	0.095	16.050	15.957	0.050	16.016	15.982	0.016	16.006	15.989	0.006	16.000	15.989	0.000
20	MAX	20.240	20.000	0.370	20.117	20.000	0.169	20.053	20.000	0.074	20.028	20.000	0.041	20.021	20.000	0.034
	MIN	20.110	19.870	0.110	20.065	19.948	0.065	20.020	19.979	0.020	20.007	19.987	0.007	20.000	19.987	0.000
25	MAX	25.240	25.000	0.370	25.117	25.000	0.169	25.053	25.000	0.074	25.028	25.000	0.041	25.021	25.000	0.034
	MIN	25.110	24.870	0.110	25.065	24.948	0.065	25.020	24.979	0.020	25.007	24.987	0.007	25.000	24.987	0.000
30	MAX	30.240	30.000	0.370	30.117	30.000	0.169	30.053	30.000	0.074	30.028	30.000	0.041	30.021	30.000	0.034
	MIN	30.110	29.870	0.110	30.065	29.948	0.065	30.020	29.979	0.020	30.007	29.987	0.007	30.000	29.987	0.000

Cont.

APPENDIX 46
PREFERRED SHAFT BASIS CLEARANCE FITS—CYLINDRICAL FITS (Cont.)

AMERICAN NATIONAL STANDARD
PREFERRED METRIC LIMITS AND FITS

ANSI B4.2-1978

Dimensions in mm.

BASIC SIZE		LOOSE RUNNING Hole C11	Shaft h11	Fit	FREE RUNNING Hole D9	Shaft h9	Fit	CLOSE RUNNING Hole F8	Shaft h7	Fit	SLIDING Hole G7	Shaft h6	Fit	LOCATIONAL CLEARANCE Hole H7	Shaft h6	Fit
40	MAX	40.280	40.000	0.440	40.142	40.000	0.204	40.064	40.000	0.089	40.034	40.000	0.050	40.025	40.000	0.041
	MIN	40.120	39.840	0.120	40.080	39.938	0.080	40.025	39.975	0.025	40.009	39.984	0.009	40.000	39.984	0.000
50	MAX	50.290	50.000	0.450	50.142	50.000	0.204	50.064	50.000	0.089	50.034	50.000	0.050	50.025	50.000	0.041
	MIN	50.130	49.840	0.130	50.080	49.938	0.080	50.025	49.975	0.025	50.009	49.984	0.009	50.000	49.984	0.000
60	MAX	60.330	60.000	0.520	60.174	60.000	0.248	60.076	60.000	0.106	60.040	60.000	0.059	60.030	60.000	0.049
	MIN	60.140	59.810	0.140	60.100	59.926	0.100	60.030	59.970	0.030	60.010	59.981	0.010	60.000	59.981	0.000
80	MAX	80.340	80.000	0.530	80.174	80.000	0.248	80.076	80.000	0.106	80.040	80.000	0.059	80.030	80.000	0.049
	MIN	80.150	79.810	0.150	80.100	79.926	0.100	80.030	79.970	0.030	80.010	79.981	0.010	80.000	79.981	0.000
100	MAX	100.390	100.000	0.610	100.207	100.000	0.294	100.090	100.000	0.125	100.047	100.000	0.069	100.035	100.000	0.057
	MIN	100.170	99.780	0.170	100.120	99.913	0.120	100.036	99.965	0.036	100.012	99.978	0.012	100.000	99.978	0.000
120	MAX	120.400	120.000	0.620	120.207	120.000	0.294	120.090	120.000	0.125	120.047	120.000	0.069	120.035	120.000	0.057
	MIN	120.180	119.780	0.180	120.120	119.913	0.120	120.036	119.965	0.036	120.012	119.978	0.012	120.000	119.978	0.000
160	MAX	160.460	160.000	0.710	160.245	160.000	0.345	160.106	160.000	0.146	160.054	160.000	0.079	160.040	160.000	0.065
	MIN	160.210	159.750	0.210	160.145	159.900	0.145	160.043	159.960	0.043	160.014	159.975	0.014	160.000	159.975	0.000
200	MAX	200.530	200.000	0.820	200.285	200.000	0.400	200.122	200.000	0.168	200.061	200.000	0.090	200.046	200.000	0.075
	MIN	200.240	199.710	0.240	200.170	199.885	0.170	200.050	199.954	0.050	200.015	199.971	0.015	200.000	199.971	0.000
250	MAX	250.570	250.000	0.860	250.285	250.000	0.400	250.122	250.000	0.168	250.061	250.000	0.090	250.046	250.000	0.075
	MIN	250.280	249.710	0.280	250.170	249.885	0.170	250.050	249.954	0.050	250.015	249.971	0.015	250.000	249.971	0.000
300	MAX	300.650	300.000	0.970	300.320	300.000	0.450	300.137	300.000	0.189	300.069	300.000	0.101	300.052	300.000	0.084
	MIN	300.330	299.680	0.330	300.190	299.870	0.190	300.056	299.948	0.056	300.017	299.968	0.017	300.000	299.968	0.000
400	MAX	400.760	400.000	1.120	400.350	400.000	0.490	400.151	400.000	0.208	400.075	400.000	0.111	400.057	400.000	0.093
	MIN	400.400	399.640	0.400	400.210	399.860	0.210	400.062	399.943	0.062	400.018	399.964	0.018	400.000	399.964	0.000
500	MAX	500.880	500.000	1.280	500.385	500.000	0.540	500.165	500.000	0.228	500.083	500.000	0.123	500.063	500.000	0.103
	MIN	500.480	499.600	0.480	500.230	499.845	0.230	500.068	499.937	0.068	500.020	499.960	0.020	500.000	499.960	0.000

APPENDIX 47

PREFERRED SHAFT BASIS TRANSITION AND INTERFERENCE FITS—CYLINDRICAL FITS

AMERICAN NATIONAL STANDARD
PREFERRED METRIC LIMITS AND FITS

ANSI B4.2-1978

Dimensions in mm.

BASIC SIZE		LOCATIONAL TRANSN. Hole K7	Shaft h6	Fit	LOCATIONAL TRANSN. Hole N7	Shaft h6	Fit	LOCATIONAL INTERF. Hole P7	Shaft h6	Fit	MEDIUM DRIVE Hole S7	Shaft h6	Fit	FORCE Hole U7	Shaft h6	Fit
1	MAX	1.000	1.000	0.006	0.996	1.000	0.002	0.994	1.000	0.000	0.986	1.000	-0.008	0.982	1.000	-0.012
	MIN	0.990	0.994	-0.010	0.986	0.994	-0.014	0.984	0.994	-0.016	0.976	0.994	-0.024	0.972	0.994	-0.028
1.2	MAX	1.200	1.200	0.006	1.196	1.200	0.002	1.194	1.200	0.000	1.186	1.200	-0.008	1.182	1.200	-0.012
	MIN	1.190	1.194	-0.010	1.186	1.194	-0.014	1.184	1.194	-0.016	1.176	1.194	-0.024	1.172	1.194	-0.028
1.6	MAX	1.600	1.600	0.006	1.596	1.600	0.002	1.594	1.600	0.000	1.586	1.600	-0.008	1.582	1.600	-0.012
	MIN	1.590	1.594	-0.010	1.586	1.594	-0.014	1.584	1.594	-0.016	1.576	1.594	-0.024	1.572	1.594	-0.028
2	MAX	2.000	2.000	0.006	1.996	2.000	0.002	1.994	2.000	0.000	1.986	2.000	-0.008	1.982	2.000	-0.012
	MIN	1.990	1.994	-0.010	1.986	1.994	-0.014	1.984	1.994	-0.016	1.976	1.994	-0.024	1.972	1.994	-0.028
2.5	MAX	2.500	2.500	0.006	2.496	2.500	0.002	2.494	2.500	0.000	2.486	2.500	-0.008	2.482	2.500	-0.012
	MIN	2.490	2.494	-0.010	2.486	2.494	-0.014	2.484	2.494	-0.016	2.476	2.494	-0.024	2.472	2.494	-0.028
3	MAX	3.000	3.000	0.006	2.996	3.000	0.002	2.994	3.000	0.000	2.986	3.000	-0.008	2.982	3.000	-0.012
	MIN	2.990	2.994	-0.010	2.986	2.994	-0.014	2.984	2.994	-0.016	2.976	2.994	-0.024	2.972	2.994	-0.028
4	MAX	4.003	4.000	0.011	3.996	4.000	0.004	3.992	4.000	0.000	3.985	4.000	-0.007	3.981	4.000	-0.011
	MIN	3.991	3.992	-0.009	3.984	3.992	-0.016	3.980	3.992	-0.020	3.973	3.992	-0.027	3.969	3.992	-0.031
5	MAX	5.003	5.000	0.011	4.996	5.000	0.004	4.992	5.000	0.000	4.985	5.000	-0.007	4.981	5.000	-0.011
	MIN	4.991	4.992	-0.009	4.984	4.992	-0.016	4.980	4.992	-0.020	4.973	4.992	-0.027	4.969	4.992	-0.031
6	MAX	6.003	6.000	0.011	5.996	6.000	0.004	5.992	6.000	0.000	5.985	6.000	-0.007	5.981	6.000	-0.011
	MIN	5.991	5.992	-0.009	5.984	5.992	-0.016	5.980	5.992	-0.020	5.973	5.992	-0.027	5.969	5.992	-0.031
8	MAX	8.005	8.000	0.014	7.996	8.000	0.005	7.991	8.000	0.000	7.983	8.000	-0.008	7.978	8.000	-0.013
	MIN	7.990	7.991	-0.010	7.981	7.991	-0.019	7.976	7.991	-0.024	7.968	7.991	-0.032	7.963	7.991	-0.037
10	MAX	10.005	10.000	0.014	9.996	10.000	0.005	9.991	10.000	0.000	9.983	10.000	-0.008	9.978	10.000	-0.013
	MIN	9.990	9.991	-0.010	9.981	9.991	-0.019	9.976	9.991	-0.024	9.968	9.991	-0.032	9.963	9.991	-0.037
12	MAX	12.006	12.000	0.017	11.995	12.000	0.006	11.989	12.000	0.000	11.979	12.000	-0.010	11.974	12.000	-0.015
	MIN	11.988	11.989	-0.012	11.977	11.989	-0.023	11.971	11.989	-0.029	11.961	11.989	-0.039	11.956	11.989	-0.044
16	MAX	16.006	16.000	0.017	15.995	16.000	0.006	15.989	16.000	0.000	15.979	16.000	-0.010	15.974	16.000	-0.015
	MIN	15.988	15.989	-0.012	15.977	15.989	-0.023	15.971	15.989	-0.029	15.961	15.989	-0.039	15.956	15.989	-0.044
20	MAX	20.006	20.000	0.019	19.993	20.000	0.006	19.986	20.000	-0.001	19.973	20.000	-0.014	19.967	20.000	-0.020
	MIN	19.985	19.987	-0.015	19.972	19.987	-0.028	19.965	19.987	-0.035	19.952	19.987	-0.048	19.946	19.987	-0.054
25	MAX	25.006	25.000	0.019	24.993	25.000	0.006	24.986	25.000	-0.001	24.973	25.000	-0.014	24.960	25.000	-0.027
	MIN	24.985	24.987	-0.015	24.972	24.987	-0.028	24.965	24.987	-0.035	24.952	24.987	-0.048	24.939	24.987	-0.061
30	MAX	30.006	30.000	0.019	29.993	30.000	0.006	29.986	30.000	-0.001	29.973	30.000	-0.014	29.960	30.000	-0.027
	MIN	29.985	29.987	-0.015	29.972	29.987	-0.028	29.965	29.987	-0.035	29.952	29.987	-0.048	29.939	29.987	-0.061

Cont.

APPENDIX 47
PREFERRED SHAFT BASIS TRANSITION AND INTERFERENCE FITS—CYLINDRICAL FITS (Cont.)

AMERICAN NATIONAL STANDARD
PREFERRED METRIC LIMITS AND FITS

ANSI B4.2-1978

Dimensions in mm.

BASIC SIZE		LOCATIONAL TRANSN. Hole K7	Shaft h6	Fit	LOCATIONAL TRANSN. Hole N7	Shaft h6	Fit	LOCATIONAL INTERF. Hole P7	Shaft h6	Fit	MEDIUM DRIVE Hole S7	Shaft h6	Fit	FORCE Hole U7	Shaft h6	Fit
40	MAX	40.007	40.000	0.023	39.992	40.000	0.008	39.983	40.000	-0.001	39.966	40.000	-0.018	39.949	40.000	-0.035
	MIN	39.982	39.984	-0.018	39.967	39.984	-0.033	39.958	39.984	-0.042	39.941	39.984	-0.059	39.924	39.984	-0.076
50	MAX	50.007	50.000	0.023	49.992	50.000	0.008	49.983	50.000	-0.001	49.966	50.000	-0.018	49.939	50.000	-0.045
	MIN	49.982	49.984	-0.018	49.967	49.984	-0.033	49.958	49.984	-0.042	49.941	49.984	-0.059	49.914	49.984	-0.086
60	MAX	60.009	60.000	0.028	59.991	60.000	0.010	59.979	60.000	-0.002	59.958	60.000	-0.023	59.924	60.000	-0.087
	MIN	59.979	59.981	-0.021	59.961	59.981	-0.039	59.949	59.981	-0.051	59.928	59.981	-0.072	59.894	59.981	-0.106
80	MAX	80.009	80.000	0.028	79.991	80.000	0.010	79.979	80.000	-0.002	79.952	80.000	-0.029	79.909	80.000	-0.072
	MIN	79.979	79.981	-0.021	79.961	79.981	-0.039	79.949	79.981	-0.051	79.922	79.981	-0.078	79.879	79.981	-0.121
100	MAX	100.010	100.000	0.032	99.990	100.000	0.012	99.976	100.000	-0.002	99.942	100.000	-0.036	99.889	100.000	-0.089
	MIN	99.975	99.978	-0.025	99.955	99.978	-0.045	99.941	99.978	-0.059	99.907	99.978	-0.093	99.854	99.978	-0.146
120	MAX	120.010	120.000	0.032	119.990	120.000	0.012	119.976	120.000	-0.002	119.934	120.000	-0.044	119.869	120.000	-0.109
	MIN	119.975	119.978	-0.025	119.955	119.978	-0.045	119.941	119.978	-0.059	119.899	119.978	-0.101	119.834	119.978	-0.166
160	MAX	160.012	160.000	0.037	159.988	160.000	0.013	159.972	160.000	-0.003	159.915	160.000	-0.060	159.825	160.000	-0.150
	MIN	159.972	159.975	-0.028	159.948	159.975	-0.052	159.932	159.975	-0.068	159.875	159.975	-0.125	159.785	159.975	-0.215
200	MAX	200.013	200.000	0.042	199.986	200.000	0.015	199.967	200.000	-0.004	199.895	200.000	-0.076	199.781	200.000	-0.190
	MIN	199.967	199.971	-0.033	199.940	199.971	-0.060	199.921	199.971	-0.079	199.849	199.971	-0.151	199.735	199.971	-0.265
250	MAX	250.013	250.000	0.042	249.986	250.000	0.015	249.967	250.000	-0.004	249.877	250.000	-0.094	249.733	250.000	-0.238
	MIN	249.967	249.971	-0.033	249.940	249.971	-0.060	249.921	249.971	-0.079	249.831	249.971	-0.169	249.687	249.971	-0.313
300	MAX	300.016	300.000	0.048	299.986	300.000	0.018	299.964	300.000	-0.004	299.850	300.000	-0.118	299.670	300.000	-0.298
	MIN	299.964	299.968	-0.036	299.934	299.968	-0.066	299.912	299.968	-0.088	299.798	299.968	-0.202	299.618	299.968	-0.382
400	MAX	400.017	400.000	0.053	399.984	400.000	0.020	399.959	400.000	-0.005	399.813	400.000	-0.151	399.586	400.000	-0.378
	MIN	399.960	399.964	-0.040	399.927	399.964	-0.073	399.902	399.964	-0.098	399.756	399.964	-0.244	399.529	399.964	-0.471
500	MAX	500.018	500.000	0.058	499.983	500.000	0.023	499.955	500.000	-0.005	499.771	500.000	-0.189	499.483	500.000	-0.477
	MIN	499.955	499.960	-0.045	499.920	499.960	-0.080	499.892	499.960	-0.108	499.708	499.960	-0.292	499.420	499.960	-0.580

APPENDIX 48
HOLE SIZES FOR NON-PREFERRED DIAMETERS

Basic Size	C11	D9	F8	G7	H7	H8	H9	H11	K7	N7	P7	S7	U7
OVER 0 TO 3	+0.120 +0.060	+0.045 +0.020	+0.020 +0.006	+0.012 +0.002	+0.010 0.000	+0.014 0.000	+0.025 0.000	+0.060 0.000	0.000 −0.010	−0.004 −0.014	−0.006 −0.016	−0.014 −0.024	−0.018 −0.028
OVER 3 TO 6	+0.145 +0.070	+0.060 +0.030	+0.028 +0.010	+0.016 +0.004	+0.012 0.000	+0.018 0.000	+0.030 0.000	+0.075 0.000	+0.003 −0.009	−0.004 −0.016	−0.008 −0.020	−0.015 −0.027	−0.019 −0.031
OVER 6 TO 10	+0.170 +0.080	+0.076 +0.040	+0.035 +0.013	+0.020 +0.005	+0.015 0.000	+0.022 0.000	+0.036 0.000	+0.090 0.000	+0.005 −0.010	−0.004 −0.019	−0.009 −0.024	−0.017 −0.032	−0.022 −0.037
OVER 10 TO 14	+0.205 +0.095	+0.093 +0.050	+0.043 +0.016	+0.024 +0.006	+0.018 0.000	+0.027 0.000	+0.043 0.000	+0.110 0.000	+0.006 −0.012	−0.005 −0.023	−0.011 −0.029	−0.021 −0.039	−0.026 −0.044
OVER 14 TO 18	+0.205 +0.095	+0.093 +0.050	+0.043 +0.016	+0.024 +0.006	+0.018 0.000	+0.027 0.000	+0.043 0.000	+0.110 0.000	+0.006 −0.012	−0.005 −0.023	−0.011 −0.029	−0.021 −0.039	−0.026 −0.044
OVER 18 TO 24	+0.240 +0.110	+0.117 +0.065	+0.053 +0.020	+0.028 +0.007	+0.021 0.000	+0.033 0.000	+0.052 0.000	+0.130 0.000	+0.006 −0.015	−0.007 −0.028	−0.014 −0.035	−0.027 −0.048	−0.033 −0.054
OVER 24 TO 30	+0.240 +0.110	+0.117 +0.065	+0.053 +0.020	+0.028 +0.007	+0.021 0.000	+0.033 0.000	+0.052 0.000	+0.130 0.000	+0.006 −0.015	−0.007 −0.028	−0.014 −0.035	−0.027 −0.048	−0.040 −0.061
OVER 30 TO 40	+0.280 +0.120	+0.142 +0.080	+0.064 +0.025	+0.034 +0.009	+0.025 0.000	+0.039 0.000	+0.062 0.000	+0.160 0.000	+0.007 −0.018	−0.008 −0.033	−0.017 −0.042	−0.034 −0.059	−0.051 −0.076
OVER 40 TO 50	+0.290 +0.130	+0.142 +0.080	+0.064 +0.025	+0.034 +0.009	+0.025 0.000	+0.039 0.000	+0.062 0.000	+0.160 0.000	+0.007 −0.018	−0.008 −0.033	−0.017 −0.042	−0.034 −0.059	−0.061 −0.086
OVER 50 TO 65	+0.330 +0.140	+0.174 +0.100	+0.076 +0.030	+0.040 +0.010	+0.030 0.000	+0.046 0.000	+0.074 0.000	+0.190 0.000	+0.009 −0.021	−0.009 −0.039	−0.021 −0.051	−0.042 −0.072	−0.076 −0.106
OVER 65 TO 80	+0.340 +0.150	+0.174 +0.100	+0.076 +0.030	+0.040 +0.010	+0.030 0.000	+0.046 0.000	+0.074 0.000	+0.190 0.000	+0.009 −0.021	−0.009 −0.039	−0.021 −0.051	−0.048 −0.078	−0.091 −0.121
OVER 80 TO 100	+0.390 +0.170	+0.207 +0.120	+0.090 +0.036	+0.047 +0.012	+0.035 0.000	+0.054 0.000	+0.087 0.000	+0.220 0.000	+0.010 −0.025	−0.010 −0.045	−0.024 −0.059	−0.058 −0.093	−0.111 −0.146

Cont.

APPENDIX 48

HOLE SIZES FOR NON-PREFERRED DIAMETERS (Cont.)

Basic Size	c11	d9	f8	g7	h7	h8	h9	h11	k7	n7	p7	s7	u7
OVER 100 TO 120	+0.400 +0.180	+0.207 +0.120	+0.090 +0.036	+0.047 +0.012	+0.035 0.000	+0.054 0.000	+0.087 0.000	+0.220 0.000	+0.010 −0.025	−0.010 −0.045	−0.024 −0.059	−0.066 −0.101	−0.131 −0.166
OVER 120 TO 140	+0.450 +0.200	+0.245 +0.145	+0.106 +0.043	+0.054 +0.014	+0.040 0.000	+0.063 0.000	+0.100 0.000	+0.250 0.000	+0.012 −0.028	−0.012 −0.052	−0.028 −0.068	−0.077 −0.117	−0.155 −0.195
OVER 140 TO 160	+0.460 +0.210	+0.245 +0.145	+0.106 +0.043	+0.054 +0.014	+0.040 0.000	+0.063 0.000	+0.100 0.000	+0.250 0.000	+0.012 −0.028	−0.012 −0.052	−0.028 −0.068	−0.085 −0.125	−0.175 −0.215
OVER 160 TO 180	+0.480 +0.230	+0.245 +0.145	+0.106 +0.043	+0.054 +0.014	+0.040 0.000	+0.063 0.000	+0.100 0.000	+0.250 0.000	+0.012 −0.028	−0.012 −0.052	−0.028 −0.068	−0.093 −0.133	−0.195 −0.235
OVER 180 TO 200	+0.530 +0.240	+0.285 +0.170	+0.122 +0.050	+0.061 +0.015	+0.046 0.000	+0.072 0.000	+0.115 0.000	+0.290 0.000	+0.013 −0.033	−0.014 −0.060	−0.033 −0.079	−0.105 −0.151	−0.219 −0.265
OVER 200 TO 225	+0.550 +0.260	+0.285 +0.170	+0.122 +0.050	+0.061 +0.015	+0.046 0.000	+0.072 0.000	+0.115 0.000	+0.290 0.000	+0.013 −0.033	−0.014 −0.060	−0.033 −0.079	−0.113 −0.159	−0.241 −0.287
OVER 225 TO 250	+0.570 +0.280	+0.285 +0.170	+0.122 +0.050	+0.061 +0.015	+0.046 0.000	+0.072 0.000	+0.115 0.000	+0.290 0.000	+0.013 −0.033	−0.014 −0.060	−0.033 −0.079	−0.123 −0.169	−0.267 −0.313
OVER 250 TO 280	+0.620 +0.300	+0.320 +0.190	+0.137 +0.056	+0.069 +0.017	+0.052 0.000	+0.081 0.000	+0.130 0.000	+0.320 0.000	+0.016 −0.036	−0.014 −0.066	−0.036 −0.088	−0.138 −0.190	−0.295 −0.347
OVER 280 TO 315	+0.650 +0.330	+0.320 +0.190	+0.137 +0.056	+0.069 0.017	+0.052 0.000	+0.081 0.000	+0.130 0.000	+0.320 0.000	+0.016 −0.036	−0.014 −0.066	−0.036 −0.088	−0.150 −0.202	−0.330 −0.382
OVER 315 TO 355	+0.720 +0.360	+0.350 +0.210	+0.151 +0.062	+0.075 +0.018	+0.057 0.000	+0.089 0.000	+0.140 0.000	+0.360 0.000	+0.017 −0.040	−0.016 −0.073	−0.041 −0.058	−0.169 −0.226	−0.369 −0.426
OVER 355 TO 400	+0.760 +0.400	+0.350 +0.210	+0.151 +0.062	+0.075 +0.018	+0.057 0.000	+0.089 0.000	+0.140 0.000	+0.360 0.000	+0.017 −0.040	−0.016 −0.073	−0.041 −0.058	−0.187 −0.244	−0.414 −0.471
OVER 400 TO 450	+0.840 +0.440	+0.385 +0.230	+0.165 +0.068	+0.083 +0.020	+0.063 0.000	+0.097 0.000	+0.155 0.000	+0.400 0.000	+0.018 −0.045	−0.017 −0.080	−0.045 −0.108	−0.209 −0.272	−0.467 −0.530
OVER 450 TO 500	+0.880 +0.480	+0.385 +0.230	+0.165 +0.068	+0.083 +0.020	+0.063 0.000	+0.097 0.000	+0.155 0.000	+0.400 0.000	+0.018 −0.045	−0.017 −0.080	−0.045 −0.108	−0.229 −0.292	−0.517 −0.580

APPENDIX 49
SHAFT SIZES FOR NON-PREFERRED DIAMETERS

Basic Size	c11	d9	f7	g6	h6	h7	h9	h11	k6	n6	p6	s6	u6
OVER 0 TO 3	−0.060 / −0.120	−0.020 / −0.045	−0.006 / −0.016	−0.002 / −0.008	0.000 / −0.006	0.000 / −0.010	0.000 / −0.025	0.000 / −0.060	+0.006 / 0.000	+0.010 / +0.004	+0.012 / +0.006	+0.020 / +0.014	+0.024 / +0.018
OVER 3 TO 6	−0.070 / −0.145	−0.030 / −0.060	−0.010 / −0.022	−0.004 / −0.012	0.000 / −0.008	0.000 / −0.012	0.000 / −0.030	0.000 / −0.075	+0.009 / +0.001	+0.016 / +0.008	+0.020 / +0.012	+0.027 / +0.019	+0.031 / +0.023
OVER 6 TO 10	−0.080 / −0.170	−0.040 / −0.076	−0.013 / −0.028	−0.005 / −0.014	0.000 / −0.009	0.000 / −0.015	0.000 / −0.036	0.000 / −0.090	+0.010 / +0.001	+0.019 / +0.010	+0.024 / +0.015	+0.032 / +0.023	+0.037 / +0.028
OVER 10 TO 14	−0.095 / −0.205	−0.050 / −0.093	−0.016 / −0.034	−0.006 / −0.017	0.000 / −0.011	0.000 / −0.018	0.000 / −0.043	0.000 / −0.110	+0.012 / +0.001	+0.023 / +0.012	+0.029 / +0.018	+0.039 / +0.028	+0.044 / +0.033
OVER 14 TO 18	−0.095 / −0.205	−0.050 / −0.093	−0.016 / −0.034	−0.006 / −0.017	0.000 / −0.011	0.000 / −0.018	0.000 / −0.043	0.000 / −0.110	+0.012 / +0.001	+0.023 / +0.012	+0.029 / +0.018	+0.039 / +0.028	+0.044 / +0.033
OVER 18 TO 24	−0.110 / −0.240	−0.065 / −0.117	−0.020 / −0.041	−0.007 / −0.020	0.000 / −0.013	0.000 / −0.021	0.000 / −0.052	0.000 / −0.130	+0.015 / +0.002	+0.028 / +0.015	+0.035 / +0.022	+0.048 / +0.035	+0.054 / +0.041
OVER 24 TO 30	−0.110 / −0.240	−0.065 / −0.117	−0.020 / −0.041	−0.007 / −0.020	0.000 / −0.013	0.000 / −0.021	0.000 / −0.052	0.000 / −0.130	+0.015 / +0.002	+0.028 / +0.015	+0.035 / +0.022	+0.048 / +0.035	+0.061 / +0.048
OVER 30 TO 40	−0.120 / −0.280	−0.080 / −0.142	−0.025 / −0.050	−0.009 / −0.025	0.000 / −0.016	0.000 / −0.025	0.000 / −0.062	0.000 / −0.160	+0.018 / +0.002	+0.033 / +0.017	+0.042 / +0.026	+0.059 / +0.043	+0.076 / +0.060
OVER 40 TO 50	−0.130 / −0.290	−0.080 / −0.142	−0.025 / −0.050	−0.009 / −0.025	0.000 / −0.016	0.000 / −0.025	0.000 / −0.062	0.000 / −0.160	+0.018 / +0.002	+0.033 / +0.017	+0.042 / +0.026	+0.059 / +0.043	+0.086 / +0.070
OVER 50 TO 65	−0.140 / −0.330	−0.100 / −0.174	−0.030 / −0.060	−0.010 / −0.029	0.000 / −0.019	0.000 / −0.030	0.000 / −0.074	0.000 / −0.190	+0.021 / +0.002	+0.039 / +0.020	+0.051 / −0.032	+0.072 / +0.053	+0.106 / +0.087
OVER 65 TO 80	−0.150 / −0.340	−0.100 / −0.174	−0.030 / −0.060	−0.010 / −0.029	0.000 / −0.019	0.000 / −0.030	0.000 / −0.074	0.000 / −0.190	+0.021 / +0.002	+0.039 / +0.020	+0.051 / +0.032	+0.078 / +0.059	+0.121 / +0.102
OVER 80 TO 100	−0.170 / −0.390	−0.120 / −0.207	−0.036 / −0.071	−0.012 / −0.034	0.000 / −0.022	0.000 / −0.035	0.000 / −0.087	0.000 / −0.220	+0.025 / +0.003	+0.045 / +0.023	+0.059 / +0.037	+0.093 / +0.071	+0.146 / +0.124

Cont.

APPENDIX 49
SHAFT SIZES FOR NON-PREFERRED DIAMETERS (Cont.)

Basic Size	c11	d9	f7	g6	h6	h7	h9	h11	k6	n6	p6	s6	u6
OVER 100 TO 120	−0.180 −0.400	−0.120 −0.207	−0.036 −0.071	−0.012 −0.034	0.000 −0.022	0.000 −0.035	0.000 −0.087	0.000 −0.220	+0.025 +0.003	+0.045 +0.023	+0.059 +0.037	+0.101 +0.079	+0.166 +0.144
OVER 120 TO 140	−0.200 −0.450	−0.145 −0.245	−0.043 −0.083	−0.014 −0.039	0.000 −0.025	0.000 −0.040	0.000 −0.100	0.000 −0.250	+0.028 +0.003	+0.052 +0.027	+0.068 +0.043	+0.117 +0.092	+0.195 +0.170
OVER 140 TO 160	−0.210 −0.460	−0.145 −0.245	−0.043 −0.083	−0.014 −0.039	0.000 −0.025	0.000 −0.040	0.000 −0.100	0.000 −0.250	+0.028 +0.003	+0.052 +0.027	+0.068 +0.043	+0.125 +0.100	+0.215 +0.190
OVER 160 TO 180	−0.230 −0.480	−0.145 −0.245	−0.043 −0.083	−0.014 −0.039	0.000 −0.025	0.000 −0.040	0.000 −0.100	0.000 −0.250	+0.028 +0.003	+0.052 +0.027	+0.068 +0.043	+0.133 +0.108	+0.235 +0.210
OVER 180 TO 200	−0.240 −0.530	−0.170 −0.285	−0.050 −0.096	−0.015 −0.044	0.000 −0.029	0.000 −0.046	0.000 −0.115	0.000 −0.290	+0.033 +0.004	+0.060 +0.031	+0.079 +0.050	+0.151 +0.122	+0.265 +0.236
OVER 200 TO 225	−0.260 −0.550	−0.170 −0.285	−0.050 −0.096	−0.015 −0.044	0.000 −0.029	0.000 −0.046	0.000 −0.115	0.000 −0.290	+0.033 +0.004	+0.060 +0.031	+0.079 +0.050	+0.159 +0.130	+0.287 +0.258
OVER 225 TO 250	−0.280 −0.570	−0.170 −0.285	−0.050 −0.096	−0.015 −0.044	0.000 −0.029	0.000 −0.046	0.000 −0.115	0.000 −0.290	+0.033 +0.004	+0.060 +0.031	+0.079 +0.050	+0.169 +0.140	+0.313 +0.284
OVER 250 TO 280	−0.300 −0.620	−0.190 −0.320	−0.056 −0.108	−0.017 −0.049	0.000 −0.032	0.000 −0.052	0.000 −0.130	0.000 −0.320	+0.036 +0.004	+0.066 +0.034	+0.088 +0.056	+0.190 +0.158	+0.347 +0.315
OVER 280 TO 315	−0.330 −0.650	−0.190 −0.320	−0.056 −0.108	−0.017 −0.049	0.000 −0.032	0.000 −0.052	0.000 −0.130	0.000 −0.320	+0.036 +0.004	+0.066 +0.034	+0.088 +0.056	+0.202 +0.170	+0.382 +0.350
OVER 315 TO 355	−0.360 −0.720	−0.210 −0.350	−0.062 −0.119	−0.018 −0.054	0.000 −0.036	0.000 −0.057	0.000 −0.140	0.000 −0.360	+0.040 +0.004	+0.073 +0.037	+0.098 +0.062	+0.226 +0.190	+0.426 +0.390
OVER 355 TO 400	−0.400 −0.760	−0.210 −0.350	−0.062 −0.119	−0.018 −0.054	0.000 −0.036	0.000 −0.057	0.000 −0.140	0.000 −0.360	+0.040 +0.004	+0.073 +0.037	+0.098 +0.062	+0.244 +0.208	+0.471 +0.435
OVER 400 TO 450	−0.440 −0.840	−0.230 −0.385	−0.068 −0.131	−0.020 −0.060	0.000 0.040	0.000 0.063	0.000 −0.155	0.000 0.400	+0.045 +0.005	+0.080 +0.040	+0.108 +0.068	+0.272 +0.232	+0.530 +0.490
OVER 450 TO 500	−0.480 −0.880	−0.230 −0.385	−0.068 −0.131	−0.020 −0.060	0.000 −0.040	0.000 −0.063	0.000 −0.155	0.000 −0.400	+0.045 +0.005	+0.080 +0.040	+0.108 +0.068	+0.292 +0.252	+0.580 +0.540

APPENDIX 50
ENGINEERING FORMULAS

Motion

S = distance (inches, feet, miles)
t = time (seconds, minutes, hours)
v = average velocity (feet per second, miles per hour, etc.
v_1 = initial velocity
v_2 = final velocity
a = acceleration (feet per second per second)

(1) $S = vt$

(2) $V(\text{avg.}) = \dfrac{v_2 - v_1}{2}$

(3) $S = \left(\dfrac{v_1 + v_2}{2}\right)\dfrac{\text{ft}}{\text{sec}}\,(t\text{ sec})$

(4) $a = \dfrac{v_2 - v_1}{t}$

(5) $S = v_1 t + \frac{1}{2}at^2$

Angular Motion

V = linear velocity
N = number of revolutions per min
θ = angular distance in radians
1 radian $= 360°/2\pi = 57.3°$
ω(omega) = average angular velocity
$\qquad = \theta/t$ (rad per sec, rev per min)
ω_1 = initial velocity
ω_2 = final velocity
α(alpha) = angular acceleration = rad per sec^2
S = length of arc
r = radius of arc
D = diameter

(6) $\theta = (\text{avg. }\omega)t$

(7) $\omega(\text{avg.}) = \dfrac{\omega_2 + \omega_1}{2}$

(8) $\alpha = \dfrac{\omega_2 - \omega_1}{t}$

(9) $\omega_2 = \omega_1 + \alpha t$

(10) $\theta = \omega_1 t + \dfrac{\alpha t^2}{2}$

(11) $V = \pi DN$ or $V = r\omega$ (ft per sec, ft per min, etc.)

(12) $\omega = \dfrac{2\pi r n}{r} = 2\pi N$

Force and Acceleration

F = force (pounds)
M = mass
a = acceleration (ft per sec^2)
g = gravitational acceleration = 32.2 ft/sec^2
W = weight (pounds)
$M = F/a = W/g$ (units of mass in slugs)

Work

W = work (ft · lb)
F = force (lb)
d = distance
$W = Fd$

Power

W = work
t = time

1 horsepower $= 550\,\dfrac{\text{ft} \cdot \text{lb}}{\text{sec}}$

Avg. power $= \dfrac{W}{t} = \dfrac{\text{ft} \cdot \text{lb}}{\text{sec}}$ or $\dfrac{\text{ft} \cdot \text{lb}}{\text{min}}$ etc.

Kinetic Energy

W = weight
V = velocity
g = 32.2 ft per sec^2
K.E. $= WV^2/2g$

APPENDIX 51
GRADING GRAPH

This graph can be used to determine the individual grades of members of a team and to compute grade averages for those who do extra assignments.

The percent participation of each team member should be determined by the team as a whole (see Chapter 2 problems).

Example: written or oral report grades

Overall team grade: 82

Team members $N = 5$	Contribution $C = \%$	$F = CN$	Grade (graph)
J. Doe	20%	100	82.0
H. Brown	16%	80	76.4
L. Smith	24%	120	86.0
R. Black	20%	100	82.0
T. Jones	20%	100	82.0
	100%		

Example: quiz or problem sheet grades

Number assigned: 30
Number extra: 6
                   —
Total   36

Average grade for total (36): 82

$$F = \frac{\text{No. completed} \times 100}{\text{No. assigned}} = \frac{36 \times 100}{30} = 120$$

Final grade (from graph): 86.0

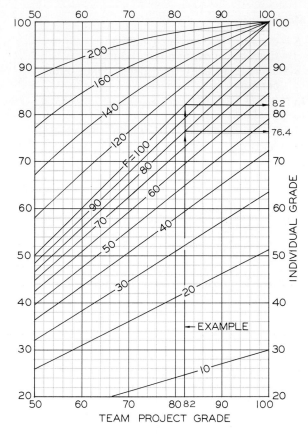

Fig. A51.–1. Grading graph.

Index

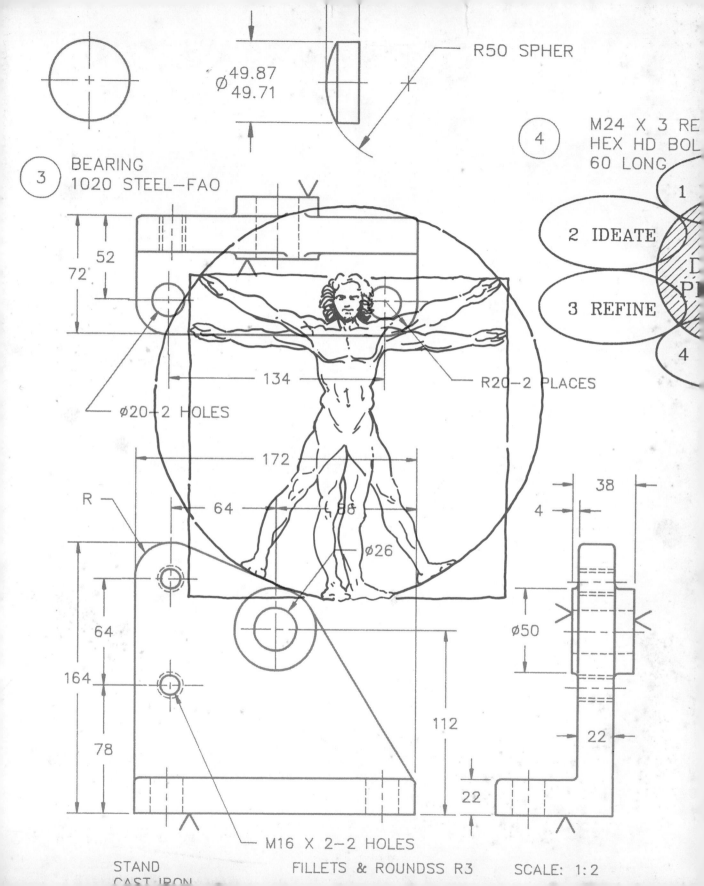